Springer Proceedings in Physics **295**

Indexed by Scopus

The series Springer Proceedings in Physics, founded in 1984, is devoted to timely reports of state-of-the-art developments in physics and related sciences. Typically based on material presented at conferences, workshops and similar scientific meetings, volumes published in this series will constitute a comprehensive up to date source of reference on a field or subfield of relevance in contemporary physics. Proposals must include the following:

- Name, place and date of the scientific meeting
- A link to the committees (local organization, international advisors etc.)
- Scientific description of the meeting
- List of invited/plenary speakers
- An estimate of the planned proceedings book parameters (number of pages/articles, requested number of bulk copies, submission deadline).

Please contact:

For Americas and Europe: Dr. Zachary Evenson; zachary.evenson@springer.com
For Asia, Australia and New Zealand: Dr. Loyola DSilva; loyola.dsilva@springer.com

H. Paul Urbach · Huilin Jiang
Editors

Proceedings of the 7th International Symposium of Space Optical Instruments and Applications

ISSOIA 2022, 21–23 Oct, Beijing, China

Springer

Editors
H. Paul Urbach
Optics Research Group
Delft University of Technology
Delft, The Netherlands

Huilin Jiang
Changchun University of Science and Technology
Changchun, China

ISSN 0930-8989 ISSN 1867-4941 (electronic)
Springer Proceedings in Physics
ISBN 978-981-99-4097-4 ISBN 978-981-99-4098-1 (eBook)
https://doi.org/10.1007/978-981-99-4098-1

This Springer imprint is published by the registered company Springer Nature Singapore Pte Ltd.
The registered company address is: 152 Beach Road, #21-01/04 Gateway East, Singapore 189721, Singapore

Contents

A 3D Adaptive Sequential Model Based Dim Small Moving Target Detection Method

Jinyan Gao[1](✉), Luyuan Wang[1], Weiwei Liu[1], and Zhongshi Pan[2]

[1] Institute of Spacecraft System Engineering, Beijing 100094, China
gaojinyan15@nudt.edu.cn

[2] Institute of Remote Sensing Satellite, Beijing 100094, China

Abstract. For dim small moving target detection in image sequences, we propose a new spatial and temporal dim small moving target detection method. First, a long multi-temporal median filter is proposed to adaptively detect small targets with different sizes and speeds. Second, a short median filter in temporal profile is designed to prevent the dim small target merged in complex background especially in low-lying areas. Therefore, the relatively movement information and intensity profile difference information is sufficiently exploited by our proposed sequence detection method. To demonstrate the robustness and efficiency of our proposed dim small moving target detection method, experiment on a real sequence is tested and the signal to clutter ratio gain is used to test the performance of the proposed method, which exhibit our method can handle complex backgrounds.

Keywords: small moving target detection · complex background · space surveillance

1 Introduction

With high penetration and detectability, infrared imaging has been widely applied in remote sensing, aerospace, surveillance, and other fields. As the imagery distance is usually far, the target appears as a small speckle in infrared images, lacking of available image information and seriously influenced by complex background and inevitable noises during imaging process, which brings great challenges to target detection.

Numerous small IR target detection methods have been designed over the last two decades. These methods can be categorized into two classes, the single frame small target detection methods and the multi-frame small target detection methods. However, single frame small target detection methods [1–6] are difficult to effectively extract targets with low signal-to-clutter ratio in the scene, which often brings problems of low detection rate and high false alarm rate in the detection results.

Many multi-frame detection methods have been proposed to detect dim small moving targets using temporal profile analysis. These temporal profile analysis methods [7–10] have achieved good performance in speckle noise and static broken cluster removal. The temporal profile of pixels affected by a point target has a pulse-like shape. The width of the pulse is inversely proportional to the speed of the target, while the intensity of

H. P. Urbach and H. Jiang (Eds.): ISSOIA 2022, SPPHY 295, pp. 1–7, 2023.
https://doi.org/10.1007/978-981-99-4098-1_1

the pulse is related to the strength of the target. A representative method using temporal profile analysis is the Temporal Variance Filter (TVF) [8], which was realized by iterating the variance of each pixel in an image sequence. This method is easy of implementation through simple iteration steps, but since its detection results usually appear as short trajectories, it is difficult to achieve accurate centroid position of an target in every frame.

Besides, The Connected Line of the Stagnation Points (CLSP) method [10] for dim small moving target detection was achieved by extracting the minimum value of each pixel in an image sequence between adjacent frames on its temporal profile, then the minimum value of each pixel was used as the stagnation point, and then it was subtracted from the temporal profile of the pixel in the original image sequence, and finally, the energy pulse signal could be obtained which was generated by a target passing through the pixel. However, the principle of this method is simple, but the implementation of the algorithm in engineering is complicated, and it is seriously affected by background noise.

In this paper, inspired by the temporal profile analysis of each pixel on the image plane, we proposed a three-dimensional adaptive sequential model based dim small moving target detection method, the three-dimensional data information includes two-dimensional data information in the spatial domain and one-dimensional data information in the temporal domain. It is composed with image pre-processing in spatial, a long multi-temporal median filter, and a short temporal median filter, the long multi-temporal median filter is used to detect small targets with different sizes and speeds, the short temporal median filter is used to enhance dim targets in the low-lying areas. Experiments shows that our proposed method achieves promising target extraction results.

The rest of this paper is organized as follows. Section 2 discusses the characteristics of dim small targets and backgrounds. Section 3 describes the model of our proposed method. In Section 4, experiments are conducted to test our proposed method. The paper is concluded in Section 5.

2 The Characteristics of Targets and Backgrounds

Dim small moving targets have been detected for many years, and it is still a difficult problem until now, we may ask why the small target is so difficult to detect. We will analyze this problem from two aspects of target and background.

For the characteristic of targets, first, since the detection distance is long, the target only occupies a few pixels on an image plane. Second, as the reflection, refraction, sensor aperture diffraction's effect, the target has a point spread characteristic in the focal plane. Third, the intensity and shapes of a small target may be changed since different weather, season, and time.

For the characteristic of background, first, the cirrus clouds with ice crystals are easy to cause sunlight reflections, broken cloud, and cloud edges drift over time. Second, strong evolving clouds may lead to a low local contrast of a target with its surrounding backgrounds. Third, cloud edge and speckle noise are the most important causes for false alarms in dim small moving target detection, and for a surveillance system, any false alarm prediction will decrease the system's confidence level. Our purpose is to minimize the false alarm rate as much as possible.

3 The Proposed Method

We propose a long multi-temporal median filter and a short temporal median filter, the long multi-temporal median filter is used to detect small targets with different sizes and speeds, the short temporal median filter is used to enhance dim targets in the low-lying areas. The block diagram of our proposed method is shown in Fig. 1.

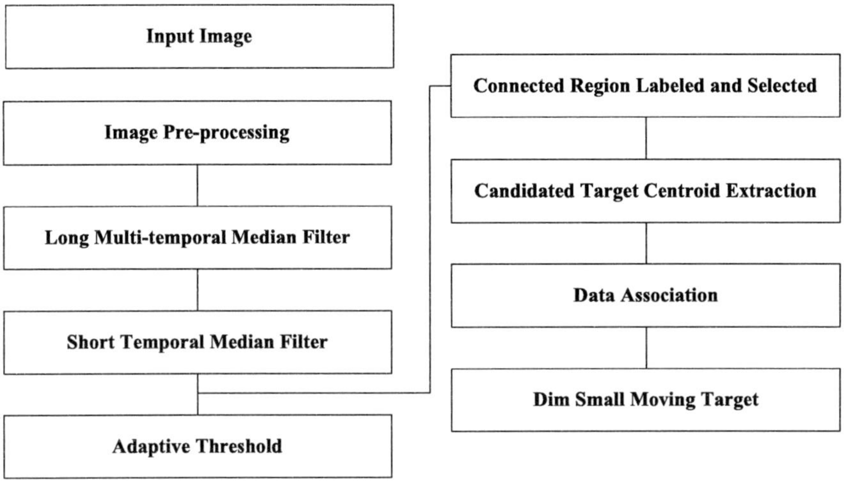

Fig. 1. The block diagram of our proposed method.

First, the image pre-processing methods are used to eliminates streak pixels and isolated noise in spatial. Second, a long multi-temporal median filter is proposed to adaptively detect small targets with different sizes and speeds. Third, a short temporal median filter is performed to detect targets in low-lying areas. Forth, an adaptive threshold segmentation method is used to furtherly remove residual false alarm clutters, and a binarized image is obtained after the image segmentation. Fifth, the connected domains are labeled and furtherly selected to extract the connected domains which satisfy target features. Finally, the centroid coordinates of candidate targets are extracted by a weighted grayscale method and the data association is performed to obtain the trajectory of targets.

3.1 The Long Multi-temporal Median Filter

A triple long-time-domain filtering template is designed to adapt the target size in the original image frame to enhance the target while suppressing its neighborhood backgrounds and noise. Specifically, the gray value of each pixel in sliding window is sorted in a descending order, the median value in sliding window is then chosen as the output of our filter, and the average of all the median filter in different length is calculated as the output of our proposed method finally. The mathematical model of our proposed long time domain filter can be modeled as follows:

$$\begin{aligned}&FilterResult_1(i,j,k)\\&=\{median\{A(i,j,k),A(i,j,k-w_t),A(i,j,k+w_t),A(i,j,k-(w_t-1)),A(i,j,k+(w_t+1))\}\\&+median\{A(i,j,k),A(i,j,k-(w_t-1)),A(i,j,k+(w_t+1)),A(i,j,k-(w_t-2)),A(i,j,k+(w_t+2))\}\\&+median\{A(i,j,k),A(i,j,k-(w_t-2)),A(i,j,k+(w_t+2)),A(i,j,k-(w_t-3)),A(i,j,k+(w_t+3))\}\}\end{aligned}$$

where (i, j) means the center of the sliding window, k means the kth frame in the image sequence. w_t = means the distance between the filtered pixels and the central pixel, and it is empirically determined with 4. $A()$ means the gray value of the filtered pixel in the sliding window. $FilterResult_1$ means the output of the long multi-temporal median filter.

3.2 The Short Median Filter

The short median filter in temporal profile is designed to prevent the dim small target merging in complex background especially in low-lying areas. The mathematical model of the short median filter can be modeled as follows:

$$\begin{aligned}FilterResult_2(i,j,k)=\{&A(i,j,k)-0.75*median\{A(i,j,k-1),A(i,j,k),A(i,j,k+1)\}\\&-0.25*median\{A(i,j,k-1),A(i,j,k+1)\}\}\end{aligned}$$

where (i, j) means the center of the sliding window, k means the kth frame in the image sequence. $FilterResult_2$ means the output of our proposed short median filter.

Assume that the three elements in the current short filter template window are [a b c]. The elements in the sliding window can be sorted in the following three situations.

Case 1: when the three pixels [a b c] in the filtering window are sorted as a < b < c, the area passed by the sliding window has the characteristics of the background area, and the output result of the triple time-domain filtering template is near zero.

Case 2: when the three pixels [a b c] in the filtering window are sorted as b < a < c, the area passed by the sliding window has the central pixel lower than the neighboring pixels. In this situation, the area passed by the sliding window has the low-lying feature, and if a target appears in the region, it is dim and faint. The output result of the triple time-domain filtering template is positive, which can enhance the target in low-lying areas.

Case 3: when the three pixels [a b c] in the filtering window are sorted as a < c < b, the area passed by the sliding window has saliency small target feature, which appears the gray level of the central area is higher than that of the neighboring pixels. The output result of the triple time-domain filtering template is negative. However, considering the target enhancement effect of above long multi-temporal median filter, the weakening effect on a target in this case can be negligible.

Above all, the proposed long and short median combined filter can be modeled as follows:

$$\begin{aligned}&FilterResult(i,j,k)\\&=\{median\{A(i,j,k),A(i,j,k-w_t),A(i,j,k+w_t),A(i,j,k-(w_t-1)),A(i,j,k+(w_t+1))\}\\&+median\{A(i,j,k),A(i,j,k-(w_t-1)),A(i,j,k+(w_t+1)),A(i,j,k-(w_t-2)),A(i,j,k+(w_t+2))\}\\&+median\{A(i,j,k),A(i,j,k-(w_t-2)),A(i,j,k+(w_t+2)),A(i,j,k-(w_t-3)),A(i,j,k+(w_t+3))\}\}\\&-1.5*\{A(i,j,k)-0.75*median\{A(i,j,k-1),A(i,j,k),A(i,j,k+1)\}\\&\qquad-0.25*median\{A(i,j,k-1),A(i,j,k+1)\}\}\end{aligned}$$

4 Experiments

Experiments are conducted on a real image sequence, qualitative analysis and quantitative analysis are performed to test the performance of the proposed method.

1) Evaluation Metrics: SCR is used to measure the difficulty of dim small moving target detection. It can be defined as:

$$SCR = \frac{\mu_t - \mu_b}{\sigma_b}$$

where μ_t is the average grayscale value of a target region, and μ_b and σ_b are the average grayscale value and standard deviation of a local surrounding background region, respectively. From the above definition, we can see that higher values of SCR means that small targets can be detected more easily.
2) Qualitative and quantitative results: As shown in Fig. 2, first, we can see that the target is completely submerged in the backgrounds, and it is difficult to distinguish in the original image. Second, after processing by the proposed combined filter, it can be seen the target has been enhanced significantly and the background has been suppressed effectively. Third, after adaptive threshold segmentation, while the target area is preserved, there are some false alarms remain in the segmentation results. Fourth, after the centroid extraction and data association, the target trajectory is finally obtained on the image plane. As shown in Fig. 3, the statistical results of signal-to-clutter (SCR) ratio show that the proposed method can detect the target with SCR above 1.

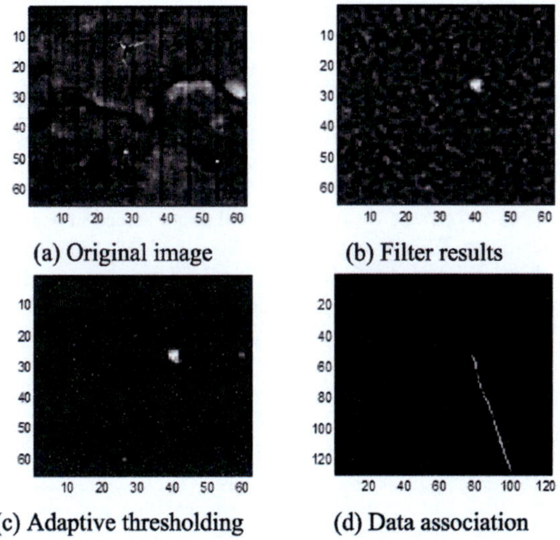

Fig. 2. Qualitative analysis of target detection results.

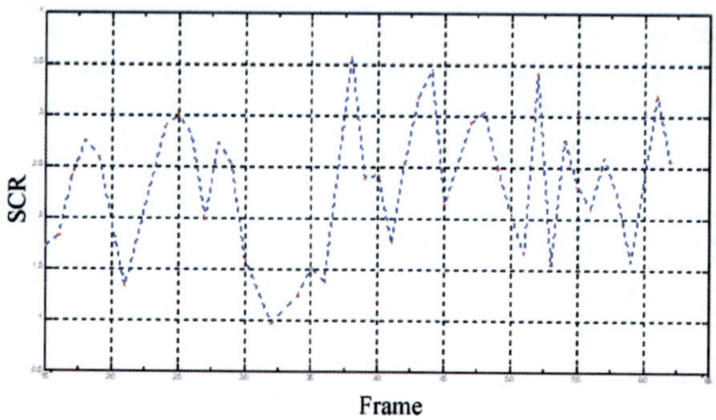

Fig. 3. Quantitative analysis of target detection results.

5 Conclusion

This paper has proposed a three-dimensional adaptive sequential model based dim small moving target detection method. Initially, a long multi-temporal median filter is proposed to suppress backgrounds and noise in high frequency while enhance targets with different velocities. Afterwards, a short time domain filter is proposed to enhance targets in low-lying areas. Then, an adaptive threshold segmentation method is used to furtherly remove residual false alarm clutters. In the next part, the connected domains are furtherly selected to confirm the connected domains that satisfy target features. Finally, the centroid coordinates of the suspected target are extracted to obtain the trajectory of a target. Experiments show that our proposed method achieves promising target extraction results.

References

1. Chen, D.L., Zhang, L., Huang, H.: Extracting small flying airplane with spatially accurate and temporally consistent foreground modeling. IEEE Trans. Geosci. Remote Sens. **60**, 1–12 (2021)
2. Zhu, Q., Zhu, S., Liu, G., Peng, Z.: Infrared small target detection using local feature-based density peaks searching. IEEE Geosci. Remote Sens. Lett. **19**, 6507805 (2022)
3. Lu, R., Yang, X., Li, W., Fan, J., Li, D., Jing, X.: Robust infrared small target detection via multidirectional derivative-based weighted contrast measure. IEEE Geosci. Remote Sens. Lett. **19**, 7000105 (2020)
4. Zhang, T., Peng, Z., Wu, H., He, Y., Li, C., Yang, C.: Infrared small target detection via self-regularized weighted sparse model. Neurocomputing **420**, 124–148 (2021)
5. Gao, J., Guo, Y., Lin, Z., et al.: Robust infrared small target detection using multiscale gray and variance difference measures. IEEE J. Sel. Top. Appl. Earth Obs. Remote Sens. **11**(12), 5039–5052 (2018)
6. Zhao, M., Li, W., Li, L., et al.: Three-order tensor creation and tucker decomposition for infrared small-target detection. IEEE Trans. Geosci. Remote Sens. **60**, 1–16 (2021)
7. Gao, C., Wang, L., Xiao, Y., et al.: Infrared small-dim target detection based on Markov random field guided noise modeling. Pattern Recogn. **76**, 463–475 (2018)
8. Gu, J., Su, T., Wang, Q., et al.: Multiple moving targets surveillance based on a cooperative network for multi-UAV. IEEE Commun. Mag. **56**(4), 82–89 (2018)

9. Gao, J., Lin, Z., Guo, Y., et al.: TVPCF: a spatial and temporal filter for small target detection in IR images. In: 2017 International Conference on Digital Image Computing: Techniques and Applications (DICTA), pp. 1–7. IEEE (2017)
10. Liu, D., Li, Z., Wang, X., Zhang, J.: Moving target detection by nonlinear adaptive filtering on temporal profiles in infrared image sequences. Infrared Phys. Technol. **73**, 41–48 (2015)

Remote Sensing Image Lossy Compression Based on JPEG with Controlled Visual Quality

Fangfang Li[1,2], Sergii Kryvenko[1], and Vladimir Lukin[1](✉)

[1] National Aerospace University, 17 Chkalova Street, Kharkov 61070, Ukraine
v.lukin@khai.edu

[2] Nanchang Hangkong University, Fenghe South Avenue 696, Nanchang 330063, China

Abstract. Remote sensing image compression plays an important role in Earth observation applications. Lossy compression techniques are efficient means to reduce the size of images, although the distortions are unavoidable. Consequently, the image quality should be controlled carefully. This paper discusses this topic with its application to JPEG. Using visual quality metrics, we show that JPEG is able to provide compression performance not worse than many advanced coders for invisible distortions and/or if images have complex content. We show that mean square error (MSE) can be provided by a two-step approach and predicted with high accuracy using a simple neural network. Such a neural network employs QF as one of the inputs and several parameters characterizing image complexity as other inputs. Different sets of input parameters are analyzed, and we show that even four input parameters produce a quite accurate prediction of MSE and, respectively, PSNR, with errors less than 1.5 dB with low computation load. This is shown using images used in training and for verification and validation image sets. Then, it is easy to determine a needed QF to provide a desired quality.

Keywords: Remote Sensing · Lossy Compression · Image Quality

1 Introduction

Nowadays, remote sensing (RS) has become a useful tool for numerous applications, including agriculture, forestry, ecology, meteorology [1, 2], etc. An improved spatial resolution of imagers and an increased number of observations, this has led to an enormous number of acquired images that have to be stored, processed, transferred via communication lines, classified, and disseminated [3]. Then, a necessity to compress RS data arises [3–5].

As it is known [6–8], there are lossless and lossy techniques of image compression (where near-lossy methods are sometimes identified as a special group [8]). Lossless compression does not introduce distortions into compressed data, but it is unable to provide a relatively large compression ratio (CR), and, besides, CR cannot be controlled. Due to this, its practical use is restricted. In turn, lossy compression is able to produce quite large values of CR with its possible variation. However, distortions are introduced,

H. P. Urbach and H. Jiang (Eds.): ISSOIA 2022, SPPHY 295, pp. 8–19, 2023.
https://doi.org/10.1007/978-981-99-4098-1_2

and they can lead to negative effects such as bad (inappropriate) visual quality and reduction of classification accuracy [3, 9]. Then, a certain compromise between a produced CR and a provided quality of compressed RS images should be attained [3, 9–11].

Providing a desired trade-off is not the only requirement that has to be satisfied in practice. Often a method to be applied should satisfy other requirements and restrictions. For example, a desired quality has to be provided quickly and with an acceptable computational load, especially on board of small satellites [3, 12]. Another requirement could be to provide pre-requisites for efficient classification where introduced losses should not lead to a (sufficient) reduction in classification accuracy [4, 9, 13]. In this sense, recent studies show that a considerable reduction of classification accuracy starts when distortions due to lossy compression become visible [9]. Thus, image visual quality is of prime importance.

Finally, one more requirement might concern the basic algorithm of image compression. In some cases, this can be a specialized coder [5, 14]. In other practical applications, the use of some standard means is desired [15, 16]. A typical representative of standards is JPEG [17] which, being quite "old," is still very widely used (especially for compressing color and three-channel RS data) and studied [15, 18]. One more reason for using JPEG is that its performance is comparable to other modern coders for high-complexity images and if introduced distortions are invisible or can hardly be noticed [18, 19]. Recall that here we consider just this case (in fact, advantages of modern coders compared to JPEG appear themselves for the quite low quality of compressed images and/or if an image to be compressed is of low complexity).

There are several main approaches to providing a desired quality of compressed images [18–21]. The first one employs iterative procedures [19] where multiple compression/decompression, metric calculation, comparison to a desired value, and the corresponding changing of the parameter controlling compression (PCC) is performed. An advantage of this procedure is that it usually allows accurate providing of the desired value of the used metric. A drawback is that the number of iterations is unclear at the beginning of compression, which can be quite large. This restricts the practical application of iterative procedures, especially on-board. A second way [20] presumes a brief analysis of distortion statistics in a limited number of image blocks with a further prediction of introduced distortions and appropriate PCC settings for compression. This method allows speedy processing without any decompression, but its accuracy in providing a desired quality is limited. Besides, it is not applicable to all possible quality metrics. Finally, a third approach (called two-step) presumes off-line getting of the average rate-distortion curve (RDC). Then, for a particular image to be compressed, one uses this curve data at the first step of compression for setting the initial PCC. Then, the image is decompressed, the metric value is calculated, and PCC is corrected using the average RDC. After this, the image is finally compressed.

The advantages of the latter approach are that it is considerably faster than iterative procedures and more accurate than the prediction approaches. The two-step approach has been successfully tested for different coders and metrics [21–23]. However, it has not been tested for JPEG yet. Recall that JPEG has a specific PCC called quality factor (QF) that varies from 1 to 100 and allows changing compressed image quality in very wide limits. In addition, since JPEG is widely used in different customer devices as well as in other imaging systems, the speed of providing a desired quality for JPEG is worth paying special attention to.

Based on this brief analysis, in this paper, we focus on testing the two-step approach of reaching a desired quality of compressed images for JPEG. In this case, specific features of using the two-step approach are discovered. In addition, we carry out a preliminary analysis showing that a very simple neural network is able to predict the mean square error of distortions introduced by JPEG, where the quality factor is one of the input parameters, while other simple parameters describe image properties from different viewpoints.

2 Rate/Distortion Curves for JPEG and Their Brief Analysis

JPEG peculiarities and properties have been intensively studied at the stage of its design [17]. Nowadays, we have it in a more or less "fixed" form being implemented in hardware and software [24], where one loads an image to be compressed, sets QF, and gets a compressed image [25]. Usually, the quantization Table for a given QF is pre-calculated and stored where its values are integers coming from QF50 as the starting point. The rate/distortion curve in the form of peak signal-to-noise ratio (PSNR) on QF then has a specific form where examples for the most interesting interval of QF values for several color images are presented in Fig. 1.a (JPEG version available at MATLAB has been used).

In addition, in Fig. 1.b, we give the RDCs as dependences of the visual quality metric PSNR-HMA on QF. Note that PSNR-HMA [26] takes into account several peculiarities of the human vision system, namely, a larger sensitivity to distortions in low spatial frequencies than in higher spatial frequencies and a masking effect. Similar to PSNR, PSNR-HMA values are expressed in dB, where larger values relate to better visual quality. PSNR-HVS-M, about 41 dB, is considered to be the threshold of distortion invisibility [27]. Note that for the metric PSNR, the distortion invisibility threshold is about 36 dB [27].

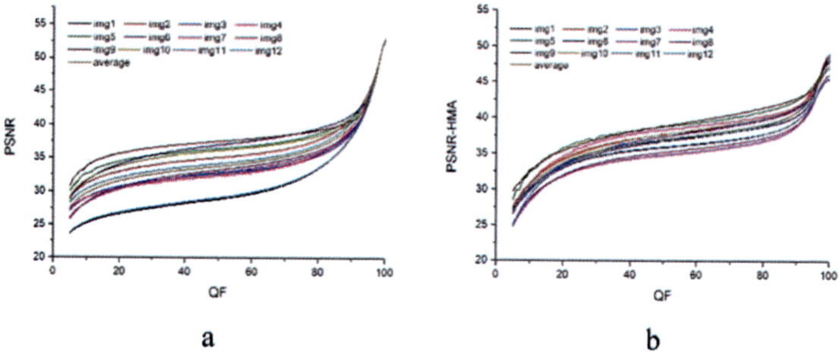

Fig. 1. RD curves for twelve test images compressed by JPEG

Keeping these properties in mind, analysis of plots for a set of color images (some of them are represented in Fig. 2) shows the following:

1) Rate-distortion curves behave individually depending on image properties; for the same QF, PSNR values can differ by up to 10 dB whilst the PSNR-HMA values have a smaller divergence - up to 5 dB, and this happens for QF about 70;
2) This means that by setting a fixed QF, one gets a sufficiently different quality of compressed images, whereas for complex structure images as, e.g., images in Figs. 2.d, the quality according to the metric PSNR is worse; however, according to the metric PSNR-HMA which characterizes the image visual quality more adequately the conclusions can slightly differ;
3) Within the limits from QF about 25 till 80, the curves behave almost linearly, and the curves are almost "parallel" to each other (this will be important for the two-step approach considered in the next Section); note that just this range of QF corresponds to the main range of interest for image compression – from PSNR > 36 dB and PSNR-HMA > 41 dB (practically invisible distortions) to PSNR from 30 dB to 36 dB and PSNR-HMA from 32 dB to 41 dB (visible but not annoying distortions) and to PSNR < 30 dB and PSNR-HMA < 32 dB (low quality of compressed images).

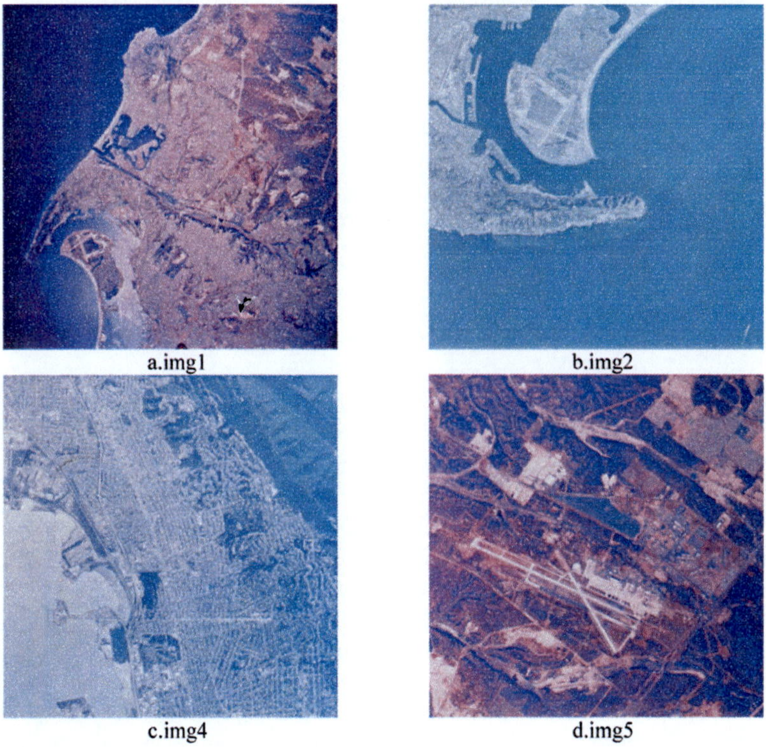

a.img1 b.img2 c.img4 d.img5

Fig. 2. Some images for which the RDCs in Fig. 1 have been obtained

Let us explain one more reason why JPEG has attracted our attention. Table 1 contains data for three standard test images of different complexity (Baboon is the most complex, and Peppers is the simplest). Three values of CR are considered – 5, 10, and 20. For CR = 5, according to PSNR, JPEG is worse than AGU [27] and BPG [28] for all three images, but JPEG is better or, at least, not worse according to visual quality metrics PSNR-HVS-M (https://ponomarenko.info/psnrhvsm.htm) and FSIM [29]. For CR = 10, where distortions are seen for the image Baboon and not seen for two other test images, JPEG is still better according to visual quality metrics. And only for CR = 20, JPEG is obviously worse than its two counterparts. Therefore, JPEG is comparable to modern coders if CR for a given image is such that introduced distortions are not noticeable or can hardly be seen. Recall that this is the case of our main interest.

Table 1. Comparison of performance for three DCT-based coders

CR	Image	AGU			BPG			JPEG		
		PSNR	PSNR-HVS-M	FSIM	PSNR	PSNR-HVS-M	FSIM	PSNR	PSNR-HVS-M	FSIM
5	Baboon	33.3	39.0	0.99	34.2	41.0	0.992	29.1	47.6	0.994
	Barbara	42.8	49.3	0.998	43.6	51.6	0.998	37.9	53.5	0.998
	Peppers	41.4	47.0	0.997	42.9	49.8	0.998	37.7	56.5	0.998
10	Baboon	28.4	31.1	0.969	29.0	32.8	0.973	25.4	34.6	0.974
	Barbara	37.7	40.9	0.992	38.7	42.919	0.994	31.9	41.3	0.989
	Peppers	37.3	41.2	0.992	38.7	43.772	0.995	35.5	44.9	0.995
20	Baboon	25.2	26.1	0.936	25.3	26.813	0.937	22.8	24.7	0.918
	Barbara	33.3	34.1	0.977	34.2	35.665	0.981	26.9	30.9	0.954
	Peppers	34.8	37.1	0.985	36.1	39.453	0.989	33.0	35.8	0.979

3 Two-step Approach for JPEG and its Accuracy

Given the analysis in the previous section, it is necessary to conduct experiments on the JPEG coder to test whether the two-step method works well for this coder, in particular, to study the method's accuracy for remote sensing images.

First, some remote sensing images have been chosen as the basic image set (partly shown in Fig. 2), which contains both complex images and simple images in terms of structure. Second, a serial JPEG compression/decompression is conducted to get the visual quality data corresponding to each QF (5~100), then the average RDC can be drawn based on these data. Third, the two-step compression method is utilized for 12 other remote sensing images (of the same size) with the average RDC. During the validation experiment, PSNR metrics are employed, and two desired visual quality values are pre-set, namely 30 and 35 dB, which roughly correspond to the obvious distortions and hardly visible distortions, respectively. The statistical data are shown in Table 2.

Here VAR_{init} denotes the variance of PSNR values provided in the first step (i.e., using fixed QF_{init} obtained from the average RDC), and VAR_{pro} is the one in the second step (where the individual QF_{cor} is set for each particular mage), as well as $Mean_{pro}$. The data shown in Table 2 suggest that the two-step method of JPEG works well in the

Table 2. Statistical result of JPEG coder for remote sensing images according to PSNR (dB).

Desired visual quality	VAR_{init}	VAR_{pro}	$Mean_{pro}$
30 dB	2.572	0.215	29.804
35 dB	2.148	0.862	35.697

considered conditions since the variance after the second step compression has dropped by approximately an order of magnitude.

Let us show some particular examples of how the two-step method works. The collected data are presented in Table 3. Here, $PSNR_{init}$ is PSNR calculated after compression with QF_{init} and decompression) and $PSNR_{prov}$ is PSNR provided after compression using the corrected QF_{cor}.

Table 3. Illustration of two-step method operation for $PSNR_{des} = 30$ dB

Test image	QF_{init}	$PSNR_{init}$	QF_{cor}	$PSNR_{prov}$	Time (s)
Im13	11	29.2755	14	29.8388	0.5912
Im14	11	28.2157	18	29.3472	0.5911
Im15	11	29.2199	14	29.7703	0.5899
Im16	11	29.9546	11	29.9546	0.5736
Im17	11	28.4708	17	29.9655	0.5944
Im18	11	28.1298	18	29.2377	0.5703
Im19	11	30.0611	11	30.0611	0.5677
Im20	11	31.5521	5	29.0448	0.5656
Im21	11	28.4156	17	29.6065	0.5872
Im22	11	30.3228	10	30.0752	0.5651
Im23	11	33.7058	5.5	30.8081	0.5752
Im24	11	29.9366	11	29.9363	0.5614

Analysis shows the following:

1) After the first step, there are several values of $PSNR_{init}$ that are already very close to $PSNR_{des} = 30$ dB; in particular, this happens for Im16, Im19, and Im24; then, $QF_{cor} = QF_{init}$ and the second step is not needed since, after the first step, we already have the compressed image;
2) After the first step, there are also several images with $PSNR_{init} > PSNR_{des}$ (e.g., images Im20 and Im23) and images with $PSNR_{init} < PSNR_{des}$ (e.g., Im14 and Im18), for which $|PSNR_{init} - PSNR_{des}|$ is larger than 1 dB, i.e., it is desired to improve the accuracy of providing $PSNR_{des}$ at the second step;

3) In all such cases, $|PSNR_{prov} - PSNR_{des}|$ after the second step occurs to be smaller than $|PSNR_{init} - PSNR_{des}|$ and in all cases $|PSNR_{prov} - PSNR_{des}|$ is smaller than 1 dB, i.e., quite high accuracy is ensured; recall that difference in images compressed with two PSNR values that differ by less than 0.5 dB cannot be noticed whilst the difference about 0.8 dB can be hardly noticed.

Thus, the two-step procedure, in general, fulfills its task, but, in most cases, it requires two steps, i.e., two compressions and one decompression. The time of QF calculation before the second step compression is also recorded in the sixth column. This time is approximately equal to the time of JPEG compression.

In general, this can be done quite quickly if the image size is not large. However, the question is, can this be done faster?

4 Neural Network-Based Predictor

It already been noticed that the performance of image lossy compression depends on image complexity, where the term complexity can be treated in different ways. In particular, we have used image entropy to develop the adaptive two-step method [23], where images are first classified into three groups, and then different average rate-distortion curves are used for setting the initial value of the parameter that controls compression and its correction. Meanwhile, entropy is not the only parameter that is able to characterize image complexity. In addition to entropy, the authors of [30] propose to apply four other complexity descriptors: edge ratio, contrast, energy, and correlation. Each of these parameters can be calculated quite quickly and characterizes image complexity from different viewpoints.

Certainly, image compression performance for JPEG also depends on QF. Thus, our idea is that QF, together with a certain subset of parameters characterizing image complexity, can be useful in the prediction of MSE of introduced distortions (and, equivalently, PSNR). There are different options for the joint processing of several input parameters in solving prediction (output approximation) tasks [31]. This can be some regression, support vector machine, or neural network (NN) based approaches. Below, we consider the latter approach since it has demonstrated itself to be efficient in similar applications. Since we have a limited number of inputs (maximally, six) and one output (that produces predicted MSE or its analog), the neural network structure does not need to be complex. In particular, we have used basic neural network fitting for which there exist standard means for its training [32]. The Levenberg-Marquardt algorithm is utilized for training the NN, where 30 neurons are set. The image we used in training is the same as the ones used in the two-step method.

Let us characterize the accuracy of prediction by goodness-of-fit parameter R that should approach unity if the prediction is absolutely perfect. Besides, we can get the distribution of MSE prediction errors.

One example of the obtained results is given in Fig. 3. Here, Target means the axis of true values, while Output denotes the predicted value. As one can see, R values are very close to unity, and all points (shown as circles) place very close to the fitted line.

Moreover, R values are almost the same for training, test, and validation sets (as can be expected, a slightly smaller value is observed for the validation set), and, as a

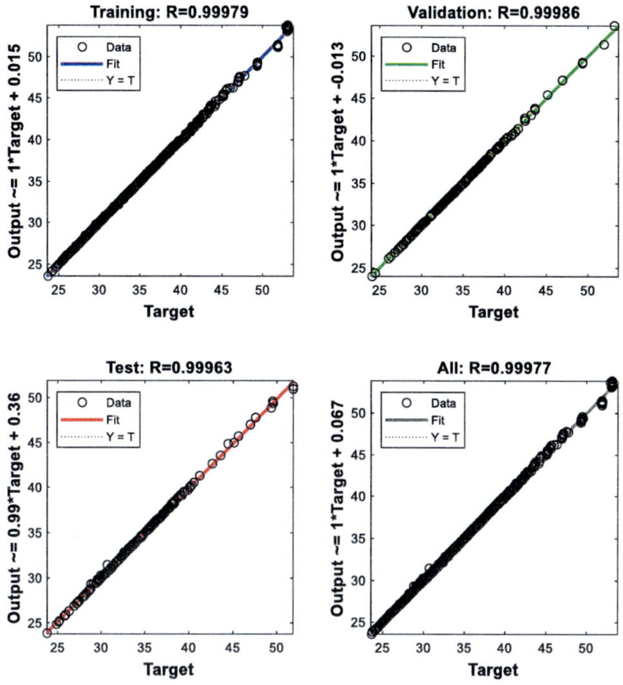

Fig. 3. Example of presenting the results for analysis carried out for the set of four input parameters, namely QF, Entropy, Contrast, and Correlation

result, R is large for all considered images (see the plot All). Slightly larger variations of scatter-plot point positions are observed for larger MSE values (smaller QF values).

It is also worth analyzing the prediction errors. The histogram of errors is shown in Fig. 4. As one can see, mainly the errors are smaller than 1.0, i.e., prediction accuracy is very high (less than a few percent of the predicted value). The larger errors are observed for larger MSE. This means that PSNR is predicted with errors less than 0.5 dB, i.e., with high accuracy acceptable in practice.

Based on the analysis of the prediction efficiency, it is worth using it to replace the first compression in the standard two-step method. Consequently, we proposed a novel hybrid method that combined the standard two-step method with an NN-based prediction algorithm. The test experiment is also conducted on the validation set, the same as in the previous section. The results of the desired PSNR equal 30 dB are shown in Table 4.

Here, the $PSNR_{pre}$ denotes the Neural Network prediction of the image to be compressed for the initial QF. As one can see, the prediction of visual quality corresponding to the initial QF is around the real value (in Table 3, third column) generally since the mean value is 31.801 dB whilst the real mean value is 29.77 dB. However, the variance differs a lot, and the prediction variance is 92.0935 dB^2 compared to the 2.572 dB^2 in real compression; the reason is that there is a large prediction error on special image #23, which has a simple structure. The provided PSNR is closer to the desired value

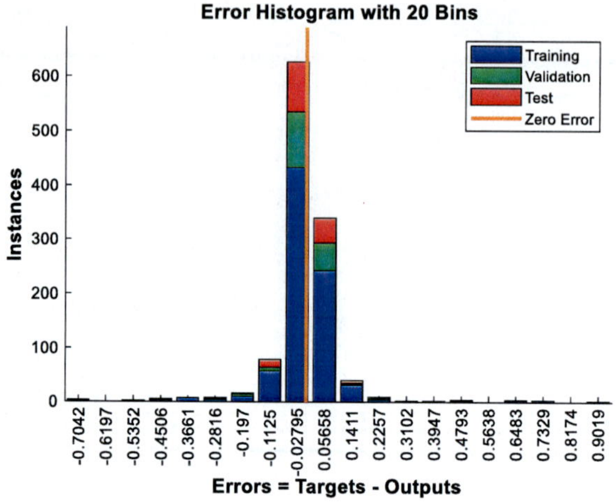

Fig. 4. Histogram of predicted MSE values for the set of input parameters as in Fig. 3.

Table 4. Illustration of the novel hybrid method operation for $PSNR_{des} = 30$ dB

Test image	*PSNRpre*	*PSNRpro*	*Time(s)*
Im13	34.0696	27.2019	0.14137
Im14	27.2715	29.0789	0.14651
Im15	36.482	27.2982	0.13439
Im16	33.4057	27.2312	0.12228
Im17	26.2258	29.7615	0.16783
Im18	22.3994	28.9737	0.13615
Im19	27.2902	31.0946	0.17625
Im20	28.4903	32.4862	0.12685
Im21	25.8657	29.4474	0.13032
Im22	29.2181	31.2715	0.1263
Im23	59.5496	30.8081	0.12385
Im24	31.3381	30.9357	0.12543
VAR	92.0935	3.0875	
MEAN	31.801	29.632	0.1381

after the correction of QF. The variance is decreased to 3.0875 dB^2, which is 1/30 of the prediction ones.

The result makes clear that the hybrid method has considerable accuracy and strong robustness, even for the special image. The prediction accuracy needs to be further improved definitely, and one possible solution is to increase the training samples in the neural network. Another advantage of the hybrid method is the improvement of time efficiency, the time has dropped to nearly 1/4 of that of the previous two-step method, and it is sufficiently smaller than the time of JPEG compression.

5 Conclusions

It is shown that PSNR for JPEG applied to color or three-channel RS images can vary in rather wide limits for the same QF. Then, it is often needed to provide a desired quality of compressed images by individual and proper setting of QF for each particular image taking into account its properties. We show that one way is to use the two-step procedure that has been earlier proposed and tested for other lossy coders. The two-step procedure performs rather well for typical PSNR values of 30 and 35 dB when introduced distortions are either not annoying or invisible. In addition, we provide prerequisites for even faster reaching a desired quality by exploiting the simple structure NN that uses a set of simple parameters characterizing image complexity. Further optimization of the input parameter set and methodology of training is needed.

References

1. Li, D., Shao, Z., Zhang, R.: Advances of geo-spatial intelligence at LIESMARS. Geo-Spat. Inf. Sci. **23**(1), 40–51 (2020)
2. Zhang, B., et al.: One-two-one networks for compression artifacts reduction in remote sensing. J. Photogramm. Remote Sens. **145**, 184–196 (2018)
3. Christophe, E.: Hyperspectral data compression tradeoff. In: Prasad, S., Bruce, L., Chanussot, J. (eds.) Optical Remote Sensing. Augmented Vision and Reality, vol. 3. Springer, Heidelberg (2011). https://doi.org/10.1007/978-3-642-14212-3_2
4. Ozah, N., Kolokolova, A.: Compression improves image classification accuracy. In: Meurs, M.J., Rudzicz, F. (eds.) Advances in Artificial Intelligence. Canadian AI 2019. Lecture Notes in Computer Science, vol. 11489. Springer, Cham (2019). https://doi.org/10.1007/978-3-030-18305-9_55
5. Penna, B., Tillo, T., Magli, E., Olmo, G.: Transform coding techniques for lossy hyperspectral data compression. IEEE Trans. Geosci. Remote Sens. **45**(5), 1408–1421 (2007)
6. Hussain, A.J., Al-Fayadh, A., Radi, N.: Image compression techniques: a survey in lossless and lossy algorithms. Neurocomputing **300**, 44–69 (2018)
7. Balasubramanian, R., Ramakrishnan, S.S.: Wavelet application in compression of a remote sensed image. In: 2013 the International Conference on Remote Sensing, Environment and Transportation Engineering (RSETE 2013), pp. 659–662 (2013)
8. Syahrul, E.: Lossless and nearly-lossless image compression based on combinatorial transforms. Université de Bourgogne (2011)
9. Lukin, V., et al.: Lossy compression of multichannel remote sensing images with quality control. Remote Sens. **12**(22), 3840 (2020)
10. Lukin, V., Zemliachenko, A., Krivenko, S., Vozel, B., Chehdi, K.: Lossy compression of remote sensing images with controllable distortions. Satellite Information Classification Interpretation, pp. 27–43. Intech Open, London (2018)
11. Blau, Y., Michaeli, T.: Rethinking lossy compression: the rate-distortion-perception tradeoff. In: International Conference on Machine Learning, pp. 675–685. PMLR (2019)

12. Valsesia, D., Magli, E.: A novel rate control algorithm for onboard predictive coding of multispectral and hyperspectral images. IEEE Trans. Geosci. Remote Sens. **52**(10), 6341–6355 (2014)
13. Zabala, A., Pons, X.: Impact of lossy compression on mapping crop areas from remote sensing. Int. J. Remote Sens. **34**(8), 2796–2813 (2013)
14. Ayoobkhan, M.U.A., Chikkannan, E., Ramakrishnan, K.: Lossy image compression based on prediction error and vector quantisation. EURASIP J. Image Video Process. **2017**(1), 1–13 (2017)
15. Zabala, A., Pons, X.: Effects of lossy compression on remote sensing image classification of forest areas. Int. J. Appl. Earth Obs. Geoinf. **13**(1), 43–51 (2011)
16. Radosavljević, M., et al.: Lossy compression of multispectral satellite images with application to crop thematic mapping: a HEVC comparative study. Remote Sens. **12**(10), 1590 (2020)
17. Wallace, G.K.: The JPEG still picture compression standard. Commun. ACM **34**(4), 30–44 (1991)
18. Stojanović, N.M., Pavlović, B.Z., Bondžulić, B.P., Petrović, V.S.: Picture-wise just noticeable difference prediction model for JPEG image quality assessment. Vojnotehnicki glasnik/Mil. Tech. Cour. **70**(1), 62–86 (2022)
19. Zemliachenko, A., Lukin, V., Ponomarenko, N., Egiazarian, K., Astola, J.: Still image/video frame lossy compression providing a desired visual quality. Multidimension. Syst. Signal Process. **27**(3), 697–718 (2016)
20. Krivenko, S., Zriakhov, M., Lukin, V., Vozel, B.: MSE and PSNR prediction for ADCT coder applied to lossy image compression. In: 2018 IEEE 9th International Conference on Dependable Systems, Services and Technologies (DESSERT), pp. 613–618. IEEE, Piscataway (2018)
21. Li, F., Krivenko, S., Lukin, V.: Atwo-step approach to providing a desired visual quality in image lossy compression. In: 2020 IEEE 15th International Conference on Advanced Trends in Radioelectronics, Telecommunications and Computer Engineering (TCSET), pp. 502–506. IEEE, Piscataway (2020)
22. Krivenko, S., Li, F., Lukin, V., Vozel, B., Krylova, O.: Prediction of visual quality metrics in lossy image compression. In: 2020 IEEE 40th International Conference on Electronics and Nanotechnology (ELNANO), pp. 478–483 (2020)
23. Proskura, G., Rubel, O., Lukin, V.: On classifier learning methodologies with application to compressed remote sensing images. Radioelectron. Comput. Syst. (3), 174–189 (2022)
24. Bradski, G.: The openCV library. Dr. Dobb's J.: Softw. Tools Prof. Program. **25**(11), 120–123 (2000)
25. Jiang, Y., Pattichis, M.S.: JPEG image compression using quantization table optimization based on perceptual image quality assessment. In: 2011 Conference Record of the Forty Fifth Asilomar Conference on Signals, Systems and Computers (ASILOMAR), pp. 225–229 (2011)
26. Ponomarenko, N., Ieremeiev, O., Lukin, V., Egiazarian, K., Carli, M.: Modified image visual quality metrics for contrast change and mean shift accounting. In: 2011 11th International Conference the Experience of Designing and Application of CAD Systems in Microelectronics (CADSM), pp. 305–311 (2011)
27. Ponomarenko, N., Lukin, V., Egiazarian, K., Astola, J.: DCT based high quality image compression. In: Kalviainen, H., Parkkinen, J., Kaarna, A. (eds.) SCIA 2005. LNCS, vol. 3540, pp. 1177–1185. Springer, Heidelberg (2005). https://doi.org/10.1007/11499145_119
28. Albalawi, U., Mohanty, S.P., Kougianos, E.: A hardware architecture for better portable graphics (BPG) compression encoder. In: 2015 IEEE International Symposium on Nanoelectronic and Information Systems, pp. 291–296. IEEE, Indore (2015)
29. Zhang, L., Zhang, L., Mou, X., Zhang, D.: FSIM: a feature similarity index for image quality assessment. IEEE Trans. Image Process. **20**(8), 2378–2386 (2011)

30. Zhang, Y., Zhang, Z., Wang, X., Wang, X., Ge, J., Bian, F.: An adaptive infrared image preprocessing method based on background complexity descriptors. In: 2018 Eighth International Conference on Instrumentation & Measurement, Computer, Communication and Control (IMCCC), pp. 344–349 (2018)
31. Rubel, O.S., Lukin, V.V.: An improved prediction of DCT-based image filters efficiency using regression analysis. Inf. Telecommun. Sci. **1**, 30–41 (2014)
32. Zhu, L., Dai, W., Wang, X.: Tool wear prediction based on Canonical Correlation Analysis and Neural Network fitting method. In: 2020 Global Reliability and Prognostics and Health Management (PHM), pp.1–5. IEEE, China (2020)

GaoFen-1 Performance and Status After Nine Years on Orbit

ChunLing Lu, Qian Yu(✉), and Yuan Yuan

DFH Satellite Co., Ltd., Beijing 100094, China
{chunl1501,yu39170}@sina.com

Abstract. Gaofen-1 is a new generation long-life small remote sensing satellite in China. It has been observing the earth for more than nine years after being launched into 645 km sun-synchronous orbit on Apr. 26, 2013. It is worth mentioning that Gaofen-1 is a novel satellite with 2 high resolution cameras (PMS1 and PMS2) and 4 wide swath cameras (WFV1 to WFV4). Two PMS cameras can provide PAN 2 m and 4 bands 8 m images with 69 km total swath. 4 WFV cameras can acquire 4 bands 16 m images with 842 km total swath. During 9 years in orbit, what changes in the performance of remote sensors and the satellite health status have been concerned. This paper writes from satellite evolution of orbit, main subsystems of platform, and so on. Some important events and operations are mentioned. Some improvements of payload are introduced. In particular, the yaw 90° radiometric calibration (side-slither radiometric calibration), TDI stage and gain adjusting, and focus adjustment to ensure the imaging quality are shown. The 9-year change of radiometric calibration coefficient from the ground system is introduced. At last, in order to ensure the illumination condition of remote sensor imaging, the experience of satellite orbit inclination maneuver to changing the local descending node time (>10:00AM) is introduced. All efforts are made to ensure that the satellite has a good condition. All the analysis results show that the satellite works in good status and its excellent performance shows that it will continue to running well in the future.

Keywords: high resolution · wide swath · image quality monitoring · side-slither radiometric calibration · optical remote sensing satellite

1 Introduction

GaoFen-1 carried two high resolution cameras (PMS) and four wide swath cameras (WFV). PMS with PAN 2 m GSD, MS 8 m GSD four bands from the visible to near infra-red, 69 km total swath. WFV with 16 m GSD four bands from the visible to near infra-red, 842 km total swath. It has a 5~8-year design life, and now it has already been running on orbit for 9 years. In 645 km sun-synchronous orbit, every day the spacecraft orbits the earth about 14.75 circles, suffers high and low temperature variations, charges and discharges about 48453 circles in 9 years. It had gone through 9 summer solstices and winter solstices in the low earth orbit. Platform and payloads are all running stably in space environment. All subsystems' conditions are very good and have realized perfect

H. P. Urbach and H. Jiang (Eds.): ISSOIA 2022, SPPHY 295, pp. 20–29, 2023.
https://doi.org/10.1007/978-981-99-4098-1_3

capability without component faults. Many technical papers and characteristics and typical applications of this satellite are shown in reference [1–3].

2 Key Events of GaoFen-1

1) At the beginning of the mission, the satellite camera's focal plane was adjusted to make images clearer; Then, based on images' grey histograms, camera's Time delay and integration (TDI) stages and gain were adjusted to widen the greyscale dynamic range distribute. Then, the side-slither radiometric calibration was completed by yawing the satellite 90° [1–3]. Later, temperature and power supply setting were adjusted according to the system demands.
2) The "Early Extended Mission" [4] phase is from 2014 to 2016. The satellite operates orbit maintenance about once every 2 months since Apr. 2014; The focal plane is adjusted every one or two year [5]; it also conducts ground radiometric calibration and synchronous observing. The calibration coefficient can be submitted every year.
3) The "Extended Mission" is from 2017 to 2018. The spacecraft not only was uploaded the information for three new receiving stations, but also extended imaging time (from 80 min to 120 min) and storage time (from 20 min to 25 min) since May. 2017 (Fig. 1).
4) The "Life Extending Mission" is from 2018 until now. In June 2022, the inclination maintenance was operated by performing the inclination maneuver to keep the satellite's descending node time over 10:00 a.m. Now all components function well and run stably without switching to backups.

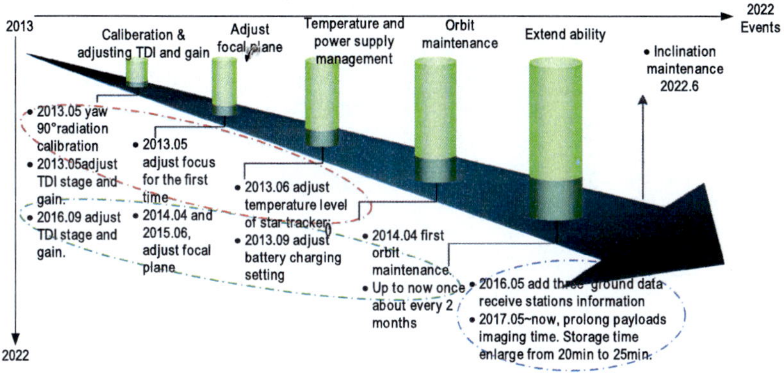

Fig. 1. Key events of the satellite in orbit

3 Performance and Status of GaoFen-1

3.1 Orbit

Two types of orbit maneuvers to ensure the safety of the satellite: ①In-plane maintenance [6]: maintain altitude; Collision avoidance or risk mitigation. ②Inclination maintenance: orbit inclination maneuver.

In-plane Maintenance

GaoFen-1 initial altitude was 1.68 km higher than the reference orbit [5]. In order to avoid the fuel consumption, GaoFen-1 dragged and reduced altitude naturally rather than orbit maneuver to reduce the altitude in the first year. In Apr. 2014, the first orbit maneuver was conducted to maintain the 645 km reference orbit altitude. After that, orbit maintenance has been conducted every two or three months in order to increase semi-major axis (SMA) and maintain the target phase point on reference orbit (Fig. 2). If the SMA is below the reference orbit caused by drag, the nadir track will drift in the positive phase direction. The relationship between relative SMA, relative phase and spacecraft drag is shown in Fig. 3 [7]. Orbit maintenance in fact is altitude maintenance.

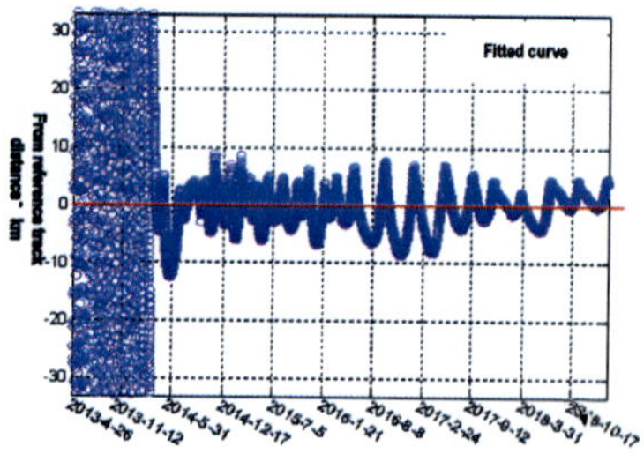

Fig. 2. Satellite orbit track statistics

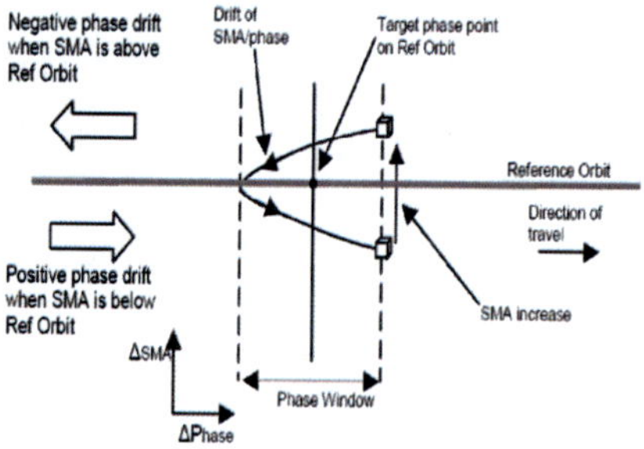

Fig. 3. Satellite orbit maneuver trajectory

Orbit maintenance ensures the nadir track always bias ±10 km of reference track, as shown in Fig. 3. By ~±35°roll, 2 m&8 m images can revisit any area in 4days. Four WFVs' 16 m images can cover global in 4days without roll. All sensors' imagers have good revisit and return characteristics.

Collision avoidance or risk mitigation maneuvers are executed if the risk of a collision or close approach to space debris. This maneuver is equivalent to perform orbit maintenance before the bias period.

Inclination Maintenance

The descending trend of nine years' orbit inclination curve and the local descending node time (LDNT) are shown in Fig. 4. In June 2022, the local descending node time (LDNT) was varied lower than 10:00AM. This caused the solar array current to drop 9 A lower than the pick 47 A, and the power was reduced by 275 W (lower 20%). These phenomena were caused by the solar perturbation which brought the lower inclination and lower LDNT. An effective solution was to raise the inclination. After 11 days inclination maneuver, delta-inclination (Δi) was raised by 0.28°, 18 kg of fuel was consumed and 12 kg of fuel was left. Figure 4 shows that the inclination and LDNT increases after inclination maintenance. The effect was that in the next 6 years, there would be no inclination maintenance, and LDNT will gradually drift to the peak of 10:30AM in 2025 (Fig. 5). Good solar illumination brought the solar array current increasing from 40.59 to 43 A. All efforts had achieved the expected results.

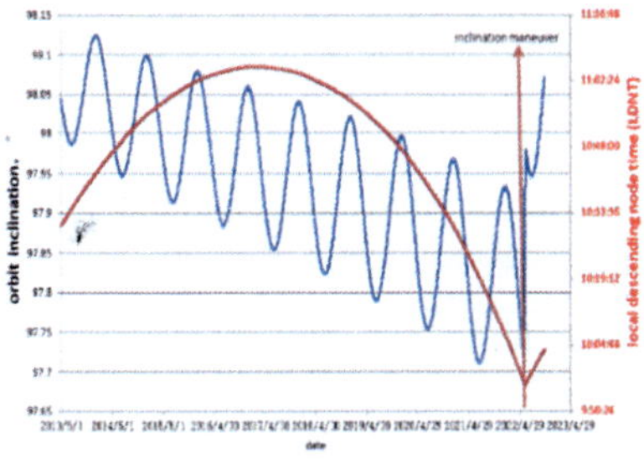

Fig. 4. Inclination and local descending node time (LDNT) curve

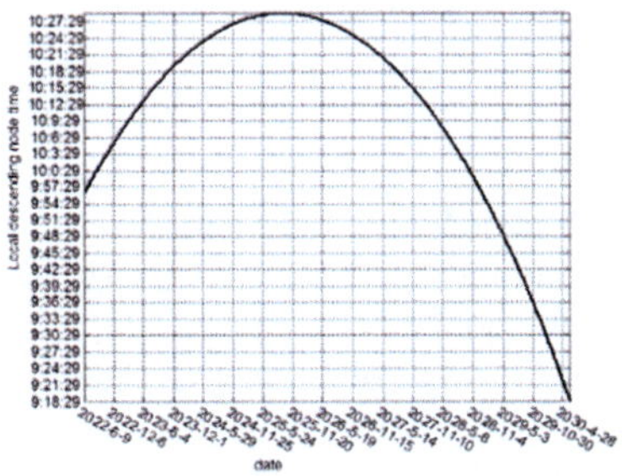

Fig. 5. Future 6-year's LDNT curve

3.2 Temperature

The temperature of the electronic equipment, which is controlled by passive thermal control, is very stable within more than 12 °C margin. The temperature of the equipment is higher near winter solstice than near summer solstices. In the first year the temperature trend raised 1.0~1.4 °C, then slightly raised about 4.05 °C in 9 years. The sun constant affects the temperature with a variation of about 3 °C, as shown in Fig. 6. All phenomena are related to the degradation of the thermal control coatings.

The temperature of cameras are controlled by active thermal control with stable curves (Fig. 6). A stable temperature environment ensures the stable image quality.

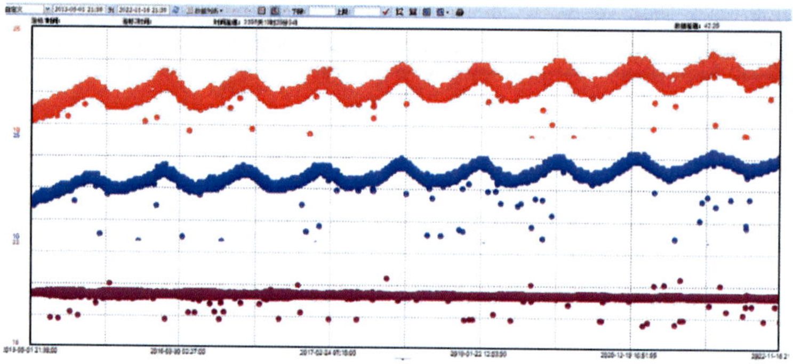

Fig. 6. The temperature of passive and active thermal control equipment

3.3 Position Precision and Attitude Control Accuracy

Attitude and orbit control subsystem (AOCS) utilizes gyros and star trackers to realize attitude determination. It also utilizes momentum wheels for attitude control. GaoFen-1 can point nadir or off-nadir. GPS receiver performs positioning, orbit determination and automatically provides time synchronization pulse. Position precision and attitude control accuracy are given in Table 1. High precision of positioning and attitude stability make images have high geometric accuracy.

3.4 Relative Radiometric Calibration

GaoFen-1 adopts an attitude yaw of 90° to realize relative radiometric calibration using a side-slither technique. In this configuration, each detector on the focal plane is positioned parallel to the ground-track direction thereby exposing each detector to the same segment of the ground [8]. This method enables both pixel-to-pixel and chip-to-chip differences can be detecting quickly with no need for a large calibration site.

The side-slither calibration method is shown in Fig. 7(a) and 7(b). In order to have a different gray level, a side-slither calibration site with high-middle-low gray is chosen rather than a radiometric flat-field. The side-slither derives detector correction parameters

Table 1. Position precision and attitude control accuracy

No.	Items	requirement	In orbit test (nadir)	In orbit test (off nadir)
1	GPS receiver position precision (rms)	≤10 m	≤7 m (rms)	
2	Roll, Pitch, Yaw point accuracy (3σ)	≤0.1°	≤0.0004°	≤0.003°
3	Roll, Pitch, Yaw attitude stability (3σ)	≤0.0005°/s	≤0.0002°/s	≤0.0003°/s

by using the histograms method (Fig. 7(c)) [9]. More gray stage lines indicate that wider distribution histograms for each pixel can be obtained [10, 11]. A significant improvement in image correction was achieved. Through this method, each detector's non-uniformity and average band response also can be obtained efficiently.

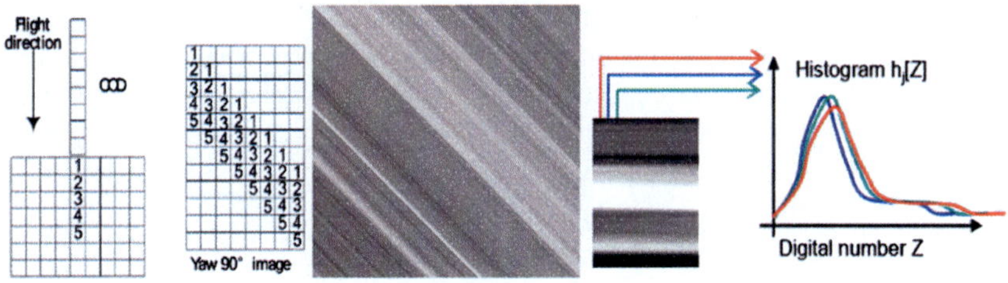

(a) (b) (c)

Fig. 7. (a) principle of side-slither calibration by yaw 90°; (b) raw side slither image; (c) histograms method.

3.5 TDI Stage and Gain Adjustment

PAN Arrays: TDI Stages 6, 12, 24, 32, 48; MS Arrays: B, G, R, NIR; TDI Stages 04, 08, 16, 24, 32. In Sep. 2016, TDI stages and gain were adjusted (Table 2). The effect was very obvious. The grey distribution and dynamic range of imagery were widened. The ratio of change was about 1.58 times.

Table 2. Adjust TDI stages and Gain during in orbit

Band	Pre-launch	2013.5.11	2016.9.27	Ratio of Change (2016 vs. 2013)
	TDI stages & gain			
P	TDI 12–0 dB	TDI 06–4 dB	TDI 12–2 dB	Increase 1.58 times
B1	TDI 24–2 dB	TDI 24–0 dB	TDI 24–4 dB	Increase 1.58 times
B2	TDI 16–0 dB	TDI 08–4 dB	TDI 16–2 dB	Increase 1.58 times
B3	TDI 08–2 dB	TDI 08–0 dB	TDI 12–0 dB	Increase 1.5 times
B4	TDI 04–2 dB	TDI 04–0 dB	TDI 04–4 dB	Increase 1.58 times

3.6 Absolute Radiometric Calibration

Two methods are used to determine the absolute radiometric calibration of the sensor. Pre-launch laboratory calibration and after-launch ground site calibration. The former provides initial values, while the latter provides updated values year by year. In order to calculate the radiance, the sensor's relative spectral response curves are shown in Fig. 8. The gain coefficient of absolute radiometric calibration not only presents the relationship between incident radiance and detector digital output, but also show radiometric resolution by $W{\cdot}m^{-2}{\cdot}sr^{-1}{\cdot}\mu m^{-1}/DN$. 10 bit quantization improves the radiometric resolution much more than 8bit quantization. All pixels are working without failure pixel.

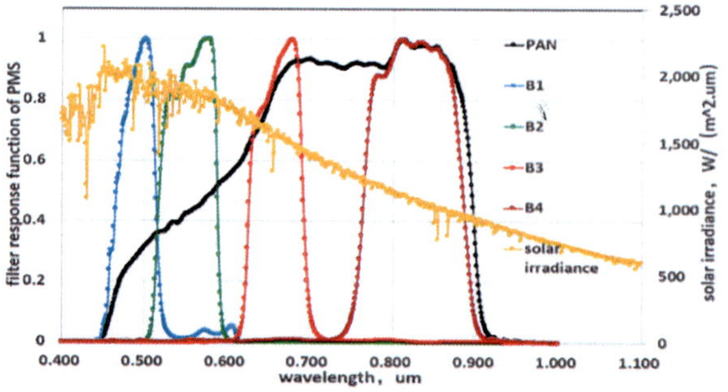

Fig. 8. PMS1 relative spectral response and solar irradiance

In Sep. 2016, the sensor's TDI stages and gain were adjusted for increasing average gray level and enlarge the dynamic range of the histograms. Therefore, the gain coefficient of ground absolute radiometric calibration can be seen changing dramatically in 2017 (Fig. 9). The lower the gain coefficient, the higher the radiometric resolution (Fig. 10).

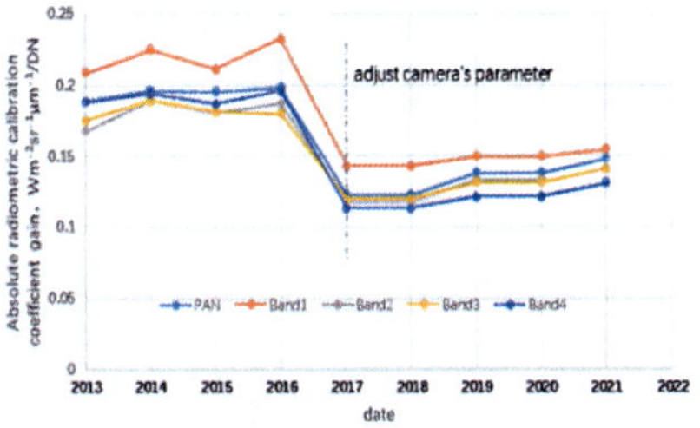

Fig. 9. The gain of PMS1 absolute radiometric calibration.

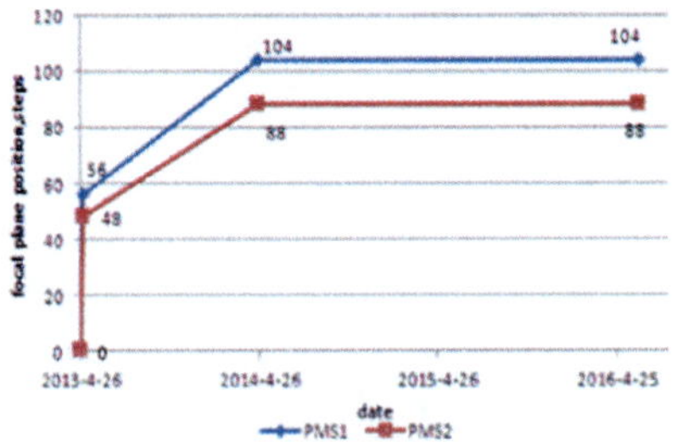

Fig. 10. Focal plane position shift

3.7 Focus Calibration

Over the 9 years' trend analysis of the platform and subsystem, GaoFen-1 works steadily and has excellent performance in orbit indicating the superiority of its design. The orbit inclination maneuver effect is good, the solar array power generation current increases, and the local descending node time rises slowly. The peak will reach 10:30 in 2025, and the inclination will not be maintained for at least six years in the future. The remaining fuel is 12 kg, meeting the requirements of future orbit maintenance. The ground illumination meets the imaging requirements. The focus, TDI stages and gain of cameras are adjusted whenever necessary. The image quality of the cameras are stable. All equipment works well without any reversion to backups. The mission has exceeded its original 5 years' lifetime and there is full confidence to running longer on orbit. It will benefit other missions for the years to come.

3.8 Image Quality Monitoring

We use SongShan calibration site to realize image quality monitoring, such as spatial resolution, MTF, band response, SNR and radiometric resolution. Gradually take photos of SongShan target site, any change can be find from the series of target images. The different years' images of painted concrete radial edge target and edge response of traps are shown in Fig. 11. The radial edge target is a 96 m radial target, 112° angle, which can meet the requirement of measuring <4 m spatial resolution. The edge response of traps is 35 m * 35 m. The reflectance of white and black target is 60% and 5% [11].

Using these target, the PAN spatial resolution is 2.35 m (2 m GSD). From the reflective tarp, the edge response curve and MTF can be obtained. The average PAN MTF is 0.112. The different years' target images (Fig. 11) show this satellite has stably and superiority image quality.

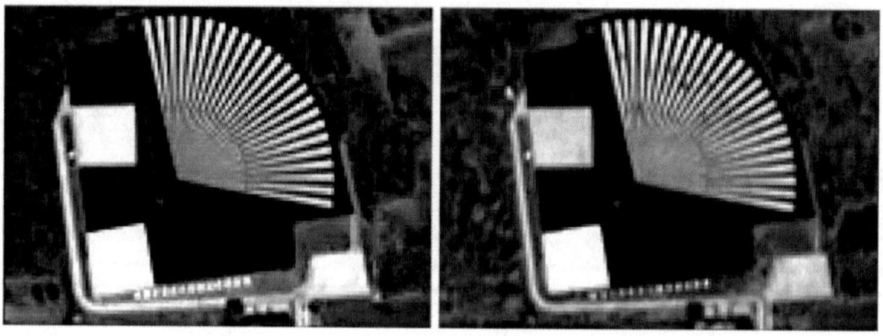

(a) 2015.11.2 PAN1; (b) 2016.5.29 PAN1

(c) 2017.12.24 PAN1; (d) 2018.10.11 PAN2

Fig. 11. PAN images of SongShan calibration site

4 Summary

Over the 9 years' trend analysis of the platform and subsystem, GaoFen-1 works steadily and has excellent performance in orbit indicating the superiority of its design. The orbit inclination maneuver effect is good, the solar array power generation current increases, and the local descending node time rises slowly. The peak will reach 10:30 in 2025, and the inclination will not be maintained for at least six years in the future. The remaining fuel is 12 kg, meeting the requirements of future orbit maintenance. The ground illumination meets the imaging requirements. The focus, TDI stages and gain of cameras are adjusted whenever necessary. The image quality of the cameras are stable. All equipment works well without any reversion to backups. The mission has exceeded its original 5 years' lifetime and there is full confidence to running longer on orbit. It will benefit other missions for the years to come.

References

1. Lu, C., Bai, Z., Li, C., Lv, D.: GF-1 satellite high resolution & wide swath mission design and technology characteristics. In: Proceedings of the International Astronautical Congress, IAC, vol. 5, pp. 3586–3590 (2014)
2. Lu, C., Bai, Z.: Characteristics and typical applications of GF-1 satellite. In: 2015 IEEE International Geoscience and Remote Sensing Symposium, Milan, Italy, November 2015, pp. 1246–1249 (2015)
3. Bai, Z.: GF-1 satellite design and technology characteristics. Spacecraft Engineering **23** (2014)
4. Middleton, E.M.: The earth observing one (EO-1) satellite mission: over a decade in space. IEEE J. Sel. Top. Appl. Earth Obs. Remote Sens. **6**(2), 243–256 (2013)
5. Lu, C., Bai, Z.: GF-1 Satellite on orbit test and evaluation. The Small Satellites & Services Symposium (4S), Majorca, Spain
6. Ihlen, V.: Landsat 7 (L7) Data Users Handbook, LSDS-1927 Version 2.0, pp. 16–17 (November 2019)
7. Gebbie, J., Pollard, M., Kadhem, H., Boland, L.: Spacecraft constellation deployment for the Rapid eye earth observation system, IAC-09-B4.3.11, 60th IAC, 12–16 October 2009, Daejeon, Republic of Korea (2009)
8. Anderson, C., Naughton, D., Brunn, A., Thiele, M.: Radiometric correction of RapidEye imagery using the on-orbit side-slither method. In: Proceedings of SPIE, vol. 8180 (2011)
9. Kubik, P., Pascal, V.: AMETHIST: a method for equalization thanks to histograms. In: Sensors, Systems, and Next-Generation Satellites VIII, Proceedings of SPIE 2004, vol. 5570 (2004)
10. Henderson, B.G., Krause, K.S.: Relative radiometric correction of QuickBird imagery using the side-slither technique on orbit. SPIE **5542**, 426–433 (2004)
11. Lebegue, L., Greslou, D.: Pleiades-HR image quality commissioning foreseen methods, pp. 1675–1676. IEEE (2010)
12. Li, Z.Z., Xu, W., Fu, Q., Min, X., Zhang, L.M.: Building and application for China SongShan artificial target site. J. Atmos. Environ. Opt. **9**(2), 81–89 (2014)

An Infrared Ship Detection Algorithm Based on VIMI Infrared Multispectral Images

Li Yan(✉)

Beijing Institute of Space Mechanics and Electricity, Beijing, China

Abstract. The Visual and Infrared Multispectral Imager (VIMI) on the GF-5 Satellite and the Hyperspectral Observation Satellite had a wide spectral range (0.45 μm~12.5 μm). Images with 12 bands from visible spectrum to thermal infrared spectrum could be obtained simultaneously based VIMI. It can monitor the maritime traffic in ports and related sea areas. A modified Infrared ship detection algorithm based on multiple infrared bands images was proposed in this paper. This algorithm was divided into two steps: Sea-Land segmentation; ship detection. The basic principle of infrared multispectral target recognition is the quantitative comparison and analysis of the spectral vectors and the standard spectral database. A modified algorithm was proposed in this paper to detect the ships. In the modified algorithm, a combination spectral operator was constructed with the spectral angle mapper and the spectral information entropy. This operator took use of their information complementary and creates their advantages to overcome the disadvantages of traditional SAM information divergence. Finally, the effectiveness of this algorithm was proved by the experiments with VIMI infrared images.

Keywords: VIMI · infrared · ship detection

1 Overview

GF-5 Satellite was the only satellite which could get high spectral resolution and high space resolution in the High resolution Special project in China. GF-5 was launched on May 9, 2018. It was the first hyperspectral and full spectrum satellite that could realize comprehensive observation of the atmosphere and the land. It was also the highest spectral resolution satellite, the highest quantitative performance satellite, the most detection means satellite, the longest design life satellite in China. The Hyperspectral Observation Satellite was the follow-up satellite of the GF-5 Satellite.The Visual and Infrared Multispectral Imager (VIMI) was the main payload on the GF-5 Satellite and the Hyperspectral Observation Satellite. VIMI had a wide spectral range (0.45 μm~12.5 μm). Images with 12 bands from visible spectrum to thermal infrared spectrum could be obtained simultaneously based VIMI [1]. The thermal infrared spectrum with split-window of retrieving land surface temperature on VIMI could get 40 m images. Its spectrum was set as follows (Table 1).

H. P. Urbach and H. Jiang (Eds.): ISSOIA 2022, SPPHY 295, pp. 30–37, 2023.
https://doi.org/10.1007/978-981-99-4098-1_4

Table 1. The Spectrum of VIMI

Band	Spectral range	GSD
B1	442.6 nm–515.6 nm	20 m
B2	513.0 nm–586.3 nm	
B3	621.2 nm–685.0 nm	
B4	761.8 nm–882.6 nm	
B5	1.551 μm–1.779 μm	
B6	2.067 μm–2.353 μm	
B7	3.465 μm–3.897 μm	40 m
B8	4.769 μm–4.969 μm	
B9	7.963 μm–8.440 μm	
B10	8.409 μm–8.893 μm	
B11	10.446 μm–11.315 μm	
B12	11.413 μm–12.495 μm	

VIMI could obtain 12 bands images from visible spectrum to thermal infrared spectrum simultaneously. The port environment and large ships sailing on the sea or entering the port could be monitored all the day based on VIMI. Ships are important transport carriers at sea. Ship detection has a wide application prospect in civil field. It could monitor the maritime shipping and traffic in specific ports and related sea areas. It could also cooperate with the safety management department to monitor and combat the illegal activities such as smuggling and piracy [2].

In this paper, a modified Infrared ship detection algorithm based on multiple infrared bands images was proposed. The effectiveness of this algorithm was proved by the experiments with VIMI infrared images.

2 Ship Detection

In this paper, ship detection mainly included two steps: Sea-Land segmentation and target detection.

2.1 Sea-Land Segmentation

In a typical remote sensing image with the port, the gray distribution of the sea area was relatively uniform. The temperature of the sea area in daytime is generally lower than the land. The situation was reversed at night. The distribution of land features was complete. It included some areas similar to the sea areas, such as small rivers, vegetation, lakes, etc. According to these characteristics of the port images, the sea-land segmentation process of infrared images in this paper was shown as the figure below [3, 4] (Fig. 1).

The maximum between-cluster variance method proposed by the Japanese scholar OTSU in 1979 was an adaptive threshold determination method, also known as OTSU. It

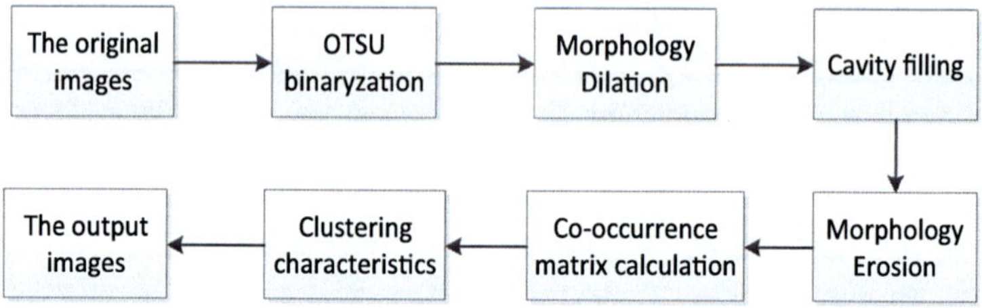

Fig. 1. The Sea-Land segmentation flow chart

was a kind of threshold segmentation method. The maximum variance between different pixel classes was calculated as the segmentation threshold [5, 6].

This method reduced the intra-class variance as much as possible, and the sum of various intra-class variances can be calculated as formula (1).

$$\sigma_w^2(t) = w_1(t)\sigma_1^2(t) + w_2(t)\sigma_2^2(t) \tag{1}$$

In the above formula, $w_i(t)$ was the proportion of class i divided by the threshold t in the whole image; σ_i^2 was the variance of the class i; $\sigma_i^2(t) = (\mu_i - \mu)^2$, μ was the mean value, μ_i was the mean value of class i.

2.2 Target Detection

Six bands images from Medium wave infrared to thermal infrared spectrum could be gotten simultaneously based on VIMI. The target recognition algorithm based on multi-spectrum infrared images was used in this paper.

The basic principle of multispectral target recognition was the quantitative comparison and analysis of spectral vectors between the multispectral image data and standard spectral database. Based on the emissivity spectrum of materials, the object types could be identified by matching and comparing the waveforms or features. Spectral matching worked as follows: based on the spectral database, standard spectral curves of materials with certain characteristics were used as templates to compare with target image pixels, and the most similar pixels were identified and labeled [7, 8].

Among the existing multispectral target recognition algorithms, the most representative algorithms were the Spectral Angle Mapper method (SAM) and the Spectral Information Divergence method (SID). The SID described the difference between two spectrum. In the SAM algorithm, the correlation coefficients of the test spectrum and the reference spectrum were calculated to judge the similarity between the two spectral vectors. SAM had a high probability of false detection because of the same result could be got for some similar spectral vectors. A modified SAM algorithm was proposed in this paper to avoid this defect. In the modified algorithm, a combination spectral operator was constructed with the spectral angle mapper and the spectral information entropy [9].

Among the existing multispectral target recognition algorithms, the most representative ones were the Spectral Angle Mapper (SAM), and the Spectral Information Divergence (SID).

SAM was a basic target recognition algorithm. The Angle between spectral vectors was calculated to measure the similarity between the spectrum of the pixel to be identified and the spectrum of the known target. Whether the target was contained in the pixel could be determined based on the angle. But, a same angle result could be got by two similar spectral vectors. A high probability of false detection would be caused by this [10].

In the SID algorithm, the spectral vectors were taken as random vectors. The similarity between two random vectors was analyzed based on mathematical statistics theory.SID described the information difference between two spectra. The similarity between two spectra was greater when the value of SID was closer to 0.

Therefore, the following combined spectral identification operator was constructed to improve the classical spectral Angle matching operator. The improved combination algorithm was as follows:

$$CM(S, Y) = \alpha SAM(S, Y) + (1 - \alpha)\, SID(S, Y) \tag{2}$$

In the above formula, α was the adjustment coefficient, $0 < \alpha < 1$, $S = (s_1, s_1, \cdots, s_m)^T$ was the reference spectra, $Y = (y_1, y_1, \cdots, y_m)^T$ was the test spectra, $CM(S, Y)$ was the combinatorial spectral operator of the reference spectra and the test spectra. $SAM(S, Y)$ was the SAM spectral operator of the reference spectra and the test spectra. $SID(S, Y)$ was the SID spectral operator of the reference spectra and the test spectra. The adjustment coefficient α could be got by the minimum error standard.

Based on the combined operator, the ship could be detected by the formula (2).

$$CM(S(j), Y(i)) \begin{matrix} \geq \mu & then & C_i = 0 \\ < \mu & hen & C_i = 1 \end{matrix}, \quad i = 1, 2, \cdots, M \tag{3}$$

In the above formula, μ was the threshold of the target detection, C_i was the target detection result, $C_i = 1$ indicated that the target $Y(i)$ and the spectrum $S(j)$ were the same material.

3 Experiment

3.1 Sea-land Segmentation

The effectiveness of the sea-land segmentation algorithm was proved through experiment based on the VIMI infrared images. TianJin Port image of VIMI was used here. TianJin Port was a very important port in China. The container throughout of the TianJin Port exceeded 10 million TEUs in 2021.

The sea-land segmentation result was shown in Fig. 2. The segmentation results showed the overall outline of the coast and the ship. It were not affected by the factors such as illumination changing.

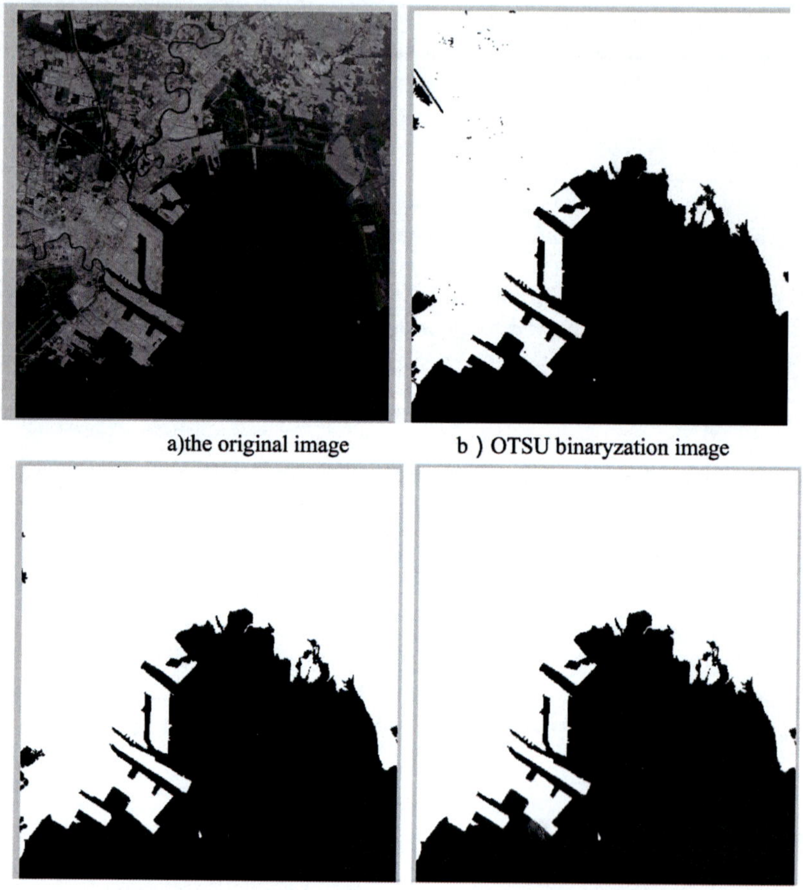

a)the original image b) OTSU binaryzation image

c)Morphology erosion image d) the final image

Fig. 2. Sea-land segmentation images

3.2 Target Detection

Based on the sea-land segmentation result of the VIMI infrared images, the ship detection experiment were showed as follows. Ships were detected with a 92% accuracy rate (Figs. 3, 4 and 5).

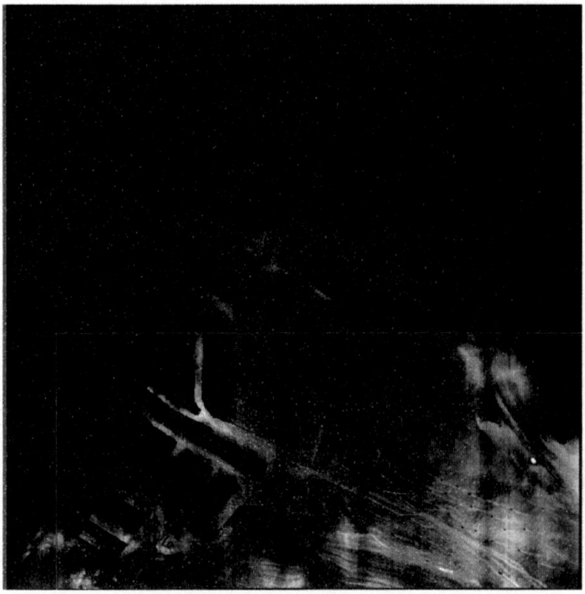

Fig. 3. The result of the combined spectral operator

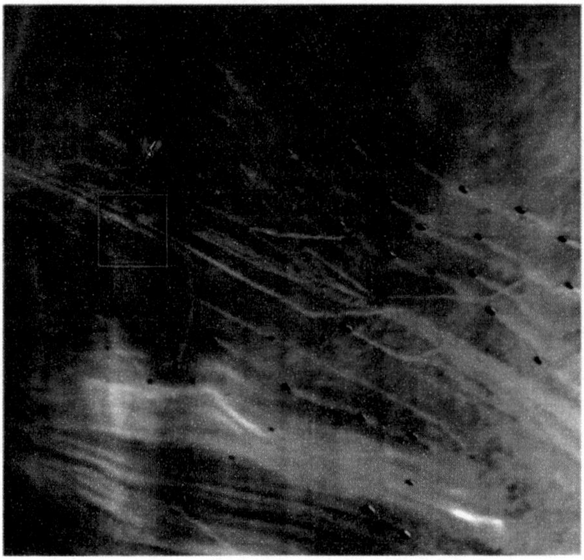

Fig. 4. The partial enlarged image

Fig. 5. The result of the ship detection

4 Conclusion

Currently, ships detection were mostly realized based on SAR or VNIR images. In this paper, the infrared spectral information was used firstly to the ship detection. In this paper, a ship target detection algorithm based on VIMI multi-spectrum spectral imager infrared image was proposed. Finally, the effectiveness of this algorithm was proved by the experiments with VIMI infrared images.

References

1. Sun, Y., Jiang, G., et al.: F-5 satellite: overview and application prospects. Spacecr. Recover. Remote Sens. **39**(3), 1–13 (2018)
2. Zhao, S., Liu, S., et al.: Potential application analysis of F-5 satellite in ecology and environment fields. Spacecr. Recover. Remote Sens. **39**(3), 115–120 (2018)
3. Xiao, H., Xu, Q., Hu, L.: A sea-land segmentation algorithm based on gray smoothness ratio. In: 4th International Workshop on Earth Observation and Remote Sensing Applications, pp. 117–121 (2016)
4. Liu, Q., Xie, X., Qi, Y., et al.: Study of acquisition method of ROI in infrared image of ship targets near shore. Infrared **36**(1 1), 41–46 (2015)
5. Zhou, H., Chen, L., Zhuang, Y., et al.: Infrared remote sensing image sea-land segmentation algorithm based on multi-scale integral image reconstruction. In: National Symposium on Signal and Intelligent Information Processing and Applications, vol. 53, no. S2, pp. 322–325 (2017)
6. Wang, W., Fu, Y., et al.: Research on infrared on-orbit ship target detection method. Mod. Electron. Tech. **41**(14), 92–95 (2018)

7. Vafadar, M., Ghassemian, H.: Anomaly detection of hyperspectral imagery using modified collaborative representation. IEEE Geosci. Remote Sens. Lett. **15**(4), 577–581 (2018)
8. Kim, S., Song, W.J., Kim, S.H.: Robust ground target detection by SAR and IR sensor fusion using Adaboost based feature selection. Sensors **16**(7), 1–27 (2016)
9. Zou, N.: Wake Detection and Ship Recognition by Data Fusion in Infrared Image, pp. 9–12. Huazhong University of Science and Technology, ShangSha (2018)
10. Zou, N., Tian, J.: Research on multi feature fusion infrared ship wake detection. Comput. Sci. **45**(11A), 172–175 (2018)

Building Change Detection on High-Resolution Imagery with a Multi-task Semantic Change Detection Method

Zhipan Wang[1](✉), Xiang Liao[2], Huan Liu[3], and Hua Zhang[3]

[1] Sun Yat-Sen University, Shenzhen 528406, Guandong Province, China
wang749195@outlook.com
[2] MagicCubesat Technology Co.LTD., Shenzhen, Guandong Province, China
[3] Changsha University of Science and Technology, Changsha, Hunan Province, China

Abstract. Building change detection from high-resolution images plays an important role in landcover classification and the analysis of urban spatial information. With the development of deep learning (DL) in remote sensing, DL has become the mainstream method in the building change detection field. However, the existing building change detection methods still only can detect building change, but there are few studies investigating whether the building is increased or decreased, in other words, they can only capture the spatial change of buildings, but can't get the temporal change of buildings. To solve this problem, this manuscript proposes a new simple but effective multi-task learning building change detection method, which can capture building increase or decrease on high-resolution images at the same time. First, we modify the DenseNet121 as our backbone to extract deep semantic features, mainly because of DenseNet has a strong feature extraction ability, but the computation cost is relatively low. Second, we design a differencing structure to detect building increases or decreases at the same time with a new loss function, namely Tanimoto loss. A large image with very high resolution is selected to validate the effectiveness and robustness of our method, on the one hand, the result indicates that our proposed building change detection method can detect building increases or decreases accurately, and the quantitative precision shows that the Accuracy is over 90%, on the other hand, the model design idea of our building change detection model, which can provide some new insights to other research fields, such as object detection on high-resolution images, change detection of other objects.

Keywords: Building Extraction · Multi-task Learning · High-resolution Images

1 Introduction

Using remote sensing images to monitor large-area building change is an effective way for developing countries, such as illegal building construction monitoring. Deep learning methods, get an impressive result than traditional machine learning methods due to their strong self-learning ability [1]. Thus, building change detection with deep learning methods has become the mainstream method [2]. However, the full application ability of deep learning was not fully investigated.

H. P. Urbach and H. Jiang (Eds.): ISSOIA 2022, SPPHY 295, pp. 38–46, 2023.
https://doi.org/10.1007/978-981-99-4098-1_5

Building change detection can be divided into two categories, binary change detection (BCD) and semantic change detection (SCD) [3]. Binary change detection, which can only detect building change in bitemporal images, the whole model structure is similar to the semantic segmentation model, that is considers the change detection task as a pixel-level or object-level segmentation task [4]. SCD method, can not only detect building change but also obtain segmentation results. A classical SCD method is the semantic feature-constrained change detection method (SFCCD) [3], it can get three product types simultaneously, they are the former image segmentation result, the latter image segmentation result, and the change detection result.

The training dataset can influence the final detection accuracy heavily. There are excellent open-sourced datasets for building change detection, such as NJDS [3] and WBDS [4], Many of them are labeled from Google Earth images or other commercial satellite images. However, there are some limitations in these datasets. For instance, building increase or decrease information cannot be obtained. Building increases and decreases are important for illegal building construction investigations. Consequently, we need to develop a method, which can get the building to increase and decrease. In this paper, we develop a building semantic change detection method, which was trained based on the existing building semantic change detection dataset, such as NJDS [3] and WBDS [4]. Moreover, in order to help other researchers to investigate building change, we have developed an easy-used building change detection software, namely, Easy-Building Change.

The main contribution of this paper is as follows:

1. A multi-task building semantic change detection model is proposed, which can obtain building increase and decrease simultaneously.
2. The principle of our multi-task semantic change detection, can provide new insight ideal for other tasks, such as forest change detection, and road change detection.

2 Materials and Methods

2.1 Data

In this manuscript, we use WBDS dataset to train our building semantic change model. First, we clip the large image into small blocks, such as a small image with fixed size (512 $\times$ 512 pixels), then we mix these small patches, and split the whole dataset into three sub-datasets, they are train dataset, valid dataset, and test dataset, their ratio is 4:4:3, finally, we use the three sub-datasets to train our deep learning model. The detailed information about this dataset is as follows (Table 1):

Table 1. Building change detection dataset comparison

Open dataset	Spatial resolution	Number of samples (512 × 512 pixels)	Data description
NJDS [3]	0.3	617	In their paper, this dataset is open-sourced, but we cannot download their dataset
WBDS [4]	0.2	1902	This dataset was generated by Wuhan University, it is open-sourced and free to use

A large-area study area is chosen to validate the effectiveness of our deep learning method. The two different temporal images are processed from the SuperView satellite, which is captured in Jiangsu Province, China, and the spatial resolution of the two images is 0.5 m, and the image size is 16978 × 14955 pixels. The RGB true color images are as follows (Fig. 1):

(a) (b)

Fig. 1. Study area. (a) 2014 image. (b) 2022 image.

2.2 Change Detection Based on Multi-task Learning

Detecting building increase and decrease at the same time, which can be defined as a multi-class classification task [5]. Though it is simple and easy to accomplish, the detection accuracy of the relatively small number category may be unsatisfactory. Consequently, in this paper, we define building increase and decrease detection as a multi-task learning question, the whole architecture is as follows:

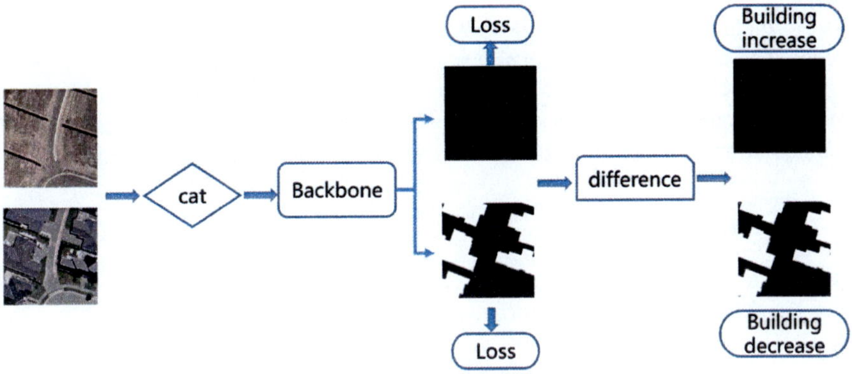

Fig. 2. Building change detection model

There are two main parts in our deep learning model, the backbone, and the semantic change detection head. DenseNet121 [6] is chosen as our backbone to extract deep semantic features, though there are many other excellent deep feature models, such as ResNet101 [7], ResNext101 [8], DenseNet101 and so on, the computation efficiency is not better. Considering the balance between efficiency and accuracy, we use DenseNet121 as the backbone. Then, we use a difference operation to get building increase or decrease, this operation is very simple, it is equal to basic math "subtract" operation.

The most important part of our building change detection model is the semantic change detection head, inspired by YOLOX [9], we use a decoupled head to detect building semantic change. When the abundant deep semantic feature of the building is extracted, two independent convolution module is utilized to extract building increase and decrease, respectively. From Fig. 2, the whole architecture is a multi-task learning model, it is simple and effective.

2.3 Loss Function

The loss function plays an important role in the deep learning method. Cross-entropy [10] is the most popular loss function in semantic segmentation [11], object detection, or instance segmentation. However, the classical cross-entropy function cannot solve the class-unbalance problem, in order to address this question, weighted cross-entropy is proposed [12], which is an effective and robust way to alleviate unbalanced class. Other excellent loss functions, such as focal loss [13], and OHEM [14], also give an interesting but effective way to help solve the class-unbalanced problem. Nonetheless, most of them need hyper-parameter setting, in other words, more prior knowledge is needed. Therefore, we introduce a new loss function, namely, Tanimoto loss [15], which can optimize classification and prediction tasks at the same time, and it also doesn't need any other parameter setting, moreover, small objection detection or segmentation with Tanimoto loss shows a robust result. The detailed information can be seen in [15].

2.4 Accuracy Assessment

We quantitatively evaluate the model performance using the three metrics, Accuracy, Recall, and F1-score [3], whose formulas are:

$$\text{Accuracy: } accuracy = (TP + TN)\,(Tp + TN + FP + FN) \tag{1}$$

$$\text{Recall: } recall = TP/(TP + FN) \tag{2}$$

$$\text{F1 - score: } F1 = 2TP/(2TP + FN + FP) \tag{3}$$

Here, TP indicates that the predicted result is a change pixel and the ground truth pixel is also a change pixel. TN indicates that the predicted pixel is the background value and the ground truth pixel is also the background value. FP indicates that the pixel in the ground truth image is changed but is predicted as the background value. FN indicates that the pixel in the ground truth image is the background value but is predicted as changed.

2.5 Results

The building change detection result of the study area can be seen in the following figure (Fig. 3).

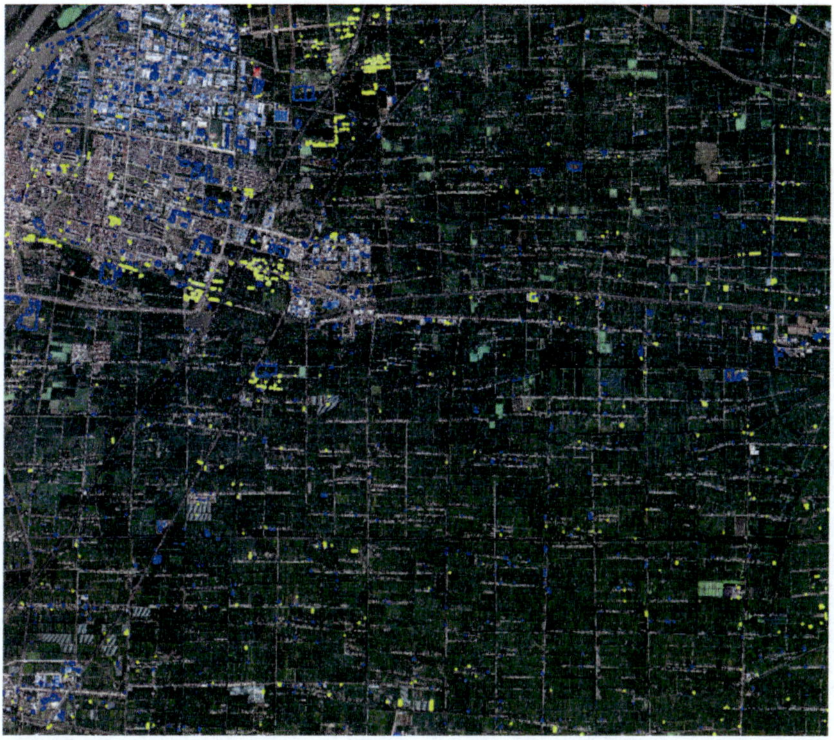

Fig. 3. Building change detection result. (The blue color and yellow color are building increase, building decrease, respectively)

The above figure indicates that building increase usually appears in the center of the city, while building decrease appears in the edge of the city or suburbs. In addition to the upper-left of the whole study area, building increases or decreases show a uniform distribution in other regions. A few small patch areas are chosen to describe the detailed information about building semantic change (Fig. 4).

We can see that, our change detection model can detect building increase and decrease accurately, whether it is a small building change or a large building change, our method can detect accurately outline the changed boundary.

The quantitative accuracy assessment is in Table (Table 2).

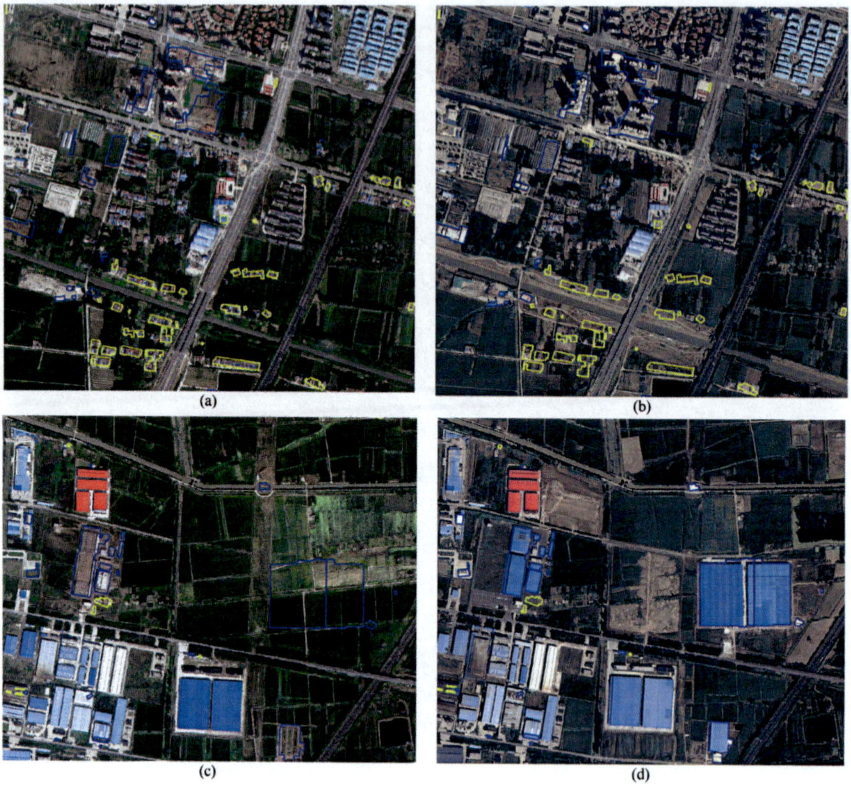

Fig. 4. Detailed building semantic change comparison. (The blue color and yellow color are building increase, building decrease, respectively). (a) and (c) is the former time-phase image, (b) and (d) is the latter time-phase image.

Table 2. Accuracy assessment

Accuracy	Recall	F1-score
0.9219	0.8193	0.8094

We use ArcGIS10.6 to statistic the area of building increase and decrease, the result is as follows (Table 3):

Table 3. Building increase and decrease area

Building increase (km^2)	Building decrease (km^2)
1.528	0.3987

We can see that the area of building increase is relatively faster than the decrease, one main reason is as the economic development of Jiangsu Province, the area of residential and industrial also increased.

2.6 Discussion

Though our deep learning model can detect building change in very high-resolution images, the detection accuracy on other median resolution images may be unsatisfactory, for example in 10m spatial resolution images. Another limitation of our method is it can detect building change with only RGB images, while in multispectral images, such as with an auxiliary near-red band, it cannot use the full spectral information, thus our next step is to use the multispectral band to train our multi-task deep learning model.

In order to help other researchers to capture building semantic change quickly, we also designed a building change detection software, which is user-friendly and high-performance. The graphical user interface is as follows (Fig. 5):

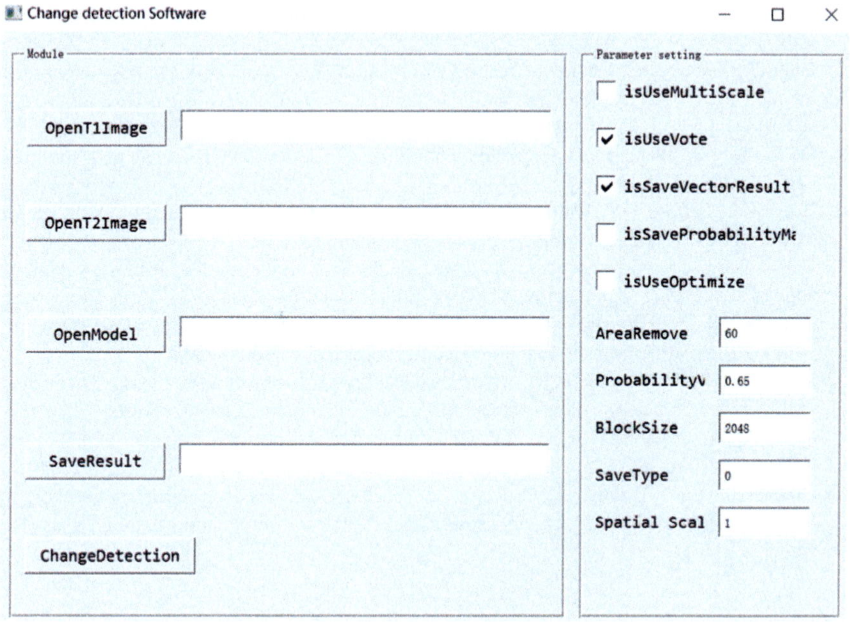

Fig. 5. Building change detection software

If users want to use this software freely, please contact the author of this manuscript. This software was developed on CUDA11.3, thus, on a client computer, users must install CUDA11.3 to get high-performance computing.

References

1. LeCun, Y., Bengio, Y., Hinton, G.: Deep learning. Nature **521**(7553), 436–444 (2015)
2. Liu, Y., Pang, C., Zhan, Z., et al.: Building change detection for remote sensing images using a dual-task constrained deep siamese convolutional network model. IEEE Geosci. Remote Sens. Lett. **18**(5), 811–815 (2020)
3. Shen, Q., Huang, J., Wang, M., et al.: Semantic feature-constrained multitask siamese network for building change detection in high-spatial-resolution remote sensing imagery. ISPRS J. Photogramm. Remote. Sens. **189**, 78–94 (2022)
4. Ji, S., Shen, Y., Lu, M., et al.: Building instance change detection from large-scale aerial images using convolutional neural networks and simulated samples. Remote Sens. **11**(11), 1343 (2019)
5. Dong X, Qian L, Guan Y, et al. A multiclass classification method based on deep learning for named entity recognition in electronic medical records. In: 2016 New York Scientific Data Summit (NYSDS), pp. 1–10. IEEE (2016)
6. Zhu, Y., Newsam, S.: Densenet for dense flow. In: 2017 IEEE International Conference on Image Processing (ICIP), pp. 790–794. IEEE (2017)
7. Szegedy, C., Ioffe, S., Vanhoucke, V., et al.: Inception-v4, inception-resnet and the impact of residual connections on learning. Thirty-first AAAI Conference on Artificial Intelligence (2017)
8. Pant, G., Yadav, D.P., Gaur, A.: ResNeXt convolution neural network topology-based deep learning model for identification and classification of Pediastrum. Algal Res. **48**, 101932 (2020)
9. Ge, Z., Liu, S., Wang, F., et al.: Yolox: exceeding yolo series in 2021. arXiv preprint arXiv: 2107.08430 (2021)
10. Botev, Z.I., Kroese, D.P., Rubinstein, R.Y., et al.: The cross-entropy method for optimization. In: Handbook of Statistics. vol. 31, pp. 35-59. Elsevier (2013)
11. Garcia-Garcia, A., Orts-Escolano, S., Oprea, S., et al.: A review on deep learning techniques applied to semantic segmentation. arXiv preprintarXiv:1704.06857 (2017)
12. Aurelio, Y.S., de Almeida, G.M., de Castro, C.L., et al.: Learning from imbalanced data sets with weighted cross-entropy function. Neural Process. Lett. **50**(2), 1937–1949 (2019)
13. Lin, T.Y., Goyal, P., Girshick, R., et al.: Focal loss for dense object detection. In: Proceedings of the IEEE International Conference on Computer Vision, pp. 2980–2988 (2017)
14. Shi, F., Qian, H., Chen, W., et al.: A fire monitoring and alarm system based on YOLOv3 with OHEM. In: 2020 39th Chinese Control Conference (CCC), pp. 7322–7327. IEEE (2020)
15. Diakogiannis, F.I., Waldner, F., Caccetta, P., et al.: ResUNet-a: a deep learning framework for semantic segmentation of remotely sensed data. ISPRS J. Photogramm. Remote. Sens. **162**, 94–114 (2020)

NM-LinkNet: Cloud Detection from Remote Sensing Images with Non-local Operation and Multi-scale Feature Aggregation

Yongshi Jie[1], Anzhi Yue[2,3], Naijian Wang[4], Yan Wang[4], Xuejie Xu[5], Ding Ding[6], Wei Tan[1], Hongyan He[1], and Kun Xing[1](✉)

[1] Beijing Institute of Space Mechanics and Electricity, China Academy of Space Technology, Beijing, China
xingkunfeixiang@163.com

[2] Aerospace Information Research Institute, Chinese Academy of Sciences, Beijing, China

[3] National Engineering Laboratory for Integrated Aero-Space-Ground-Ocean Big Data Application Technology, Xian, China

[4] Bureau of Geophysical Prospecting INC., China National Petroleum Corporation, Zhuozhou, Hebei Province, China

[5] Department of Engineering Physics, Tsinghua University, Beijing, China

[6] Powerleader Computer Systems Co., Ltd, Shenzhen, Guangdong Province, China

Abstract. Cloud detection is an important preprocessing process in remote sensing applications. Cloud detection methods have developed from traditional methods to deep learning methods which are widely used at present. However, current cloud detection methods still have limitations in the detection of thin clouds and broken clouds, mainly because the thin clouds are sparsely distributed and the size of broken clouds is relatively small. To solve the above difficult problems, this paper proposes a cloud detection network NM-LinkNet, which combines non-local operation and multi-scale feature aggregation, to improve the detection ability of thin clouds and broken clouds. Based on LinkNet50, NM-LinkNet uses non-local operation to obtain the long-distance context information of sparse distributed thin clouds to enhance the features of thin clouds. The multi-scale feature aggregation module designed in this paper is used to extract the features of clouds of different scales to improve the detection ability of small broken clouds. The experimental results on SPARCS dataset show that the IoU and F1 of NM-LinkNet proposed in this paper reach 87.50% and 93.33% respectively, which exceeds the other five mainstream deep learning methods in quantitative data and visualization results.

Keywords: cloud detection · remote sensing · deep learning · non-local operation · multi-scale feature aggregation

1 Introduction

With the rapid development of remote sensing technology, optical remote sensing image plays an important role in semantic segmentation [1], object detection [2], change detection [3] and other remote sensing application tasks. However, more than 66% of the

H. P. Urbach and H. Jiang (Eds.): ISSOIA 2022, SPPHY 295, pp. 47–55, 2023.
https://doi.org/10.1007/978-981-99-4098-1_6

earth's surface area is covered by clouds [4], which has a negative effect on remote sensing surface information extraction. Optical satellite sensors cannot penetrate the clouds during imaging, which leads to the pollution of the acquired remote sensing images by clouds, results in the loss or weakening of the features of ground targets and limits the further remote sensing applications. Therefore, as an important preprocessing work, cloud detection has extremely important practical significance for remote sensing applications. However, the difficulty of cloud detection is to detect thin clouds and broken clouds. Thin clouds are sparsely distributed and relatively transparent compared with thick clouds, and the size of broken clouds is usually small, which makes cloud detection task challenging.

So far, researchers have developed a variety of cloud detection methods, which are mainly divided into threshold-based methods, machine learning-based methods and deep learning-based methods.

Threshold-based methods mainly construct spectral features from multi-band images, and then set a specific threshold to distinguish cloud pixels from non-cloud pixels. For example, ACCA proposed by Irish et al. [5] is an automatic cloud coverage evaluation algorithm for Landsat-7. The core idea of this algorithm is to combine the information of multiple spectral bands and use decision tree and threshold function to determine whether each pixel is cloud. The automatic cloud and shadow detection algorithm developed by Huang et al. [6] takes clear forest pixels as a reference, separates the cloud and reference pixels in the spectral-temperature space, estimates the height of the cloud on the basis of the cloud extraction results, and determines the position of the shadow in combination with the sun illumination. The FMask method proposed by Zhu et al. [7] sets multiple thresholds according to the spectral features of clouds in different bands of Landsat images, thus realizing automatic detection of clouds and cloud shadows. The threshold-based method has two disadvantages: first, it needs to obtain spectral feature information from multiple bands, which is limited by optical sensors and has poor universality in different satellite images. Second, the effect of this method is poor in complex background images, because there may be ground objects similar to clouds in the background, such as snow and ice, which makes it difficult to identify the threshold method [8].

The main content of the cloud detection algorithm based on machine learning is to label a large number of cloud pixels and non-cloud pixels as training samples, manually design some cloud features to build a feature set, and then use the feature set to train the classifier. Li et al. [9] proposed a cloud detection method based on SVM classifier, texture features, average gradient and other features. Bai et al. [10] integrated various features such as spectral features, texture features and NDVI, and used SVM classifier to carry out cloud detection. Shao et al. [11] designed a cloud detection method based on auto encoding network and fuzzy function. However, if machine learning methods are to be used to achieve better detection results, researchers are required to manually design better features.

Deep learning methods use deep convolutional neural networks to automatically learn target features from a large number of samples, which significantly improves the accuracy of image information extraction. In the field of Cloud detection, Mohajerani et al. [12] proposed Cloud-Net, which adopted a U-shaped encoder-decoder structure and achieved good results on Landsat 8 datasets. Guo et al. [13] proposed a Cloud detection network called Cloud-AttU based on UNet [14] and attention mechanism. Liu et al. [15] improved the feature extraction network of D-LinkNet [16] and applied it in the cloud detection task. Although existing deep learning methods have significantly improved the accuracy of cloud detection, thin cloud and broken cloud detection is still a difficult problem.

The novel cloud detection network NM-LinkNet proposed in this paper adopts the encoder-decoder structure, introduces the non-local operation (NL) in the skip connection to capture the long-distance context information, and adds the multi-scale feature aggregation (MFA) module designed in this paper in the decoding process to aggregate the feature information of different scales in each stage of the decoder. The experimental results on SPARCS dataset show that NM-LinkNet improves the detection ability of thin clouds and broken clouds, and the IoU and F1 are 87.50% and 93.33%, respectively, which are better than many mainstream algorithms.

2 Method

2.1 Network Structure

The network structure of NM-LinkNet is shown in Fig. 1. NM-LinkNet uses LinkNet50 [17] as the basic network and uses the encoder-decoder structure. The encoder is the left branch of the network structure, which is composed of ResNet50 [18]. It uses a deep convolutional network to extract low-level features containing spatial details and high-level deep features rich in semantic information. NM-LinkNet adds non-local operations at each level of skip connection to aggregate long-distance context information at each stage. The function of the decoder is to gradually restore the features to their original size. Different from the original LinkNet decoder, this paper designs a multi-scale feature aggregation module in the decoding process to fuse the features of different scales in different stages of the decoder. The input of NM-LinkNet is an image with a dimension of $384 \times 384 \times 3$, and the output is a prediction result of $384 \times 384 \times 1$. Here, the prediction result is displayed with a color image for the convenience of display.

2.2 Non-local Operation

Compared with thick clouds, thin clouds are more transparent and sparsely distributed, which brings challenges to thin cloud detection. In order to deal with the problem that thin clouds are difficult to detect, this paper introduces non-local operation [19] into the network model to calculate the correlation between each pixel, obtain the context information of thin clouds, and enhance the features of thin cloud pixels with sparse distribution. The structure diagram of non-local operation is shown in Fig. 2. The dimension of the input feature is $H \times W \times 2C$. Firstly, the input features are passed through three 1×1 convolutional layers to reduce the number of channels by half, and three features with dimensions of $H \times W \times C$ are obtained, which are respectively represented by u, v and g. Then the dimensions of feature v and g are converted to $HW \times C$, and the dimensions of features u are transformed and transposed to $C \times HW$. Then, the two-dimensional feature of $HW \times HW$ is obtained by matrix multiplication of the feature

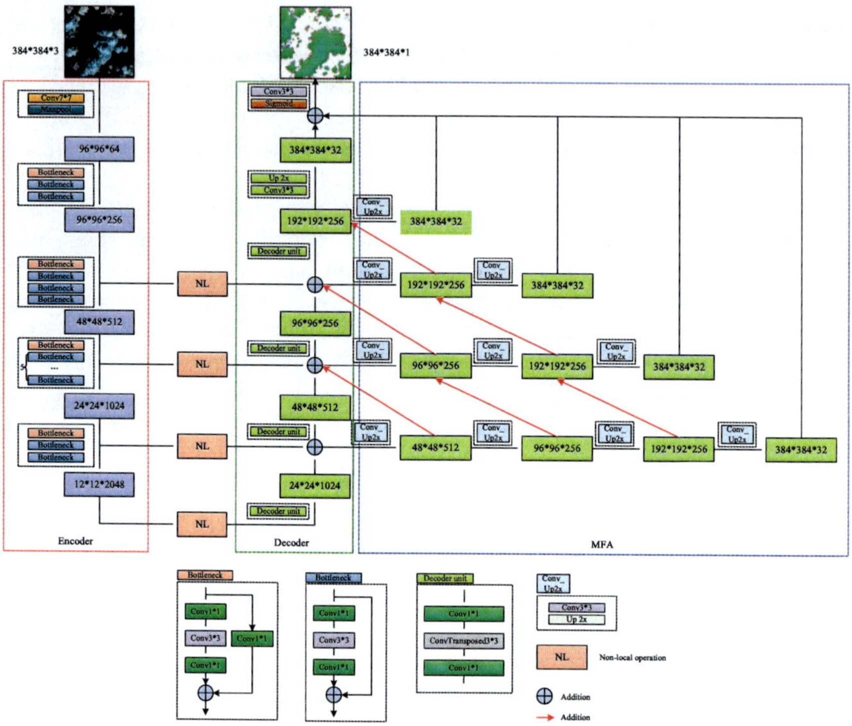

Fig. 1. Network structure of NM-LinkNet.

v and u after dimensional transformation, which represents the similarity between each pixel. The similarity matrix was normalized by softmax function, and then the matrix was multiplied with the feature g after dimension transformation, and then the feature with dimension H × W × C was generated after dimension transformation. Finally, it goes through 1 × 1 convolutional layer to double the number of channels, and then adds the input features, so as to finally get the output features of non-local operation.

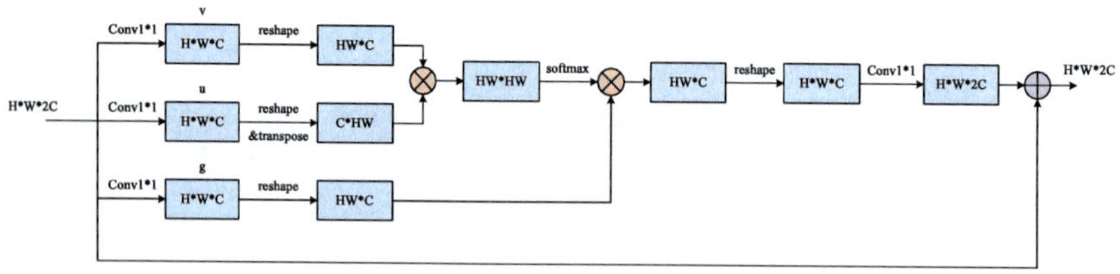

Fig. 2. Structure of non-local operation.

2.3 Multi-scale Feature Aggregation

The size of broken clouds in the image is generally small due to non-aggregation, and the detection ability of common methods for small broken clouds is often poor. To solve this problem, this paper proposes the MFA module to extract features of different scales to enhance the detection accuracy of the network for small broken clouds. The structure of the module is shown in Fig. 1. In the decoder of the original LinkNet, the output features of the decoding units of each stage, after being added with the features of the skip connection, will be used as the input of the next decoding unit. The output feature of the last decoding unit is then processed by up-sampling layer and convolution layer to get the prediction result. Different from LinkNet, the proposed NM-LinkNet incorporates the features of different scales at different stages of the decoder. In order to better preserve the spatial details of the features, MFA adopts a stepwise up-sampling method for the features of each stage of the decoder, and adds the features with the same dimension in adjacent stages to achieve the full fusion of features of different scales. Finally, MFA adds the $384 \times 384 \times 32$ features obtained after up-sampling at each stage of the decoder, and then obtains the cloud prediction result after up-sampling layer, convolution layer and sigmoid function.

3 Experiment

3.1 Dataset

In this paper, experiments are carried out on SPARCS dataset. The SPARCS dataset contains 80 Landsat8 images with 1000×1000. In this paper, 80 original images are seamlessly cropped into 720 384×384 image tiles, and the data are divided according to the ratio of 7:1:2. Finally, the training set contains 504 image tiles, the validation set contains 72 image tiles, and the testing set contains 144 image tiles.

3.2 Experimental Setup

In this paper, Pytorch 1.8.0 [20] is used to complete all the experiments. The hardware environment of the experiments is NVIDIA GeForce RTX 3060 (12G), and the operating system is Windows 10. The epochs and bach size of the experiments are 200 and 4, respectively. The optimizer is SGD, initial learning rate is 0.01 and the learning rate reduction strategy is "poly". The loss function used in the experiments is binary cross-entropy (BCE), and the predicted threshold is 0.5. In addition, this paper adopts random horizontal flip, vertical flip, Gaussian blur and rotation to expand the training samples.

3.3 Evaluation Metrics

In this paper, IoU and F1, which are commonly used in cloud detection, are used to evaluate the performance of each method. The formulas of IoU and F1 are as follows. The meaning of IoU is the ratio of intersection and union of prediction image and label image. F1 is an index to comprehensively evaluate Precision and Recall, which is often used in remote sensing semantic segmentation tasks. True Positive (TP) indicates the number of pixels that are correctly predicted to be the cloud. False Positive (FP) indicates the number of pixels that are predicted to be cloud, but are predicted incorrectly. False Negative (FN) indicates the number of pixels that are incorrectly predicted as the background.

$$\text{IoU} = \frac{\text{TP}}{\text{TP+FP+FN}} \tag{1}$$

$$\text{F1} = \frac{2 \times \text{Precision} \times \text{Recall}}{\text{Precision} + \text{Recall}} \tag{2}$$

$$\text{Precision} = \frac{\text{TP}}{\text{TP+FP}} \tag{3}$$

$$\text{Recall} = \frac{\text{TP}}{\text{TP+FN}} \tag{4}$$

3.4 Result Analysis

In order to verify the effectiveness of NM-LinkNet, this paper conducted comparative experiments with FCN [21], UNet [14], LinkNet50 [17], DeepLabv3+ [22] and Cloud-Net [12] on SPARCS datasets. The quantitative comparison results are shown in Table 1. The accuracy of NM-LinkNet is the highest, with IoU and F1 reaching 87.50% and 93.33%, respectively, which exceeds other comparison methods. The IoU and F1 of NM-LinkNet are 5.53% and 3.24% higher than that of FCN, and 0.77% and 0.44% higher than that of LinkNet50, respectively. Compared with the basic network, the reason for the accuracy improvement of NM-LinkNet is the addition of non-local operation and multi-scale feature aggregation module, which improves the detection ability of thin clouds and broken clouds, thus improving the performance on SPARCS dataset.

Table 1. Comparison of results of different methods on SPARCS dataset (%)

Method	IoU	F1
FCN	81.97	90.09
Cloud-Net	85.03	91.91
DeepLabv3 +	85.11	91.95
UNet	85.39	92.12
LinkNet50	86.73	92.89
NM-LinkNet	**87.50**	**93.33**

In addition to quantitative comparison, this paper also compares the methods in terms of qualitative visualization results, as shown in Fig. 3. Thin clouds in the first column of images are sparsely distributed and relatively transparent. FCN detected only a small fraction of thin clouds. Compared with other methods, NM-LinkNet detected the most thin cloud regions. Similarly, the images in the second and third columns are also thin clouds. NM-LinkNet obtains long-distance context information through non-local operation, so it has a better detection effect on thin clouds. The fourth to sixth columns contain images and results of broken clouds. FCN has many missed detection, and Cloud-Net, DeepLabv3+, UNet and LinkNet50 also have different degrees of missed detection. However, NM-LinkNet has a stronger ability to detect broken clouds due to

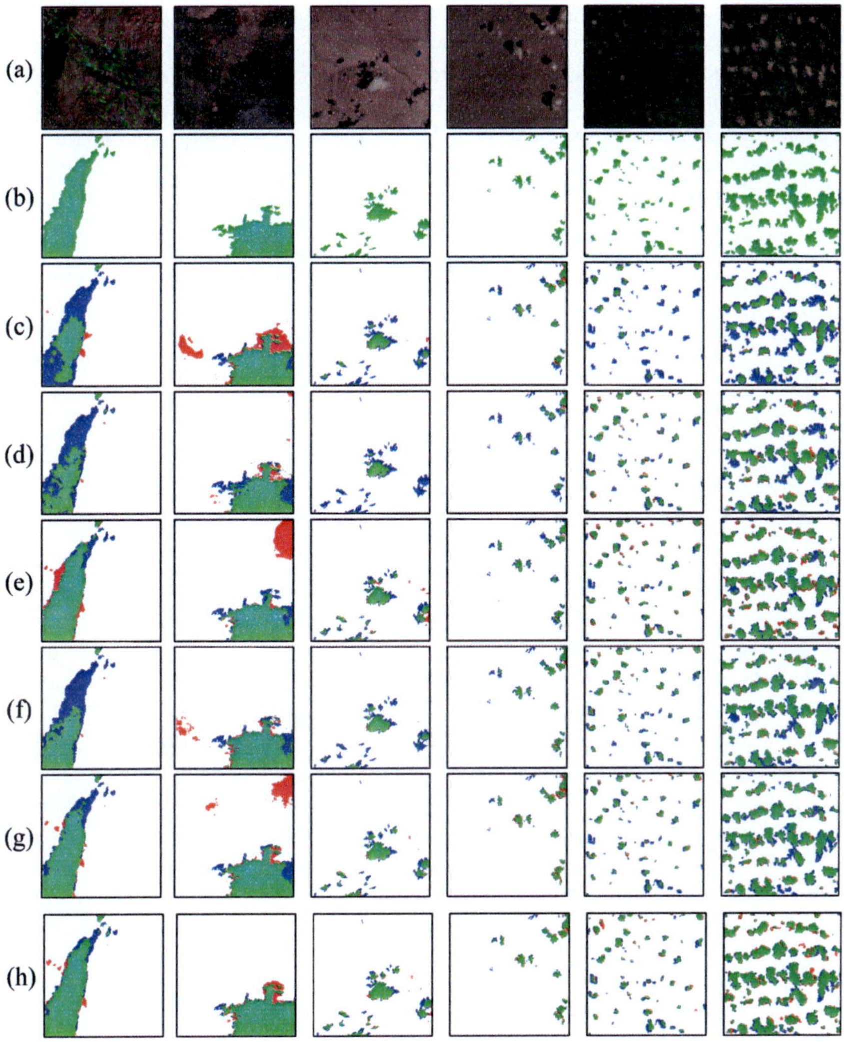

Fig. 3. Visualization results of each method on SPARCS dataset. (a) Image; (b) Label; (c) FCN results; (d) Cloud-Net results; (e) DeepLabv3 + results; (f) UNet results; (g) LinkNet50 results; (h) NM-LinkNet results; Green: TP; White: TN; Blue:FN; Red:FP.

the full aggregation of multi-scale features using the MFA module, with relatively less missed detection.

4 Discussion

In order to verify the effectiveness of each improved module, the ablation experiment of NM-LinkNet is carried out in this paper. In this paper, non-local operation and MFA are added to LinkNet50 successively to evaluate the performance of the model. The results of the ablation experiment are shown in Table 2. The IoU and F1 of LinkNet50 are 86.73% and 92.89%, respectively. After adding non-local operation on the basis of LinkNet50, IoU and F1 are improved by 0.07% and 0.04%, respectively. After adding the MFA module, IoU and F1 increased by 0.7% and 0.4%, respectively. After adding non-local operation and MFA module respectively, the accuracy of the network is improved, indicating that these two modules can improve the accuracy of cloud detection.

Table 2. Results of ablation experiment (%)

Method	IoU	F1
LinkNet50	86.73	92.89
LinkNet50 + NL	86.80	92.93
LinkNet50 + NL + MFA	87.50	93.33

5 Conclusion

In order to solve the problem that thin clouds and broken clouds are difficult to detect, a new cloud detection network NM-LinkNet is proposed in this paper. NM-LinkNet adopts encoder-decoder structure and introduces non-local operation to obtain long-distance context information, so as to improve the detection ability of sparse thin clouds. In addition, a multi-scale feature aggregation module is designed to fully aggregate the features of different scales of broken clouds, so as to improve the detection ability of small broken clouds. The experimental results on the public SPARCS dataset show that NM-LinkNet outperforms many mainstream methods, and improves the detection ability of thin and broken clouds, indicating that NM-LinkNet is effective for improving the accuracy of cloud detection.

Funding. This research was supported by the National Natural Science Foundation of China (42050202).

References

1. Yuan, X., Shi, J., Gu, L.: A review of deep learning methods for semantic segmentation of remote sensing imagery. Expert Syst. Appl. **169**, 114417 (2021)

2. Cheng, G., Han, J.: A survey on object detection in optical remote sensing images. ISPRS J. Photogramm. Remote. Sens. **117**, 11–28 (2016)
3. Asokan, A., Anitha, J.: Change detection techniques for remote sensing applications: a survey. Earth Sci. Inf. **12**(2), 143–160 (2019)
4. Zhang, Y., Rossow, W.B., Lacis, A.A., et al.: Calculation of radiative fluxes from the surface to top of atmosphere based on ISCCP and other global data sets: Refinements of the radiative transfer model and the input data. J. Geophys. Res. Atmos. **109**(D19) (2004)
5. Irish, R.R., Barker, J.L., Goward, S.N., et al.: Characterization of the Landsat-7 ETM+ automated cloud-cover assessment (ACCA) algorithm. Photogramm. Eng. Remote. Sens. **72**(10), 1179–1188 (2006)
6. Huang, C., Thomas, N., Goward, S.N., et al.: Automated masking of cloud and cloud shadow for forest change analysis using Landsat images. Int. J. Remote Sens. **31**(20), 5449–5464 (2010)
7. Zhu, Z., Woodcock, C.E.: Object-based cloud and cloud shadow detection in Landsat imagery. Remote Sens. Environ. **118**, 83–94 (2012)
8. Qiu, S., He, B., Zhu, Z., et al.: Improving Fmask cloud and cloud shadow detection in mountainous area for Landsats 4–8 images. Remote Sens. Environ. **199**, 107–119 (2017)
9. Li, P., Dong, L., Xiao, H., et al.: A cloud image detection method based on SVM vector machine. Neurocomputing **169**, 34–42 (2015)
10. Bai, T., Li, D., Sun, K., et al.: Cloud detection for high-resolution satellite imagery using machine learning and multi-feature fusion. Remote Sens. **8**(9), 715 (2016)
11. Shao, Z., Deng, J., Wang, L., et al.: Fuzzy autoencode based cloud detection for remote sensing imagery. Remote Sens. **9**(4), 311 (2017)
12. Mohajerani, S., Saeedi, P.: Cloud-Net: an end-to-end cloud detection algorithm for Landsat 8 imagery. In: IGARSS 2019–2019 IEEE International Geoscience and Remote Sensing Symposium, pp. 1029–1032. IEEE (2019)
13. Guo, Y., Cao, X., Liu, B., et al.: Cloud detection for satellite imagery using attention-based U-Net convolutional neural network. Symmetry **12**(6), 1056 (2020)
14. Ronneberger, O., Fischer, P., Brox, T.: U-net: Convolutional networks for biomedical image segmentation. In: Navab, N., Hornegger, J.., Wells, W.M., Frangi, A.F. (eds.) MICCAI 2015. LNCS, vol. 9351, pp. 234–241. Springer, Cham (2015). https://doi.org/10.1007/978-3-319-24574-4_28
15. Liu, G., Wang, G., Bi, W., et al.: Application of improved D-LinkNet model in cloud detection of domestic satellite image. Bull. Surv. Mapp., (11): 54–58, 64 (2021)
16. Zhou, L., Zhang, C., Wu, M.: D-LinkNet: LinkNet with pretrained encoder and dilated convolution for high resolution satellite imagery road extraction. In: Proceedings of the IEEE Conference on Computer Vision and Pattern Recognition Workshops, pp. 182–186 (2018)
17. Chaurasia, A., Culurciello, E.: LinkNet: exploiting encoder representations for efficient semantic segmentation. IEEE Vis. Commun. Image Process. (VCIP) **2017**, 1–4 (2017)
18. He, K., Zhang, X., Ren, S., et al.: Deep residual learning for image recognition, In: Proceedings of the IEEE Conference on Computer Vision and Pattern Recognition, pp. 770–778 (2016)
19. Wang, X., Girshick, R., Gupta, A., et al.: Non-local neural networks. In: Proceedings of the IEEE Conference on Computer Vision and Pattern Recognition, pp. 7794–7803 (2018)
20. Paszke, A., Gross, S., Massa, F., et al.: Pytorch: An imperative style, high-performance deep learning library. In: Advances in Neural Information Processing Systems, vol. 32 (2019)
21. Long, J., Shelhamer, E., Darrell, T.: Fully convolutional networks for semantic segmentation. In: Proceedings of the IEEE Conference on Computer Vision and Pattern Recognition, pp. 3431–3440 (2015)
22. Chen, L., Zhu, Y., Papandreou, G., Schroff, F., Adam, H.: Encoder-decoder with atrous separable convolution for semantic image segmentation. In: Proceedings of the European Conference on Computer vision, pp. 801–818 (2018)

Design and Performance of the SOLAR-Induced Fluorescence Spectrometer of BISME

Yazhou Jing(✉), Pengfei Duan, Bicen Li, and Weigang Wang

BISME, Beijing, China
15810106739@163.com

Abstract. At the end of 2020, the Chinese government proposed the goals of "carbon peaking and carbon neutrality", which quickly became the focus of the times. The quantitative remote sensing needs of "carbon monitoring" are very urgent.

In the field of vegetation remote sensing, the solar-induced fluorescence remote sensing is a new type of remote sensing technology developed rapidly in recent years, and is the most breakthrough research frontier in the field of vegetation remote sensing in the past decade. Vegetation chlorophyll fluorescence has obvious advantages in photosynthesis detection, so it will play a very important role in vegetation monitoring and carbon cycle application.

This paper introduces the design scheme and performance of a spectrometer for the solar-induced fluorescence, including the task analysis, requirements, configurations and results of the instrument. This spectrometer was developed by the Beijing Institute of Space Mechanics and Electricity, and has been successfully launched in August 2022, entering the stage of in-orbit service.

Keywords: Carbon peaking and carbon neutrality · Solar-induced fluorescence remote sensing · Spectrometer · Design scheme and performance

1 Summary

Subsequent paragraphs, however, are indented.Vegetation fluorescence detection is to obtain the spatially continuous vegetation fluorescence signal by means of ultra high spectral resolution detection under the condition of weak reflection signal, analyze the photosynthesis capacity of land vegetation, and further answer and explain the carbon absorption capacity of forests and other vegetation and the growth of agricultural crops [1–4]. The task of the Sun Induced Fluorescence Spectrometer (hereinafter referred to as the Spectrometer) is to obtain the sun induced vegetation fluorescence remote sensing data, so as to accurately map the spatial and temporal distribution of regional/global vegetation fluorescence, and meet the needs of global carbon source, carbon sink quantitative monitoring, and forest vegetation productivity assessment [5–8].

This spectrometer was developed by Beijing Institute of Space Mechanics and Electricity, and has been successfully launched in August 2022, entering the orbit service phase.

H. P. Urbach and H. Jiang (Eds.): ISSOIA 2022, SPPHY 295, pp. 56–65, 2023.
https://doi.org/10.1007/978-981-99-4098-1_7

2 Task Analysis

Photosynthesis of plants is the most basic biochemical process on the earth. Plants absorb carbon dioxide in the air, convert it into organic matter, and release generated oxygen into the air at the same time. Photosynthesis of green plants in terrestrial ecosystems is an important part of the earth's carbon absorption, and also an important way for the earth's ecosystem to achieve carbon neutralization [9, 10].

Chlorophyll fluorescence spectrum is narrow, only nanometer wide, and its energy is very weak in the reflected energy of vegetation, only about 0.5% to 2% of which is emitted in the form of fluorescence. Therefore, the spectrometer needs the detection ability of high spectral resolution, high signal to noise ratio and high stability [11, 12].

3 Indicator System

The design requirements of the hyperspectral detector should aim at the spectrum and radiation characteristics of chlorophyll fluorescence, combine the satellite orbital height and observation requirements, and comprehensively consider the spectral range, spectral resolution, spatial resolution, swath, signal-to-noise ratio and dynamic range of the detector. Through analysis and simulation, the indicator system of the spectrometer is shown in the following Table 1.

Table 1. The design Requirements of the spectrometer

Serial No	Project	Performance index requirements
1	Spectral range	670 nm–780 nm
2	spectral resolution	≤0.3 nm
3	Spectral sampling interval	≤0.15 nm
4	GSD	≤1.5 km
5	SNR	≥200 (input radiance 10 mWm–2 sr-nm-1)
6	Dynamic range of radiance	10– 350 mWm-2sr-1nm-1
7	System MTF (static)	≥0.2 (Nyquist frequency)
8	width of cloth	≥30km
9	Keystone/Smile Distortion	Less than 0.3 pixels
10	Stray light	≤1%
11	Polarization sensitivity	≤1%
12	Radiometric calibration accuracy	On board absolute radiometric calibration accuracy≤5%
13	Spectral calibration accuracy	$\leq \Delta\lambda / 10$

4 Instrument Configurations

4.1 System Composition

The spectrometer consists of three single component, including optic-structure, signal processor and power unit. The system composition and functional block is shown in the following figure (Fig. 1).

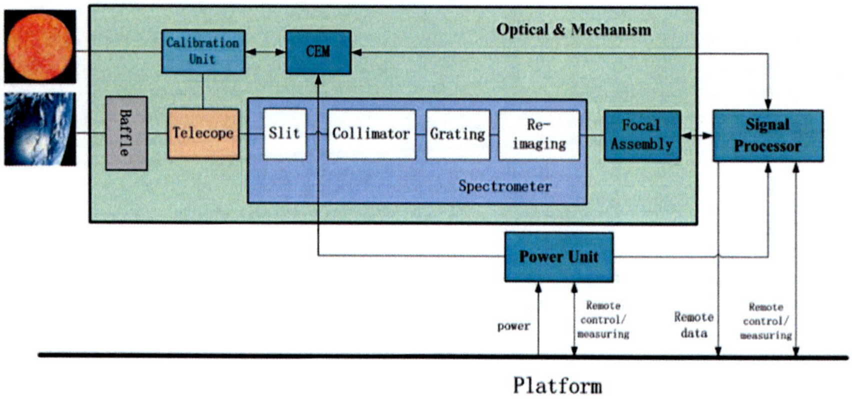

Fig. 1. Function and Composition Block of the spectrometer

4.2 Working Principle

The spectrometer adopts push scan imaging mode. The optical axis is perpendicular to the flight direction, the along orbit direction is the spectral dimension, and the through orbit direction is the spatial dimension. The input signal enters the spectrometer through the optical slit after passing through the telescope system. The transmitted collimation system expands the beam into parallel light and shines on the grating to disperse into fine spectrum. The spectrum is obtained by imaging on the focal plane detector through the convergence system. The analog signal of the focal plane detector is collected, amplified and converted into digital signal by A/D, and then processed and formatted. It is transmitted to the ground through the satellite data transmission system (Fig. 2).

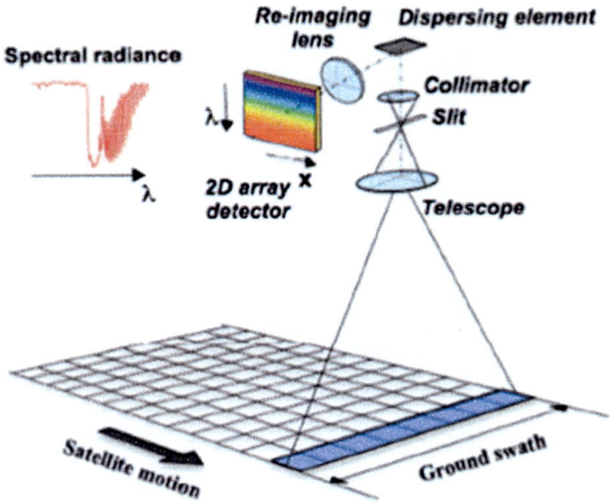

Fig. 2. Schematic diagram of the working principle of the spectrometer

4.3 Operating Mode

The spectrometer mainly has four working modes: launch mode, standby mode, observation mode and calibration mode, as shown in the following table (Table 2).

Table 2. Working Mode Design

Serial No	Project	Performance index requirements
1	Launch	To the orbit, power-off, mechanism locked
2	Standby	Waiting to work, thermal control, power-off
3	Activation	Observing, power-on
4	Calibration	Calibration on orbit: Solar calibration & Background calibration

The on orbit working mode of the spectrometer is shown in the figure below (Fig. 3).

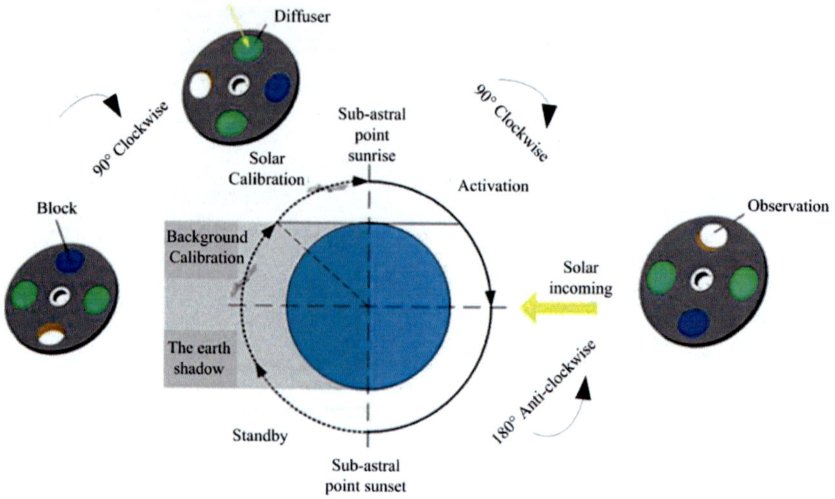

Fig. 3. On orbit operation mode of the spectrometer

4.4 Optical System Design

4.4.1 Optical System Composition

The order of the spectrometer optical path is as follows: radiation incident to the ground field of view or calibrated sunlight incident, then the energy enters the main optical and mechanical components, and finally converges on the detector. According to the function, the optical mechanical components can be divided into calibration components, telescope system components, spectrometer system components and focal plane components. The calibration components include calibration mechanisms and calibration controllers. The spectrometer system components include slit and field mirror components, collimation components, grating components and imaging components according to the detection order. As shown in the figure below (Fig. 4).

4.4.2 Optical System Design Results

In the optical system design of the spectrometer, in addition to the fine optical simulation, the influence of stray light and polarization on the imaging quality is fully considered. After establishing the simulation model of stray light, through multiple ray tracing, check the path of stray light, modify the structure that causes the stray light, and adopt the polarization suppression design of the double depolarizer, which ensures the detector's high-precision detection of weak target signals.

The comparison between the design results of the optical system of the spectrometer and the index system is shown in the following table (Table 3).

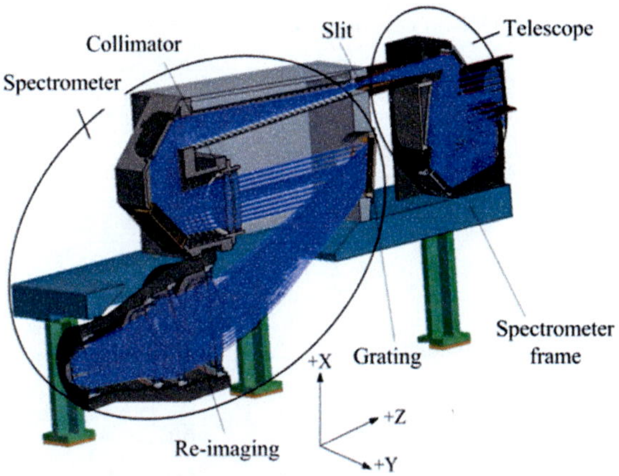

Fig. 4. Composition of main optical mechanical components

Table 3. Optical System Design Results

Indicator items	Index requirements	Design results
Working spectrum section	670 nm–780 nm	670 nm–780 nm
Spectral resolution	0.3 nm	0.3 nm
Spectral sampling interval	0.1 nm/pixel	0.1 nm/pixel
Spectral sampling rate	≥2.8 pixel ≤3.2 pixel	2.8 pixel–3.08 pixel
Focal length (rail crossing/space direction)	185 mm	185 mm
F	2.7≤(F through * F along) 1/2≤2.8	2.78
Field angle	2ω=3.9°	3.9
Optical system transmissivity	≥0.65	0.7
MTF	≥0.9@1.85 lp/mm	0.98@1.85 lp/mm
MTF	≥0.9@7.4 lp/mm	0.96@7.4 lp/mm
Keystone	≤0.2 pixel	0.10 pixel
Smile	≤0.2 pixel	0.15 pixel

4.5 Optical Mechanical Structure Composition

The opto mechanical structure of the detector is composed of the base plate, calibration unit, optic-structure and thermal control unit. The detailed composition is shown in the table below (Table 4).

Table 4. Optical and Mechanical Structure Composition

Serial No	Components	
1	Base Plate	
2	Calibration Unit	Frame
3		Mechanism
4		Diffusers
5		Cover
6	Optic-structure	Telescope
7		Frame
8		Silt
9		Collimator
10		Grating
11		Re-imaging
12	Thermal Control	CCD heat-pipe
13		Focal plane heat-pipe
14		Radiator
15		LMI, heaters, etc
16	Focal assembly	

4.6 Thermal Control Design

The spectrometer has high requirements for spectral resolution, stability and signal-to-noise ratio. The spectrometer system uses a plane reflective grating, which has high requirements for temperature gradient, temperature level and temperature stability. At the same time, the spectrometer is a high signal-to-noise ratio detector for weak signals, and reducing noise is one of the main means to improve the signal-to-noise ratio. Noise is mainly affected by device temperature noise and readout noise, so reducing device temperature is the main means to suppress device noise. The spectrometer operates in the sun synchronous orbit with an orbital period of about 90 min. Each circle has to go in and out of the earth's shadow, and the external heat flow conditions are complex, which makes the thermal control design of the spectrometer more difficult. Therefore, the spectrometer adopts the following thermal control measures.

a. Integrated design of indirect thermal control and structure. The spectrometer has been optimized in structure, and a series of auxiliary structures have been added. By controlling the temperature of such auxiliary structures, a good thermal environment is provided for key components to form an indirect temperature control and meet the temperature stability requirements;
b. Combination of passive thermal control and active thermal control, and use mature technologies as far as possible, such as thermal control coatings, multi-layer thermal

insulation materials, thermal insulation pads, thermal conductive fillers, film electric heaters and other measures;

c. Compensated heating method is adopted to ensure the temperature stability of key components and reasonably allocate heating power.

4.7 On Orbit Calibration Scheme

The improvement of data accuracy depends on the high-precision radiometric calibration and spectral calibration of the instrument. In addition to providing complete instrument calibration parameters through laboratory calibration, accurate and reliable in orbit calibration is also required to eliminate the inconsistency between heaven and earth caused by space environment. The key to achieve high-precision calibration on orbit is to select appropriate calibration scheme and stable and reliable calibration source. The spectrometer uses the sun as the radiation calibration source, and achieves full aperture, full field of view, and full optical path absolute radiation calibration through the diffuser.

The on orbit calibration mode of the spectrometer includes solar calibration and circuit background calibration.

4.7.1 Solar Calibration Mode

The solar calibration shall be carried out at an appropriate time between the satellite's departure shadow area and the subsatellite point before the daytime. Turn the diffuser to the front of the main lens of the optic-structure, and the solar calibration lasts for about 36s. The solar calibration is performed once a day, and the primary diffuser and standby diffuser switching is performed once a month.

4.7.2 Circuit Background Calibration Mode

The circuit background calibration is carried out in the satellite shadow area. Turn the dark background block to the front of the main lens of the optic-structure, and the dark background calibration lasts for about 36s. The dark background calibration is conducted once a day.

5 System Verification

5.1 Validation Items

The spectrometer has been fully tested and verified in the laboratory. In order to ensure the validity and accuracy of the verification, all verification items are carried out under vacuum conditions. The verification items include radiometric calibration, spectral calibration, stray-light suppression test, polarization sensitivity test, BSDF test, etc. The test results provide a good data basis for the calibration and use of the spectrometer in orbit.

5.2 Validation Results

The laboratory test results of the spectrometer are summarized in the following table (Table 5).

Table 5. Test Results of the spectrometer

Indicator items	Requirements	Configurations	Compliance
Spectrum	670–780 nm	670–780 nm	C
Spectral Resolution	≤0.3 nm	0.3 nm	C
SSI	≤0.15 nm	0.1 nm	C
Swath	≥30 km	34 km	C
SSD	≤1.5 km	1.5 km	C
SNR	≥200	513.5	C
Radiation Calibration	absolute≤5%	4.81%, on orbit	C
	relative≤3%	2.2% in lab	C
Spectrum Calibration	<1/10*FWHM	<1/50*FWHM, on orbit & in lab	C

6 Conclusion

The spectrometer is a precision optical instrument for sun induced vegetation fluorescence detection. The system adopts the technical scheme of pushbroom imaging and grating dispersion light splitting. It has carried out comprehensive task analysis, fine system design and full test verification. All indicators meet and exceed the requirements of the detection task, and has the detection capabilities of high spectral resolution, high signal to noise ratio and high stability. After being officially put into operation, the spectrometer will play a very important role in vegetation monitoring and carbon cycle applications [13, 14].

References

1. Sun, Y., et al.: OCO-2 advances photosynthesis observation from space via solar-induced chlorophyll fluorescence. Science **358**(6360), eaam5747 (2017)
2. Frankenberg, C. et al.: New global observations of the terrestrial carbon cycle from GOSAT: patterns of plant fluorescence with gross primary productivity. Geophys. Res. Lett. **38**(17) (2011)
3. Damm, A., et al.: Far-red sun-induced chlorophyll fluorescence shows ecosystem-specific relationships to gross primary production: an assessment based on observational and modeling approaches. Remote Sens. Environ. **166**, 91–105 (2015)
4. Joiner, J., et al.: Global monitoring of terrestrial chlorophyll fluorescence from moderate-spectral-resolution near-infrared satellite measurements: methodology, simulations, and application to GOME-2. Atmos. Measur. Tech. **6**(10), 2803–2823 (2013)
5. Du, S., et al.: Retrieval of global terrestrial solar-induced chlorophyll fluorescence from TanSat satellite. Sci. Bull. **63**(22), 1502–1512 (2018)
6. Coppo, P., Taiti, A., Pettinato, L., Francois, M., Taccola, M., Drusch, M.: Fluorescence imaging spectrometer (FLORIS) for ESA FLEX mission. Remote Sens. **9**(7), 649 (2017)
7. Drusch, M., et al.: The fluorescence explorer mission concept-ESA's earth explorer 8. IEEE Trans. Geosci. Remote Sens. **55**, 1273–1284 (2016)

8. Guanter, L., et al.: Developments for vegetation fluorescence retrieval from spaceborne high-resolution spectrometry in the O2-A and O2-B absorption bands. J. Geophys. Res. **115**(D19) (2010)
9. Ma, Y., Liu, L., Chen, R., et al.: Generation of a global spatially continuous TanSat solar-induced chlorophyll fluorescence product by considering the impact of the solar radiation intensity. Remote Sens. **12**(13), 2167 (2020)
10. Coppo, P., Pettinato, L., Nuzzi, D., et al.: Instrument predevelopment activities for FLEX mission. Opt. Eng. **58**(7), 075102 (2019)
11. Abdon, S., et al.: Digital correction of residual straylight in FLEX images. In: SPIE Proceedings of the International Conference Space Optics ICSO, Chania, Greece (2018)
12. Taiti, A., Coppo, P., Battistelli, E.: Fluorescence imaging spectrometer optical design. Proc. SPIE **9626**, 96261N (2015)
13. Meroni, M., et al.: Remote sensing of solar-induced chlorophyll fluorescence: review of methods and applications. Remote Sens. Environ. **113**, 2037–2051 (2009)
14. Du, S., Liu, X., Chen, J., Liu, L.: Prospects for Solar-induced chlorophyll fluorescence remote sensing from the SIFIS Onboard the TECIS-1 satellite. J. Remote Sens. 2022, Article ID 9845432, 9 pages (2022)

Unified Approach to Inshore Ship Detection in Optical/radar Medium Spatial Resolution Satellite Images

Mykhailo O. Popov[1], Sergey A. Stankevich[1], Valentyn V. Pylypchuk[1], Kun Xing[2(✉)], and Chunxiao Zhang[2]

[1] Scientific Centre for Aerospace Research of the Earth, National Academy of Sciences of Ukraine, Kyiv 01054, Ukraine
[2] Beijing Institute of Space Mechanics and Electricity, Beijing 100190, China
st@casre.kiev.ua

Abstract. High-resolution optical or radar images are usually used as input data, while ship detection is based mainly on certain kinds of features, namely geometric or spectroradiometric ones. Image processing and analysis algorithms usually differ significantly for optical and radar ones. This circumstance complicates the software for ship detection and requires high-qualified experts' participation. The authors propose an approach to ship detection involving a combination of specific geometric and radiometric features, which is quite possible to extract from a medium spatial resolution multiband satellite image. An essential advantage of the approach is the unification of algorithms for calculating features for both optical and radar satellite images.

The proposed approach is instantiated as the following step-by-step procedure:

In the first step, a mask of the water surface is acquired, where search targets, i.e. ships, can be located. Further processing is carried out within this mask only.

In the second step, anomalies of the water surface, which can be considered as candidates on ships, are detected by geometric features and separately by multidimensional spectroradiometric features. Then, the detected anomalies are presented in a unified probabilistic form. In the third step, the partial maps of geometric and spectroradiometric anomalies are fused into a single joint probability of locked targets using a modified Bayesian rule. Finally, in the fourth step, the decision on the detection is made by comparing the fused probability value with a specified threshold.

Evaluation of the proposed approach, conducted over actual medium resolution Sentinel 1 radar satellite images and Sentinel 2 optical ones, demonstrated fair performance.

Keywords: ship detection · optical/radar satellite image · geometric and spectroradiometric features · probability fusion

H. P. Urbach and H. Jiang (Eds.): ISSOIA 2022, SPPHY 295, pp. 66–79, 2023.
https://doi.org/10.1007/978-981-99-4098-1_8

1 Introduction

UN statistics show that about 90% of the world's cargo is transported by sea [1]. Therefore, many vessels differ in their type, dimensions, tonnage and other specifications inside the coastal maritime zones of port cities.

The management of such ships' conglomerates is impossible without their identification. The Automatic Identification System (AIS) [2] ensures the sea vessels' identification, including those located inside the coastal zone. The AIS provides an identification number from the International Maritime Organization (IMO) for each vessel registered. The vessel's transponder transmits information on the ship's type, current location, heading, speed, etc. [2, 3].

However, AIS solves the problem of the identification of marine vessels only partially. First, according to the protocol, not all vessels are subject to mandatory registration in their register; for example, vessels with a length of fewer than 15 m and a gross tonnage of fewer than 15 tons are not obliged to transmit information to the AIS. Second, in maritime practice, many vessels prefer to be in the so-called "dark" mode without registration in the AIS. Such mode is practiced by vessels trying to hide the fact of their illegal activities, for example, engaged in smuggling operations, drug trafficking, illegal immigration, illegal fishing, performing intelligence, etc.

Meanwhile, prompt and complete information about ships inside the coastal zone is crucial for many organizations and departments, particularly port administrations, border and customs authorities, migration services, security and intelligence agencies, etc. The fact is that inside the coastal zone, along with the problem of identifying ships, there are other urgent tasks, such as identifying the pollution of seawater and other artificial violations of the environment, detection of illegal fishing, smuggling, migrants' transportation, etc. However, the possibilities of solving such problems using AIS only are very limited. Therefore, attempts are being made to develop systems that complement the AIS capabilities [4]. For example, to obtain complete information about ships, their movement and dangerous cargo, the European Maritime Safety Agency (EMSA) established its own SafeSeaNet information system for the movement of ships monitoring [5].

According to experts, the most promising solution is the global system for monitoring the operational situation in coastal marine areas development using satellite technologies [6, 7]. Such systems, which use both optical and radar satellite imagery, are actively investigated [8–11].

The authors propose an approach to ship detection inside the coastal sea area by the medium spatial resolution satellite images analysis. Both optical and radar images can be used for this purpose. Implementation of this approach can provide ships of various sizes detection and allow a selective search of unidentified vessels by length or shape ellipticity.

Further, the paper is presented as follows. Section 2 analyzes the state of the problem and considers the study's motivation. Section 3 describes the methodology of the proposed approach. Section 4 provides a reasoned treatment of the technique. Section 5 presents the results of experimental studies. Section 6 discusses the results obtained and ways to improve the approach developed. Conclusions and future research directions are formulated in the final section.

2 Related Work and Motivation

Remote sensing satellites equipped with on-board optical or radar imagers are used to acquire operational information about the objects inside the coastal region [4, 8–11].

The optical imagers are mainly VNIR/SWIR multispectral systems (the operating wavelengths from 0.4 to 2.5 μm) of high or medium spatial resolution and thermal infrared sensors [12].

The best results in ship detection are achieved using multispectral satellite images of high spatial resolution. Analysis of such images with science-intensive techniques such as knowledge-based methods [13], object-based image analysis [14], and deep learning [15] provide the contours of object extraction, very accurate geometric measurements, automatic interpretation of situations within the sea area, etc.

Some difficulties arise because high spatial resolution satellite imagery is a commercial product of stiff cost. In addition, the image analysis methods listed above require large training samples, but this condition is challenging to accomplish.

Therefore, some recent research and developments are aimed at utilizing medium spatial resolution multispectral imagery, particularly acquired by the Resourcesat-2/2A [16] or Sentinel-2A/2B [17] satellites' on-board imagers. Such imagers' swath must be wide, providing a large area survey with sufficiently high productivity. Also, medium spatial resolution satellite images are usually in the public domain.

Thermal imagers operating in the infrared bands make it possible to detect local temperature anomalies, which are one of the important identifying features of the desired objects [15, 18, 19]. However, information about thermal anomaly alone is not enough to make a decision about the object's presence.

The disadvantage of optical imagers is dependence on the time of day and cloudiness. This disadvantage limits the satellite monitoring possibilities. So, satellite synthetic aperture radar (SAR) imaging technology has quickly developed during the last decades. SAR sensor operates in centimeter/decimeter wavelength bands, hence it can acquire images in any daypart regardless of clouds, rainfall, or atmospheric haze.

The recent developments in the field of satellite systems for radar monitoring of the marine area and detection of various objects are described, in particular, in the [20, 21] papers. In addition, algorithms for processing and analyzing radar imagery are constantly being improved [22–24].

However, several studies [25–27] have noted the following SAR disadvantages:

- the quality of the images acquired and the number of false alarms dependence on local wind gusts, surf waves, storm cells and other natural effects that cause a strong backscattering clutter;
- the restrictions on the ship detection made of dielectric materials such as wood or glass fibre;
- sophisticated SAR design, significant power consumption and high cost.

From the previous, one of the possible ways to improve the satellite monitoring of the coastal zones is the joint operation of optical and radar systems. This approach is proposed, in particular, in the [14, 28]. In [14], object-oriented vessel detection is developed based on the joint analysis of an optical and radar image pair. Both images of the pair must have a close spatial resolution – either high (WorldView and COSMO-SkyMed pair) or medium (Sentinel 2 and Sentinel 1 pair). In [28], it is proposed to engage a 4-band (0.45–0.9 μm) Kompsat-2 camera, 4 m spatial resolution, and Sentinel-1 *C*-band (5.405 GHz) SAR, 10 m spatial resolution for vessels detection in coastal zones. Also, the AVIRIS hyperspectral sensor (224 spectral bands inside the range from 0.4 to 2.5 μm) can be used for this purpose [29].

A comparative analysis of the optical and radar satellite images of marine zones, carried out in the [25, 27], shows the following:

1) The sea surface statistics in images of both types are mainly preserved. For example, if an adequate model of sea surface state is the Rayleigh distribution, then this model remains applicable to images of the sea surface too;
2) The ships' signatures in both types of images form similar compact areas that contrast with the surrounding water. The size of such an area depends on the shiple size and spatial resolution of the imaging system.

Thus, despite significant differences in the principles of operation of optical and radar systems, the images generated by them have similar statistical properties, and marine object detection is based on the analysis of the same features. Reasoning from this, the same algorithms for the efficient detection of vessels in satellite images of both types can be ensured.

The [25] paper describes a two-stage unified algorithm for ship detection in optical and SAR spaceborne images. In the first stage, potential vessel candidates are detected using a discrete wavelet transform (DWT) and constant false alarm rate (CFAR) detector [30]. In the second stage, the false objects are excluded using multiscale reasoning on the contours of the targets.

The [25] method's disadvantages are: the processing procedure is made concerning two constraints: the detection rate should be 100%, and the false alarm rate should be as low as possible. Both of these limitations in practical data processing are difficult to meet.

Therefore, our research proposes a more realistic and no less efficient than the currently known solutions approach to inshore ship detection in optical/radar medium spatial resolution satellite images. This approach involves the satellite multipolarization radar image (SMRI) or satellite multispectral optical image (SMOI) as input data on the coastal area.

3 Methodology

The key issues of object detection are the choice of informative features and methods for their determination, as well as the decision rule.

3.1 Objects Detection Features

In this study, the object geometry (size and shape) and spectroradiometric distinction (optical contrast or radar backscattering) are assumed as the most informative features of an object.

The planned image of a ship in satellite image must have the properties of compactness and convexity. However, due to noise, interference, and other distorting factors, the real image of the ship usually does not retain these properties, which negatively affects the accuracy of measuring the geometric properties.

To extract ship shapes from the satellite image, we used the Local Binary Patterns (LBP) method [31].

The statistical signal separation [32, 33] was used to map spectroradiometric anomalies (contrast or backscattering) in satellite images.

3.2 Local Binary Patterns

The LBP was originally designed in 1996 for texture description [34], but later this method was successfully used in vessel classification problems [35]. An LBP modification used in current research can be applied to multiband images. For example, an SMRI may consist of two bands with different polarizations. SMOI, as a rule, includes several band images acquired simultaneously in corresponding optical spectral bands.

The satellite image LBP processing is a sequential pixel-wise procedure. Any unity pixel representing dryland and artificial objects is considered a growth point, to which neighbouring unity pixels are appended or rejected according to specific rules. Thus, an array object as a candidate for the ship is stacked. After that, any other unity pixel that is not included in the previously formed object is selected as initial, and so on. In this way, the set of candidates for the ship is obtained.

3.3 Spectroradiometric Anomaly Mapping

The rich information content of spectroradiometric anomaly on the seawater surface is noted in a number of studies [15, 18, 23]. The reasons for such features can be the ship's reflectance, emissivity, backscattering, etc.

Currently, the CFAR algorithm is most often used to detect spectroradiometric anomalies in the sea zone [30]. This algorithm provides anomaly detection with a fixed false alarm probability but does not guarantee the value of correct detection. Our approach differs from the CFAR because it considers the full probability of error, not just the false alarm probability. In addition, the decision threshold is not yet specified because it will be applied later – after data fusion with other sources.

3.4 Probability Fusion and Decision Making

Let's assume that object detection depends on the N probability values of N probabilities $p_n, n = 1, ..., n$. The decision on object detection is denoted as $d = 1$ (detected) or $d = 0$ (no object). The decision is made using probabilities p_n.

If the probabilities p_n are known, then, according to the modified Bayes rule proposed in [36], the posterior detection probability can be calculated as

$$P^{ps}(p_1, ..., p_N) = \frac{P^{pr} \cdot \psi(p_1, ..., p_N|1)}{P^{pr} \cdot \psi(p_1, ..., p_N|1) + (1 - P^{pr}) \cdot \psi(p_1, ..., p_N|0)} \quad (1)$$

where $\psi(\cdot)$ is a joint probability density of detection, P^{pr} is a prior probability of detection.

If the probabilities $p_1, \ldots, p_N$ are mutually independent, then

$$\psi(p_1, \ldots, p_N|d) = \prod_{n=1}^{N} \psi(p_n|d), \quad d \in \{0, 1\} \tag{2}$$

Substituting (2) into (1) gives

$$P^{ps}(p_1, \ldots, p_N) = \left(1 + \left(\frac{P^{pr}}{1 - P^{pr}} \cdot \frac{\prod_{n=1}^{N} \psi(p_n|1)}{\prod_{n=1}^{N} \psi(p_n|0)}\right)^{-1}\right)^{-1} \tag{3}$$

where the $\frac{\prod_{n=1}^{N} \psi(p_n|1)}{\prod_{n=1}^{N} \psi(p_n|0)}$ term is the likelihood ratio.

In [36], it is proposed to consider the p_n probability as a random variable with a beta distribution. Then the distribution density $\psi(\cdot)$ of p_n $(0 \le p_n \le 1)$ can be determined by equation [37]:

$$\psi(p) = \frac{p^{\alpha-1}(1-p)^{\beta-1}}{B(\alpha, \beta)} \tag{4}$$

where $B(\alpha, \beta)$ is a beta function $B(\alpha, \beta) = \frac{\Gamma(\alpha)\cdot\Gamma(\beta)}{\Gamma(\alpha+\beta)}$; here $\Gamma(\cdot)$ is a gamma function; α and β are parameters, $\alpha > 0$, $\beta > 0$.

The [36] assumption is due to two considerations. First, the density of the beta distribution is defined on the interval [0, 1], which coincides with the probability range. Second, due to the presence of two parameters α and β, the beta function can be easily transformed through these parameters, which allows controlling the probability "weights" in the likelihood ratio.

Under the beta distribution with fixed the α and β parameters, it is possible to estimate the posterior probability of object detection $(d = 1)$ by (3) with known probabilities p_n.

To do this, the symmetric beta likelihood model [36] is applied, wherein $\alpha > \beta$:

$$\frac{\psi(p_n|1)}{\psi(p_n|0)} = \frac{p_n^{\alpha_n-1}(1-p_n)^{\beta_n-1}}{p_n^{\beta_n-1}(1-p_n)^{\alpha_n-1}} = \left(\frac{p_n}{1-p_n}\right)^{\alpha_n-\beta_n} \tag{5}$$

Then the likelihood ratio will be rewritten as:

$$\frac{\prod_{n=1}^{N} \psi(p_n|1)}{\prod_{n=1}^{N} \psi(p_n|0)} = \prod_{n=1}^{N} \left(\frac{p_n}{1-p_n}\right)^{\alpha_n-\beta_n} \tag{6}$$

Equation (6) significantly simplifies the calculation of the posterior probability of object detection. Taking the $P^{pr} = 0.5$ substituting (6) into Eq. (3) gives

$$P^{ps}(p_1, \ldots, p_N) = \left(1 + \prod_{n=1}^{N} \left(\frac{p_n}{1-p_n}\right)^{\beta_n-\alpha_n}\right)^{-1} \tag{7}$$

The decision about the object's presence is made by comparing the (7) value with a certain threshold. Bayesian theory recommends the threshold, which ensures minimum risk over the entire decision set [38]. However, other recommendations regarding the threshold value may be considered too.

4 Proposed Approach

Let's assume that input data are a satellite image (SMRI or SMOI) of the coastal zone, and the following parameters are known: $L_{\max}$ is the maximum allowable length of ships under detection; P_{thr} is the threshold value of detection probability.

The proposed approach is instantiated as the following step-by-step procedure (Fig. 1).

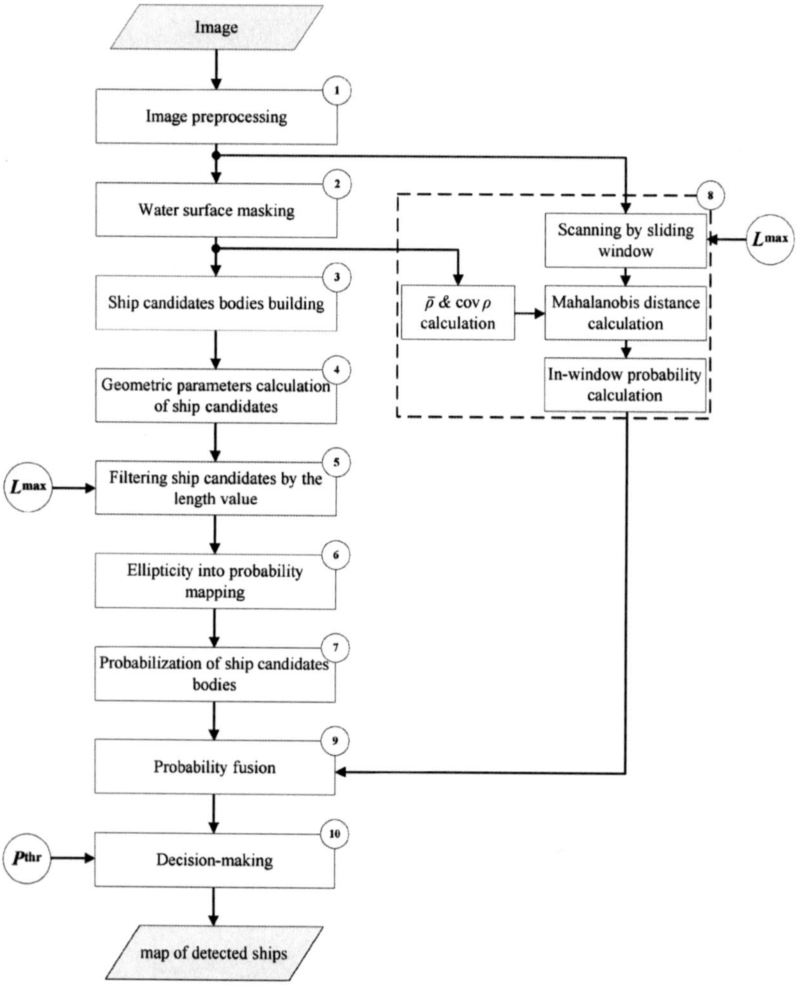

Fig. 1. A general flowchart of the proposed approach

Step 1: *Image preprocessing.*

The image preprocessing includes geometric, radiometric and atmospheric correction, conversion of raw DN pixels into land surface reflectance for optical images or into backscattering coefficients for radar ones, speckle noise filtering, other image improvement operations and removal or mitigation of environmental effects such as clouds, shadows, rainfall, haze etc.

Step 2: *Water surface masking.*

A pixel mask of the water surface is created. This is done by one of the known algorithms of unsupervised image classification, for example, *k*-nearest neighbor (*k*NN) or IsoData [39, 40].

Step 3: *Ship candidates' bodies detection.*

The compact objects, i.e. the bodies of ships' candidates, are formed by pixel-wise image processing with the LBP algorithm.

Step 4: *Geometric parameters estimation of ship candidates' bodies.*

The geometric dimensions of each ship candidate – length L and width B – are measured over compact objects, and then the ellipticity of each object is calculated as

$$\varepsilon = \frac{L^2 - B^2}{L^2 + B^2} \tag{8}$$

Step 5: *Filtering ship candidates by the length value.*

All potential ship candidates whose length L is greater than the threshold L_{thr} are rejected.

Step 6: *Ellipticity into probability converting.*

The entire domain of ellipticity values is transformed into the probability axis. The normal distribution $\mathbb{N}(\mu_\varepsilon, \sigma_\varepsilon)$ is used as the transform operator, where μ_ε and σ_ε are the mean value and standard deviation of ellipticity, respectively. The values of μ_ε and σ_ε can be determined, for example, by statistical analysis of data on the geometric dimensions of various types of ships stored in the AIS database.

Step 7: *Probabilization of ship candidates' bodies.*

Probabilization means that the probability value for the ellipticity of the candidate object is assigned to all pixels of that object's body.

Step 8: *Sea surface spectroradiometric anomaly probabilistic mapping.*

This step estimates the spatial distribution of spectroradiometric anomaly probabilities within the sea surface (see block 8 in Fig. 1).

Each input image pixel (SMRI or SMOI) is represented by a vector of backscattering or spectral reflectances $\langle\rho\rangle = \{\rho_1, \rho_2, ..., \rho_m\}$, where m is the number of polarization/spectral bands. Also, for each band, a minimum reflectance threshold $\tau_{\min}$ is assigned, in comparison with which allows to pre-separate the water from other objects within the scene. Finally, the statistical characteristics – the means vector $\overline{\rho}$ and the covariance matrix cov ρ are calculated over the extracted water mask.

The scanning with a sliding window, the size of which is matched with the expected target size L is carried out within a given area of the sea surface. The Mahalanobis statistical distance value b^2 between the in-window pixels and the background is calculated at each position of the sliding window:

$$b^2 = \Delta\rho^{\mathrm{T}} \times (\mathrm{cov}\rho)^{-1} \times \Delta\rho \tag{9}$$

where $\Delta\rho$ is the means vector difference between the in-window sample and background [41].

The anomaly detection probability p will be [42]:

$$p \cong 1 - \Phi\left(-\frac{b^2}{2\sqrt{b^2 + \frac{m}{n} + \frac{m}{w}}}\right) \quad (10)$$

where n is the number of samples in the background, w is the number of pixels in the sliding window, $\Phi(x) = \frac{1}{\sqrt{2\pi}}\int_{-\infty}^{x} e^{-\frac{z^2}{2}}\,dz$ is the cumulative distribution function of the standard normal distribution [43].

With increasing sample size $n \gg m$, the anomaly detection probability (10) asymptotically tends to [44]:

$$p \cong e^{-\frac{b^2}{8}} \quad (11)$$

The Eq. (11) estimates the anomaly detection probability of size specified by the sliding window size in each pixel of the input multilayer image and thereby obtains a probability map of the ship candidates.

Step 9: *Probability fusion.*

So, in steps 7 and 8, partial probabilities of ship candidates by LBP and spectroradiometric anomalies were obtained. Then, in step 9, these two probabilities are pixel-by-pixel merged into a common a posteriori probability P^{ps} by Eq. (7). The output is the spatial distribution of the ship's presence probability in the observed marine area image.

Step 10: *Decision-making.*

A final decision is made on each object detected. In particular, the decision can be made by comparing the P^{ps} probability with the threshold P_{thr}: the detected object is a ship if $P^{ps} > P_{thr}$.

Step 11: *Output a map of the ships detected.*

5 Experiments

The developed ship detection technique was tested using satellite radar and optical images over the coastal zone in the vicinity of Quindao (China) seaport. Radar images were acquired by the on-board *C*-band SAR of the Sentinel 1A satellite. Optical images were acquired by an on-board multispectral instrument of Sentinel 2A/2B satellites.

Dual polarization (VH and VV) IW GRDH radar data products [45] of 10 m resolution and four bands VNIR (0.443–0.665 μm) Level 2A optical images [46] of 10 m resolution too were used.

A typical test sample of objects in the image included 80–90 vessels of various types. Examples of radar and optical images used in testing the technique for ship detection are shown in Fig. 2.

The probabilization was performed using the normal distribution law. The ship geometry data were captured from the vessels database of AIS (https://www.vesselfinder.com/vessels). The ship ellipticity was calculated by (8) for all ships registered, and then the mean value and standard deviation of this feature were estimated.

Probability maps estimated by radar and optical images are shown in Fig. 3 and Fig. 4 correspondingly.

a *b*

Fig. 2. Examples of optical and radar images: *a* – Sentinel 1A dual polarization false color composite radar image. *b* – Sentinel 2A VNIR bands natural color composite optical image

a *b* *c*

Fig. 3. Ship detection probability mapping by Sentinel 1A radar image: *a* – Probability map derived from the object's geometry. *b* – Probability map of spectroradiometric anomalies. *c* – Map of fused probabilities P^{ps} of ship detection

The resulting probability of ship detection P^{ps} was estimated with (7), while the difference between the beta distribution parameters was $\alpha - \beta = 2$.

The precision and the false alarm probability were used as a quantitative measure of the developed technique's performance. The experiments demonstrated the following ship detection performance: for radar images, the precision is 0.874, the false alarm probability is 0.125; for optical images, the precision is 0.855, and the false alarm probability is 0.145.

a b c

Fig. 4. Ship detection probability mapping by Sentinel 2B optical image: *a* – Probability map derived from the object's geometry. *b* – Probability map of spectroradiometric anomalies, *c* – Map of fused probabilities P^{ps} of ship detection

6 Discussion

In general, the test evaluation demonstrates the good efficiency of the proposed approach for ship detection within the sea area using both radar and optical imagery.

The proposed approach can detect ships of various sizes, but, on the other hand, by choosing appropriate thresholds of the ship length and the ellipticity value, the approach can provide the adaptive selection of vessels detected.

At the same time, a comparative analysis of satellite images, as well as the detection probability estimates obtained in experiments, clearly shows that the satellite optical and radar images' features impose some limitations on the detection performance, viz. Both the precision and the false alarm probability.

The ship detection probability in the optical image is affected by local statistical inhomogeneity (fragmentation) of the sea surface. The fragmentation makes it difficult to detect spectroradiometric anomaly, which is an important feature of an artificial object. In addition, the issue of filtering clouds and cloud shadows in satellite optical images remains open.

In radar images, the adequate display of the object's actual shape is complicated: firstly, by the strong corner re-scattering from the ship's deck and, secondly, by filtering the speckle noise inherent in radar imagery.

During the experiments, quite a few cases were revealed when the false alarms differed in optical and radar images of the same sea scene. The main reason is the unequal sensitivity of optical and radar systems to natural phenomena. For optical images, the false alarm sources are clouds and whitecaps. They have a high reflectance compared to the sea surface and can be close to vessels. The main false alarm sources for SAR data are local wind effects, breaking waves, storm cells, and all high local backscatter striking.

Various ways can be used to improve ship detection's precision within the marine area. For optical imagers, this may be the optimal selection of spectral bands. For radar systems, this is an extended use of multipolarization capabilities. However, the most promising way to improve detection precision is optical and radar data fusion.

7 Conclusion

The paper proposes and explores a new approach to inshore ship detection in optical/radar medium spatial resolution satellite images. The methodology grounded in the proposed approach is described, and a step-by-step procedure for one's implementation is provided.

The approach was tested on actual satellite optical and radar images. The results demonstrated the realism and fair performance of the approach developed.

Future research of the authors is planned to be directed to the engineering implementation of the developed approach, as well as to the algorithm optimization for the objects feature extraction, probability fusion and decision-making.

References

1. Stalker, P. (ed.): Review of Maritime Transport 2021. United Nations Publications, New York (2021)
2. Metcalfe, K., et al.: Using satellite AIS to improve our understanding of shipping and fill gaps in ocean observation data to support marine spatial planning. J. Appl. Ecol. **55**(4), 1834–1845 (2018). https://doi.org/10.1111/1365-2664.13139
3. Proud, R., Browning, P., Kocak, D.M.: AIS-based mobile satellite service expands opportunities for affordable global ocean observing and monitoring. In: Proceedings of the MTS/OTS OCEANS'2016 Conference, pp. 1–8. IEEE, Monterey (2016). https://doi.org/10.1109/OCEANS.2016.7761069
4. Heiselberg, H., Stateczny, A.: Remote sensing in vessel detection and navigation. Sensors **20**(20), 5841 (2020). https://doi.org/10.3390/s20205841
5. EMSA. Vessel traffic monitoring in EU waters (SafeSeaNet). http://www.emsa.europa.eu/ssn-main.html. Accessed 18 Nov 2022
6. Chintoan-Uta, M., Silva, J.R.: Global maritime domain awareness: a sustainable development perspective. WMU J. Marit. Aff. **16**(1), 37–52 (2016). https://doi.org/10.1007/s13437-016-0109-5
7. Reggiannini, M., et al.: Remote sensing for maritime prompt monitoring. J. Mar. Sci. Eng. **7**(7), 202 (2019). https://doi.org/10.3390/jmse7070202
8. Corbane, C., Najman, L., Pecoul, E., Demagistri, L., Petit, M.: A complete processing chain for ship detection using optical satellite imagery. Int. J. Remote Sens. **31**(22), 5837–5854 (2010). https://doi.org/10.1080/01431161.2010.512310
9. Shajini, V.S., Kumar, K.M.: Analysis of ship detection techniques in remote sensing image. Int. Res. J. Eng. Technol. **5**(11), 832–835 (2018)
10. Kanjir, U., Greidanus, H., Oštirc, K.: Vessel detection and classification from spaceborne optical images: a literature survey. Remote Sens. Environ. **207**, 1–26 (2018). https://doi.org/10.1016/j.rse.2017.12.033
11. Li, B., Xie, X., Wei, X., Tang, W.: Ship detection and classification from optical remote sensing images: a survey. Chin. J. Aeronaut. **34**(3), 145–163 (2021). https://doi.org/10.1016/j.cja.2020.09.022
12. Cheng, G., Han, J.: A survey on object detection in optical remote sensing images. ISPRS J. Photogramm. Remote. Sens. **117**, 11–28 (2016). https://doi.org/10.1016/j.isprsjprs.2016.03.014

13. Mazzarella, F., Arguedas, V.F., Vespe, M.: Knowledge-based vessel position prediction using historical AIS data. In: Proceedings of International Conference on Sensor Data Fusion: Trends, Solutions, Applications (SDF 2015), pp. 1–6. IEEE, Bonn (2015). https://doi.org/10.1109/SDF.2015.7347707
14. Aiello, M., Vezzoli, R., Gianinetto, M.: Object-based image analysis approach for vessel detection on optical and radar images. J. Appl. Remote Sens. **13**(1), 014502 (2019). https://doi.org/10.1117/1.JRS.13.014502
15. Shao, J., Yang, Q., Luo, C., Li, R., Zhou, Y., Zhang, F.: Vessel detection from nighttime remote sensing imagery based on deep learning. IEEE J. Sel. Top. Appl. Earth Observations Remote Sens. **14**, 12536–12544 (2021). https://doi.org/10.1109/JSTARS.2021.3125834
16. Ghosh, S., et al.: On-board ship detection for medium resolution optical sensors. Sensors **21**(9), 3062 (2021). https://doi.org/10.3390/s21093062
17. Štepec D., Martinčič T., Skočaj, D.: Automated system for ship detection from medium resolution satellite optical imagery. In: Proceedings of the MTS/OTS OCEANS'2019 Conference, pp. 1–10. IEEE, Seattle (2019). https://doi.org/10.23919/OCEANS40490.2019.8962707
18. Miao, R., Jiang, H., Tian, F.: Robust ship detection in infrared images through multiscale feature extraction and lightweight CNN. Sensors **22**(3), 1226 (2022). https://doi.org/10.3390/s22031226
19. Kiruba, K.: A survey on ship detection in remote sensing images. J. Anal. Comput. **XII**(I), 2861 (2019)
20. Ouchi, K.: Recent trend and advance of synthetic aperture radar with selected topics. Remote Sens. **5**(2), 716–807 (2013). https://doi.org/10.3390/rs5020716
21. Li, J., Xu, C., Su, H., Gao, L., Wang, N.: Deep learning for SAR ship detection: past, present and future. Remote Sens. **14**(11), 2712 (2022). https://doi.org/10.3390/rs14112712
22. Shu, G., Chang, J., Lu, J., Wang, Q., Li, N.: A novel method for SAR ship detection based on eigensubspace projection. Remote Sens. **14**(14), 3441 (2022). https://doi.org/10.3390/rs14143441
23. Iervolino, P., Guida, R., Lumsdon, P., Janoth, J., Clift, M., Minchella, A., Bianco, P.: Ship detection in SAR imagery: a comparison study. In: Proceedings of International Geoscience and Remote Sensing Symposium (IGARSS 2017), pp. 2050–2053. IEEE, Fort Worth (2017). https://doi.org/10.1109/IGARSS.2017.8127384
24. Chang, Y.-L., Anagaw, A., Chang, L., Wang, Y.C., Hsiao, C.-Y., Lee, W.-H.: Ship detection based on YOLOv2 for SAR imagery. Remote Sens. **11**(7), 786 (2019). https://doi.org/10.3390/rs11070786
25. Jubelin, G., Khenchaf, A.: A unified algorithm for ship detection on optical and SAR spaceborne images. Proc. SPIE **9244**, 924415 (2014). https://doi.org/10.1117/12.2067154
26. Leng, X., Ji, K., Zhou, S., Xing, X.: Ship detection based on complex signal kurtosis in single-channel SAR imagery. IEEE Trans. Geosci. Remote Sens. **57**(9), 6447–6461 (2019). https://doi.org/10.1109/TGRS.2019.2906054
27. Yang, C.-S., Kim, T.-H.: Integration of SAR and AIS for ship detection and identification. Proc. SPIE **8372**, 83720A (2012). https://doi.org/10.1117/12.920359
28. Park, K.-A., Park, J.-J., Jang, J.-C., Lee, J.-H., Oh, S., Lee, M.: Multi-spectral ship detection using optical, hyperspectral, and microwave SAR remote sensing data in coastal regions. Sustainability **10**(11), 4064 (2018). https://doi.org/10.3390/su10114064
29. Vane, G., Green, R.O., Chrien, T.G., Enmark, H.T., Hansen, E.G., Porter, W.M.: The airborne visible/infrared imaging spectrometer (AVIRIS). Remote Sens. Environ. **44**(2–3), 127–143 (1993). https://doi.org/10.1016/0034-4257(93)90012-M
30. Gandhi, P.P., Kassam, S.A.: Analysis of CFAR processors in nonhomogeneous background. IEEE Trans. Aerosp. Electron. Syst. **24**(4), 427–445 (1988). https://doi.org/10.1109/7.7185

31. Ojala, T., Pietikäinen, M., Mäenpää, T.: Multiresolution gray-scale and rotation invariant texture classification with local binary patterns. IEEE Trans. Pattern Anal. Mach. Intell. **24**, 971–987 (2002). https://doi.org/10.1109/TPAMI.2002.1017623
32. Cardoso, J.-F.: Blind signal separation: statistical principles. Proc. IEEE **86**(10), 2009–2025 (1998). https://doi.org/10.1109/5.720250
33. Stankevich, S.A., Gerda, M.I.: Small-Size target's automatic detection in multispectral image using equivalence principle. Cent. Eur. Res. J. **6**(1), 1–9 (2020)
34. Ojala, T., Pietikäinen, M., Harwood, D.: A comparative study of texture measures with classification based on featured distributions. Pattern Recogn. **29**(1), 51–59 (1996). https://doi.org/10.1016/0031-3203(95)00067-4
35. Zhao, H., Wang, Q., Huang, J., Wu, W., Yuan, N.: Method for inshore ship detection based on feature recognition and adaptive background window. J. Appl. Remote Sens. **8**(1), 083608 (2014). https://doi.org/10.1117/1.JRS.8.083608
36. Haigang, S., Zhina, S.: A novel ship detection method for large-scale optical satellite images based on visual LBP feature and visual attention model. Int. Arch. Photogrammetry Remote Sens. Spat. Inf. Sci. **XLI-B3**, 917–921 (2016). https://doi.org/10.5194/isprs-archives-XLI-B3-917-2016
37. Knysztofowicz, R., Long, D.: Fusion of detection probabilities and comparison of multisensor systems. IEEE Trans. Syst. Man Cybern. **20**(3), 665–677 (1990). https://doi.org/10.1109/21.57281
38. Krishnamoorthy, K.: Handbook of Statistical Distributions with Applications. Chapman & Hall/CRC, New York (2006). https://doi.org/10.1201/9781420011371
39. Ash, R.B.: Basic Probability Theory. Dover Publications, New York (2008)
40. Duda, R.O., Hart, P.E., Stork, D.G.: Pattern Classification, 2nd edn. Wiley-Interscience, New York (2001)
41. Kobayashi, Y.: Effects of numerical errors on sample Mahalanobis distances. IEICE Trans. Inf. Syst. **E99.D**(5), 1337–1344 (2016). https://doi.org/10.1587/transinf.2015EDP7348
42. Mešalkin, L.D., Serdobol'skiy, V.I.: Classification errors in the case of multidimensional distributions. Theory Probab. its Appl. **XXIII**(4), 772–781 (1978). https://doi.org/10.1137/1123090
43. Gallego, G., Cuevas, C., Mohedano, R., García, N.: On the Mahalanobis distance classification criterion for multidimensional normal distributions. IEEE Trans. Signal Process. **61**(17), 4387–4396 (2013). https://doi.org/10.1109/TSP.2013.2269047
44. Fukunaga, K.: Introduction to Statistical Pattern Recognition, 2nd edn. Academic Press, San Diego (1990). https://doi.org/10.1016/C2009-0-27872-X
45. Schmidt, K., Schwerdt, M., Miranda, N., Reimann, J.: Radiometric comparison within the Sentinel-1 SAR constellation over a wide backscatter range. Remote Sens. **12**(5), 854 (2020). https://doi.org/10.3390/rs12050854
46. Szantoi, Z., Strobl, P.: Copernicus Sentinel-2 calibration and validation. Eur. J. Remote Sens. **52**(1), 253–255 (2019). https://doi.org/10.1080/22797254.2019.1582840

Research on Reinforcement Method of Optical Cable Assembly Used in Space Optical System

Chen Xiaolin[1(✉)], Zhou Yao[2], Zhang Lei[1,2], and Liu Tao[1,2]

[1] Avic Jonhon Optronic Technology Co, Ltd., Luoyang, China
chenxiaolin@jonhon.cn
[2] China Academy of Space Technology, Beijing, China

Abstract. With the rapid increase ofm the amount of data transmitted by spacecraft, optical fiber communication has become an ideal solution for spacecraft big data transmission due to its high bandwidth, small size, light weight, strong electromagnetic interference resistance, and high confidentiality. This article analyzes several typical structures of the optical cable components inside and outside the cabin, respectively. In view of the bending radius of the optical cable assembly and the insufficient radiation resistance, a reinforcement scheme is proposed to effectively improve the aerospace reliability of the optical cable assembly, and the application scenarios of optical cable assemblies with various structures are proposed.

Keywords: Aerospace · Optical Cable Assembly · Reinforcement Method

1 Introduction

The data transmission system used in traditional spacecraft, such as 1553B bus and 100 Mbit Ethernet, has some shortcomings, such as large volume and weight, short transmission distance, low transmission rate and poor anti-interference ability. With the continuous improvement of spatial resolution, spectral resolution, temporal resolution and quantization depth of spacecraft such as the new generation of satellites, the high resolution spectral image data has increased dramatically. At present, the microwave data transmission technology and other electrical signal transmission technologies commonly used in spacecraft communication can no longer meet the requirements of large data volume and high-speed transmission. The optical fiber communication network technology has the advantages of bandwidth, small volume, light weight, strong anti-electromagnetic interference, and high confidentiality. It is an ideal solution for large data volume and high-speed transmission of spacecraft. Optical cable assembly will replace cable assembly and become a new data transmission channel connecting spacecraft equipment.

The optical fiber communication system is composed of a variety of optoelectronic devices and optical elements, of which optical fiber and cable are the key media for optical signal transmission. However, compared with cables, optical cables have shortcomings such as insufficient stiffness, strict requirements for bending radius, and poor radiation

H. P. Urbach and H. Jiang (Eds.): ISSOIA 2022, SPPHY 295, pp. 80–91, 2023.
https://doi.org/10.1007/978-981-99-4098-1_9

resistance. In particular, the harsh radiation environment in the space will have different degrees of radiation effects on optical fibers, resulting in increased loss.

This article analyzes several typical structures of the optical cable assembly inside and outside the cabin. To solve the problems such as insufficient stiffness, bending radius and radiation resistance of optical cable assembly, find solutions to improve the use reliability of optical cable components in space environment.

2 Typical Aerospace Optical Cable and Optical Cable Assembly Structure

2.1 Typical Optical Cable Structure

At present, there are two types of aerospace optical cables: single-core optical cable and multi-core optical cable. The single-core optical cable structure is shown in Fig. 1. The multi-core optical cable is shown in Fig. 2.

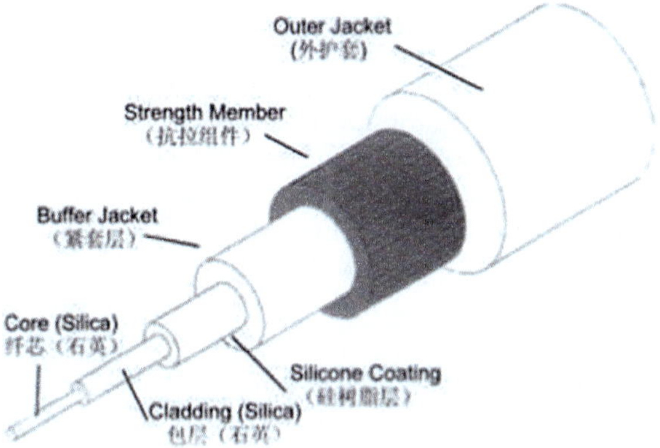

Fig. 1. Single-core optical cable structure schematic diagram

Single-core optical cable is usually composed of outer jacket, strength member, tight sleeve layer, coating, cladding and fiber core, among which coating, cladding and fiber core constitute radiation resistant optical fiber.

Multi-core optical cables are usually composed of outer jacket, strength member, spiral armor tubes, internal protective sleeve, and multiple anti-radiation fiber optic fiber.

As the optical cable used in aerospace environment needs to meet the requirements of thermal vacuum gas release and radiation resistance, the outer jacket, internal protective sleeve and tight sheath of optical cable are made of fluoroplastic materials.

Multi-core optical cable has good resistance to pressure and bending due to the existence of spiral armor tube. The single core optical cable has the disadvantages of poor compression and bending resistance due to its small diameter and thin thickness of each component material.

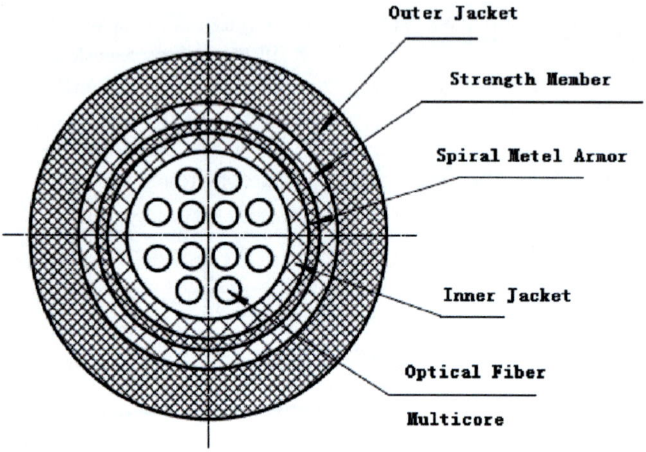

Fig. 2. Multi-core optical cable structure schematic diagram

2.2 Typical Optical Cable Assembly Structure

Optical cable assembly are mainly composed of connectors and optical cables. For multi-core optical cable assembly, the typical structure is shown in Fig. 3. Due to the existence of the spiral armor tube of the multi-core optical cable, its own elasticity can effectively support the weight of the connector. After installation and laying, the optical cable at the end of the connector can meet the requirements of the minimum bending radius of the optical cable under the effect of its own elasticity.

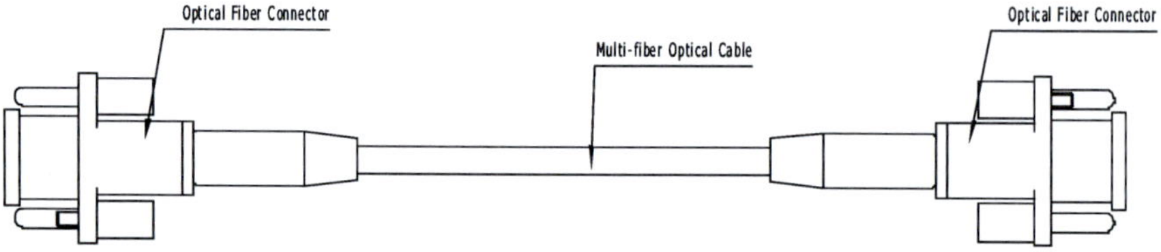

Fig. 3. Structural diagram of typical multi-core optical cable assembly

The structure of the optical cable assembly composed of traditional multiple single-core optical cables is shown in Fig. 4, and its production method is to use brocade silk sheath or nylon sheath and other materials for clustering. During the actual production, installation and laying process, it was found that due to the narrow installation space of the aircraft, when the installation personnel installed, the connector sagged under its own gravity, driving the optical cable will cause about 180° bending, and the actual bending radius is far less than the minimum allowable bending radius of the optical cable. After the connector is installed, when the optical cable at the tail of the connector is naturally sagging, there is also a condition that the optical cable is less than the minimum bending radius. Therefore, for the aerospace optical cable assembly made of single core optical cable, it is necessary to improve the traditional bunching method and conduct bending reinforcement to make the optical cable assembly meet the minimum bending radius requirements during installation and laying.

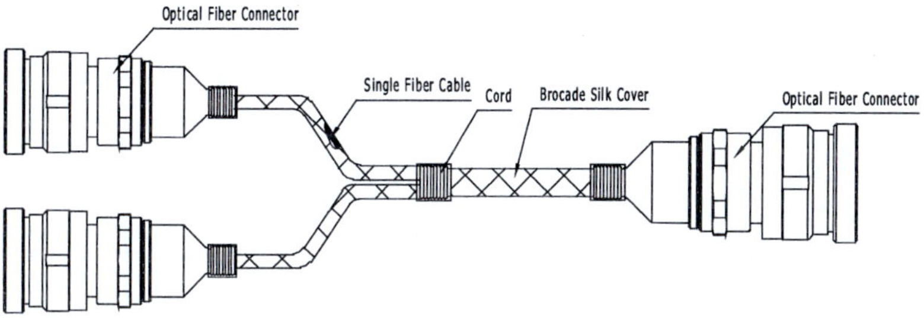

Fig. 4. Structural diagram of typical single-core optical cable assembly

3 Radiation Effect of Optical Fiber and Cable

The role of γ-rays on optical fiber mainly includes photoelectric effects, Compton effects, electronic effects and scattering effects. In addition, there are fluorescence effect and Cerenkov effect. When γ radiation acts on the fiber, the absorption loss of fiber will increase, the fiber refractive index will change. The macro-effect of γ-rays on the electromagnetic effect of optical fiber, whether it is to increase the loss of optical fiber absorption or change the refractive index of the fiber, will cause optical fiber loss to increase. This optical fiber loss due to radiation is called radiation induced loss.

Radiation induced loss [1] can be explained from two aspects: first, radiation changes the color center of optical fiber; Second, the radiation changes the refractive index distribution of fiber cladding and core diameter.

Color center refers to the absorption of specific spectrum caused by impurities, atomic defects, irregular arrangement of atoms or carriers bound around some missing or impurity atoms in optical fibers. The color center corresponds to a series of electronic energy levels. The existence of these energy levels is easy to generate resonance absorption and electronic transition for the transmission photons, which is shown by the specific absorption spectrum of the transmission photons in the optical fiber. Under the action of ionizing radiation, the chemical bond of the optical fiber breaks, resulting in the generation of new defect atoms or irregularly arranged atoms, increasing the concentration of defect atoms, which also leads to the corresponding increase of optical fiber absorption loss. The life of color centers is very short, and their recombination will prevent new color centers from continuing to form. Therefore, within a certain radiation dose range, the radiation induced loss of optical fiber has a linear relationship with the dose, but with the increase of the dose, the radiation induced loss of optical fiber will show a saturated trend. And the general rule is that the absorption coefficient of long wavelength is small, and the absorption coefficient of short wavelength is large. Figure 5 [2] shows that in 1.0×10^5 rad (Si)/h dose rate irradiation, the attenuation loss of signal light at different wavelengths changes with the irradiation dose. It can be seen that the loss shows a saturated trend with the increase of dose, indicating that the number of absorption defects is limited. Once the defects are exhausted, no new absorption center will be formed. The radiation damage of 980 nm wavelength is far greater than that of other wavelengths, because the absorption of near-infrared light zone is mainly caused by the tail of ultraviolet absorption and the strong absorption at 630 nm. Therefore, the overall trend of infrared zone is that the shorter the wavelength, the more serious the damage.

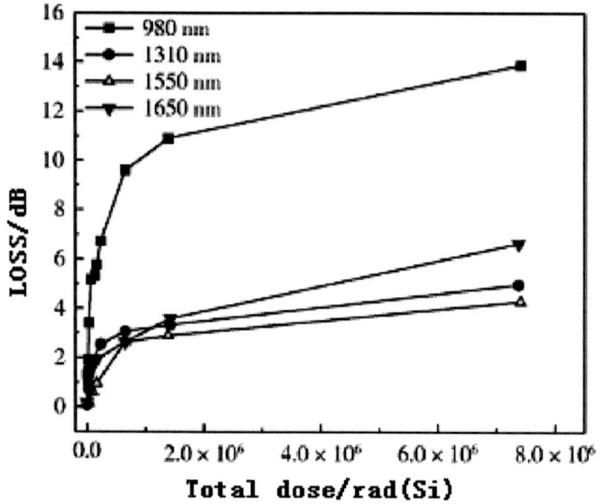

Fig. 5. The loss of the fiber as a function of total dose (dose rate 1.0×10^5 rad(Si)/h)

In general, for medium and low dose rate levels, the radiation induced loss increases linearly with the increase of total dose with a slow trend. At a higher dose rate, with the increase of total dose, the radiation induced loss increases at a faster rate. Because in this case, the defect center in the optical fiber is formed faster and moves in a large range, and the defect healing requires enough time to form more absorption bands. For the same total dose, the radiation resistance of quartz fiber at low dose rate is better than that at high dose rate, which is mainly due to the difference in the recovery effect after irradiation in the two cases.

Liu Fangxin [3] et al. used a Co^{60} radiation source, and the radiation dose rate was controlled at 50–3 $\times$ 10^4 rad/min, the radiation resistance of three kinds of optical fibers is studied. A large number of experiments have confirmed that when the radiation dose is less than 200 rad, the radiation induced loss is linear with the dose. In the range of 200–10^5 rad with moderate radiation dose, the relationship between them is power exponential. When the dose is greater than 105 rad, the change of radiation induced loss tends to be flat and saturated with the increase of radiation dose. The approximate relationship between radiation induced loss and dose of 850 nm optical fiber is given in combination with relevant literature.

$$\alpha_D = \begin{cases} \alpha D & (D \leq 200\,\text{rad}) \\ kD^b & (200\,\text{rad} < D \leq 10^5\text{rad}) \\ c & (D > 10^5\, rad) \end{cases} \quad (1)$$

in formula (1): α_D is radiation induced loss, D is dose, a,b, c, k are constants, which are related to the material and composition of the optical fiber.

Radiation changes the color center of optical fiber and also changes the refractive index. The change of refractive index is mainly caused by the change of dielectric constant, which is ultimately caused by the change of atomic energy level. The change of optical fiber refractive index affects the selection ability of optical fiber to waveguide mode, resulting in the corresponding increase of waveguide loss. Therefore, the radiation induced loss of optical fiber is the result of two mechanisms: the change of color center and the change of refractive index.

It can be seen from the electromagnetic field distribution in the fiber that the light wave power of single-mode fiber is mainly concentrated in the fiber core diameter, while the light wave mode field of multi-mode fiber is relatively dispersed. Under the same irradiation conditions, the absorbed radiation dose of single-mode fiber is relatively small, so the radiation induced loss of single-mode fiber is lower than that of multi-mode fiber.

The test data shows that the radiation induced loss of single-mode optical fiber after being irradiated by 18 Mrad (Si) at 1310 nm/1550 nm, which is commonly used, is $\leq$2 dB/100 m. However, after the multi-mode OM3 optical fiber is irradiated by 100 Krad (Si) at the commonly used 850 nm wavelength, the radiation induced loss is $\leq$15 dB/100 m. At the same time, fluoropolymer materials such as the outer sheath, tight sleeve layer and coating layer in the optical cable will become brittle after being exposed to high dose irradiation outside the cabin, and their protection to the optical fiber will be reduced.

It can be seen that under the high dose irradiation environment outside the spacecraft, multimode optical cables need to be irradiated and reinforced to reduce the radiation induced loss and the damage to the polymer protective layer caused by irradiation.

4 Reinforcement Method of Optical Cable Components

4.1 Bending Resistance and Reinforcement Method of Optical Cable Assembly

Bending Rreinforcement Principle. Fiber winding molding technology is a kind of processing technology to wrap fiber material evenly on the core mold surface as far as possible. By controlling the movement of fiber and core mold, the fiber can completely cover the core mold surface and form the thickness to meet the strength requirements. Fiber winding process can be divided into two categories according to its forming method: outer winding method and inner winding method. The outside winding method is to wind fibers on the outer surface of the core mold. The geometry of the fiber winding member formed in this way is determined by the shape of the outer surface of the core mold. The fiber distribution is reasonable and anisotropic during the winding process, and the finished product after winding has a good strength-to-weight ratio. As the direction, level and quantity of fiber arrangement can be determined according to the bearing requirements during winding, it is easy to achieve equal strength design [4]. Fiber optic cable anti-bending reinforcement is the use of outer winding method in the dry winding process, the use of ring winding method, the cable bundle core mold, in the cable bundle outside the fixed Angle winding multi-layer fiber products, so that the cable bundle has a higher strength and stiffness, to meet the bending radius is not less than the minimum allowable bending radius of the cable requirements. As shown in Fig. 6.

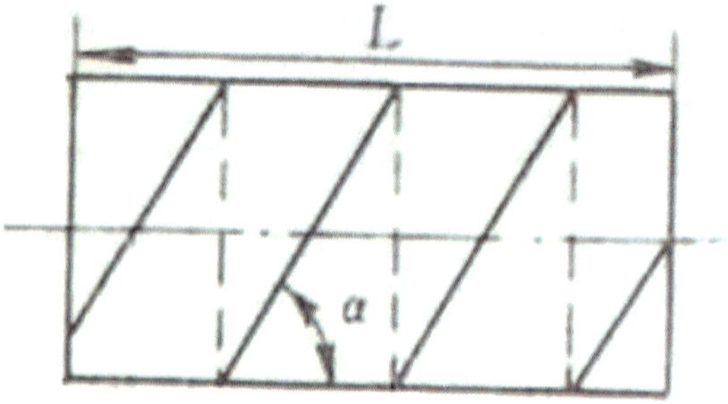

Fig. 6. Schematic diagram of the outer ring winding method

Implementation Method of Bending Reinforcement. The purpose of bending reinforcement of optical cable assembly is to make the optical cable have certain rigidity and toughness, and ensure that the bending radius of the optical cable assembly is not less than the minimum allowable bending radius of the optical cable during the whole installation process and during the operation of the aircraft.

The reinforcement method is to use polyimide tape to wind and fix the optical cable and the rear end of the connector tail cover, and wind and fix all optical cables in the optical cable assembly according to the predetermined path, then use anti atomic oxygen flame retardant cloth to wind from the rear end of the connector tail cover, and completely cover the polyimide tape according to the predetermined path. One layer of polyimide tape and anti atomic oxygen flame retardant cloth. During winding, the polyimide tape and the anti atomic oxygen flame retardant cloth shall be inclined at an angle of about 45° so that the rear end can compress half of the front end winding. The optical cable assembly made by this reinforcement method significantly improves the bending resistance of the optical cable, and can ensure that the optical cable assembly always maintains the minimum bending radius requirement of bending radius greater than 30 mm under the effect of its own stiffness. Figure 7 shows the comparison of the bending radius of the unreinforced and reinforced optical cable assemblies under natural conditions. The bending radius of the fiber optic cable assembly assembled with nylon sheath before reinforcement is about 20 mm–25 mm in the natural state, and the bending radius of the fiber optic cable assembly after reinforcement is about 50 mm in the natural state. And because the raw materials used are non-metallic materials, the same bending resistance effect as the armor layer can be obtained, while effectively reducing the weight of the optical cable components. Bending resistant and reinforced optical cable components have been widely used in the signal transmission system of the space station.

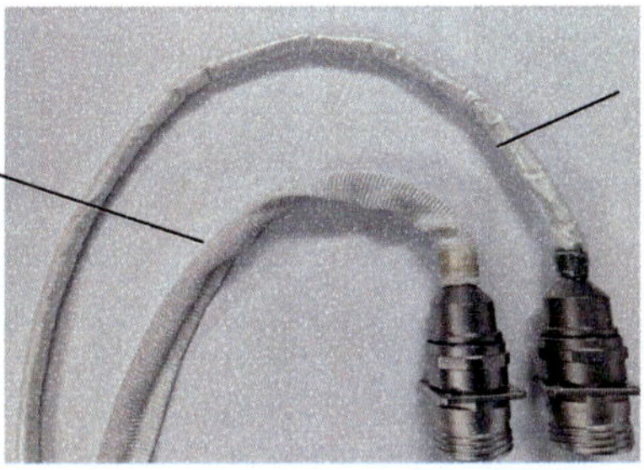

Fig. 7. Comparison of bending radius before and after reinforcement

4.2 Irradiation Strengthening Method of Optical Cable Assembly

Radiation Hardening Principle. In the harsh space radiation environment, radiation effect seriously affects the operation of equipment, and space radiation protection is essential. In order to ensure the reliability of spacecraft, a variety of anti-radiation measures are usually adopted, including anti-radiation reinforcement and radiation shielding. The shielding method of covering the surface of radiation sensitive devices with metal material or applying protective layer before packaging is a common radiation reinforcement method in China, which has achieved good results [5].

In the 1970s, the United States began to study the performance of packaging in terms of radiation, and the package-level shielding method using heavy metal as shielding layer began to be widely used in semiconductor devices [6]. The main effects of shielding on space radiation are to reduce the energy of primary particles, generate and subsequently absorb secondary radiation, and change energy spectrum, etc. [7]. The low atomic number shielding material can not prevent the transmission of charged particles, but can effectively reduce the motion velocity of charged particles. The high ordinal number shielding material can play a good role in preventing charged particles from passing through, but the energy of Bremsstrahlung radiation produced is also very high [8]. Therefore, for the selection of radiation shielding materials, metals with smaller atomic numbers should be selected as far as possible, while metals with larger atomic numbers should be avoided to avoid more obvious secondary radiation.

For optical cable assemblies used outside spacecraft, the total radiation dose will generally reach more than 10^6 rad (Si). Under the irradiation of this dose, the fluoropolymer sheath material of the optical cable will become brittle, reducing the protection effect on the optical fiber, and the radiation induced loss of multi-mode optical fiber will increase dramatically. There is a risk of communication interruption caused by excessive radiation induced loss. The radiation reinforcement of optical cable is mainly through adding

Fig. 8. Structural diagram of stainless steel metal hose

metal shielding layer outside the optical cable to reduce the space radiation energy of the optical cable.

Irradiation Reinforcement Method. The irradiation reinforcement method of optical cable assembly is to protect the optical cable with stainless steel metal hose. The stainless steel metal hose is made of stainless steel SUS304, and the common specification is 3 mm to 20 mm outside diameter. The structure of the stainless steel metal hose is shown in Fig. 8. It is made of double buckle.The stainless steel metal hose has excellent flexibility, corrosion resistance, high temperature resistance, wear resistance, tensile and bending resistance. Stainless steel metal hose and connector tail accessories are usually fixed with crimping structure. As shown in the Fig. 9 below.It can not only effectively attenuate the energy of space radiation, but also improve the overall tensile, bending and compressive properties of optical cable components.

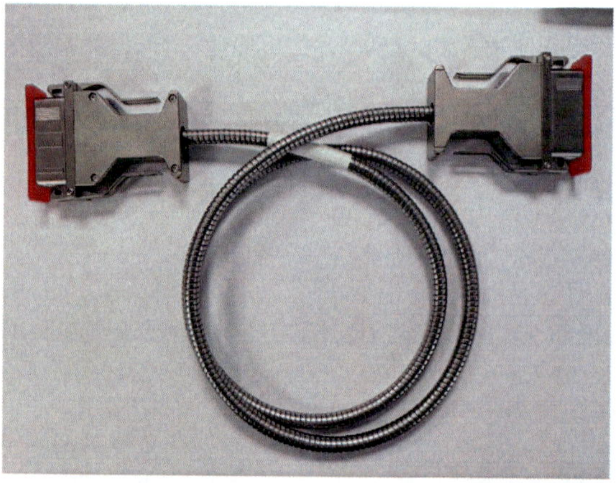

Fig. 9. Schematic diagram of radiation hardened optical cable assembly

Take 3 mm stainless steel metal hose as an example, and its main specifications are shown in the table below (Table 1).

When calculating the radiation dose behind different absorbing or shielding materials, the usual practice is to convert all materials to equivalent aluminum thickness by appropriate density correction [7]. Calculation formula based on equivalent aluminum shielding thickness:

$$\rho = M/S \tag{2}$$

In formula (2):

Table 1. Main parameters of 3mm stainless steel metal hose

Project	Parameter
Material	Stainless steel SUS304
Diameter	External diameter φ 2.9 $\pm$ 0.1 mm, internal diameter φ 1.8 $\pm$ 0.1 mm
Working temperature	−100°C–200°C
Wall thickness	0.2 mm
Compressive strength	1000N
Tensile strength	200N

ρ- Mass surface density of material, g/cm^2.
M - mass of large area of material, g.
S - actual area of material, cm^2.

$$H = \rho/0.27 \tag{3}$$

In formula (3):
H-equivalent aluminum shielding thickness, cm.

According to the calculation, the equivalent aluminum shielding thickness of the stainless steel metal hose is 0.578 mmAl. Taking the space station as an example, the total radiation dose after 10 years of operation is 9 Mrad (Si). After the stainless steel metal hose is used for reinforcement, the total radiation dose of the internal optical cable can be reduced to about 80.4 Krad (Si), which can meet the requirements that multimode optical cables can still work normally after irradiation.The optical cable assembly with this irradiation hardening method has been applied in space station, "Zhurong" rover and other spacecraft extravehicular environment.This radiation hardening method can be widely used in the extravehicular environment of spacecraft.

5 Test and Verification

5.1 Test and Verification of Anti-bending Reinforced Optical Cable Components

Mechanical tests such as sinusoidal vibration (frequency: 10–2000 Hz, acceleration: 147 m/s^2), random vibration (frequency: 10–2000 Hz, power spectral density: 0.4 G^2/Hz, root mean square value of total acceleration: 23.1G), bending (±180°, bending for 100 times) and other mechanical tests were carried out on the optical cable assembly after bending reinforcement. The test results are shown in Table 2.

The insertion loss has no obvious change after various mechanical tests, and the optical performance is good. The anti-atomic oxygen flame retardant cloth in the optical cable assembly shall be free of looseness and detachment. The test results show that the anti-bending reinforced optical cable assembly meets the requirements of spacecraft use.

Table 2. Test results of anti-bending reinforced optical cable assembly

Result	Project				
	Before test	Sinusoidal vibration	Random vibration	Bending at the seal of optical cable	Required value
Insertion loss (dB)	0.06–0.19	0.05–0.24	0.06–0.24	0.06-0.27	≤0.6

5.2 Test and Verification of Irradiation Hardened Optical Cable Assembly

The radiation test shall be carried out by selecting one anti-radiation reinforced optical cable assembly and one un-reinforced optical cable assembly respectively. The length of the optical cable is 1m. The optical cable adopts the anti-radiation multimode 12-core optical cable with the outer sheath of ETFE material (the radiation induced loss index of the optical cable is: ≤15 dB/100 m@850 nm after the total irradiation dose of 100 Krad (Si)). The total irradiation dose test of 9Mrad (Si) was conducted for two groups of samples at a dose rate of 50 rad/s. The test results are shown in Table 3.

Table 3. Comparison results of irradiation test

Result	Project			
	Before test	Unreinforced optical cable assembly	Reinforced optical cable assembly	Required value
Insertion loss (dB)	0.31–0.61	Some light paths are blocked	0.41–0.76	≤1
Appearance	Good appearance	The ETFE sheath of the optical cable is broken	Good appearance	/

The test results show that the ETFE sheath material has hardened and brittle broken after the irradiation test of the optical cable assembly without radiation reinforcement, and the protection of the internal optical fiber has been lost, and the optical fiber of some optical paths has broken. However, the insertion loss of the anti-irradiation reinforced optical cable assembly after the test is qualified, the optical cable has good flexibility, and no fiber breakage occurs. The radiation induced loss generated is about 0.1 dB–0.15 dB, which meets the requirements of the radiation induced loss index of the optical cable, and proves that the anti-irradiation reinforcement measures have significant effects.

6 Conclusion

In this paper, the typical structures of spacecraft optical cable assemblies are analyzed, and the existing problems and risks of conventional optical cable assemblies are identified. Aiming at the bending radius of optical cable components, polyimide tape and anti atomic oxygen flame retardant cloth are used for winding reinforcement, which effectively improves the rigidity of optical cable components. In view of the insufficient radiation resistance of the extravehicular optical cable, the stainless steel metal hose is used for radiation reinforcement to effectively reduce the total radiation dose borne by the optical cable. The test results show that the above reinforcement methods can effectively improve the long-term reliability of optical cable components in the aerospace environment, and provide support for the extensive promotion and application of optical cable components in spacecraft.

References

1. Liu, F., An, Y., Wang, Q., et al.: Effects of gamma-ray radiation on optical fibers. High Power Laser Part. Beams **22**(7), 1668–1671 (2010)
2. Wang, X., Zhang, C., Jin, J., et al.: Radiation induced attenuation effect on special optical fibers applied in space. Infra. Laser Eng. **40**(12), 2516–2520 (2011)
3. Liu, F., Zhang, C., Li, Z., et al.: Research on γ–ray radiolytic loss. J. Chin. Univ. Sci. Technol. **24**(3), 390–393 (1994)
4. Leng, X.: Filament winding principle. Shandong Science and Technology Press (1990)
5. Zhao, K., Wen, C., Sang, J., et al.: Passive radiation protection technology for space electronic products. Space Electron. (3), 105–111 (2021)
6. Jingyang, L., Heran, Z.: Research on packaging radiation hardening technology. Microprocessor **2**, 21–25 (2021)
7. Holmes-Siedle, A., Adams, L.: Handbook of radiation effects, p. 479, Oxford Science Publications, **Oxford** (1994). ISBN0–19–856347–7
8. Yang, X., Feng, B., Yang, Z., et al.: Shielding analysis of high-energy electrons by multi-layer metal materials - Application of MULASSIS program. Rare Metals **31**(4), 481–485 (2007)

A Method of Cloud Detection in Remote Sensing Image Based on FPGA

Wu Panfeng[1,2], An Jian[1(✉)], Wu Baolin[2], Bai Hongyang[3], Wang Yunsen[1], and Liu Zhiyuan[1]

[1] Shandong Institute of Space Electronic Technology, Yantai 264000, China
anjianwork@163.com
[2] Harbin Institute of Technology, Harbin 150001, China
[3] Nanjing University of Science and Technology, Nanjing 210094, China

Abstract. In order to obtain effective information from remote sensing images and meet the demand of efficient extraction of effective information from massive data, this paper presents a cloud detection method of remote sensing image based on FPGA. In this method, the average, variance, average gradient and entropy of image blocks in remote sensing images are calculated by using the high parallelism of FPGA, and the cloud detection of image blocks is quickly completed by relying on the above four features. The application of this method has two advantages. Firstly, the cloud detection method can effectively acquire the effective area without clouds in remote sensing images, and further improve the efficiency of remote sensing image application and processing. Secondly, the highly parallel computing of FPGA can greatly improve the execution efficiency of cloud detection tasks. Compared with the traditional embedded CPU, the implementation of cloud detection by FPGA has increased the task execution speed by more than 9 times. The method can support real-time cloud detection of remote sensing satellites in orbit, avoid the waste of storage resources, bus resources and computing resources caused by cloud cover, improve the task execution efficiency, and further greatly improve the usability of remote sensing satellites. This method has high application value in the field of on-orbit processing of remote sensing images.

Keywords: Remote sensing images · FPGA · Real-time cloud detection

1 Introduction

Remote sensing is a non-contact detection technology based on aerial photography to obtain remote information, developed in the sixties of last century. With the rapid development of information technology and sensors, remote sensing has become a new field of science and technology, and is entering a new stage that can provide a variety of earth observation remote sensing data quickly and timely. Although visible light remote sensing images have achieved the expansion of high resolution in orbit, it is almost impossible to obtain clear and cloudless large area from remote sensing images because of climate and other factors. At the same time, the weather phenomena in the earth's

H. P. Urbach and H. Jiang (Eds.): ISSOIA 2022, SPPHY 295, pp. 92–104, 2023.
https://doi.org/10.1007/978-981-99-4098-1_10

atmosphere are very rich, and cloudy atmospheric phenomena are very common due to the formation of water vapor condensation in the atmosphere. In China, for example, cloud cover accounts for 1/2 of the weather in the north (take Beijing as an example), while in the south, the cloud cover is higher in the subtropics and tropics. Therefore, how to remove the cloud interference is of great significance to improve the on-orbit processing efficiency of the sensitive information of the remote sensing image and enhance the accuracy and validity of the remote sensing data [1].

In recent years, scholars have proposed a variety of cloud detection algorithms for remote sensing images. For example, the detection algorithm based on image spatial features [2] generally regards cloud detection as a two-classification problem between cloud layer and underlying surface, and extracts gray features, fractal dimension and texture features of cloud regions. The classifier is used to classify the image blocks to distinguish between cloud and cloud-free regions. This kind of cloud detection method can be subdivided into threshold method [3, 4], texture analysis method [5, 6] and statistical method. However, this kind of method has high complexity and poor real-time performance, so it is usually used in the image cloud detection scene of ground execution. With the continuous development of artificial intelligence technology, cloud detection methods based on deep learning have become the mainstream research direction, including FCN [7], Unet, SegNet [8] and so on. Compared with classical methods, intelligent processing methods are basically data-driven, can better extract cloud features, and have higher detection accuracy. However, the deep-learning-based method needs a lot of computing resources, and it is difficult to deploy in orbit.

At present, the common high-performance computing processing platforms in the aerospace field at ground and abroad contain DSP, GPU, FPGA and heterogeneous SoC [9]. Among them, FPGA has numerous parallel computing units and the main frequency is much lower than other platforms, so it can achieve high-performance computing under the condition of lower power consumption, which can meet the application conditions of on-board embedded environment, and has application potential [10]. Therefore, this paper proposes a remote sensing cloud detection algorithm based on FPGA architecture, which can be used for real-time cloud detection of remote sensing image in orbit, has the characteristics of low power consumption and easy to implement, and can greatly improve the application efficiency of remote sensing satellite system.

2 Algorithm Construction of Cloud Detection

2.1 Feature Extraction of Cloud Image

In the remote sensing images captured by satellites, due to the diversity of clouds and the complexity of background features, it is usually necessary to determine the cloud image area by a combination of multiple features, so as to increase the distinguishability between cloud image and background features in feature space. At the same time, in order to satisfy the hardware realizability and better real-time requirements, the computational complexity of the selected features is not suitable to be too complex (Fig. 1).

Because of the obvious gray feature difference between cloud image and ground object in remote sensing image, this paper finally selects the following four image features as the basis of cloud image region discrimination in cloud judgment algorithm.

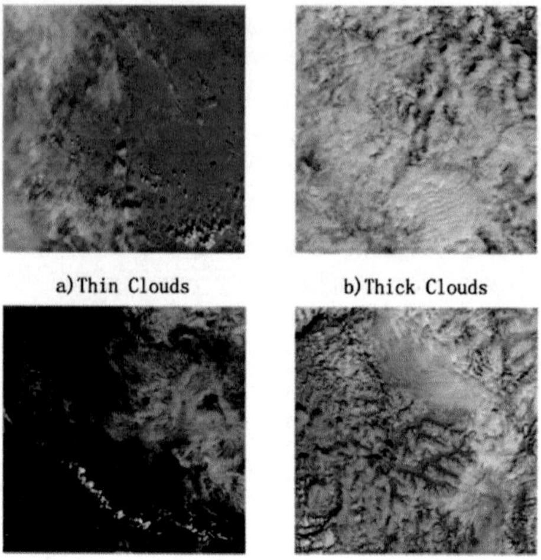

a) Thin Clouds b) Thick Clouds

c) Ocean Background d) Forest Background

Fig. 1. Diversity of remote sensing cloud images

(1) mean value of Pixel block

$$\overline{x} = \frac{\sum_{i=1}^{n} x_i}{n} \tag{1}$$

(2) Variance value of Pixel block

$$S2 = \frac{\sum_{i=1}^{n} (x_i - \overline{x})^2}{2} \tag{2}$$

(3) average gradient of Pixel block

$$G_{xy} = \frac{1}{M \times N} \sum_{i=1}^{M} \sum_{j=1}^{N} \sqrt{\frac{\left(\frac{\partial f}{\partial x}\right)^2 + \left(\frac{\partial f}{\partial y}\right)^2}{2}} \tag{3}$$

(4) Entropy of Pixel block

$$H = -\overline{x}\left(\sum_{i=0}^{255} p_i \, Inp_i\right) \tag{4}$$

In the formula (1) ~ (4) above, x_i is the pixel value of the image block, n is the total number of pixels in the image block, $\overline{x}$ is the pixel mean value of the image block, S^2 is the variance of the image block, M is the number of pixels of the image block in the x direction, N is the number of pixels of the image block in the y direction, G_{xy} is the average gradient of the image block, p_i is the probability that the pixel value is i in the image block, H is the entropy of the image block.

2.2 Classification and Judgment of Cloud Image Features

To determine the feature classification of cloud images by using feature vectors, it is necessary to manually select 10 kinds of cloud and ground images with obvious features from the image database of the images to be tested, and calculate the feature vectors C_i of these cloud image blocks and G_i of ground image blocks. These feature vectors will be used as the key parameters to determine the accuracy of whether the image block to be tested is a cloud occlusion region. Therefore, when preparing the feature vector of cloud image block and ground object image block, we should take into account the practical application and reduce the false alarm rate as much as possible. In the extreme case, there can be some criteria for missing alarm (that is, the coexistence of cloud and figure should be identified as the area of figure) to screen the clustering centers of cloud vector and ground vector.

The feature space composed of four feature quantities $[\overline{x}, S^2, G_{xy}, H]$ is a four-dimensional Euclidean space. The distance function $S(x, y)$ can be used to calculate the Euclidean distance between the feature vector x and the feature vector y in the feature space. Then the minimum Euclidean distance classifier is used to quickly classify the cloud image region and the feature region.

For example, first calculate the minimum Euclidean distance $S_{min}(T, C_i)$ between the feature vector T of the image block to be tested and the cloud vector C_i, and the minimum Euclidean distance $S_{min}(T, G_i)$ of the feature vector T and the feature vector G_i of the image block to be tested. Then compare the size of $S_{min}(T, C_i)$ and $S_{min}(T, G_i)$, if the former is small, the image block to be tested is the cloud image region, otherwise the image block to be tested is the figure area.

To sum up, the specific steps of the cloud decision algorithm are as follows:

(1) The cloud region vector and ground vector of many kinds of images are calculated, the clustering centers of cloud region vector and ground vector are screened, and 10 known cloud feature vectors and 10 known feature vectors are obtained.
(2) The measured image is divided into blocks to form a m × m pixel image block as the basic discrimination unit.
(3) Calculate and extract the four features needed for each image block to form a feature vector $[\overline{x}, S^2, G_{xy}, H]$
(4) Calculate the extracted feature vector and the minimum Euclidean distance between the 10 known feature vectors under cloud and ground classification.
(5) Compare the minimum Euclidean distance under the two categories calculated above, and take the classification of the smaller one as the discriminant result of the image block to be tested and final export.

3 Design of FPGA Processing Architecture for Cloud Detection Model

According to the processing process of the model above, the processing architecture of the cloud detection model on FPGA is designed as follows (Fig. 2):

In the image above, after the remote sensing image data is input into the FPGA, the first module will buffer and preprocess the image data. After the color is converted to

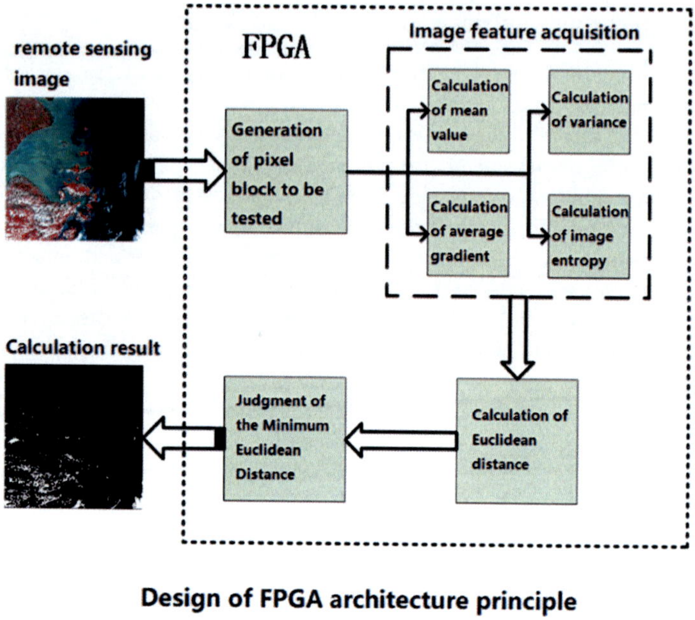

Fig. 2. FPGA architecture design diagram

grayscale, 4 × 4 pixels are extracted sequentially to generate the image unit block to be processed. The detailed extraction method is shown in Sect. 3.1. The second module parallel calculates four image features for the acquired image unit block to form the feature vector $[\overline{x}, S^2, G_{xy}, H]$. The detailed extraction method is shown in Sect. 3.2. The third module is responsible for parallel calculation of the Euclidean distance between the feature vector output of the second module and the cloud feature vector and the ground feature vector. The detailed implementation method can be found in Sect. 3.3. The last module makes a quick comparison and judgment of the Euclidean distance of the output of the third module, obtains the minimum Euclidean distance, and discriminates the measured image unit block and outputs the discriminant results.

3.1 Acquisition of 4 × 4 Pixel Image Block to be Measured

The acquisition of image blocks will occupy the buffer resources of FPGA. Considering the utilization of various resources on FPGA, 4 × 4 pixel image blocks are selected as the basic computing unit of the cloud decision processing algorithm in this paper. This paper uses Shift RAM IP to check the input data for buffering. Set the buffer length to the length of a row of data, and a Shift RAM IP core can buffer a row of data. Firstly, the image data is converted into gray value by the grayscale processing module, and then it is buffered by three levels of Shift RAM IP, plus a row of data currently being input to read four rows of data at the same time. Send the image data from clock to Shift RAM IP. When the current four rows of data are fully input, as shown below (Fig. 3):

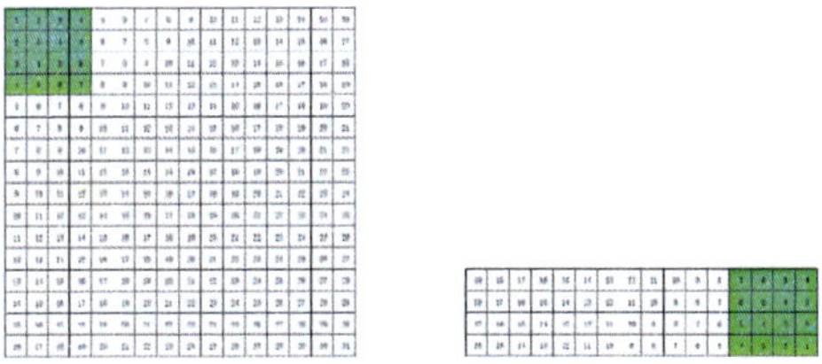

Fig. 3. Shift buffer logic diagram 1

The pixel block is obtained after the first clock as shown in the following figure (Fig. 4):

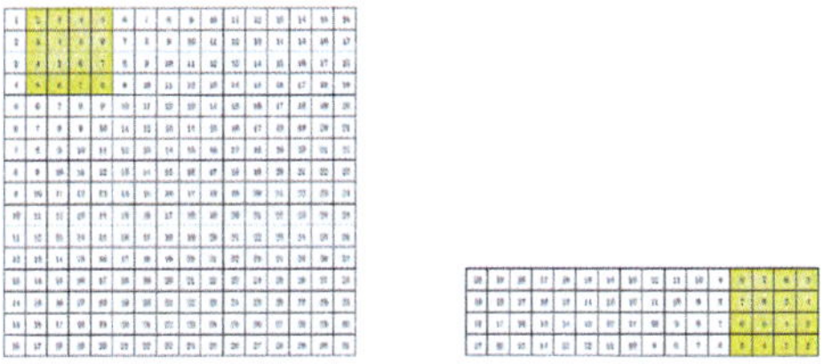

Fig. 4. Shift buffer logic diagram 2

The pixel block is obtained after the second clock as shown in the following figure (Fig. 5):

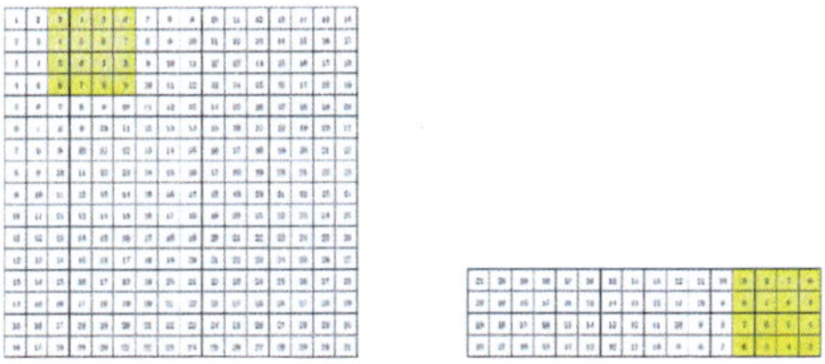

Fig. 5. Shift buffer logic diagram 3

Get the pixel block after the third clock as shown in the following figure (Fig. 6):

The pixel block is obtained after the fourth clock as shown in the following figure (Fig. 7):

After obtaining four rows of data, according to the needs of the algorithm, the four rows of data are read every four pixel clock cycles, and the values of the four pixels are $p_{11}, p_{12}, p_{13}, p_{14}, p_{21} \ldots\ldots p_{44}$ respectively. The 4×4 pixel blocks generated in this way will not overlap, which can reduce the unnecessary amount of computation.

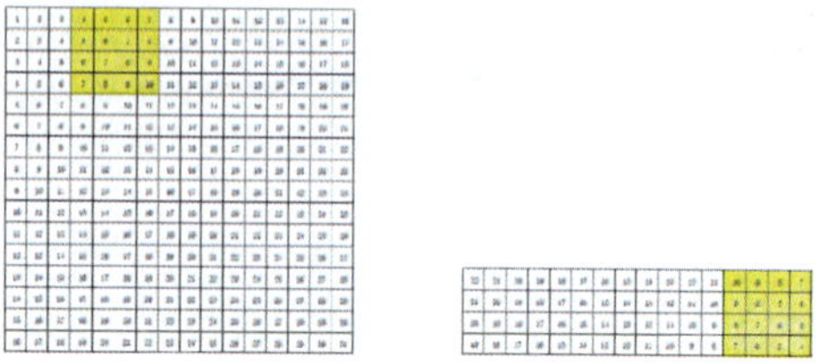

Fig. 6. Shift buffer logic diagram 4

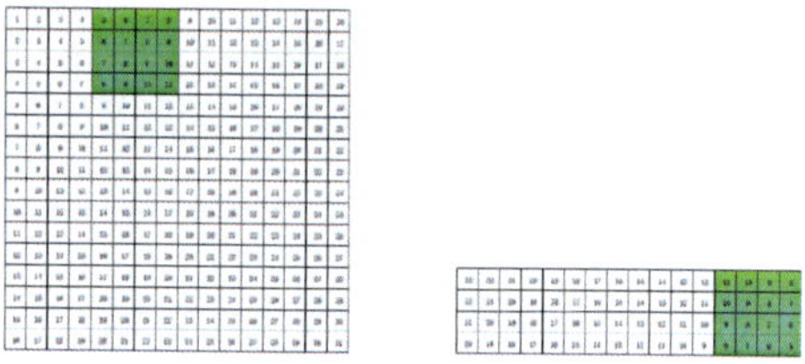

Fig. 7. Shift buffer logic diagram 5

3.2 Extract Image Eigenvalues

Cloud detection algorithm needs to calculate and extract four eigenvalues of mean, variance, average gradient and image entropy of the image block to be tested. The realization of mean, variance and average gradient is more common, so it is not introduced in detail. In view of the probability of each pixel value needed in the calculation formula of image entropy, the method of subtraction is used to calculate the probability.

(1) Subtraction module

Using Verilog hardware description language to design a subtraction module, the function is to subtract the pixel value p_{test} and the pixel $p_{11}, p_{12}, p_{13}, p_{14}, p_{21} \ldots\ldots p_{44}$ in the image block to be tested in turn, and record the number of times when the result is 0 as *counter*. The pixel p_{11} is then used as a p_{test} input subtraction module to obtain an output $counter_{11}$. Making use of the characteristics of FPGA parallel processing, pixel p_{12}, pixel p_{13}….. Pixel p_{44} are used as p_{test} input subtraction module in turn, and $counter_{12} \ldots\ldots counter_{44}$ is obtained at the same time.

(2) Statistical module

In the previous step, the possible value of the output *counter* of the subtraction module is any value from 1 to 16. The statistical module is also designed by using Verilog hardware description language. Record the number of times with a median value of 1 in the above 16 *counter* calculated by the statistical module as N_1, and the number of other values as $N_2, N_3 \ldots N_{16}$ in turn. Thus, the value of N_i divided by i is the probability of occurrence of pixel values. Finally, the calculation formula of image entropy H is changed as follows:

$$H = -\overline{x}\left(\sum\nolimits_{i=0}^{255} p_i \, Inp_i\right) \tag{5}$$

3.3 Cloud Layer Discrimination

After obtaining the feature vectors of the image blocks to be tested, the Euclidean distances between them and 10 cloud feature vectors and 10 ground feature vectors need to be calculated and compared to get their minimum Euclidean distance. In this paper, the corresponding comparison module is designed, and the minimum Euclidean distance can be obtained in less clock cycle. The specific design of the process of the comparison module is as follows (Fig. 8):

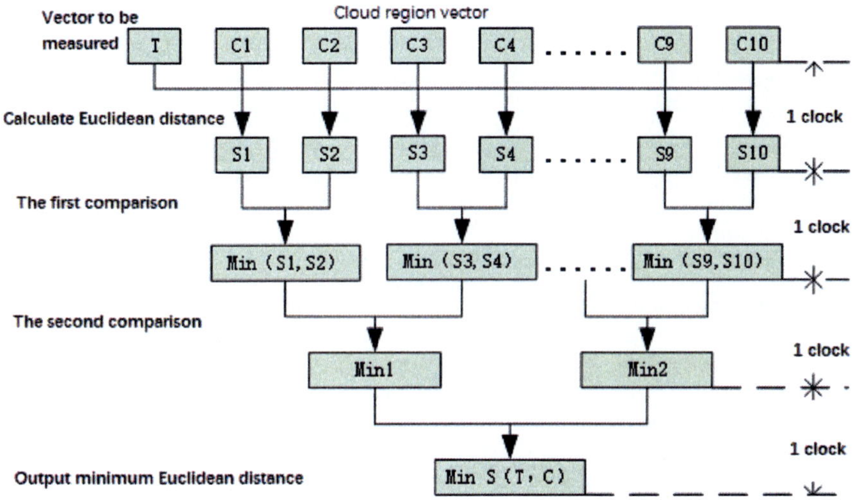

Fig. 8. Flow chart of minimum Euclidean distance calculation

As shown in the figure above, it takes a total of 4 clock cycles to compare the vector T under test with the minimum Euclidean distance $S_{min}(T, C_i)$ of the cloud feature vector. At the same time, the minimum Euclidean distance $S_{min}(T, G_i)$ between the vector to be tested and the feature vector can be obtained by making use of the advantage of FPGA parallel computing. In the last step, the final discriminant result can be output by comparing the size of $S_{min}(T, C_i)$ and $S_{min}(T, G_i)$.

4 Analysis of Test Results

ZYNQ7020 is selected as FPGA in this design. Some 10K*10K cloud remote sensing images are segmented, and a total of 1000 images are obtained for cloud detection testing. Image size is 1K*1K pixels.

4.1 Analysis of Test Results of Buffer Processing Module

The solution of this paper is to use three cache modules, each of which sets a buffer length of 1 row of data, plus the data being input, so as to complete the first step of

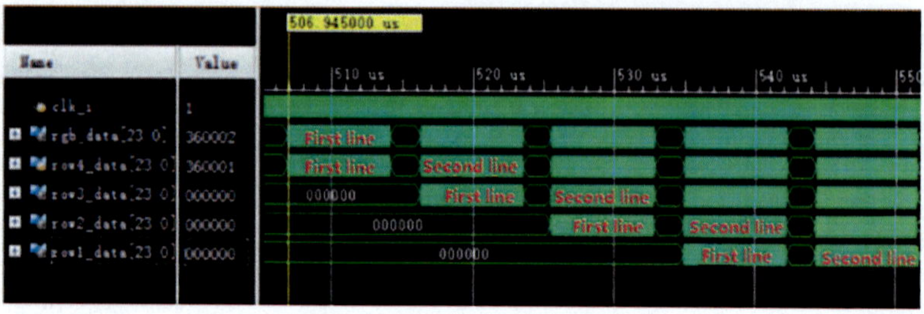

Fig. 9. Multi-line buffer waveform

reading 4 rows of data at the same time. The waveform capture diagram is shown in the following figure (Fig. 9):

From the observation of the above picture, we can see that the data can be read once for every 4 rows of buffer, which avoids the reading of the repetitive region. When reading four rows of data, four image data are read from each row in turn to form an image block to be tested.

4.2 Analysis of the Results of Image Feature Module

(1) Calculation of Mean and Variance of Image Block

The mean value of the image block is the easiest to calculate. Through the observation of the variance calculation formula, we can find that the variance can only be calculated when the mean value is known, so the two features are calculated and extracted together. The waveform is shown in the following figure (Fig. 10):

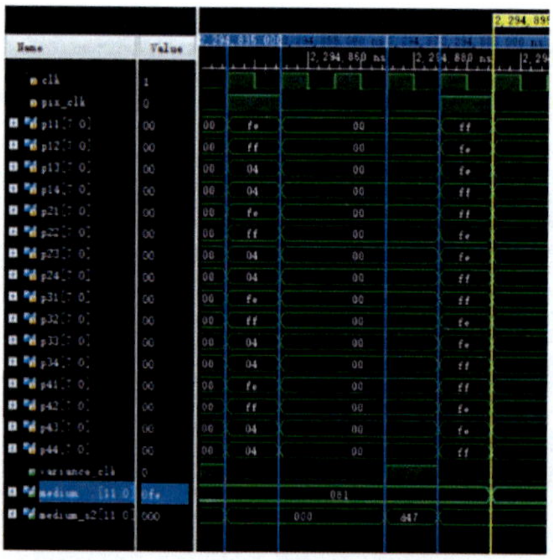

Fig. 10. The results of calculating the mean and variance of image blocks

It can be observed from the above picture that it takes only one pixel clock cycle to calculate the average image block and three clock cycles to calculate the image block variance.

(2) Average gradient calculation of image blocks

When calculating the average gradient of the image block, the gradient of the image block in the X and Y direction is calculated in parallel, and then the average gradient is calculated, so as to reduce the clock cycle. The waveform is shown in the following figure (Fig. 11):

Fig. 11. Calculate the average gradient of the image block

As shown from the image above, it takes 2 pixel clock cycles to calculate the average gradient of the image block.

(3) Entropy calculation of image block

The waveform for calculating the entropy of the image block is shown in the following figure (Fig. 12):

In the process of calculating image block entropy, parallel computing is adopted in many places, so it can be seen from the above picture that it takes 2 pixel clock cycles to calculate image block entropy.

According to the calculation results of the above-mentioned modules for calculating image features, there are some differences in the length of the clock cycle, the shortest one takes only one clock cycle, and the longest takes three clock cycles. Therefore, a buffering process is needed to achieve the synchronization and consistency of the output timing of various image features.

4.3 Analysis of the Effect of Cloud Judgment

The discriminant waveform of the cloud judgment module is shown in the following figure (Fig. 13):

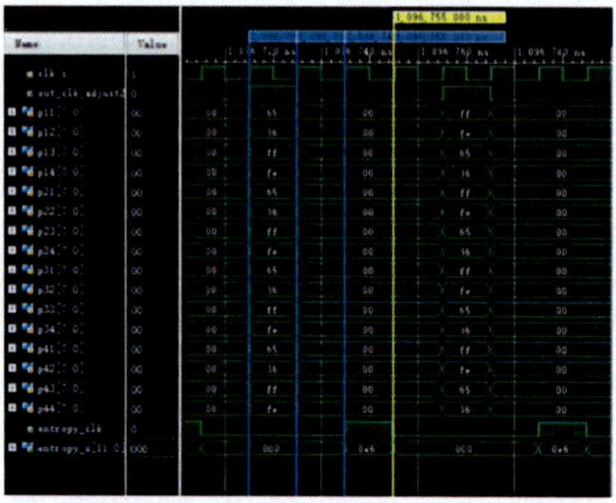

Fig. 12. The result of calculating image block entropy

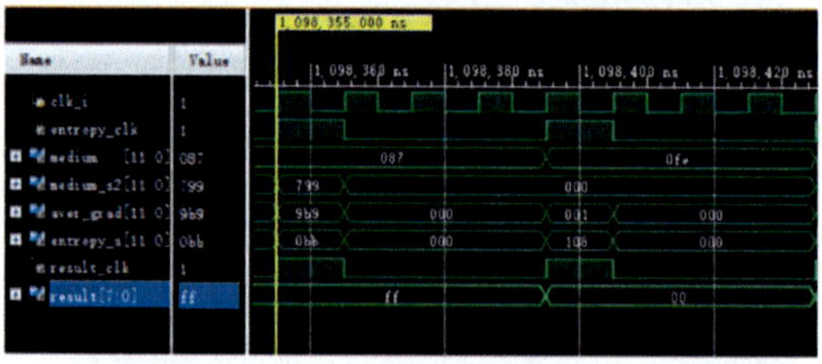

Fig. 13. Cloud judgment computing result

From the observation of the above image, it can be concluded that after the image features are extracted, it takes a total of 4 clock cycles to obtain the discrimination results. The work of full cloud detection for each 4 $\times$ 4 pixel block takes a total of more than a dozen FPGA clocks.

Using this cloud detection method, the cloud detection results of large cloud-containing remote sensing images is shown in the following figure (Fig. 14).

From the cloud detection results, it can be seen that the cloud detection method has high detection accuracy for thick clouds, and there is a low probability of misjudgment for thin clouds. The misjudgment belongs to the false alarm misjudgment, that is, in the case of the coexistence of cloud and ground, the thin cloud area would be misjudged as the ground area, which can ensure the maximum preservation of the ground area.

4.4 Summary

Through the analysis, through the calculation of the characteristic values of the mean, variance, average gradient and entropy of the input image block, and compared with the eigenvalues of the cloud image, the cloud decision function of the image block can be completed in more than ten clocks. To implement the above functions, FPGA resource usage is shown in the following table. Only the cloud detection function of this article is run in ZYNQ7020, and the power consumption is about 4.2W (Table 1).

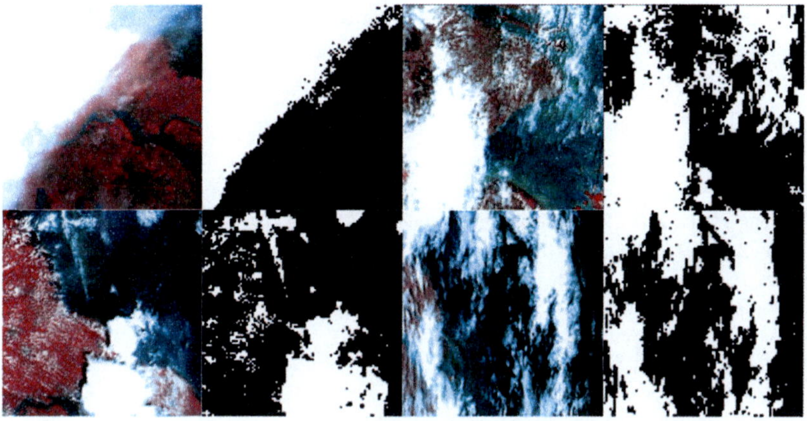

Fig. 14. Samples of cloud

Table 1. FPGA resource usage table

Resource name	Total resources	Utilization rate
LUT	53200	14%
LUTRAM	4.9M	1%
FF	106400	2%
DSP	220	57%
BUFG	32	9%

Through the actual measurement of 1000 1K*1K pixels images, the cloud judgment speed of this method is about 100 FPS, which is close to the prediction speed, and the time consuming for a single image is about 10 ms. Compared with the 100 ms needed by FT2000 dual-core processor to run cloud decision program, 90 ms required by DSP to run the cloud decision program [11], the cloud detection method described in this paper can improve the processing speed by more than 9 times.

5 Conclusions

In this paper, a remote sensing image cloud detection method based on FPGA is proposed. By calculating the value of block mean, variance, gradient and entropy after image segmentation, the Euclidean distance between the pre-obtained cloud feature vector and the ground feature vector, the cloud detection of image block could be completed efficiently.

This paper expounds and designs the cloud detection method of remote sensing image based on FPGA from three aspects: the construction of cloud detection algorithm, the FPGA architecture design of cloud detection model and the analysis of test results. The test result is that the processing time of a single 1K*1K pixels image is about 10ms. The cloud detection processing speed of this method is more than 9 times that of the traditional embedded processor.

References

1. Zheng, H., Li, Z., Li, Z.: Real-time cloud detection that uses data parallel and pipeline. In: 2011 4th International Congress on Image and Signal Processing.Shanghai: 2011 4th International Congress on Image and Signal Processing, pp. 1681–1684 (2011)
2. Pengfei, L., Limin, D., Huachao, X., et al.: A cloud image detection method based on SVM vector machine. Neuro Comput. **169**, 34–42 (2015)
3. Poli, G., Adembri, G., Gherardelli, M., et al.: Dynamic threshold cloud detection algorithm improvement for AVHRR and SEVIRI images. In: 2010 IEEE International Geoscience and Remote Sensing Symposium, pp. 4146–4149 (2010)
4. Arai, K.: Thresholding based method for rainy cloud detection with NOAA/AVHRR data by means of Jacobi Itteration method. Int. J. Adv. Res. Artif. Intell **5**, 21–27 (2016)
5. Cao, Q., Zheng, H., Li, X.-S.: A method for detecting cloud in satellite remote sensing image based on texture. Acta Aeronaut. et Astronaut. Sin. **28**(3), 661 (2007)
6. Mauniga, M., Padma, S.: Automatic cloud detection and weather forecasting using gradient based SPS. Imp. J. Interdiscip. Res. **2**(7) (2013)
7. Mohajerani, S., Saeedi, P.: Cloud-Net: an end-to-end cloud detection algorithm for Landsat 8 imagery. In: IGARSS 2019–2019 IEEE International Geoscience and Remote Sensing Symposium, pp. 1029–1032 (2019)
8. Lu, J., Wang, Y., Zhu, Y., et al.: P_Segnet and NP_Segnet: New neural network architectures for cloud recognition of remote sensing images. IEEE Access **7**, 87323–87333 (2019)
9. George, A.D., Wilson, C.M.: Onboard processing with hybrid and reconfigurable computing on small satellites. Proc. IEEE **106**(3), 458–470 (2018)
10. Raissouni, N., El Adib, S., Sobrino, J., et al.: Towards LST split-window algorithm FPGA implementation for cubesats on-board computations purposes. Int. J. Remote Sens. **40**(5–6), 2435–2450 (2019)
11. Sui, Y.: Tensor based cloud detection and onorbit real-time processing of multispectral images. Beijing: University of Chinese Academy of Sciences (Changchun Institute of Optics, Fine Mechanics and Physics, Chinese Academy of Sciences) (2019)

Multi-mode Detection System with the Same Aperture for Space Target

Ke Wang(✉), Xiaoyu Shang, and Jianing Hu

Beijing Institute of Spacecraft System Engineering, Beijing 100094, China
wangke315@mails.ucas.ac.cn

Abstract. With the dramatic increase in the number of space debris and spacecraft operating in orbit, the space environment has become more and more complex in recent years. The space target detection system play an important and critical role in such a critical situation. A common problem of space target space-based detection payloads is single detection mode, which cannot detect important space target in multiple dimensions, thus making it difficult to adapt to changes in the space environment. In this paper, a multi-mode co-aperture space target detection system is proposed, which consists of a visible-near infrared zoom detection subsystem, a long-wave infrared detection subsystem and a LiDAR detection subsystem. The visible-near infrared detection module is used to receive the reflective beam of the target; the long-wave infrared detection module is used to receive the signal of the target's spontaneous radiation; LiDAR detection module is used to supplement the defect of optical passive detection module. Because this system has the capability of multi-dimensional integrated detection, it can guarantee the accuracy and timeliness of target detection by space-based detectors in complex space environments under the premise of existing satellite resource constraints, thus effectively enhancing the space environment adaptation capability of space-based detectors. It is of great importance for the effective use and in-depth exploration of extraterrestrial space.

Keywords: Multi-mode detection system · Optical payload · Common aperture

1 Introduction

With the rapid development of space technology, a large number of space-based detection devices for remote sensing imaging, communication, atmospheric sounding, and many other fields have been developed and sent to space in recent years around the world. At the same time, the amount of space debris has increased dramatically with the large number of human space activities [1–3]. In 2009, a Russian satellite collided with a U.S. satellite. The collision produced more than 600 pieces of space debris. Currently, more than 23,000 pieces of space debris larger than 10 cm in diameter are regularly tracked and numbered by the U.S. Air Force [4]. The dense distribution of space debris in space orbit is one of the main threats to the safe operation of spacecraft. Therefore, in order to improve the operational lifetime of spacecraft in orbit, it is necessary to enhance sensing and identification capabilities of it to be able to take appropriate protection or evasion measure in time when facing threats [5, 6].

H. P. Urbach and H. Jiang (Eds.): ISSOIA 2022, SPPHY 295, pp. 105–113, 2023.
https://doi.org/10.1007/978-981-99-4098-1_11

Most of the current space-based detection systems for space target have the disadvantage of a single detection mode. This means that these detection systems are poorly adapted to the space environment and may even fail in some specific space environments: when the space target is in the backlight environment, the detection instrument will be blinded by strong light; for low-light space, such as shadow area, light energy received by conventional passive optical detection instrument is insufficient to detect the space target quickly. In addition, the constraint of focal length and detector pixel size make the detection instrument have a fixed resolution. The factor also causes some difficulties in space target identification and threat prediction. A solution to above problems is to install multiple detection devices on the space-based platform, so that the space vehicle has a multidimensional detection capability, such as the Japanese Hayabusa II [7]. However, it is obvious that the method is not universal. Because satellite resources are often limited, the installation of multiple types of detection equipment is extremely demanding on the performance of the spacecraft.

Based on the above discussion, it is extremely important to solve the problem of poor space environment adaptability of space-based detector under the existing satellite resource constraints. Therefore, a multi-mode co-aperture detection system is proposed in this paper. The system consists of a visible-near infrared (VIS-NIR: 0.4–0.9 μm) continuous zoom detection subsystem, a long-wave infrared (LWIR: 8–14 μm) detection subsystem, and a LiDAR (1064 nm) detection subsystem. Among them, the VIS-NIR zoom detection subsystem has a continuous zoom function, which can achieve high-resolution detection of space target at different distances in a well-illuminated environment. For space target in low-light space or backlight environment, the LWIR detection subsystem is used to receive the thermal radiation signal of the target so as to detect it effectively. When space targets in poorly illuminated environments do not have thermal radiation characteristics, LiDAR detection subsystem based on active detection mode is used to measure their spatial distances to make up for the deficiencies of passive optical detection [8–10]. By the integrated and synergistic work of the three detection modes mentioned above, it is possible to guarantee that the resource-limited sensing spacecraft has a high-performance sensing capability for space targets in complex environments.

2 Design of the Multi-mode Detection System

The multi-mode detection system with the same aperture uses a Ritchey-Chretien telescope to simultaneously receive VIS-NIR reflected or scattered signals, LWIR spontaneous radiation signals and LiDAR echo signal. And then the received signals are introduced into their respective detection subsystem by using a two-way color beam splitter. The schematic diagrams and main parameters of the multi-mode co-aperture detection system are shown in Fig. 1 and Table 1, respectively.

For the above detection modes, the wavelength ranges of VIS-NIR, LWIR and LiDAR are 0.4–0.8 μm, 8–14 μm and 1064 $\pm$ 5 nm, respectively. Among them, the VIS-NIR mode senses the reflected or scattered beam of the target. The LWIR mode detects the spontaneous radiation characteristics of the target, which is used to compensate for the defect of VIS-NIR detection in the ground shadow region or backlight environment. The LiDAR mode is used for long-range real-time sensing of space target without thermal radiation properties, which is in poorly illuminated environments.

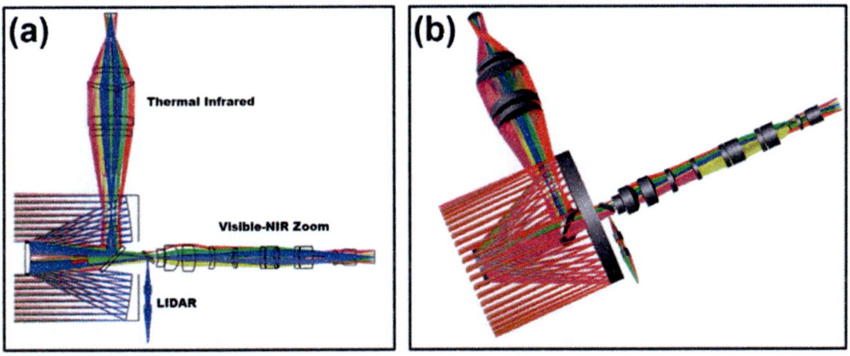

Fig. 1. Schematic diagrams of the multi-mode detecting system

Table 1. Main parameters of multi-mode detection system with the same aperture.

Parameter	VIS-NIR	LWIR	LiDAR
Wavelength range (μm)	0.4–0.9	8–14	1.064
Focal length (mm)	600/1200	800	1000
Field of view (°)	1.6° @f 600 mm; 0.8° @f 1200 mm	1.4°	0.26°
F-number	3 @f 600 mm; 6 @f 1200 mm	4	5

2.1 VIS-NIR Zoom Detection Subsystem

In space target detection, the resolution needs to be adjusted to different tasks, so the detection subsystem should have zoom function. For the VIS-NIR zoom detection subsystem shown in Fig. 2, the movement of the lens group enables the doubling of the focal length (The focal length is expanded from 600 mm to 1200 mm), thus doubling the resolution.

If a detector with an image element size of 10 μm is used, the effective adjustment range of the angular resolution of the zoom subsystem is 0.0083 mrad– 0.01667 mrad. Figure 3 shows the MTF of the zoom subsystem at room temperature, where the working focal lengths of (a)–(c) are 600 mm, 900 mm and 1200 mm, respectively. Which indicates that the image generated by the subsystem has high quality contrast.

Considering that the operating temperature of the space optical detection system is generally controlled to 20 ± 5 °C by the thermal control system, Fig. 4 shows the MTF of the zoom subsystem at Nyquist frequency within the normal operating temperature range. It can be seen that the MTF of the VIS-NIR zoom detection subsystem in a temperature range from 15 °C to 25 °C are greater than 0.7 at all working focal lengths. The simulation result also demonstrates that the imaging performance of the subsystem working at normal operating temperature is good.

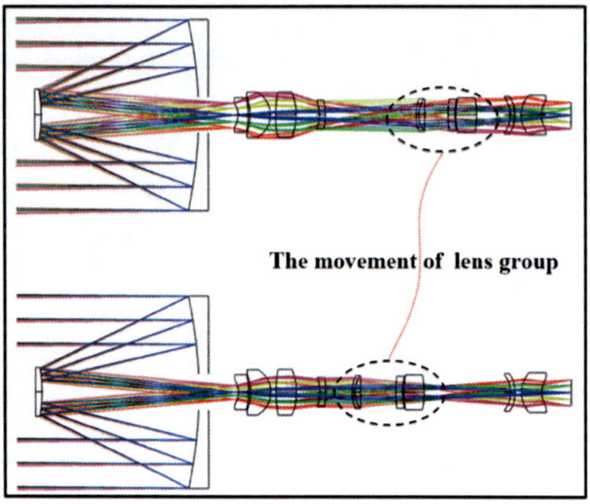

Fig. 2. VIS-NIR zoom detection subsystem

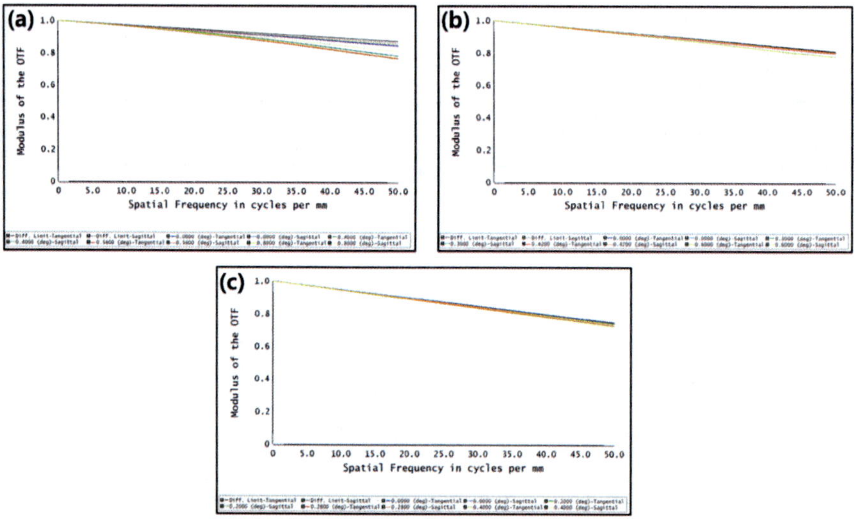

Fig. 3. MTF of the VIS-NIR zoom detection subsystem at different working focal lengths: (a) 600 mm, (b) 900 mm, (c) 1200 mm

The spot diagrams of the zoom subsystem at different focal lengths are shown in Fig. 5, where the focal lengths of (a)–(c) are 600 mm, 900 mm and 1200 mm, respectively. And the black circles in the picture are Airy spot. It can be seen that the dot column plot sizes of all fields of view are close to the diffraction-limited spots. And the RMS radius of the diffuse spots are all below 3 μm, which is smaller than the size of the detector pixels (10 μm). In other words, the subsystem has a high-quality focusing effect.

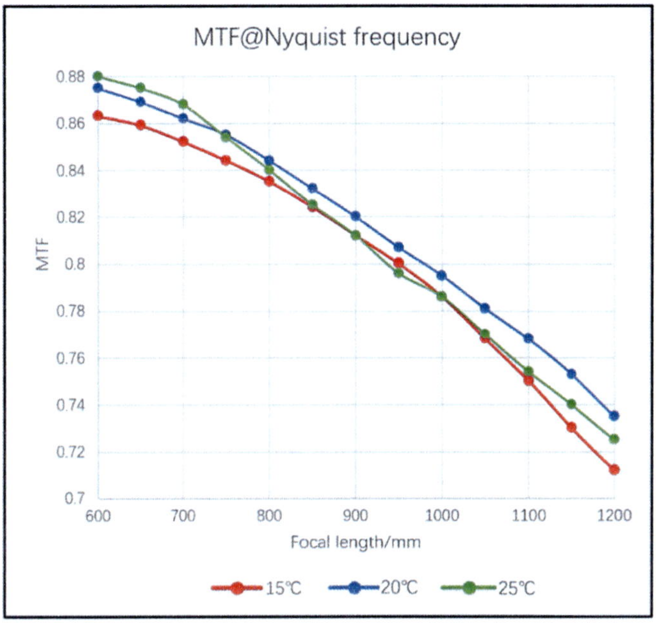

Fig. 4. MTF of the VIS-NIR zoom detection subsystem working at normal operating temperature range

2.2 LWIR Detection Subsystem

LWIR detection subsystem is characterized by low transmission material and low transmittance. Therefore, the number and thickness of lenses should be minimized. In addition, because the refractive index of long wave materials varies greatly with temperature, the effect of temperature on performance of the detection subsystem should be considered in the design process. It should be noted that for LWIR subsystem, cooling of the subsystem is required to reduce the thermal noise of the system. Figure 6 shows a LWIR detection subsystem with a focal length of 800 mm, which uses a cooled standard detector (The pixel size and the F number of the detector are 30 μm and 4, respectively). As shown in the figure, the subsystem consists of a reflective front system and a relay system consisting of four lenses. In order to minimize the number of lenses and increase the transmission rate of the subsystem, the lens surface types of the relay system are all aspherical.

The MTF of the LWIR subsystem is shown in Fig. 7(a). As shown in the figure, the MTF in the full field of view are close to the diffraction limit. Thus the image generated by the subsystem has high quality contrast. The spot diagram of the subsystem is shown in Fig. 7(b). Which shows that the image spot size of all fields of view are smaller than the Airy spot, so the subsystem has good focusing characteristics.

In the design and implementation of this subsystem, the detector is always set to the best focal plane position at both low and high working temperatures by the directional movement of the two lenses close to the focal plane. Figure 8 shows the MTF of the

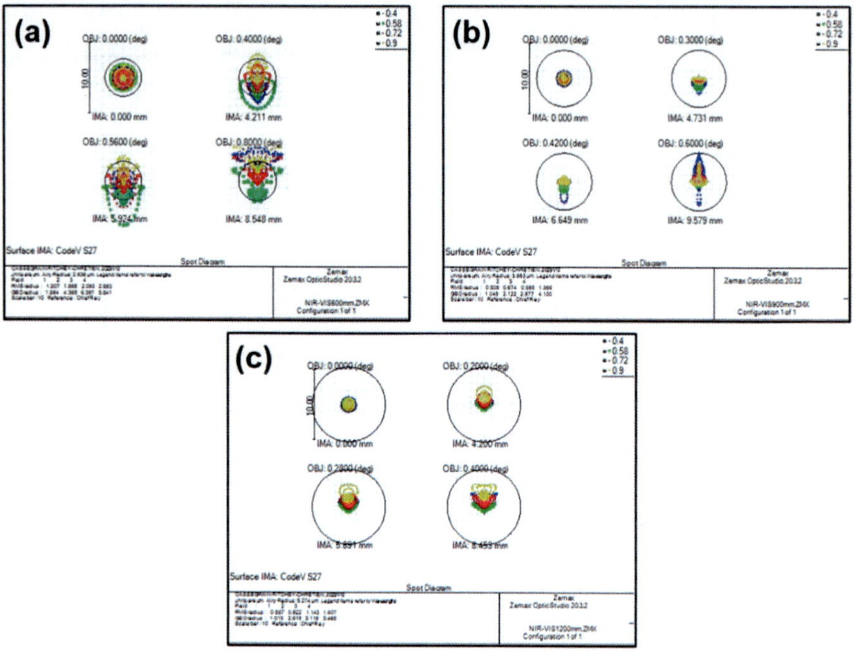

Fig. 5. Spot diagrams of the VIS-NIR zoom detection subsystem at different focal lengths: (a) 600 mm, (b) 900 mm, (c) 1200 mm

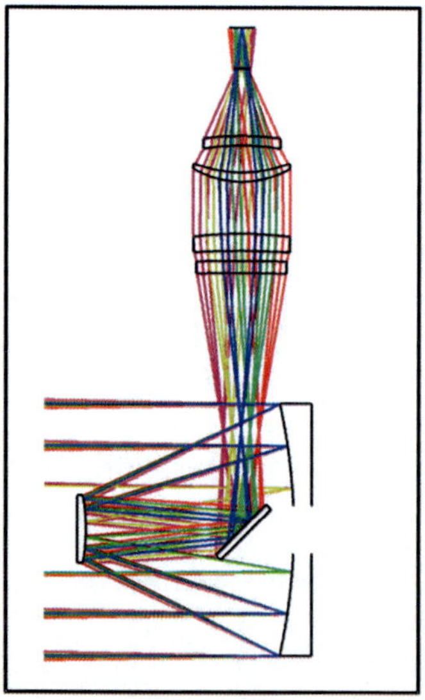

Fig. 6. LWIR detection subsystem

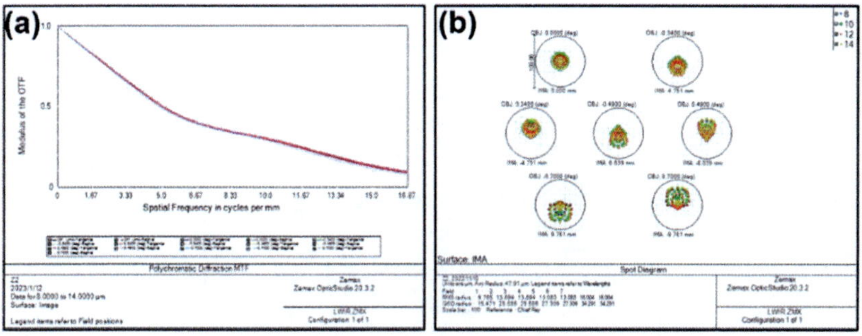

Fig. 7. Image quality of the LWIR detection subsystem, (a) MTF, (b) spot diagram

subsystem working at 15 °C and 25 °C. It can be seen that the MTF at both low and high working temperatures are close to the diffraction limit. The result means that thermal aberration of the subsystem is greatly eliminated.

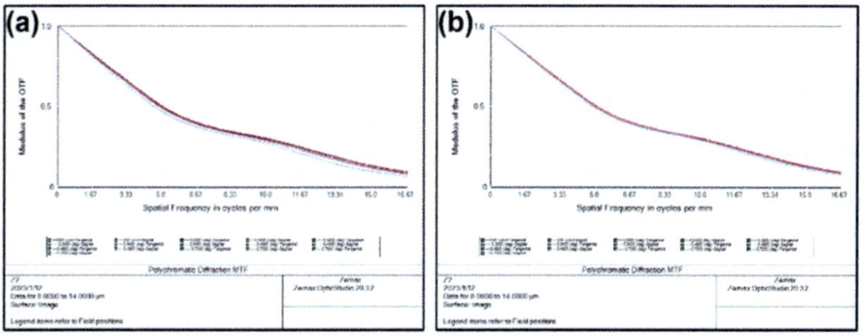

Fig. 8. MTF of the LWIR detection subsystem at both low and high working temperatures, (a) 15 °C (b) 25 °C

2.3 LiDAR Detection Subsystem

Unlike optical passive detection, the active detection method of LiDAR is used to detect the distance information of the cold target in poorly illuminated environments. In addition, the 3D profile of the target can also be identified by laser-scanning. The LiDAR detection subsystem is shown in Fig. 9.

Considering the shift characteristic of laser wavelength under vary temperature and the half-height width of laser spectrum, the laser spectral bandwidth is selected as 10 nm. The chromatic aberration of the laser echo signal can be disregarded due to the narrow band range of the LiDAR. In addition, the detector of the LiDAR detection subsystem is a photodiode with a photosensitive surface size of 4.6 mm × 4.6 mm. Figure 10(a) and (b) show the fraction of enclosed energy of the LiDAR detection subsystem and the position distribution of the laser spot on the photosensitive surface of the detector at different fields of view. Which indicates that the laser echo signal can be well received.

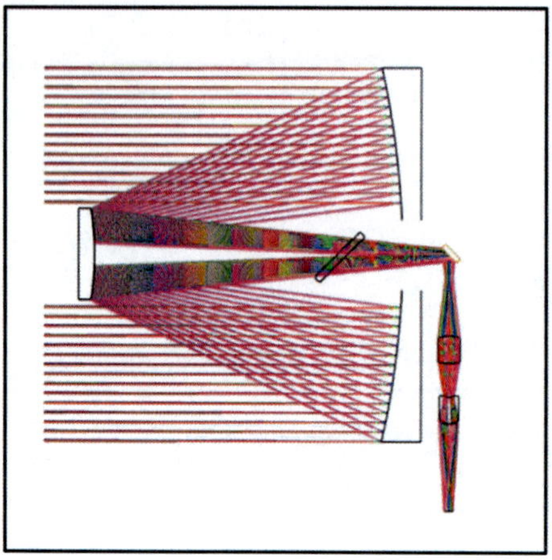

Fig. 9. LiDAR detection subsystem

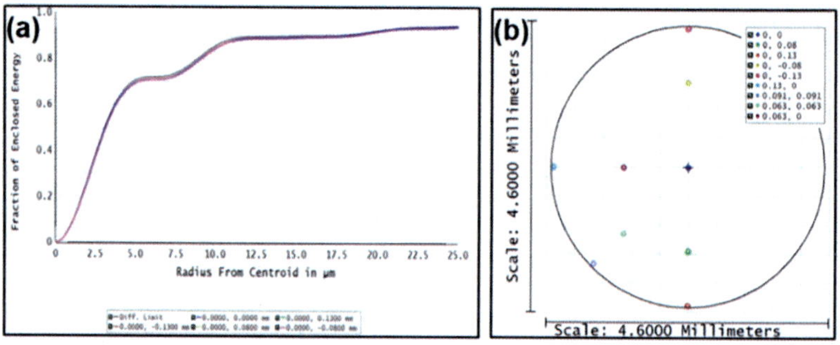

Fig. 10. (a) Fraction of enclosed energy of the LiDAR detection subsystem, (b) The laser spot on the photosensitive surface of the detector at different fields of view

3 Conclusion

In this paper, an integrated multi-mode co-aperture detection system with multi-dimensional detecting capability is proposed. The simulation models of the optical subsystems are established, including a VIS-NIR zoom detection subsystem, a LWIR detection subsystem, and a LiDAR detection subsystem. Among them, the VIS-NIR zoom detection subsystem is able to achieve a high contrast variable angular resolution of 0.0083 mrad–0.01667 mrad (CCD with a pixel of 10 μm) by zoom adjustment. The LWIR detection subsystem with a focal length of 800 mm can effectively detect space target in low-light space or backlight environment. The instantaneous FoV of a single detection of the LiDAR subsystem with a focal length of 1000 mm is 0.26°, which can be used for scanning to obtain distance information of space object or three-dimensional mapping. Because this system has the capability of multi-dimensional integrated detection, it can guarantee the accuracy and timeliness of target detection by space-based detectors in complex space environments under the premise of existing satellite resource constraints, thus effectively enhancing the space environment adaptation capability of space-based detectors. It is of great importance for the effective use and in-depth exploration of extraterrestrial space.

References

1. Li, C., Ouyang, Z., Du, H.: Space debris and space environment. Quat. Sci. **22**, 540–551 (2002)
2. Walker, R., Martin, C.E.: Cost-effective and robust mitigation of space debris in low earth orbit. Adv. Space Res. **34**(5), 1233–1240 (2004)
3. Huo, J., Qin, D., Xianfeng, Q.I.: Study on general situation of space debris. J. Acad. Equip. Command Technol. (2007)
4. Lal, B., Balakrishnan, A., Caldwell, B.M., et al.: Global trends in space situational awareness (SSA) and space traffic management (STM). Institute for Defense Analyses Washington DC (2018)
5. Freiwald, A., Freiwald, J.: Range-gated laser and ICCD camera system for on-orbit detection of small space debris. Proc. SPIE Int. Soc. Opt. Eng. **2214**, 116–123 (1994)
6. Dong, J., Hu, Q., et al.: Space debris detection with space-based avalanched photodiode laser radar. Chin. Opt. Lett. **5**, S176–S179 (2007)
7. Tsuda, Y., Yoshikawa, M., Abe, M., et al.: System design of the Hayabusa 2-Asteroid sample return mission to 1999 JU3. Acta Astronaut. **91**, 356–362 (2013)
8. Ying, W., Jianming, H., Yu, L., et al.: Simulation of LiDAR imaging for space target. Infrared Laser Eng. **45**(9), 906002–0906002 (2016)
9. Nocerino, A., Opromolla, R., Fasano, G., et al.: LiDAR-based multi-step approach for relative state and inertia parameters determination of an uncooperative target. Acta Astronaut. **181**, 662–678 (2021)
10. Dong, J., Hu, Q.: Detecting space debris with LiDAR. Laser Technol. **31**, 185–187 (2007)

Applications of High-Power Semiconductor Lasers in Marine Environmental Monitoring

Dongxin Xu[1,2], Yana Zhu[1,2], Guojun Liu[1,2], Yi Qu[1,2(✉)], Zhibin Zhao[1,2], Hao Chen[1,2], Cheng Cheng[1,2], and Zaijin Li[1,2]

[1] College of Physics and Electronic Engineering, Hainan Normal University, Haikou 571158, China
quyihainan@126.com

[2] Hainan Provincial Key Laboratory of Laser Technology and Optoelectronic Functional Materials, Innovation Center of Hainan Academician Team, Haikou 571158, China

Abstract. The protection and utilization of marine resources is an important development strategy in China. It is of great significance to the healthy development of marine environment to monitor the sea targets round the clock timely and effectively, to perform measurement of wind power and direction, Ranging at sea, extraction of wave-making resistance information, and water pollution detection, etc. High power semiconductor lasers have the advantages of compact, light weight, high efficiency, good tunability, etc., and have become the ideal light source for ocean environment monitoring photoelectric systems such as wind detection radar, through fog range finder, wave monitoring, heavy metal detection instruments, and satellite-borne ocean altimeter and underwater imaging system for marine pollution control. This paper focuses on high power semiconductor lasers' applications in the field of marine environmental monitoring. The development status and prospects of the respective semiconductor lasers used for these applications have also been summarized and analyzed.

Keywords: Semiconductor Lasers · Ranging · Detection · Marine Environmental monitoring

1 Introduction

Oceans are the cradle of human civilization and also the important carrier of human sustainable development, having great impacts on modern and future human life and survival. The effectiveness of the oean's development, utilization, and protection determines the quality of human survival and development. With the continuous development and utilization of marine resources, the protection of the marine environment has also attracted worldwide attention. Marine environmental monitoring can normalize the supervision of the ocean and provide emergency response to disasters such as sudden marine red tides and marine pollution. It is utterly important to comprehensively, timely, and accurately collect and master the laws and trends of ocean movement, help people protect marine resources, make rational use of marine resources, and ensure the sustainable development of marine resources.

H. P. Urbach and H. Jiang (Eds.): ISSOIA 2022, SPPHY 295, pp. 114–128, 2023.
https://doi.org/10.1007/978-981-99-4098-1_12

Laser technology has been dubbed the golden technology of the 21st century, of which the high-power semiconductor laser is the favorite, being small size, light weight, high efficiency, and easily tunable, which is an ideal light source for various marine environmental monitoring photoelectric systems. Semiconductor lasers have developed rapidly in many application fields since they were invented more than half a century ago. With the continuous emergence of various wavelengths and high-power semiconductor lasers, they have been applied in many fields, such as medicine, military, industrial processing, human life and work, and natural sciences. It has become the common consensus to actively develop and utilize rich and excellent marine resources, develop marine ecological environment monitoring, and improve the public's awareness of marine ecological security. High power semiconductor lasers have been widely used in the field of marine environmental monitoring.

2 High Power Semiconductor Laser

Semiconductor laser, also known as laser diode, is a device that uses high crystal quality semiconductor materials as working media to generate laser through the process of stimulated radiation. Its working principle is to realize the non-equilibrium charge carrier population inversion either between the energy bands (conduction band and valence band) of the semiconductor material, or between the energy band of the semiconductor material and the impurity (acceptor or donor) energy level through a certain excitation mode. When a large number of electrons in the population inversion state are combined with holes, the stimulated emission dominates, which generates laser radiation. Semiconductor lasers are developing rapidly, with a wide wavelength range, simple packaging, compact, all-solid, reliable and easy to modulate. The application of semiconductor lasers has covered the whole field of optoelectronics and become the core technology of optoelectronics. Semiconductor lasers are widely used in ranging, radar, communication, simulation weapons, warning, guidance, fuze, aerospace, etc.

In 1962, the first generation of homojunction-injected semiconductor lasers in GaAs was born in the United States [1]. In the following year [2], a heterojunction semiconductor laser was successfully developed at the Joffe Institute of the Soviet Academy of Sciences. In 1967, IMB grew AlGaAs on GaAs using liquid phase epitaxy (LPE) technology. In 1970, the first AlGaAs/GaAs single heterojunction laser was successfully studied at Bell Laboratories in the United States. In the 1980s, the theory of energy band engineering has been proposed and great strikes have been made in new epitaxial technologies for the growth of crystal semiconductor materials such as molecular beam epitaxy (MBE), metal organic chemical vapor deposition (MOCVD), liquid phase epitaxy (LPE) and chemical beam epitaxy (CBE), accompanied by breakthroughs in material passivation technology, heat deposition technology, and LD structure design, including single-mode lasers, distributed feedback lasers, laser arrays, and the simultaneous development of complementary technologies, all these contribute to the realization of high power output of semiconductor lasers.

So far, high power semiconductor laser single emitter CW power has exceeded 70 W, pulse output power has exceeded 100 W; LD bar CW power exceeds hundreds of watts, Quasi continuous wave (QCW) output exceeds kilowatts; The quasi-continuous output power of the LD stack and the optical fiber combined laser already exceeds ten thousand watts. Multi-unit combination of the array laser light source through the LD stack enables power output above megawatt level and multiwavelength megawatt laser emission. The development of high power semiconductor lasers makes the downstream laser technology develop rapidly. In the field of downstream laser technology, semiconductor lasers are used for pumping doped crystals (diode-pumped solid-state lasers) or doped fibers (fiber lasers) with high conversion efficiency, long life, good stability and excellent beam quality. With thermal management, spectrum control, output beam control, and optical beam shaping element process improvement, high frequency or continuous megawatt high power semiconductor lasers have been developed. Commercial products will also be widely used in various fields (Fig. 1).

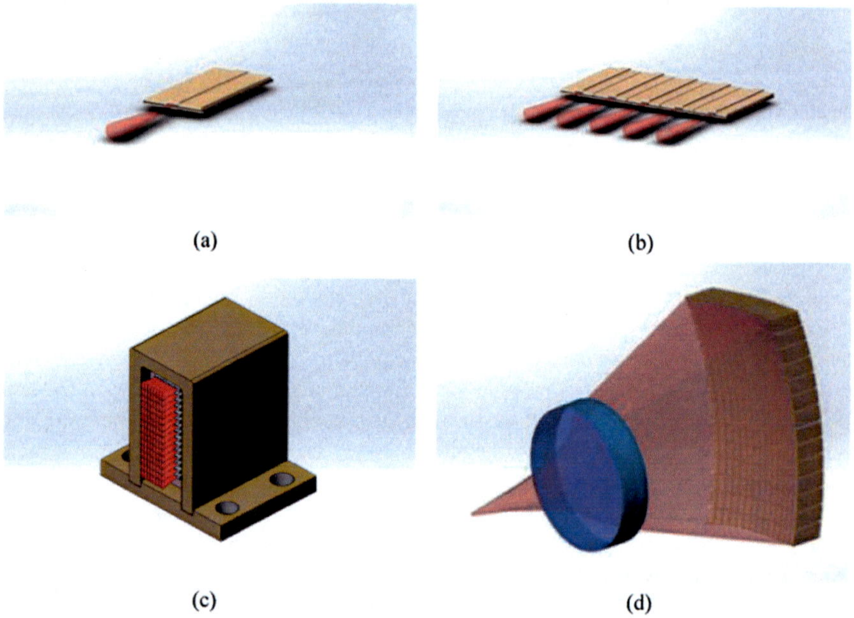

Fig. 1. Various forms of semiconductor lasers. **(a)** Hundred watt class single emitter LD. **(b)** Kilowatt level LD Bar. **(c)** Ten thousand watt level LD stack unit. **(d)** Megawatt level LD stack array.

3 High Power Semiconductor Lasers for Wind Measurement Radar

The wind measurement radar, or Anemometer radar, is a radar device used to measure the wind direction and wind speed in the air. Using the Doppler shift principle of laser, the wind speed and direction information is obtained by measuring the frequency change of light wave reflection in the air when it meets the aerosol particles of wind motion and calculating the vector wind speed and direction data of the corresponding height. The wind measurement radar detection system mainly consists of the laser system, transmitting and receiving system, heterodyne detection system and digital signal processing system. Using semiconductor laser atmospheric acquisition sensor, combined with a variety of data processing and servo control system, remote sensing detection of atmospheric wind field at different heights can be achieved, fast and accurately and with high resolution, based on data from the measured space of three-dimensional atmospheric wind field information, these atmospheric wind field data can significantly improve the accuracy of numerical weather forecasting for military and civilian purposes (Fig. 2).

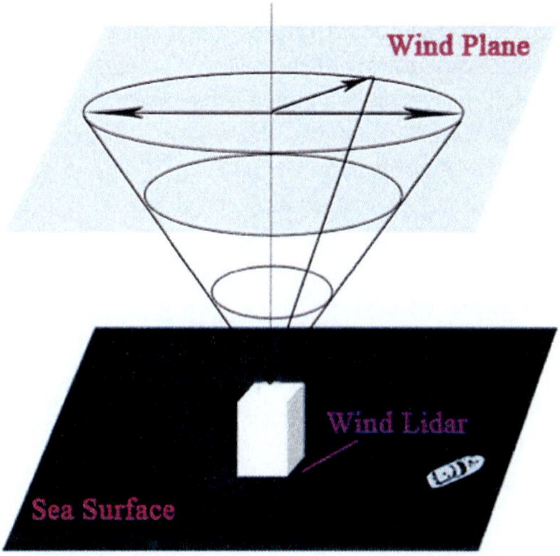

Fig. 2. Schematic Diagram of Semiconductor Laser Wind Measuring Radar.

Since the first coherent continuous wave wind laser radar (lidar) based on 10.6 μm CO_2 lasers was developed in 1970 [3], Doppler lidar technology has been developed soon afterwards. In 2000, Karlsson et al. [4] first released an all-fiber CW coherent wind lidar. In 2004, QinetiQ Company of the United Kingdom developed a commercial all fiber continuous wave coherent wind lidar, which can measure at 10 customized locations within the range of 10–200 m. It has played a significant role in wind energy fields such as wind resource assessment and wind farm planning. In 2014, the WindVISION wind measuring radar developed by Windar Photonics of Denmark measured the wind turbine wake, gust, turbulence and other data by transmitting four continuous wave laser beams focused at a point within 0–100 m and with a certain angle, providing data protection for the optimization of the wind power field systems.

The introduction of high power semiconductor lasers has led to a renewed interest in the use of wind measurement radar once composed of solid state lasers. The development of high power semiconductor laser wind measurement radar has a positive impact on human life and ocean monitoring. The use of semiconductor light wind measurement radar can avoid natural dangers, study the natural laws of the marine environment, and utilize marine resources, etc. High power semiconductor laser wind measurement radar can be used in clear sky, which can provide high precision and high resolution real-time atmospheric wind field information, and is the international recognized advanced technology for measuring wind field. The traditional microwave technology is applicable to a certain range of wavelengths, and the monitoring effect is better in cloudy and rainy conditions, and the detecting capability will be greatly hampered when encountering clear weather. Semiconductor lidar system can achieve high resolution wind speed measurement in small scale, combined with the scanning module, fixed elevation angle wind speed scanning measurement can be achieved, so that the wind speed scanning observation of near-surface wind field can also be achieved. High-power semiconductor laser-pumped solid-state lasers have replaced CO_2 lasers and flash lamp-pumped solid-state lasers, providing significant improvements in compactness, efficiency and lifetime. With the continuous improvement of laser power level, the distance of laser wind measurement radar system also comes farther and farther, and the expansion of output wavelength and detection capability, the wavelength of laser wind measurement radar is also developing in the direction of human eye safety. 1.55 μm wave band is located in the atmospheric absorption window, so this 1.55 μm laser band, with low absorption and less fiber loss in the atmosphere, has become the mainstream of laser wind measurement radar band, for example, with 500 μJ pulse energy, the action of the distance of 16 km can be reached.

High power semiconductor laser wind measurement radar is a new atmospheric remote sensing equipment, which can realize small scale, high precision and high resolution non-contact measurement and spatial scale wind velocity gradient measurement. Three-dimensional atmospheric wind field remote sensing observation can solve the problem of low level wind shear detection which seriously affects the ocean wind direction monitoring, maritime navigation safety and offshore areas. It can be widely used in meteorology, environmental protection, national defense, airport and other fields. It can be called the "downwind eye of the sea".

4 Application of High Power Semiconductor Lasers in Through-Frog Rangefinder

Lasers have been used for laser ranging for many years, and laser rangefinders should be able to have fog-transparent ranging capabilities in harsh environments with low visibility such as at sea level. The advantages of laser ranging are high measurement accuracy, high resolution and anti-interference capability, and it is a ranging technology that uses laser as the information carrier to measure the distance to the target. It can be classified into pulsed laser ranging and phased laser ranging according to different laser signal processing methods. Due to the promising application of laser ranging and the development of semiconductor laser devices, pulsed laser ranging technology is developing in the direction of miniaturization, high accuracy, multifunctionality, solid-state components and human eye safety. The high power semiconductor laser as the light source of laser ranging technology is small in size, light in weight, simple in structure and high in efficiency, and is the preferred light source for fog penetrating rangefinders (Fig. 3).

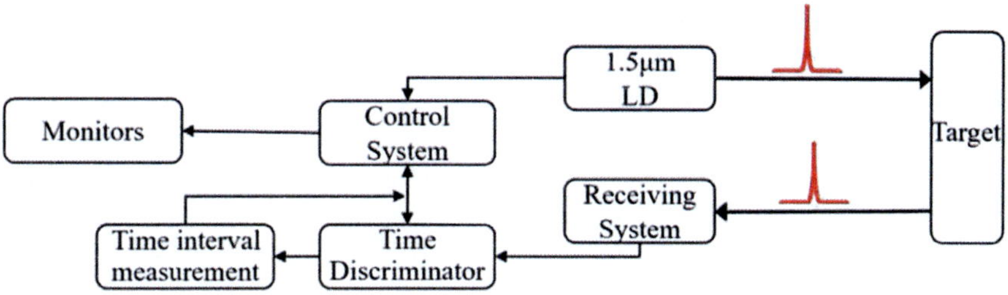

Fig. 3. Schematic diagram of 1.5 μm LD ranging principle.

In 1960, the world's first ruby laser rangefinder at 694 nm was applied in the US military since its birth [5]. The early ruby solid laser rangefinder was large in size, low in efficiency and high in power consumption, and was soon replaced by the second generation 1064 nm near-infrared neodymium laser rangefinder [6]. Although the second generation laser rangefinder has greatly improved in volume, power consumption, efficiency and other performance, it does not meet the human eye safety standards. The rapid development of the third generation of laser rangefinders with eye safety and high integration has broken the limitations of the second generation of laser rangefinders, rapidly expanded the application scope of laser rangefinders, and more companies have also increased their efforts in the development of laser rangefinders. In the field of portable eye safe laser rangefinders, Tuyade and Danergy in the United States, Bosch and Olfa in Germany, Nikon in Japan, etc. have made great achievements. Leica and Bosch have obvious advantages in the direction of hand-held short-range laser ranging. Leica's S910 handheld laser rangefinder can directly measure the distance between two points, and can directly measure the area of irregular objects and any angle of space. In 2001, Swiss Research Institute used micro optics system to effectively shape the beam of high-power laser diode, and the overall beam efficiency increased by 28%. In 2011, Tuya developed the SP800 telescopic rangefinder using human eye safe semiconductor laser. The military 905 nm semiconductor laser range finder launched by Nukang Company in Canada has a measuring distance of 4 km and a ranging accuracy of ±1 m. Its new product LP20K laser rangefinder uses semiconductor pumped 1570 nm solid-state laser, with a maximum operating distance of 20 km; The measuring distance of the range finder of the company can reach 1200 m, and the measuring distance of small targets can reach 450 m.

At present, 1530 nm Erbium glass laser which meets the safety standards of human eye is used in our country, and its measuring distance can be up to 6 km. Using a high power 1570 nm laser pumped by a semiconductor laser, the measuring range of large hulls, buildings or mountains over 20 km can be realized. 1.5 μm laser has high transmittance, can achieve fog penetration, and good concealment, and can effectively reduce the interception or interference by the other side. Eye safe long-distance laser fog ranging system at 1.5 μm can be called "sea perspective eye".

5 High Power Semiconductor Lasers for Ocean Wave Monitoring

Protection and utilization of Marine resources is an important development strategy in our country. As the main means of Marine traffic and transportation, the performance of ships plays a key role in the efficiency of Marine resources development and utilization. However, the wave-making resistance produced in the process of moving has an serious effect on the sailing speed and shipping efficiency of the ship. Therefore, in order to measure wave-making resistance timely and effectively, it is more urgent to observe and measure the waves. In addition, the development and utilization of ocean wave energy also requires real-time and accurate observation of waves (Fig. 4).

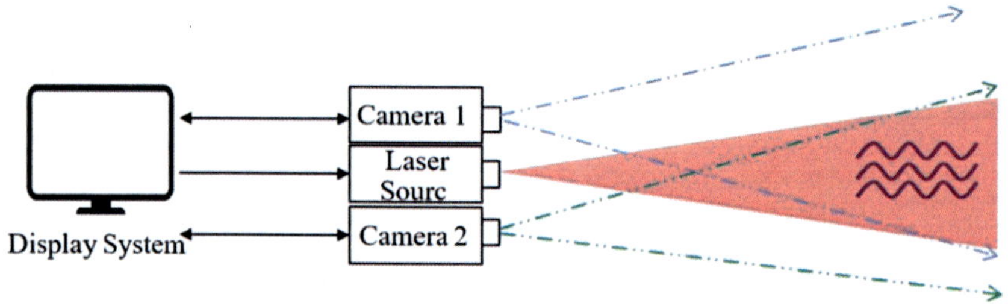

Fig. 4. Schematic diagram of semiconductor laser wave monitoring principle.

In the 1970s, laser technology was applied as a new light source and observation tool for wave monitoring. In 1973, Tober et al. [7] used a 632.8 nm He-Ne laser as a light source to calculate the slope of waves by observing the spatial position change of the beam after refraction of the underwater laser beam through the wave surface. In 2004, Wang Juncheng et al. carried out wave measurements with a solid-state laser. Using ship-based laser wave measurement instrument, there was no instrument with the same principle used at home as abroad at the time. What were used is a laser wave measurement device installed on the oil platform [8]. Computer vision-based motion target detection technology was applied to wave detection at an early stage, and the United States developed the EVS (Erdman Video System) system in 1998, which combined a video camera and a high-resolution camera to achieve coastal protection, beach management, and marine vessel monitoring. Australia's Coast Watch developed the Coastal COMS (Conditions Monitoring System) system in 2008 for coastal monitoring needs, mainly for beach visitor detection, shoreline extraction and offshore sports vessel monitoring. The Center for Shoreline Protection Technology at Atlanta University in the United States has developed a marine video monitoring system for marine traffic monitoring.

Since 2006, the National Marine Technology Center has conducted in-depth research on the nearshore monitoring technology, and made remarkable achievements in ship target detection and coastal monitoring. The application of moving target detection technology in the monitoring of sea moving target (wave) not only cuts short the development cycle of wave monitoring software, reduces the difficulty of software development, but also can be quickly applied to the real life, which is helpful for the subsequent analysis of the height of the wave climbing, event alarm and so on.

With the development of laser technology, small size, light weight, high efficiency and tunable semiconductor laser light source will be gradually extensively used in the field of wave monitoring. The ocean wave parameters measurements in the target sea area have been readily realized, based on high definition through-frog infrared night vision technology and binocular stereo vision technology, and combined with digital image processing technology, in which a high power semiconductor laser is used as the light source and a binocular stereo vision system as the acquisition platform. With two high-resolution cameras, the system can respond to the infrared band, complete the acquisition of all-weather wave video pictures. The wave height, period and direction information of the ocean waves in the target sea area can be calculated from the captured wave video. Wave eigenvalues are extracted by synchronous contrast processing of two-sided video images, and the inversion algorithm is used to complete the three-dimensional reconstruction of wave information. Wave video preprocessing is used to extract image frames and calculate the inter-frame difference from them to form a three-channel image containing static and dynamic wave height information. A model capable of detecting wave height is built, and a training dataset is constructed from the image set after video preprocessing. The network model is trained to detect different wave heights, and the training model is used to detect the instantaneous wave heights of nearshore waves with high accuracy. Using continuous wave high power fiber coupled semiconductor laser over 30 W as auxiliary illumination, day and night sea area monitoring of 3 km can be achieved. With the binocular wave measuring system, the wave elevation range is 0 ~ 20 m, with the measurement error of $\pm 10\%$, the accuracy of wave period measurement is 0.5 s, and the accuracy of wave direction measurement is $\pm 10°$.

High power semiconductor laser night vision and ocean wave monitoring system can be used for inter-island lookout, waterway management, maritime management, marine dynamic monitoring, fishery law enforcement, aquaculture safety, hydrological monitoring, port and dock management, offshore mine, offshore power generation device perimeter prevention, etc. High-power semiconductor lasers are truely the "eye of the sea nighthawk".

6 High Power Semiconductor Lasers for Marine Heavy Metal Detection

With the accelerated global industrialization, the problem of heavy metal pollution in water bodies is becoming more and more serious, posing a great threat to human health and ecological environment. Therefore, there is an urgent need for a method that can perform rapid and sensitive detection of hazardous metal elements in aqueous solutions. At present, the commonly used analytical methods for elemental detection in aqueous solutions include inductively coupled plasma emission spectrometry, atomic absorption spectrometry, atomic fluorescence spectrometry and electrochemical analysis, etc. Although these methods are more sensitive, the detection process is time-consuming and requires complex pretreatment of the sample, which is difficult to meet the practical needs of rapid on-site detection. Laser-induced breakdown spectroscopy was first proposed by Brech and Cross et al. in 1962, and can be used for qualitative and quantitative determination of elemental components in solution samples.

Laser Induced Breakdown Spectroscopy (LIBS) is a new method in spectroscopic analysis applied to trace element analysis. The basic principle is that a laser excites the measured substance to a plasma state, and based on the position and intensity of the spectral lines of light emitted from the plasma, the analysis obtains valid information about the composition and concentration of the measured element. It has many advantages over atomic absorption spectroscopy, liquid chromatography and other detection methods, such as no pretreatment of samples, short analysis time, multi-element analysis, no additional contamination of samples, and no restriction of sample morphology, and therefore has a wide range of applications in solid, liquid and gas fields. This method not only has the advantages of fast detection and no or less sample pretreatment, but is also one of the few techniques that can perform real-time, in situ, non-contact detection, and is particularly suitable for monitoring toxic metal elements in industrial wastewater. Despite the obvious advantages of LIBS over other methods, the conventional single-pulse LIBS technique for trace metal element detection and analysis in water has some problems, such as low detection sensitivity and poor signal reproducibility. This can be partly attributed to the properties of water bodies, which are known to have a higher thermal conductivity and are difficult to be compressed compared to air. Therefore, underwater LIBS is a spectroscopic technique that uses short pulses of high energy laser light focused on the sample surface to excite the sample to the plasma state, generating atomic emission spectra and then analyzing the elemental content of the substance. This technique has the following advantages: easy to meet the requirements of online detection, easy and fast processing and analysis, and no need to add any processing and analysis reagents.

LIBS technology is widely used in the detection of various substances, and the exploration of marine detection is also being carried out. LIBS technology can not only detect various heavy elements in the ocean, but also determine the growth environment of marine organisms by detecting them, and determine the global climate change during their growth cycle by detecting shells. The full use of LIBS technology can obtain more information about the ocean and provide support for people to develop and protect marine resources. Research on the detection of metal pollution in the ocean has mainly focused on Cu, Pb, Zn, Fe, Mn, Cr, Hg and other elements. Although trace elements such as Mn, Cu and Zn in human body are essential for human life activities, but when the concentration of these heavy metal elements exceeds a certain range, it turns harmful to human body (Fig. 5).

Overseas, the LIBS technique has developed rapidly since 1962. In 1995, Dana et al. used LIBS to measure the salt concentration in seawater aerosol droplets, determined the plasma temperature, and made a standard curve of 589 nm for Na and 656.3 nm for Na as a standard for determining the Na concentration in seawater, and this method can determine the Na concentration in the range of 100 ppm to 10000 ppm. In 2011, Man et al. used the LIBS technique to determine the concentrations of four elements, K, Mg, Ca, and Al, to classify 11 sea salts by attribution, and compared the results obtained with the ICP-AES method to demonstrate the reliability of the LIBS method for sea salt classification, and to demonstrate that the production method of sea salt can be determined by the spectral intensity of Al. In 2017, Luan Kunxiang et al. of the First Institute of Oceanography, State Oceanic Administration of China, used an underwater

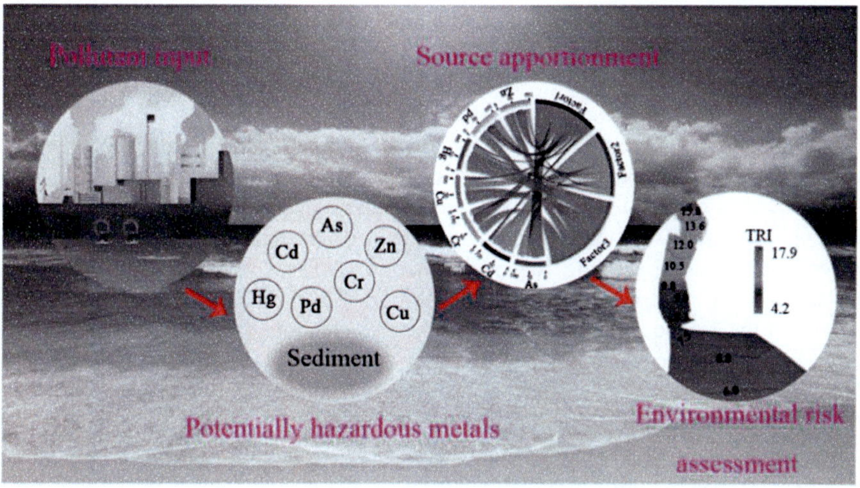

Fig. 5. Marine Heavy Metal Pollution Detection.

intelligent robot to sample our sea area and combined UV-Vis spectrophotometer with Kalman filter algorithm to achieve real-time monitoring of a variety of heavy metal elements to detect pollution in our sea area [9]. In 2004, A. De Giacomo et al. used a double-pulse laser to act inside seawater to detect the signal of metal elements in it. Wang Chuanhui and Cui Deifeng from Anhui Normal University and China Ocean University tried to analyze the metal elements in aqueous solutions of $AlCl_3$ and $CuSO_4$, and the detection limit reached 23 ppm.

High-power semiconductor lasers can be used to create miniature heavy metal detectors that can perform real-time metal element analysis at sea. The method and system for measuring the content of heavy metals in water based on laser induced breakdown spectroscopy can be called "the golden eye of underwater knowledge", which has the ability to comprehensively improve the environmental supervision of marine heavy metals.

7 Applications of High Power Semiconductor Laser in Satellite Ocean Altimetry System

The impact of human activities on the atmospheric and oceanic systems has been increasing, and global temperature rise, sea level rise, and global climate extremes have become hot issues of global concern. The development of satellite-based lidar altimetry technology has provided a new data source for scientific exploration of polar regions, measurement of global land topography and sea surface height measurement, and a new perspective and means for ocean altimetry.

Compared with the traditional microwave altimetry, the high-powered semiconductor laser satellite altimetry system has the advantages of wide coverage, more concentrated pulse energy, high detection accuracy, and the ability to obtain three-dimensional information. The satellite-based laser altimetry system provides a new observation method to obtain large spatial scale and high spatial and temporal resolution observation data, which greatly expands the depth and breadth of ocean research. The accuracy of decimeter level provides a new way and favorable conditions for the measurement of sea surface height, sea ice height and sea ice thickness in polar regions.

Satellite laser altimetry system and microwave altimetry system have the same basic principle and both are satellite-carried system. For microwave altimetry system, the sensor is microwave radar, its operating frequency band is microwave, while laser altimetry is an optical remote sensor, its operating frequency band is green and near infrared light, so the laser altimeter operating frequency is many times higher than that of the microwave altimeter. High frequency means that the laser pulse will not penetrate deeply into the surface of the ground and snow and ice surface, which not only can reduce the altimetry error, but also can be well applied in the polar region containing snow and ice. Satellite laser altimeter adopts active detection method, its basic working principle is to obtain the target information by accurately detecting the time experienced by the emitted laser pulse from the emission to the earth's surface and reflected back to the receiver, and the shape of the returned full waveform. Then, based on the precise orbit and attitude data, the precise elevation data and geographic location information of the laser foot point can be determined (Fig. 6).

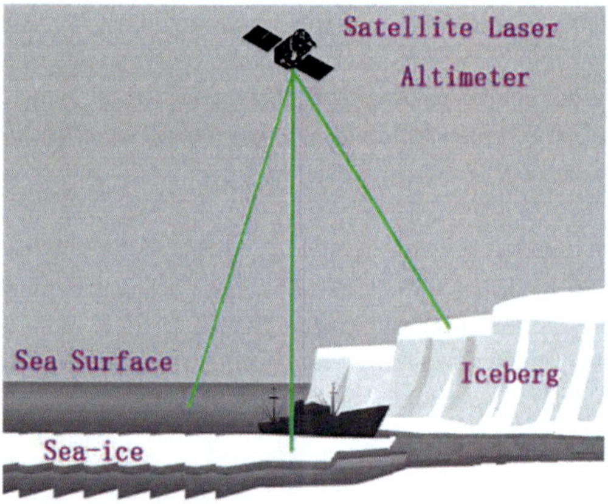

Fig. 6. Schematic Diagram of Satellite Laser Altimeter.

The United States was the first country to use satellites for altimetry. On May 14, 1973, the National Aeronautics and Space Administration of the United States launched the first satellite Skylab with altimetry performance. Its orbit running height is 425 km, the orbit inclination is 50°, the altimetry accuracy is 85–100 cm, and its working life is only about three months. As the Skylab is the first satellite used for height measurement, it is the result of the principle experiment of satellite height measurement, marking the official launch of the height measurement satellite. On April 9, 1975, Ceos-3 was launched from a U.S. Air Force base with an orbital altitude of 840 km, an orbital inclination of 115°, a high accuracy of 25–50 cm, and an operational lifetime of 2–3 years and a repeat cycle of 23 days. Compared with Skylab, Ceos-3 not only increases the working lifetime, its orbital height, orbital inclination and height measurement accuracy have also been optimized and improved. During the 3 years of normal operation of Ceos-3, measurements of parameters such as Earth's level, wind speed, waves, ice and ocean currents can be made, providing data-based support for oceanography and geodynamics. On June 28, 1978, the U.S. launched the ocean satellite Seasat, with an orbital altitude of 800 km, an orbital inclination of 108°, a high accuracy of 20–30 cm, a working life of 106 days, and a repeat cycle of 3 days (pre) and 17 days (post). Its purpose is to measure the level of the ocean, wind speed, waves and currents. The launch of the ocean satellite Seasat is a historic milestone in oceanography. On March 12, 1985, the United States Navy Base launched Geosat, a geodetic satellite that provides geodynamics with accurate data of the earth's gravity field covering the whole world, and can analyze the sea surface shape, sea ice, level and other parameters. Geosat has a track running height of 800km, a running track inclination of 108°, a height measuring accuracy of 10-20cm, a working life of 4 years, a repeat period of 17 days. The first European remote sensing satellite, ERS-1, was launched on July 17, 1997, with an orbital altitude of 785 km, an orbital inclination of 98.5°, a high accuracy of 10 cm, and an operational lifetime of 3 years. Its outstanding feature is that it can change its own orbital altitude and change its own operation period. This remote sensing satellite ERS-1 can detect sea surface topography, circulation, sea ice, and it can detect 90% of the world's oceans. On August 10, 1992, the National Aeronautics and Space Administration (NASA) and the French Space Agency (ASA) jointly launched the Topex/Poseidon ocean topography test satellite, which enables more accurate measurements of the ocean surface, currents and tides. The orbital altitude of this satellite is 1336 km, the orbital inclination angle is 66°, the measurement accuracy is 6cm, and the working life is 4 years [10, 11].

Compared with foreign countries, China's independent research and development of altimetry satellite started later [12]. On August 16, 2011, Taiyuan Satellite Launch Center carried the "Ocean II (HY-2A)" satellite on a Long March 4B launch vehicle and successfully launched it into space. HY-2A [13] is China's first independently developed altimetry satellite applied to the ocean power environment, which can detect sea surface height, waves, currents, sea surface temperature and other ocean information parameters. During 2018–2021, China successfully launched the ocean power satellite HY-2B/2C/2D, HY-2 series altimetry satellites. The measurement accuracy of the satellites is 5-8cm, the effective wave height measurement accuracy is 0.4 m, the sea surface wind speed measurement accuracy is 1.5 m/s, the wind direction accuracy is 19.5°, the sea surface temperature measurement accuracy is 1 K, and the wet tropospheric path delay error is <20 mm. so far, 23 altimetry satellites have been launched worldwide.

The high power semiconductor pumped solid laser altimeter carried by the satellite can measure the elevation of the sea surface, sea ice and land surface, draw the sea and land topographic map, detect the slope, reflectivity and roughness of the ice surface and sea surface, and retrieve the vegetation canopy height and biomass. Satellite altimetry has played an important role in polar scientific research, providing more abundant observation data for polar ice and snow surface elevation, sea ice density, sea ice freeboard, height and other aspects of research. The satellite laser altimeter can be called the "eye of the sea".

8 High Power Semiconductor Lasers in Marine Underwater Imaging Systems

The development of marine resources puts higher requirements on underwater imaging technology. At present, there are two ways of underwater imaging, optical imaging and acoustic imaging. Acoustic imaging can transmit to a long distance but with limited resolution, mostly used for seafloor topographic survey. Sonar equipment has high resolution but expensive. Optical imaging form is intuitive, more convenient for human eye observation, with the advantages of fast transmission, high resolution, real-time, etc., is the most suitable imaging method for underwater environmental monitoring.

The irregularity of suspended particle motion in the ocean causes multiple reflections and severe backscattering of seawater for optical signals, which results the light wave transmission with certain amounts of loss and strong noise, and seriously affects the imaging distance and quality of underwater optical imaging systems [14]. In order to suppress the strong backscattering of the optical signal by seawater and obtain the ideal imaging quality, researchers at home and abroad mainly use underwater laser imaging techniques such as distance selective pass and simultaneous scan imaging. The simultaneous scanning technique can effectively reduce the overlapping area between the receiver field of view and the illuminated water body, so that the backward and forward scattered light entering the receiver can be reduced to the minimum, and effectively improve the imaging quality and extend the observation distance. Underwater laser imaging plays an important role in the observation and identification of marine targets, compared with the traditional means of sonar, laser imaging has significant advantages in terms of image visualization, resolution and visibility, affected by the backward scattering of laser signals by suspended particles in the water. In order to improve the signal-to-noise ratio and imaging clarity of underwater laser imaging equipment, it is needed to use the distance selection pass synchronization control technology: using pulsed laser and image intensifier CCD camera, to separate the scattered light and the reflected light of the target in the process of laser propagation, so that the radiation pulse reflected back by the observed target just reaches the camera and images are formed within the selective time of ICCD camera (Fig. 7).

The development of foreign underwater laser imaging technology is mainly focused on the improvement of the traditional camera lens, color balance filter, pressure hole, in addition, the introduction of high-resolution camera has also greatly improved the technology. At this stage, the quality of the photos taken underwater is often not high, so the impact caused by the physical properties of water has become the focus of research on underwater laser imaging. Research on underwater laser imaging systems in the United States began several decades ago and was not reported for reasons of confidentiality. At present, the more recognized underwater laser imaging system is a synchronous scanning underwater laser imaging system used by the U.S. Navy [15], which uses an argon ion laser as the laser source, and the field of view of this system can reach 70°, the imaging distance range is 2.5–45 m, and the resolution can reach 25 mm for imaging distance of 30 m. There are several models of foreign underwater Laser imaging system being introduced, mainly used in underwater video recording, target detection and other purposes. Such systems can also be deployed underneath ships and vessels to capture two-dimensional images of the underwater. When the ship is underway, the

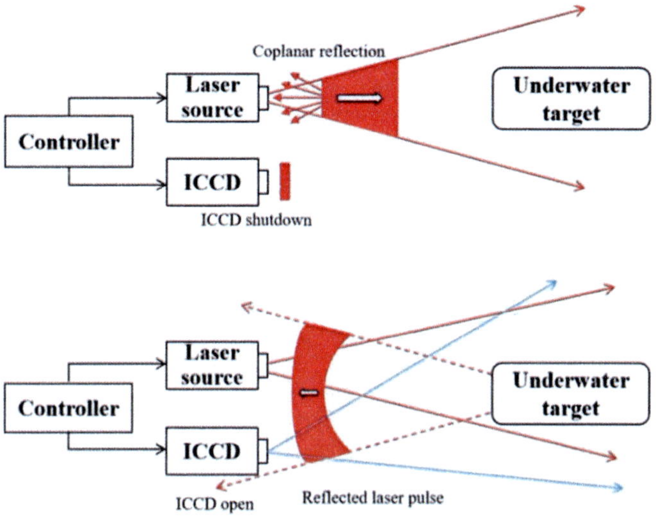

Fig. 7. Schematic diagram of underwater laser imaging principle.

two-dimensional image of the seabed will be formed by the system, the underwater laser imaging and detection distance can reach the attenuation length, the image clarity and resolution are very high. In addition, there are reports on the development of prism control laser technology system with fast rotation characteristics, which has been assembled in the military submarine.

The laser underwater imaging system can monitor submarine cables and underwater pipelines for easy detection and maintenance, and can search for submarine and mine and other underwater military targets. In the search and observation of moving targets at the same time can also achieve tracking, ranging and video recording. At present, underwater laser imaging technology has entered the practical stage, and has received the attention of the military departments of various countries.

The underwater imaging system uses a 532 nm green laser light source combined with polarization technology [16], which can be used to simulate imaging in the turbid marine environment. By combining the 532 nm laser active illumination with polarization technology, the acquired polarized image at a certain polarization angle is richer in image information and clearer in imaging quality compared to the conventional imaging raw image. This underwater imaging method can improve the resolution of the target through high contrast image information, and can provide new ideas for ocean underwater observation, worthy of being the "laser eye under the sea".

9 Conclusions

In addition to ocean wind-detecting radar, through-frog rangefinder, ocean wave monitoring device, heavy metal detector, satellite range measurement system, and underwater imaging system, the high-power semiconductor laser is also the preferred light source for underwater communication system, marine oil spill measurement fluorescence system, seabed mapping system, ocean chlorophyll concentration measurement system, marine airborne and shipborne monitoring multi-function radar system, etc. There is no doubt that high-power semiconductor lasers are the eyes of the ocean.

With the rapid development of computer science and technology, microelectronics integration technology, material chemistry science and technology, semiconductor packaging technology, precision processing, and manufacturing technology, intelligent big data fusion technology, and aerospace technology, high-power semiconductor lasers can not only be widely used in the field of marine monitoring, but will infiltrate into every corner of human activities. It is the "ideal light" and "wisdom eye" for us to look forward to and enjoy beautiful life, protect ecology, maintain peace, and explore the universe.

References

1. Hall, R.N., Fenner, G.E., Kingsley, J.D., et al.: Coherent light emission from GaAs junctions. Phys. Rev. Lett. **9**(9), 366 (1962)
2. Kroemer, H.: A proposed class of hetero-junction injection lasers. Proc. IEEE **51**(12), 1782–1783 (1963)
3. Huffaker, R.M., Jelalian, A.V., Thomson, J.A.L.: Laser-Doppler system for detection of aircraft trailing vortices. Proc. IEEE **58**(3), 322–326 (1970)
4. Karlsson, C.J., Olsson, F.A., Letalick, D., et al.: All-fiber multifunction continuous-wave coherent laser radar at 1.55 μm for range, speed, vibration, and wind measurements. Appl. Opt. **39**(21), 3716–3726 (2000)
5. Lai, H.H., Li, J.Q., Dong, T.: Study on the light source for long range laser ranging in unmanned aerial vehicles. In: Proceedings of the 5th China UAV Congress, Beijing (2014)
6. Chai, J.H., Liu, N.: Analysis of long range laser ranging in fog. In: Photon Science and Technology Innovation and Industrialization - Photon Science and Technology Innovation Forum in Yangtze River Delta, Anhui, vol. 9999, pp. 1–13 (2006)
7. Tober, G., Anderson, R.C., Shemdin, O.H.: Laser instrument for detecting water ripple slopes. Appl. Opt. **12**(4), 788–794 (1973)
8. Wang, J.C., Hou, G.L., Liu, Y., et al.: Research on shipboard wave measurement instrument based on laser technology. Ocean Technol. **23**(04), 14–17 (2004)
9. Luan, K.X., Ma, X.D.: The intelligent underwater vehicle detects the status of heavy metal pollution in sea. Ship Science and Technology **39**(16), 187–189 (2017)
10. Zhai, G.J., Huang, M.T., Ouyang, Y.Z., et al.: Principle and application of satellite altimetry. Hydrogr. Surv. Charting **22**(01), 57–62 (2002)
11. Jin, T.Y., Liu, W.X., Jiang, W.P., et al.: Research progress on global sea level change determined by satellite altimetry. J. Geomatics **47**(03), 1–8 (2022)
12. Wang, R., Xu, H., Zhang, H., et al.: Characteristics and application of HY-2 Satellite. Chin. Aerosp. **10**, 7–11 (2012)
13. Zhang, Q.J., Zhang, J., Zhang, H., et al.: The study of HY-2A satellite engineering development and in-orbit movement. Strateg. Study CAE **15**(07), 12–18 (2013)
14. Zhao, Y., Gai, Z.G., Zhao, J., et al.: Research on range - gated underwater laser imaging technology. Logistics Eng. Manage. **36**(07), 269–270 (2014)
15. Acharekar, M.A.: Underwater laser imaging system (ULIS). In: Detection and Remediation Technologies for Mines and Minelike Targets II, vol. 3079, pp. 750–761. SPIE, (1997)
16. Wang, W.T., Feng, W.W., Zhu, Z.L., et al.: Research on underwater imaging technology of ocean Ranch based on laser polarization technology. J. Optoelectron.·Laser **32**(06), 581–586 (2021)

A General Recognition Approach for Formation Change of Group Targets

Dan Wang(✉), Yongfu Wang, Ke Li, Yufei Huang, and Luyuan Wang

China Academy of Space Technology (CAST), Beijing Institute of Spacecraft System Engineering (ISSE), Beijing, China
wangdan_ict_hit@163.com

Abstract. The carrier strike groups (CSGs) are the major components of the national navy. The surveillance of CSG formation changes and other behaviors is critical to identify the other party's behavioral intentions and win the first opportunity in the battle field. As the spread range of a CSG is large, satellite remote sensing imagery is the most effective way to capture the CSG behavior change. However, due to the difference of satellite imaging angle of view, the apparent characteristics of similar formation may vary greatly, which may lead to the decline of formation change recognition performance. Inspired by the fact that a CSG formation is usually composed of multiple basic units, which appears a strong spatio-temporal relationship between each other, this paper proposes a graph model based method to identify the formation changes of group targets. Firstly, the graph model method is proposed to model the spatial structure relationship of the group targets. Secondly, multi-scale convolution kernel features and spatio-temporal graph convolution features are calculated to capture the formation sequential changes. Thirdly, a graph based, deep recursive convolution neural network model is learned and used to recognize formation changes. In our experiments, the formation transformation refers to the sequential transformation between two of seven specified formations. Experimental results show that the graph model and the spatio-temporal graph convolution feature representation are efficient and robust, to describe formation trajectory of the simulated data. Meanwhile, the recursive convolution neural network model greatly improves the recognition performance of target formation change, which can provide necessary technical support for correct decision-making and occupying the initiative in the battlefield.

Keywords: Group Targets · Graph Model · Deep Learning

1 Introduction

Group behavior analysis and recognition has attracted more and more attention from researchers [1–4]. With the continuous maturity of computer vision technology, the research on this topic has been deepened [5–7]. Group targets can be divided into two categories: one is dense target groups, such as crowds gathered in various occasions such as shopping malls; the other is sparse target group, such as athletes on the stadium,

H. P. Urbach and H. Jiang (Eds.): ISSOIA 2022, SPPHY 295, pp. 129–137, 2023.
https://doi.org/10.1007/978-981-99-4098-1_13

carrier battle group on the sea, aircraft group in the air, satellite group in space, etc. This paper focuses on the behavior analysis of sparse target groups, especially the formation change of groups. Taking the carrier strike group as an example, it is a typical sparse target group, which can be seen as composed of some discrete points. The change of array formation often indicates the change of a certain battlefield situation, and is the basis for identifying behavior intention of the enemy and winning the battlefield first opportunity [8, 9]. Consequently, it is of great application value to study the identification method of array formation changes.

2 Related Works

The CSGs are the main organization of national navy. By monitoring the CSG with satellites, identifying its formation and change behavior, it can provide decision-making support to military commanders, and even determine the outcome of a campaign. Due to the wide spread range of CSGs, the most effective way to obtain the navigation of the opposite CSGs in the actual battlefield is satellite imagery. However, the research work on predicting formation changes based on satellite detection results is relatively limited.

Deng et al. [9] proposed MVC (Multi view Point Context) descriptor based on Archimedes spiral for ship formation recognition, which introduced a probability density function of observation points, and proposed a similarity measurement method to directly identify ship formation. However, the descriptors are lack of spatial and temporal structure information. The paper [10] regarded the formation as a scene in two-dimensional space, in which the known formation is assumed as a template, and the tested ones can be recognized by calculating the scene similarity with the template. The method requires target locations and less target positioning accuracy, but has poor deformation resistance. The works in [1] and [11] mainly focus on the design of linear formation. In [11], an algorithm for recognition of warship formation has been studied based on Hough transform technology. When the target information is seriously polluted, the improved K-means clustering algorithm is further used to cluster the local peaks of the accumulation matrix obtained by Hough transform. Finally, the parameters of the formation are accurately extracted, according to the peak clustering results. However, limited by the design of line detection, its adaptability is not strong. In addition, some researchers use graph models composed of a series of key points to identify object types, and even array formation.

Generally, there are several basic types of fleet formation, based on unit structures of fleet groups. By analyzing the basic units in the formation and inferring the formation composition, the goal of identifying the formation can be achieved. It is difficult to find a unified approach to recognize group targets, when they are rotated, scaled and deformed. Thus some scholars have utilized deformation ability of graph models to build target recognition algorithms. In this paper, a graph based deep network learning method is proposed. By combining the graph model and deep convolution neural networks, we proposed an efficient and robust graph feature representation method for feature extraction and analysis of large-scale formation trajectory data. Robust dynamic graph prediction method is also used to achieve dynamic real-time prediction and analysis of formation trajectory data.

In this paper, the typical sparse group of target carrier strike group is taken as an example to study two aspects: (1) Extraction both of static features and dynamic spatio-temporal changes of the carrier strike group formation; (2) Identification and analysis of aircraft carrier strike group formation change. Since the ship formation is usually composed of multiple basic units, there is a strong spatial relationship between ships. Therefore, a graph model based on spatiotemporal trajectory data is constructed to describe the spatio-temporal correlation and dependency between the formation change data, which has a strong anti deformation ability. By establishing the recursive convolution neural network (RCNN) model of dynamic graph, the problem of joint modeling of temporal motion and spatial structure is solved, and the recognition accuracy of group target formation changes is improved.

3 Proposed Method

3.1 Graph Model Construction of Group Targets

Given a remote sensing image, a triplet can be constructed according to the ship group attributes in the image, which represents the graph model $\mathcal{G} = (\mathcal{V}, \mathbf{X}, \mathbf{A})$ where $\mathcal{V}$ represents the set of nodes, $\mathbf{X}$ represents the node features and $\mathbf{A}$ represents the adjacency matrix. In the node set $\mathcal{V} = \{v_1, \ldots, v_n\}(|\mathcal{V}| = n)$, we assume each ship as a node and extract features according to the node attributes, including the ship position, speed, dwell time etc. Formally, $\mathrm{x}_i = f(x, v_i) \in \mathbb{R}^{n\times d}$, where x_i represents the i-th node, f represents the feature extractor, and d represents the data dimension. By concatenating node features, we can get the feature matrix composed of all node features: $\mathrm{X} = [\mathrm{x}_1^{\mathrm{T}}; \mathrm{x}_2^{\mathrm{T}}; \ldots;\mathrm{x}_{\mathrm{n}}^{\mathrm{T}}] \in \mathbb{R}^{\mathrm{n}\times \mathrm{d}}$. For simplicity, $\mathbf{X}_i$ and x_i will not be distinguished later, which represent the i-th row in characteristic matrix or i-th node features, with the same meaning.

Assume the edge set ε consists of the edges connecting all nodes in graph $\mathcal{G}$. The edge construction depends on the spatial, semantic or other potential influence factors between nodes. We formally define the edge between v_i and v_j node as $e_{ij} = \lessdot_\epsilon(\mathcal{K}(v_i, v_j))$, where $\lessdot_\epsilon(\cdot)$ is the indicator function with the threshold ϵ, and $\mathcal{K}$ is the similarity measurements, such as Euclidean distance. For example, the edge value is 1, if the distance $\mathcal{K}$ between v_i and v_j is greater than the threshold value ϵ, otherwise it is 0. All connection node pairs form an edge set $\varepsilon = \{(v_i, v_j)|e_{ij} = 1, v_i, v_j \in \varepsilon\}(|\varepsilon| = m)$, which can also be represented as an unweighted adjacency matrix $\mathbf{A} \in \mathbb{R}^{\mathrm{n}\times \mathrm{n}}$,

$$A = (e_{ij})_{n\times n}, e_{ij} = \begin{cases} 1, (v_i, v_j) \in E; \\ 0, \mathit{otherwise}. \end{cases}$$

In addition, for the relationship with edges connected, we use features to enhance edge information. The most common method is to introduce weighted edges. The higher the correlation between two nodes, the greater the corresponding edge weight. We can use measurement function $\mathcal{K}$ to define $\mathrm{A_{ij}} = e_{ij} \times \mathcal{K}(v_i, v_j)$. For example, only the adjacent edges have non-zero weights, so the adjacency matrix $\mathbf{A}$ not only records the adjacency of nodes, but also describes the correlation degree of adjacent nodes.

Based on adjacency matrix **A**, Laplacian matrix can be constructed as a matrix representation of node relationships in the graph. Given a graph model $\mathcal{G} = (\mathcal{V}, \mathbf{X}, \mathbf{A})$, its Laplace matrix can be defined as

$$L = D - \mathbf{A},$$

where D is the degree matrix of the graph, expressed as a diagonal matrix $D = diag[d_1, d_2, \ldots, d_n]$ and $d_i = \sum_j A_{ij}$.

3.2 Spatial-Temporal Graph Convolution (STGC) Model Based Group Target Formation Change Recognition

In this paper, the ship formation is modeled as an undirected attribute graph $\mathcal{G} = (\mathcal{V}, \mathrm{A}, \mathrm{X})$, with a set of N vertices $\mathcal{V} = \{v_n\}_{n=1}^{N}$, a weighted adjacency matrix $\mathbf{A} \in \mathbb{R}^{N \times N}$ is, and a signal matrix on the vertices $\mathbf{X} \in \mathbb{R}^{N \times d}$. So the formation graph sequence with the length of T can be represented as $(\mathcal{G}_1, \mathcal{G}_2, \ldots, \mathcal{G}_T)$, where $\mathcal{G}_k = (\mathcal{V}_k, A_k, X_k)$ represents the formation on the k-th slice. Next, we will introduce the Spatial Temporal Graph Convolution (STGC) model for dynamic graph sequence analysis.

3.2.1 Multiscale Graph Convolution Kernel

Since the image has a grid structure, the receptive field of standard convolutional neural network (CNN) on the image can be easily defined as a local rectangular region. So convolution filtering on regular structured data is easy to operate. On the contrary, it is relatively difficult to construct convolution kernels on irregular structured graphs, mainly because homogeneous graph structures correspond to the same filtering response. Inspired by graph theory, with the help of the adjacency matrix **A**, which represents the connection relationship between the vertices of a graph. A_k accurately records the vertices that can be reached by k step by step connection on the graph. Therefore, the receptive field of k neighborhood $\psi_k(\mathbf{A})$ can be constructed by defining k order polynomial of **A**. The simplest strategy is adopted here, $\psi_k(\mathbf{A}) = \mathbf{A}_k$. In practice, the Laplacian matrix L of the graph can be used instead of **A** to avoid being affected by the scale of matrix norm in recursive reasoning. Therefore, for the local receptive field with a scale of K, the following multi-scale convolution filtering is defined:

$$Z = \mathcal{G} * f = \sum_{k=0}^{K-1} \psi_k(L) X V_k,$$

where $\psi_k(L)$ represents the receptive field of k scale, $V_k \in \mathbb{R}^{d \times d\prime}$ is the corresponding signal transformation matrix. The vertex information in all the k scale receptive fields is weighted and synthesized by calculating $\psi_k(L)X$. Consequently, for the graph, its filtering response is homogeneous and invariant.

3.2.2 Spatio-Temporal Graph Convolution

Base on the above multi-scale graph convolution kernel, and further inspired by the design concept of Auto Regressive Moving Average (ARMA) model, the following spatio-temporal graph convolution model is built:

$$Y_{t+1} = \sum\nolimits_{k=0}^{K_1-1} \psi_k(L) Y_t W_k + X_t V_0,$$
$$O_{t+1} = Y_{t+1} + \sum\nolimits_{k=1}^{K_2-1} \psi_k(L) X_t V_k,$$

where, $\psi_k(\cdot)$ is the receptive field function of k-th scale, $\{\mathbf{W}_k \in \mathbb{R}^{d' \times d'}, \mathbf{V}_k \in \mathbb{R}^{d \times d'}\}$ is the signal transformation matrix of the k-th scale,. In the above model, $\mathbf{Y} = [\mathbf{y}_1, \mathbf{y}_2, \ldots, \mathbf{y}_{d'}]$ and $\mathbf{O} = [\mathbf{o}_1, \mathbf{o}_2, \ldots, \mathbf{o}_{d'}]$ can be regarded as hidden state and output state respectively. Along the time dimension, the signal Y_{t+1} is regressed recursively by the local convolution kernel in the above equation, so that the dynamic changes can be serialized and encoded.

The output signal O_{t+1} in the above equation integrates the spatial graph convolution signal and the dynamic sequential signal, and each output signal $\mathbf{o}_i$ depends on all input signals $\{\mathbf{x}_1, \mathbf{x}_2, \ldots, \mathbf{x}_d\}$. In particular, when signals are independent and the spatial-temporal convolutional filter operates on the signal channel respectively, the dynamic graph convolutional model can be written as:

$$Y_{t+1} = \sum\nolimits_{k=0}^{K_1-1} \psi_k(L)\, Y_t diag(w_k) + X_t diag(v_0),$$
$$O_{t+1} = Y_{t+1} + \sum\nolimits_{k=1}^{K_2-1} \psi_k(L)\, X_t diag(v_k),$$

where $\mathbf{w}_k = [w_{k1}, w_{k2}, \ldots, w_{kd}]^{\mathrm{T}}$ and $\mathbf{v}_k = [v_{k1}, v_{k2}, \ldots, v_{kd}]^{\mathrm{T}}$ are mapping parameters, and w_{ki} and v_{ki} is related to k-th scale of the i-th signal. Note that it is assumed that the number of output dimensions is the same as the one of the input. In the case of signal independence, it is easy to expand to $d' \neq d$.

3.2.3 Construction of Deep Graph Convolution Neural Network Model

The above recursive convolutional model can be easily extended to the deep structures. Specifically, the recursive model is used as a basic layer to stack into a multi-layer network architecture, where the output signal **O** of the bottom layer is used as the input of the top layer. Formally,

$$Y_{t+1}^{(l)} = \sum\nolimits_{k=0}^{K_1-1} \psi_k(L) Y_t^{(l)} W_k^{(l)} + O_t^{(l-1)} V_0^{(l)},$$
$$O_{t+1}^{(l)} = Y_{t+1}^{(l)} + \sum\nolimits_{k=1}^{K_2-1} \psi_k(L)\, O_t^{(l-1)} V_k^{(l)},$$

where l represents the number of network layers. With the increasing number of layers, the receptive field scale of the convolution kernel can become larger, so the top layer can extract more global information. In other words, the recursive model is regarded as a basic neural network layer, and a multi-layer network structure is formed by stacking more layers. The output signal of the current layer can be used as the input signal of the next layer. As the number of network layers increases, the receptive field of convolution kernels becomes larger, so that the top layer of the network can capture global information. The framework is shown in the figure below (Fig. 1).

Different basic features in the graph are extracted by constructing multi-channel local convolution kernels, while higher level features are extracted by designing multiple

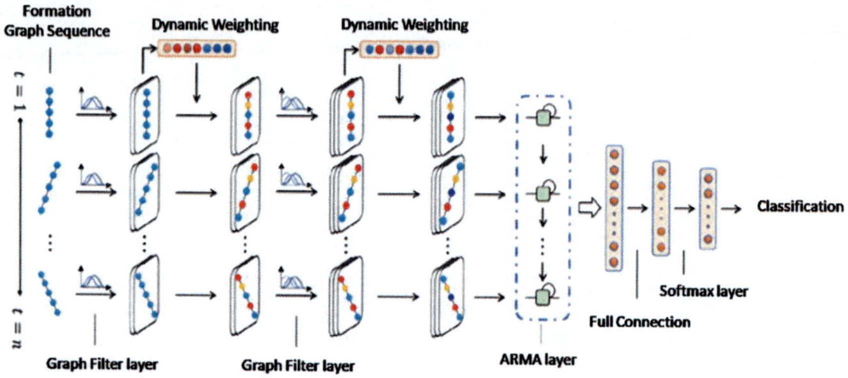

Fig. 1. Spatio-Temporal Dynamic Graph Network Framwork

convolutional layers. Gradient descent method is adopted to learn the model parameters. By using k-neighborhood local filters, the learning complexity is independent of the number of input samples and only depends on $\boldsymbol{O}(k)$. We consider of the maximum pooling operation, local average pooling, or constructing a binary tree according to some rules, to merge some similar vertices for down sampling. Finally, with a full connection layer, the learned feature representation is mapped into the sample label space, to classify the traveling trajectory.

4 Experiments

4.1 Dataset

The formation change in this paper refers to the sequence transformation between two of seven formations, thus there are 42 transformation types of sequences. Each sequence has 100 frames; the pixel difference between adjacent frames is the average difference of the formation distance between the two ends of the sequence. During the experiments, 4200 sequences of training set and 4200 sequences of test set are simulately generated, that is, 100 sequences of each formation change in training set and test set respectively. Among them, several formation transformation diagrams are shown in Fig. 2.

4.2 Comparisons

The proposed STGC model is experimentally compared with the latest methods. The experimental results are shown in Table 1. Among them, P-LSTM is a model for emotion classification based on short-term memory recurrent neural network. Comparatively, STGC has greatly improved the recognition accuracy (5.5%). Cov-RP proposed a new region feature descriptor, the covariance matrix feature, to represent the region of interest, which is applied to target detection and texture classification. Ker-RP proposes an open framework, which uses the kernel matrix on feature dimension as a universal representation, and explores a more appropriate representation based on symmetric positive

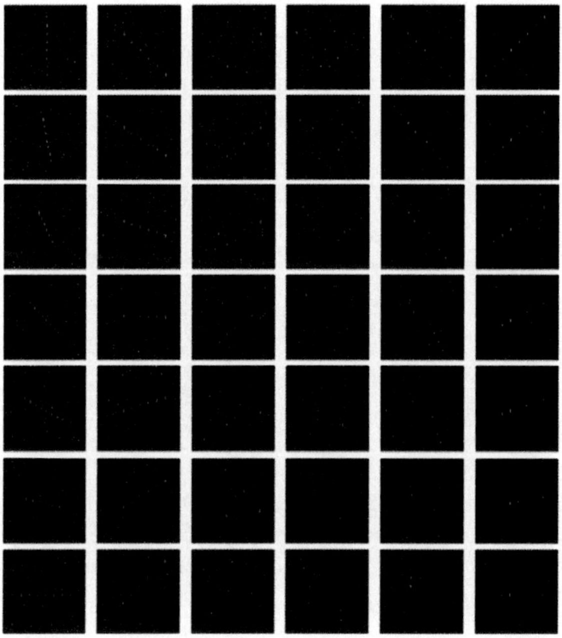

(a) (b) (c) (d) (e) (f)

Fig. 2. Example of formation change. From left to right, transforms are shown in columns. (a) from columns to rows, (b) from right echelon to left echelon, (c) from V-shaped echelon to wedge-shaped echelon, (d) from V-shaped echelon to interlaced echelon, (e) from right echelon to interlaced echelon, (f) from left echelon to rows.

definite matrix, to obtain better recognition performance. RSR-ML can deal with high dimensional symmetric positive definite matrices, by constructing a low dimensional and more discriminative symmetric positive definite manifold. Thus it adopts standard orthogonal projection to model the mapping from high-dimensional symmetric positive definite manifold to low dimensional positive definite manifold, to improve the recognition accuracy.

Table 1. Comparison results of STGC model

Methods	Recognition rates
P-LSTM	92.6%
Cov-RP	92.4%
Ker-RP	95.2%
RSR-ML	96.8%
STGC	**98.1%**

Based on the above experiments, we extract 75%, 50% and 25% sequence fragments from each complete sequence as a new test set and use the complete sequence fragments in the training set. Comparisons are shown below, which realizes the sequence type prediction, according to incomplete sequence frames.

Table 2. Recognition rate of observed sequence data with different proportion

Proportion of observed sequences	Recognition rate
25%	87.5%
50%	92.4%
75%	97.8%
100%	98.1%

Table 2 shows, it will bring a great reduction 9.6% on the accuracy of the final prediction, when using only the first 25% of the full sequence, while it will bring less impact on the results, when using more than 75% of the full sequence. It demonstrates the robustness of the proposed model.

5 Conclusion

This paper proposes an effective graph structured modeling method for the formation change of group targets, based on the deep graph network model. By comprehensively considers the spatio-temporal data information of the dynamic change of group targets, we focus on the problem of efficient and robust graph feature representation. A spatio-temporal graph convolution method is proposed to extract features from large-scale formation trajectory data, which greatly improves the discrimination ability and accuracy of the deep graph network. Finally the dynamic identification and analysis of formation change data of ship group targets and experimental results demonstrate the effectiveness of the proposed method.

References

1. Bing, Q., Bin, X.: Research on automatic recognition of warship formation based on template matching. Comput. Simul. **23**(9), 4–6 (2006)
2. Lan, T., Wang, Y., Yang, W., Ro-binovitch, S.N., Mori, G.: Discriminative latent models for recognizing contextual group activities. IEEE Trans. Pattern Anal. Mach. Intell. **34**, 1549–1562 (2012)
3. Zhang, Y., Dong, S., Bi, K.: Recognition algorithm of ship formation based on Hough transform and clustering. J. Mil. Ind. **37**(4), 648–655 (2016)
4. Deng, Z., Vahdat, A., Hu, H., Mori, G.: Structure inference machines: recurrent neural networks for analyzing relations in group activity recognition. In: IEEE International Conference on Computer Vision and Pattern Recognition (2016)

5. Ibrahim, M.S., Mori, G.: Hierarchical relational networks for group activity recognition and retrieval. In: European Conference on Computer Vision (2018)
6. Wu, J., Wang, L., Wang, L., Guo, J., Wu, G.: Learning actor relation graphs for group activity recognition. In: IEEE International Conference on Computer Vision and Pattern Recognition (2019)
7. Azar, S.M., Atigh, M.G., Nickabadi, A., Alahi, A.: Convolutional relational machine for group activity recognition. In: IEEE International Conference on Computer Vision and Pattern Recognition (2019)
8. Xiaochun, Z.: Analysis of US aircraft carrier formation and anti submarine capability. Ship Sci. Technol. **35**(9), 143–148 (2013)
9. Deng, C., Cao, Z., Xiao, Y., Chen, Y., Fang, Z., Yan, R.: Recognizing the formations of CVBG based on multiviewpoint context. IEEE Trans. Aerosp. Electron. Syst. **51**(3), 1793–1810 (2015)
10. Shouquan, D., Wei, S.: Formation recognition algorithm based on spatial direction similarity. Fire Command Control **35**(11), 167–169 (2010)
11. Leng, H., Guan, Q., Wu, X.: Recognition of maritime formation linetype based on domain knowledge. Ship Sci. Technol. **35**(2), 103–106 (2013)

Check for updates

The Technology of Aerospace Monitoring of Changes in the Ecological Condition of the Environment Using the Statistical Criteria Method with Virtual Standards, on the Example of Mining and Industrial Territories of Ukraine

O. D. Fedorovskyi, A. V. Khyzhniak, A. YU. Porushkevych(✉), L. P. Lishchenko, and T. A. Yefimenko

State Institution "Scientific Centre for Aerospace Research of the Earth of the Institute of Geological Science of the National Academy of Sciences of Ukraine", Olesia Honchara Str., 55-B, Kyiv 01054, Ukraine

adfedorovsky@ukr.net, aporushkevich@gmail.com

Abstract. The article describes the technology of determining changes in geosystems of the environment based on the satellite images using the method of statistical criteria with virtual standards on the example of the study of mining areas. The modification and adaptation of the statistical criterion was carried out, virtual benchmarks were formed for four categories of ecological changes of the territory (natural, anthropogenically modified, anthropogenic and technogenic). The developed technology was tested on the mining and industrial areas of Nikopol and Kryvyi Rih of Ukraine.

1 Introduction

There is a well-known technology for recognizing and classifying images of objects (or areas of the territory) on satellite images, which makes the conclusion about the classification of the studied object into one or another class is based on the maximum value of the membership function, which is formed by the parameters or one object of the object taken as a standard, or by a compromise set of parameters obtained from averaging the parameters of several standard objects.

The difference and advantage of the proposed technology lies in the fact that in the case of the study of the territory with different geological and landscape conditions, objects of different geological and geophysical nature and dimensions are simultaneously investigated thanks to virtual standards that can be formed as data from satellite images, as well as numerical informative features of standard objects.

H. P. Urbach and H. Jiang (Eds.): ISSOIA 2022, SPPHY 295, pp. 138–147, 2023.
https://doi.org/10.1007/978-981-99-4098-1_14

2 Purpose of the Study

The purpose of the study is to justify the technology of aerospace monitoring of the determination of changes in the ecological state in geosystems of the environment by the method of statistical criteria with virtual standards on the example of the study of the mining and industrial territories of Ukraine.

The calculation of the above parameters is performed by software created on the basis of a statistical criterion (Arhipov, Glazunov, & Hizhnjak, 2018), when several objects of different classes with different sets of values of informative features are taken as standards. The classification of each investigated object is performed by calculating the probability value of its relation to each of the reference object classes presented on the aerospace image, and belonging to a specific class is determined by the maximum value of the probability of its relation to a specific reference object of the corresponding class, which involves a priori input of informative features of all references and automatic calculation of the probability values of the relation of each studied object to the available reference objects.

3 Method

This method provides the possibility of simultaneous classification of researched objects of different classes according to different standards when they are arbitrarily located in the analysis area, which allows for greater objectivity and reliability in the recognition and classification of objects on images.

Geomonitoring of the ecological state with the quantitative assessment of the impact of man-made load through its reflection on aerial images and ground observation simultaneously on several standards, each of which represents the corresponding level of load is relevant. In order to achieve this, the "virtual standards" technology was chosen, which is used to obtain greater opportunities in the formation of standard parameters and increase the accuracy of measurements (Ostapiv, Pindus, Chekhovskyi, & Klochko, 2016). A similar technology is used for images recognition, and for more justified formation of standards thanks to the selection of different levels and load parameters, in accordance with the informative features of the studied objects. Thus, the virtual benchmark of the maximum load level consists of the maximum values of man-made or anthropogenic load indicators. Similarly, the standard of minimum load level consists of minimum values. Accordingly, standards of intermediate load values are made. The number of standards levels is determined by the expert depending on the task (Fedorovskij & Lishhenko, 2003).

The methodology of the statistical criterion with virtual standards that implements the proposed method are the follows: empirical data indicate that the values of the spectral brightness $\overline{\xi}_q = (\xi_{q1}, \ldots, \xi_{qk})$ of a specific section of the studied area under the number q change from section to section in a random manner, taking values around their average values. In accordance with this, we will consider the vector $\overline{\xi}_q$ to be random, the coordinates of which are random normally distributed values.

Correspondence of the specified distributions to normality is determined by sample values using, for example, the χ^2 Pearson's test, or the Kolmogorov's test, or the Smirnov's test.

Average values $D_{q,k}$ and root mean square $\sigma_{q,k}$ values Bare calculated according to standard formulas:

$$D_{q,k} = \frac{1}{N} \sum_{n=1}^{N_q} \xi_{q,k,n}$$

$$\sigma_{q,k} = \pm \sqrt{\frac{1}{N_q - 1} \sum_{n=1}^{N_q} (\xi_{q,k,n} - D_{q,k})^2}$$

where N_q is the sample volume for determining the unknown parameters of a random variable $\xi_{q,k}$, and $\xi_{q,k,n}$ are its sample values.

The function $P_{q,k}(t_k)$, which is the density of the distribution of a random variable $\xi_{q,k,n}$, has the form:

$$P_{q,k}(t_k) = \frac{1}{\sqrt{2\pi \sigma_{q,k}}} e^{-\frac{(t_k - D_{q,k})^2}{2\sigma_{q,k}^2}}$$

And the multidimensional density $P_q(\bar{t})$ of the vector distribution $\bar{\xi}_q$ is defined as a product

$$P_q(\bar{t}) = \left(\frac{1}{2\pi}\right)^{\frac{K}{2}} \prod_{k=1}^{K} e^{-\frac{(t_k - D_{q,k})^2}{2\sigma_{q,k}^2}} \frac{1}{\sigma_{q,k}}$$

The specified value of the vector $\overline{X} = (X_1, \ldots, X_k)$ is considered as a sample value of some random vector $\bar{\xi}_q$ with an unknown value q that needs to be determined. This corresponds to the fact that the area under consideration corresponds to the number q.

The purpose of the study is to specify a criterion that will allow with a high degree of confidence to determine the number q based on the given value of the vector $\overline{X}$. At the same time, it should be known for certain that the area being investigated corresponds to the numbers $1, \ldots, Q$.

This means that from the total number Q of competing hypotheses H_q, one should choose the one that most corresponds to the set $\overline{X}$.

Finally, we get a criterion in the form:

$$C_q = \frac{P_q(\overline{X})}{\sum_{r=1}^{Q} P_r(\overline{X})}$$

An algorithm was developed for the further study of man-made load on mining and industrial areas (Fig. 1).

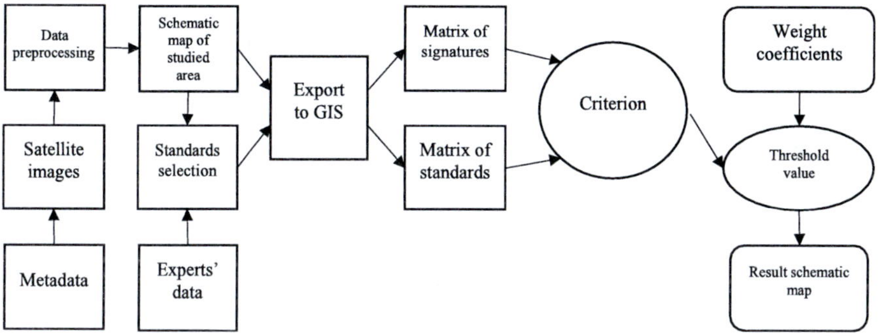

Fig. 1. Algorithm for further determination of changes in the ecological state of mining and industrial territories.

The developed technology was implemented to assess the man-made load on the territory of the Nikopol Mining and Industrial District. The Nikopol area was chosen due to the fact that this area has undergone significant transformations and man-made loads under the influence of economic and mining activities. Since the 2000s, the intensive quarrying of manganese ore raw materials and its beneficiation has decreased significantly. However, this zone requires thorough study and monitoring in connection with the significant deterioration of the environment - lithosphere, soils, air, underground and surface waters, biocenoses and human living conditions. To assess the man-made load in the Nikopol mining and industrial territory, the probability of the ratio of the spectral informative features of the studied objects to the spectral informative features of each reference object present in the aerospace image is calculated (Khyzhniak, Fedorovskyi, & Yefimenko, 2021) (Fig. 2).

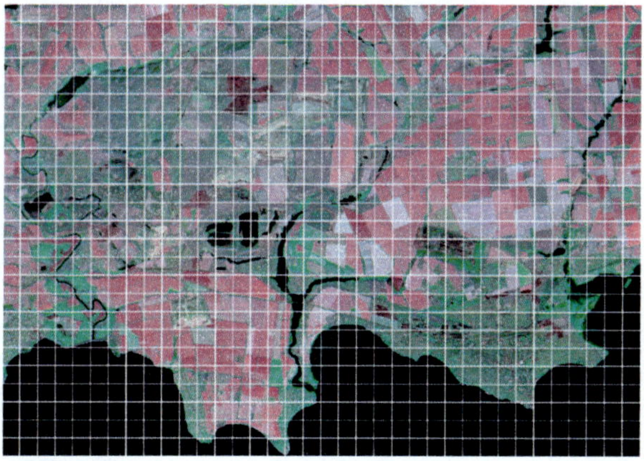

Fig. 2. Research territory of the Nikopol mining district. Landsat 8 space image (OLI/TIRS) for August 31, 2018 (channel combination 4–5-1)

Multispectral space images from Landsat 8 (OLI/TIRS) on August 31, 2018, were used to prepare input data set. For all used channels (1 (0.435–0.451), 2 (0.452 μm – 0.512 μm), 3 (0.533 μm – 0.590 μm), 4 (0.636 μm – 0.673 μm), 5 (0.851 μm – 0.879 μm), 6 (1.566 μm – 1.651 μm), 7 (2.107 μm – 2.294 μm) and 9 (1.363 μm – 1.384 μm)) atmospheric correction was applied.

For the correct operation of the program, reference areas were selected on the space image, 10 reference areas for each degree of load. Figure 3 shows examples of selected benchmark sites for I, II, III and IV levels of man-made load, respectively.

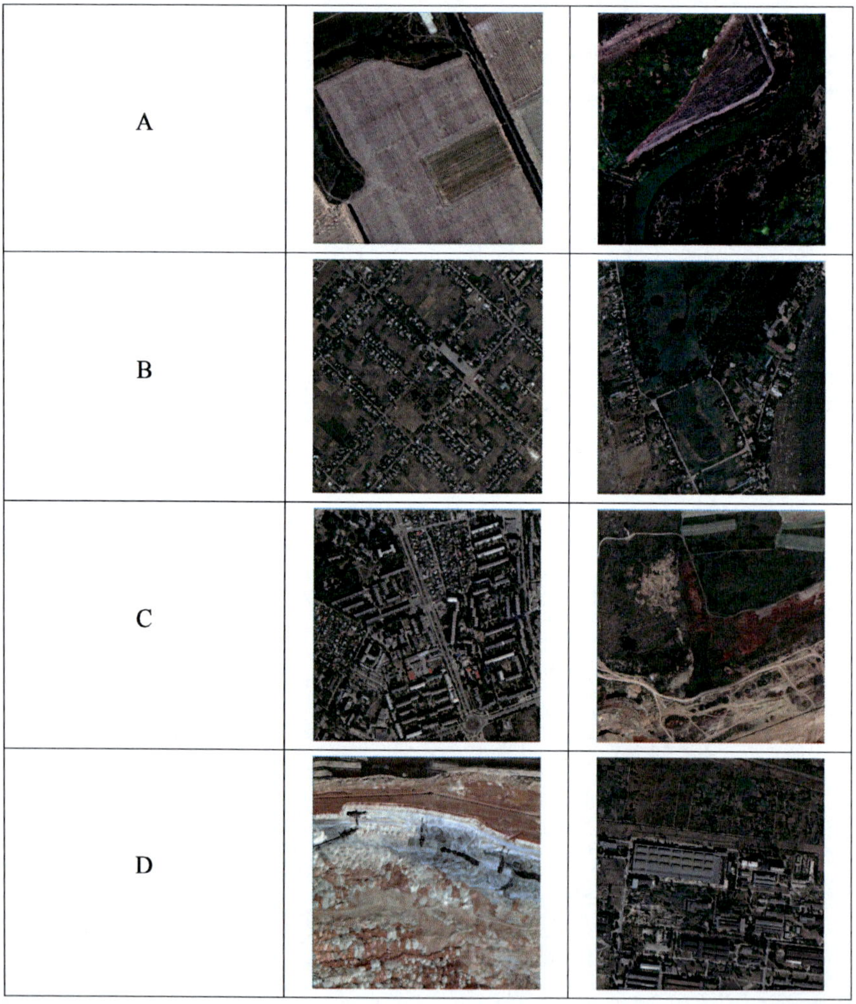

Fig. 3. Examples of reference areas selected on the space image: a) 1st degree corresponds to the natural state; b) 2nd degree of man-made load c) 3rd degree of man-made load; d) 4th degree of man-made load.

Since the study area is heterogeneous and its structure is quite complex, each of the load levels can include several different landscape complexes, with different percentage ratios to the total sample area. For example, areas dominated by natural landscape complexes (LC) (which form the primary background geoecological conditions for the functioning of the LC, strengthen the development of negative natural and natural-anthropogenic processes, including environmental pollution) were assigned to I - "Light", to a small extent there are anthropogenically modified and almost or completely absent anthropogenic and technogenic. II - "Medium" refers to reference areas mainly with anthropogenically modified landscape complexes (characterized by insignificant variability of individual components of the landscape, when the action of anthropogenic influence is stopped, they are restored and return to the natural evolutionary path of the development of the landscape), to a moderate extent there are natural and anthropogenic, and man-made almost absent. Up to III - "Moderate" - mainly anthropogenic LCs (characterized by the introduction of artificial formations into the LC, which changes the external image of the LC, its physiognomic characteristics and internal dependencies between the components of the LC), and a small percentage are anthropogenically modified and man-made, and up to IV - "Severe» level of load - man-made (landscapes that have undergone fundamental changes in certain natural components, even to their complete destruction, formed in favorable and unfavorable natural conditions), with a small percentage of the presence of 3 other classes of landscape complexes.

In order to systematize the input data, a project in geo-information system of the area under study was created based on the received data of remote information and expert assessments.

Further, on the basis of the obtained data, matrices of informative features were created for the studied plots and reference plots. This process includes the formation of feature values in the form of electronic tables, each row of which contains the results of measurements of all types of remote information, and the columns provide feature values for each of the 900 sites.

This stage of the work was automated using a software specially created on the basis of a statistical criterion.

The final product of this process is the determination of the class of the studied objects based on the maximum value of the probability of their belonging to a specific reference object, the corresponding class, and the creation of a map scheme characterizing the man-made load of the Nikopol mining and industrial district. The results are presented in the form of maps on Fig. 4 in relative gradations of brightness: light color - natural, light gray - light load, gray - medium and black - heavy load, respectively.

Kryvyi Rih is a large industrial city, the center of the Kryvyi Rih iron ore basin, the most important raw material base of metallurgy in Ukraine. The Kryvyi Rih industrial region plays a leading role in the economy of Ukraine and is the main raw material base for the development of ferrous metallurgy, is of strategic importance for the economic independence and security of the state.

There are about 5,000 sources of atmospheric air pollution in Kryvyi Rih. Geologically, the city is located in the central part of the Ukrainian Crystalline Massif and is characterized by a complex geological structure of the crystalline foundation, the structure of which includes dislocated metamorphic and ultrametamorphic worlds of the

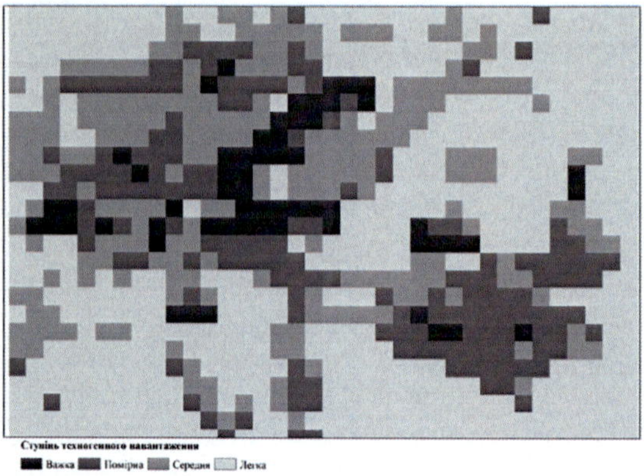

Fig. 4. Map scheme in relative gradations of brightness according to four levels of man-made load, which characterizes the geoecological state of the Nikopol mining district (obtained on the basis of a statistical criterion).

Proterozoic, complicated by fault zones and covered for the most part by sedimentary rocks of the Cenozoic age.

Fig. 5. The territory of research of mining and industrial territory of the Kryvyi Rih industrial district. Sentinel 2 space image for April 20, 2018 (combination of channels 2-3-4)

Below the result of using the method based on a statistical criterion for assessing the anthropogenic-technogenic load on 115 plots with an area of 500 × 500m each of the mining and industrial territory of the Kryvyi Rih industrial region.

Evaluation of man-made load was carried out according to the following informative features: Gravika, Magnitka, Spektr - data of the average value of the brightness of a pixel of optical space images, data from temperature maps, data obtained from radar space images for four classes (degrees) of load: natural - 0 points, light - 3 points, medium - 6 points and difficult - 9 points.

In order to systematize the input data, a geo-information system of the site project was created, which is investigated based on the data of remote information and expert assessments. In addition, in order to obtain the values of all types of features at individual points, a grid with 115 sections (Fig. 5) was drawn on the selected research territory, similar to the example for the Nikopol mining and industrial territory.

The calculation of the value of the probability of assigning the values of the informative features of each researched object to the values of the informative features of each reference object present in the space image was automated, similar to the example with the Nikopol mining area (Table 1).

Table 1. Benchmarks for four categories (classes) of man-made load (0, 3, 6, 9) based on three informative features (Gravika, Magnitka, Spektr).

Signatures\Standard	0	3	6	9
Gravika	−0,50	−1,60	−1,95	4,40
	−0,95	1.50	3,50	4,91
	−1,50	2,00	4,70	5,10
	−1,95	3,25	5,15	5,45
Magnitka	310	−1000	390	−4500
	360	200	415	−8653
	390	420	−2500	−9500
	415	−123	−3500	−10504
Spektr	1200	1280	1320	1220
	1220	1300	1345	1265
	1240	1350	1380	1290
	1260	1370	1418	1600

4 Final Product

The final product of this process is the determination of the class of the studied objects according to the maximum value of the probability of their belonging to a specific reference object, the corresponding class and the creation of a map scheme characterizing the ecological state of the mining and industrial territory of the Kryvyi Rih industrial region. The results are presented in the form of maps in Fig. 6 in points and relative gradations of brightness: light color (0) - natural, light gray (3) - light load, gray (6) - medium and black (9) - heavy load, respectively.

0	3	0	0	6	3	6	6	6	0	6
6	6	0	0	0	0	6	6	6	9	9
9	9	3	3	3	3	6	6	6	0	0
0	0	6	3	3	3	0	6	0	0	0
0	0	6	6	6	9	9	9	9	9	3
3	3	3	0	9	3	9	9	0	6	6
6	3	0	0	0	0	0	0	0	6	6
6	9	9	9	9	9	6	6	3	0	3
0	3	3	6	6	6	0	0	0	0	0
0	0	0	0	0	6	6	0	0	0	0

3	0	0	0	3	6	0	0	0	0	0
0	0	0	0	0	3	3	9	9	9	9
9	9	6	3	3	3	3	3	3	3	0
0	0	0	0	0	0	0	0	0	0	0
0	3	6	6	6	9	9	9	9	6	6
3	3	3	0	3	3	0	9	0	6	0
6	3	0	0	0	0	0	0	0	3	6
6	6	9	9	9	9	6	6	9	9	9
3	3	3	0	0	0	0	0	0	0	0
0	0	0	0	3	0	0	0	0	0	0

Fig. 6. Maps for assessing the ecological state of mining and industrial areas of the Kryvyi Rih industrial region: a) the map is presented in points according to the degree of load according to expert assessments; b) map scheme in points and relative gradations

5 Conclusions

1. Based on the example of assessing the ecological state of mining and industrial areas of the Nikopol and Kryvyi Rih industrial regions, methodological foundations of the technology for determining anthropogenic-technogenic changes in geosystems of the environment using aerospace information and statistical criterion, which involves a priori input of informative features of all standards and automatic calculation of probability values, have been developed the ratio of each investigated object to all presented reference objects.
2. Virtual standards were created based on space images and numerical values of informative features for four classes of anthropogenic-technogenic load: natural - 0 points, light - 3 points, medium - 6 points, and heavy - 9 points. The number of standard levels is determined by the expert depending on the problem.
3. The proposed method was substantiated by determining the correspondence of the results obtained visually with ground observations and using the method of statistical criteria based on space images. For this, the correlation coefficient was calculated, which was 0.7, which is acceptable for practical express assessments of man-made or anthropogenic load.
4. The research perspective is the adoption of the developed methodology for use in other industries, for example, agriculture, water management, urban management.

References

Arhipov, A.I., Glazunov, N.M., Hizhnjak, A.V.: Jevristicheskij kriterij raspoznavanija klassa po spektral'nym jarkostjam [Heuristic criterion for class recognition by spe Kibernetika i Sistemnyj Analiz **54**(1), 105–110 (2018)

Fedorovskij, A.D., Lishhenko, L.P.: Landshaftno-sistemnyj podhod pri ocenke geojekologicheskoj situacii v regione [Landscape-system approach in assessing the geoecological situation in the region]. Dopovidi Natsionalnoi Akademii Nauk Ukrainy (11), 37–40 (2003)

Khyzhniak, A.V., Fedorovskyi, O.D., Yefimenko, T.A.: Kontseptsiia formuvannia metodolohii aerokosmichnoho heomonitorynhu dlia vyrishennia zavdan pryrodokorystuvannia [The concept formation of the aerospace geomonitoring methodology of the decision of problems in natural resource]. Ukrainian J. Remote Sens. **8**(4), 4–11 (2021). https://doi.org/10.36023/ujrs.2021.8.4.203

Ostapiv, V.V., Pindus, N.M., Chekhovskyi, S.A., Klochko, N.B.: Virtualni etalony, yak zasib pidvyshchennia tochnosti vymiriuvan ["Virtual standard" as a means of improving the measurement accuracy]. Syst. Obrobky Informatsii **6**(143), 108–11. (2016)

Geometric Positioning Sensitivity Analysis of Spaceborne Laser Altimeter

Qiu panghe[1,2], Yue Chunyu[1,2], Wang Chunhui[1,2], and Tao Yuliang[1,2](✉)

[1] Beijing Institute of Space Mechanics and Electricity, Beijing 100094, China
1371280479@qq.com

[2] Key Laboratory for Space Laser Information Perception Technology of CAST, Beijing 100094, China

Abstract. The spaceborne laser altimeter is widely used in forestry, polar ice cap elevation measurement and other fields due to its advantages of high ranging accuracy. During the launch and in-orbit operation of the satellite, the laser altimeter will generate a number of systematic errors, so it is necessary to use the on-orbit geometric positioning error analysis to check and avoid errors when design the system. The spaceborne laser altimeter will be affected by various errors from ranging, attitude, GPS positioning, etc. in the process of positioning the ground, and these errors will affect the positioning results to varying degrees. The working principle of the laser altimeter is introduced, and the geometrical positioning error model of the system is deduced. It is found that the measurement data of the laser altimeter is greatly affected by the angle measurement error, and the error sources are mutually influenced. In order to improve the measurement accuracy of the altimeter and make more effective use of the measurement data of the spaceborne laser altimeter, this paper makes a comprehensive analysis of the positioning error from the geometric positioning model of the system. The results show that the influence of different error sources on the measurement results is different, and this analysis can provide a basis for the design and application of subsequent space laser altimeter measurements.

Keywords: laser altimeter · laser ranging · positioning error · angle measuring error

1 Introduction

Spaceborne laser altimeter provides a new way for Digital Surface Model (DSM) acquisition, and its ranging accuracy is comparable high. However, the angle measurement error has become the main obstacle to the application of the spaceborne laser altimeter because it will have an impact of 100 m or even kilometers in the X and Y directions. Therefore, in order to further improve the effectiveness of the spaceborne laser altimeter measurement data, how to improve the measurement accuracy of the altimeter is of great significance [1–12].

H. P. Urbach and H. Jiang (Eds.): ISSOIA 2022, SPPHY 295, pp. 148–158, 2023.
https://doi.org/10.1007/978-981-99-4098-1_15

We starting from the geodetic coordinates under the earth reference framework, this paper first studies the error propagation law of each outer azimuth element during the positioning of the satellite position and the spaceborne laser altimeter. Then, the main sources of positioning errors and their effects are analyzed to provide reference for the design and application of spaceborne laser altimeter.

2 Altimeter Laser Footing Geometric Positioning Model

Laser footprint position is a three-dimensional coordinate position under the reference frame centered on the center of mass of the earth. Spaceborne laser altimetry systems usually use GPS to determine the orbital position of satellites and combine the satellite attitude obtained by star sensitivity to determine the propagation direction of the laser beam [13–18]. After knowing the direction of laser propagation, the distance measured by the laser is used to calculate the vector pointed to the surface by the satellite, and the two vectors can be added to obtain the vector from the center of the earth to the foot point of the surface, and finally the geodetic coordinates of the foot point are found.

Further analysis of the geometric positioning principle of the spaceborne laser altimeter yields Fig. 1 [6, 19–21], where S is the satellite centroid, and the coordinates in the spatial coordinate system are (X_s, Y_s, Z_s), and P is the observed ground point, and the coordinates are (X, Y, Z) in the spatial coordinate system, Δref is the distance from the satellite center of mass to the zero point of the laser measurement, L is the distance measurement, (α_1, β_1) is the pointing angle of the vector between the satellite center of mass and the zero point of the laser measurement and the X axis and Z axis, respectively, and (α_2, β_2) is the pointing angle of the laser altimeter exit light and the X axis and Z axis respectively, (φ, ω, k) is the outer azimuth element of the laser altimeter. In order to describe the vector relationship more clearly, Fig. 2 is the local vector relationship diagram of Fig. 1. The laser altimeter Earth observation equation is

$$\begin{bmatrix} X \\ Y \\ Z \end{bmatrix} = \begin{bmatrix} X_s \\ Y_s \\ Z_s \end{bmatrix} - \begin{bmatrix} a_1\ a_2\ a_3 \\ b_1\ b_2\ b_3 \\ c_1\ c_2\ c_3 \end{bmatrix} \begin{bmatrix} rcos\beta_1 \cos\alpha_1 + Lcos\beta_2 cos\alpha_2 \\ rcos\beta_1 sin\alpha_1 + Lcos\beta_2 sin\alpha_2 \\ rsin\beta_1 + Lsin\beta_2 \end{bmatrix} \tag{1}$$

where $(\mathrm{a}_1, \mathrm{a}_2, \mathrm{a}_3, \mathrm{b}_1, \mathrm{b}_2, \mathrm{b}_3, \mathrm{c}_1, \mathrm{c}_2, \mathrm{c}_3)$ is the directional cosine parameter of the outer azimuth element of the laser altimeter.

$$\begin{cases} a_1 = cos\varphi cosk - sin\varphi sin\omega sink;\ a_2 = -cos\varphi sink - sin\varphi sin\omega cosk;\ a_3 = -sin\varphi cos\omega; \\ b_1 = cos\omega sink;\ b_2 = cos\omega cosk;\ b_3 = -sin\omega; \\ c_1 = sin\varphi cosk + cos\varphi sin\omega sink;\ c_2 = -sin\varphi sink + cos\varphi sin\omega cosk;\ c_3 = cos\varphi cos\omega; \end{cases} \tag{2}$$

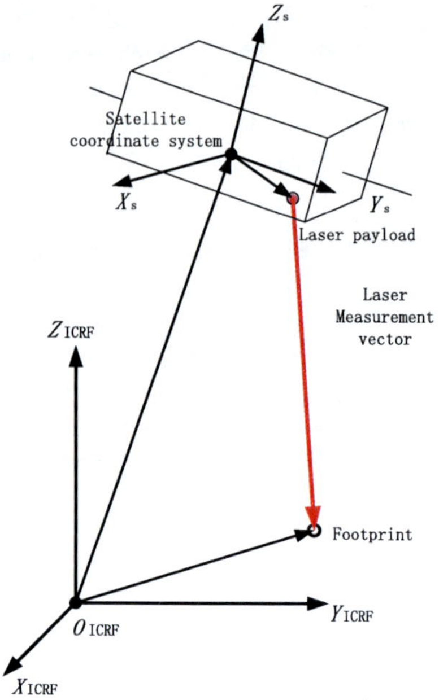

Fig. 1. Principle of geometric positioning of laser altimeter

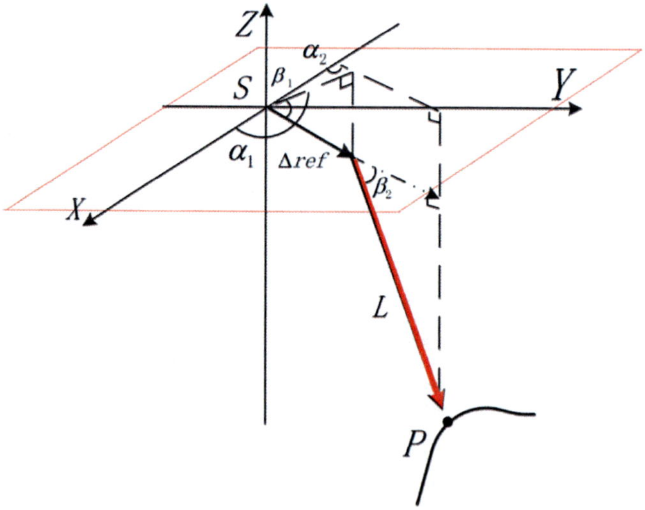

Fig. 2. Geometric positioning vector relation diagram of laser altimeter

3 Spaceborne Laser Altimeter Positioning Error Model and Analysis

3.1 Positioning Error Propagation Model

In order to obtain the outer azimuth element error of the error source (ΔX_s, ΔY_s, ΔZ_s, $\Delta\varphi$, $\Delta\omega$, Δk), where ΔX_s, ΔY_s, ΔZ_s is the positioning error in the spatial coordinate system, and $\Delta\varphi$, $\Delta\omega$, Δk is the angular error component of the three orthogonal directions of pitch, roll and yaw. The above errors are mainly errors generated by instruments such as star sensitivity and gyroscope when measuring the attitude of satellites when the satellite is in orbit. The distance error from the center of mass of the laser altimeter satellite to the zero point of the laser measurement Δref, hereinafter abbreviated as Δr. Height measurement error ΔL. The vector between the satellite center of mass and the zero point of the laser measurement is $\Delta\alpha_1$, $\Delta\beta_1$ with the pointing angle error of the X axis and the Z axis respectively, and the pointing angle error $\Delta\alpha_2$, $\Delta\beta_2$ of the laser altimeter exit light and the X and Z axes. Taking these 12 error corrections as unknowns, it can be obtained by linearly expanding Eq. (1) with Taylor's formula

$$\begin{bmatrix} X \\ Y \\ Z \end{bmatrix} = \begin{bmatrix} X_0 \\ Y_0 \\ Z_0 \end{bmatrix} + \begin{bmatrix} \frac{\partial X}{\partial X_s} & \frac{\partial X}{\partial Y_s} & \frac{\partial X}{\partial Z_s} & \frac{\partial X}{\partial \varphi} & \frac{\partial X}{\partial \omega} & \frac{\partial X}{\partial k} & \frac{\partial X}{\partial \alpha_1} & \frac{\partial X}{\partial \beta_1} & \frac{\partial X}{\partial \alpha_2} & \frac{\partial X}{\partial \beta_2} & \frac{\partial X}{\partial r} & \frac{\partial X}{\partial L} \\ \frac{\partial Y}{\partial X_s} & \frac{\partial Y}{\partial Y_s} & \frac{\partial Y}{\partial Z_s} & \frac{\partial Y}{\partial \varphi} & \frac{\partial Y}{\partial \omega} & \frac{\partial Y}{\partial k} & \frac{\partial Y}{\partial \alpha_1} & \frac{\partial Y}{\partial \beta_1} & \frac{\partial Y}{\partial \alpha_2} & \frac{\partial Y}{\partial \beta_2} & \frac{\partial X}{\partial r} & \frac{\partial X}{\partial L} \\ \frac{\partial Z}{\partial X_s} & \frac{\partial Z}{\partial Y_s} & \frac{\partial Z}{\partial Z_s} & \frac{\partial Z}{\partial \varphi} & \frac{\partial Z}{\partial \omega} & \frac{\partial Z}{\partial k} & \frac{\partial Z}{\partial \alpha_1} & \frac{\partial Z}{\partial \beta_1} & \frac{\partial Z}{\partial \alpha_2} & \frac{\partial Z}{\partial \beta_2} & \frac{\partial X}{\partial r} & \frac{\partial X}{\partial L} \end{bmatrix} \begin{bmatrix} \Delta X_s \\ \Delta Y_s \\ \Delta Z_s \\ \Delta\varphi \\ \Delta\omega \\ \Delta k \\ \Delta\alpha_1 \\ \Delta\beta_1 \\ \Delta\alpha_2 \\ \Delta\beta_2 \\ \Delta r \\ \Delta L \end{bmatrix} \tag{3}$$

where (X_0, Y_0, Z_0) is the coordinate value obtained from ground observations, as an approximation of (X, Y, Z).

The error propagation coefficient of each error correction number is

$$\frac{\partial X}{\partial X_s} = 1, \quad \frac{\partial X}{\partial Y_s} = 0, \quad \frac{\partial X}{\partial Z_s} = 0;$$

$$\begin{aligned} \frac{\partial X}{\partial \varphi} = {} & r(sin\varphi coskcos\beta_1 cos\alpha_1 + cos\varphi sin\omega sinkcos\beta_1 cos\alpha_1 - sin\varphi sinkcos\beta_1 sin\alpha_1 \\ & + cos\varphi sin\omega coskcos\beta_1 sin\alpha_1 + cos\varphi cos\omega sin\beta_1) \\ & + L(sin\varphi coskcos\beta_2 cos\alpha_2 + cos\varphi sin\omega sinkcos\beta_2 cos\alpha_2 \\ & - sin\varphi sinkcos\beta_2 sin\alpha_2 + cos\varphi sin\omega coskcos\beta_2 sin\alpha_2 + cos\varphi cos\omega sin\beta_2); \end{aligned}$$

$$\frac{\partial X}{\partial \omega} = r(sin\varphi cos\omega sinkcos\beta_1 cos\alpha_1 + sin\varphi cos\omega coskcos\beta_1 sin\alpha_1 - sin\varphi sin\omega sin\beta_1) \\ +L(sin\varphi cos\omega sinkcos\beta_2 cos\alpha_2 + sin\varphi cos\omega coskcos\beta_2 sin\alpha_2 \\ -sin\varphi sin\omega sin\beta_2);$$

$$\frac{\partial X}{\partial k} = r(cos\varphi sinkcos\beta_1 cos\alpha_1 + sin\varphi sin\omega coskcos\beta_1 cos\alpha_1 + cos\varphi coskcos\beta_1 sin\alpha_1 - sin\varphi sin\omega sinkcos\beta_1 sin\alpha_1) \\ +L(cos\varphi sinkcos\beta_2 cos\alpha_2 + sin\varphi sin\omega coskcos\beta_2 cos\alpha_2 + cos\varphi coskcos\beta_2 sin\alpha_2 - sin\varphi sin\omega sinkcos\beta_2 sin\alpha_2);$$

$$\frac{\partial X}{\partial \alpha_1} = r(a_1 cos\beta_1 sin\alpha_1 - a_2 cos\beta_1 cos\alpha_1); \quad \frac{\partial X}{\partial \alpha_2} = L(a_1 cos\beta_2 sin\alpha_2 - a_2 cos\beta_2 cos\alpha_2);$$

$$\frac{\partial X}{\partial \beta_1} = r(a_1 sin\beta_1 cos\alpha_1 + a_2 sin\beta_1 sin\alpha_1 - a_3 cos\beta_1);$$

$$\frac{\partial X}{\partial \beta_2} = L(a_1 sin\beta_2 cos\alpha_2 + a_2 sin\beta_2 sin\alpha_2 - a_3 cos\beta_2)$$

$$\frac{\partial X}{\partial \mathrm{r}} = -a_1 cos\beta_1 cos\alpha_1 - a_2 cos\beta_1 sin\alpha_1 - a_3 sin\beta_1;$$

$$\frac{\partial X}{\partial L} = -a_1 cos\beta_2 cos\alpha_2 - a_2 cos\beta_2 sin\alpha_2 - a_3 sin\beta_2;$$

$$\frac{\partial Y}{\partial X_s} = 0, \quad \frac{\partial Y}{\partial Y_s} = 1, \quad \frac{\partial Y}{\partial Z_s} = 0, \quad \frac{\partial Y}{\partial \varphi} = 0;$$

$$\frac{\partial Y}{\partial \omega} = r(sin\omega sinkcos\beta_1 cos\alpha_1 + sin\omega coskcos\beta_1 sin\alpha_1 + cos\omega sin\beta_1) \\ +L(sin\omega sinkcos\beta_2 cos\alpha_2 + sin\omega coskcos\beta_2 sin\alpha_2 + cos\omega sin\beta_2);$$

$$\frac{\partial Y}{\partial k} = r(cos\omega sinkcos\beta_1 sin\alpha_1 - cos\omega coskcos\beta_1 cos\alpha_1) + L(cos\omega sinkcos\beta_2 sin\alpha_2 - \\ cos\omega coskcos\beta_2 cos\alpha_2);$$

$$\frac{\partial Y}{\partial \alpha_1} = r(b_1 cos\beta_1 sin\alpha_1 - b_2 cos\beta_1 cos\alpha_1); \quad \frac{\partial Y}{\partial \alpha_2} = L(b_1 cos\beta_2 sin\alpha_2 - b_2 cos\beta_2 cos\alpha_2);$$

$$\frac{\partial Y}{\partial \beta_1} = r(b_1 \sin\beta_1 cos\alpha_1 + b_2 \sin\beta_1 sin\alpha_1 - b_3 cos\beta_1);$$

$$\frac{\partial Y}{\partial \beta_2} = L(b_1 \sin\beta_2 cos\alpha_2 + b_2 \sin\beta_2 sin\alpha_2 - b_3 cos\beta_2);$$

$$\frac{\partial Y}{\partial r} = -b_1 cos\beta_1 cos\alpha_1 - b_2 cos\beta_1 sin\alpha_1 - b_3 sin\beta_1;$$

$$\frac{\partial Y}{\partial L} = -b_1 cos\beta_2 cos\alpha_2 - b_2 cos\beta_2 sin\alpha_2 - b_3 sin\beta_2;$$

$$\frac{\partial Z}{\partial X_s} = 0; \frac{\partial Z}{\partial Y_s} = 0; \frac{\partial Z}{\partial Z_s} = 1;$$

$$\frac{\partial Z}{\partial \varphi} = r(-cos\varphi coskcos\beta_1 cos\alpha_1 + sin\varphi sin\omega sinkcos\beta_1 cos\alpha_1 + cos\varphi sinkcos\beta_1 sin\alpha_1 + sin\varphi sin\omega coskcos\beta_1 sin\alpha_1 + sin\varphi cos\omega sin\beta_1) + L(-cos\varphi coskcos\beta_2 cos\alpha_2 + sin\varphi sin\omega sinkcos\beta_2 cos\alpha_2 + cos\varphi sinkcos\beta_2 sin\alpha_2 + sin\varphi sin\omega coskcos\beta_2 sin\alpha_2 + sin\varphi cos\omega sin\beta_2);$$

$$\frac{\partial Z}{\partial \omega} = r(cos\varphi sin\omega sin\beta_1 - cos\varphi cos\omega sinkcos\beta_1 cos\alpha_1 - cos\varphi cos\omega coskcos\beta_1 sin\alpha_1) + L(cos\varphi sin\omega sin\beta_2 - cos\varphi cos\omega sinkcos\beta_2 cos\alpha_2 - cos\varphi cos\omega coskcos\beta_2 sin\alpha_2);$$

$$\frac{\partial Z}{\partial k} = r(sin\varphi sinkcos\beta_1 cos\alpha_1 - cos\varphi sin\omega coskcos\beta_1 cos\alpha_1 + sin\varphi coskcos\beta_1 sin\alpha_1 + cos\varphi sin\omega sinkcos\beta_1 sin\alpha_1) + L(sin\varphi sinkcos\beta_2 cos\alpha_2 - cos\varphi sin\omega coskcos\beta_2 cos\alpha_2 + sin\varphi coskcos\beta_2 sin\alpha_2 + cos\varphi sin\omega sinkcos\beta_2 sin\alpha_2);$$

$$\frac{\partial Z}{\partial \alpha_1} = r(c_1 cos\beta_1 sin\alpha_1 - c_2 cos\beta_1 cos\alpha_1);$$

$$\frac{\partial Z}{\partial \alpha_2} = L(c_1 cos\beta_2 sin\alpha_2 - c_2 cos\beta_2 cos\alpha_2);$$

$$\frac{\partial Z}{\partial \beta_1} = r(c_1 \sin\beta_1 cos\alpha_1 + c_2 \sin\beta_1 sin\alpha_1 - c_3 cos\beta_1);$$

$$\frac{\partial Z}{\partial \beta_2} = r(c_1 \sin\beta_2 cos\alpha_2 + c_2 \sin\beta_2 sin\alpha_2 - c_3 cos\beta_2);$$

$$\frac{\partial Z}{\partial r} = -c_1 cos\beta_1 cos\alpha_1 - c_2 cos\beta_1 sin\alpha_1 - c_3 sin\beta_1;$$

$$\frac{\partial Z}{\partial L} = -c_1 cos\beta_2 cos\alpha_2 - c_2 cos\beta_2 sin\alpha_2 - c_3 \sin\beta_2;$$

3.2 Positioning Error Propagation Coefficient Analysis

Considering the sub-star point ranging, without side swing, the set distance measurement value is equal to the orbital height of 500 km (L = 500km, $\varphi = \omega = k = 0$). Taking a certain model 03 star laser altimeter as an example, the coordinates of the zero point measured by the laser altimeter under satellite coordinates are x = −1600 mm y = 1300 mm z = −1600 mm.The influence of laser altimeter altimetry vector direction angles α_2 and β_2 on positioning accuracy was analyzed.

As can be seen from Fig. 3(a), the effect of the change of β_2 on the direction of the X-axis is on the order of 10^{-11} and is therefore negligible. As can be seen from Fig. 3(b), the change in the Z direction is parabolic tendency, and the sensitivity coefficient is 0 when β_2 takes 90°, which means that when the laser exit direction is completely perpendicular to the ground, the Z direction accuracy reaches the highest. Therefore wr chooses β_2=90° as the ideal angle between the laser pointing and the XOY plane, but it is difficult to reach 90° in the actual design and manufacture of the system, so the design is infinitely close to

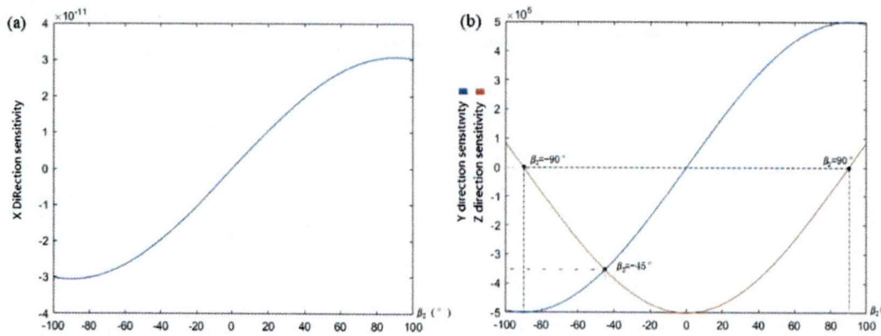

Fig. 3. **(a)** The change of sensitivity in X direction with β_2; **(b)** Change of sensitivity in Y and Z directions with β_2.

90°, 89.89° is taken as the consideration value, and 88°is used as the comparison value to analyze the influence of α_2 on the sensitivity of X, Y, and Z directions, as shown in Fig. 4. It can be seen from the figure that when the laser exit direction is infinitely close to the Z axis (when pointing to the sub-star point,$\alpha_2 = 90°$), the influence of positioning accuracy in the X and Y directions is about 0(the situation that the outgoing light is on the YOZ plane). Moreover, the change of α_2 does not affect the positioning accuracy of the Z direction, so $\alpha_2 = 90°$ is selected as the reference value in this paper to further analyze the sensitivity coefficient of the remaining variables in the three directions.

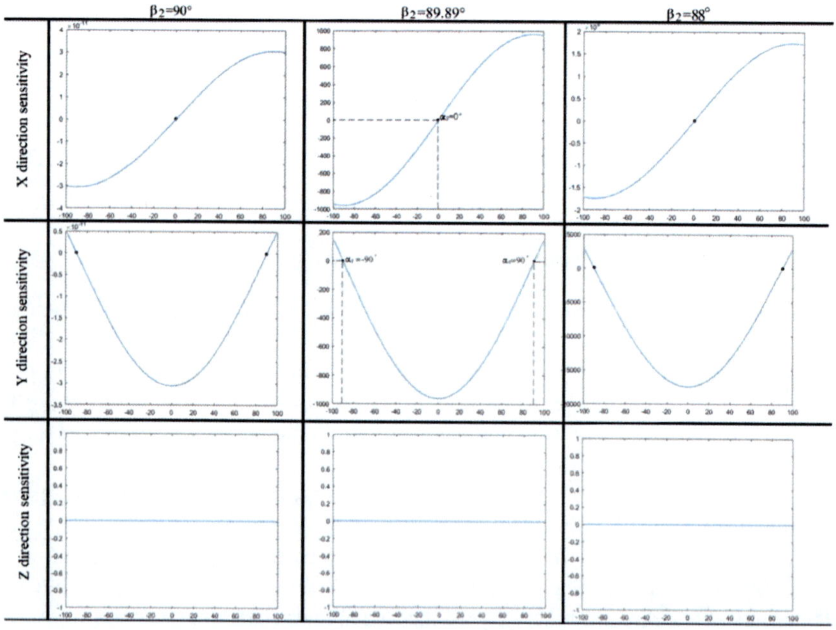

Fig.4. Variation of sensitivity in X, Y and Z directions with α_2

A certain model 03 star laser altimeter main and standby two beams and Z axis angle design is $\pm 0.11°(\beta_2 = 90° \pm 0.11°)$. The sensitivity coefficient of the laser altimeter's omnidirectional positioning accuracy was analyzed, as shown in Table 1. At this time, the error propagation coefficient analysis of each error variable is shown in Table 1.

Table 1. Table of propagation system of each error variable taking a certain type of laser altimeter as an example ($\alpha_2 = 90°, \beta_2 = 90° - 0.11°$).

Orientation sensitivity	Propagation coefficient of each error variable											
	X_S	Y_S	Z_S	φ	ω	k	α_1	α_2	β_1	β_2	r	L
X	1	0	0	500000	0	958.3305	−1.6	959.9305	1.0089	0	−0.4982	0
Y	0	1	0	0	500000	−1.3	−1.3	0	−1.2418	500000	0.6131	−0.0019
Z	0	0	1	−1.3	−958.3305	0	0	0	−2.0616	−959.9305	−0.6131	−1

It can be seen that the positioning error in X direction is mainly related to the laser altimeter's outgoing light and the X-axis direction Angle α_2, pitch Angle φ and yaw Angle k. The positioning error in Y direction is mainly related to the side roll Angle ω and the pointing Angle β_2 between the laser altimeter and the Z axis. The accuracy of Z direction is mainly affected by the attitude Angle φ, ω, and the pointing Angle β_2, and is mainly affected. It can be seen from the above analysis that the z-direction error propagation coefficient of the laser altimeter is smaller than that of the XY direction, which can achieve higher measurement accuracy. And the laser positioning accuracy is affected by attitude, orbit, laser installation, ranging and other factors, and the laser altimeter error is mainly affected by the installation Angle of the laser altimeter after orbit, because the above errors are random errors, it is difficult to avoid, so the pointing measurement system must be set up to correct the measurement data of the satellite-borne laser altimeter. In order to further analyze the specific impact of sensitivity on positioning error, the following mainly analyzes the impact of the above elements on the positioning accuracy of the satellite-borne laser altimeter.

The outer azimuth element of the laser altimeter (ΔX_s, ΔY_s, ΔZ_s) is calculated by the orbit parameters of the satellite platform, and the measurement accuracy can reach the decimeter level ($\Delta X_S = \Delta Y_S = \Delta Z_S = 0.1$ m). The outer azimuth element takes 1 "for k. When the influence of changing $\Delta\varphi$, $\Delta\omega$, $\Delta\alpha_2$, $\Delta\beta_2$ on positioning accuracy is discussed separately, the error of the other two parties is tentatively 1 ". Assuming the pointing Angle error between the laser altimeter's outgoing light and the X axis and Z axis $\Delta\alpha_2 = 1''$, $\Delta\beta_2 = 1''$, the laser ranging error is ΔL=0.3m, $\Delta r = 0.01$m, $\Delta\alpha_1 = 1''$, $\Delta\beta_1 = 1''$. According to the law of error propagation, the positioning error of the laser altimeter is:

$$\begin{bmatrix} \Delta X \\ \Delta Y \\ \Delta Z \end{bmatrix} = \begin{bmatrix} \sqrt{\left(\frac{\partial X}{\partial X_S}\right)^2(\Delta X_S)^2 + \left(\frac{\partial X}{\partial \varphi}\right)^2(\Delta\varphi)^2 + \left(\frac{\partial X}{\partial \omega}\right)^2(\Delta\omega)^2 + \left(\frac{\partial X}{\partial k}\right)^2(\Delta k)^2 + \left(\frac{\partial X}{\partial L}\right)^2(\Delta L)^2 + \left(\frac{\partial X}{\partial r}\right)^2(\Delta r)^2 + \left(\frac{\partial X}{\partial \alpha_1}\right)^2(\Delta\alpha_1)^2 + \left(\frac{\partial X}{\partial \alpha_2}\right)^2(\Delta\alpha_2)^2 + \left(\frac{\partial X}{\partial \beta_1}\right)^2(\Delta\beta_1)^2 + \left(\frac{\partial X}{\partial \beta_2}\right)^2(\Delta\beta_2)^2} \\ \sqrt{\left(\frac{\partial Y}{\partial Y_S}\right)^2(\Delta Z_S)^2 + \left(\frac{\partial Y}{\partial \varphi}\right)^2(\Delta\varphi)^2 + \left(\frac{\partial Y}{\partial \omega}\right)^2(\Delta\omega)^2 + \left(\frac{\partial Y}{\partial k}\right)^2(\Delta k)^2 + \left(\frac{\partial Y}{\partial L}\right)^2(\Delta L)^2 + \left(\frac{\partial Y}{\partial r}\right)^2(\Delta r)^2 + \left(\frac{\partial Y}{\partial \alpha_1}\right)^2(\Delta\alpha_1)^2 + \left(\frac{\partial Y}{\partial \alpha_2}\right)^2(\Delta\alpha_2)^2 + \left(\frac{\partial Y}{\partial \beta_1}\right)^2(\Delta\beta_1)^2 + \left(\frac{\partial Y}{\partial \beta_2}\right)^2(\Delta\beta_2)^2} \\ \sqrt{\left(\frac{\partial Z}{\partial Z_S}\right)^2(\Delta Z_S)^2 + \left(\frac{\partial Z}{\partial \varphi}\right)^2(\Delta\varphi)^2 + \left(\frac{\partial Z}{\partial \omega}\right)^2(\Delta\omega)^2 + \left(\frac{\partial Z}{\partial k}\right)^2(\Delta k)^2 + \left(\frac{\partial Z}{\partial L}\right)^2(\Delta L)^2 + \left(\frac{\partial Z}{\partial r}\right)^2(\Delta r)^2 + \left(\frac{\partial Z}{\partial \alpha_1}\right)^2(\Delta\alpha_1)^2 + \left(\frac{\partial Z}{\partial \alpha_2}\right)^2(\Delta\alpha_2)^2 + \left(\frac{\partial Z}{\partial \beta_1}\right)^2(\Delta\beta_1)^2 + \left(\frac{\partial Z}{\partial \beta_2}\right)^2(\Delta\beta_2)^2} \end{bmatrix} \quad (4)$$

The influence of $\varphi, \omega, \alpha_2and\beta_2$ accuracy on the positioning accuracy of laser altimeter is shown in Table 2.

In Table 2, the laser altimeter is sensitive to the change of observation error of attitude angle φ, ω, β_2 in the X and Y directions. When directly positioning and the

Table 2. Table of positioning accuracy affected by error ($\beta_2 = 89.89°$)

Error source	$\Delta\varphi$			$\Delta\omega$			$\Delta\alpha_2$			$\Delta\beta_2$		
Error magnitude	1″	3″	5″	1″	3″	5″	1″	3″	5″	1″	3″	5″
ΔX	2.43	7.27	12.12	2.43	2.43	2.43	2.43	2.43	2.43	2.43	2.43	2.43
ΔY	3.43	3.43	3.43	3.43	7.67	12.36	3.43	3.43	3.43	3.43	7.67	12.36
ΔZ	0.34	0.34	0.34	0.34	0.42	0.54	0.34	0.34	0.34	0.34	0.42	0.54

error of $\Delta\varphi$, $\Delta\omega$, $\Delta\alpha_2$, $\Delta\beta_2$ is above 1", it can only meet the requirements of small-scale mapping above 1:50,000, which meet the requirements of laser altimetry of a certain model 03 star.

With the increase of pitch the error of angle $\Delta\varphi$, the X direction positioning error of the laser altimeter increases, while the Y direction positioning error remains unchanged. With the increase of the error of rolling angle $\Delta\omega$ or the pointing angle $\Delta\beta_2$ which is the angle between the laser exit light and the Z axis, the positioning error of the laser altimeter in the Y direction increases, while the positioning error in the X direction remains unchanged. With the increase of the error of the pointing angle $\Delta\alpha_2$ which is angle between the laser exit light and the X axis, the X and Y directions basically do not change, and only when the system design is set to deviate from 90°, the positioning error in the X direction will be slightly affected with the change of $\Delta\alpha_2$.

4 Conclusion

Starting from the working principle of laser altimeter, this paper constructs a tight geometric model of the spaceborne laser altimeter by using parameters such as attitude angle, orbital positioning, pointing angle, etc., and analyzes the main error sources to show that the positioning accuracy in the X and Y directions is mainly affected by the accuracy of attitude angle and laser altimeter pointing angle. Based on the error propagation coefficient, the following conclusions can be drawn:

1) Whether the positioning accuracy of the spaceborne laser altimeter X and Y directions is under ideal circumstances(the pointing angle of the altimetry vector is infinitely close to 90°), or under the general case that the pointing angle of the altimetry vector deviates from 90°, the positioning accuracy is mainly affected by the observation accuracy of the pitch angle φ, roll angle ω, and pointing angle β_2 of altimetry vector, and the degree of influence is positively correlated with the orbital height. The positioning accuracy in the Z direction is mainly affected by the roll angle ω and the pointing angle β_2, and the degree of influence is positively correlated with the track height.

2) Through simulation analysis, it can be seen that the geometric positioning of the spaceborne laser altimeter requires high accuracy of external azimuth elements and pointing angles, and the positioning accuracy in the Z direction is greater than that in the X and Y directions.

Finally, the Z direction accuracy of the spaceborne laser altimeter is less affected by the angle measurement reason for the laser ranging accuracy is generally high, and it can be corrected by other means. However, the accuracy of satellite attitude and orbit determination is very high, the error propagation coefficient of attitude angle and laser exit light and the direction angle of the X axis and Z axis is very large, which is hundreds of thousands of times the error propagation coefficient of laser ranging. Therefore, the accuracy of the X, Y direction of the spaceborne laser altimeter is seriously affected by the angle measurement on the whole, and it is necessary to consider the parameters when installing the spaceborne laser altimeter. In order to further improve the measurement accuracy of the spaceborne laser altimeter, improving the angle measurement accuracy has become an indispensable priority.

References

1. Yu, J., Chen, Q., Pan, Z.: Overview of time interval measurement of pulsed Lidar. Laser Infrared (2001)
2. Yu, H.: Review on time-of-flight measurement of pulsed laser radar. Laser Infrared (2001)
3. Kostamovaara, J., Maatta, K., Koskinen, M., et al.: Pulsed laser radars with high-modulation-frequency in industrial applications (1992)
4. Yoder, J.A., Aiken, J., Swift, R.N., et al.: Spatial variability in near-surface chlorophyll a fluorescence measured by the Airborne Oceanographic Lidar (AOL). Deep Sea Res. Part II **40**(1), 37–53 (1993)
5. Fan, C., Li, J., Wang, D., et al.: Application of laser altimeter satellite ICESAT in geoscience research. Geodesy Geody. **25**(2), 4 (2005)
6. Zwally, H.J., et al.: ICESat's laser measurements of polar ice, atmosphere, ocean, and land. J. Geody. **34**, 405–445 (2002)
7. Weiping, J., Jiancheng, L., Zhengtao, W.: Determination of global mean sea surface WHU2000 by combining various altimetry data [J]. Chin. Sci. Bull. **47**(15), 5 (2002)
8. Hu, M.: Theory and Application of Modern Geodesy. Surveying and Mapping Press (2003)
9. Seeber, G.: Satellite Geodesy. Walter de Gruyter, Berlin (2003)
10. Herring, T.A., Quinn, K.J.: The algorithm theoretical basis document for the atmospheric delay correction to GLAS laser altimeter ranges, vol. 8 (2012)
11. Renxiang, W.: Photogrammetric satellite camera without ground control point in China. Space Return Remote Sens. **29**(3), 6–9 (2008)
12. Chunyu, Y., Yongchao, Z., Yuliang, T.: Analysis on satellite photogrammetry assisted by satellite-based laser altimeter. Space Return Remote Sens. **34**(4), 6 (2013)
13. Jianfeng, J., Jiao Wenhai, H., Hu, G.: Principle of satellite laser altimetry and its ranging accuracy. Surv. Mapp. Tech. Equip. **2**(7), 46–48 (2005)
14. Simard, M., Rivera-Monroy, V.H., Mancera-Pineda, J.E., et al.: A systematic method for 3D mapping of mangrove forests based on shuttle radar topography mission elevation data, ICEsat/GLAS waveforms and field data: application to Cienaga Grande de Santa Marta, Colombia. Remote Sens. Environ. **112**(5), 2131–2144 (2008)
15. Wang, X., Gong, P., Zhao, Y., et al.: Water-level changes in China's large lakes determined from ICESat/GLAS data[J]. Remote Sens. Environ. **132**, 131–144 (2013)

16. Simard, M., Neumann, M., Buckley, S.: Validation of the new SRTM digital elevation model (NASADEM) with ICESAT/GLAS over the United States. In: 2016 IEEE International Geoscience and Remote Sensing Symposium (IGARSS). IEEE (2016)
17. Schutz, B.E.: Laser footprint location (geolocation) and surface profiles. GLAS ATBD **3**(10), 16–18 (2002)
18. Dang, Y., Kim, C.M., Cheng, Y.: Principle and Application of Global Navigation Satellite System. pp. 26–32. Surveying and Mapping Press (2007)
19. Hanjiang, W., Pengfei, C.: Principle and application of ICESAT/GLAS laser allocator. Sci. Surv. Mapp. **30**(5), 33–35 (2005). (in Chinese)
20. Chunyu, Y., Hongyan, H., Yunfei, B., et al.: Geometric positioning error propagation analysis of spaceborne laser altimeter. Aerosp. Return Remote Sens. **35**(2), 6 (2014)
21. Yu, J., Sun, S.: Some ideas on the development of satellite photogrammetry observation technology. In: The 23rd National Space Exploration Academic Exchange Conference, Xiamen, pp. 1–6 (2010)

Improvement of the Information Technology Basis for Aerospace Monitoring of the Sea Shelf for the Search for Hydrocarbon Deposits in the Optical Spectral Range

Oleksandr Fedorovskyi, Olga Sedlerova(✉), and Anna Khyzhniak

Scientific Center for Aerospace Research of the Earth of the Institute of Geological Sciences of the National Academy of Sciences of Ukraine, Olesia Honchara Street, Kyiv 01054, Ukraine
sedlerova@casre.kiev.ua

Abstract. The paper presents an analysis of satellite images of temperature anomalies over the Schmidt and Golitsyn fields. An analysis of the state of the aquatic environment under various hydrological and hydrophysical conditions has been carried out. Firstly, with hydrology close to isotherm and intense fluid flow, this can be an airlift process, i.e., the removal of cold deep waters by a fluid flow to the sea surface. Second, if there are density gradients in the stratified water column, migrating fluids also cause the formation of internal waves propagating towards the water surface and interacting with it. Thirdly, in hydrology with a pronounced thermocline, the flow of fluids, reaching it, causes fluctuations in density and temperature in it, propagating from the area of disturbance to the free water surface in the form of internal waves (IW). The frequency of these waves is known as the Brent-Waisell frequency. For further detailed study of the process of formation of temperature anomalies on the free water surface over hydrocarbon deposits, the result of modeling this process obtained earlier at different times from the board of ships and offshore platforms in the Black, Japanese and Barents Seas was used. After processing and analyzing the results of marine physical modeling of the formation of temperature anomalies on the sea surface, a unique statistical material was obtained in different marine areas, hydrology and seasons. Among other things, several dozens of different Haralik parameters were tested and the five most informative of them were selected for further use: Contrast, Sum Variance, Sum Entropy, Entropy and Difference Entropy. As a result of the research, new technological methods for searching and predicting explosive deposits were proposed.

Keywords: aerospace monitoring · multispectral satellite imagery data · structural and textural parameters · hydrocarbon deposits

H. P. Urbach and H. Jiang (Eds.): ISSOIA 2022, SPPHY 295, pp. 159–175, 2023.
https://doi.org/10.1007/978-981-99-4098-1_16

1 Introduction

Interest in the study of hydrophysics and hydrodynamics of sea surface temperature and its distribution in the surface layer of water are due to two main factors. First, it is energy-mass exchange between the atmosphere and the ocean. The spatial distribution of temperature is important in the simulation of large-scale processes in climate models, which is essential for weather forecasting and assessments of long-term climate changes. The second factor is interest in the hydrodynamics of the near-surface layer of water, which can be used as one of the informative features when searching for hydrocarbon deposits in the shelf zone. The water surface for aerospace monitoring is the natural integrator of information, which allows you to detect processes occurring not only on the sea surface, but also in the water column and bottom space. Information about the temperature of the surface layer of the sea is carried by its infrared thermal radiation, the intensity of which is related to the usual thermodynamic temperature by the well-known Stefan-Boltzmann law [1].

The purpose of the research is to reveal the hydrophysical and geological features of the sea surface in the optical spectral range, which form informative features for aerospace geomonitoring of the search for hydrocarbon deposits and to propose new technological methods for assessing the oil and gas prospects of areas of the sea shelf.

2 Background and Related Works

2.1 Studying the Heterogeneous State of the Marine Environment

The first experiments on remote studies of the processes occurring in the layer between the sea and the atmosphere were conducted by Mac Allister with the help of a radiometer developed by him [2]. It was experimentally established that a specific boundary layer with a size of several millimeters and a temperature gradient of several degrees is formed in the surface layer. The temperature from the surface of the water increases linearly to a depth of several millimeters, and then monotonically falls to the temperature of the deep sea horizons. This hydrophysical formation was called the skin layer. The thickness of this boundary layer and the temperature difference between its boundaries mainly depend on the local conditions that determine the flow of heat through the water-atmosphere boundary and the characteristics of turbulence in the upper layer of the sea.

The question of the physical essence of fluid-conducting structures is currently quite complex and ambiguous. First of all, this is due to the entry into the water environment through the atmosphere-water interface of solar radiation and the atmosphere's own radiation. In turn, the water environment forms its own flow of long-wave radiation. In addition, the near-surface layer is affected by convective flows occurring in it, turbulence, internal and surface waves, hydrology, wind effects, evaporation, precipitation, cloud cover, currents, and surfactants.

It is known that hydrocarbons can be discharged into the water environment through fault zones of various geological structures, forming bubbles and mud volcanoes. Gas sources on the seabed are due to significant loosening and high fluid conductivity of rocks, processes of vertical migration of formation and deep, both gaseous and liquid fluids. Under the latter, linear zones of rock densification are considered, which are fixed by distance methods in the form of lineament zones and nodes of their intersection. The movement of formation and deep fluids, which has a discrete or batch mode, in the zone of the geodynamic node, is associated with a change in pressures, both in the fluid-conducting structure itself and in the adjacent zones [3].

Figure 1 shows a diagram of high-frequency sonar of gas "torches" (according to Yevhen Shnyukov). The parameters of gas "torches" depend on the structure of the geofluidodynamic structure, pressure, volume and composition of migrating gases. A high density of gas sources is characteristic of most of the Black Sea, which can be considered as a consequence of active degassing of the subsoil of the region due to the stretching of the lithosphere as a result of mantle diapirism and the processes of the expansion of the planet at the present stage [4].

The mechanism of passage of the migratory flow of hydrocarbon fluids through the water column and its interaction with the water surface is determined by a number of factors, including the type of hydrology, which depends on the season and water area. Figure 2 shows hydrological sections by depth, which were obtained in different water areas and at different times of the year.

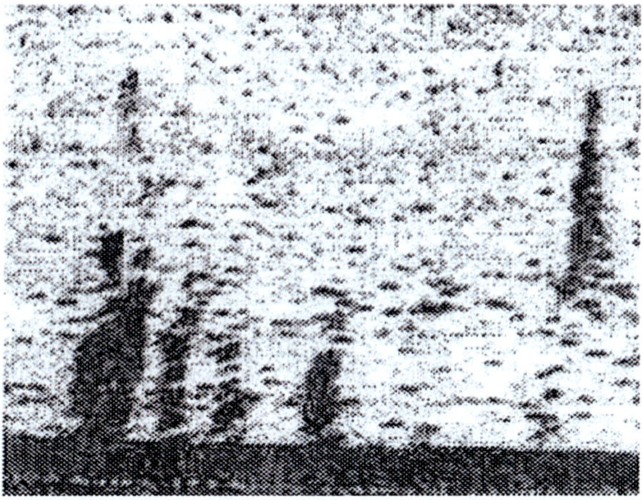

Fig. 1. Diagram of high-frequency sonar of gas "torches"

Figure 3 shows shadow photos of the marine environment in different hydrological conditions. Photographing was carried out using a Tepler camera when the equipment was submerged to a depth of up to 100 m.

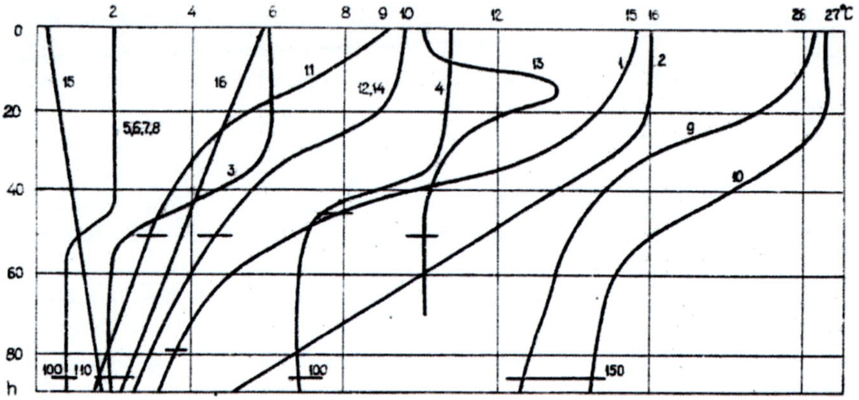

Fig. 2. Hydrological sections by depth, which were obtained in different water areas and at different times of the year

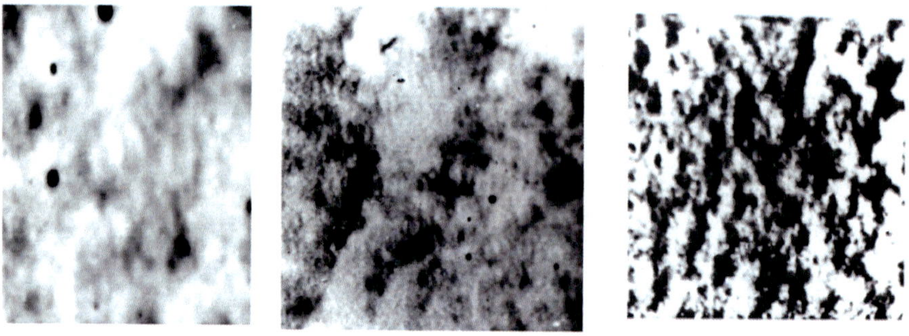

a b c

Fig. 3. Examples of different conditions of the marine environment at deep horizons: isotherm or weakly stratified (a), stratified (b), thermocline (c)

2.2 Methods and Results of Studying Specific Temperature Changes

In Fig. 4 provides examples of the state of the marine environment at deep horizons under external excitation, which are noticeably different depending on hydrological conditions, changes in temperature, salinity, underwater currents, the structure of the water environment is different. In the Black Sea, the thermocline is usually located at a depth of about 15–25 m. The disturbance that occurs in it generates density and temperature fluctuations that propagate from the area of disturbance in the form of internal waves (IW).

With hydrology close to isotherm and intense flow of fluids, this can be an airlift process - the rise of cold deep waters to the sea surface by a fluid flow. Secondly, in the presence of density gradients in the stratified water column, migrating fluids also cause the formation of internal waves that propagate to the water surface and interact with it. Thirdly, in hydrology with a pronounced thermocline, the flow of fluids, reaching it, causes density and temperature fluctuations in it, which spread from the disturbed area to the free water surface in the form of internal waves (IW).

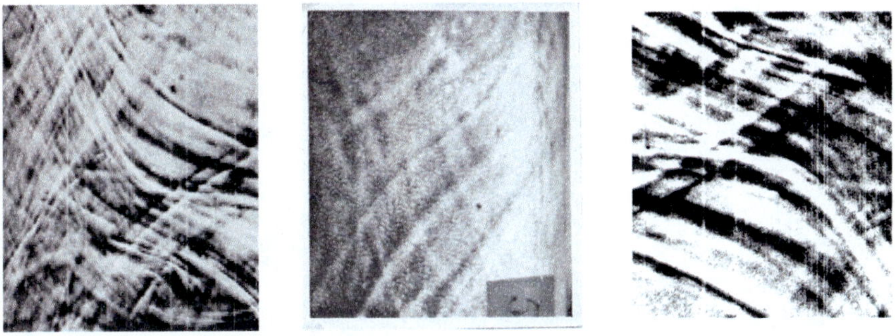

Fig. 4. Examples of shadow images of changes in the texture of the marine environment at deep horizons after external excitation

The frequency of these waves is known as the Brent-Vaisail frequency, and its inverse value (the wave period) serves as the fundamental time scale that determines oscillatory movements in a stratified water medium. The resulting HCs, reaching the water surface, cause changes in the hydrophysical characteristics of the near-surface layer of water, which are reflected on the sea surface as temperature changes, and in the near-surface layer as changes in the temperature gradient. The results of the interaction of emerging deep hydrodynamic processes (DHP) with the near-surface water layer are naturally reflected on the free water surface as temperature anomalies recorded on space images in the infrared spectral range [5].

Figure 5 shows a characteristic temperature profile in the near-surface layer of the liquid [5].

In most cases, under natural conditions, the surface temperature is lower than the temperature below the lying layers of water, which first increases with depth, and then begins to monotonously fall to the temperature of deep horizons. The reason for this phenomenon, with rare exceptions, is that the water surface gives off heat to the atmosphere by radiant and turbulent heat transfer.

To fulfill the condition of heat balance, it is necessary to have a flow compensating for these losses to the surface from the lower water layers. In addition, the near-surface layer is affected by convective flows occurring in it, turbulence, internal and surface waves, hydrology, wind effects, evaporation, precipitation, cloud cover, currents, and surfactants. As a result, an unstable stratification occurs in the surface layer, which can lead to convective movements, the upper layers of water fall into the depth of the liquid, forming cold thermals. The process of formation of such thermals is periodic in nature.

Two forms of free convective flows from the water surface are known: a jet with a spherical front and flows in the form of vertical thermals.

Figure 6 shows the results of laboratory experimental studies of microconvection in the near-surface layer of water with a size of the order of 0.5 cm (shadow photographs of a vertical section of the water surface), in which a dynamic picture of the process

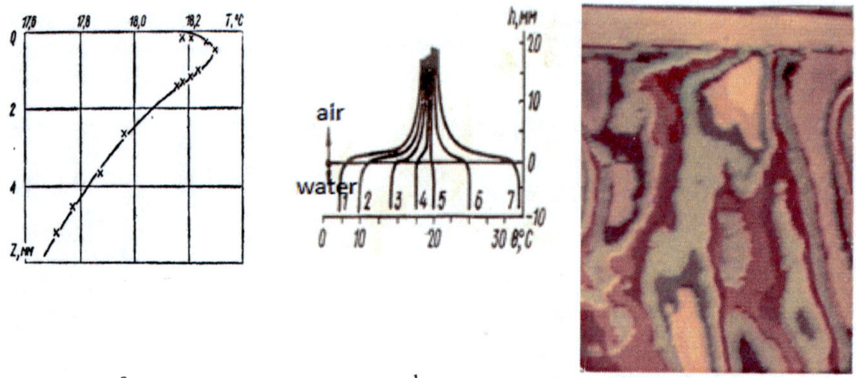

a b c

Fig. 5. Characteristic temperature profile in the near-surface liquid layer: a) idealized temperature profile for daytime in summer with weak winds; b) dependence of surface temperature profiles according to Heisler; c) vertical section of the near-surface liquid layer

of formation and development of a cold skin layer near the surface of the water–air separation is presented [6]. The shadow photos show the successive development of cold thermals (1, 2 and 3) in the form of drops breaking off from the water surface. At the same time, areas of increased density are depicted in lighter tones. The formed thermals and their subsequent evolution within 6 s are indicated by dark arrows in the photographs. The speed of movement of the thermals is several millimeters per minute with the size of the thermals of the order of 0.5 cm. In the obtained photos, you can observe the temperature inversion in the upper layer of the studied volume of liquid. At the same time, water cooled by evaporation and radiation "flows" down through these layers, forming a thin surface film. The contrast of the image reflects temperature differences in the near-surface layer, the value of which is proportional to the brightness of the image. Figure shows the formation and movement of cold thermals from the surface into the depth of the liquid during a natural process.

Figure 7 shows shadow pictures of the development of microconvection in the near-surface layer during forced cooling of the water surface. It was found that water in the surface layer is located in thin unstable layers caused by convective processes.

With the appearance of cloudiness, the initial stationary state of the cold laminar layer, ensured by the balance of heat flows, was disturbed. The effective radiation decreased, which caused an increase in the temperature of the skin layer and an increase in evaporation, which is accompanied by the return of heat to the atmosphere. The latter is compensated by the inflow of heat from the homogeneous layer into the cold sublayer before the onset of a new quasi-stationary state caused by the lack of inflow of direct solar radiation. After the passage of the cloud, the temperature of the skin layer remains above its initial state for some time until the complete disappearance of cloudiness, when the thermal parameters of the skin layer begin to asymptotically approach their initial values. Thus, with the appearance of cloudiness, an increase in the temperature of the water surface was recorded and substantiated.

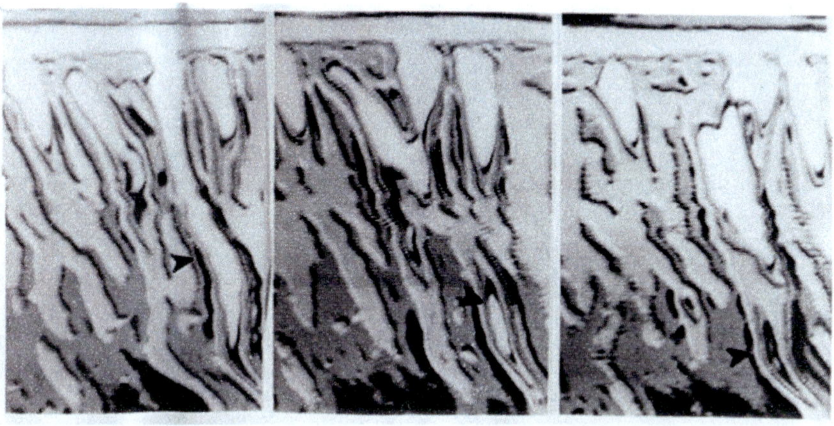

Fig. 6. Results of laboratory experimental studies of microconvection in the near-surface layer of water with a size of the order of 0.5 cm

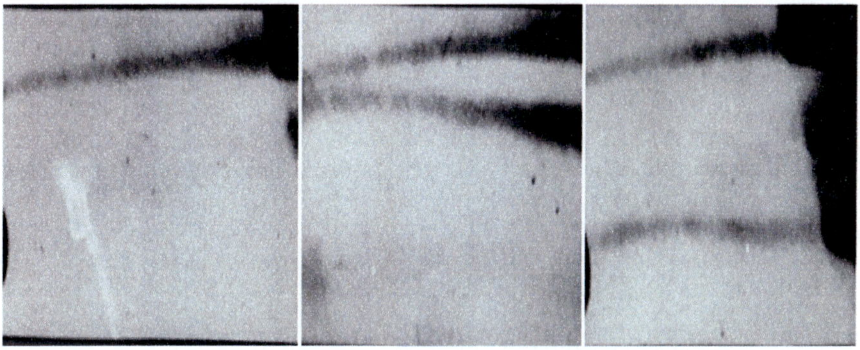

Fig. 7. Results of laboratory experimental studies of microconvection in the near-surface layer of water with a size of the order of 0.5 cm during forced cooling of the water surface

The cold skin is preserved at a wind speed of up to 10 m/s, and the recovery time of the skin layer depends on many external factors. After its destruction due to the collapse of waves and other factors, the recovery time of the skin can be tens of seconds. Therefore, it can be assumed that the existence of a cold skin layer is a rather stable phenomenon.

The volume of water in the thermocline, shifted up or down, will be affected by forces acting in the direction of the reverse shift. Therefore, the perturbation that occurs in the thermocline generates density and temperature fluctuations that propagate from the perturbation region. Convergence and divergence processes caused by internal waves affect the distribution of films of surfactants, oil, and organic liquids that are always present on the surface of the ocean, leading to the formation of characteristic "streaks" that can be seen from the air or from space.

3 Forecasting the Oil and Gas Prospects of Areas Based on Various Characteristics

Analysis of the location of the gas torches on the seabed with the attachment of the latter to the appropriate space images made it possible to determine that they are confined to high -density of liniment and cause thermal anomalies on the sea surface, which certify the presence and hydrocarbon deposits. The physical basis of the formation of geo-infection features of underwater landscapes is the transmission of information on the intensity of the stress-deformed state of the lithosphere and fluidogeoodynamic processes passing from depth to the earth's surface [7].

Direct and indirect geoindicative signs of underwater landscapes are distinguished. Direct signs are related to the physiognomic features of underwater landscapes, which shine through everywhere in the water column. Indirect geoindicative features associated with hydrodynamic processes occurring in the water column and on the surface of the sea, where the main role is played by the hydrological and hydrophysical factor.

Thus, in shallow water conditions, the neotectonic active structure associated with the Schmidt field (Fig. 8) is clearly visible in the relief of the seabed, where the lighter phototone corresponds to the area in the relief located above the Schmidt field. This is triggered by the neotectonic factor in the formation of an optical anomaly on the aerospace image of the sea surface, through the thickness of the water the light rocks of the seabed.

Figure 9 shows a fragment of a space image of the northwestern shelf of the Black Sea (NOAA satellite channel 4 (10.3–11.3 μm)) from the temperature anomaly over the Golitsyn hydrocarbon field (according to O.Yu. Kotlyar). From Fig. 9 we see that there are both cold (black spots) and warm (light spots) anomalies on the sea surface, that is, it can be assumed that both processes of surface temperature anomaly formation are present: bubbles and internal waves [8] (Tables 1 and 2).

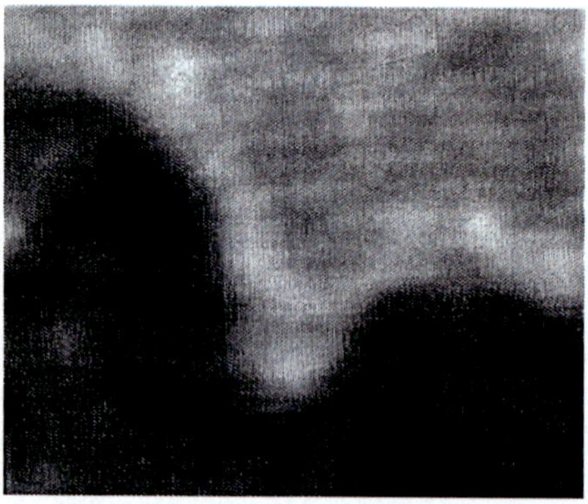

Fig. 8. Fragment of a space image of an optical anomaly over the Schmidt hydrocarbon deposit

Table 1. The value of the Haralik parameters over the Schmidt hydrocarbon deposit

Haralik parameters	Anomaly	Background
1. Contrast	1133	155
2. Sum Variance	8,1	0,8
3. Sum Entropy	0,12	0,08
4. Entropy	13,9	5,0
5. Difference Entropy	0,7	0,5

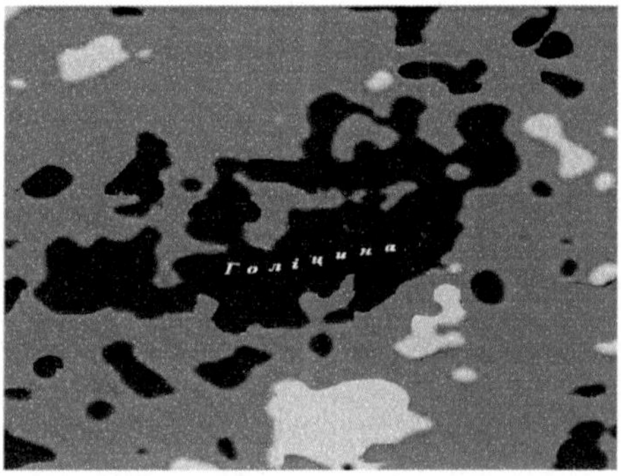

Fig. 9. Temperature anomaly over the Golitsyn hydrocarbon deposit. (according to O. Yu. Kotlyar)

Table 2. Values of Haralik parameters over the Golitsyn hydrocarbon deposit

Haralik parameters	Anomaly	Background
1. Contrast	4,88	0,01
2. Sum Variance	288	0,37
3. Sum Entropy	4,24	0,41
4. Entropy	5,18	0,42
5. Difference Entropy	1,93	0,08

In this connection, interest arose in researching the mechanism of formation of optical anomalies and hydrothermal processes that occur at the boundary of the sea surface-atmosphere separation for the subsequent use of the results of theoretical studies in the development of methods for deciphering space images and detecting oil and gas deposits.

The surface of the sea and the boundary layer of the atmosphere adjacent to it are in a close thermal and dynamic interaction, which occurs through radiant and contact heat transfer, as well as evaporation.

For a near-surface layer of water of the order of 1 cm (with tolerances), the heat balance equation has the form:

$$q = R + Q + i, \tag{1}$$

where R is the balance of long-wave radiation; Q - contact heat exchange of the sea surface with the atmosphere; i- heat consumption for evaporation.

In connection with the difficulties of determining the total heat flow by means of calculations, in practice, the method of its direct measurement along the temperature gradient is used [9].

$$q = kd/dZ, \tag{2}$$

k is the coefficient of thermal conductivity of the near-surface skin layer of water Z.

The value of k is easily calculated within the upper two-centimeter layer of water

$$k = k_0(1 + vZ), \tag{3}$$

k_0 is the molecular value of the thermal conductivity coefficient (Z = 0), v is the rate of change of k (600 m–1).

Therefore, the task of determining the total heat flow from the sea surface is reduced to measuring the temperature gradient in the surface layer of water.

The water surface - atmosphere system is modeled by the liquid - vapor system, the thermodynamic equilibrium between which is disturbed by external radiation. A heat-conducting liquid obeying the Navier–Stokes law is chosen as a model. The complexity of the mathematical description lies in the task of non-equilibrium boundary conditions that must be satisfied by the hydrothermodynamic parameters of the two media. There are a number of temperature boundary layer models. However, all of them are based on the assumption that its destruction occurs periodically and the influence of phase transitions on the structure of the thermal boundary layer is not taken into account.

In a number of cases, it is possible to obtain a pronounced dependence of the effective radiation depth Z_{ef} and the thermal state of the water. The mathematical model for Z_{ef} after a number of transformations looks like this

$$Z_{ef} = \frac{\int_{\lambda_1}^{\lambda_2} \varepsilon(\lambda)(dm_{eS}/dZ)\alpha e(\lambda)^{-1} d\lambda}{\int_{\lambda_1}^{\lambda_2} \varepsilon(\lambda)(dm_{eS}/dZ) d\lambda}, \tag{4}$$

where ε(λ) is the spectral emission coefficient, me is the energy luminosity of the black body, αe(λ) is the spectral absorption coefficient.

To calculate the effective depth of radiation in the case of a narrow spectral interval, H. Macalister derived the following expression for Z_{ef}.

$$Z_{ef} = 1/\alpha e(\lambda), \tag{5}$$

Thus, for the spectral intervals of 2.0–2.4 μm and 3.5–4.0 μm, the depths at which the temperature was measured are 0.5 mm and 0.06 mm, respectively.

By successively measuring the radiation of water in different spectral ranges, it is possible to determine the water temperature at different depths and, as a result, knowing the value of Z_{ef}, calculate the temperature gradient. The latter can be used as an informative sign of the presence of hydrocarbon deposits during space remote sensing monitoring of the shelf zone of the Azov-Black Sea basin and determination of temperature gradient anomalies in the near-surface layer of the sea surface.

For a further detailed study of the process of formation of temperature anomalies on a free water surface, the results of an experiment previously performed at different times in the Black Sea from the ship OS-6 (Fig. 10), as well as continued in the Barents Sea and the Sea of Japan, were used. Modeling was performed by disturbing the deep horizons with the help of a submerged self-propelled model (as a simulator of hydrocarbon deposits), and the registration of emerging anomalies was carried out as a result of their sequential crossing by a ship with optical highly sensitive scanning equipment with parameters: spectral range 3.5–5.2 μm, spatial resolution on the water surface is 1–2 m.

Fig. 10. A ship with an installed thermal imager

Figure 11 shows fragments of an IR image of the temperature background of the sea surface (a) and temperature anomalies disturbed on the water surface (b, c, d, f) by a submerged mobile model and obtained as a result of their successive crossing by the ship as the anomalies develop from 30 min (b) up to 3 h (f) respectively.

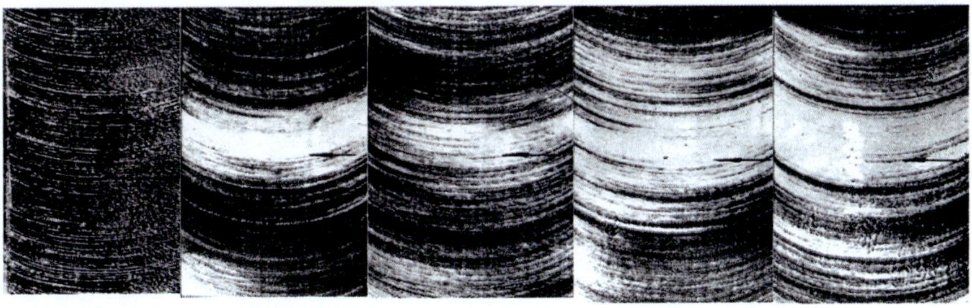

a b c d f

Fig. 11. Fragments of an IR image of the temperature background of the sea surface (a) and temperature anomalies on the water surface (b, c, d, f)

Arrows in Fig. 11 shows the direction of movement of the hydrodynamic process model. The arc-shaped image of temperature anomalies is related to the relationship between the scanning speed and the speed of the ship carrying the thermal imager.

The obtained IR images of anomalies were processed to obtain informative features of the background and signal. For this, the structural and textural characteristics of the water surface were used by calculating the values of the Haralik parameters from infrared images (Fig. 11). Entropy (H) was determined as the most informative parameter of Haralik [10].

$$H = \sum_{i=1}^{N} \log(1/p_i) \cdot p_i, \tag{6}$$

where, p_i is the probability of a pixel value of a digital image, $i = 1 \ldots N$ is the number of a pixel value of a digital image (Fig. 12).

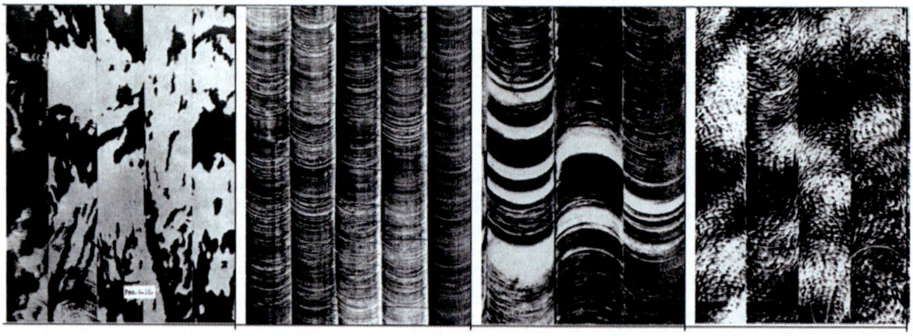

Fig. 12. Temperature anomalies on the surface of the Black, Barents, Japanese and Mediterranean seas

Table 3. Value of entropy at different times of development of temperature anomalies on the water surface by the submerged moving hydrodynamic process model

Images	a	b	c	d	f
Time of development of anomalies	background	30 min	1.5 h	2 h	3 h
Entropy	8,21	8,68	8,82	8,86	8,92

Table 3 shows that the background has a minimum entropy value, and with the appearance of an anomaly and the life time of the anomaly, the entropy value increases to the maximum value, after which, as the temperature wake of the model dissipates, it begins to fall to the background values.

where

$$p(i,j) = \frac{x(i,j)}{\sum_{i=1}^{N} \sum_{j=1}^{N} x(i,j)}$$

$$\mu_{x+y} = \sum_{k=2}^{2N} k \times p_{x+y}(k)$$

Table 4. Haralik parameters and corresponding formulas

N	Sings	Formulas
1	Contrast	$\sum_i \sum_j (i-j)^2 p(i,j)$
2	Sum Variance	$\sum_{k=2}^{2N} \left(k-\mu_{x+y}\right)^2 p_{x+y}(k)$
3	Sum Entropy	$-\sum_{k=2}^{2N} p_{x+y}(k) \log p_{x+y}(k)$
4	Entropy 1	$-\sum_{i=1}^{N} \sum_{j=1}^{N} p(i,j) \log p(i,j)$
5	Entropy 2	$-\sum_{k=0}^{N-1} p_{x-y}(k) \log p_{x-y}(k)$

$$p_{x+y}(k) = \sum_{i=1}^{N} \sum_{j=1}^{N} p(i,j),\ i+j=k$$

$$p_{x-y}(k) = \sum_{i=1}^{N} \sum_{j=1}^{N} p(i,j),\ |i-j|=k$$

As a result of the retrospective analysis and processing of archival materials of physical modeling of the process of formation of temperature anomalies of natural and man-made origin on the water surface, several dozen different parameters of Haralik were checked and the five most informative of them were selected for further use: Contrast, Sum Variance, Sum Entropy, Entropy, Difference Entropy (Table 4).

A comprehensive method of prediction of oil and gas perspective areas on shelf is tested, which, taking into account the geological features of the bottom of the marine shelf, consists of four stages [11].

At the first stage of the research, visual contrast-analog decoding of space images is performed, which is based on the property of remote data to reflect the deep geological structure. At the second stage, structural and geomorphological studies of the bottom relief are carried out based on the analysis of bathymetric and geomorphological maps and their transformations. This stage consists of numerous methods of analyzing the modern topography of the bottom in order to highlight anomalies in the structure, which are geoindicators of deep faults and blocks active at the latest tectonic stage, and to determine the relative neotectonic activity of the latter. To assess the relative neotectonic activity of the blocks, morphometric constructions (maps of horizontal and vertical dismemberment, gradient zones) are carried out; data on the modern geodynamics of the surrounding land, analysis of modern sediment accumulation and granulometric composition of bottom sediments are included. The comparative tectonic analysis is carried out at the third stage. The basis of this analysis is the visual comparison of the obtained analytical maps with a priori geological and geophysical information. Lineaments are identified with tectonic faults and disturbances in the sedimentary cover revealed by geophysical data, classification of lineament zones into confirmed and unconfirmed ones is carried out. By the comparison method, an analysis of the ratio of structural plans of the Meso-Cenozoic is carried out and structures with inherited development are revealed. In automatic mode, by integrating remote sensing data and geological-geophysical data, an analysis of the heredity of the territory's development is performed [12]. At the fourth stage, an assessment of the oil and gas prospects of areas of the sea shelf is carried out on the basis of neotectonic, structural-geomorphological, structural-geological criteria. Further analysis consists of determining the rank of oil and gas prospective areas according to the main characteristics obtained by decoding of space images, neotectonic and morphometric analysis, structural and tectonic data. The characteristics of each zone are compiled taking into account the rank of oil and gas prospective areas, which were evaluated by rating according to a set of criteria. Establishing a priority rating is the basis for determining the priority areas for exploration for oil and gas.

The main factors affecting the formation and redistribution of oil and gas deposits are geofluidodynamic and neotectonic. The result of the action of these deep processes is manifested in the contours of the modern relief and landscape as a whole. Thus, when identifying zones promising for oil and gas on the sea shelf, there is a study of geofluidodynamic processes and neotectonic movements. Geofluidodynamic processes due to the impact on landscapes, water environment and sea surface are quite clearly recorded on remote sensing materials as temperature anomalies.

In the proposed method of assessing the oil and gas prospects of areas on the sea shelf [13], as an additional informative sign of hydrocarbon deposits, changes in the temperature gradient of the near-surface layer of water under the influence of hydrocarbon fluids are used by calculating the values of radiation values in two spectral ranges (far and near infrared) based on the data of multispectral space surveys, then calculate the effective radiation depth for the given spectra and, based on the obtained data, determine the radiation gradient in the near-surface layer of water, compare the calculation results with the parameters of the reference areas, and make a predictive assessment of the presence of gas deposits.

In order to increase the probability of prediction and the effectiveness of the search for explosives, as an additional informative sign of the presence of explosive deposits, use changes under the influence of hydrocarbon fluids of structural and textural parameters, namely "entropy". The novelty and physical essence of the proposed method is that the structural and textural features of temperature anomalies allow calculating the size, shape, orientation of the relative location of the spatial component images of temperature anomalies. The increase in the probability of prediction and the effectiveness of the search for explosives is due to the fact that the structural-textural feature on the sea surface is a more stable unmasking informative feature, which, despite seasonal weather variability, maintains a qualitative and quantitative assessment throughout the time of searching for explosives. To substantiate the proposed method, modeling of the formation of temperature anomalies on the water surface in the experimental basin was performed, which confirmed the information content of the function of "entropy" to decipher optical anomalies on the images of the water surface.

4 Modeling by the Method of Analytical Networks

To determine the most promising oil and gas area, mathematical modeling was applied by creating a corresponding mathematical model. As an example, consider the model (Fig. 13), created on the basis of the analytical network method (ANM) [14].

The choice of this method is due to the fact that ANM allows you to process more diverse and complex structures taking into account the dependencies between levels and feedback between level elements, thereby achieving greater objectivity and reliability in decision-making. On the basis of ANM, the problem is structured in the form of a network model, on the basis of which the relationships between the proposed alternatives and the generalized criteria for choosing the most promising oil and gas site are determined. For this purpose, three search areas (# 1, # 2 and # 3) of the Caspian shelf of Turkmenistan were selected, which received the highest score at the first stage according to the objective function.

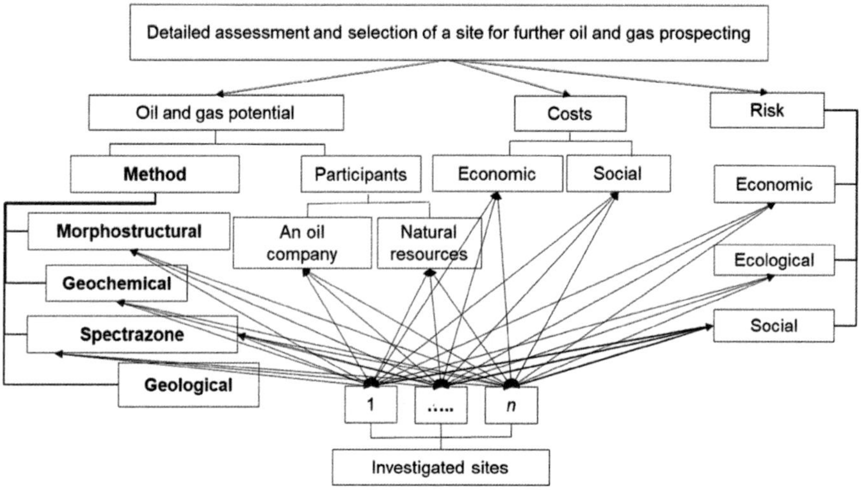

Fig. 13. Network model of relationships between proposed alternatives and generalized selection criteria for determining the most promising oil and gas area

On the created network model and interpretation of the results, calculations conducted on the basis of the Super Decision software product (SP) were performed. Calculations were carried out in three stages:

Step 1. Experts in the field of the subject are determined the priorities by which each of the three regions will be evaluated. The assessment was carried out on an intensity scale from 1 to 9.
Next, in the SP Super Decision, relationships between criteria and alternatives are built and expert evaluations are made for each.
Step 2. The supermatrix and the limit matrix of the elements of relationships are calculated. The most promising oil and gas areas are determined according to separate generalized criteria ("Oil and gas prospects", "Costs", "Risks").
Step 3. Depending on the task, in our case, it is an assessment of the oil and gas potential of prospective areas, a formula is determined, according to which the final calculation will be carried out taking into account the generalized criteria and expert assessments based on the priorities of the selected criteria. In this case, the standard additive (probabilistic) formula was used.

As a result of a detailed study based on the ANM and according to the final ranking data, the following estimates of the oil and gas prospects of three selected areas of the Caspian shelf of Turkmenistan were obtained: # 1–0.31; #2–0.39 and #3–0.30.

5 Conclusion

1. Hydrophysical processes occurring in the near-surface layer of water are caused primarily by the entry into the water environment through the atmosphere-water separation of solar radiation and the action of hydrocarbon fluids that create bubbles and internal waves, which are reflected on the sea surface by the appearance of temperature anomalies that are registered in optical (infrared) range.
2. Between the hydrophysical and hydrodynamic processes in the near-surface layers of water at the water-atmosphere interface, on the one hand, and fluid-dynamic processes, the geological structure of the lithosphere, and underwater landscapes, on the other, there is a relationship that must be taken into account when detecting hydrocarbon deposits based on remote sensing space shooting.
3. Important factors and conditions for obtaining high-quality space images of the sea shelf for effective search for hydrocarbons are the time and season of shooting, cloud cover, sea temperature, presence of wind and fog, density of phytoplankton and salinity of sea water.
4. A mathematical model created on the basis of the analytical networks method (ANM) is used to determine the most oil and gas -perspective area. As a result of a detailed study on the basis of ANM and, according to the final ranking data, the assessments of oil and gas per prospection of the three selected sections of the Caspian shelf of Turkmenistan were obtained.

References

1. Lyalko, V.I., Popov, M.O., Fedorovskyi, O.D.: Multispectral methods of remote sensing of the Earth in nature management problems, Kyiv, Naukova dumka (2006)
2. Mc Alister, E.D. Measurement of total heat flow from the sea surface. Appl. Opt. **5b**, 188–201 (1964)
3. Break, V.M., Pererva, V.M., Kostyna, T.I.: Geofluid dynamic structures of the lithosphere and diapirism. Rep. NAS Ukraine **2**, 131–136 (2002)
4. Shnyukov, E.F., Starostenko, V.Y., Kobelev, V.P.: Geological studies. Geof. J. **26**(4), 116–132 (2004)
5. Akimov, E.A., Stanichny, S.V., Polonsky, A.B.: Using SEVIRI scanner data to estimate the temperature of the surface layer of the Black Sea. Mor. Hydrophys. J. **6**, 37–46 (2014)
6. Fedorovsky, A.D., Nikyforovych, E.I., Prykhodko, N.A.: Transport processes in gas-liquid systems, Kyiv, Naukova dumka (1988)
7. Kolodiy, V.V., Kolodiy, I.V.: Fluid dynamics of the shelf part of the North-Black Sea oil and gas-bearing basin. Geol. Ukraine **1**, 41–44 (2004)
8. Fedorovskyi, O.D., Khyzhniak, A.V., Filimonov, V.Y.: Justification of the dual-use of aerospace geomonitoring of the offshore shelf: exploration of hydrocarbon deposits and "highlighting" the marine situation. Space Sci. Technol. **27**(129), 38–44 (2021)

9. Nikyforovych, E.I., Fedorovsky, O.D.: Hydrothermodynamics of the near-surface liquid layer. Visn. Acad. Sci. Ukrainian SS **11**, 15–21 (1984)
10. Haralick, R.M.: Statistical and structural approaches to texture. Proc. IEEE **67**(5), 786 (1979)
11. Yefimenko, T.A., Sedlerova, O.V.: Geoindicative signs of neotectonic processes in the Azov-Black Sea region, problems of geodynamics and oil and gas carrying capacity of the Black Sea-Caspian region. In: 5th International Conference "Crimea-2003", 8–13 September, pp. 269–272 (2003)
12. Sedlerova, O.V.: Geological interpretation of the data integration multispectral satellite imagery and geological and geophysical data for predict oil and gas zones at the regional level (in the example of the north west shelf of the Black sea). Ukr. J. Remote Sens. **7**, 47–57 (2015)
13. Patent of Ukraine No. 108696, Lyalko V.I., Fedorovsky O.D., Yakymchuk V.G., Sokolovska A.V., Vorobyev A.I. Method of forecasting gas deposits on the sea shelf. 05/25/2015
14. Saati, T.: Decision-Making with Dependencies and Reverse Connections: Analytical Networks. Librokom, Moscow (2009)

Research and Comparison on Remote Sensing Image Restoration Methods Under the Influence of Microvibration

Jingyu Liu[1](✉), Yufu Cui[1], Yu Wang[2], Hongyan He[2], and Chunrui Zhang[2]

[1] DFH Satellite Co. Ltd, Beijing 100094, China
liujingyubetter@163.com
[2] Beijing Institute of Space Mechanics and Electricity, Beijing 100190, China

Abstract. With the improvement of resolution of remote sensing satellite, the effect of microvibration on the imaging quality of space camera is becoming more and more obvious. In this paper, different restoration methods of remote sensing image under the influence of microvibration are studied and compared. Firstly we analyze the exposure rule and the imaging principle of space camera when the imaging device is moving relative to the target. And a simulation degradation model of space camera imaging under microvibration is established based on the principles. And the simulated degraded image is obtained by simulating the model. Then, a non-blind restoration algorithm of remote sensing image under the influence of microvibration is proposed. The degraded image is restored and processed. The indexes of the restored image are improved by more than 25% and the proposed method is effective. Besides, four kinds of other restoration algorithms are also used to restore the degraded image. And we can get that the proposed non-blind restoration algorithm is better than all four kinds of other restoration algorithms about 33% by comparing the results. Finally, several suggestions are put forward in order to provide reference for remote sensing image restoration technology under the influence of microvibration.

Keywords: Microvibration · Imaging Model · the Non-Blind Restoration Algorithm · Effect Comparison · Development Suggestions

1 Introduction

In recent years, satellite high-resolution imaging technology has been greatly developed. And the resolution of remote sensing satellites has been improved from tens of meters to less than 1 m. China's commercial remote sensing satellites have also entered the sub-meter era, in which SuperView-1 satellite's resolution can reach 0.5 m. But at the same time, it should be recognized that with the improvement of image resolution, the factors which were not important under the previous low resolution conditions will also have a great impact on image quality now. For example, microvibration is one of the important factors affecting the image quality of high spatial resolution remote sensing satellite.

H. P. Urbach and H. Jiang (Eds.): ISSOIA 2022, SPPHY 295, pp. 176–190, 2023.
https://doi.org/10.1007/978-981-99-4098-1_17

At present, some scholars have made some progress in the study of the effects of microvibration. Zhang Bowen and Hu Yongli applied the idea of integrated modeling to analyze the effect of microvibration on high-resolution space camera, as in [1]. Fan Chao studied that TDICCD camera had a higher requirement for satellite flutter amplitude, and with the increase of number of stages, the flutter effect was particularly prominent, as in [2]. Xu Shiwen analyzed the effect from the angle of image shift, as in [3]. Deng Qin of Zhejiang University has established flutter degradation model for the array and TDICCD, and carried out simulation and verification, as in [4]. Zhao Guangxu of Southwest Jiaotong University has studied the influence and compensation of platform flutter on the quality of high-resolution remote sensing images with the support of 973 subprojects, as in [5]. Shi Junxia has used fast array CCD to detect the vibration parameters, as in [6]. Li Haoyang has studied the influence of low-frequency sinusoidal vibration model, as in [7]. Zhuang Xuxia and Li Bo of Harbin Institute of Technology and Peng Xujiao of National Defense University have also analyzed the influence of platform vibration on the image quality from different angles, as in [8–10]. Fu Zhongliang and Xu Zhihai of Zhejiang University [11] and Liao Yongzhong of Central South University [12] also made relevant research on moving image restoration technology. Tong Xiaohua of Tongji University [13, 14], Zhao Hongzhi [15]and Cao Qipeng [16] of DFH Company made some research on microvibration detection and compensation method.

At present, the restoration methods of motion blurred images include improving hardware conditions and software algorithm processing. And the software algorithm processing mainly includes non-blind restoration algorithm and blind restoration algorithm. The non-blind restoration algorithm reconstructs the PSF (Point Spread Function) through vibration parameters which are the prior knowledge. The blind restoration iterative algorithm directly solves the PSF through multiple iterations, without prior knowledge for vibration parameters. Generally, the non-blind restoration algorithm is better than the blind restoration algorithm. In this paper, we mainly combine the imaging process and principle of TDICCD camera to establish a degradation model, and simulate the degraded image. Then, a non-blind restoration algorithm for remote sensing image under the influence of microvibration is proposed. The degraded image is restored and each index of the restored image is improved to a certain extent. At the same time, several different restoration algorithms are used to process the degraded image. The results of non-blind restoration and blind restoration are compared and the proposed non-blind restoration algorithm has better effects.

2 Simulation Model

The camera exposure imaging process can be understood as an integral process in which the detector receives the intensity of light. The process is described by (1) as follows.

$$g(x_0, y_0) = \frac{1}{T}\int_0^T f(x_0, y_0)dt \quad (1)$$

In (1), (x_0,y_0) is the imaging plane coordinates, T is the exposure time, $g(x_0,y_0)$ and $f(x_0,y_0)$ are the image function and object function respectively. If the imaging device is moving relative to the target, the exposure process is described as (2).

$$g(x_0, y_0) = \frac{1}{T}\int_0^T f[x_0 + x(t), y_0 + y(t)]dt \tag{2}$$

The $x(t)$and $y(t)$are motion functions.

TDICCD increases the sensitivity of the system by time delay integration and multi-level energy accumulation. The detailed working principle diagram is shown in Fig. 1.

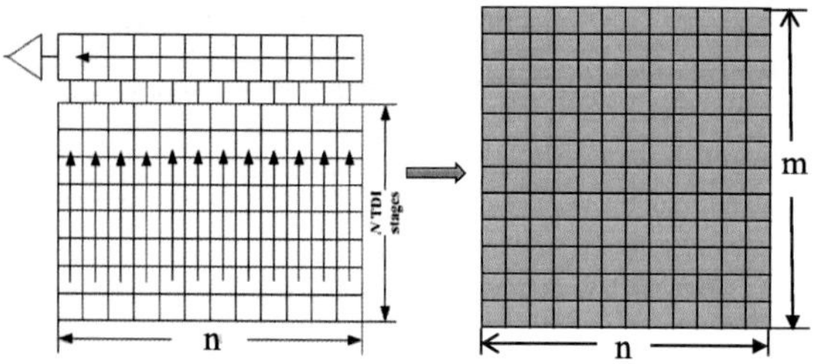

Fig. 1. Working principle diagram of TDICCD

TDICCD is a plane array CCD. There are N × n pixels in total. N represents the number of stages, and n represents the number of pixels in the vertical direction of TDI. Generally n is much larger than N. The right side is the obtained remote sensing image whose size is m×n. TDICCD uses pushing-broom imaging method to output data line by line. In the pushing-broom process, the charge is transferred and accumulated between different stages of CCD. And the charge transfer rate is consistent with the image scanning rate, so that the charges from different stages of CCD imaging the same scene can be accumulated and output to get a row of scene images. At the same time, the ground objects can be imaged and output line by line by pushing broom to get the remote sensing image on the right side.

Combining with the exposure imaging rule when the imaging device is moving relative to the target and the imaging process of TDICCD, the imaging rule of TDICCD under the influence of microvibration can be analyzed and studied. The analysis shows that the time range when the k_{th} stage of TDICCD is scanning to the i_{th} line of image is [(i + k-2) T, (i + k-1) T]. And the magnitude of the corresponding microvibration in this integral time can be known, so that the effect of microvibration can be considered in the integral process. The simulation process is shown as (3).

$$g(x_0, y_0) = \sum_{k=1}^{N}\int_{(i+k-2)T}^{(i+k-1)T} f[x_0 + x(t), y_0 + y(t)]dt \tag{3}$$

T is the imaging line period, and *x(t), y(t)* are the influence function of microvibration. And the simulated degradation image under the influence of microvibration can be calculated by (3). For discrete digital images, the corresponding pixel gray value is calculated by bilinear interpolation method when the sampling point does not fall on the center of the original input image pixels. Simpson integral method is used to calculate the integral operation in each line period T. As shown in (4), a and b are the upper and lower bounds of the integral operation, f (x) is the original integral function, and P (x) is the interpolation polynomial function.

$$\int_a^b f(x)dx \approx \int_a^b P(x)dx = \frac{b-a}{6}[f(a) + 4f(\frac{a+b}{2}) + f(b)] \tag{4}$$

3 Simulation Results

According to the simulation model in Sect. 2, a clear original image of St. Paul's commercial street with a resolution of 0.3 m is simulated on the platform of MATLAB. The size of the street image is 128 × 512. The parameters of the satellite and camera are shown in Table 1 and the parameters of the microvibration are shown in Table 2. The simulation results are shown in Fig. 2, and Fig. 3 is the local enlarged drawing at the circle of Fig. 2.

Table 1. Parameters of satellite and camera

Size of horizontal pixel/um	Size of vertical pixel/um	Resolution/m	Orbit height/km	Number of stage
6	6	0.3	600	48

Table 2. Parameters of microvibration

Amplitude/um	Frequency/Hz	Initial phase/°	Angle with vertical flight direction/°
20	50	0	15

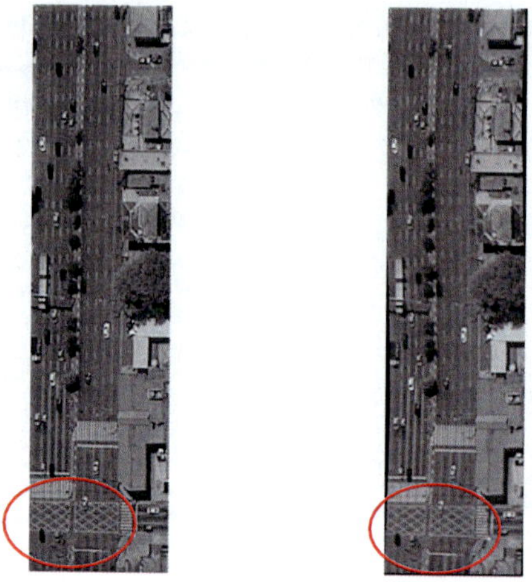

(a) (b)

Fig. 2. Simulation results. (a) the original image (b) the simulated degraded image.

(a) (b)

Fig. 3. Local enlarged drawing. (a) the original image (b) the simulated degraded image.

The contrast of the original image is 374.64 and the clarity is 1803.2. The contrast of the degraded image is 233.31 and the clarity is 837.69. The MSE of degraded image is 854.87, the PSNR is 18.81 and the SSIM is 0.5871. Compared with the original image, the quality of the degraded image under the influence of microvibration decreases, which is consistent with the theoretical analysis.

4 Image Restoration

4.1 Proposed Non-Blind Restoration Algorithm

In this paper, a non-blind restoration algorithm is proposed to restore the degraded image.

Uniform linear motion is a simple form of motion. The calculation of motion fuzzy PSF under uniform linear motion has been studied [17]. It is assumed that the current image only moves uniformly in a straight line in the X direction with a given equation $X(t) = V_0*t = at/T$. When t equals T, the displacement reaches the maximum value a. The PSF of uniform linear motion can be obtained as (5).

$$h(x) = 1/a, 0 < x < a \tag{5}$$

For complex motion, it can be considered as the combination of multiple uniform linear motions shown as in Fig. 4. And Eq. (5) can be improved and extended.

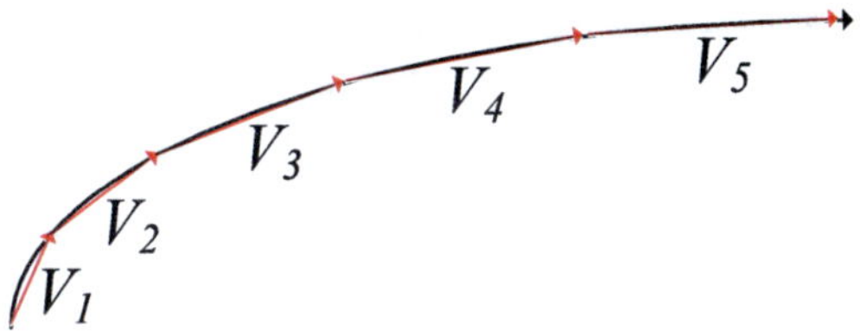

Fig. 4. Complex motion is considered as the combination of multiple uniform linear motions.

Equation (5) can be transformed to (6).

$$h(x) = \frac{1}{a} = \frac{1}{T \bullet V_0} \tag{6}$$

And Eq. (6) can be improved and extended to (7) for complex motion, where V(t) is variable speed. Since there is a corresponding relationship between X and t, V(t) can be transformed to V(x).

$$h(x) = \frac{1}{T \bullet V(\mathrm{t})} = \frac{1}{T \bullet V(x)} \tag{7}$$

Considering that the relative motion may be alternating motion, that is, it passes through the same position many times in the exposure process such as sinusoidal motion. The Eq. (7) should be further improved to (8). $V_i(x)$ is the speed when the imaging device passes through the corresponding position for the i_{th} time. The subscript q is proposed because the imaging device may move back and forth. q is total times for moving back and forth.

$$h(x) = \sum_{i=1}^{q} \frac{1}{T \bullet V_i(x)} \tag{8}$$

The PSF for two-dimension is extended to be calculated by (9).

$$h(x, y) = \sum_{i=1}^{q} \frac{1}{T \bullet V_i(x, y)} \tag{9}$$

The application of non-blind restoration algorithm should determine prior knowledge firstly. Because of the randomness of in-orbit microvibration, the prior knowledge can be obtained by some microvibration measurement methods.

The traditional concept usually considers the correlation between speed and time. The core of the non-blind restoration algorithm proposed in this paper is to establish the relationship between position and speed, and then calculate the PSF. In this paper, the common sinusoidal motion is taken as an example for calculation and analysis. For other types of working conditions, similar methods can also be used for calculation (Tables 3 and 5).

Table 3. Related parameters

Amplitude/um	Frequency/Hz	Phase/°	Exposure time/ms	Size of pixel/um
24	100	0	5	6*6

According to the above method, the following equations can be obtained for sinusoidal vibration.

$$\Rightarrow V(t) = \omega A \cos(\omega t + \varphi) \tag{10}$$

$$\Rightarrow (\frac{x}{A})^2 + (\frac{V}{\omega A})^2 = 1 \tag{11}$$

$$\Rightarrow V(x) = \omega\sqrt{A^2 - x^2} \tag{12}$$

(13) can be obtained by substituting those parameters.

$$V(x) = 200\pi\sqrt{(24/6)^2 - x^2} \tag{13}$$

The results are obtained as Fig. 5, Fig. 6 and Table4.

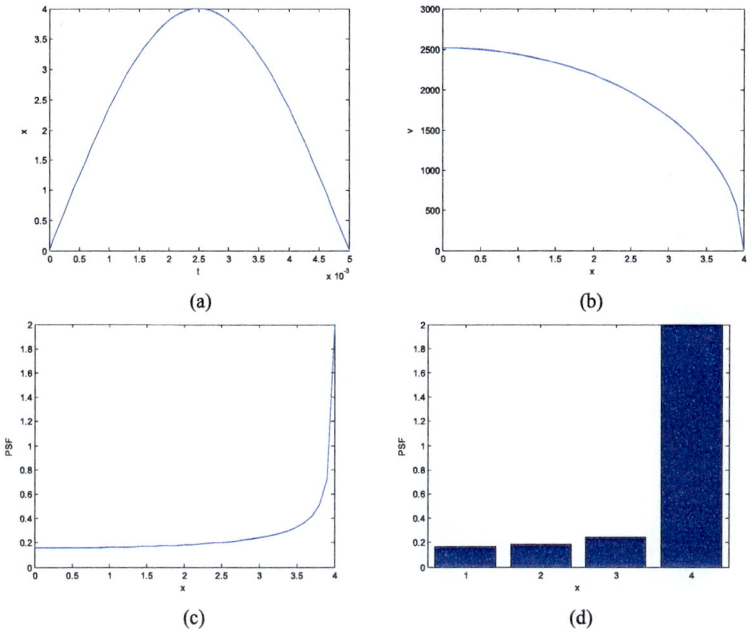

Fig. 5. The implementation steps of obtaining the fuzzy process PSF. (a)relation between velocity and time(b) relation between velocity and displacement (c)continuous PSF (d)discrete PSF

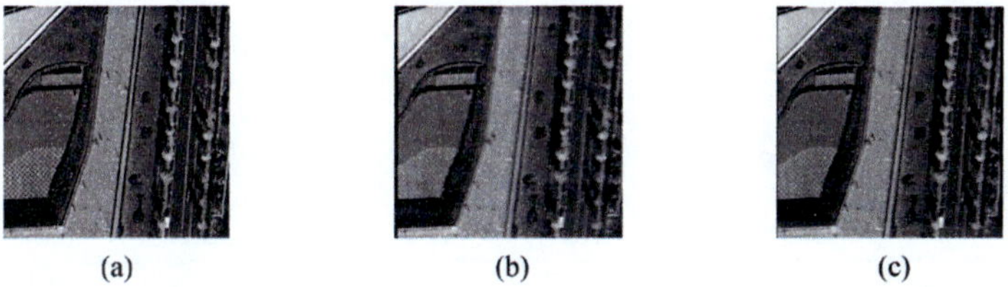

Fig. 6. Simulation results of non-blind restoration algorithm. (a)the original image; (b)the degraded image; (c)the restored image

Table 4. Index comparison between original, degraded and restored images

Parameters	MSE	PSNR	SSIM	Clarity	Contrast
the original image	–	–	–	1834.1	401.31
the degraded image	934.03	18.43	0.7389	338.64	116.31
the restored image	242.31	24.29	0.9320	540.09	163.99

As can be seen from Fig. 6 and Table 4, compared with the degraded image, the non-blind restored image has a certain improvement in each index by more than 25%, and the method is effective.

4.2 Other Restoration Algorithms

A Non-blind Restoration Algorithm Based on Laplace Gradient Prior Knowledge. The non-blind restoration algorithm based on Laplace gradient prior knowledge uses a single exponential distribution to constrain every element in the fuzzy kernel shown as (14).

$$h(x) = \lambda \exp(-\lambda x) \tag{14}$$

λ represents the power exponent of the model (Fig. 7).

Fig. 7. Restoration images with Laplace algorithm

Table 5. Restoration indexes with Laplace algorithm

Index	Contrast	Clarity	MSE	PSNR	SSIM
Restored image	119.63	367.82	881.31	18.68	0.7482

Direct Iterative Blind Deconvolution Algorithm
The direct iterative blind deconvolution algorithm is a common method among blind restoration algorithms. It is an algorithm based on fast Fourier transform. The main conversion formula is as follows.

$$G(u, v) = F(u, v) \bullet H(u, v) + N(u, v) \tag{15}$$

For this algorithm, 100, 200 and 500 iterations were performed respectively, and the restored image and index results are shown in Fig. 8 and Table 6.

Iterative Blind Restoration Algorithm Based on Coefficient Tradeoff
Based on the direct iterative algorithm in the above section, the parameters for tradeoff can increase the convergence of the algorithm.

(a) 100 iterations (b) 200 iterations (c) 500 iterations

Fig. 8. Restoration images with different iterations

Table 6. Restoration indexes with different iterations

Iteration times	Contrast	Clarity	MSE	PSNR	SSIM
100	334.38	2555.7	1309.6	16.96	0.6253
200	166.24	633.1	1218.9	17.27	0.6523
500	142.58	472.47	1163.3	17.47	0.6667

Three different constants of α are selected for simulation. The algorithm all converges within 10 iterations and converges to the same result. The corresponding parameters α and β are shown in Table 7, the restored image is shown in Fig. 9, and the index results are shown in Table 8.

Table 7. Results of parameter α and β

α	0.3	0.6	0.8
β	0.3	0.15	0.05

Table 8. Restoration indexes of coefficient tradeoff algorithm

Index	Contrast	Clarity	MSE	PSNR	SSIM
Restored image	82.06	213.42	1214.9	17.29	0.7048

Fig. 9. Restoration image of coefficient tradeoff algorithm

Simulated Annealing Algorithm

In the blind image restoration algorithm, the restoration algorithm is transformed into a function optimization problem, and the simulated annealing algorithm is used to solve the optimal solution. In this problem, we find f and h to minimize the cost function Q in formula (16).

$$Q(f,h) = \sum_{i,j} [f(i,j) * h(i,j) - g(i,j)]^2 / \sum_{i,j} [g(i,j)]^2 \tag{16}$$

The simulated annealing algorithm is used to solve the problem. The annealing temperature curve is shown in Fig. 10, the optimization function Q curve is shown in Fig. 11, the recovered image is shown in Fig. 12, and the recovered image index results are shown in Table 9. It can be seen from the curves in Fig. 10 and Fig. 11 that when the outer loop is about 40 times, the calculation results of the algorithm have converged and the curve tends to be stable.

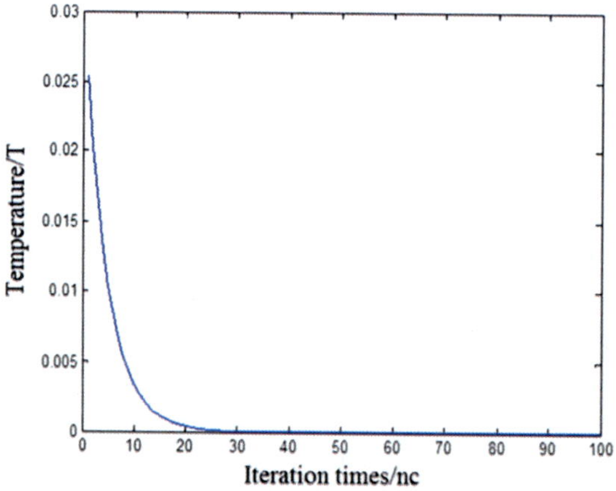

Fig. 10. Annealing temperature curve

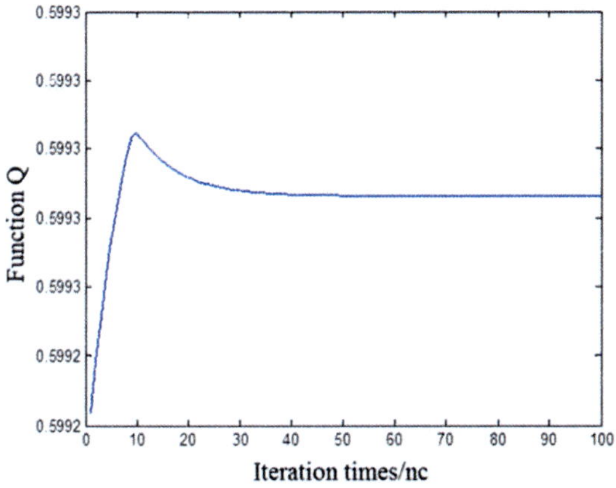

Fig. 11. Curve of optimization function Q

Fig. 12. Restoration image of simulated annealing algorithm

Table 9. Restoration indexes of simulated annealing algorithm

Index	Contrast	Clarity	MSE	PSNR	SSIM
Restored image	540.82	4359.7	1417.4	16.62	0.6998

4.3 Restoration Algorithm Comparison

In this chapter, a non-blind restoration algorithm is proposed. At the same time, four different other restoration algorithms are used for image restoration. The comparison of the indexes of the five restoration methods is shown in Table 10.

It can be seen from the comparison between the restored images of each algorithm in this chapter that the visual effect of the restored images of the proposed non-blind restoration algorithm is better than that of other restoration algorithms. In terms of image indexes, MSE, PSNR and SSIM of proposed non-blind restored images are much better than those of other restored images, and the indexes are better by 33%. And clarity and contrast of some blind restored images appear to be larger because of ringing effect, which leads to wavy stripes in the restored images. These two indexes of such images are not meaningful for reference. In addition, some indexes of the image restored by

Table 10. Comparison of five kinds of restoration algorithm

Different images		Contrast	Clarity	MSE	PSNR	SSIM	Calculation time/s
Original image		401.31	1834.1	—	—	—	—
Degraded image		116.31	338.64	934.03	18.43	0.7389	—
Proposed Non-blind restoration algorithm		163.99	540.09	242.31	24.29	0.932	0.24
Laplace algorithm		119.63	367.82	881.31	18.68	0.7482	0.26
Direct Iterative Blind Deconvolution Algorithm	100	334.38	2555.7	1309.6	16.96	0.6253	1.45
	200	166.24	633.1	1218.9	17.27	0.6523	1.84
	500	142.58	472.47	1163.3	17.47	0.6667	3.95
Iterative Blind Restoration Algorithm Based on Coefficient Tradeoff		82.06	213.42	1214.9	17.29	0.7048	0.98
Simulated Annealing Algorithm		540.82	4359.7	1417.4	16.62	0.6998	33.6

some blind restoration algorithms are even worse than those of the degraded image, and the restoration effect is poor. At the same time, the blind restoration algorithm involves iterative operation, and its calculation time is 4 times or 16 times even 140times as the non-blind restoration algorithm for the same size image. Considering visual effect, image evaluation indexes and calculation time of image restoration, the proposed non-blind restoration algorithm is superior to other restoration algorithms.

5 Conclusion

In this paper, we mainly study the image restoration method under the influence of microvibration. Firstly, by combining the exposure imaging rule when the imaging device and the target have relative motion and the imaging principle of TDICCD camera, the imaging model of TDICCD camera under microvibration is established, and the degraded image is obtained through simulation. Then, a non-blind restoration algorithm of remote sensing image under the influence of microvibration is proposed. The degraded image is restored, and each index of the restored image is improved by more than 25%. Finally, four other restoration algorithms are used to process the degraded image, and the results of the five methods are compared. The proposed non-blind restoration algorithm has a better image restoration effect by more than 33%.

According to the relevant research on image restoration technology under the influence of microvibration, the following suggestions are put forward in order to have some reference significance in this direction.

(1) In view of the improvement of image quality under the influence of microvibration, it is suggested that the mechanical analysis and suppression from the front end should be combined with the image analysis and processing from the back end. And the best effect can be achieved by combining two approaches.
(2) When conducting microvibration experiments on the ground, it is recommended to measure the relevant parameters of microvibration. At the same time, the target image shall be taken by the camera payload accordingly. And the corresponding vibration data and image data shall be archived. The model establishment and restoration method can be studied by ground experiment data.
(3) At present, some satellites have acceleration measuring devices to measure microvibration data on orbit. And other satellites will be also installed by such devices to obtain microvibration data with time tag on orbit. At present, these data are not fully utilized. It is recommended to make full use of these data in the future, so as to study the model establishment and recovery methods better.

References

1. Zhang, B.W., Wang, X.Y., Hu, Y.L.: Integrated analysis on effect of micro-vibration on high resolution space camera imaging. Spacecraft Recovery Remote Sens. **33**(2), 60–66 (2012)
2. Fan, C., Li, Y.C., Yi, H.W.: Influence analysis of buffeting on image quality of TDICCD camera. Acta Photonica Sinica **36**(9), 1714–1717 (2007)
3. Xu, S.W., Yao, X.C., Fu, L.: An analysis for the influence of image method. Opto-Electr. Eng. **26**(4), 60–63 (1999)
4. Deng, Q.: Research of motion estimation algorithm based on digital images. Zhejiang University, Hangzhou (2013)
5. Zhao, G.X.: The camera platform flutter analysis of the influence of high-resolution optical remote sensing imaging. Southwest Jiaotong University, Chengdu (2015)
6. Shi, J.X., Xue, X.C., Guo, Y.F.: Effect of satellite vibration on imaging quality of TDICCD camera and compensation method. Opto-Electr. Eng. **37**(12), 11–16 (2010)
7. Li, H.Y., Liu, Z.J., Xu, P.M.: Influence of satellite Jitter on imaging quality of space-born remote sensing systems. Comput. Simul. **29**(1), 25–27 (2012)
8. Zhuang, X.X.: The effect analysis and simulation of platform motion on image quality of spaceborn TDICCD camera. Harbin Institute of Technology, Hei longjiang (2011)
9. Li, B.: Analysis of random vibration on image quality degradation of TDICCD camera. Harbin Institute of Technology, Hei longjiang (2013)
10. Peng, X.J.: Image degradation modeling due to satellite platform vibration and blur parameter identification. National University of Defense Technology, Changsha (2013)
11. Fu, Z.L.: Restoration of two-dimensional motion blurred image. Zhejiang University, Hangzhou (2010)
12. Liao, Y.Z., Cai, Z.X., He, X.H.: Fast algorithm for motion blurred image blind deconvolution. Opt. Precis. Eng. **21**(10), 2688–2694 (2013)
13. Tong, X.H., Ye, Z., Liu, S.J.: Essential technology and application of jitter detection and compensation for high resolution satellites. Acta Geod. Cartographica Sin. **46**(10), 1500–1508 (2017)
14. Tong, X.H., Ye, Z., Liu, S.J.: Jitter detection for high resolution satellites based on phase correlation with local frequency analysis. Acta Geod. Cartographica Sin. **47**(10), 1346–1352 (2018)

15. Zhao, H.Z., Huang, J.J., Cao, Q.P.: Method of measuring image offset caused by onboard micro-vibration. Spacecraft Eng. **23**(5), 35–40 (2014)
16. Cao, Q.P., Liu, J.Y., Ye, Z.: Micro-vibration measurement method of high-resolution space camera based on target image. J. Nanjing Univ. Aeronaut. Astronaut. **51**(S), 83–87(2019)
17. Zhao, Y.Y., Yuan, Y., Su, L.J.: Point spread function estimation of blurring due to uniform linear motion in arbitrary direction. Chin. J. Lasers **39**(8), 1–7 (2012)

Status and Progress of Research on HgCdTe Photovoltaic Infrared Detectors

Chen Xin(✉), Chen Yong, Lv Wei Dong, Zhang Jiu Shuang, and Chen Lei

Beijing Institute of Space Mechanics & Electricity, Beijing 100094, China
920353389@qq.com

Abstract. In the field of infrared remote sensing, cadmium mercury telluride photovoltaic infrared detectors play a crucial role in the performance of the entire infrared camera optical system. HgCdTe photovoltaic detectors have the advantages of high quantum efficiency and low dark current. This paper firstly describes the working mechanism of the HgCdTe photovoltaic detector. When the pn junction is exposed to light, the electron-hole pairs generated in the p and n regions break the equilibrium of the original pn junction, and the electrons generated in the p region and the holes generated in the n region move under the action of the internal electric field, generating a photogenerated current. The dark current model is then analysed. The dark current originates from four main mechanisms, diffusion currents in the n and p regions (I_{diff}), complex currents generated in the depletion region (I_{gr}), trap-assisted tunneling currents (I_{tat}), and direct tunneling currents in the band (I_{bbt}). Finally, the current status and development direction of HgCdTe detectors at home and abroad are introduced.After the development of TDI small line array detectors and condensed surface array technology, large format, very long wave and multi-colour detection are still the current development direction of HgCdTe detectors.

Keywords: HgCdTe photovoltaic detector · dark current · thin film growth

1 Introduction

Materials used for infrared detectors in recent years are HgCdTe, InSb, InGaAs, Si:X, QWIP and InAs/GaSbT$_2$ SL, of which HgCdTe is a ternary compound, an alloy of CdTe and HgTe ratios [1]. It is an ideal infrared detector material with a large adjustable range, and the forbidden band width can cover an energy range of 0.1–1.0eV with the change of material components, which can develop infrared detectors with high quantum efficiency, long carrier life and high operating temperature [2].The indirect composite (Shockley Read Hall, SRH) of HgCdTe has a long carrier lifetime and low dark current, making it ideal for room temperature operation. Mercury cadmium telluride ($Hg_{1-x}Cd_xTe$, x is the material component) detectors, which came into being with the invention of HgCdTe materials, are now technologically well developed, high performance detectors and occupy a central position in aerospace infrared remote sensing applications.

The working mechanism of the photovoltaic detector is that when the pn junction is illuminated, the electron-hole pairs generated in the p and n regions break the equilibrium

H. P. Urbach and H. Jiang (Eds.): ISSOIA 2022, SPPHY 295, pp. 191–206, 2023.
https://doi.org/10.1007/978-981-99-4098-1_18

of the original pn junction, and the electrons generated in the p region and the holes generated in the n region move under the action of the internal electric field, generating photogenerated currents, so the photovoltaic detector has a very high optical sensitivity and high detection rate [3].

2 Theoretical Foundations

Photovoltaic detectors are devices based on the photovoltaic effect. The photovoltaic effect is present in pn junctions formed by the same semiconductor material, in heterojunctions formed by two semiconductors with different forbidden band widths and in Schottky junctions formed by metal-semiconductor contacts.

When the device works in the open circuit state, the pn junction type photoelectric device produces photovoltaic effect, this working state is called photovoltaic working mode; if the device works in the reverse bias state, no light when the pn junction resistance is very large, the current is very small; when there is light, the junction resistance becomes small, the current becomes large.

In the absence of light, the current flowing through the pn junction can be expressed as follows

$$I = I_0 e^{qV/\beta kT} - I_0 \tag{1}$$

where k is the Boltzmann constant, T is the thermodynamic temperature, $I_0 e^{qV/\beta kT}$ denotes the diffusion current of the majority carriers into the other region and I_0 denotes the drift current of the minority carriers in the junction region (reverse saturation current); v is the voltage across the p-n junction.

In the presence of light and an applied voltage of V, the current flowing through the p-n junction can be expressed as follows

$$I = I_0(e^{qV/\beta kT} - 1) - I_p \tag{2}$$

where I_p is the photocurrent, the direction of which is opposite to the p-n junction current, and its magnitude is related to the irradiation power, irradiated area, quantum efficiency and other factors.

If the pn junction is short-circuited and V is 0, the short-circuit photocurrent is obtained as

$$I = -I_p \tag{3}$$

If the pn junction is disconnected and I is 0, the open-circuit optical voltage is obtained as

$$V_{oc} = \frac{kT}{q} \ln(\frac{I_p}{I_0} + 1) \tag{4}$$

The current of the p-n junction increases rapidly with the applied voltage at the applied forward voltage, and with the applied reverse voltage at the pn junction, the current tends to a saturation value I_0; at the same applied voltage, the current increases with the increase of the light intensity, as Fig. 1 shown. The specific mechanism is that when the pn junction is irradiated by light, the energy of photons hυ is greater than the forbidden band width E of the material$_g$, the pn junction produces numerous electron-hole pairs. p-zone depletion layer produces electron-hole pairs, photogenerated electrons move to the n-zone under the action of the internal electric field, so the n-zone accumulates negative charge, while the p-zone depletion layer also produces electron-hole pairs, photogenerated electrons can diffuse to reach the depletion layer, and then reach the n-zone under the action of the internal electric field The photogenerated electrons can diffuse to the depletion layer and then reach the n-region under the action of the internal electric field. Similarly, electron-hole pairs are generated in the n-region depletion layer and outside the depletion layer under infrared radiation, and photogenerated holes reach the p-region under the action of the internal electric field, which accumulates a positive charge. The resulting photogenerated electron-hole forms a current opposite to the internal electric field, producing an electric potential called the photogenerated electric potential, an effect known as the photovoltaic effect.

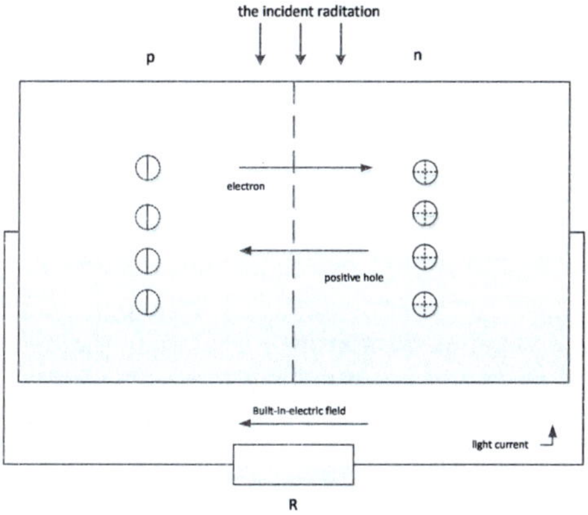

Fig. 1. Schematic diagram of photogenerated current of p-n junction [4]

2.1 Photovoltaic Model

Photovoltaic infrared detectors can be divided into tabletop and flat infrared detectors according to their structure, as Fig. 2 shown here. The photovoltaic detector is formed by interconnecting a photovoltaic sensitive element array chip with a silicon readout circuit. The main function of the photovoltaic sensitive element array chip is to convert the incident infrared radiation signal into an electrical signal, while the silicon readout circuit stores, converts and outputs the collected electrical signal, simplifying the signal processing process by direct coupling between the two [5]. The direct coupling of the two simplifies the signal processing process.

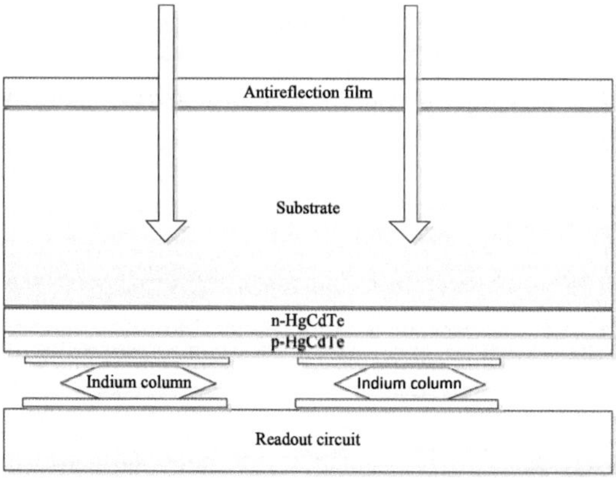

Fig. 2. Schematic diagram of the structure of a photovoltaic MCT infrared detector

2.2 Dark Current Characteristics

Dark currents in photovoltaic devices can be divided into four mechanisms: diffusion mechanisms, compound generation mechanisms, tunneling mechanisms and surface leakage mechanisms.

In recent years, the magnitude of the dark current has been reduced by in situ doping techniques. In situ doped p-on-p heterojunction materials can avoid the damage introduced by ion injection and can effectively reduce the p-n junction depletion region as well as the surface generation-composite current, and the defect-assisted tunneling centre density; through the reduction of the energy band structure, the direct chance of tunneling can be greatly reduced, and the tunneling current is reduced accordingly. The reduction of the energy band structure can greatly reduce the chance of direct tunneling and correspondingly the tunneling current. The p + -on-n double-layer heterojunction material prepared by Tian Zhen et al. of the North China Institute of Optoelectronics Technology and obtained by wet etching, bench etching and passivation processes for CdTe p + -on-n long-wave heterojunction bench-type devices can effectively overcome the problems of low oligon lifetime and have lower dark current values in the long-wave and very long-wave bands [6].

Temperature is the main factor affecting the dark current. Shanghai Institute of Technology and Physics will p-type HgCdTe material x component 0.49, through the MBE method to grow to obtain Si-based HgCdTe epitaxial film material, after cleaning, etching, ion injection to form n-type area after the surface growth of 400 nm, ZnS film as a passivation layer, and then through the lithography process to form p/n electrode, after the test, obtained at room temperature, the dark current is mainly diffusion current and generate compound current, and the dark current of conventional structure A is the largest [7]; at low temperature, the dark current is mainly manifested as tunneling current, and the dark current of conventional structure A is the smallest instead.

The dark current is also extremely closely related to the bias voltage at positive bias, the dark current is dominated by the generated compound current at low temperature, and the diffusion current increases as the temperature rises, determined by both the generated compound current and the diffusion current; near zero bias, at 60 K, the trap-assisted tunneling current dominates, and at 80 K, it is determined by both the trap-assisted tunneling current and the generated compound current, and the contribution of the generated compound current increases as the temperature rises, to 110 K when the generated The complex current dominates the dark current at reverse bias, with the trap-assisted tunneling current dominating the dark current and the direct tunneling current having a weak effect on the dark current. pn junction voltage-current characteristics are shown in the following Fig. 3. The voltage-current characteristics of the pn junction are shown below. The horizontal coordinate is the detector bias voltage V_{DET} and the vertical coordinate is the detector current I. The graph shows that the detector bias voltage will have a direct effect on the detector output current, if the bias voltage is too large at both ends, the detector will break down and not work properly. The detector current is mainly composed of dark current I_{dark} and photogenerated current I_{light}. When the bias voltage at both ends of the detector $V_{DET} = 0$, the dark current $I_{dark} = 0$. In order to reduce the dark current generated by the detector as much as possible and thus reduce the power consumption of the circuit, the readout circuit unit provides a stable zero bias voltage for the detector which should be the optimal result to be achieved by the design [8].

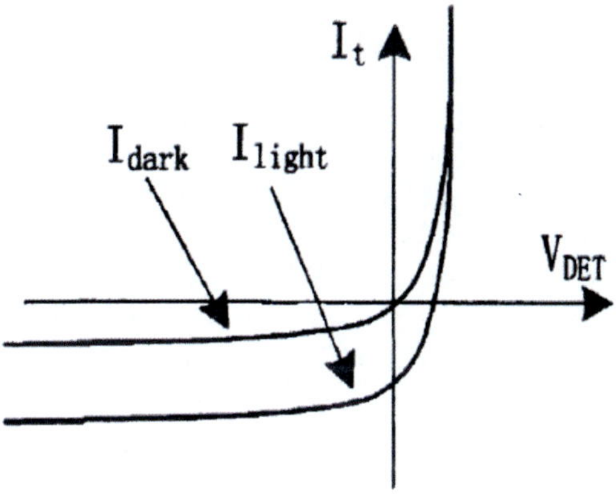

Fig. 3. HgCdTe PN junction I-V characteristic curve [9]

3 Film Growth Methods

The substrate materials available for thin film growth are CdTe, sapphire (Al_2O_3), GaAs, Ge, Si, etc. $Cd_{l\text{-}y}$ Zn_y Te (CZT) crystals are currently recognized as a good substrate for lattice matching with HgCdTe, and the difficulty in preparing thin film materials lies in the difficulty in obtaining the ideal CdZnTe substrate material. Research has been devoted to alternative substrates such as PACE (Producible Altemative to CdTe for Epitaxy)-I(HgCdTe/CdTe/Sapphire), PACE-II (HgCdTe/CdTe/GaAs) and PACE-III (HgCdTe/CdTe/Si).

Thin film growth methods include liquid phase epitaxy (LPE) growth, molecular beam epitaxy (MBE) liquid phase epitaxy, metal organic compound vapour deposition (MOCVD) and pulsed laser deposition (PLD), while LPE growth of HgCdTe has matured, making it possible to produce HgCdTe on an industrial scale. The MBE or MOCVD method can produce complex HgCdTe multilayer heterojunction materials, which are currently necessary for the growth of two-colour and multi-colour devices. PLD is a new thin film preparation technique that has shown its unique superiority in the deposition of infrared narrow forbidden band films.

3.1 LPE Method [10]

The LPE technique for growing HgCdTe films is currently the most mature and widely used technique for fabricating HgCdTe films, requiring a strictly lattice-matched substrate. Continuous production is also possible as the same masterbatch can be reused several times. HgCdTe liquid-phase epitaxial films are usually grown on CdTe or CdZnTe substrates using the Te or Hg-rich masterbatch method, using H2 as the carrier gas. There are several methods such as slider, dip, tilt and rotary, all with growth temperatures around 500 °C. A closed-tube low-temperature heat treatment of the grown primary wafers in an Hg-saturated atmosphere is usually used to tune the stoichiometric ratio of the film and improve the electrical properties of the film in order to make it better suited to the requirements of infrared devices.

The HgCdTe 2D arrays are made from LPE as-grown P + -on-n heterojunction structure on CdZnTe substrate as shown in Fig. 4. The narrow band-gap base layer (n-) is doped with indium, and the wide band-gap cap layer (P +) is doped with arsenic. Processing steps include mesa formation, surface passivation, p-metal and n-metal contacts, under-bump metal, In-bump, epoxy backfill, hybridization, backside polishing, and AR coating [11].

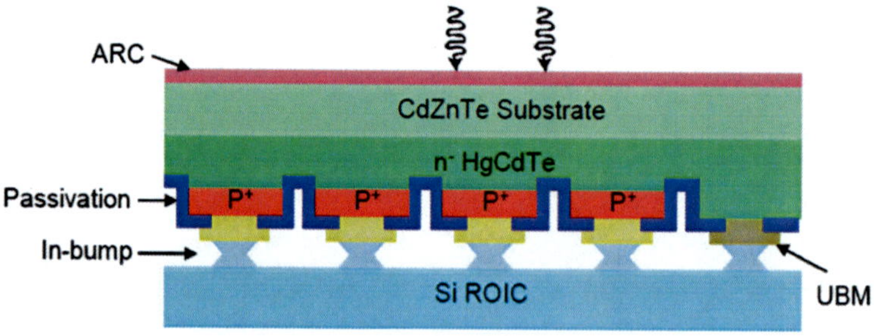

Fig. 4. Cross-section of HgCdTe FPA structure.

3.2 MBE Method

The MBE method can compensate for the difficulties of LPE in heterojunction HgCdTe materials and is the core technology for the preparation of third generation HgCdTe focal plane devices, and the component ratios as well as the doping concentration can be very easily regulated using the MBE method. During MBE growth, indium (In) is typically injected as the sender for n-type doping and arsenic (As) is injected as the recipient for p-type doping. The use of buffer layers is essential when the MBE method uses lower cost Si, Ge or GaAs substrates, and the use of lattice-matched CdZnTe buffer layers can reduce the problem of secondary lattice matching. Overall, the quality of the HgCdTe material reaches the level of the material produced by the LPE method, regardless of whether the material is grown as a single layer or as a multilayer [12]. Figure 5 shows a simplified schematic of a vertical MBE system (the normal to the substrate wafer surface is vertical) where sources of Hg, CdTe, and Te are evaporated (sublimated) towards the substrate with their beam fluxes adjusted to give the correct mole fraction for the epitaxial HgCdTe layer.

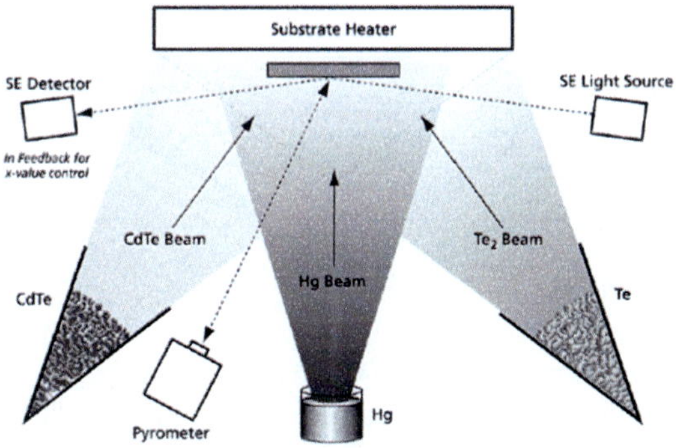

Fig. 5. Simplified schematic of the HgCdTe MBE process

MBE technology allows growth at temperatures below 200 °C, which is lower than LPE (around 500 °C) and MOCVD (350 °C), thus reducing the impact of undesired thermal activation phenomena such as diffusion, improving the control of the stoichiometric ratios in the growth layer and reducing the number of undesired activation phenomena [13]. The control of stoichiometric ratios and the reduction of uncontrollable metal sublattice doping vacancies are therefore reduced. Due to the slow growth rate, the thickness of the layer can be precisely controlled and very thin films can be prepared. It is also possible to change the composition or doping of the layer by varying the molecular beam source, which provides flexibility. However, the MBE method has the disadvantage of high input, low output and difficulty in achieving mass production when used for the preparation of HgCdTe. Moreover, it can only be carried out in the lower temperature range, which does not guarantee the chemical proportion of the material, and the higher Hg air pressure required for preparation is not allowed by the MBE equipment. Figure 6 shows a photo of MWIR HgCdTe on 8-inch Si wafer that Raytheon made. The wafer surface appears specular across the entire area, indicating good quality. They grew six 8-inch-diameter MWIR HgCdTe wafers on Si substrates and measured within-wafer and wafer-to-wafer variations of key parameters such as cut off wavelength, HgCdTe thickness, macrodefect density including voids and microvoids, and arsenic doping uniformity.

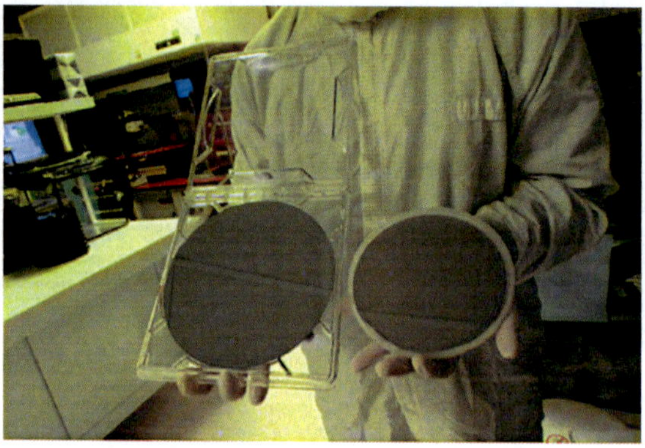

Fig. 6. HgCdTe epitaxial layer on 6-inch and 8-inch wafers [14].

3.3 MOCVD Method

MOCVD is a method for the epitaxial growth of semiconductor compounds by decomposing the compounds using heat. Initially, the preparation of HgCdTe films by MOCVD was achieved using dimethyl cadmium (DMCd) and diethyl tellurium (DETe) at a substrate temperature of 410 °C. However, this method has major drawbacks, mainly in the large difference between the pyrolysis temperatures of DMCd and DETe, with DETe having a much higher pyrolysis temperature. Figure 7 shows the scheme of the gas delivery installation and the reactor of AIX 200 MOCVD system adapted for HgCdTe epitaxial growth. An arsenic precursor,tris-dimethy laminoarsenic (TDMAAs), has been used for the acceptor doping and it provides control over doping concentration range between $10^{14}$and 5×10^{17} cm^{-3}.An iodine precursor, ethyl iodide (EI), has been used for donor doping and it provides control over doping range from 10^{14} to $1 \times$ 1018 cm^{-3}. The special partition in the reactor separates As precursor from DMCd one to prevent premature reaction [15].

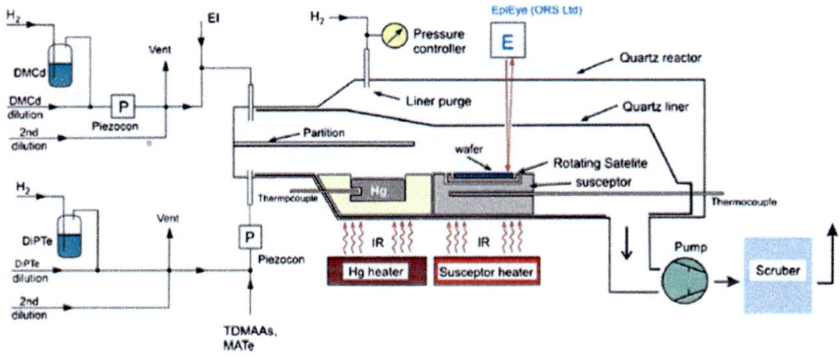

Fig. 7. The scheme of the gas delivery installation and the reactor in AIX 200 MOCVD system adapted to HgCdTe growth [15].

3.4 PLD

PLD is a new thin film preparation technology developed in the late 1980s, which is a high power pulsed laser beam generated by a pulsed laser focused on the surface of the target material to generate high temperature and fusion on the surface of the target material, and further generate high temperature and high pressure plasma ($T \geq 10^4$ K) [16], this plasma directed local expansion emission and deposition on the substrate and thin films are formed. Initially pulsed laser deposition was successfully used for the preparation of high temperature superconductors of polyoxides represented by Yb_2 Cu_3 $O_{7\text{-}x}$. Subsequently, it has gradually been used for the preparation of thin films of multiple compounds, organic films, more difficult to synthesise materials and nanoparticles. It is characterised by high deposition rates, low substrate temperatures and the fact that the film components can be very close to those of the target material.

4 Current Situation at Home and Abroad

The development of detectors has gone through four stages, as shown in Fig. 8 below. After the first generation of line array detectors, the second generation of photovoltaic multiple line array detectors and two-dimensional surface array infrared focal plane detectors, and the third generation of large-scale, highly integrated, high-resolution, low-cost and lightweight detectors, the era of fourth generation detectors with more wavelength bands, optical polarisation information and optical phase information was ushered in. At the same time, research into artificial microstructures, artificial intelligence and neuromorphic computing is gradually emerging, advancing the development of optical modulation of infrared radiation, on-chip intelligence and chip-level integration of traditional infrared photodetectors. Many new concepts, technologies and devices have emerged that are highly cutting-edge and forward-looking [17].

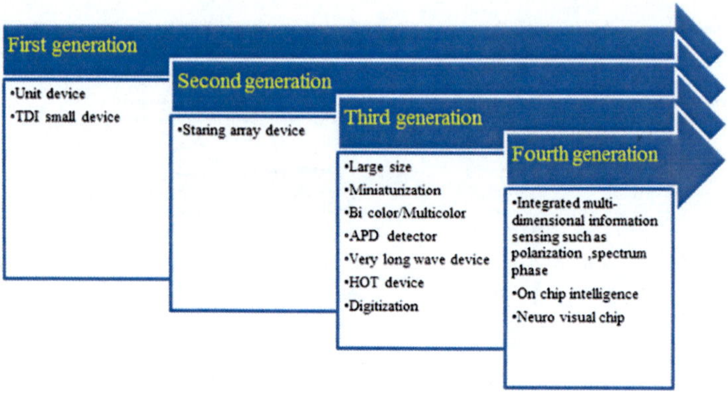

Fig. 8. History and future trends of infrared detectors [18–21]

4.1 Status of Development Abroad

The main foreign companies researching HgTe infrared detectors are the French Sofradir company, the American Teledyne company, the American Raytheon company, the British SELEX company and so on. In the US, UK, France and other developed countries, monochromatic HdCdTe infrared focal plane technology has basically reached a mature stage in terms of application. The development level of long-wave HgCdTe focal plane devices has reached a scale of 640 × 512 pixel, and medium and short-wave devices have reached a scale of 2048 × 2048 pixel. The most advanced infrared focal plane detectors used for satellite detection abroad have reached 1024 × 1024 pixel, 2048 × 2048 pixel and 4096 × 4096 pixel scale.

Raytheon in USA has developed a number of large surface array detectors for astronomical applications to meet the demanding requirements of low background science applications. The arrays developed range in size from 40 x 16 to 8000 x 8000, with image element sizes ranging from 61 um to 8 um [22]. The high cost of making large CdZnTe substrates and the difficulty of the fabrication process led Raytheon to use Si materials as substrates from 2000 onwards. The advantage of this substrate material is that it is very low cost and very easy to fabricate in large sizes. Currently, HgCdTe materials can be grown on 6in size Si substrates using molecular beam epitaxy to produce 4k × 4k image size of 20 um or 8k × 8k meta-detector arrays of 10 um image size [22]. The HgCdTe material was grown on a substrate for the preparation of 4k × 4k or 8k × 8k meta-detector arrays with an image size of 20 um or 10 um. To solve the problem of the different expansion coefficients of Si readout circuits and HgCdTe detectors, Raytheon has adopted a new cooling method using a sandwich connection to allow the readout circuits to shrink at the same rate as HgCdTe when cooled.

Teledyne in USA has developed a 4k × 4k HgCdTe shortwave infrared detector with an image element spacing of 15um using the MBE method, which is mounted on a telescope for space and astronomical applications. It has a very low dark current ($0.01e^-$/s/image-element) and a high quantum efficiency (70% to 90%) with an effective image rate of > 97%. Teledyne also manufactures high fractional rate focal plane arrays with an array size of 4096 × 4096 and an image element spacing of 10 um, using a Si-based p-i-n structure [23]. The detector has a Si-based p-i-n structure.

Sofradir in France has been developing detectors since the 1990s, and a 15 um centre pitch mid-wave 1280 × 1024 pixel IR focal plane detector is available on the shelf, which is the smallest centre pitch mid-wave CdTe IR focal plane detector available today and supports SXGA high resolution high performance applications. A 2 K × 2 K low-noise NIR device with a dark current of less than $0.5e^{-}$ s/element and a quantum efficiency of 70% has been developed for space applications, using CdZnTe as the substrate material and HgCdTe epitaxial materials prepared by MBE and LPE methods [24]. The DAPHNIS-HD MW is a pre-projected 1280 × 720 meta-facet CdTe IR detector from Sofradir, France, which is not yet a shelf product. It supports 16:9 HD and can be used in a variety of aerospace, naval and land-based detection applications. The detector assembly adds non-uniform correction and blind source replacement to the circuitry for easy processing of data.

Selex in UK has been working on the preparation of HgCdTe infrared focal plane devices using MOCVD epitaxy technology with an n-on-p table top bilayer component heterostructure, which eliminates crosstalk between devices. The EAGLE type long-wave 640 × 512 element HgCdTe focal plane array components have been developed and are now available on the shelf, with a centre spacing of 24 um × 24 um, a noise equivalent temperature difference of 24 mK, an operating band of 8–10 μm, a blind element rate of less than 1%, a response rate non-uniformity of less than 10%, an operating temperature of −45 °C + 70 °C, and a Stirling cooler [26]. The operating temperature of the module is −45 °C + 70 °C, suitable for Stirling coolers.

4.2 Status of Domestic Development

There are few units in China engaged in the research of HgCdTe infrared detectors, mainly including Kunming Institute of Physics, the 11th Research Institute of China Electronics Technology and Shanghai Institute of Technology and Physics. As China started late in developing HgCdTe infrared focal plane technology, there is still a big gap compared to foreign countries, but the strong support from the state has made great progress in CdHg infrared technology in recent years.

The 11th Research Institute of China Electronics Technology's long-wave gaze focal plane detectors include 256 × 256 (30 um pitch), 320 × 256 (30 um pitch), 640 × 512 (25 um pitch), 1280 × 1024 (25 um pitch), 640 × 512 (15 um pitch). 1280 × 1024 (10 um pitch) will be developed in 2020 [26].

The Kunming Institute of Physics started the development of CdTe medium wave focal plane detector in 2006, and after four years of technical research, mass production of CdTe medium wave 320 × 256 infrared focal plane detector components was achieved in 2010, and the annual production level of hundreds of components was reached in 2013 [27]. In 2013, it reached the level of several hundred sets per year. By continuing to promote the research of non-intrinsic doping of HgTe materials and the improvement of surface passivation and dry etching in the device process, HgTe medium wave focal plane detector components (analogue/digital) of 640 × 512 (15 um, 20 um, 25 um) and 1280 × 1024 (15 um) sizes have been released one after another.

5 Direction of Development

As a typical representative of the third generation of infrared detectors, large-field HgCdTe infrared detectors are currently being developed and used on a large scale, but they are also facing many problems. Due to the limited area of high quality HgCdTe epitaxial substrate CdZnTe, large surface array HgCdTe materials are very expensive and the manufacturing cost is a barrier to the advancement of HgCdTe infrared detectors to a wider market; the defect density of HgCdTe materials available for large area Si substrates is still high and the oligon lifetime is still short, which makes its dark current and noise affect its performance. This makes it impossible to achieve optimal performance due to its dark current and noise. In general, HgCdTe IR focal plane detectors are currently evolving in the following areas.

5.1 Large Size of the Surface Array

In recent years, large surface array infrared detector technology continues to mature, large surface array infrared detection in remote sensing, military, meteorological detection and other fields of application, with the image resolution and target recognition capability needs to improve, increasing the detector size technology continues to develop. Existing monolithic detectors cannot fully meet the needs of astronomy, satellite remote sensing, etc. Focal plane stitching methods are also required to obtain larger detector arrays. For the growth of large-scale arrays, the epitaxial layer flatness as well as the material components must be strictly controlled, and the maximum deviation of the cut-off wavelength can be controlled within 0.18 um (for a detector chip with a cut-off wavelength of 9 um).

The main challenge when producing large scale arrays is to increase the number of effective pixels. Each step of the process, including substrate and material growth as well as etching, interconnection and passivation, can result in defective pixels and only very mature process lines can end up with a high effective pixel rate. AIM's large scale detector arrays are above 99%, while mid-wave devices can reach 99.96% at 80K and SELEX super array stitching chips are up to 99.6% [28, 29]. The SELEX super array stitching chip has an effective pixel rate of up to 99.6%.

Detectors can be optically or mechanically stitched together to achieve large scale focal plane arrays, the largest focal plane array in the world today is Teledyne's 147 megapixel detector, shown in Fig. 4. The detector assembly consists of 35 2k × 2k HgTe IR detectors stitched together [30]. Foreign countries in the large surface array and long line array splicing, and mastered the relevant key technology, the product has also been used in the relevant remote sensors and astronomical telescopes. China is a late starter in stitching and needs to step up its efforts in the implementation of key technologies [31].

Fig. 9. Teledyne's 35 2k × 2k CdTe stitching arrays [32]

5.2 Long Wave/Very Long Wave Infrared Detection

With the commercialization of remote sensing satellites, long wave/VLW detectors will be the trend of the future [33]. Very Long Wave Infrared Radiation (VLWIR) with wavelengths greater than 14 μm contains information on the molecular content of atmospheric H_2O, CO_2, etc. and the temperature distribution of the atmosphere, and has an important role in proposing high sensitivity for low-temperature targets below 200 K. The detection of VLWIR is an essential part of obtaining spectral information in the entire infrared band. International research has been successful in the laboratory 15–17 μm of good performance of the very long wave detector, such as the French cut-off wavelength 15.1 μm 384 × 288 detector, the United States cut-off wavelength ≥ 17 μm (80 K) detector [34]. In China, 14 μm detectors are being developed. In China, 14 μm and 16 μm cut-off wavelength CdM and II-SLs infrared focal plane detectors are being developed, and some VLWIR detectors have been adopted by space projects such as the Feng Yun meteorological satellite.

VLWIR detector materials have a very narrow forbidden band width (Eg less than 0.1eV) to obtain a larger cut-off wavelength. HgCdTe materials have been the material of choice for VLW devices due to their proven application in the IR band. However, the uniform growth of HgCdTe has been a challenge for VLW device manufacturing processes, as a 1% change in the Cd component x can lead to a change in the cut-off wavelength of the image element greater than 0.4 μm. Compared to MBE, LPE grown HgCdTe has a flat epitaxial layer and a uniform component distribution and is more commonly used in VW device fabrication processes.

5.3 Multi-colour Detection

Multi-colour detection is accurate and can effectively improve the device's ability to extract and identify target features in complex environments. Imaging techniques using multi-colour devices can include the spectral and spatial information of the detection target, significantly improving the ability to distinguish the truth from the target and also enhancing the target's resistance to interference. Mercury cadmium telluride enabled the preparation of two-colour infrared detectors in the 1970s [36, 37] Thepreparation of multi-colour infrared detectors, which require a ring-hole process and multilayer epitaxy, is complex and the duty cycle and quantum efficiency of the photosensitive elements limit the development of large-scale applications.

The MW-MW ($\lambda c = 4.6\ \mu m$, $5.5\ \mu m$) dual-band detectors produced in France have a quantum efficiency of > 50%, a NETD below 20mK and an effective pixel rate of 99.8% at 98K. For MW-LW ($\lambda c = 5.6\ \mu m$, $10\ \mu m$) detectors, effective image rates of 99.8% (MW) as well as 98.5% (LW) are achieved. Two-colour arrays are available in 640 x 480 format with an image centre distance of 24 μm, while a 20 μm centre distance array is also under development [35] A 20 μm centre distance array is also under development. Raytheon USA has developed a 3-layer n-p-n structured MW-LW ($\lambda c = 5.1\ \mu m$, 9.6 μm@81K) two-colour IR detector on Si with an effective pixel rate of 99.98% (MW) and 98.7% (LW) at 640 × 480. Compared to the device of 2009, the main advantage is the increase in the effective pixel rate [2, 36].

6 Summary

HgCdTe photovoltaic detectors have high output impedance, low power consumption and good impedance matching with CMOS readout circuits, and have wide application prospects in space remote sensing, laser detection, agricultural detection, hyperspectral imaging, and military fields. It is foreseeable that HdCdTe photovoltaic infrared detectors will continue to dominate the field of infrared detection, ultra-large scale, very long wave infrared detection, multi-colour detection will further develop, structural coupling will lead the development of infrared detection to multi-dimensional, data cubic detection direction, the formation of integrated optical imaging system, reduce the complexity and cost of optical systems. In the future, the development of MBE and MOVPE will allow for more complex and accurate HgCdTe multilayer structures, while relying on improvements in various parts of the process to increase the degree of accurate processing will greatly facilitate the development of a new generation of HgCdTe infrared detectors in the direction of cutting-edge hot technologies.

References

1. Guiying, Z.: A review of the hybrid focal plane of mercury cadmium telluride infrared. Infrared Technol. **06**, 1–4 (1984)
2. Sun, I.: A review on the development of visible-shortwave infrared cadmium mercury telluride focal plane detectors. In: Proceedings of the 2019 Symposium on Infrared, Remote Sensing Technology and Applications and Interdisciplinary Forum (2019)
3. Ding, R., Chen, L., Sun, S., et al.: Research progress on infrared focal plane devices with cadmium mercury telluride for space. In: Proceedings of the 11th National Conference on Molecular Beam Epitaxy (2015)
4. Boliang, C., Xiangyang, L.: Infrared Imaging Detectors for Space Applications, pp. 42–45.Science Press, Beijing (2016)
5. Cui, A.: Atomic layer deposition passivation technology for cadmium mercury telluride micro-tablet chips. University of Chinese Academy of Sciences (Shanghai Institute of Technology Physics, Chinese Academy of Sciences) (2021)

6. Tian, Z., Song, S., Wang, X., et al.: Study on the preparation of mercury cadmium telluride p-on-n long-wave heterojunction detector materials. Laser Infrared **48**(6) (2018)
7. Yue, T., Yin, F., Hu, X.N.. Dark current characterization of silicon-based HgCdTe photovoltaic devices. In: Proceedings of the 3rd National Symposium on Advanced Focal Plane Technology (2007)
8. Zhang, W.: Reliability techniques of large surface array infrared focal plane detectors with cadmium mercury telluride. University of Chinese Academy of Sciences (Shanghai Institute of Technology Physics, Chinese Academy of Sciences) (2021)
9. Li, X.: Research on key technologies of infrared focal plane array readout signal processing circuit design. Tianjin University (2010)
10. Ye, Z.H., Chen, Y.Y., Zhang, P.: A review of frontier technologies for cadmium mercury telluride infrared detectors. Infrared **35**(2) (2014)
11. Yuan, H., Apgar, G., Kimet, J., et al.:Infrared Technology and Applications XXXIV. . In: Proceedings Volume (2008)
12. Reddy, M., Jin, X., Lofgreen, D.D., et al.: Demonstration of high-quality MBE HgCdTe on 8-inch wafers. J. Electr. Mater. **48**(10), 6040–6044 (2019)
13. Chorier, P., Tribolet, P., Destefanis, G.: From visible to infrared: a new approach. In: Proceedings of the SPIE [6206–01], Orlando (2006)
14. Zhang, J., Chattaraj, S., Lu, S., Madhukar, A.: Mesa-top quantum dot single photon emitter arrays: growth, optical characteristics, and the simulated optical response of integrated dielectric nanoantenna-waveguide systems". J. Appl. Phys. **120**(24), 243103 (2016)
15. Madejczyk, P., Gawron, W., Martyniuk, P., et al.: MOCVD grown HgCdTe device structurefor ambient temperature LWIR detectors. Semicond. Sci. Technol. **30**(10) (2013)
16. Liu, M.: Structural characterization of pulsed laser deposition (PLD) mercury cadmium telluride (HgCdTe) thin film materials (2009)
17. Hu, W., Li, Q., Chen, X., et al.: Infrared photoelectric detectors with transformative features. J. Phys. **68**(12) (2019)
18. Ye, Z.H., Li, F.H., Wang, J.D., et al.: Frontier hotspots and changing trends of infrared photodetectors. J. Infrared Millimeter Waves **41**(1) (2022)
19. Yu, S., Zhu, X., Zhou, L., et al.: Development of mega-pixel mid-wave infrared focal plane assembly. Space Return Remote Sens. **37**(4) (2016)
20. Cai, J.F., Zhao, M.G., Hong, G.T.: Pulse tube cooler for the Gaofen-4 satellite gaze camera. Space Return Remote Sens. **37**(4) (2016)
21. Qiu, M.P., Ma, W.P.: Optical stitching method of detector for space infrared push-scan imaging system. Space Return Remote Sen. **40**(6) (2019)
22. Holland, A.D., Beletic, J., Starr, B., et al.: RVS large format arrays for astronomy. In: SPIE, vol. 9915, p. 99152x (2016)
23. Wen, H.E., Cong, W.A.N.G., Zhen, T.I.A.N., et al.: Research on the performance tuning technology of mercury cadmium telluride infrared detectors. Infrared **42**(12), 6–14 (2021)
24. Yanlan, L., Da, G., Zhen, L., et al.: Current status and trends of research on large-size mercury cadmium telluride materials. Laser Infrared **52**(8), 1204–1210 (2022)
25. Zhen, T., Shufang, S., Yanlei, X., et al.: Study on the preparation of very long-wave mercury cadmium telluride infrared detectors. Laser Infrared **52**(10), 1527–1531 (2022)
26. Jiaojiao, Q.I., Xiaoyu, F.E.N.G., Yanguan, C.H.E.N., et al.: Progress in the development of 10 μm pitch long-wave 1280$\times$1024 HgCdTe detectors. Infrared **43**(2), 1–6 (2022)
27. Han, F.Z., Zhou, L.J., Yuan, S., Wu, J.: HgCdTe infrared focal plane detectors from R&D to production. Infrared Technol. **36**(04), 271–274 (2014)
28. He, L., Chen, L., Chen, J.X., et al.: Overview of infrared focal plane research at Shanghai institute of technology and physics. In: Proceedings of the Tenth National Conference on Molecular Beam Epitaxy, pp. 3–4 (2013)

29. Dorn, M., McMurtry, C., Pipher, J., et al.: +A mono- lithic 2k×2k LWIR HgCdTe detector array for passively cooled space missions. In: SPIE, vol. 10709, p.1070907 (2018)
30. ubaldo, L., Taalat, R., Berthoz, J., et al.: Latest Improvements on Long Wavep-on-nHgCdTe Technology at Sofradir. In: SPIE, vol. 10177, p. 101771E (2017)
31. Sprafke, T., Beletic, J.W.: Infrared focal plane arrays for space applications. OPN JUNE 22–27 (2008)
32. Lv, W.D., et al.: Study on the splicing technology of infrared detectors for space use. Infrared Technol. **44**(10), 999–1008 (2022)
33. Finger, G., Beletic, J.W.: Review of the state of infrared detectors for astronomy in retrospect of the June 2002 workshop on scientific detectors for astronomy. In: Proceedings of SPIE, vol. 4841, pp. 839–852 (2003)
34. Shi, M., Ling, L.: Status and trends of development of large surface array mercury cadmium telluride infrared focal plane array. J. Arms Equipment Eng. **38**(6) (2017)
35. Rogalski, A., Antoszewski, J., Faraone, L.: Third generation infrared photodetector arrays. J. Appl. Phys. **105**(9), 091101–1–091101–44 (2009)
36. Vilela, M.F., et al.: High-performance M/LWIR dual-band HgCdTe/Si focal-plane arrays. J. Electron. Mater. **42**(11), 3231–3238 (2013)

Starting from the Sampling Imaging System, A Comprehensive Review on the Remote Sensing Image Super-Resolution Technology

Lin Lan and Chunling Lu(✉)

DFH Satellite CO., LTD., Beijing 100094, China
chun11501@sina.com

Abstract. High resolution is the eternal pursuit of remote sensing satellites. At present, the highest resolution of American lock-eye satellite KH-12 has reached 10 cm, but due to the limitation of the diffraction limit of the remote sensing satellite optical system, the development of a higher resolution remote sensing satellite has encountered a bottleneck. The satellites in orbit, such as Pleiades-NEO, have found another way to super-resolve 30 cm images to 15 cm image products using super-resolution technology based on a convolutional neural network. This method of restoring low-resolution images to high-resolution images through image super-resolution techniques has attracted a lot of attention since the middle and late 20th century. The physical continuous graphic signal is sampled and quantized into discrete digital arrays by CCD or CMOS camera, while the diffraction limit of the optical system and many other factors of degradation also exacerbated this resolution degradation phenomenon. This paper writes from the optical sampling imaging system to induce the model of remote sensing image degradation and analyze the causes of resolution degradation from the source; Further, this paper investigates the image super-resolution techniques of about the last two decades and categorizes them into two categories: traditional algorithm-based, and learning-based. This paper analyzes in detail the key algorithms in the history of super-resolution, and focuses on today's deep learning-based algorithms, clarifying the problems targeted by each type of algorithm, analyzing their design ideas and implementation principles, and how these algorithms can be adapted for super-resolution algorithms for remote sensing images. Then this paper compares the basic features of the main remote sensing image super-resolution algorithms and their advantages and disadvantages. It also introduces the current more widely used super-resolution effect evaluation metrics. Finally, this paper lists and analyzes the latest and most complete various remote sensing super-resolution datasets publicly available on the Internet, looks forward to the possible future development trends, and points out that joined with imaging systems on satellites, unsupervised learning, and multi-source remote sensing image fusion are the development directions of future remote sensing image super-resolution technologies.

Keywords: Remote Sensing Image · Super-Resolution · Degradation Model · Deep Learning

H. P. Urbach and H. Jiang (Eds.): ISSOIA 2022, SPPHY 295, pp. 207–228, 2023.
https://doi.org/10.1007/978-981-99-4098-1_19

1 Introduction

Since the 21st century, high-resolution remote sensing satellites have become a hot spot for competition among space powers, and the current highest resolution of the U.S. lock-eye satellite KH-12 has reached 10 cm [1]. However, the cost and risk of increasing the resolution by reducing the orbital altitude and increasing the focal length of the optical system are very huge. And large remote sensor equipment cannot be equipped in the current popular nanosatellites, because of its small size and small weight limitation, which means it is a luxury for remote sensing nanosatellites to obtain high-resolution images directly. Therefore, a super-resolution (SR) image processing technique, which restores low-resolution (LR) images into high-resolution images (HR) at the software level, has gradually come into view. We counted the number of papers on remote sensing images super-resolution on four bibliometrics (ScienceDirect, IEEE Xplore, CNKI database, and Wanfang database) in the last 20 years, as shown in Fig. 1. It can be seen that this software-level algorithm has received more and more attention and research. Harris [2] is the first to propose the concept of super-resolution in the 1960s, and most of the early techniques used simple interpolation, such as nearest-neighbor interpolation, bilinear interpolation, and bicubic interpolation [3]. Into the 1990s, pioneering methods such as multi-frame super-resolution based on sub-pixel shifts [4] and multi-sensor image fusion [5] were also proposed, and then super-resolution techniques began to develop rapidly. Some super-resolution models based on probability theory, transform domain, machine learning, and artificial neural networks were proposed and applied.

After 2008, super-resolution methods based on sparse coding became popular. Then 2012, because of the significant increase in hardware computing power, methods based on deep artificial neural networks gained widespread attention because of their powerful multi-layer feature learning and representation capabilities, And super-resolution techniques are evolved into models such as SRCNN [6] based on convolutional neural networks, SRGAN [7] based on generative adversarial networks, attention mechanisms, and unsupervised models. Early super-resolution techniques were applied to natural images before remote sensing image super-resolution (RSISR) techniques were developed. For single frame image super-resolution (SISR), the techniques for natural and remotely sensed images are very similar, with the possible difference that remote sensing images have more noise, as well as lower resolution and less texture information. For the super-resolution of multi/hyperspectral images with multiple frames, there are more changes to the SISR technique. In this paper, we will start from the sampling imaging of optical remote sensing images, introduce the degradation model. Then we use the degradation model as traction, discuss the mainstream SISR algorithms, and make some comparisons, as well as the implementation of SR on remote sensing images. Finally, we introduce the metrics to measure the processing results and the latest and most comprehensive remote sensing datasets, summarize the SR technology for remote sensing images and make an outlook.

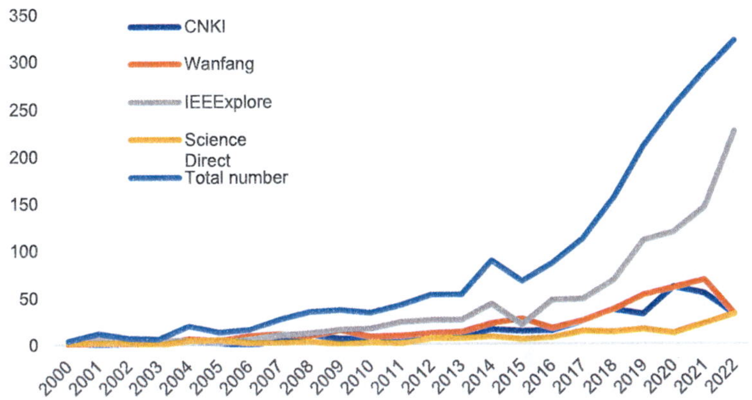

Fig. 1. RSISR literature quantity trend chart

2 Sampling Imaging System and Degradation Model

Super-resolution originated from the study of natural images and was later extended to remote sensing images. However, remote sensing images contain rich types of features, many degradation factors such as sampling, deformation, degradation clarity and noise on the imaging link as well as ground artifacts caused by cloud cover, terrain undulation, haze and other lighting changes, which make the semantic information of remote sensing images much more complex than natural images, and thus the super-resolution of remote sensing images is more difficult.

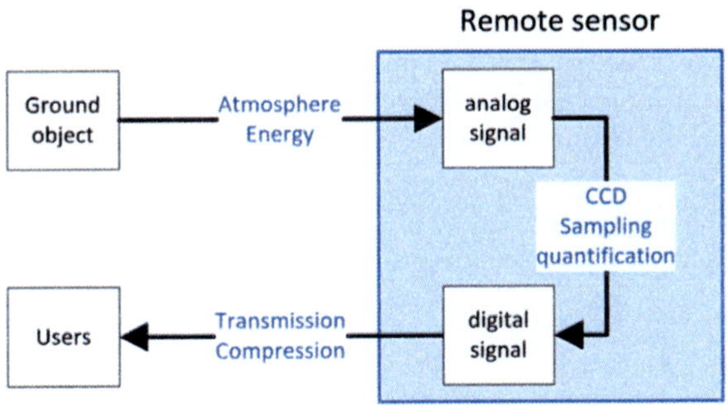

Fig. 2. Remote sensing image imaging link and degradation process

The remote sensing imaging link can be briefly summarized as follows: the target irradiance is captured by the remote sensor (CCD camera) in the sky after being disturbed by atmospheric and energy factors, then transmitted to the ground by the remote sensor through compression, and finally displayed and image processed on the ground for other applications. Among them, the continuous graphic signal of nature is quantified into discrete digital array by CCD camera sampling, this process directly degrades the continuous clear image into an image composed of one pixel, if the GSD of remote sensing satellite is 1 m, it represents the degradation of a square meter area of the ground into the gray value of one pixel point, which greatly reduces the resolution, coupled with the atmospheric interference of the earth, energy attenuation, noise, etc. impact, further causing blurring of the image, making the image degraded from high resolution to low resolution. This process is shown in Fig. 2.

In the actual image super-resolution process, we do not reconstruct the digital signal into an analog signal, but generate more pixels to make the image clearer, that is, to enlarge (recover) a low-resolution image **LR** into a high-resolution image **HR**, the common magnification is 2, 3, 4, 8, etc. The formula can be expressed as $Y = DBX + N$, where $\boldsymbol{Y}$ is the low-resolution image we observe, $\boldsymbol{X}$ is the high-resolution image we want to obtain, D is a down-sampling operation, B is a blurring filter, N is the noise. After integrating D, B and N into $\boldsymbol{D_m}$, the formula can be further abstracted to $\boldsymbol{Y} = \boldsymbol{D_m X}$, where $\boldsymbol{D_m}$ represents the degradation model. The essence of the super-resolution algorithm is to find a suitable degenerate model $\boldsymbol{D_m}$ to ensure that the recovered $\boldsymbol{X}$ is consistent with the input $\boldsymbol{Y}$.

3 Traditional Algorithms

The traditional implementation idea of super-resolution is to reconstruct HR images based on a custom degradation model with a priori constraints. In this paper, according to the traditional algorithm development lineage and highlights, we focus on analyzing only four types of these algorithms: interpolation method (with bicubic interpolation [3] as an example), iterative back-projection method (IBP) [8], convex set projection method (POCS) [9] and maximum a posteriori probability method (MAP) [10].

3.1 Bicubic Interpolation (BC)

The bicubic interpolation (BC) method proposed by Keys in 1981 [3] is the most commonly used interpolation method in two-dimensional space. The core idea is that the pixel value of a certain interpolated point P is obtained by weighting the pixel values of the surrounding 16 sampled points, and the 16 nearest neighboring points of point P are selected by their relative positions as in Fig. 3. Keys constructed a bicubic function to calculate the weights of the surrounding 16 points as follows.

$$W(x) = \begin{cases} x = (a+2)|x|^3 - (a+3)|x|^2 + 1 & \text{for } |x| \le 1 \\ y = a|x|^3 - 5a|x|^2 + 8a|x| - 4a & \text{for } 1 < |x| < 2 \\ z = \quad 0 & \text{otherwise} \end{cases} \tag{1}$$

Here a is generally taken as −0.5. After getting to the weights, we just need to weight up the pixel values of these 16 points, and the formula for interpolation is as follows.

$$f(x, y) = \sum_{i=0}^{3} \sum_{j=0}^{3} f(x_i, y_j) W(x - x_i) W(y - y_j) \tag{2}$$

It is important to note that the pixel values obtained by weighting are restricted to 0 to 255. Since the bicubic interpolation uses 16 points and a smoother cubic function, the computation is more complicated than the previous nearest-neighbor interpolation and bilinear interpolation, but the generated images are smoother and have better details.

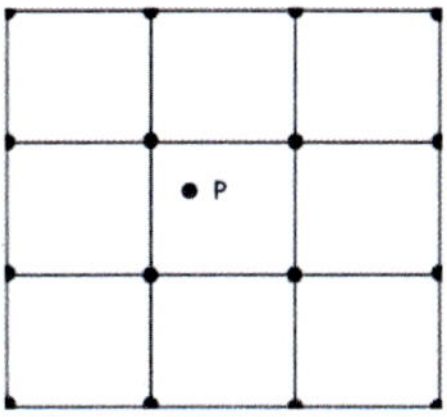

Fig. 3. Diagram of the bicubic interpolation method

3.2 Iterative Back-Projection (IBP)

Iterative back-projection (IBP) [8] was proposed by Irani and Peleg in 1991, and the core idea is to iteratively eliminate the residuals between the observed LR image and the simulated LR image to recover multiple LR images into one HR image. Which customizes the degradation model H and the inverse degradation model H^{BP}. The iterative formula is as follows

$$\hat{f}^{n+1} = \hat{f}^{n} - \lambda \sum_{i=1}^{p} H^{BP}\left(\hat{y}_i^n - y_i\right) \tag{3}$$

where: n is the number of iterations. $\hat{f}^{n+1}$ and $\hat{f}^{n}$ denote the super-resolution images obtained from the $(n + 1)_{th}$ and n_{th} iterations. H^{BP} is the inverse projection operator, which can be obtained by synthesizing the matrix of translation, rotation, lucidity reduction, down-sampling and noise. $\hat{y}_i^n$ is the simulated LR image, and $\hat{y}_i^n = H\hat{f}^n$; p is the number of frames of the LR image used for reconstruction. λ is the projection relaxation factor.

The advantage of IBP is that the algorithm is simple and intuitively easy to understand, but it is very sensitive to high frequency noise. The method has no unique solution because the problem is inherently ill-posed. The method has some difficulties in choosing H^{BP} H^{BP}, and it is more difficult to add a priori constraints.

3.3 Projections onto Convex Sets (POCS)

In order to solve the drawback of the IBP algorithm which is difficult to use a priori information, in 1989, Stark et al. [9] applied the projections onto convex sets (POCS) method to super-resolution. In this theory, constraints or prior knowledge (such as positivity, energy boundedness, observation consistency, and smoothness) can be defined as convex sets in vector space (convex sets), and the solution space of the super-resolution reconstruction problem is formed by intersecting these convex sets.

For m prior knowledge, there will be m corresponding closed convex sets C_i, $i = 1, 2, \cdots m$. Stark and Oskoui define a convex set projection operator P_i for each convex set C_i. Then the high-resolution image is $f \in C_0 = \bigcap_{i=1}^{m} C_i P_i$, where C_0 is a nonempty closed convex set, and the iterative formula for the high-resolution image f_{k+1} is as follows.

$$f_{k+1} = T_m T_{m-1} \cdots T_1 f_k\,,\ k = 1, 2 \cdots\cdots \tag{4}$$

where $T_i = (1-\lambda_i)I + \lambda_i P_i$, and $0 < \lambda_i < 2$ is the relaxed projection operator, meaning f_{k+1} weakly converges to a feasible solution at the C_0. Any element in the intersection set C_0 is one that satisfies all the prior knowledge or constraints, so the feasible solution obtained by the POCS method is generally not unique.

The POCS algorithm has remarkable features such as simple and direct and powerful image prior knowledge embedding capability, but it also has some disadvantages as follows: 1) non-uniqueness of the solution, 2) dependence on the initial estimate, order of each projection operator, 3) higher computational complexity, but less than MAP algorithm.

3.4 The Maximum a Posteriori Probability (MAP)

The maximum a posteriori probability (MAP) method belongs to statistical estimation methods. The MAP algorithm in super-resolution reconstruction [10] refers to the maximization of the posterior probability of the occurrence of a high-resolution image given a known sequence of low-resolution images. The ideal high-resolution image is A and the observed low-resolution image is B. According to Bayes' probability theorem, the posterior probability of the synthesized high-resolution image is $P(A|B) = P(A) \cdot P(B|A) / P(B)$. Where $P(B|A)$ is the conditional probability of a sequence of low-resolution images given an ideal high-resolution image; $P(A)$ is the prior probability of an ideal high-resolution image. In comparing the posterior probability $P(A|B)$, P(B) is the same and can be deleted, so the optimization equation is $A_{MAP} = \arg\max_{A}\left[\log P(A) + \log P(B|A)\right]$. The conditional probability $P(B|A)$ generally uses a Gaussian model. The prior probability $P(A)$ uses different models in different algorithms, such as the Markov random field model and the Gibbs random field model.

4 Learning-Based Algorithms

The learning-based method is a data-driven approach, where the learned data is divided into labeled data (supervised learning) and unlabeled data (unsupervised learning). Ideally, the larger the number of dataset and model, the more information can be learned, and the better the final result can be. Of course, the actual learning, due to hardware and time constraints, the experimenter should choose the right amount of data, as well as optimize the model size. Learning-based super-resolution algorithms are essentially based on big data, learning the mapping of low-resolution images to high-resolution images, or learning an inverse degradation model. In this section, the learning-based single-frame image super-resolution algorithms are divided into four methods based on sparse coding, based on CNN, based on GAN and based on attention, of which the former belongs to traditional machine algorithms and the latter three belong to deep learning algorithms. In addition, since unsupervised is the development trend of deep learning, we also analyze the current single-frame image super-resolution algorithms based on unsupervised deep learning.

4.1 Based on Sparse Coding

Sparse Coding (SC). The sparse coding (called sparse representation also) wants to express most or all of the original signal with a linear combination of fewer basic signals, just as we can remember a person's face based on a small number of features. Where the original signal is an N*1-dimensional vector that can be flattened from a 2D image patch which in general is dense, i.e., most elements are not 0. These basic signals are called atoms, one atom is a vector (N*1), K atoms form an under-determined dictionary D (N*K), and the under-determined dictionary means K > > N. In general the bases of the under-determined bases are redundant, making the image representation under the under-determined basis more sparse than the determined orthogonal basis, and the clean part of the image can be linearly represented using a small number of non-zero sparse representation coefficients.

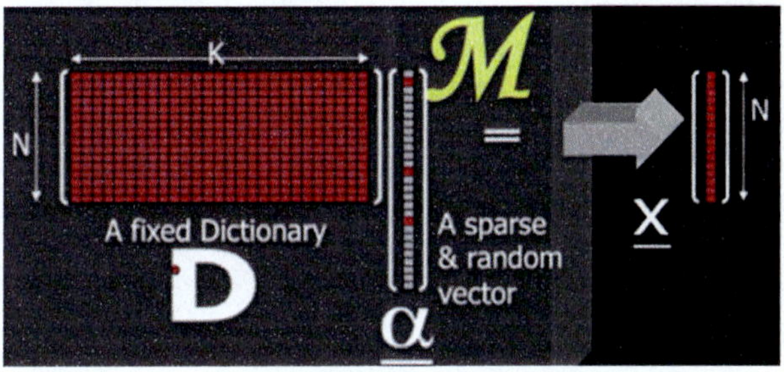

Fig. 4. Sparse representation

A sparse coding means finding a coefficient vector α_i (K*1) and a dictionary matrix D (N*K) such that $D\alpha_i$ as close as possible to x_i and α_i is as sparse as possible, α_i is the sparse representation of x_i, and the formula is $D\alpha_i = x_i$, as shown in Fig. 4. Expressed as an optimization problem, the simplest form of the sparse coding is

$$\min_{D,\alpha_i} \sum_{i=1}^{m} \|x_i - D\alpha_i\|_2 + \lambda \sum_{i=1}^{m} \|\alpha_i\|_2 \tag{5}$$

where x_i is the i_{th} sample, D is the dictionary matrix, α_i is the sparse representation of x_i, and λ is a penalty factor greater than 0 (used to make α_i more sparse). The specific methods for finding b and solving the dictionary D are not expanded in detail due to space limitations.

Application of Sparse Coding in Image Super-Resolution. The super-resolution technique based on sparse coding can be understood from the perspective of degradation model, let X represent **HR** image and Y represent **LR** image, both X and Y can be used for sparse coding, corresponding to under-determined dictionaries as D_H and D_L, then $D_H\alpha_X = X, D_L\alpha_Y = Y$; if $\alpha_X = \alpha_Y$, then $X = (D_H)^{-1}D_L Y = D_m Y$, $(D_H)^{-1}$ is the pseudo-inverse, D_m is the degradation model we are looking for. Yang et al. [11] did just that, and Yang first proposed a super-resolution reconstruction method based on sparse coding in 2008. Yang establishes the underdetermined dictionaries D_L and D_H for high- and low-resolution image patches, respectively, while assuming that the sparse representation α of the low-resolution image patches is the same as the sparse representation α of the corresponding high-resolution image patches, and further synthesizes D_L and D_H as D, transforming the problem into a solution Eq. 6 For better results, in addition to the standard steps mentioned above, subsequent global constraints, etc., are often required. In the testing stage, we synthesize HR image using D_H and the sparse representation α of LR image, as shown in Fig. 5.

Zheng et al. [12]. First applied the sparse coding to the super-resolution of remote sensing images. For the problem that remote sensing images contain more noise, he proposed that the previous super-resolution technique of early denoising may bring more interference to the later super-resolution, so he used K-SVD [13] and OMP [14] to solve the sparse coefficients while suppressing the noise. Further, Dong et al. [15] generated multiple sub-dictionaries by clustering image patches so that a given patch of images could select sub-dictionaries adaptively, and imposed adaptive regularization constraints on the optimization equations to make the super-resolution reconstruction more accurate; Zhang et al. [16] used the initial sparse dictionary and the residual sparse dictionary to significantly improve the resolution of remote sensing images, with the former dictionary reconstructing LR images as the initial HR image, and the latter dictionary repairs the detail information lost in the initial HR image.

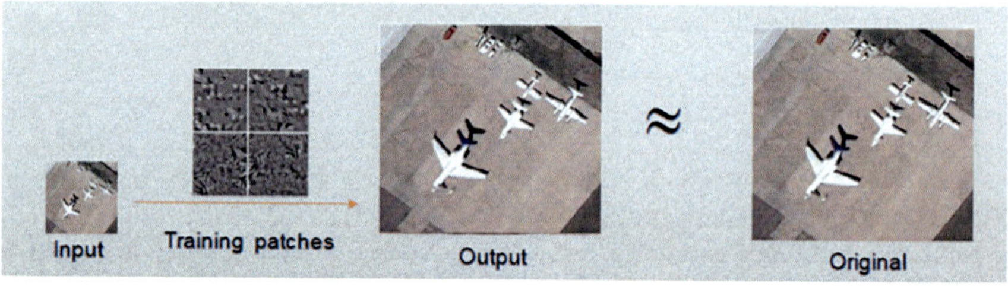

Fig. 5. Sparse coding for super-resolution reconstruction

4.2 Based on CNN

In order to solve the problem of losing spatial information by expanding images into vectors for processing in previous fully connected neural networks, convolutional neural networks use 2-D convolution operators to process images in 2-D directly (Fig. 6), which not only retains spatial information but also reduces the parameters of the network, and at the same time, the randomness of rounding off information also makes CNNs enter the overfitting state more slowly and improves the generalization performance of CNNs. CNNs containing 2-D convolution operator, pooling, dropout, and batch-normalization processes have opened a new era of deep learning image processing.

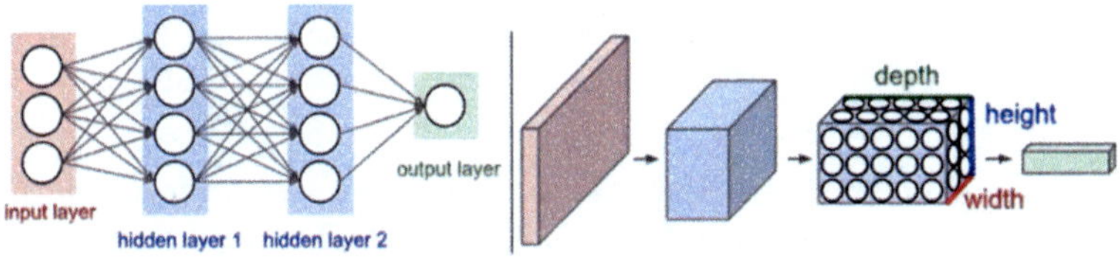

Fig. 6. Fully connected neural networks vs. convolutional neural networks

SRCNN. Dong et al. [5] proposed SRCNN (Super Resolution Convolutional Neural Network) in 2015, which kicked off the convolutional neural network for super-resolution. The network structure of SRCNN is shown in Fig. 7, where a LR_image (1*256*256) is input, which passes through convolution_kernel_1 (1,9,9,64) to get feature_map_1 (64*248*248), then passes through convolution_kernel_2 (64*1*1*32) to get feature_map_2 (32*248*248), and finally passes through convolution_kernel_3 (32*5*5 *1) to output the HR_image (1*244*244). When input is a color image, the input image is the Y channel in the YCbCr color space; and the input low-resolution image is pre-interpolated by bicubic interpolation and scaled up to a high-resolution size. The experiment shows that the high-resolution image recovered by CNN is significantly better than generated by bicubic interpolation.

The whole process from data to learning is: original HR image $\mathbf{X} \rightarrow$ bicubic interpolation down-sampling $\rightarrow$ LR $\mathbf{Y_1} \rightarrow$ bicubic interpolation up-sampling $\rightarrow$ immature HR $\mathbf{Y_2} \rightarrow$ CNN $\rightarrow$ generated HR $\mathbf{Y}$. Where the CNN network learns the best convolutional kernel parameters in CNN by optimizing the mean square error of $\mathbf{X}$ and $\mathbf{Y}$ by SGD method. In fact, CNN has a great similarity with the idea of sparse coding, as shown in Fig. 7. Convolution_kernel_1 is like a dictionary of low-resolution image patches, transforming each low-resolution image patch into a representation on the feature map, although not sparse. Convolution_kernel_2, on the other hand, represents a nonlinear mapping that maps the low-resolution representation into a representation of the high-resolution image patch. Convolution_kernel_3 is a dictionary of high-resolution image patches, which eventually reconstructs the representations of the high-resolution patches into a high-resolution image. However, these are implicitly implemented through the convolution kernel.

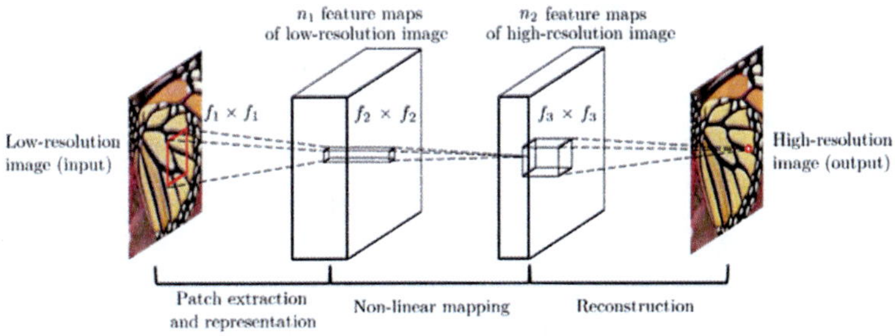

Fig. 7. The framework of SRCNN [5]

super-resolution is only applied on the luminance channel (Y channel in YCbCr color space). so c = 1 in the first/last layer, and performance (e.g., PSNR and SSIM) is evaluated on the Y channel.

Remote Sensing Images Super-Resolution on CNN. Since remote sensing images are missing local detail information, and the common deep CNN super-resolution methods using large perceptual fields tend to ignore local information, Lei et al. [17] designed a Local-Global Combined-Network (**LGCnet**) to address this problem. The network is designed to learn the multi-scale representation of remote sensing images by combining the convolution results of different layers to better perform RSISR.

Pan et al. [18] proposed a method based on residual dense networks (Residual DBPN, **RDBPN**) inspired by the Deep Back-Projection Networks (DBPN) proposed by Haris et al. [19]. The method adds dense skip connections to the DBPN projection units to construct global and local residuals, and provides information for high magnification through feature reuse, thus making it show better performance at high magnification.

4.3 Based on GAN

Deep learning models can be divided into discriminative models and generative models. Due to the invention of algorithms such as back propagation (BP), discriminative models (e.g., BP networks, CNNs, RNNs, etc.) have been developed rapidly. However, the development of generative models was slow until 2014, when Goodfellow et al. [20] proposed the most successful generative model, the Generative Adversarial Network (GAN), and generative models exploded.

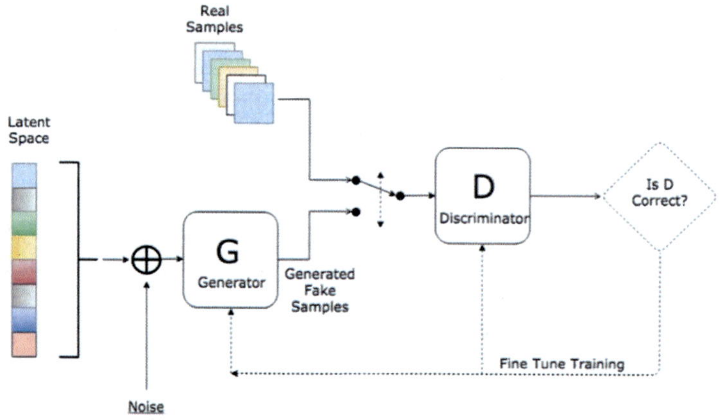

Fig. 8. The framework of GAN network

GAN consists of a generator and a discriminator, as shown in Fig. 8. Latent Space is a random hidden variable of fixed size, often obeying Gaussian distribution, and the **G**enerator tries to generate fake samples from the hidden variable. The **D**iscriminator learns the real samples to obtain the ability to identify the real and fake samples. In this way, through the continuous learning and confrontation between the **D**iscriminator and the **G**enerator, the **G**enerator is finally able to generate fake samples that are highly similar to the real samples, causing the **D**iscriminator to be unable to judge, which means that the **G**enerator has successfully modeled the process of producing real samples.

SRGAN. GAN network is applied by **SRGAN** [7] for super-resolution algorithm reconstruction for the first time. SRGAN added ResNet to the Generator, which enables to train deeper network and improve the network accuracy. SRGAN proposed perceptual loss function, which can generate images more in line with human visual perception and the loss is an effective guarantee for the algorithm. The perceptual loss can be split into content loss and adversarial loss, and the content loss is a great innovation. The previous MSE loss can improve the PSNR, but it also loses some high-frequency information, resulting in blurred images. The content loss is calculated on the VGG's feature map by passing the true and false samples through the VGG network, which can greatly improve the blurring problem caused by MSE loss.

Remote Sensing Images Super-Resolution on GAN. The degradation process of remote sensing images often contains more noise, and the original GAN-based method is more sensitive to noise, which generates high-frequency noise independent of the input image. In response, Jiang et al. [21] proposed **EEGAN** (Edge-Enhanced GAN) from the perspective of edge enhancement. The Edge-Enhanced Sub-Network (EESN), constructed by Laplacian operator, fuses the SR reference images and their edges to generate HR remote sensing images with clear edges, so as to alleviate the problem of blurred edges in remote sensing images.

It is found through statistical analysis that there are more low-frequency components (flat regions) in the remotely sensed images than in the natural images. When using GAN for RSISR, it is difficult for the Discriminator in the network to determine whether these low-frequency regions are generated from the real HR remote sensing images, which leads to the quality of the generated HR images to be affected. In this regard, Lei et al. [22] designed a Coupled-Discriminate GAN network (**CDGAN**). The two-channel network in the coupled discriminator joins the features extracted from the real HR image and the generated HR image and input them into the subsequent layers, and a dedicated coupled loss function is constructed to update the network parameters. The model improves the GAN-based image SR method in processing low-frequency image regions with blurred resolution.

4.4 Remote Sensing Images Super-Resolution on Attention

Although CNNs have made significant achievements in deep learning, there are some problems, such as sliding weight windows applied to all spatial channels or spectral channels of the feature map equally when performing convolutional computation, and this uniform computation makes it difficult to extract the part of features that need special attention. As a result, attention mechanisms came into the vision of researchers, Hu et al. [23] proposed the Squeeze-and-Excitation network SENet based on channel attention module (**CAM**), and Woo et al. [24] proposed the spatial attention module (**SAM**)-based convolutional block attention module (CBAM), like Fig. 9. On the whole, CAM means generating a weight vector MC with the same number of channels as the input feature map, and the size of the weights corresponds to whether a channel in the feature map is more important. SAM means generating a weight matrix MS with the same height and width dimensions as the input feature map, and the size of the weights corresponds to whether a pixel in the feature map is more important. These two attention models can be used independently or mixed, and more attention models have been developed later. Also the attention mechanism can be used not only in CNNs but also in other deep learning networks, and later it has been developed into the transformer [25] module which is very popular nowadays.

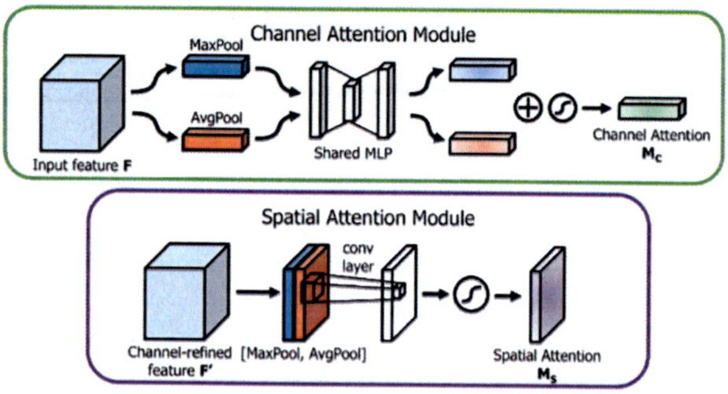

Fig. 9. Channel Attention Module and Spatial Attention Module [24]

The Attention module was first introduced into image super-resolution processing by Zhang et al. [26], who proposed a deep residual channel attention network (**RCAN**). Zhang utilized the channel attention module to adaptively adjust the channel features. Later, based on the idea of RCAN [26], Huat et al. [27] proposed **RSRCAN** network to introduce channel attention module into remote sensing image super-resolution. Huat pointed out that the current deep learning super-resolution model is difficult to train due to its own complexity and the lack of important training data, which becomes an important limitation for satellite remote sensing image super-resolution. Moreover, most CNN-based super-resolution algorithms for remote sensing images default to equal importance of all features extracted from LR input images, which may lead to a lack of flexibility in analyzing the presence of different types of features in remote sensing images. Moreover, remotely sensed image data suffers some degradation during its acquisition, and this usually introduces a lot of noise and variability in the data (in addition to the rich low-frequency information), so Huat uses the channel attention module to focus on the surface features that require finer HR detail by enhancing the high-frequency information of the image and suppressing the low-frequency information, thus allowing the model's to learn more about the mapping relationships between the high-frequency components. Since texture information varies greatly with different remote sensing images, but most SR methods use the same learning model for all scenes, the existing SR methods have poor generalization capability. To be able to adaptively adjust the network according to the input images, Jia et al. [28] proposed a multi-attentive remote sensing super-resolution reconstruction network (**MA-GAN**). The main body of MA-GAN [28] contains three modules: a pyramidal convolutional residual dense (PCRD) block, an attention-based up-sampling (AUP) block, and an attention-based fusion (AF) block. The PCRD block combines multi-scale convolution and channel attention to automatically learn and scale the residuals for better representation, the AUP block uses pixel attention to perform arbitrary scale up-sampling, and the AF block uses branch attention to integrate the up-sampled low-resolution images with high-level features.

4.5 Unsupervised Algorithms

Supervised deep neural networks often require HR-LR image pairs during training, and the LR images used are generally obtained from HR images by custom downsampling algorithms (e.g., bicubic interpolation). However, in reality, high-resolution remote sensing images are usually difficult to obtain, and there are still differences between the images obtained by custom degradation and the actual low-resolution remote sensing images.

Haut et al. [29] constructed a CNN-based deep generative network (**ANDGN**) for RSISR from an unsupervised perspective. The method first uses a CNN to expand the random noise to the target HR dimension, then downsamples [30] the generated HR results to obtain the generated LR image. The loss is calculated by generated LR image and original LR image and minimized through iterations until the final desired HR remote sensing image is generated. The unsupervised aspect of ANDGN are reflected in the fact that only LR images are used. Wang et al. [31] proposed an unsupervised learning network **CycleCNN** for SR of remote sensing images based on CycleGAN [32], which contains a cyclic network composed of image degradation network and image super-resolution network. The HR image selected in the paper is a panchromatic image with GSD of 1 m/pixel in GaoFen-2, and the HR image is single-frame of multispectral image with GSD of 4 m/pixel in GaoFen-2. The LR image undergoes the cycle process of super-resolution and degradation in the network, and the HR image undergoes the cycle process of super-resolution and degradation in the network. Wang then constructs cycle loss and identity loss functions to enable the degraded network to degrade more realistic low-resolution images and improve the performance of the super-resolution network. The unsupervised nature of CycleCNN is reflected in the fact that no HR-LR image pairs are used. Considering that high-resolution remote sensing images are difficult to obtain, Zhang et al. [33] in 2022 proposed an unsupervised visible image-guided remote sensing image super-resolution network (**UVRSR**) by guiding low-resolution (LR) remote sensing images through HR visible natural images, which was successfully established.

5 Comparison and Discussion

5.1 Comparison of Algorithms

The comparison of traditional algorithms is shown in Table 1, and it can be seen that since the bicubic interpolation method is a typical mathematical violence interpolation method, which is relatively fixed and not good for improvement, although it is simple and intuitive. The later appearing super-resolution algorithms (IBP, POCS, MAP, etc.) have more data, theoretical support and more priori information making them much superior to the bicubic interpolation method in terms of image super-resolution, which is now basically used to appear as a comparison method in the paper. Specifically, the IBP algorithm obtains the inverse projection operator based on a custom degradation model, which is used for iterative optimization to obtain super-resolved images. However, this prevents the introduction of a priori information such as, for example, positivity, energy boundedness, observation consistency, and smoothness, which is compensated by the MAP and POCS methods, where the MAP method goes further and uses the statistical laws of a large amount of data to hyper-segment the image, and also lays the foundation for later data-driven models based on them.

Table 2 shows the improvement of learning-based super-resolution techniques for remote sensing images compared to bicubic interpolation on PSNR and SSIM, using data from the original papers of the relevant methods. SC in Table 2 is the traditional machine learning algorithm, and the others are deep learning algorithms. Overall, different learning-based algorithms have slight advantages at different magnifications or different evaluation metrics, for example, RDBPN, MA-GAN and UVRSR still have good results at 8 times. The latter emerged methods have better image quality compared to the former usually, but it also tends to mean that their networks also become more complex and require more hardware resources. Specifically for deep learning-based super-resolution methods, the network in the algorithm only targets a specific magnification and the size of the input image. If the image size is larger than the set network input size, it is often super-resolved by cropping and then stitched into a large super-resolved image. If the image size is smaller than the set network input size, it is generally not recommended to select this network for super-resolution, or to up-sample the magnification to the network input size first. In the case of end-to-end pure convolutional networks, in principle, images of various sizes can be input, but in practice, it is limited by hardware resources as well as image quality requirements. Different networks need to be trained for different magnifications, which means that deep learning-based methods have high hardware requirements and are currently difficult to be used directly for Onboard processing.

It shows the performance of the previously mentioned methods on the road test set of UC Merced with a super-resolution factor of 4 and an objective evaluation metric of PSNR, with data and images from the papers of ANDGN [27] and RSRCAN [29] in Fig. 10. ANDGN is the unsupervised method and the remaining six are supervised methods. From the figure, we can roughly see that the super-resolution techniques have been developed over the years and the results that can be presented are getting better and better. For the current public dataset, the supervised methods may yield somewhat better results than the unsupervised ones, but unsupervised methods are still the future direction of development.

Table 1. Comparison of traditional super-resolution algorithms

method	bicubic interpolation	IBP	MAP	POCS
Applicable theory	Single	Limited	Lots of	Limited
priori information	no	no	Prior probability density function	Convex sets, simple and efficient
Super-resolution optimal solution	Existing and unique	Not unique, cannot be constrained a priori	Existence and uniqueness, MAP optimal estimation	Not unique, within the intersection of constrained sets
Optimization method	No iteration	Iterative projection method	Standard iterative algorithm	Iterative projection method
Convergence speed	Fast	Slow	Slow	Slow
Convergence stability	High	Relatively low	High	Relatively low
Calculation volume	Relatively low	Relatively high	High	Relatively high
Complexity	Medium	Definition of the projection operator	Optimization under non-convex a priori information	Definition of the projection operator
Noise reduction capability	Relatively low	Relatively low	High	Relatively low
Image smoothing control capability	No	Relaxation factor coefficient	A priori model, regularization factor	Relaxation factor coefficient
Edge preservation capability	Medium	Relaxation factor coefficient	Medium	High
Number of pictures needed / piece	1	$> = 1$	$> = 1$	$> = 1$

5.2 Evaluation Metrics

There is not yet a unified metric for evaluating super-resolution-generated images, because it is not like object detection, image segmentation, etc., where only two options of right or wrong are evaluated on pixels that have already existed. Super-resolution is achieved by adding pixels, and whether the added pixels satisfy the requirements is often subjective and objective criteria are difficult to unify. Therefore, new evaluation metrics have also been continuously proposed by researchers, and the commonly used metrics are as: 1) Mean Square Error (**MSE**). MSE is the average of the difference between the

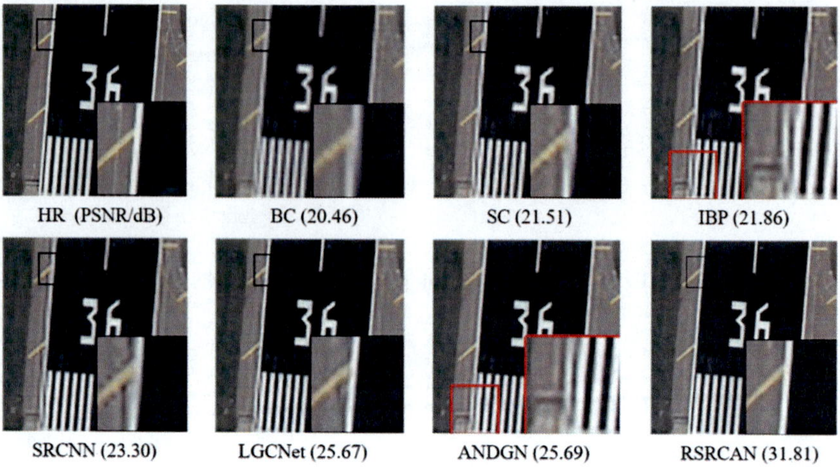

Fig. 10. PSNR of the UC Merced road test image considering a 4 × scaling factor. [27, 29]

pixel values of two images, the smaller the value, the smaller the difference between the two images, but it focuses too much on the comparison of individual pixels, ignoring the overall visual perception of the image. 2) Normalized mean square error (**PSNR**). PSNR is one of the most commonly used image quality evaluation metrics, which combines MSE with the dynamic range of the whole image, and has a more holistic sense compared to MSE. The larger the PSNR value, the more similar the image is. 3) Structural similarity (**SSIM**) [35], another popular objective assessment metric, is a global perceptual model using luminance, contrast, and structure. The higher the SSIM value, the more similar the two signals are. 4) Natural Image Quality Evaluator (**NIQE**) [36], NIQE neither seeks a priori information about distorted images nor relies on any human opinion scores, and a smaller NIQE value indicates better visual quality. 5) Perceptual Index (**PI**) [37], pi value represents the subjective perceptual quality of an image and is also a popular evaluation metric. Often, the lower the pi value, the better the perceptual quality of the image, which is the opposite of the PSNR value. 6) Learned Perceptual Image Patch Similarity (**LPIPS**) [38], LPIPS is more consistent with human perception than traditional methods (than PSNR, SSIM). a lower value of LPIPS indicates that the two images are more similar, and vice versa, the greater the difference. In addition, other metrics, which will not be continued in this paper.

5.3 Datasets

The common datasets of deep learning-based methods for super-resolution reconstruction of remote sensing images are as follows.

1) UC Merced [39]. It is a dataset used to study land use, with a total of 2,100 images containing 21 categories of scenes, 100 images per category, and the pixel size of each image is 256 × 256.
2) RSCNN7 [40]. A total of 2800 remote sensing images from 7 different scene categories, each category contains 400 images.

Table 2. Learning-based methods relative to bicubic linear interpolation for PSNR and SSIM enhancement

method	dataset	metrics	enlarge factor			
			x2	x3	x4	x8
SC [34]	Google Earth	PSNR	–	3.99%	–	–
		SSIM	–	3.80%	–	–
LGCNet [17]	UC Merced	PSNR	8.84%	6.63%	14.15%	–
		SSIM	5.19%	7.95%	22.50%	–
RDBPN [18]	UC Merced	PSNR	–	–	9.64%	5.15%
		SSIM	–	–	16.94%	16.10%
EEGAN [21]	UC Merced	PSNR	14.14%	14.15%	13.38%	–
		SSIM	3.73%	7.59%	11.14%	–
CDGAN [22]	UC Merced	PSNR	–	–	1.95%	–
		SSIM	–	–	–	–
RSRCAN [27]	UC Merced	PSNR	11.74%	10.20%	8.69%	–
		SSIM	5.77%	11.48%	14.60%	–
ANDGN [29]	UC Merced RSCNN7 WHU-RS19	PSNR	8.75%	–	6.87%	–
		SSIM	5.12%	–	11.69%	–
CycleCNN [31]	GaoFen-2	PSNR	–	–	4.66%	–
		SSIM	–	–	1.27%	–
MA-GAN [28]	NWPU-RESISC45	PSNR	11.59%	–	12.16%	10.80%
		SSIM	5.31%	–	16.64%	14.73%
UVRSR [33]	UC Merced	PSNR	7.08%	–	4.74%	18.17%
		SSIM	–	–	–	–

3) NWPU-RESISC45 [41]. From Northwestern Polytechnic University, it contains a total of 31,500 images divided into 45 scene categories with 700 images in each category, and the pixel size of each image is 256 × 256.
4) Kaggle open source dataset [42]. This dataset consists of more than 1000 VHR aerial photographs collected in southern California, USA. It contains 350 images for training and 1370 images for testing.
5) Sentinel-2 dataset, which is one of the Sentinel series, the data is free of charge, and the main payload is a multispectral imager with 13 bands in the spectrum from 0.4 to 2.4 m, covering visible, near-infrared and short-wave infrared, and this dataset is increasingly used as a complement to Landsat in the field of Earth observation.
6) AID [43]. This dataset was released by Wuhan University in 2012. The data source is Google Earth and includes 30 types of remotely sensed scenes such as parks, airports, mountains, and churches. Each type has 200 to 400 images, and all images are 600 × 600 pixels in size. The spatial resolution is 0.58 m/pixel

7) PatternNet. This dataset was published by Wuhan University in 2018. The data source is Google Maps and includes 38 types of remotely sensed scenes, such as forests, highways, railroads, shipyards, and soccer fields. There are 800 images for each category. The size of all images is 256 × 256 pixels with a spatial resolution of 0.064.7m/pixel.
8) WHU-RS19 [44]. WHU-RS19 contains 19 different land classes. Fifty VHR images for each category were obtained from Google Earth with a size of 600 × 600 pixels.
9) IEEE Data Fusion Contest (DFC) 2019 [45]. The DFC (2019) dataset consists of 2783 multi-date satellite images captured by the WV-3 satellite for training and 50 images for testing. The size of the sample image blocks is 1024 × 1024 pixels.
10) SpaceNet dataset (AWS2018) [46]. SpaceNet is another large-scale satellite image dataset, acquired exclusively from the VHR WV-3 satellite.
11) GeoEye-1 dataset. GeoEye-1 satellite is a commercial satellite launched by the United States on September 6, 2008 from Vandenberg Air Force Base, California, to acquire quad-band (blue, green, red, and near-infrared) multispectral images with a spatial resolution of 0.41 m and a spatial resolution of 1.65 m.
12) SPOT-6 dataset. SPOT-6 satellite was launched on September 9, 2012, with a spatial resolution of 1.5 m for panchromatic images and 6 m for multispectral images, including blue, green, red and near infrared.
13) Gaofen2 dataset. Launched on August 19, 2014, the Gaofen 2 satellite has a spatial resolution better than 1 m. It is equipped with two cameras, HR for 1 m panchromatic imaging and 4 m multispectral imaging.
14) DOTA [47]. DOTA is a large-scale benchmark dataset based on aerial imagery generated for target detection tasks. The dataset contains 2806 images collected from Google Earth, GF-2 and JL-1 satellites and aerial images provided by CycloMedia B.V.
15) Multi-sensor Remote Sensing Dataset (MSRSD) [48]. This dataset consists mainly of VHR satellite images acquired by Pleiades 1A/1B, GeoEye-1, QuickBird2, WV-2, WV-3, and DEIMOS satellites and most of them are publicly available. MSRSD includes satellite images from seven different satellites, from different geographic locations and various landscape conditions to simplify the transferability of the model globally and universality, it is a common and rich dataset.

6 Conclusion

Super-resolution processing has been a hot topic in the field of image processing, especially in the field of remote sensing, which has very important application value. This paper reviews image super-resolution processing techniques in remote sensing, categorizing them into traditional methods and learning-based methods, among which, deep learning-based methods are the hot topic nowadays, and more pages are spent on that part in this paper. To facilitate the introduction of RSISR techniques, many methods are categorized in this paper, but there are actually many methods that do not belong to one of these categories, and they integrate many classes of methods as a sub-method to form one method that works better. For example, SWCGAN [49], which integrates CNN, GAN, and Swin Transformer [50] together. For single image super-resolution (SISR), the techniques based on natural images are easy to migrate to remote sensing images, but because the resolution of remote sensing images is much lower than that of natural images, thus causing much less detailed texture information, these images also contain more noise, so blindly performing remote sensing image super-resolution may result in a large amount of false information, which instead reduces the image quality. Therefore, in order to improve the quality of remote sensing image super-resolution, it is suggested that the original resolution of single frame satellite remote sensing image utilized by super-resolution technique is often above 2 m, and preferably higher than 1 m. In addition, during this literature research, the author considers the following main development directions of optical remote sensing image super-resolution technology.

1) Research on the evaluation metric of super-resolution algorithm. As mentioned in Sect. 5.1, super-resolution is intuitively expressed as an increase in the number of pixels, while whether the image meets the requirements after the increase of pixels often needs to be evaluated by the human eye, and objective evaluation metrics cannot be formed yet. Therefore, the image quality evaluation metrics applicable to super-resolution processing still need further research.
2) Research on unsupervised learning of super-resolution reconstruction method for remote sensing images. It is difficult to obtain remote sensing images with different resolutions in the same scene, so there is a lack of training samples, and the low-resolution samples in current supervised algorithms are usually obtained by artificial down-sampling, which cannot simulate the image degradation model well and can hardly cope with the super-resolution reconstruction tasks of some actual scenes, so unsupervised super-resolution reconstruction models of remote sensing images have important research values.
3) Research on the special neural network structure for remote sensing image characteristics. Although the existing methods of super-resolution reconstruction of remote sensing images based on deep learning have achieved good results, remote sensing images are characterized by scale diversity, viewpoint specificity, multi-direction and high background complexity, so it is still a question worth exploring how to use existing technologies (such as attention mechanism, multi-scale feature fusion, reinforcement learning, etc.) to build a more efficient and super-resolution network adapted to the characteristics of remote sensing images problem.
4) Jointly with the imaging system on the satellite. For example, the algorithm proposed by CNES to enhance the resolution of PAN SPOT5 using a quincunx sampling mode, which uses two CCD detectors shifted in the focal plane in order to acquire two images with a resolution of 5 m, resulting in a synthetic resolution of about 2.5 m images [53]

References

1. Lin, C.: Analysis of Electronic intelligence safeguards during the gulf war. National Air Intelligence Center Wright-Patterson AFB OH (1996)
2. Harris, J.L.: Diffraction and resolving power. JOSA **54**, 931–936 (1964)
3. Keys, R.: Cubic convolution interpolation for digital image processing. IEEE Trans. Acoust. Speech Sig. Process. **29**, 1153–1160 (1981). https://doi.org/10.1109/TASSP.1981.1163711

4. Peleg, S., Keren, D., Schweitzer, L.: Improving image resolution using subpixel motion. Pattern Recogn. Lett. **5**, 223–226 (1987)
5. Nguyen, N., Milanfar, P., Golub, G.: A computationally efficient super resolution image reconstruction algorithm. IEEE Trans. Image Process. **10**, 573–583 (2001)
6. Dong, C., Loy, C.C., He, K., Tang, X.: Image super-resolution using deep convolutional networks. http://arxiv.org/abs/1501.00092 (2015)
7. Ledig, C., et al.: Photo-realistic single image super-resolution using a generative adversarial network. http://arxiv.org/abs/1609.04802 (2017)
8. Irani, M., Peleg, S.: Improving resolution by image registration. CVGIP: Graph. Model. Image Process. **53**, 231–239 (1991). https://doi.org/10.1016/1049-9652(91)90045-L
9. Stark, H., Oskoui, P.: High-resolution image recovery from image-plane arrays, using convex projections. J. Opt. Soc. Am. A. **6**, 1715 (1989). https://doi.org/10.1364/JOSAA.6.001715
10. Schultz, R.R., Stevenson, R.L.: Extraction of high-resolution frames from video sequences. IEEE Trans. Image Process. **5**, 996–1011 (1996). https://doi.org/10.1109/83.503915
11. Yang, J., Wright, J., Huang, T., Ma, Y.: Image super-resolution as sparse representation of raw image patches. In: 2008 IEEE Conference on Computer Vision and Pattern Recognition, pp. 1–8. IEEE, Anchorage, AK, USA (2008). https://doi.org/10.1109/CVPR.2008.4587647
12. Liebel, L., Körner, M.: Single-image super resolution for multispectral remote sensing data using convolutional neural networks. Int. Arch. Photogramm. Remote Sens. Spatial Inf. Sci. **XLI-B3**, 883–890 (2016). https://doi.org/10.5194/isprsarchives-XLI-B3-883-2016
13. Aharon, M., Elad, M., Bruckstein, A.: $rm K$-SVD: an algorithm for designing over complete dictionaries for sparse representation. IEEE Trans. Signal Process. **54**, 4311–4322 (2006). https://doi.org/10.1109/TSP.2006.881199
14. Tropp, J.A., Gilbert, A.C.: Signal recovery from random measurements via orthogonal matching pursuit. IEEE Trans. Inform. Theory. **53**, 4655–4666 (2007). https://doi.org/10.1109/TIT.2007.909108
15. Dong, W., Zhang, L., Shi, G., Xiaolin, W.: Image Deblurring and super-resolution by adaptive sparse domain selection and adaptive regularization. IEEE Trans. Image Process. **20**, 1838–1857 (2011). https://doi.org/10.1109/TIP.2011.2108306
16. Zhang, Y., Wu, W., Dai, Y., Yang, X., Yan, B., Lu, W.: Remote sensing images super-resolution based on sparse dictionaries and residual dictionaries. In: 2013 IEEE 11th International Conference on Dependable, Autonomic and Secure Computing, pp. 318–323. IEEE, Chengdu, China (2013). https://doi.org/10.1109/DASC.2013.82
17. Lei, S., Shi, Z., Zou, Z.: Super-resolution for remote sensing images via local-global combined network. IEEE Geosci. Remote Sensing Lett. **14**, 1243–1247 (2017). https://doi.org/10.1109/LGRS.2017.2704122
18. Pan, Z., Ma, W., Guo, J., Lei, B.: Super-resolution of single remote sensing image based on residual dense back projection networks. IEEE Trans. Geosci. Remote Sensing. **57**, 7918–7933 (2019). https://doi.org/10.1109/TGRS.2019.2917427
19. Haris, M., Shakhnarovich, G., Ukita, N.: deep back-projection networks for super-resolution. In: 2018 IEEE/CVF Conference on Computer Vision and Pattern Recognition, pp. 1664–1673. IEEE, Salt Lake City, UT, USA (2018). https://doi.org/10.1109/CVPR.2018.00179
20. Goodfellow, I.J., et al.: Generative adversarial networks. http://arxiv.org/abs/1406.2661 (2014)
21. Jiang, K., Wang, Z., Yi, P., Wang, G., Lu, T., Jiang, J.: Edge-enhanced GAN for remote sensing image super resolution. IEEE Trans. Geosci. Remote Sens. **57**, 5799–5812 (2019). https://doi.org/10.1109/TGRS.2019.2902431
22. Lei, S., Shi, Z., Zou, Z.: Coupled adversarial training for remote sensing image super-resolution. IEEE Trans. Geosci. Remote Sens. **58**, 3633–3643 (2020). https://doi.org/10.1109/TGRS.2019.2959020

23. Hu, J., Shen, L., Albanie, S., Sun, G., Wu, E.: Squeeze-and-excitation networks. http://arxiv.org/abs/1709.01507 (2019)
24. Woo, S., Park, J., Lee, J.-Y., Kweon, I.S.: CBAM: convolutional block attention module. http://arxiv.org/abs/1807.06521 (2018)
25. Vaswani, A., et al.: Attention is all you need. http://arxiv.org/abs/1706.03762 (2017)
26. Zhang, Y., Li, K., Li, K., Wang, L., Zhong, B., Fu, Y.: Image super-resolution using very deep residual channel attention networks. http://arxiv.org/abs/1807.02758 (2018)
27. Haut, J.M., Fernandez-Beltran, R., Paoletti, M.E., Plaza, J., Plaza, A.: Remote sensing image superresolution using deep residual channel attention. IEEE Trans. Geosci. Remote Sens. **57**, 9277–9289 (2019). https://doi.org/10.1109/TGRS.2019.2924818
28. Jia, S., Wang, Z., Li, Q., Jia, X., Xu, M.: Multiattention generative adversarial net-work for remote sensing image super-resolution. IEEE Trans. Geosci. Remote Sens. **60**, 1–15 (2022). https://doi.org/10.1109/TGRS.2022.3180068
29. Haut, J.M., Fernandez-Beltran, R., Paoletti, M.E., Plaza, J., Plaza, A., Pla, F.: A new deep generative network for unsupervised remote sensing single-image super-resolution. IEEE Trans. Geosci. Remote Sens. **56**, 6792–6810 (2018). https://doi.org/10.1109/TGRS.2018.2843525
30. Turkowski, K.: Filters for common resampling tasks. In: Graphics Gems, pp. 147–165. Elsevier (1990). https://doi.org/10.1016/B978-0-08-050753-8.50042-5
31. Wang, P., Zhang, H., Zhou, F., Jiang, Z.: Unsupervised remote sensing image super-resolution using cycle CNN. In: IGARSS 2019 - 2019 IEEE International Geoscience and Remote Sensing Symposium, pp. 3117–3120. IEEE, Yokohama, Japan (2019). https://doi.org/10.1109/IGARSS.2019.8898648
32. Zhu, J.-Y., Park, T., Isola, P., Efros, A.A.: Unpaired image-to-image translation using cycle-consistent adversarial networks. In: 2017 IEEE International Conference on Computer Vision (ICCV), pp. 2242–2251. IEEE, Venice (2017). https://doi.org/10.1109/ICCV.2017.244
33. Zhang, Z., Tian, Y., Li, J., Xu, Y.: Unsupervised remote sensing image super-resolution guided by visible images. Remote Sens. **14**, 1513 (2022). https://doi.org/10.3390/rs14061513
34. Zhihui, Z., Bo, W., Kang, S.: Single remote sensing image super-resolution and de-noising via sparse representation. In: 2011 International Workshop on Multi-Platform/Multi-Sensor Remote Sensing and Mapping, pp. 1–5. IEEE, Xiamen, China (2011). https://doi.org/10.1109/M2RSM.2011.5697420
35. Wang, Z., Bovik, A.C., Sheikh, H.R., Simoncelli, E.P.: Image quality assessment: from error visibility to structural similarity. IEEE Trans. Image Process. **13**, 600–612 (2004). https://doi.org/10.1109/TIP.2003.819861
36. Mittal, A., Soundararajan, R., Bovik, A.C.: Making a "completely blind" image quality analyzer. IEEE Sig. Process. Lett. **20**, 209–212 (2013). https://doi.org/10.1109/LSP.2012.2227726
37. Blau, Y., Mechrez, R., Timofte, R., Michaeli, T., Zelnik-Manor, L.: The 2018 PIRM challenge on perceptual image super-resolution (2019). https://doi.org/10.48550/arXiv.1809.07517. http://arxiv.org/abs/1809.07517
38. Zhang, R., Isola, P., Efros, A.A., Shechtman, E., Wang, O.: The unreasonable effectiveness of deep features as a perceptual metric (2018). https://doi.org/10.48550/arXiv.1801.03924. http://arxiv.org/abs/1801.03924
39. Yang, Y., Newsam, S.: Bag-of-visual-words and spatial extensions for land-use classi-fication. In: Proceedings of the 18th SIGSPATIAL International Conference on Ad-vances in Geographic Information Systems - GIS 2010, p. 270. ACM Press, San Jose, California (2010). https://doi.org/10.1145/1869790.1869829
40. Zou, Q., Ni, L., Zhang, T., Wang, Q.: Deep learning based feature selection for remote sensing scene classification. IEEE Geosci. Remote Sens. Lett. **12**, 2321–2325 (2015). https://doi.org/10.1109/LGRS.2015.2475299

41. Cheng, G., Han, J., Lu, X.: Remote sensing image scene classification: benchmark and state of the art. Proc. IEEE. **105**, 1865–1883 (2017). https://doi.org/10.1109/JPROC.2017.2675998
42. Kaggle: Kaggle Open Source Dataset. https://www.kaggle.com/c/draper-satellite-image-chronology/data
43. Xia, G.-S., et al.: AID: a benchmark data set for performance evaluation of aerial scene classification. IEEE Trans. Geosci. Remote Sens. **55**, 3965–3981 (2017). https://doi.org/10.1109/TGRS.2017.2685945
44. Sheng, G., Yang, W., Xu, T., Sun, H.: High-resolution satellite scene classification using a sparse coding based multiple feature combination. Int. J. Remote Sens. **33**, 2395–2412 (2012). https://doi.org/10.1080/01431161.2011.608740
45. Bertrand: DATA FUSION CONTEST 2019. https://ieee-dataport.org/open-access/data-fusion-contest-2019-dfc2019
46. SpaceNet: The SpaceNet Datasets. https://spacenet.ai/datasets/
47. Xia, G.-S., et al.: DOTA: A Large-Scale Dataset for Object Detection in Aerial Images. In: 2018 IEEE/CVF Conference on Computer Vision and Pattern Recognition, pp. 3974–3983. IEEE, Salt Lake City, UT (2018). https://doi.org/10.1109/CVPR.2018.00418
48. Wang, P., Bayram, B., Sertel, E.: A comprehensive review on deep learning based remote sensing image super-resolution methods. Earth Sci. Rev. **232**, 104110 (2022). https://doi.org/10.1016/j.earscirev.2022.104110
49. Tu, J., Mei, G., Ma, Z., Piccialli, F.: SWCGAN: generative adversarial network combining swin transformer and CNN for remote sensing image super-resolution. IEEE J. Sel. Top. Appl. Earth Ob. Remote Sens. **15**, 5662–5673 (2022). https://doi.org/10.1109/JSTARS.2022.3190322
50. Liu, Z., Lin, Y., Cao, Y., Hu, H., Wei, Y., Zhang, Z., Lin, S., Guo, B.: Swin Trans-former: Hierarchical Vision Transformer using Shifted Windows (2021). https://doi.org/10.48550/arXiv.2103.14030. http://arxiv.org/abs/2103.14030

Research on Levelness Measurement Method of Large Size Object Based on Laser Tracker and Electronic Theodolite

Ziwen Wang[1](✉), Guowei Ruan[1], Minxing Zhou[2], Deyao Hu[1], and Chenxu Duan[1]

[1] Beijing Institute of Spacecraft Environment Engineering, China Academy of Space Technology, Beijing 100094, China
1162126366@qq.com

[2] Institute of Telecommunication and Navigation Satellites, China Academy of Space Technology, Beijing 100094, China

Abstract. In view of the fact that in the current test process, the levelness measurement of the test object mostly relies on small range levelness measurement tools such as a level meter, and the levelness test process of large size test objects is tedious, and there may be loss of data details, data distortion, etc., a levelness measurement method for large size objects based on laser tracker and electronic theodolite is proposed. Through the joint measurement of laser tracker and electronic theodolite, the establishment of the horizontal coordinate system and the measurement of the levelness of the measured object are realized. Through comparative tests, the deviation between the measurement results of high-precision level gauge and the measurement results of this method is less than 0.01 mm/m, which meets the test requirements and can provide more detailed information, and can be used as a solution for levelness measurement of large objects.

Keywords: Laser tracker · Electronic theodolite · Large objects Levelness

1 Preface

At present, in the process of installation and debugging of high-precision tool machines, it is necessary to detect the levelness or verticality of various mechanical equipment, and the measurement of levelness is put forward higher and higher requirements [1–3]. The leveler is usually used as the measuring tool for levelness measurement. With the development of technology, the level gauge has developed from the traditional bubble level, quadrant, image combiner and other one-dimensional angle measuring level gauges based on the principle of level gauge to the highly sensitive inductive and capacitive level gauges based on the principle of pendulum, as well as the photoelectric level gauge [4].

During the levelness measurement, the measuring surface of the level gauge closely fits with the measured surface, and the levelness of the measured object can be obtained by reading the level gauge readings [5], as shown in Fig. 1. When the area of the object to be measured is large, due to the small size of the level gauge, the level gauge is usually used for multi position measurement, and then the levelness data of multiple positions

H. P. Urbach and H. Jiang (Eds.): ISSOIA 2022, SPPHY 295, pp. 229–236, 2023.
https://doi.org/10.1007/978-981-99-4098-1_20

are integrated as the overall levelness of the object to be measured; or use the strip truss together. During the measurement, the truss shall be fitted with the measured surface, and the level gauge shall be fitted closely with the truss to complete the measurement of the area covered by the truss. The measurement range shall be widened by moving the truss. When measuring the levelness of large objects with the help of a level gauge at multiple locations or a strip truss, the local levelness data integration is used as the overall levelness processing method, which may lead to data distortion or missing details.

Fig. 1. Electronic level

In view of the problems existing in the levelness measurement of large objects at present, a method to establish the horizontal coordinate system, and to measure the levelness of the measured object is proposed in this paper, which based on the laser tracker and electronic theodolite. The point position measurement function of laser tracker and the horizontal alignment function of electronic theodolite are applied in this method, to achieve high-precision and rapid measurement of levelness of large objects. It provides a new solution for the establishment of horizontal coordinate system and the measurement of levelness.

2 Method Introduction

The levelness measurement method of large size objects based on laser tracker and electronic theodolite comprehensively uses the laser tracker and electronic theodolite, and reflective target ball is applied to realize the combination of laser tracker and electronic theodolite, and the establishment of the horizontal coordinate system and the measurement of the levelness of the measured object are completed through the contact measurement method.

2.1 Laser Tracker

The laser tracker measurement system is composed of a laser tracking head, a controller, a user computer, a reflector, etc., and it adopts laser interferometry ranging, photoelectric detection and other technologies, which contribute high distance measurement and angle measurement accuracy [6–8]. The measuring principle is shown in Fig. 2. By simultaneously measuring distance L and horizontal angle α、 Zenith angle β, the three-dimensional coordinates of the measured object in the laser tracker measurement coordinate system are obtained [9].

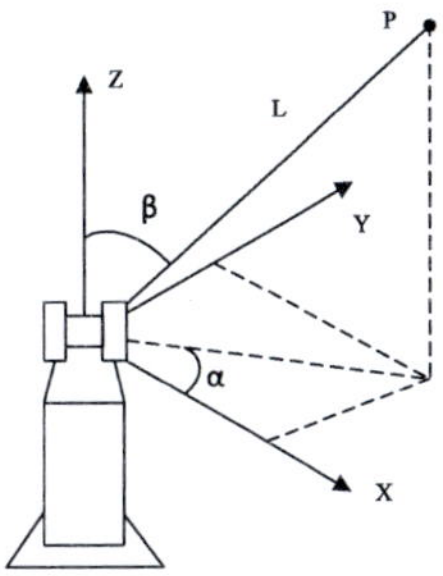

Fig. 2. Measuring principle of laser tracker

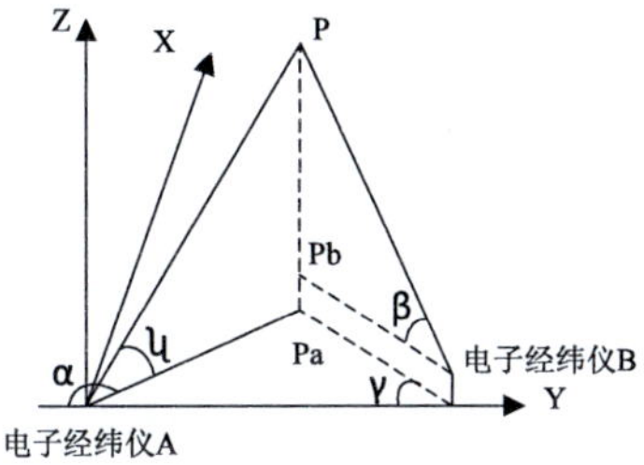

Fig. 3. Principle of electronic theodolite orientation measurement

2.2 Electronic Theodolite

The electronic theodolite is a precision angle measuring equipment integrating light, machine and electricity, which can realize automatic measurement of horizontal angle and pitch angle, and has a dual axis horizontal compensation function, and can correct the angle observation value in real time [10]. The theodolite measurement system mainly has two application scenarios. One is the system orientation measurement based on the principle of space front intersection. Through the joint construction of multiple electronic theodolites, the three-dimensional coordinates (X, Y, Z) of the measured object in the measurement coordinate system are obtained, as shown in Fig. 3. The other is the self collimation attitude measurement based on optical cube mirror. Using the self collimation function of electronic theodolite, the attitude matrix of the measured object in the measurement reference coordinate system is measured by jointly building stations with multiple electronic theodolites [11].

2.3 Establishment Method of Horizontal Coordinate System

According to the light propagation principle, when there is an angle between the light beam and the horizontal plane, that is, there is no parallel relationship with the horizontal plane, the height of the horizontal plane at any point on the light beam and its extension line is different, Hd > He > Hf, as shown in Fig. 4(a). When the light beam is parallel to the horizontal plane, any point on the light beam and its extension line are at the same horizontal height, Ha = Hb = Hc, as shown in Fig. 4(b).

Based on the above principle, this method takes the collimated light of the electronic theodolite mirror tube as the light source, precisely level the electronic theodolite, and precisely adjust its pitch angle to 90.0000°. In this state, the beam emitted by the electronic theodolite barrel is parallel to the horizontal plane. In order to specify the beam emitted from the electronic theodolite lens barrel and the point on its extension line, the reflection target ball is placed on the support with adjustable height, to make it on the extension line of the electronic theodolite optical axis. Set up a laser tracker to measure the three-dimensional coordinates of the reflection target ball center P1 under the laser tracker measurement coordinate system. Keep the elevation angle of the electronic theodolite unchanged, change its horizontal angle, correspondingly move the position

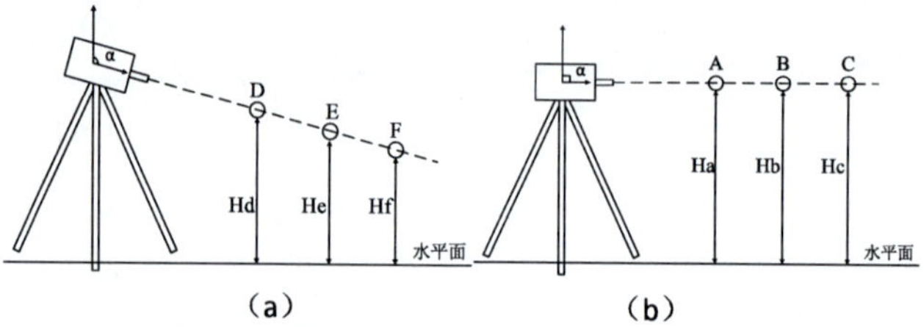

Fig. 4. Height of Horizontal Plane at Each Point on the Beam and Its Extension Line

of the reflecting target ball, and ensure that the target ball is on the extension line of the optical axis of the electronic theodolite. Measuring the central coordinates of the reflecting target ball after the position moves with the help of laser tracker (Fig. 5).

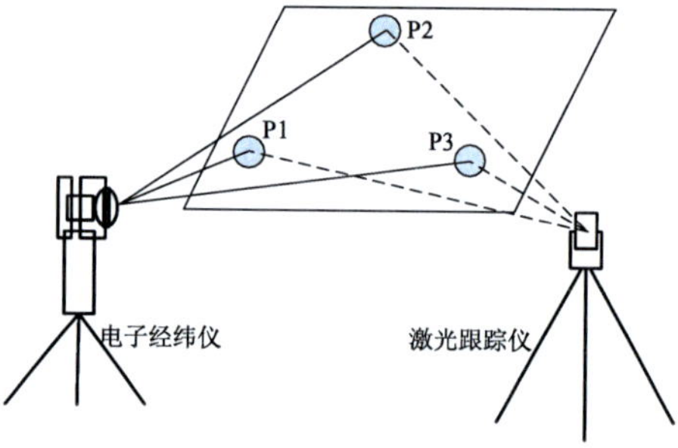

Fig. 5. Establishment method of horizontal coordinate system (taking three-point measurement as an example)

The least square method is used to fit the above collected data to obtain the surface M and its normal HM. Since the electronic theodolite has completed accurate horizontal adjustment, and the optical axis of the mirror tube is precisely located at 90.0000°, the original point Pi used to fit plane M is on the extension line of the optical axis of the mirror tube, so the fitting plane M is parallel to the horizontal plane, which can be used as the benchmark for levelness comparison, and its normal HM is perpendicular to the horizontal plane.

The plane equation is as follows:

$$z = a0x + a1y + a2 \tag{1}$$

For points in the same plane (Xi, yi, Zi) (i = 0,1,2,…, n-1), according to the principle of least squares, the sum of squares of errors is the minimum, namely:

$$S = \sum_{i=0}^{n-1} (z - zi) \tag{2}$$

In order to minimize S, the partial differential derivation of the equations is performed:

$$\left.\begin{array}{l}\frac{\partial S}{\partial a_0}=0\\ \frac{\partial S}{\partial a_1}=0\\ \frac{\partial S}{\partial a_2}=0\end{array}\right\}\tag{3}$$

Integrated:

$$\begin{bmatrix}\sum_{i=0}^{n-1}x_i{}^2\sum_{i=0}^{n-1}x_iy_i\sum_{i=0}^{n-1}x_i\\ \sum_{i=0}^{n-1}x_iy_i\sum_{i=0}^{n-1}y_i{}^2\sum_{i=0}^{n-1}y_i\\ \sum_{i=0}^{n-1}x_i\sum_{i=0}^{n-1}y_in\end{bmatrix}\begin{bmatrix}a_0\\ a_1\\ a_2\end{bmatrix}=\begin{bmatrix}\sum_{i=0}^{n-1}x_iz_i\\ \sum_{i=0}^{n-1}y_iz_i\\ \sum_{i=0}^{n-1}z_i\end{bmatrix}\tag{4}$$

The coefficient of the above equation is solved to obtain the plane equation to be measured.

Keep the position and state of the laser tracker unchanged, and use the reflective target ball to measure the point position of the characteristic point Qi (i = 1, 2, 3,…) on the surface of the object that needs to measure levelness. After the measurement is completed, Qi is taken as the original point data, and the plane fitting of the measured object is carried out to obtain the face N and its normal HN. The levelness of the measured object is the angle between plane N and the horizontal plane β. Since plane M is parallel to the horizontal plane, the levelness of the measured object is also equal to the angle between plane M and plane N γ, It is also equal to the angle between two plane normal HM and HN δ, I.eβ = γ = ε (Fig. 6).

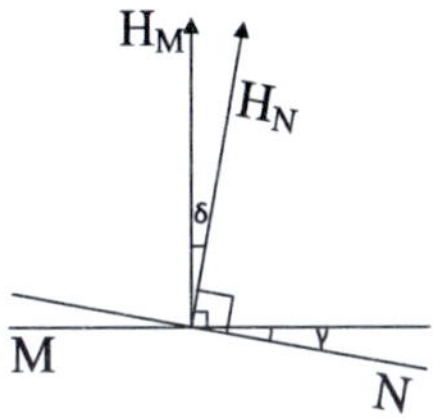

Fig. 6. Measurement of included angle of levelness

3 Test Verification

According to the levelness measurement method proposed in the above chapters, a test verification system is built as shown in Fig. 7. Erect the laser tracker and complete the preheating, erect a Leica TM5100A electronic theodolite, accurately adjust the level and adjust the elevation angle of the lens barrel to 90.0000°, erect another lifting bracket, and firmly paste a target ball socket on its tray. A reflective target ball is stably placed on the target base, and the position of the target ball bracket is moved and the height of the target ball is adjusted so that when the optical axis of the electronic theodolite is aimed at the target ball, its pitch angle is still 90.0000°. Ensure that the state of the target ball bracket remains unchanged, rotate the target ball towards the laser tracker, and measure the point position of the center of the reflection target ball under the laser tracker measurement coordinate system. Ensure that the electronic theodolite remains unchanged, adjust its azimuth angle, repeat the above operations, so that when the target ball center is on the extension line of the optical axis of the electronic theodolite, the elevation angle of the electronic theodolite is still 90.0000°. Repeat the above operation to measure at least 3 non collinear points. With the above measurement points as the source data, the datum plane and its normal are established.

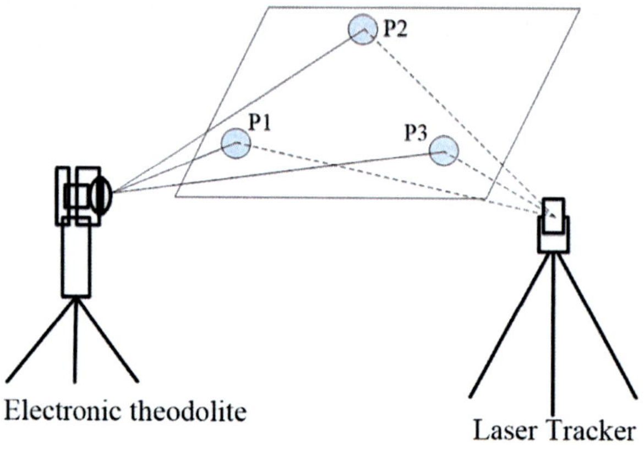

Fig. 7. Establishment method of horizontal coordinate system (Taking three-point measurement as an example)

Keep the position and state of the laser tracker unchanged, use the reflective target ball as the measuring tool to conduct contact measurement on the characteristic points of the measured object, complete the measurement of at least three non collinear points of the measured object, and complete the establishment of the measured plane and its normal. See Table 1 for the observed data (Tables 2, 3, and 4).

Table 1. Datum observation data from electronic theodolite

Num	Azimuth angle of electronic theodolite (°)	Elevation angle of electronic theodolite (°)
P1	0.0087	90.0002
P2	90.0094	90.0001
P3	270.0124	89.9998

Table 2. Datum observation data from laser tracker

Num	Datum observation data from laser tracker (mm)		
	X	Y	Z
P1	−205.0356	−882.5163	4188.0218
P2	−206.2556	−616.0406	3955.5661
P3	−200.2744	−383.6790	4222.0687

Table 3. Observation data of the measured surface from the laser tracker

Num	Observation data of the measured surface from the laser tracker		
	X	Y	Z
P1	−204.8899	−245.5163	4656.0290
P2	−206.1176	20.9594	4423.5733
P3	−199.0759	−13.1556	4922.4196

Table 4. Horizontal contrast data

Object	Levelness(°)
Result of this Method	0.003
Result Measured by Level Gauge	0.002
diviation	0.001

4 Summary

In this paper, aiming at the problems of more times of moving in the process of measuring the levelness of large objects with electronic level, data distortion and loss of details in taking local data integration as the overall levelness data, a method for measuring the levelness of large objects based on laser tracker and electronic theodolite is proposed. Through theoretical analysis and comparative test verification, for the levelness measurement of a small range of measuring objects, the levelness value measured by this method is equivalent to the accuracy measured by a high-precision level; For large-scale object levelness measurement, this method can reflect more details of the levelness of the test object, and the overall levelness performance is good. To sum up, this method provides a new solution for the levelness measurement of a large range of objects, and can provide an idea for related measurement.

References

1. Wang, D., Wei, J., Wang, C., Fan, B., Su, X.: Analysis and testing of electronic horizontal instrument. Eng. Survey. Mapp. **27**(1), 68–72 (2018)
2. Xianhui, H., Xiangjun, Y., Guorong, P.: Application developing of inclination sensor. J. Geomatics **32**(5), 36–37 (2007)
3. Liu, L.: Image detection and development of the two-dimensional Gradienter, Southeast University, pp. 2–3 (2017)
4. He, L., Wang, C., Wu, J.: The measure algorithm and implement of omnidirectional electronical Gradienter. Manuf. Inf. Eng. Chin. **38**(5), 60–62 (2009)
5. Yuhua, Z., Anmin, W., Chengjiang, W.: Application of differential capacitive sensor in intellectual electronic level gauge. J. Transducer Technol. **21**(12), 52–55 (2022)
6. Wu, T.: Research on long-rang collimation of tracking beam and dynamic measuring and control technology for laser tracker. Harbin Institute of Technology, pp. 1–3 (2016)
7. Umetsu, K., Furutnani, R., Osawa, S., et al.: Geometric calibration of a coordinate measuring machine using a laser tracking system. Meas. Sci. Technol. **16**(12), 2466 (2005)
8. Lijie, T.: The application of laser tracker in the aircraft components docking process. Dazhong Keji **15**(163), 3–4 (2013)
9. Wang, Y., Zhou, W., Wang, Z., Ding, L.: Correction method of air refractivity for laser tracking measurement system. Optoelectron. Technol. **31**(2), 83–86 (2011)
10. Krishna, R.: Improved pointing accuracy using high precision theodolite measurement. SPIE **2812**, 199–209 (1996)
11. Hongrui, M.: Study on integrated application of rangefinder and electronic theodolite and key technologies. Nanjing University of Science and Technology, pp. 4–5 (2013)

Research on TDI Mode of Area Array Camera for Space Debris Detection

Lin He(✉), Liguo Song, Xuguo Zhang, Yuchen Liu, Wenping Lei, and Pu Huang

Beijing Institute of Space Mechanics and Electricity, Beijing, China
helincoo@163.com

Abstract. In order to improve the detection capability of small-scale space debris, the method of simply increasing the camera aperture cannot meet the requirements of satellite resource constraints. In this paper, a time delay integration (TDI) imaging mode of area array camera based on point target detection is designed, and the imaging parameters are designed. Based on the TDI imaging principle of an area array detector, a calculation model of the camera's signal-to-noise ratio is given, and the effects of target motion and integration time on imaging are analyzed. The principle prototype test shows that the camera can achieve the detection capability of space debris target better than 15 magnitude, and the signal-to-noise ratio is greater than 5.

Keywords: space camera · space debris · TDI mode of area array imaging · SNR

1 Introduction

In order to improve the detection capability of small-scale space debris, the method of simply increasing the camera aperture cannot meet the requirements of satellite resource constraints. The space debris is a point target in the focal plane of the camera due to its long distance, and its visual brightness is relatively weak. The smaller the target size is, the smaller its sunlight reflection cross section area is, and the lower its radiant brightness is. In order to detect and find smaller scale space debris, if the optical aperture is increased, the weight and volume of the camera will be greatly increased. d. If the plane array staring imaging mode is adopted, the integration time can be extended, but the gimbal needs to be added to ensure the relative stability of the camera optical axis and the target, which increases the complexity of the system. The integration time is increased by time delay integration (TDI), so as to improve the detection sensitivity of the camera. Uses a smaller camera aperture to achieve the same magnitude level, effectively reducing the weight of the camera, and has a higher cost-effectiveness ratio.

This paper designs an imaging mode based on time delay integration of area array camera (area array TDI). Its advantages are that, on the one hand, the camera integration time is increased through TDI technology, which greatly improves the camera's ability to detect dim and weak space debris under the same optical aperture; On the other hand, the camera adopts the area-array output mode, which can ensure the observation arc required by the space debris orbit determination.

H. P. Urbach and H. Jiang (Eds.): ISSOIA 2022, SPPHY 295, pp. 237–245, 2023.
https://doi.org/10.1007/978-981-99-4098-1_21

2 Working Principle

Time Delay Integration (TDI) technology is often used in remote sensing cameras. Based on multiple exposures of the same ground object, the collection of light energy is greatly increased, thus obtaining higher sensitivity, wider dynamic range. The working principle of area array TDI is similar to that of linear array TDI. The difference is that the output of area array TDI is area-array image at the same exposure time, which can meet the demand for orbit determination of space targets.

The area array TDI mode is realized by using a frame transfer TDICCD structure detector. By controlling the CCD working time sequence, the relative motion speed of

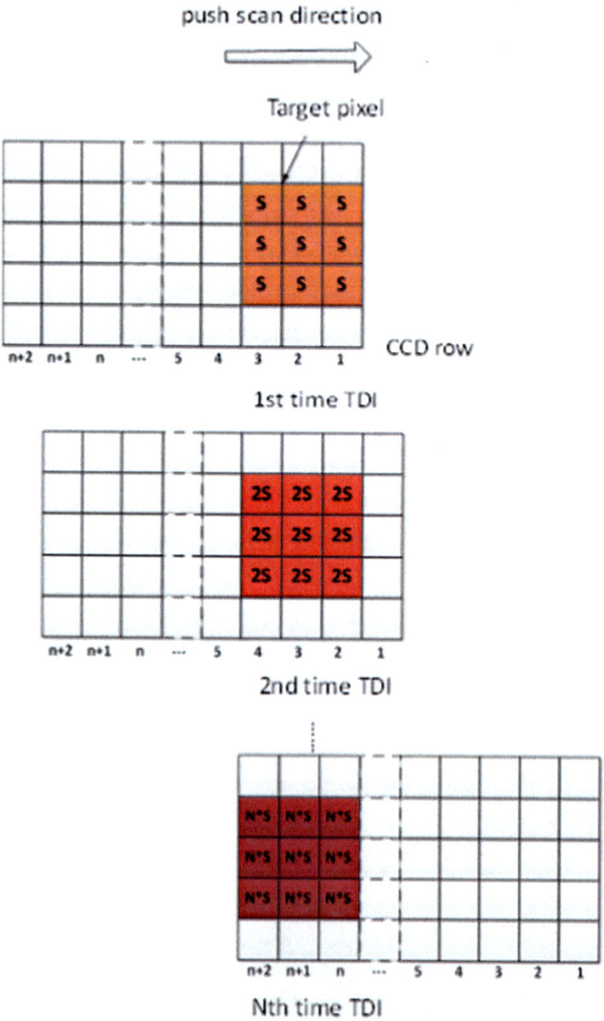

Fig. 1. Schematic diagram of TDI imaging principle of area array

the target is matched with the charge transfer speed, so that multiple exposures of the same point target are accumulated to significantly enhance the signal, and the target smear caused by the motion is eliminated. For one M × N array detector, after n times of TDI, the signal strength becomes nS, and the output image size is M × (N-n) (Fig. 1).

3 Calculation and Analysis of TDI Imaging Parameters

1) Relative motion

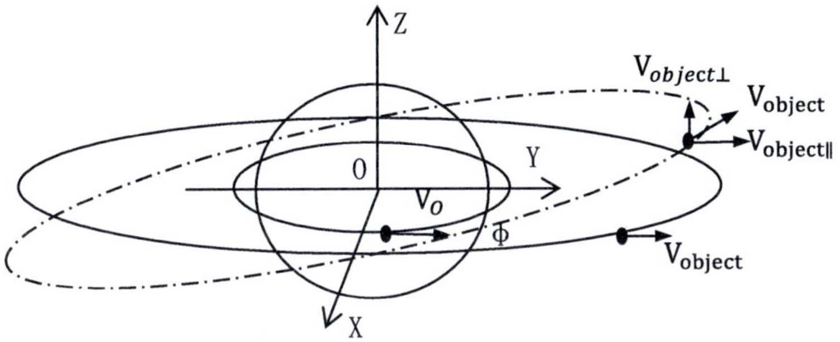

Fig. 2. Observation of different inclination space debris

Taking 600 km–900 km orbits as an example, typical inclinations are shown in Table 1. For space debris targets with different inclinations, the variation range of their east-west relative angular velocity relative to the camera is very small, which can be considered as constant 0.065°/s. With the increase of the orbital inclination, the north-south angular velocity increases. When the orbital inclination is 15°, the north-south relative angular velocity is 0.0013°/s at the maximum, which is far less than the east-west angular velocity (Fig. 2). When designing the TDI CCD circuit, ensure that the charge transfer speed strictly matches the relative angular velocity in the east-west direction. Assume that the angular size of camera pixel is a × a. The focal length of the camera is f, and the TDI row frequency of CCD is f_T needs to meet:

$$f_T = \frac{V_{object\parallel}}{a} \times f \tag{1}$$

2) Maximum integration time

Analyze the image motion velocity under the influence of the orbital velocity of space debris with different inclinations, and comprehensively calculate the reasonable integration time. In order to ensure the imaging quality, the amount of image movement in the integration time should be less than one pixel. Considering the horizontal and vertical velocity components, the exposure time shall meet the following conditions:

$$T_{exp} \leq min\left(\frac{a}{f \cdot \left(\mathrm{V}_{\mathrm{object}\parallel} - V_{O\parallel}\right)}, \frac{a}{f \cdot \left(\mathrm{V}_{object\perp} - V_{O\perp}\right)}\right) \quad (2)$$

After the horizontal relative speed is compensated by TDI through circuit design, it can be considered as $\left(V_{object\parallel} - V_{O\parallel}\right)$ is almost zero and can be ignored. Image motion is mainly determined by the relative velocity in the vertical direction. When $V_{O\perp}$ is zero, the integration time meets:

$$T_{exp} \leq \frac{a}{f \cdot \mathrm{V}_{object\perp}} \quad (3)$$

When the target orbital inclination is 0°, there is no relative angular velocity in the north-south direction, so there is no image movement. As the orbital inclination becomes larger, it is concluded from Table 2 that the relative angular velocity in the north-south direction gradually increases, and the image movement also increases accordingly. Therefore, the integration time is limited and cannot be too long, and there is a maximum integration time:

$$T_{max} = \frac{a}{f \cdot \mathrm{V}_{object\perp}} \quad (4)$$

Assume the angular resolution of the camera is 15″. In the TDI mode, the east-west relative velocity of the target is compensated to zero, and the north-south vertical image motion and maximum integration time of the typical dip space debris are calculated respectively. It can be seen from the figure that the longer the integration time is, the greater the north-south vertical image motion is. The larger the target inclination is, the greater the north-south velocity component is, and the greater the image motion is. The maximum integration time allowed by the camera when observing a 15° inclination target is 3.2 s in the extreme case, as shown in the table.

Table 1. Maximum integration time allowed for space debris targets with different inclinations

NO	orbital inclination (°)	North-south image motion (pixel)	Maximum integration time allowed (s)
	2	≤1	23.84
	4	≤1	11.93
	6	≤1	7.96
	8	≤1	5.98
	10	≤1	4.79
	12	≤1	4.05
	15	≤1	3.2

3) Detection sensitivity and signal-to-noise ratio

For the detection of space point targets, the visual magnitude (Mv) is usually used to measure the detection sensitivity of visible light cameras. According to the magnitude formula:

$$m_2 - m_1 = 2.5\log\frac{E_1}{E_2} \tag{5}$$

In the above formula, m_2 and m_1 is the magnitude of two celestial bodies respectively, while E_1 and E_2 is the irradiance of two space object respectively. Take the sun as the reference magnitude −26.5, and the sun irradiance of 0.4–0.9 μm is 763.6 W/m^2, which can calculate the irradiance E_{in} of the observation target before entrance pupil.

The point target energy is imaged on the focal plane image sensor after receiving light through the camera. After time integration through the detector, the photon signal is converted into the number of electrons to obtain the signal S:

$$\mathrm{S} = \frac{\pi D^2(1-\varepsilon)\tau_o E_{in} a^2 \eta T_{exp}}{4 S_{img} E_{ph}} \tag{6}$$

where, D is the optical aperture of the camera, ε is the camera occlusion ratio, τ_o is optical transmittance, a is pixel size, S_{img} is light spot size, η is sensor QE, T_{exp} is integration time, E_{ph} is photon energy. It can be concluded that the signal strength is linear with the integration time when other quantities remain unchanged. In the area array TDI mode, the total integration time of the camera is equal to the product of a single pixel integration time T_{line} and TDI stage N.

The calculation formula of σ_{noise} is as follows, and the unit is the electronic number e-

$$\sigma_{noise} = \sqrt{S_b^2 + (I_d \times \mathrm{T_{exp}})^2 + N_r^2 + N_q^2} \tag{7}$$

where, Sb is the number of background noise electrons, $S_b^2 = S$, Id is dark current, Nr is the readout noise, Nq is quantization noise.

The camera signal-to-noise ratio is calculated as follows:

$$\mathrm{SNR} = \frac{S}{\sigma_{noise}} \tag{8}$$

Signal to noise ratio after N times TDI integrations:

$$SNR_{TDI} = \frac{NS}{\sqrt{NS + N^2(I_d \times \mathrm{T_{exp}})^2 + N_r^2 + N_q^2}} \tag{9}$$

Assuming that the visual magnitude of space debris is 15Mv, it is generally required that the signal-to-noise ratio is greater than 5 to ensure effective detection. According to Formula (9), when the system noise includes detector noise, circuit noise and background stray light noise, it is necessary to increase the camera aperture and the integration time for improving SNR. Assume that the camera spectral band is 450 nm–900 nm, the optical transmittance is 70%, and the average quantum efficiency is 60%, the energy concentration of 3 × 3 pixels is 90%, the CCD reads out noise is 5e−, dark circuit noise is 1e −/s, and the circuit noise 6e−. Calculate the camera optical aperture and integration time required for detecting 15 magnitude, as shown in the table. Based on the above analysis, we finally design optical aperture is 150 mm, and integration time is 3 s.

Table 2. Optical aperture and integration time of camera

	Optical aperture (mm)	Integration time (s)	SNR
1	140	3.9	5.09
2	150	3.4	5.10
3	150	2.9	5.00
4	170	2.6	5.05
5	180	2.3	5.02
6	190	2.1	5.09
7	200	1.9	5.11

4 Experiments and Results

In order to verify the area array TDI imaging mode and camera design specifications, a principle prototype camera is developed. The camera lens uses an off-axis three mirror optical system without optical occlusion to obtain higher optical transmittance. The camera adopts a frame transfer type area array CCD detector, with a pixel size of 13 um, and operates in the area array TDI mode. The detection performance and signal-to-noise ratio of the camera are tested. The infinitely distant star point target is simulated by using the method of "adjustable aperture integrating sphere + long focal length collimating system". Put the camera on a high-precision turntable, and the turntable rotates at a constant speed of 0.065°/s to simulate the relative angular velocity between the satellite and the space debris. Set the CCD row cycle to 32 ms, adjust the TDI stage, and collect images in real time. Calculate the SNR through the image by selecting 50 × 50 pixels area around points target uniform background, DN value of the brightest pixel of the target signal is marked as S, background mean value is u, background root mean square noise in the area is σ_{noise}, the SNR ratio calculated according to the point source target SNR formula is

$$\mathrm{SNR} = \frac{S - u}{\sigma_{noise}} \tag{10}$$

Acquire 50 images according to the above method to calculate the SNR of each image. Then average it to obtain the SNR of the camera (Table 3). The actual measurement shows that the SNR of the camera in the area array TDI mode is 6.9, the TDI stage is 96, and the integration time is 3.072 s, which verifies the design indicators (Figs. 3 and 4).

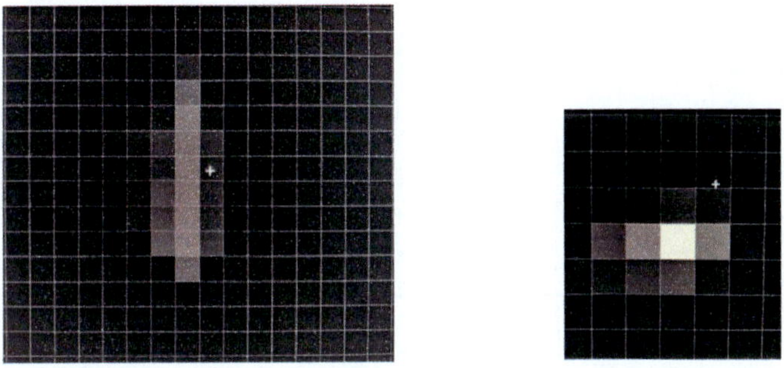

a) Moving point target imaging smear(no TDI) b) Moving point target imaging of area array TDI (8-stage TDI)

Fig. 3. Point target imaging of area array TDI mode test results

Table 3. Camera response linearity test with integration time

integration time(s)	average DN value of 50 images
0.08	310
0.15	382
0.32	685
0.64	1471
0.96	1978
1.28	2579
1.6	3217
1.92	3829
2.24	4461
2.56	5056
2.88	5694
3.0	6321

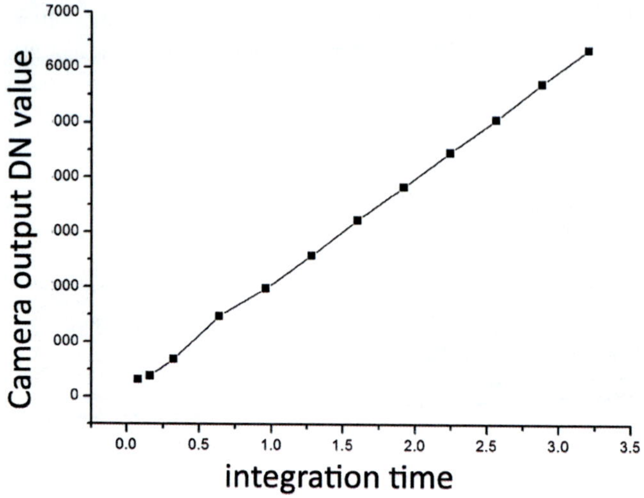

Fig. 4. Camera Response Curve

5 Concluding Remarks

In this paper, a TDI imaging mode based on area array camera is studied to observe space debris targets, which can extend the exposure time through time delay integration without increasing the optical aperture, so as to improve the detection sensitivity of the camera. Combined with the principle of area-array TDI mode, the calculation methods of the main imaging parameters are applied to the camera design, and the principle prototype of camera is developed for verification. The experimental results show that the camera can realize the detection ability of space debris, which is better than 15Mv, and the SNR is greater than 5.

References

1. Xuxia, Z., Ningjuan, R., Sisi, Z.: Characterization techniques of space objects based on point target information. Infrared Laser Eng. **45**(S1), S126001 (2016)
2. Stokes, G.H., Von Braun, C., Sridharan, R.: The space-based visible program. Lincoln Lab. J. **2**, 205–229 (1998)
3. Gaposchkin, E.M., Von Braun, C.: Space-based space surveillance with the space-based visible. J. Guid. Control. Dyn. **23**(1), 148–152 (2000)
4. eoPortal. ORS-5(Operationally Responsive Space-5)/SensorSat (2018). https://directory.eoportal.org/web/eoportal/satellite-missions/o/ors-5
5. Lianda, W.: The Obit and Detection of Artificial Satellites and Space Debris, pp. 266–293. Science and Technology of China Press, Beijing (2011)
6. Yuheng, Z., Yihua, Y.: Design of frame transfer area CCD imaging system for dark objects. Acta Astronom. Sinica **56**(3), 295–304 (2015)
7. Qingyu, M., Wei, Z., Funian, L.: Analysis on detection ability of space-based space target visible camera. Infrared Laser Eng. **41**(8), 2019–2083 (2012)

8. Guangjun, F., Zhen, M., Yingcai, L.: Design and performance analysis of standard starlight simulator. J. Appl. Opt. **31**(1), 39–42 (2010)
9. Yi, H., Ming, C., Jianfeng, X., et al.: Simulation & validation of ground-based optical—telescope imaging on target craft. Infrared Laser Eng. **48**(12), 1214002 (2019)
10. Luyao, H.A.N., Chan, T.A.N., Yunmeng, L.I.U., et al.: Research on the on-orbit real-time space target detection algorithm. Spacecraft Recov. Remote Sens. **42**(6), 122–131 (2021)
11. Wenyu, Z.: Point target imaging simulation and experiment based on temporal-spatial oversampling system. Infrared Laser Eng. **44**(10), 3124–3129 (2015)

A Micro-Satellite Design for Deep Space Exploration

Zhenhua Su, Yibing Ding(✉), Yumei Zhang, Liang Yin, Hong Zeng, Haiming Wang, Xiaojun Zhang, and Zhendong He

DFH Satellite Co., Ltd., Beijing 100094, China
dingyibing@buaa.edu.cn

Abstract. Deep space exploration is of great significance for mankind to develop and utilize space resources and explore the origin of the universe. Due to the huge consumption of human, material, financial and time costs of deep space exploration missions, therefore, both the miniaturization and high reliability requirements of aircraft are urgently required. Micro-satellite based on MEMS technology has the characteristics of small size, light weight, high integration and reliability, which is applicable to deep space exploration missions. Based on the systematic analysis of deep space exploration missions, a micro-satellite design suitable for deep space exploration is proposed. The energy supply of the aircraft, intense radiation environmental effects and protection, flight strategy and orbit optimization design, extreme high and low temperature thermal control technology and long-life autonomous operation of interplanetary crossing are analyzed in detail. The first was to elaborate the requirements of deep space exploration micro-satellites in terms of energy supply, intense radiation environmental effects and protection, flight strategy and orbit optimization design, extreme high and low temperature thermal control technology and long-life autonomous operation of interplanetary crossing. Then the paper demonstrates the difference between deep space exploration micro-satellites and near-Earth exploration satellites and provides feasible technology paths and design schemes by investigating the status of domestic and international research. The micro-satellite is finally verified to meet the deep space exploration mission, and its technical features and advantages and disadvantages are analyzed. The results indicate that the micro-satellite is capable of matching the requirements of miniaturization and high reliability for deep space exploration simultaneously and has high-cost performance. The MEMS-based micro-satellite energy supply, intense radiation protection, orbit and thermal control technology and long-life autonomous operation technology solve the problems of miniaturization and reliability of deep space exploration, laying the foundation for micro-satellites application in deep space exploration missions. The paper also discusses the development trend of micro-satellites applicable to deep space exploration.

Keywords: Deep space exploration · micro-satellite · scheme design · MEMS

H. P. Urbach and H. Jiang (Eds.): ISSOIA 2022, SPPHY 295, pp. 246–256, 2023.
https://doi.org/10.1007/978-981-99-4098-1_22

1 Introduction

Deep space exploration refers to the exploration of planets and their satellites, asteroids and comets in the solar system, as well as the planets outside the solar system. Deep space exploration is of great significance to human beings and the earth. Through the exploration of extraterrestrial space, human beings can further understand the origin and evolution of the earth and the formation of stars, which is conducive to improving human utilization of space resources and may even discover extraterrestrial life. Over the past few decades, European and American countries have launched more than 100 deep space probes, accumulating rich information and experience for deep space exploration [1]. China's deep space exploration program has also been implemented. Compared with other space activities, deep space exploration has more urgent requirements for satellite miniaturization, and the resources of mass, volume and energy are more valuable. Therefore, the miniaturization of deep space exploration satellites determines the cost performance of exploration missions [2]. During the execution of deep space exploration missions, if a certain characteristic or parameter of the deep space probe has a large deviation that exceeds the normal operating conditions, it may cause the overall system to fail, which in turn leads to mission failure [3]. As deep space exploration mission has invested huge financial, human, material and time costs, it requires high reliability of deep space exploration satellites. Micro-satellite based on MEMS technology has the advantages of small size, light weight, high integration, good reliability and satisfactory performance. It is suitable for deep space exploration and is also the future development trend [4].

Deep space exploration micro-satellites experience extremely special environment, and its design has the following characteristics: 1) Extreme high and low temperature environment, in different orbital positions will face several times the earth's illumination or a few percent of the earth's illumination changes, not only to withstand extremely high temperature, but also to withstand extremely low temperature environment; 2) In the extreme high and low temperature environment, the satellite needs to carry out orbit and attitude maneuver, so the micro-satellites are required to have sufficient strength in the extreme high and low temperature environment; 3) micro-satellites need sufficient energy supply to ensure the long-term exploration mission; 4) The radiation environment faced by deep space exploration missions is complex and harsh. The energy and flux of particles in deep space are much higher than those of the Earth's radiation belt, therefore the protection design of micro-satellites is required [5].

Based on the above analysis, this paper studies the overall design and energy supply of micro-satellites for deep space exploration, strong radiation environmental effects and protection, flight strategy and orbit optimization design, extreme high and low temperature thermal control technology and long-life autonomous operation. A micro-satellite design for deep space exploration is proposed, and the micro-satellite is verified to meet the deep space exploration mission, and its technical characteristics and advantages and disadvantages are analyzed. The rationality and effectiveness of the micro-satellite design were verified, and it can meet the requirements of miniaturization and reliability for deep space exploration and has a high-cost performance. It also provides a valuable idea for the design of miniaturized and highly reliable deep space exploration satellites and reducing the cost of deep space exploration.

2 Overall Design Based on MEMS

2.1 MEMS Technology for Deep Space Exploration

With many virtues unmatched by large satellites, micro-satellites perform an essential role in deep space exploration [6]. First, the development cycle of micro-satellites is short; then, micro-satellites have a low R&D (research and development) expense and minimal risk. MEMS technology can be understood to some extent as a design method and process means to realize system integration and miniaturization. MEMS technology based on the manufacturing technology of semiconductor devices is a powerful tool to realize small devices [7]. The emergence of MEMS technology provides an important support for the miniaturization of micro-satellites. The prominent features of MEMS devices are small mass and size, high integration, low power consumption, and batch manufacturing, thus reducing costs. On the one hand, MEMS technology can geometrically minimize the size of the satellite system to achieve the goal of miniaturization while obtaining the desirable functional performance; on the other hand, MEMS technology also makes it feasible for the incorporation and integration of microsatellite design and manufacturing [8].

MEMS technology-based microsatellites combine small size and light weight, high integration, good reliability, better performance characteristics, which is the ideal solution for deep space exploration.

2.2 Overall System Design Based on MEMS

The deep space exploration micro-satellite adopts a highly integrated, high-precision, highly reliable and lightweight micro-satellite platform based on MEMS technology. The platform service part includes thermal control module, integrated electronic module, energy module, measurement and control module, inter-satellite communication module. The satellite platform is capable to meet the requirements of deep space exploration attitude and orbit maneuver. According to the specific exploration mission, the satellite flexibly configures deep space exploration payloads such as optical observation, signal detection and scientific exploration. The overall system design is shown in Fig. 1. The deep space exploration micro-satellite will be delivered to the transfer orbit firstly by launch vehicle, then the micro-satellite's propulsion system executes orbit maneuver to enter the deep space exploration mission orbit.

2.3 An Example of a MEMS-Based Microsatellite Design

A MEMS-based microsatellite is designed with an open architecture and standard interface specifications, applying innovative design concepts and technologies such as Commercial-Off-The-Shelf devices, industrial technologies and miniaturized products. The microsatellite module is designed according to the PC104 specification, which ensures that each module unit has a certain degree of functional and structural independence and integrity. The module components include micro integrated electronic system, micro measurement and control transponder, micro navigation receiver, micro sun sensor, lightweight high precision star sensor, micro momentum wheel, micro thruster, micro magnetometer and other miniaturized components based on MEMS design.

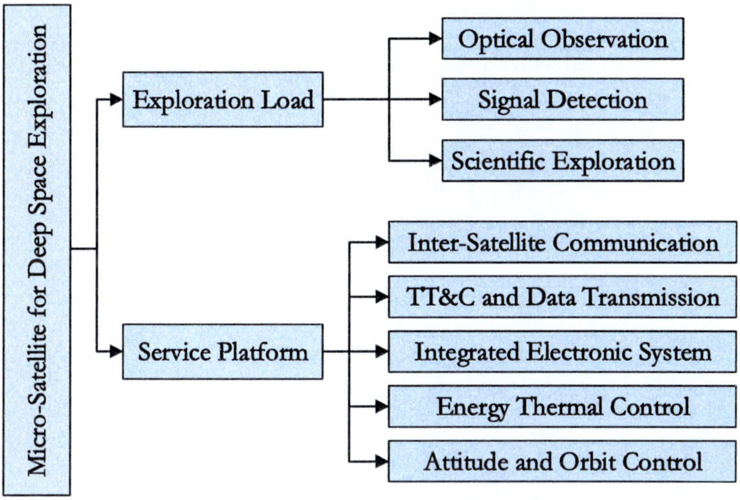

Fig. 1. Overall System Components.

The PC104 specification stack structure of the microsatellite integrated power, measurement and control, integrated electronics and other modules, realized the electronic stacking and signal processing centralization of the satellite. Beyond the stack, only the battery, RF module, antenna, control components, structural parts are retained, which greatly contributes to the integration and miniaturization of the satellite. Through actual testing, the power consumption of the whole microsatellite designed with PC104 specification stack structure has been reduced by 70%, and the weight and size have been reduced by 50%, realizing a miniaturized and highly reliable deep space exploration satellite design (Fig. 2).

In addition, MEMS technology and SOC technology are used to package the power supply system, propellant storage and supply system and thruster modules into a very small volume as shown in Fig. 3 to achieve a high functional density and work stability. The nozzle area of the micro thruster processed by MEMS technology is less than $10^{-10}m^2$, and the action time can reach millisecond level, the minimum thrust is up to mN or even μN, and the minimum impulse reaches μN/s, which meet the demands of high-precision attitude control of micro/nano-satellites.

3 Micro-Satellite Module Design

3.1 Energy Supply Design

As the space environment solar energy flux in the near-Earth orbit is large, which can reach several thousands of W/m^2, China's near-Earth exploration satellites energy supply generally use solar power generation currently. While the solar energy flux in the deep space environment is small only a few hundred W/m^2 [9]. For example, the solar energy flux of Uranus orbit is only 3 W/m^2, in such cases using the solar power will be a huge expense, and even cannot use solar energy in some extreme conditions (Fig. 4).

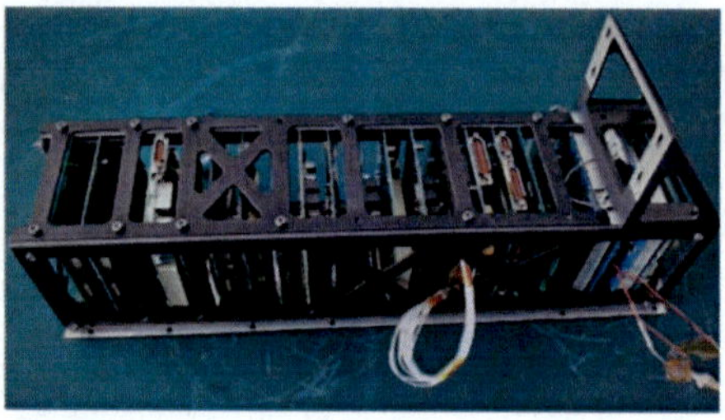

Fig. 2. Stack Structure Design based on PCI104 Specification.

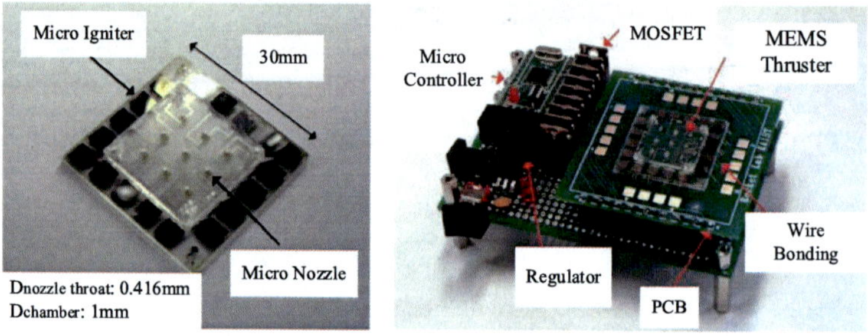

Fig. 3. MEMS Thruster and MEMS Thruster Array.

On the one hand, the interplanetary crossing phase of the deep space exploration mission is in a low temperature and low solar energy flux environment. As the distance of the probe from the sun increases, the solar radiation energy obtained continues to decrease, while the energy consumption required to maintain the stable operation of the thermal control system keeps increasing. On the other hand, the scarcity of power and energy brings constraints to its use. To solve this problem, space isotope power sources are generally used in deep space exploration missions, which can not only solve the power source issue, but also provide a continuous source of heat to sustain satellite operation using extra heat. Space isotope power technology is a power source technology using the decay heat of radioisotopes as the energy source to generate electrical energy available for satellites through a thermoelectric conversion power generation unit [10].

3.2 Strong Radiation Protection Design

The radiation environment faced by deep space exploration missions is complex and harsh. The energy and flux of particles in the radiation belts of some planets in deep space are much higher than those in the Van Allen belt. For example, the flux of low-energy protons of Jupiter is ten times higher than that of the Van Allen belt, and the flux of high-energy electrons is 2–3 orders of magnitude higher than that of the Van Allen belt. There are numerous high-energy electrons with energy exceeding 10 meV in the radiation belts, and even some electrons with energy up to GeV, In contrast, the maximum energy of electrons in the Van Allen belt does not exceed 7 meV [11].

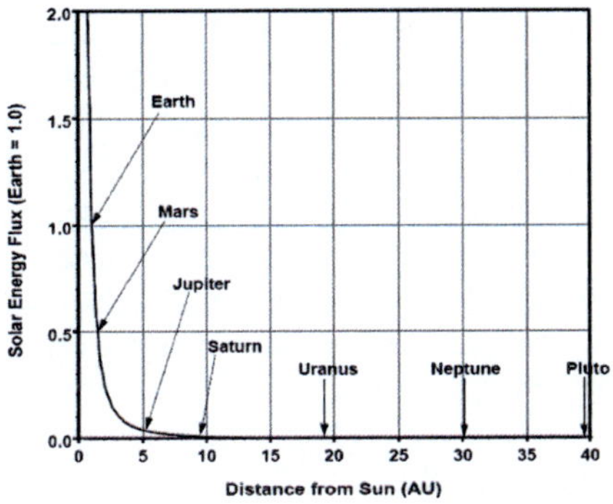

Fig. 4. Diagram of Solar Energy Flux as Function of Solar Distance

High energy, high flux particles have potential risks for satellite operation, which may lead to abnormal operation of the probe in orbit, or lead to satellite deactivation or loss of control, seriously threatening the survival of the satellite. For example, the total ionizing dose on electronic components and single event effects will cause errors in the logic circuitry and even lead to device failure. And the solar panel damage by radiation can result in output power decay, in general, the solar panel efficiency decay of about 15% per year.

The space radiation environment and its effects are one of the vital factors affecting the high reliability and long-life operation of satellites and the protection of the space radiation environment is a difficult point in the design of micro-satellites. Due to the huge difference in the energy spectrum of particles in the deep space environment, the space radiation environment and the radiation effect of high-energy particles are analyzed and studied, and the ground simulation tests are carried out to formulate the space environment protection design for deep space exploration [12].

To provide reliable environmental parameters for the design of satellite anti-radiation reinforcement, the elements that may do harm to the satellite in the space radiation environment of deep space exploration mission is studied, the applicability of each environmental model is evaluated, and the data credibility is analyzed. Through conducting the analysis of three-dimensional shielding model and the study of unified protection design of shield cover, the radiation resistance indicator requirements of electronic devices are specified. Then the indicator requirements are reduced through optimization analysis to save satellite weight resources. We have completed the electronic device protection design, solar power battery anti-radiation reinforcement and system radiation protection design of the deep space exploration microsatellite to support the deep space exploration missions.

3.3 Orbit Control Design

Compared with near-Earth exploration, the detection range of deep space exploration mission is more distant. The traditional direct transfer method based on Hohmann orbit cannot meet the mission requirements of deep space exploration in engineering due to the limited launch capacity of current launch vehicles. A new deep space transfer orbit strategy must be adopted to solve this problem. Deep space exploration missions are highly expensively, and gravity-assisted orbit transfer by natural planets has become a key technology in deep space orbit design to ensure that sufficient payloads can be carried and to maximize the scientific payoff of the mission [13]. Considering that deep space exploration satellites can change their orbits by means of the gravity when flying over large planets in the solar system, it can reduce the launch energy of deep space exploration satellites, save the cost of exploration missions and reduce fuel consumption [14]. At present, all international deep space exploration missions have adopted the orbit transfer scheme of borrowing force from large solar system planets. The borrowed flight transfer orbit involves many aspects, such as precise orbit design, launch trajectory splicing, error transfer analysis, as well as measurement and control, energy, radiation protection and other engineering constraints.

Aiming at the problems of complex deep space exploration missions, multiple mission constraints, and uncertain flight environment factors. Breakthroughs have been made in many key technologies such as deep space exploration micro-nano satellite mission planning and engineering constraint analysis, multiple leveraging planet optimization and leveraging sequence planning technology, orbit correction strategy under accurate dynamic model, continuous thrust leveraging flight orbit optimization technology and orbit evolution characteristic analysis. Combined with these techniques, the problems of multi-factor uncertainty and multi-mission flight under large system constraints are solved (Fig. 5).

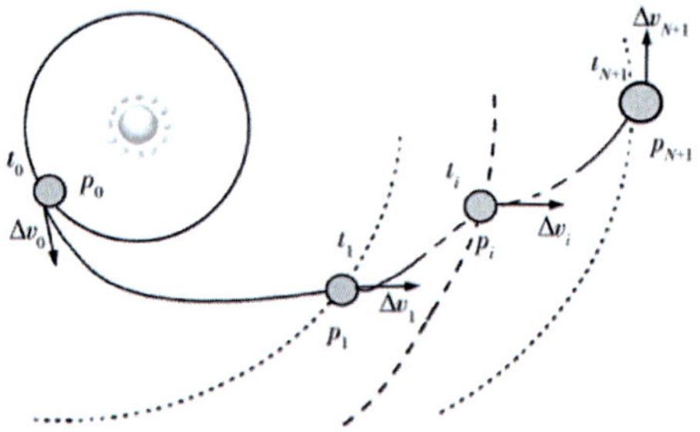

Fig. 5. Schematic diagram of multiple gravity-assisted transfer orbits.

Based on a kinetic model of gravitational assist, a powered gravitational assist scheme is proposed to apply maneuvers at the near star point and the edge of the planetary impact ball, and the maneuver pulse size is calculated considering the constraint of the probe and the leveraged planet, and the influence of gravitational assist on the orbit energy and orbit direction change is analyzed. The orbit design method based on gravity assist is studied. Two energy maps are derived and analyzed to optimize the launch matching window and find the optimal gravitational assist sequence. The deep space orbit segmentation design problem based on energy matching and conic curve splicing is analyzed. There are two main cases in terms of the effect of using planets gravity to assist in changing the orbital energy of the probe: one is to reduce the energy escape from the earth and the other is to reduce the energy difference reaching the target planet [15]. The maximum energy gain through gravitational assist is obtained when the perihelion radius vector is exactly opposite to the planetary velocity. Gravitational assist is usually applied in cases where there is an angle between the target orbit and the Earth's orbit plane when used to change the orbit direction of the probe.

3.4 Extreme High and Low Temperature Thermal Control Design

The deep space exploration mission is generally controlled by the flight trajectory with the help of celestial gravity. The satellite will withstand extreme high and extreme low temperature harsh thermal environment, for example, the solar radiation intensity near Venus is about two solar constants, while the solar radiation intensity near Jupiter is less than 0.04 solar constants. The external temperature range reaches 400 °C (−200 ~ +200 °C) [16]. Deep space exploration satellites need to accomplish heat dissipation in extreme high temperature environments and thermal insulation in extreme low temperature environments to achieve long life operation requirements.

In view of at the new space radiation environment faced by deep space exploration missions, the corresponding thermal control materials such as multi-layer thermal insulation components and fluid loop working fluids are designed, and the extreme high and low temperature adaptive thermal control scheme is adopted. The scheme adopts a highly reliable heat bus based on the fluid loop to collect and transmit the heat in the detector, so that the internal equipment of the detector is thermally coupled together, as shown in Fig. 6.

In the high radiation intensity environment, the heat dissipation system is adaptively adjusted to high radiation capacity for efficient heat dissipation. In the environment of low radiation intensity, high-efficiency heat acquisition and storage technology is used to strengthen the collection and reuse of waste heat from high-temperature equipment. At the same time, the heat dissipation system is used to adaptively adjust the low radiation capacity to carry out on-demand heat dissipation and reduce heat leakage to the outside of the satellite. The extreme high and low temperature adaptive thermal control system is adopted to realize the integrated coupling thermal control design of the whole satellite and ensure the adaptation of the satellite to the extreme ambient temperature changes from −200 to +200 °C.

3.5 Long-Life Autonomous Operation Design

Deep space exploration is characterized by extremely long exploration distance, extremely low light intensity, complex and harsh radiation and magnetic field environment, and the ground can only provide a very low satellite-ground measurement and control bandwidth [17]. As a result, it is necessary to carry out satellite long-life autonomous management design. First, the ultra-long measurement and control distance and large time delay are difficult to meet the requirements of complex task control and data return in the process of deep space exploration and cannot deal with on-orbit faults in time, which seriously threatens the safety of the satellite. Second, such as mission planning, task execution, regular self-test, fault detection and isolation, energy management, thermal control require to consume huge manpower and resources if completely

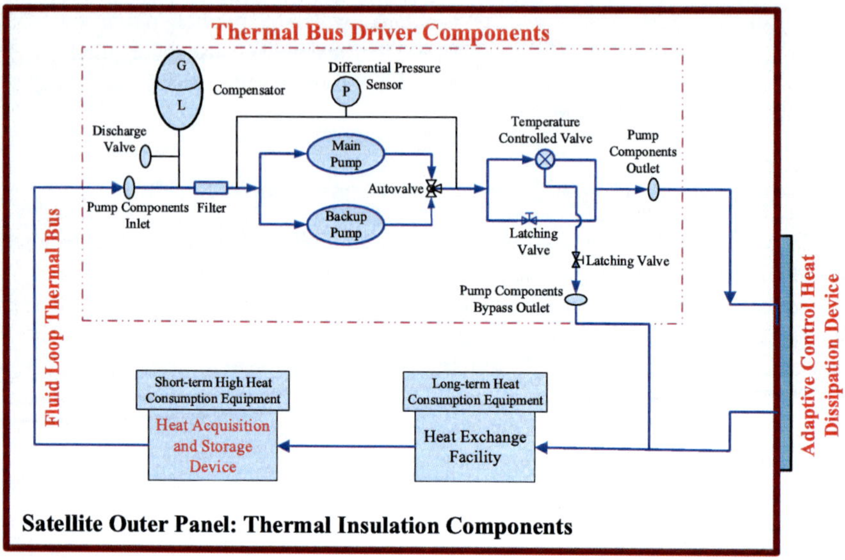

Fig. 6. Schematic diagram of extreme high and low temperature adaptive thermal control system.

dependent on the ground. In addition, decades of mission cycle also places high demands on the design life of the satellite electronics.

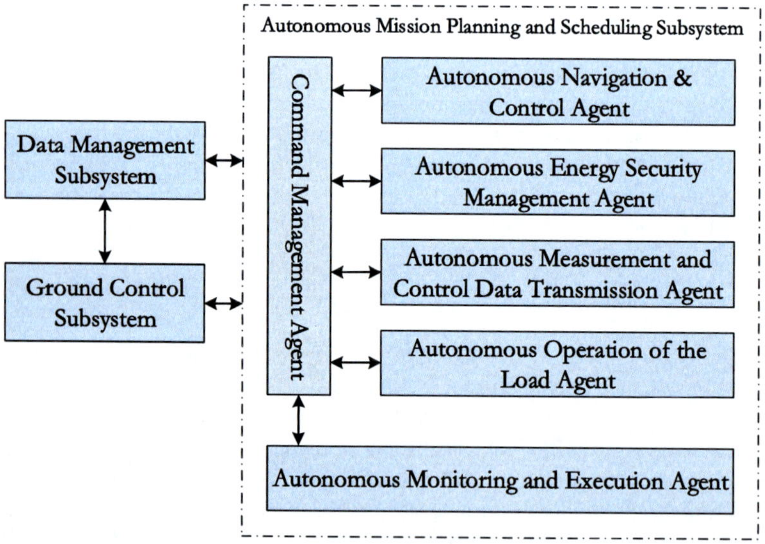

Fig. 7. Schematic diagram of autonomous mission planning and scheduling system based on multi-agent.

The research is carried out for the autonomous operation architecture design and autonomous mission planning method of deep space exploration satellite. The architecture of satellite autonomous operation based on multi-agent is adopted. The autonomous system software structure and autonomous management operation sequence are designed to realize the integrated autonomous mission management of the whole satellite [18–20]. The satellite autonomous operation management system consists of the following parts: autonomous mission planning and scheduling system, execution system, execution monitoring and fault diagnosis system. Among them, the autonomous mission planning and scheduling system performs reasoning and searching with the information in the state parameter manager; the execution system is responsible for the distribution and execution of commands; the execution monitoring and fault diagnosis system is responsible for collecting the status information of each subsystem and performing anomaly and fault monitoring. The architecture of multi-agent based autonomous mission management is shown in Fig. 7. The functional modules of the autonomous mission planning and scheduling subsystem mainly include: autonomous navigation and control agent, autonomous energy security management agent, autonomous measurement and control data transmission agent, autonomous operation of the load agent, command management agent and autonomous monitoring and execution agent, totaling 6 functional modules, which accomplish the functions and tasks of attitude and orbit information forecasting, energy management, communication management, load management and execution monitoring and so on.

4 Conclusion

Aiming at the requirements of miniaturization and reliability of satellites for deep space exploration, considering the constraints of weight, volume and energy in the process of deep space exploration, a micro-satellite based on MEMS technology is designed to meet the requirements of deep space exploration. In this paper, the overall framework of the micro-satellite is designed. Detailed and credible design is carried out in energy supply, strong radiation environment protection, flight strategy and orbit optimization design, extreme high and low temperature thermal control and long-life autonomous operation in response to the extremely special environment in deep space exploration. The design of the micro-satellite is reasonable and effective, which can meet the requirements of miniaturization and reliability in deep space exploration and has a high-cost performance. It also supplies a useful idea and method for the application of micro-satellite in the field of deep space exploration.

The future micro-satellites for deep space exploration will evolve towards smaller size, lighter weight, higher integration, better performance and higher reliability with the development of MEMS technology. By carrying multiple microsatellites in one deep space exploration project to achieve the exploration of multiple planets, as well as its increasingly low cost will accelerate the process of deep space exploration globally.

References

1. George, T.: Overview of MEMS/NEMS technology development for space applications at NASA/JPL. In: SPIE Proceedings, vol. 5116, pp. 136–148 (2003)

2. Roy, P.: MEMS technology & its application for miniaturized space system. Inert. MEMS **3**, 93–97 (2015)
3. Hunter, G., et al.: Smart sensor systems for aerospace applications: from sensor development to application testing. ECS Trans. **16**, 333–344 (2008)
4. De Rooij, N.F., et al.: MEMS for space. In: TRANSDUCERS 2009 - 2009 International Solid-State Sensors, Actuators and Microsystems Conference, pp. 17–24. IEEE (2009)
5. Katyal, A., Heena, S., Raja, M., Chaturvedi, S.K.: Application for MEMS in space. Int. Rev. Appl. Eng. Res. **4**, 313–316 (2014)
6. Tuan, L.K., et al.: University nanosatellite distributed satelllite capabilities to support TechSat 21. In: AIAA/USU Small Satellite Conference, UT (1999)
7. Momma, T., Nara, H., Osaka, T.: Applications – portable | micro hybrid power system: fuel cells/capacitors. In: Garche, J. (ed.) Encyclopedia of Electrochemical Power Sources, pp. 1–12. Elsevier (2009)
8. Silva, M.A.C., Guerrieri, D.C., Cervone, A., Gill, E.: A review of MEMS micropropulsion technologies for CubeSats and PocketQubes. Acta Astronaut. **143**, 234–243 (2018)
9. van der Ha, J.C., Stramaccioni, D.: Thermal radiation effects on deep-space trajectories. Adv. Astronaut. Sci. **136**, 1861–1880 (2010)
10. Robertson, G., et al.: Preliminary analysis: Am-241 RHU/TEG electric power source for nanosatellites (2014)
11. Roussos, E., et al.: A radiation belt of energetic protons located between Saturn and its rings. Science **362**, eaat1962 (2018)
12. Bykov, A.M., et al.: High-energy particles and radiation in star-forming regions. Space Sci. Rev. **216**, 1–37 (2020)
13. Sims, J.A., Longuski, J.M., Staugler, A.J.: $V\infty$ leveraging for interplanetary missions: multiple-revolution orbit techniques. J. Guid. Control Dyn. **20**, 409–415 (1997)
14. Sun, P., Yang, H., Li, S.: Accessibility of near-Earth asteroids and main-belt asteroids in a gravity-assisted multi-target mission. Planet. Space Sci. **182**, 104851 (2020)
15. Gaowei, T.A.N., Yang, G.A.O., Xin, Y.A.N.G.: Global search of multiple gravity assist transfer trajectories for deep space probes. Spacecr. Eng. **21**, 18–27 (2012)
16. Bonnici, M., Mollicone, P., Fenech, M., Azzopardi, M.A.: Analytical and numerical models for thermal related design of a new pico-satellite. Appl. Therm. Eng. **159**, 113908 (2019)
17. Hengbo, X., Nan, Y.: Key technologies of high speed modulator for satellite data transmission. In: 2022 IEEE 4th International Conference on Power, Intelligent Computing and Systems (ICPICS), pp. 32–35 (2022)
18. Yang, L., Li, X., Feng, C., Sun, Z., Zhang, T.: Research on autonomous health management and reconstruction technology of satellite. In: International Conference on Signal and Information Processing, Networking and Computers, pp. 115–123 (2018)
19. Xiang, S., Chen, Y.G., Li, G.L., et al.: Review on satellite autonomous and collaborative task scheduling planning. Acta Autom. Sin. **45**, 252–264 (2019)
20. Zhang, G., et al.: Summary of research on satellite mission planning based on multi-agent-system. J. Phys. Conf. Ser. **1802**, 022032 (2021)

A Method for Increasing Dynamic Range of Space Point Target Detecting System

Yang Xiaole(✉), Xing Mailing, and Shi Manli

Beijing Institute of Space Mechanics and Electricity, Beijing, China
274861918@qq.com

Abstract. Point target detecting system is detecting and measuring system, and has essential distinction with land imaging system. It needs to detect both strong and weak targets, achieve high sensitivity and large dynamic range detection. Traditional detecting system realizes large dynamic range detection by switching detector integral capacitance, However, it cannot meet the requirements of simultaneous detection for strong targets and weak targets in single frame detection. This paper discusses a space point target detecting system based on new TDI infrared detector. The infrared detector automatically implements two integration time samplings under an integration capacitor, and automatically select the output. System single imaging can detect both strong and weak targets, this scheme effectively solves the problem of space point target detection.

Keywords: Point target detecting system · TDI · Detector · Dynamic range

1 Introduction

Space infrared astronomical telescope and space-based missile early warning system are typical space point target detecting systems. Point target detecting system is detecting and measuring system, it needs to find the target and measure the target radiation intensity, so has essential distinction with land imaging system [1].

It is necessary to ensure that the weak target has a high signal-to-noise ratio, and the strong target is unsaturated. This means that it can detect targets of different intensities at the same time, and achieve high sensitivity and large dynamic range detection.

In order to achieve large field of view detection, based on the scale of the existing detector array, the imaging system of scanning mechanism and linear detector is adopted. By increasing the detector integration time, the signal-to-noise ratio can be improved, but the scanning speed will be reduced and the system frame frequency will be reduced. The TDI function is a good solution to this contradiction [2]. Therefore, the TDI infrared detector is used in the system. The introduction of the time delay integration (TDI) function can improve the signal-to-noise ratio without changing the integration time, and can eliminate the defect pixels to improve the circuit reliability.

H. P. Urbach and H. Jiang (Eds.): ISSOIA 2022, SPPHY 295, pp. 257–265, 2023.
https://doi.org/10.1007/978-981-99-4098-1_23

2 Conventional TDI Infrared Detector

Time delay integration (TDI) is developed based on the concept of multiple exposure accumulation of the same object. Its output signal is the result of multi-level charge accumulation, so it is not sensitive to the optical response non-uniformity of the device. Because the signals generated by the same target on different pixels are completely correlated in time, according to the correlation theory, these signals can be completely superimposed, and the noise is completely irrelevant due to its random fluctuation, and cannot be completely superimposed, so the TDI function improves the signal-to-noise ratio [2] (Fig. 1).

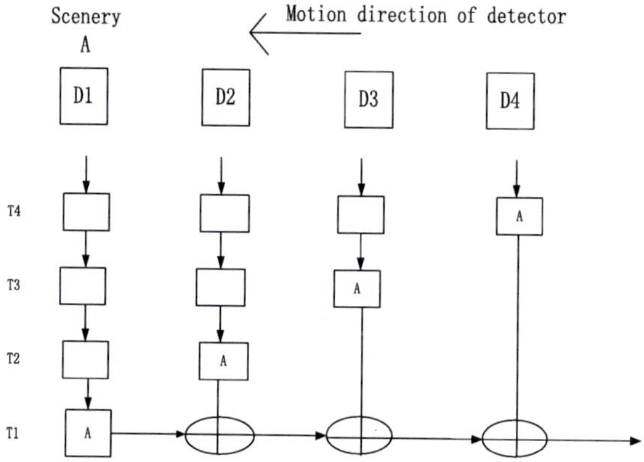

Fig. 1. TDI working principle

The output signal of the target obtained by a single pixel is V_{S0}, and the circuit noise is V_{n0}. The readout signals of N pixels with N-level TDI function are Vs1 ~ VsN, and the circuit noise is Vn1 ~ VnN. The output signal of a TDI channel is V_S, and the circuit noise is V_n. In the analysis, it is assumed that the pixel and the circuit are in good agreement, that is, the signal of each pixel and the noise of the circuit are equal, which is the case of a single pixel. Since the signal is correlated and the noise is uncorrelated [2, 3], there are:

$$V_s = \frac{V_{s1} + V_{s2} + V_{s3} + \cdots + V_{sN}}{N} = V_{S0} \tag{1}$$

$$V_n = \frac{\sqrt{V_{n1}^2 + V_{n2}^2 + V_{n3}^2 + \cdots + V_{nN}^2}}{N} = \frac{\sqrt{N}V_{n0}}{N} = \frac{V_{n0}}{\sqrt{N}} \tag{2}$$

$$SNR_N = \frac{V_s}{V_n} = \frac{V_{S0}}{\frac{V_{n0}}{\sqrt{N}}} = \sqrt{N}SNR_0 \tag{3}$$

It can be seen that the SNR of N-level TDI is $\sqrt{N}$ times that of single pixel.

An example of a TDI channel of a four - stage TDI linear infrared detector, the input stage of the D1 ~ D4 pixel readout circuit completes the conversion of the optical signal to the electrical signal. During the integration time, the Q1 of the switch tube is closed, and the output signal of the input stage is output to the TDI operation circuit through the capacitance C1 sampling and storage. At the end of the integration, the Q1 is turned on and the capacitance C1 sampling is completed. The result of 4 times integral sampling is output after TDI operation. All pixels adopt the same integral time. All pixels realize different detection dynamics by selecting different sampling capacitors. This is the imaging principle of conventional TDI detector (Fig. 2).

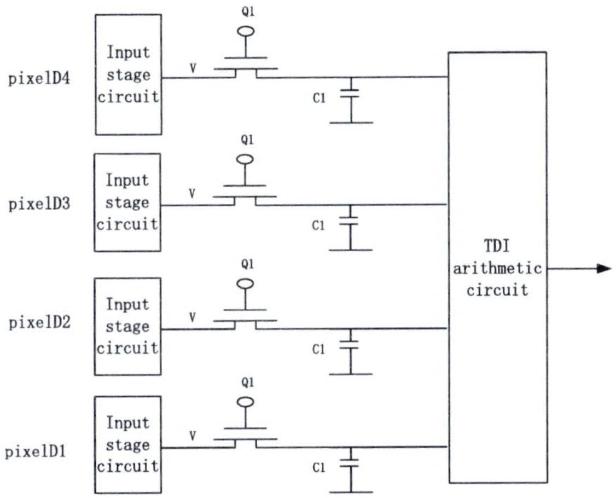

Fig. 2. TDI readout circuit schematic

The detector integration capacitance is selected according to the target energy characteristics. The full-well electron number of each capacitance determines the strongest target that the detector can detect [4, 5].

The conventional TDI infrared detector achieves large dynamic range detection by adjusting the gain when the integration time is determined. Gain adjustment is to adjust the readout circuit input stage integral capacitance, in the input stage circuit has a multi-stage integral capacitance. When the signal is strong, the photoelectric conversion generates a large number of electrons. When the number of electrons exceeds the full well number of the integral capacitor, the signal saturation is avoided by switching a larger capacitance. When the signal is weak, the photoelectric conversion produces a small number of electrons. By switching the small capacitance capacitance, a high output voltage can be obtained to improve the signal-to-noise ratio. However, the conventional TDI linear infrared detector cannot meet the needs of simultaneous detection of strong and weak targets in single frame detection by adjusting the gain [4, 5].

3 TDI Infrared Detector Based on Long and Short Integration Time

Aiming at the traditional TDI infrared detector, this paper proposes a new type of TDI infrared detector which can detect both strong and weak targets effectively in a single frame without switching integral capacitance. The detector can automatically realize the sampling of two integral time under one integral capacitance, and can automatically select the output (Fig. 3).

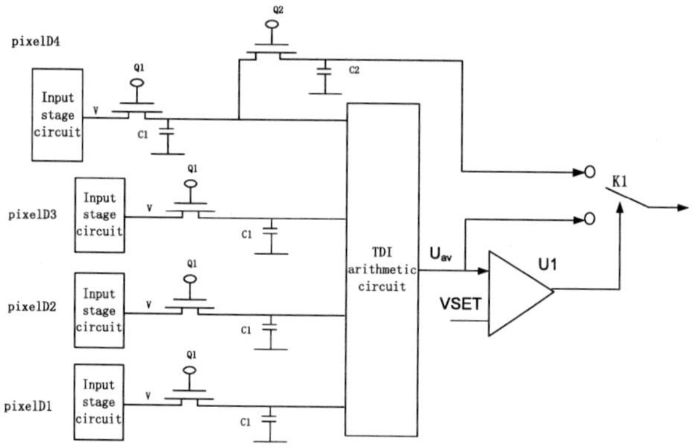

Fig. 3. New infrared detector readout circuit schematic

The normal exposure time of the detector is called 'conventional integration time, the D4 pixel in the D1-D4 pixels is a special pixel, and its 'conventional integration time' is divided into two consecutive time periods. The first time period is called 'short integration time', and the two time periods together are 'conventional integration time'. In the 'short integration time', the switch tube Q1 closed, through the capacitor C1 sample then store and output, while the switch tube Q2 closed, capacitor C2 sampling storage output. After the 'short integration time', the switch Q2 opens, the capacitor C2 ends the sampling, and the remaining integration time, Q1 continues to close, and the capacitor C1 continues to sample. After the 'conventional integration time', Q1 opens, and C1 ends the sampling. So the sampling result of capacitor C2 corresponds to 'short integration time'. The sampling result of C1 corresponds to 'conventional integration time' the same as D1 ~ D3 pixels.

The capacitance C1 sampling results of D1 ~ D4 pixels, that is, the 'conventional integration time' sampling results, are all input into the TDI operation circuit to obtain the average voltage Uav. Uav is connected to the comparator U1 and compared with the external voltage VSET (VSET is set to the saturation output voltage of the detector). When Uav is greater than VSET, that is, the detector 's' conventional integration time' output is saturated, andThe output of comparator U1 controls the switch K1 to select the sampling result of D4 pixel capacitor C2. That is, the 'short integration time' sampling result, ensures that the detector output is unsaturated. On the contrary, when Uav is less than VSET, that is, the output of the detector 's' conventional integration time' is unsaturated, and the output of the comparator U1 will control the two-choice switch K1 to select the Uav output. Which also ensures the detector output unsaturated. To prevent the comparator from noise interference, the comparator U1 is set as a hysteresis comparator.

Due to the demand of target measurement and inversion, the comparison result of each TDI channel outputs the identification bit through the shift register. Through this identification, it can be judged whether the output state of the TDI channel is short integration time output or conventional integration time output.

In the long and short integration time scheme, the signal-to-noise ratio of the single pixel output is $\sqrt{N}$ lower than that of the TDI accumulation output because it does not have the TDI function, but at this time, the signal accumulation is saturated, the signal energy is already strong, and the system SNR can still meet the requirements.

4 Response Characteristic

According to the design of the new infrared detector, the output response curve is divided into two sections [8]: conventional integration time TDI accumulation and single pixel short integration time output. The pixel theoretical radiation response characteristic curve is as follows, the abscissa is the input energy, and the ordinate is the DN value after the output voltage quantization. When the input energy is small, conventional integration time output, when the output voltage is greater than the set threshold voltage, according to the results of the comparator, short integration time output, because the cumulative energy becomes smaller, the output jumps until saturated again (Fig. 4).

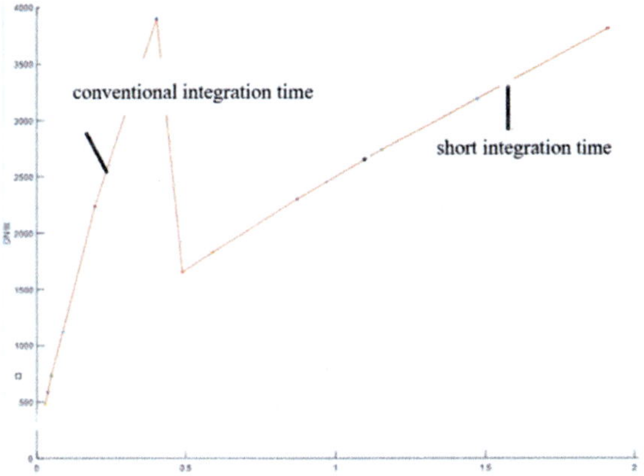

Fig. 4. New infrared detector radiation response curve

It can be seen from the figure that the conventional integration time output and the short integration time output have different response slopes and intercepts.

The conventional integral time, the radiation response equation of each pixel is:

$$DN_{TDI}(j) = L \times K_{TDI}(j) + C_{TDI}(j) \tag{4}$$

K_{TDI} and C_{TDI} are Output response slope and intercept for conventional integration time. L is the input energy.

In the output range of short integration time, the radiation response equation of each pixel is:

$$DN_S(j) = L \times K_S(j) + C_S(j) \tag{5}$$

Ks and Cs are Output response slope and intercept for short integration time.

In the use of the system, it is necessary to perform energy inversion on the detected target. The above curve is different from the radiation response curve of the conventional detection system [6, 7]. When the target energy is inverted, it is impossible to accurately know whether the target is in the conventional integration time or short integration time output state. Therefore, it is necessary to normalize the conventional integration time at the digital end of the imaging circuit, keep the response slope consistent with the short integration time output, and transform the system response curve into a single linear line. Because it is processed at the digital end, it does not affect the system SNR.

Normalized equation from conventional integration time to short integration time can be obtained by making L equal in radiation equation of conventional integration time and short integration time (Fig. 5).

$$DN_{TDI}(j) = DN_S(j) \times \frac{K_S(j)}{K_{TDI}(j)} + C_S(j) - \frac{K_S(j)}{K_{TDI}(j)} C_{TDI}(j) \tag{6}$$

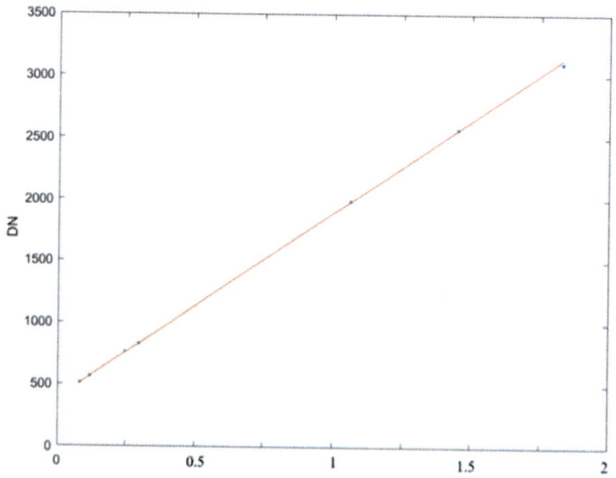

Fig. 5. New infrared detector normalized radiation response curve

5 Practical Test of Radiation Response

By placing the surface source blackbody in front of the system, the radiation response characteristics of the system using new infrared detector are tested, and the normalization coefficient and normalization effect are verified.

The blackbody is placed in front of the light entrance of the system to cover the optical lens and all detector pixels (Fig. 6).

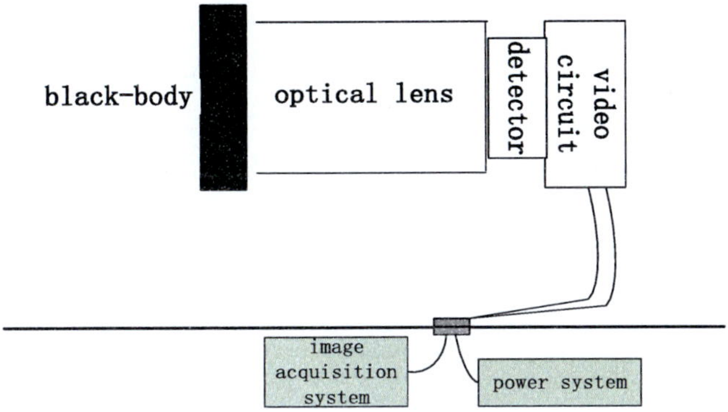

Fig. 6. Radiation response test system

In the experiment, the irradiance on the photosensitive surface of the detector is changed by changing the black body temperature to change the radiance, and the output signal voltage of the detector with different radiance input is obtained. Select a detector pixel, the measured radiation response curve is as follows (Fig. 7).

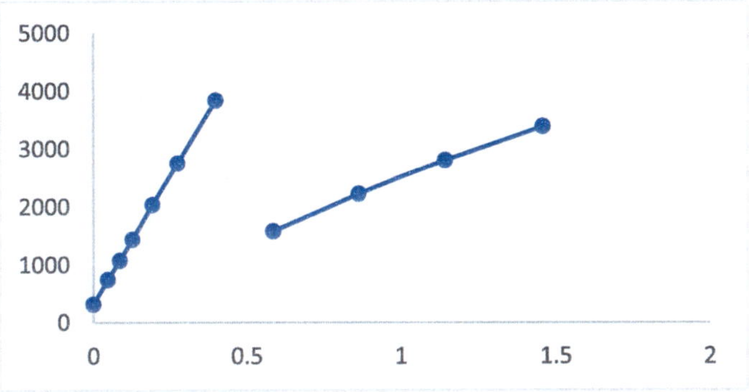

Fig. 7. Practical test of Radiation responsetest cure

The response output of all measuring points is obviously linear step. When the slope changes from large slope to small slope, it shows that with the increase of black body temperature, the radiance increases, and the output value of conventional integration time should reach saturation, which is greater than the comparator threshold. The detector outputs is in short integration time, and the pixel output is automatically switched. Consistent with the design.

According to the above test, the radiation equation is established, and the normalized coefficient and the normalized radiation calibration curve are as follows. The linear correlation coefficient is greater than 0.99, and the output has no step (Fig. 8).

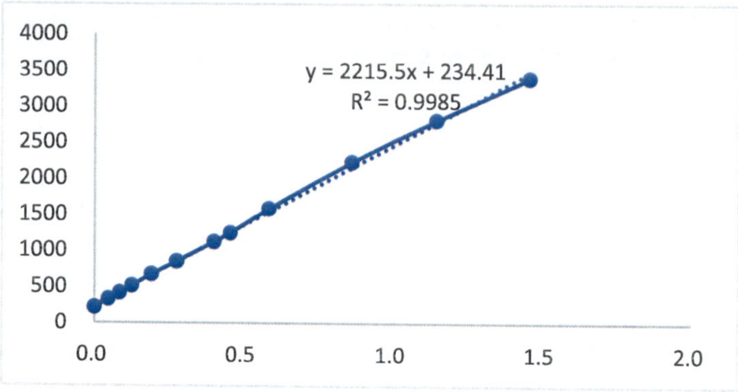

Fig. 8. Practical test of normalized radiation response curve

6 Conclusion

Based on the characteristics of the space point target detection system, this paper innovatively proposes a long and short integration time scheme, which solves the problem of weak point target detection and simultaneous detection of strong and weak targets at the same time, combined with the target inversion requirements, the system which uses the new infrared detector output response characteristics are given. The correctness of the design is verified by experimental tests. At the same time, the method of system output response reconstruction is proposed, which makes the output response monotonously linear as the conventional system, and the linear correlation coefficient is basically consistent with the conventional system.

References

1. Liang, L., Shitao, W., et al.: In-orbit radiometric calibration methods for remote sensing system to detect space infrared point target. Spacecr. Recovery Remote Sens. **2**(33), 73–80 (2012)
2. Tang, J., Lu, W.-G., et al.: 288 × 4 IRFPA ROIC with TDI. Infrared Technol. **4**(29), 206–210 (2007)
3. Han, J., Huang, B., et al.: A high sensitivity large dynamic range readout circuit. China Integrated Circ. **140**, 39–42 (2011)
4. Tao, K., Li, F., et al.: IRFPA imaging system dynamic range adaptive adjust technology. Infrared Laser Eng. **2**(37), 265–269 (2008)
5. Fang, D.: Skill analysis of readout intergrated circuits of I RFPA. Infrared Technol. **2**(26), 23–28 (2004)
6. Li, F.-W., Zhang, Y.-Q., et al.: Influence of integration time on IRFPA imaging system. J. Appl. Opt. **5**(29), 727–730 (2008)
7. Dawei, Z., Jiangyong, L., et al.: Research of the automatic-gain control method for infrared imaging system. Laser Infrared **39**(10), 1119–1122 (2009)
8. Dong, S.K., Eun, K.: Noise power spectrum measurements under nonuniform gains and their compensations. In: Physics of Medical Imaging, Proceedings of the SPIE (2016)

The Effect of Instrumental Polarization on the High Resolution Imaging of Space Interferometry

Shu Yuan(✉) and Aimin Jiang

National Astronomical Observatories, Chinese Academy of Sciences, 20A Datun Road, Chaoyang District, Beijing, China
yuanshu@bao.ac.cn

Abstract. Owe to its super high angular resolution, the space-based optical interferometry will become a unique tool for the terrestrial planet detection and characterization. In the optical interferometric system, the instrumental polarization caused by the folding mirrors in each branch plays an important role for the formation of interference fringe, which will impact the imaging quality of the system finally. In this paper, the instrumental polarization and the resulting image quality degradation for a typical Fizeau interferometer is investigated by a self-developed telescope polarization simulation tool. The purpose of this work is to show the necessity of polarization analysis and optimization of the optical system when a high-performance interferometry is pursued. We have developed a vector imaging simulation tool for telescope based on the technical of polarization ray tracing. By which, the polarized point spread functions and optical transfer functions are calculated for the target systems. The results show that the imaging quality of symmetric three- and six-apertures systems is sensitive to the instrumental polarization. An arbitrary design of optical configuration may lead to decrease of contrast of interference fringes, thus decrease the modulation transfer function in the middle and high spatial frequency regions. Generally, the influence of instrumental polarization on the imaging quality of interferometric system must be considered in the design phrase, whether for astronomical observation or remote sensing of the earth. Through careful polarization design, the impact of instrument polarization on imaging quality can be eliminated, so that the limit angular resolution can be achieved.

Keywords: Space interferometry · instrumental polarization · polarization ray tracing · contrast ratio

1 Introduction

The space-based stellar interferometry is a powerful tool for the astronomical research of exoplanet, stellar, black holes and Galaxy. By the technique of formation flying of satellites, the baseline of interferometer can extend to more than 100 m. It brings us a capability to detect and characterize the habitable terrestrial planets directly around the

H. P. Urbach and H. Jiang (Eds.): ISSOIA 2022, SPPHY 295, pp. 265–276, 2023.
https://doi.org/10.1007/978-981-99-4098-1_24

nearby stars. The challenge for such direct detection is the combination of large contrast ratio between the star and planet (~10^7 in the mid-IR; ~10^{10} in the visible) and the small angular separation (0.1arcsec for the Earth-Sun at a distance of 10pc) [1]. The method of nulling interferometry [2] has been proposed to achieve this challenge, which was pursued by the NASA's mission of Terrestrial Planet Finder [3] and the ESA's mission of Darwin [4].

Comparing to the telescope, the imaging characteristics of interferometer are much more dependent on the polarization properties of itself. Retardance effects and reorientation of the polarization coordinate frame of reference by reflections are a major cause of fringe contrast decrease [5, 6]. At the same time, the nulling interferometry requires the high symmetry of the instrumental polarization of interfering beams to obtain the high contrast of 10^{-6} [7].

In this paper, we discuss the effect of instrumental polarization on the contrast and image quality of the space interferometry. One Fizeau interferometer with three sub-apertures is given as example systems for this discuss. The method of polarization ray tracing and vectorial Fourier transform are employed to model the polarized (anisotropic) interfering fringe. The frequency-domain analysis is done for evaluating the effect on the imaging quality. In the following, we introduce the polarization component of interfering fringe for the general interferometer first. In Sect. 3, the optical configuration of the example interferometer is described. The polarization modeling method is given in Sect. 4. The simulated fringe and the corresponding frequency domain analysis are represented in Sect. 5. The final section is conclusion.

2 The Polarization Components of Fringe

The radiation emitted by a star in optical band is natural light, which may be represented as a superposition of two incoherent beams linearly polarized at right angles to each other. The interference between the x-components and y-components may then be considered separately, and the total intensity is obtained by addition of the separate intensities [8]. For a polarization free system, the two intensities are identical to each other. The addition of them does not change the profile of intensity. But, for a polarized system, the fringe is composed by four polarization components

$$I = I_{xx} + I_{yx} + I_{yx} + I_{yy}, \tag{1}$$

where the subscripts are corresponding to the input (right subscript) and output (left subscript) linear polarized components. For example, the I_{xx} represents the x-component intensity form the input x-component, and the I_{xy} represents the x-component intensity from the input y-component, and so on. The polarization component of fringe can be calculated by the following equation

$$I_{p,q} = \left| \sum\nolimits_j e^{-i\phi_j} \mathrm{ARM}_{j,p,q} \right|^2, \tag{2}$$

where the subscript $[\mathrm{p, q}] \in [\mathrm{x, y}]$, the j is the index of interfering leg, $\mathrm{ARM}_{j,p,q}$ represents the element of amplitude response matrix (ARM) for the j_{th} leg, and ϕ_j is the dephase for the j_{th} leg, and the i is imaginary unit. The ARM has the form of Jones matrix, which can be written as

$$\mathbf{ARM} = \begin{pmatrix} \mathrm{ARM}_{xx} & \mathrm{ARM}_{xy} \\ \mathrm{ARM}_{yx} & \mathrm{ARM}_{yy} \end{pmatrix}. \quad (3)$$

The ARM is the Fourier transform of the Jones pupil, which is the distribution function of Jones matrix cross the pupil. For a given optical system, the Jones pupil can be calculated by the technique of polarization ray tracing (PRT). The non-uniformity of Jones pupil is defined as polarization aberrations, which is discussed by Breckinridge [8] and Chipman [9] in detail. The method of PRT using in this work is described in Sect. 4.

3 The Example Interferometer

The example system is a typical Fizeau interferometer, which has three sub-apertures. The aperture architecture is shown by Fig. 1, which is a co-planar equilateral triangle. In this figure, the central three small circulars represent the exit pupil of the interferometer. For keeping the imaging properties of linear spatially invariant, the exit pupil of interferometer should be similar to the entrance pupil, and the reduction ratio should be equal to the reciprocal of angular magnification of the collector. Here, we defined the angular magnification as 5. The absolute size of aperture array isn't important in this work. Without loss of generality, we set the outer envelope of the three sub-apertures as 1 m, and the diameter of each collector is 360 mm.

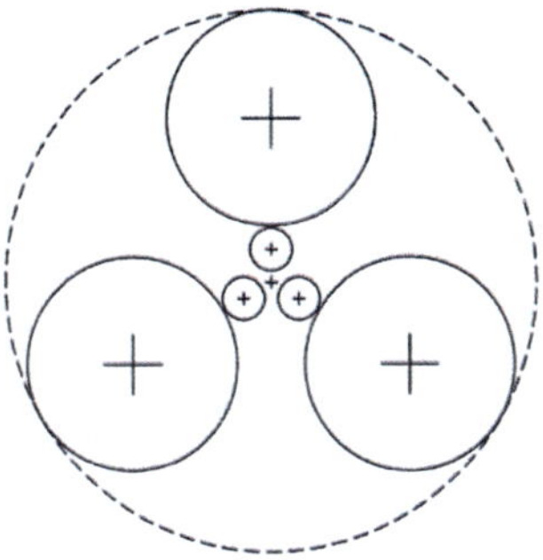

Fig. 1. Aperture architecture of the example interferometer

For the optical configuration, the interferometer is composed by collectors, folding mirrors, delay lines and combiner, as shown in Fig. 2.

The collectors are identical afocal telescopes, and both of the primary and secondary mirrors are paraboliod. As mention above, the magnification of the collector is 5. For each leg, two diagonal folding mirrors reflect the beam close to the systematic symmetric axis. The following delay line is composed by four diagonal mirrors. These mirrors share a same plane of incident (POI). The azimuth of the POI is a free parameter in scalar optical design. But, for the polarization optical design, the imaging properties are dependent to

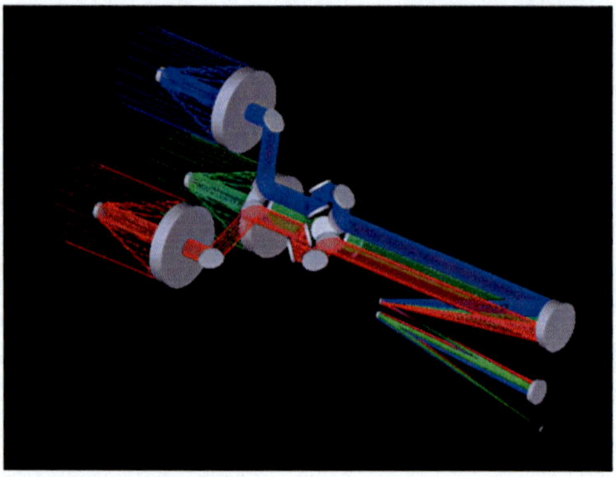

Fig. 2. Optical configuration of the example interferometer.

this parameter, which will be discussed in the follow. Here, we set the POI of the three delay lines circular symmetric to the system axis firstly. The combiner of interferometer is an off-axis three mirrors system, where the primary mirror is paraboliod, the secondary and third mirrors are ellipsoid and hyperboloid respectively. The effective focal length of the interferometer is 28000 mm, and total angular magnification is −28.75. For this design, the field of view with diffraction-limited image quality is ±2 arcmins.

4 The Method of Polarization Modeling

Polarization ray tracing, which integrates thin-film design and crystal optics into optical design, is a straightforward but nontrivial extension of conventional ray tracing [10]. PRT calculates the Jones matrix associated with an arbitrary path through an optical system. A self-developed PRT program is used to calculate the Jones pupil of the interferometer, which has been used for the polarization design and analysis of the large solar telescope [11, 12]. In the program, the Jones matrix for a given ray can be written as

$$\mathbf{J}(\mathbf{h}, \boldsymbol{\rho}, \lambda) = \mathbf{R}_{out}\mathbf{J}_N\left(\prod\nolimits_{j=1}^{N-1}\mathbf{R}_j\mathbf{J}_j\right)\mathbf{R}_{in}, \tag{4}$$

where $\mathbf{h}$ and $\boldsymbol{\rho}$ are the object and pupil coordinate respectively, λ is the wavelength; $\mathbf{J}_\mathrm{j}$ indicates the Jones matrix of the given ray at the $\mathrm{j_{th}}$ optical surface, $\mathbf{R}_\mathrm{i}$ is the rotation matrix to match the successive polarization coordinate frames (PCF); the $\mathbf{R}_\mathrm{in}$ and $\mathbf{R}_\mathrm{out}$ are the user-defined input and output PCF in the entrance pupil and the focal plane. The Jones matrix of metallic mirror in its eigen PCF is

$$\mathbf{J}_j = \begin{pmatrix} r_s & 0 \\ 0 & r_p \end{pmatrix}, \tag{5}$$

where the r_s and r_p are reflection coefficients for perpendicular and parallel polarization components respectively. Figure 3 displays the optical layout by the program, in which the input and output PCF are shown by colorful arrow. For a multiple apertures system, the corresponding axis directions of the input and output PCF should be consistent. As shown in the Fig. 3, the red arrows in the entrance are parallel to each other, which represent the directions of vertical linear polarization. The blue arrows represent the directions of horizontal linear polarization, and the green arrows represent the propagation direction of on-axis rays. By the combiner, the three beams converge on the detector plane, and a compromised output PCF is defined at the focus.

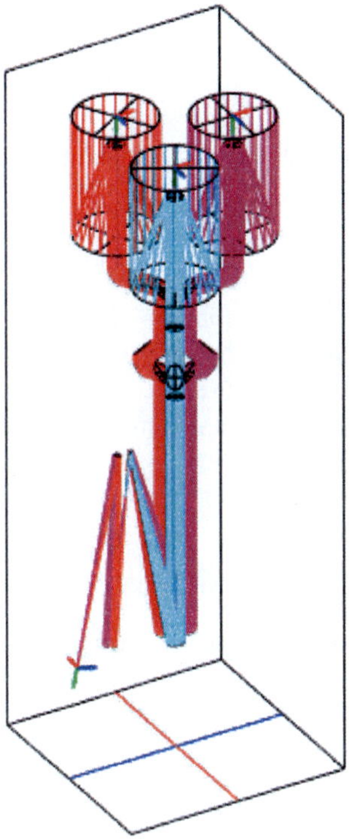

Fig. 3. The optical layout of the interferometer

The polarization properties of mirror are functions of incident angle, wavelength, film structure and the optical properties of coating material. Here, we define all mirrors in the system are coated by bare aluminum film. The effect of oxide layer on aluminum is ignored in this work. In metal optics, optical properties of metal coating are described by complex refractive index (CRI). The real and imaginary parts of it are defined as the real refractive index (n_r) and extinction coefficient (κ) of the metallic coating respectively. Both of the two quantities are functions of wavelength and coating process. In our modeling, we choose 1.0865 for the n_r and 6.8754 for the κ of aluminum coating at the wavelength of 0.5 microns [12]. By substituting the CRI into the Fresnel equations, we calculate the r_p and r_s as functions of incident angle (See Fig. 4). As shown in Fig. 4, the reflection coefficients of p and s components in the incident angle of 45° are 0.940 and 0.970 respectively, and the retardance between the two components is 168.59°.

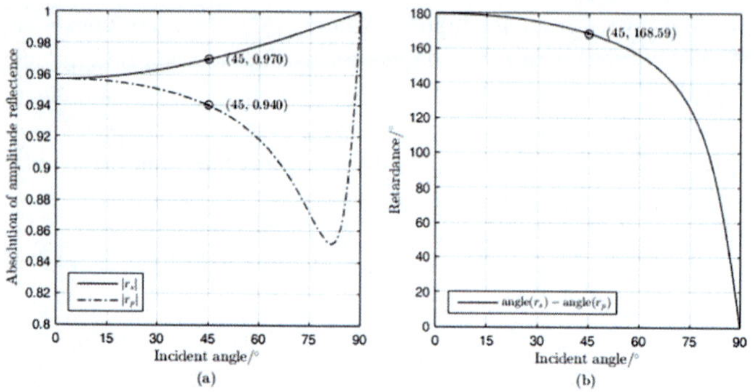

Fig. 4. Reflection coefficients of amplitude (a) and retardance (b) as functions of incident angle between 0° and 90° for the bare aluminum coating at wavelength of 0.5 microns.

Figure 5 displays the resulted polarization pupil map by the PRT program, where the output polarization states are described by the polarization ellipse. As shown in Fig. 5, the polarization properties of the rays in one sub-aperture are uniform nearly, but quite different from other sub-apertures. This feature manifests that the instrumental polarization of the example interferometer is mainly caused by folding mirrors, not the collectors and combiner.

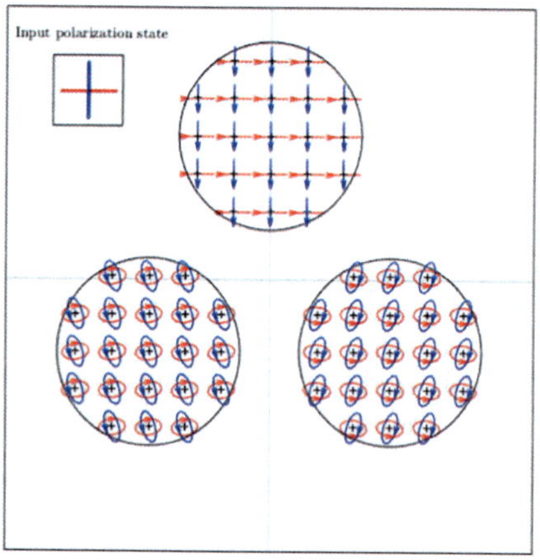

Fig. 5. The polarization pupil map results by PRT, where the blue and red polarization ellipses are corresponding to the vertical and horizontal input linear polarization respectively.

5 The Analysis of Interfering Fringe

By Fourier transform of the pupil map, and Eq. (2), we get the four intensity components of the fringe. The four intensities can combine into two intensities, which are $I_x = I_{xy} + I_{xx}$ and $I_y = I_{yy} + I_{yx}$. The two intensities are corresponding to the two orthogonal linear polarization components at the detector. Experimentally, the two intensities can be observed by inserting a linear polarizer in front of the detector, and rotating it to the horizontal and vertical azimuths. Figure 6 displays the interfering intensities for the x-polarization (horizontal) component, the y-polarization (vertical) component, superposition of the two (actual) and the ideal fringe without considering the polarization effect. Observing the two polarization intensities in Fig. 6, we can find that their maximum have an obvious bias from each other. This effect decreases the contrast of actual fringe from the ideal one.

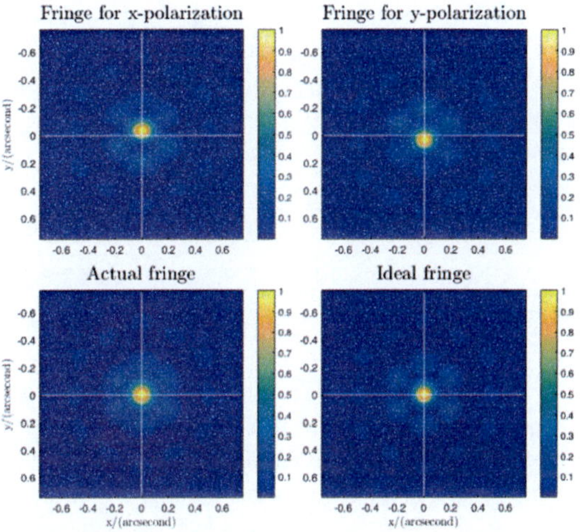

Fig. 6. The fringe intensities for the x-polarization (left upper), the y-polarization (right upper), the actual (left lower) and the ideal cases (right lower).

Figure 7 shows the intensities along an x-axis cross-section through the actual and ideal fringe in $\log_{10}$ scale in Fig. 6. The polarization effect smooth out some of the fringe minimum, and decreases the contrast ratio of the fringe dramatically. Table 1 presents the intensities of first five minima of the actual and ideal fringe, where the locations of the minima are marked in Fig. 7. Except the 3rd and 5th minima, the intensities of other minima for actual case are much higher than the value for ideal case in 3 orders of magnitude. This result indicates that the instrumental polarization will be an important factor for the high contrast imaging of space interferometers.

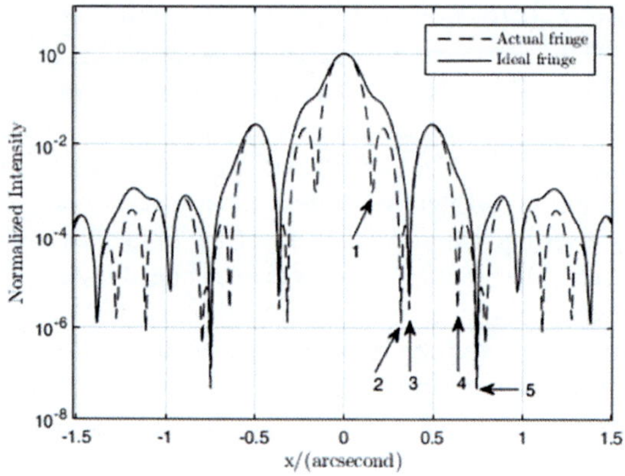

Fig. 7. Cross sections through the $\log_{10}$ I intensities for the actual and ideal fringe along the x-direction.

Table 1. The first five minima of the actual and ideal fringe

Order of fringe minimum	1	2	3	4	5
Location/(arcsec)	0.1595	0.3191	0.3663	0.6381	0.7445
Ideal intensities	9.07×10^{-4}	1.26×10^{-6}	2.50×10^{-6}	2.52×10^{-6}	4.60×10^{-6}
Actual intensities	0.1167	0.00444	5.43×10^{-6}	0.00221	1.68×10^{-7}

At the same time, the decreasing of contrast also leads to degradation of imaging quality of the interferometer, which can be described quantitively in the spatial frequency domain. Figure 8 displays the ideal modulation transfer function (MTF) which does not consider the polarization effect and the actual MTF which considers the effect. By comparison, we find that there is a remarkable decrease in the middle and high frequency of the MTF. Figure 9 shows the section of the MTF, whose frequency coordinate is normalized by the cut-off frequency. For comparison, the MTF of single sub-aperture is also presented in the figure. The actual MTF reduce by half from the ideal MTF in the frequency band beyond the normalized frequency of 0.45, which just the cut-off frequency of single sub-aperture case.

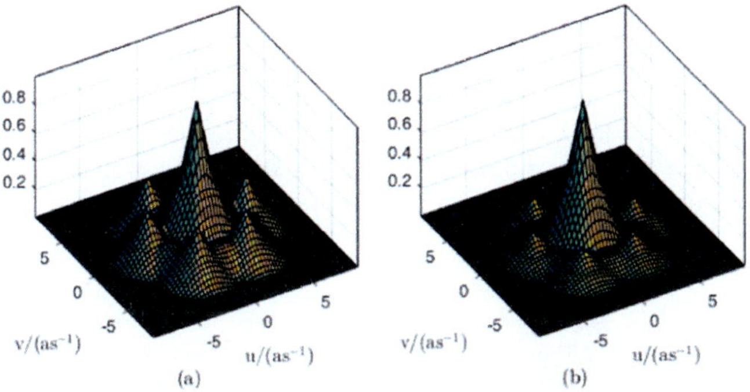

Fig. 8. The ideal MTF (a) and actual MTF (b) of the example interferometer

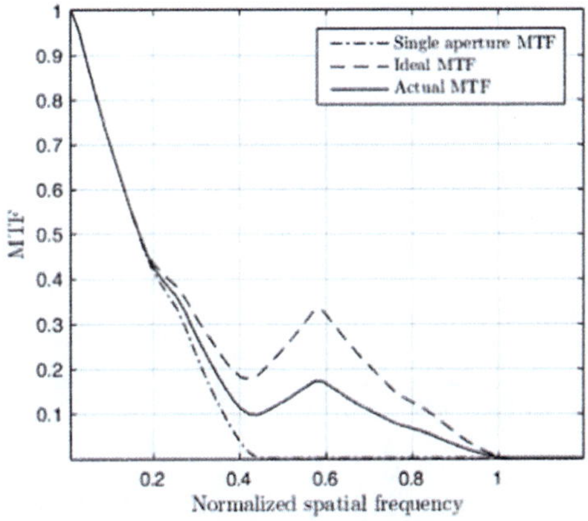

Fig. 9. The cross sections of the MTF with the normalized frequency coordinates.

6 Discussion

As the result mentioned above, the folding optics are the primary source of instrumental polarization of interferometers. Can we minimize the polarization effect on the performance of interferometer by optimizing its optical configuration? The easiest way to realize such idea is to rotate the POI of delay lines to parallel to each other, which is shown in Fig. 9. This operation does not affect the phase of each leg, but makes the polarization properties of each leg more identical. The resulted interfering intensities are displayed in Fig. 10. Obviously, both of the actual fringe and its polarization components are much more consistent with the ideal fringe than the previous. This improvement is demonstrated further by the cross sections of logarithm intensities of fringes in Fig. 12. Although, the profile of actual fringe is closed to the ideal profile, the residual effect of instrumental polarization still fills the gap of minima up. The result indicates that, if this system is used for nulling interferometry, the contrast can't achieve the expected level in visible band. Figure 13 presents the cross sections of updated MTF, where the actual MTF is almost identical to the ideal one. This result suggests that the parallel configuration of delay lines reduce indeed the polarization effect on the systematic image quality. Generally, the above discussion brings us an idea that it is the difference, not the degree, of instrumental polarization among interfering legs causes the degradation of imaging quality of interferometer. And the azimuth configuration of delay lines plays an important role in reducing the polarization effect (Fig. 11).

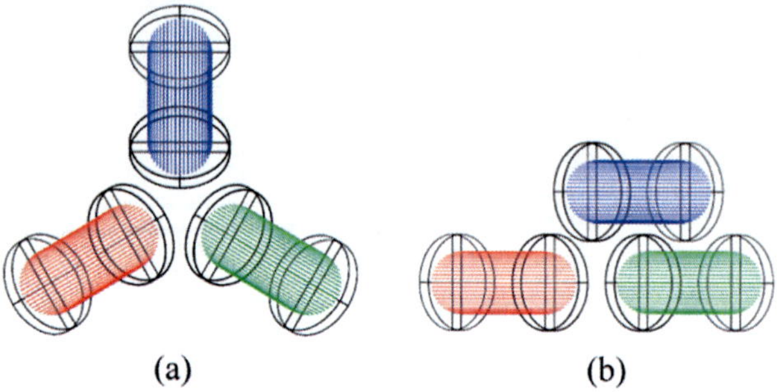

Fig. 10. The azimuth configuration of the delay lines: (a) circular-symmetry (b) parallel

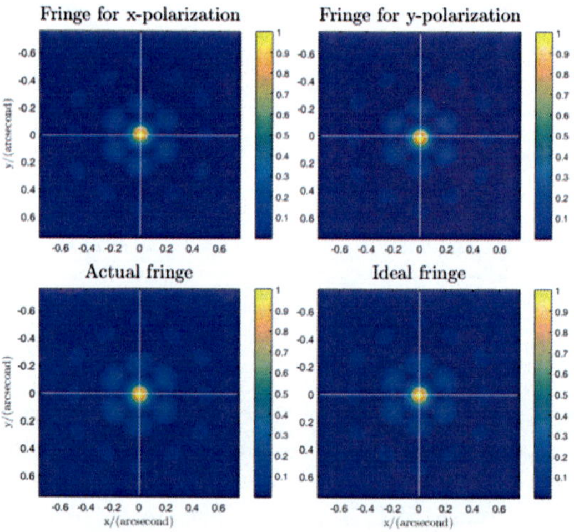

Fig. 11. The fringe intensities for the parallel configuration of delay lines.

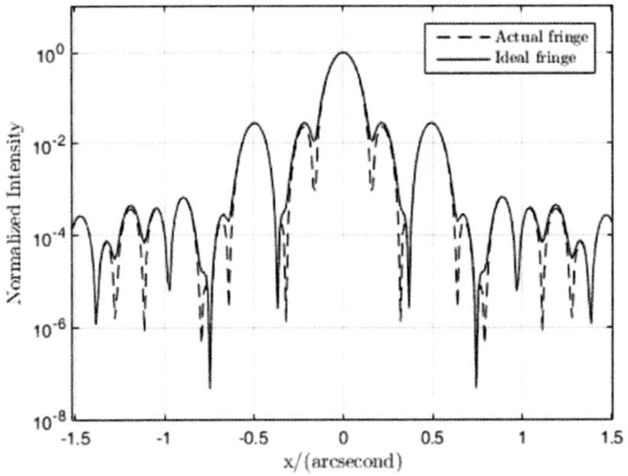

Fig. 12. Cross sections through the $\log_{10}$ I intensities for the parallel configuration of delay lines.

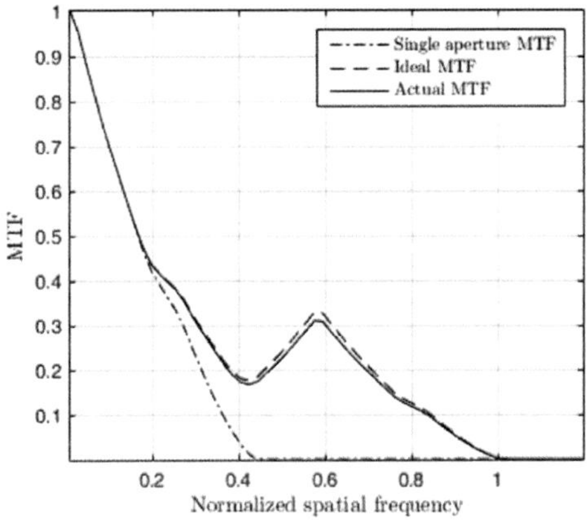

Fig. 13. The MTF for the parallel configuration of delay lines.

7 Conclusion

In this work, we investigate the polarization effect of space-based staller interferometer on the imaging quality and contrast. Due to the complication of optical configuration of interferometer, the polarization ray tracing is an essential tool for implementing the polarization analysis of interferometric system. The image performance of interferometer is relevant not the surface and alignment accuracy of the optics only, but also the optical properties of mirror coating, and geometric configuration of folding mirrors. The azimuth of delay line is a free design parameter for scalar optical design, but an important parameter for polarization optical design of the interferometer. For the given example interferometer, the MTF in middle and high frequency band decrease by half, and the contrast increase three orders of magnitude with the ideal case, when the delay lines are set to be circular-symmetry to the systematic axis. But, if the delay lines are set to be parallel to each other, the polarization effect on image quality is suppressed nearly at all. The contrast of fringe, which is very important for exoplanet detection, isn't improved remarkably for the configuration. Because the azimuth of folding mirrors, which connect the collectors to combiner, are still circular-symmetry. To eliminate the difference of instrumental polarization of this part between different interfering legs is a new subject for polarization-optimized design of the interferometer. At this point, the linear sub-apertures array may be a perfect solution for the high contrast imaging of interferometer. It is certain that the polarization optimized design is very essential for the space interferometer, which pursues the high resolution and contrast. In the future, many kinds of coherent imaging system, such as interferometer or deployable synthetic aperture system will appear in space, not only for the scientific, but also for the applied, even the commercial goal. Polarization effect will be an important design factor for these systems.

References

1. Lay, O.P.: Systematic errors in nulling interferometers. App. Opt. **43**(33), 6100–6122 (2004)
2. Bracewell, R.N.: Detecting nonsolar planets by spinning infrared interferometer. Nature **274**, 780–781 (1978)
3. Coulter, D.: NASA's terrestrial planet finder mission: the search for habitable planets. In: Fridlund, M., Henning, T. (eds.) Towards Other Earths: DARWIN/TPF and Search for Extrasolar Planets, ESA SP-539, Noordwijk, Netherlands, pp. 47–54 (2003)
4. Karlsson, A., Kaltenegger, L.: The technology of DARWIN. In: Fridlund, M., Henning, T. (eds.) Towards Other Earths: DARWIN/TPF and Search for Extrasolar Planets, ESA SP-539, Noordwijk, Netherlands, pp. 41–46 (2003)
5. Beckers, J.: Polarization effects in astronomical spatial interferometry. In: Chipman, R. (ed.) Polarization Consideration for Optical Systems II, Proceeding of SPIE, pp. 380–390 (1990)
6. Beckers, J.: Instrumental factors affecting the fringe contrast in optical interferometers. APSS **171**, 333–339 (1990)
7. Chazelas, B., Valette, C., Rabbia, Y., et al.: Study of the polarization effects in a nulling interferometer: design consquences. In: Aime, C., Vakili, F. (eds.) Direct Imaging of Exoplanets: Science & Techniques, Proceedings IAU Colloquium No. 200, pp. 251–254 (2005)
8. Breckinridge, J., Lam, W., Chipman, R.: Polarization aberrations in astronomical telescopes: the point spread function. PASP **127**, 445–468 (2015)
9. Born, M., Wolf, E.: Principle of Optics, 7th edn. Cambridge University Press, Cambridge (1999)
10. Chipman, R., Lam, W., Breckinridge, J.: Polarization aberration in astronomical teescopes. In: J. Shaw, D. LeMaster (eds.) Polarization Science and Remote Sensing VII, Proceedings of SPIE, vol. 9613, p. 96130H (2015)
11. Chipman, R.: Polarization analysis of optical systems. OptEn **28**(2), 090–099 (1989)
12. Yuan, S., Fu, Y., Di Varano, I., et al.: Simulation of high spatial resolution polarimetry with the Chinese giant solar telescope. In: Gandorfer, A., Lagg, A., Raab, K. (eds.) Proceedings of the 9th Solar Polarization Workshop SPW9 (2020). https://doi.org/10.17617/2.3194856
13. Yuan, S., Fu, Y., Jin, Z.: Polarization modeling for the main optics of Chinese giant solar telescope. In: Riva, M. (ed.) Proceedings of SPIE, vol. 10012, p. 1001205 (2016)
14. Rakic, A., Djurisic, A., Elazar, J., Magewski, M.: Optical properties of metallic films for vertical-cavity optoelectronic devices. Appl. Opt. **37**(22), 5271–5283 (1998)

Satellite Photometric Characteristics Modeling and Simulation Influenced by Attitude Changes

Yuzhe Li(✉), Limin Wu, and Xuguo Zhang

Beijing Institute of Space Mechanics and Electricity, Beijing 100094, China
18235228909@163.com

Abstract. The photometric data of space objects have important application value in the inversion of shape, attitude, surface characteristics, and operation state of space objects. In order to solve the problems that the fewer photometric data samples of on-orbit satellites and low time resolution, which is difficult to meet the demand of attitude behavior characteristic inversion, this paper established a box-wing satellite model according to the optical scattering characteristics of satellites, combined with their shape, orbit, attitude, material, and other parameters. Simulation analysis on the influence of attitude changes on photometric characteristics for GEO targets is conducted to obtain the influences of different factors on the photometric curve of space objects such as size, attitude, BRDF model, and the location of the measuring station. By simulation, the problem that the measured photometric properties of on-orbit objects are insufficient to support the inversion of target characteristics is solved. The simulation results proved that the photometric model established in this paper can effectively reflect the influence of attitude changes on the photometric properties, which can provide corresponding reference for the subsequent satellite attitude inversion work.

Keywords: Photometric characteristic · Satellite attitude · Characteristics inversion · Space objects

1 Introduction

The number of space targets is increasing annually, which not only makes orbital resources very scarce but also greatly increases the risk of spacecraft collisions. Space monitoring is one of the elements of space situational awareness and is an important support for space situational awareness [1, 2]. Photometric observation of space targets has important applications in tracking, identification, and classification of space targets.

As the number of resident space objects continues to grow, the need to detect and predict changes in active space objects becomes increasingly crucial. The space target photometric data plays an important role in estimating the state of objects [3–5]. The target photometric curves reflect the changes in the magnitude of reflected light from the object's surface to the observer and contain information about the geometric structure between the sun, the object, and the observer. In addition, the photometric curves reflect various types of information about the attitude, shape, surface properties, and health status of the space target [6, 7].

H. P. Urbach and H. Jiang (Eds.): ISSOIA 2022, SPPHY 295, pp. 277–288, 2023.
https://doi.org/10.1007/978-981-99-4098-1_25

Kelecy T et al. [8] used the Cook-Torrance model to simulate photometry of ground-based optical telescopes; Rush K A et al. [9] used the Ashikhmin-Shirley model and ray-tracing model to simulate the photometry of rectangular, cylindrical and box-wing space targets. Wang Xia et al. [10] proposed an Optical Scattering Cross Section (OCS) observation model based on a modified Phong model. Ying Wu et al. [11] used a mathematical analysis based on radiation theory to model solar radiation and solar radiation reflected from the Earth and the atmosphere.

Data acquisition by ground-based optical telescopes can be affected by the constraints of observing equipment, weather, visibility period, and other observing conditions. Trying to acquire photometric data of space targets requires a lot of time and equipment resources, while the various information in the photometric profiles indicates that the number of state variables used for estimation is often large. The current observations of space targets still suffer from insufficient samples and temporal resolution, and the estimation of space target states based on photometric data is still a challenging problem [12].

This study aims to establish the mapping relationship between attitude changes of space targets and photometric characteristics, to obtain the influence of different factors such as size, attitude, BRDF model parameters, and geographic location of the station on the photometric curve of space targets, and to solve the difficult problem that the measured photometric characteristics of on-orbit targets are insufficient to support the inversion of target characteristics through simulation.

2 Theoretical Basis

The Bidirectional Reflectance Distribution Function (BRDF) characterizes the optical properties of the object surface and the photometric curve is usually expressed using the Bidirectional Reflectance Distribution Function [13], which describes the light intensity ratio between irradiance and radiant brightness.

2.1 Bidirectional Reflection Distribution Function (BRDF)

This paper uses the BRDF model of Lambertian scattering and specular reflection, which has been widely used in the study of light curves.

BRDF describes the ratio of light intensity between irradiance L_i and reflected irradiance L_r as:

$$f_r = \frac{L_r}{L_i} \tag{1}$$

The geometric structure of the reflected light on the facet is shown in Fig. 1, where s, v and n are the sun direction vector, observer direction vector and normal direction vector of the facet, respectively, θ_i and θ_r are the angle between s and n, v and n.

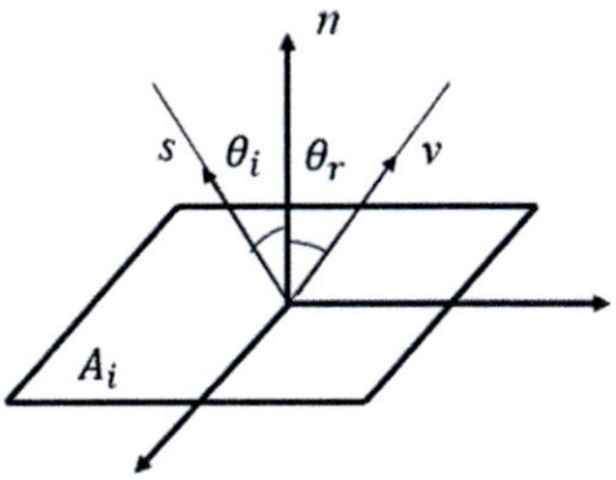

Fig. 1. Light Reflection on a facet.

2.2 Characteristic Description of Space Target Attitude

The space objects are assumed to be rigid and have constant surface parameters. The position of an object is represented by an Earth-Centric Inertial (ECI) coordinate system with the X-axis pointing to the vernal equinox, the Z-axis corresponding to the Earth's rotation axis, and the Y-axis constituting a right-handed coordinate system. The fixed coordinate system of space objects takes the center of mass as the origin, and the fixed axis corresponds to the main axis of inertia. The motion equations of the space target are as follows [14]

$$\ddot{r} = -\frac{\mu}{r^3}r \tag{2}$$

$$\mathrm{q} = \frac{1}{2}\begin{bmatrix} 0 & \omega_z & -\omega_y & \omega_x \\ -\omega_z & 0 & \omega_x & \omega_y \\ \omega_y & -\omega_x & 0 & \omega_z \\ -\omega_x & -\omega_y & -\omega_z & 0 \end{bmatrix}\begin{bmatrix} q_1 \\ q_2 \\ q_3 \\ q_4 \end{bmatrix} \tag{3}$$

r is the position vector of the space object in the inertial system, μ is the earth's gravitational constant, q is the quaternion, and $\omega = \left[\omega_x, \omega_y, \omega_z\right]^T$ is the angular velocity.

3 Satellite Modeling

This paper studies a box-wing model consisting of a satellite body (box) and two solar panels (wings), as shown in Fig. 2. The relative surface angle used for photometric curve simulation is shown in Fig. 3. In the simulation, the surface of the space target is decomposed into multiple small facets. The orbital model uses the TLE (Two-Line-Orbital element), and the quaternion is used to describe the attitude of the space target. The BRDF photometric observation model is established.

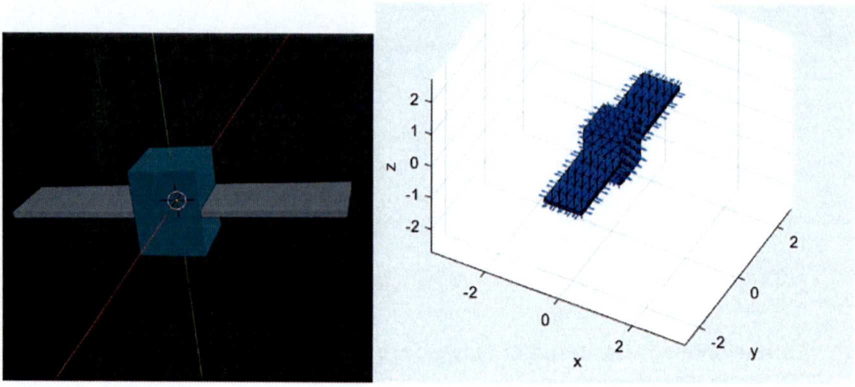

Fig. 2. Box-wing satellite model

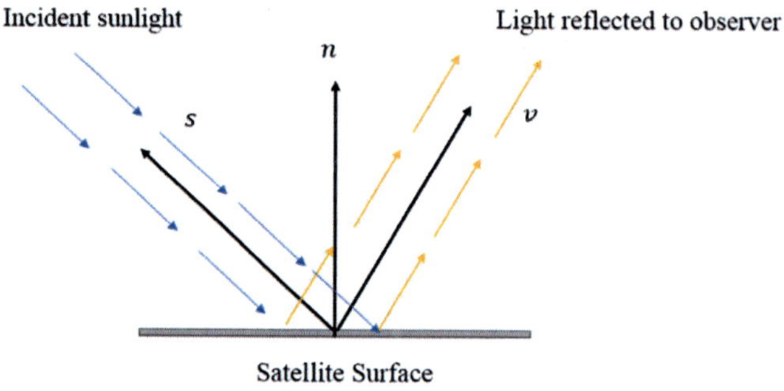

Fig. 3. Relative surface angles for light curve simulation

Photometric curves are continuous photometric data obtained from ground observations. The relative magnitude of the brightness of the i th facet is described as follows.

$$m_{app,i} = m_{sun} - 2.5\log_{10}\frac{f_{obs}}{r_{obs}^2} \tag{4}$$

where, $m_{sun} = -26.7$, r_{obs} is the distance between the space target and the observer.

$$f_{obs} = c_{total} f_i A(n^T \cdot v) \tag{5}$$

In Eq. (5), $c_{total} = dc_d + sc_s$, d and s are the coefficients of diffuse and specular reflection, c_d and c_s are the target surface diffuse and specular reflection parameters, A is the surface area, and the function f_i is expressed as follows

$$f_i = \begin{cases} 0 & if\ s^T \cdot n \leq 0 \\ s^T \cdot n & elsewhere \end{cases} \tag{6}$$

4 Photometric Curve Simulation

The photometric curve of a space target is a coupled function of the size, attitude, surface material and position of the space target, therefore, in the simulation process, all the above influencing factors need to be taken into account to improve the reliability and accuracy of the simulation data. The spatial target simulation model described in this paper takes into account the relevant influencing factors and simulates the influence of different parameters on the photometric curve through a Matlab program. The exact simulation flow of the photometric curve is shown in Fig. 4.

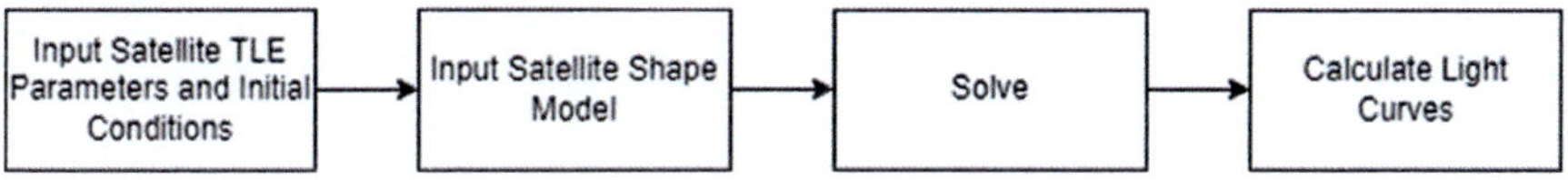

Fig. 4. Flow of the photometric curve

This section shows the simulation results of the box-wing model. It is assumed that the orbital position and attitude of the satellite are known and the position of the observation station is 130°20′ E and 33°56′N. In the simulation, each surface of the box and the solar wing is divided into 100 small facets for high-fidelity calculation of photometric curves. By using the formulae of the photometric observation model given in Sect. 2, the simulation obtains the influence of different factors such as size, attitude, BRDF model parameters, and geographic location of the station on the photometric profile of the space target.

4.1 Influence of Size Parameters on Photometric Curves

This section discusses the influence of size parameters on the photometric curve, ensuring that under the same orbit, the same attitude mode, and the same reflection conditions, the photometric curve changes of box-wing models of different sizes can be analyzed. Here, the GEO orbit is selected with a Two-Line-Orbital element as shown in Table 1, the attitude is a three-axis stabilization mode with the z-axis pointing to the sky bottom and the x-axis pointing to the velocity direction, the reflection conditions are constant, i.e. $C_d = 0.2$, $C_s = 0.8$, and the dimensional parameters of the space target are set to three cases as shown in Table 2. By the formula of the photometric model given in Sect. 3, the photometric curves shown in Fig. 5 are obtained.

Table 1. The orbital elements of a GEO.

1	29272U	06033A	19052.49283631	−.00000367	+00000-0	00000-0	0	9997
2	29272	000.0538	281.8322	0002158	050.2267	090.8544	1.002711420	20932

Table 2. Size parameters of space objects

Parameter	Box	Wing
Size 1	1.0 m × 1.0 m × 1.3 m	2.0 m × 1.0 m × 0.1 m
Size 2	2.0 m × 2.0 m × 2.6 m	4.0 m × 2.0 m × 0.2 m
Size 3	4.0 m × 4.0 m × 5.2 m	8.0 m × 4.0 m × 0.4 m

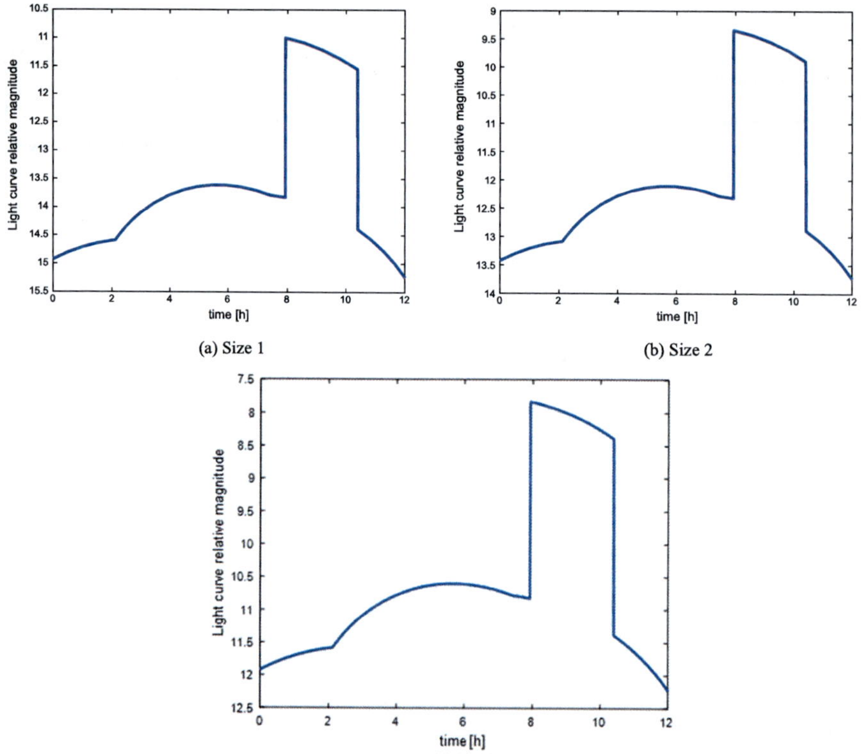

(a) Size 1

(b) Size 2

(c) Size 3

Fig. 5. Photometric simulation results of different size parameters

As can be seen from Fig. 5, changing the size has almost no effect on the shape of the luminosity curve, but only on its brightness value. When the size is doubled, the magnitude value changes by about 1.5 magnitudes.

4.2 Influence of Attitude Parameters on Photometric Curves

This section discusses the influence of attitude parameters on the photometric curve and analyzes the changes of the photometric curve with different attitude parameters under the same size, the same orbit, and the same reflection conditions. Here, the GEO orbit is selected with Two-Line-Orbital element as shown in Table 1, and the reflection conditions are unchanged, that is $C_d = 0.2$, $C_s = 0.8$ and the box size is 4.0 m × 4.0 m × 5.2 m.

There are six types of attitudes:

(a) The body z-axis points to the nadir and the body x-axis points to the velocity direction.
(b) The body z-axis points to the nadir and the body x-axis is at an angle of 15° to the inertial x-axis.
(c) The body z-axis points to the nadir and the body x-axis is at an angle of 30° to the inertial x-axis.
(d) The body z-axis points to the nadir and the body x-axis is at an angle of 45° to the inertial x-axis.
(e) The body z-axis points to the nadir and spins around the body y-axis at a rate of 0.05 revs/s.
(f) The body z-axis points to the nadir and spins around the body y-axis at a rate of 0.1 revs/s.

The photometric curves calculated by the photometric model are shown in Fig. 6. It can be seen from the Fig that the influence of attitude on the luminosity curve is very large. Through Fig. 6(a)–(d), the photometric curves of attitude 1, 2, 3, and 4 are similar in shape, but there is a difference in the peak position. The observer observes a time difference on the same surface, resulting in a difference in the time when the peak is generated.

From Fig. 6(e) and (f), it can be seen that although attitude 5 and 6 are both rotating states, the rotation rates are different, and the luminosity curves are also different. The faster the rotation speed, the more obvious the change of luminosity.

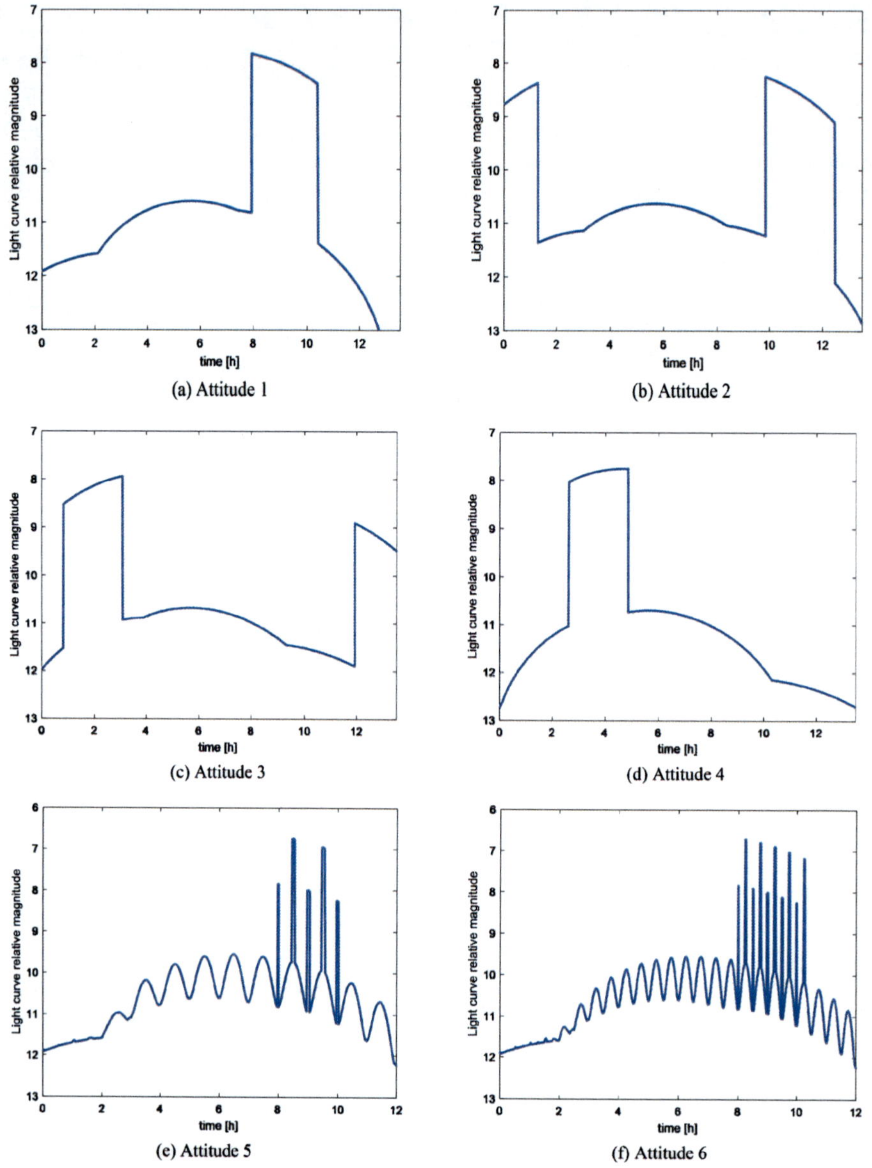

Fig. 6. Photometric simulation results of different attitude modes

4.3 Influence of BRDF Model Parameters on Photometric Curves

In this section, the effect of BRDF model parameters on the photometric curve is discussed, and the variation of the photometric curve under different reflection conditions is analyzed under the same shape, the same orbit, and the same attitude mode. Here, the GEO orbit is selected with Two-Line-Orbital element as shown in Table 1. The box size is 4.0 m × 4.0 m × 5.2 m, and the attitude is a three-axis stabilization mode, with the body z-axis pointing to the nadir and the x-axis pointing to the velocity direction. We compared the changes of the simulated photometric curves of different reflection parameters (C_d, C_s), and the results are shown in Fig. 7. It can be seen from Fig. 7 that the reflection parameters determine the degree of bending of the photometric curve and also the magnitude of the peak, the larger the specular reflection parameters, the more obvious the peak phenomenon and the greater the degree of bending of the luminosity curve, which indicates that the reflection parameters are also related to the peak phenomenon.

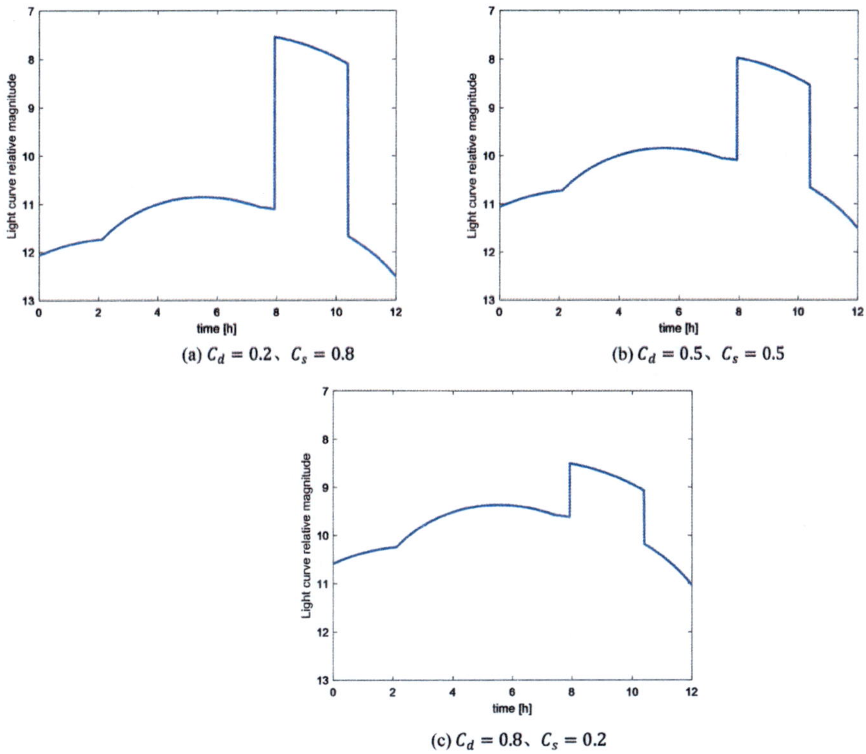

Fig. 7. Photometric simulation results of different reflection parameters

4.4 Influence of Location of the Station on Photometric Curves

In this section, by comparing the luminosity curves observed at different longitudes of the stations, we analyze the changes of the luminosity curves by the geographic locations of different stations. The orbit is a geosynchronous orbit (GEO), and the Two-Line-Orbital element is shown in Table 1. The attitude control mode is a three-axis stabilization mode, the z-axis of the body points to the nadir, and the x-axis points to the direction of the movement speed. The size of the box is 4.0 m × 4.0 m × 5.2 m, and the longitude is at intervals of 15°. The results are shown in Fig. 8.

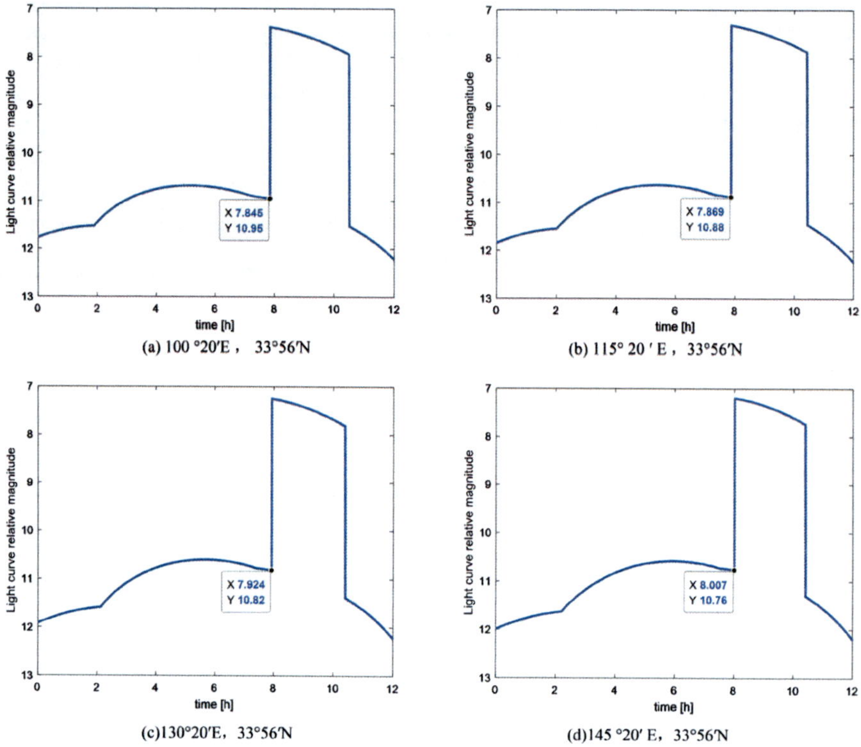

Fig. 8. Photometric simulation results of different station longitudes

It can be seen from Fig. 8 that there is a time shift when observing space targets at the same latitude and different longitude stations, and the difference in the size of the peaks is not large. It can be inferred that the difference in longitude will lead to a certain time difference in the observed peak phenomenon, and the time difference is in the level of several minutes. The time difference is caused by the geometric relationship of the space target with respect to the station and the Sun. In order to observe the peak phenomenon, the geometric relationship of the sun-target-station needs to satisfy certain specular reflection conditions.

5 Conclusion

Aiming at the problem of state estimation of photometric curve data, this paper establishes a satellite box-wing model, selects a reasonable BRDF model and establishes a photometric curve simulation model through the calculation formula of photometric model. We discuss the effects of size, attitude, BRDF model parameters, and location of the station on the photometric curve. Through the comparative analysis, it is pointed out that these influencing factors have an impact on the photometric curve. The change of the size parameter affects the change of the simulated luminosity. When the size is doubled, the brightness increases by about 1.5 magnitudes; the attitude parameter has a great influence on the photometric curve, and the shape of the photometric curve can indirectly reflect the attitude mode of the space target; BRDF model parameter is related to the peak phenomenon. The larger the specular reflection parameter (Cs) is, the more obvious the peak phenomenon is. The longitude of the station determines the time of the peak position. The influences of the above factors are consistent with the actual observation. The simulation results prove that the shape of the photometric curve is, to some extent, mainly determined by the attitude and rotation state of the space target, and the photometric curve simulation method pro-posed in this paper can provide corresponding technical support for the subsequent photometric curve attitude inversion work.

References

1. Kennewell, J.A., Vo, B.N.: An overview of space situational awareness. In: Proceedings of the 16th International Conference on Information Fusion, pp. 1029–1036. IEEE (2013)
2. Bobrinsky, N., Del Monte, L.: The space situational awareness program of the European space agency. Cosmic Res. **48**(5) (2010)
3. Yoshimura, Y., Matsushita, Y., Arakawa, R., et al.: Light curve approximation using an attitude model of solar sail spacecraft. J. Guid. Control. Dyn. **43**(10), 1960–1966 (2020)
4. Matsushita, Y., Arakawa, R., Yoshimura, Y., et al.: Light curve analysis and attitude estimation of space objects focusing on glint. In: 1st International Orbital Debris Conference (IOC), Sugar Land, Texas, USA (2019)
5. Yoshimura, Y., Matsushita, Y., Arakawa, R., et al.: Combining photometric data formulation with attitude model of a spinning solar sail spacecraft. In: First International Orbital Debris Conference, vol. 2109, p. 6082 (2019)
6. Spurbeck, J., Jah, M., Kucharski, D., et al.: Satellite characterization, classification, and operational assessment via the exploitation of remote photoacoustic signatures. In: Advanced Maui Optical and Space Surveillance Technologies Conference (AMOS) (2018)
7. Spurbeck, J., Jah, M.K., Kucharski, D., et al.: Near real time satellite event detection, characterization, and operational assessment via the exploitation of remote photoacoustic signatures. J. Astronaut. Sci. **68**(1), 197–224 (2021)
8. Kelecy, T., Gerber, E., Akram, S., et al.: Characterization of resident space object states using functional data analysis. J. Astronaut. Sci. **69**(2), 627–649 (2022)
9. Rush, K.A.: An application of the unscented Kalman filter for spacecraft attitude estimation on real and simulated light curve data (2020)
10. Wang, X., Zhang, Y.S., Xu, C., et al.: An improved Phong model for GEO target attitude estimation based on space-based optical observations. J. Photon. **49**(1), 0112004 (2020)
11. Wu, Y., Yang, L., Fan, J.Y., et al.: Modeling and simulation of visible light scattering characteristics of spatial targets. J. Harbin Univ. Technol. **14**(2), 82–85 (2009)
12. Matsushita, Y., Hanada, T., Yoshimura, Y., et al.: Attitude estimation of space objects using glint. In: 43rd COSPAR Scientific Assembly. Held 28 January-4 February, vol. 43, pp. 2191 (2021)

13. McInnes, C.R.: Solar Sailing: Technology, Dynamics and Mission Applications. Springer, Cham (2004)
14. Tsuda, Y., Saiki, T., Funase, R., et al.: Generalized attitude model for spinning solar sail spacecraft. J. Guid. Control. Dyn. **36**(4), 967–974 (2013)

Check for updates

Design of On-Orbit Radiation Calibration Scheme of Optical Remote Sensing Systems

Xiaoman Li(✉), Libing Jin, Minlong Lian, and Huiting Gao

Beijing Institute of Space Mechanics and Electricity, Beijing 100094, China
Lixiaoman_bisme@163.com

Abstract. For quantitative application of space remote sensing, strict radiation calibration before launch, to determine the initial spectral radiation response characteristics of remote sensing, due to space particle radiation, pollution, components aging will cause the degradation of remote sensing performance, to change the spectral radiation response characteristics, therefore, regularly for radiation calibration, meet the requirements of long-term quantitative application of remote sensing data. When the earth-observation remote sensing system with NIR and SWIR infrared spectrum works, the radiation received by the remote sensing device is greatly affected by the sun. For the remote sensing device with large field of view greater than 100 degrees, multi-perspective calibration is required to reduce the calibration error. Generally, external scene calibration sources (such as large desert area on the ground, the sun, the moon, the cryogenic space, etc.) or built-in calibration devices (such as lamps, silicon carbide sticks, blackbody, etc.) could be selected to reduce the calibration uncertainty. When the infrared remote sensing system uses blackbody for radiometric calibration, by using accurate blackbody radiation model and highly stable blackbody emission performance, the uncertainty of radiation calibration could be reduced. The designing of infrared optical remote sensing was introduced, and the method of calibration accuracy analysis was introduced in the paper.

Keywords: Remote sensing system · On-orbit Radiation Calibration · Blackbody · Calibration Accuracy · Quantitative Application

1 Introduction

Infrared remote sensing could be used not only in civil fields such as meteorology and hydrology, resource exploration, agricultural survey, but also in astronomical observation science applications, with a wide range of applications [1]. At present, scientific research requires high accuracy of quantitative inversion of space infrared remote sensors.

Due to a series of factors such as non-uniformity of detector chip substrate material, detector doping level, detector bias voltage, readout circuit noise level, and inconsistency between multiple parallel output channels, spatial infrared remote sensor has non-uniformity in focal plane array response. In addition, the image illumination uniformity of optical system design, stray radiation of optical system, and changes in operating temperature of optical system, Will affect the stability of the output. Before launching, the

H. P. Urbach and H. Jiang (Eds.): ISSOIA 2022, SPPHY 295, pp. 289–297, 2023.
https://doi.org/10.1007/978-981-99-4098-1_26

laboratory vacuum radiation calibration shall be strictly carried out, and the ground pixel by pixel calibration shall be completed to obtain the parameter difference between pixels, so as to determine the initial spectral radiation response characteristics of the remote sensor [2]. During the orbit flight of the remote sensor, due to the radiation of space particles, pollution, aging of components, etc., the performance of the remote sensor will be degraded, thus changing the spectral radiation response characteristics. Therefore, it is necessary to regularly calibrate the remote sensor on satellite radiation to meet the user's requirements for long-term quantitative application of remote sensing data. Generally, external scene calibration sources (such as large desert area on the ground, the sun, the moon, the cryogenic space, etc.) or built-in calibration devices (such as lamps, silicon carbide sticks, blackbody, etc.) could be selected to reduce the calibration uncertainty.

When the infrared remote sensing system uses blackbody for radiometric calibration, by using accurate blackbody radiation model and highly stable blackbody emission performance, the uncertainty of radiation calibration could be reduced. The uncertainty of radiation calibration of space remote sensing instruments has reached the best level of 2%–3%. At the same time, it is easy to achieve full aperture, full optical path and full field of view absolute radiation calibration, such as MODIS, AATSR SPECTRA and etc.all use blackbody for on orbit calibration. The long-term decay of blackbody emissivity in orbit could be monitored by other calibration methods. For example, ASTRO-F could observe stars during its orbit, and it uses the sensitivity of far-infrared detector and the brightness of point source to carry out the absolute radiation calibration of stars; SPIRIT III also carried out absolute radiometric calibration using stars in orbit. The high resolution satellite IKONOS has chosen the scheme of using stars for radiometric calibration, and the error between it and the calibration scheme of NASA team is 2.5%.

2 Radiation Calibration Scheme Design

The high temperature and low temperature area source blackbody were both used as the standard source for absolute radiometric calibration. The stability of area source blackbody will affect the calibration accuracy [3]. Because the radiation of stars is relatively stable, the calibration results of blackbody are corrected by observing stars. It is a better method of satellite calibration for infrared remote sensor [4].

There are many factors to be considered in the design of radiometric calibration scheme, including the working spectrum, calibration accuracy, life, reliability and constraints from spacecraft and aerospace optical remote sensors.

2.1 Blackbody Calibration

The blackbody surface is designed with V-shaped slots at a certain angle and the surface is anodized and blackened, which could ensure that the blackbody has good radiation uniformity and high radiation emissivity [5, 6].

The onboard blackbody design parameters of a remote sensing camera are as follows:

1) Blackbody temperature: could cover the target and scene equivalent blackbody temperature, adjustable;
2) Effective size in bold: could cover the entrance pupil of the camera);

3) Normal emissivity: ≥0.98;
4) Temperature uniformity: ≤0.2K;
5) Temperature control accuracy of blackbody surface: ± 0.1K;
6) Temperature measurement accuracy: ≤0.1K.

The high-temperature blackbody component is mainly composed of blackbody, heating sheet, thermal insulation gasket and thermal insulation gasket. Aluminum alloy 2A12 with good thermal conductivity, light weight and high hardness is selected to be processed into high-temperature black body, and thermal insulation gasket is processed from polyimide. Titanium alloy screws with poor thermal conductivity are used to connect the high-temperature blackbody and the scanning base.

The low-temperature blackbody component is mainly composed of blackbody, semiconductor cooler, fixing block (used to fix semiconductor cooler and heat pipe), heat pipe, thermal insulation gasket and thermal insulation gasket. The low-temperature black body is made of aluminum alloy 2A12 with good thermal conductivity, light weight and high hardness. The connection between the low-temperature black body and the scanning base and the fixing block is made of titanium alloy screws with poor thermal conductivity.

Optimization of high precision temperature control measures for blackbody:

1) The enclosure is arranged outside the blackbody component, and the temperature is controlled through the radiation of the enclosure to meet the temperature requirements. The external surface of the enclosure is arranged with an active temperature control heating circuit; The inner surface of the outer cover is blackened to enhance the radiation heat transfer between the outer cover and the blackbody component. The outer surface of the outer cover is covered with multi-layer thermal insulation materials to reduce radiation heat leakage. The outermost side of the multi-layer is black polyimide carburized film;
2) The two sides of the blackbody are sandwich structures, and the middle layer could be a 0.5 mm-thick-graphite film to improve the temperature uniformity in the area of the optical path. The back of the blackbody is covered with multi-layer thermal insulation materials to reduce heat leakage after cutting into the optical path;
3) Thermal insulation installation between blackbody and supporting structure to reduce mutual influence of temperature;
4) The blackbody support structure is arranged with an active temperature control heating circuit, and the support structure and the camera main frame are installed with thermal insulation.

Through the thermal design and thermal analysis of the high and low temperature blackbody, the temperature level and uniformity of the blackening side of the high and low temperature blackbody that affect the radiometric calibration accuracy are shown in the following Table 1 (Fig. 1).

2.2 Solar Calibration of Diffuse Reflective Plate

The solar diffuser has been widely used in high-precision absolute the solar radiation calibration. Using accurate solar radiation model and highly stable solar diffuser reflection performance, could achieve high calibration accuracy.

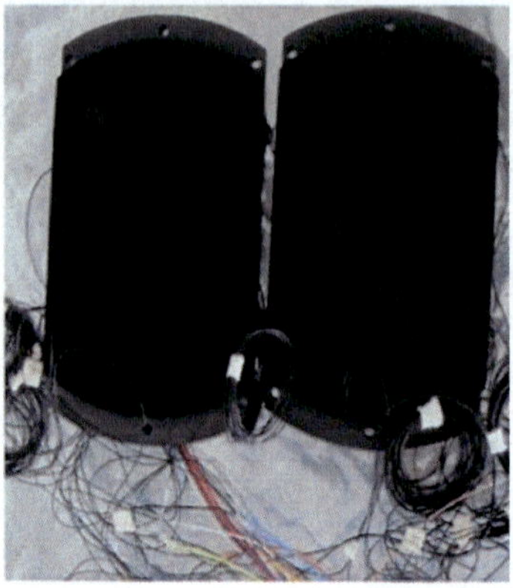

Fig. 1. High temperature blackbody component and cryogenic blackbody component

Table 1. Calculation results of high temperature blackbody and cryogenic blackbody

ITEMS	Temperature Stability	Temperature Uniformity
High temperature blackbody	<0.1 °C/min	0.3 °C
Low temperature blackbody	<0.1 °C /min	0.3 °C

1) Design of the diffuse reflection plate components.

   Reflection plate adopts high temperature fusion PTFE diffuse reflective plate with superior optical reference characteristics and certain spatial adaptability, could be approximated as ideal Lambert; and slightly backscattering; its BRDF precision 2%, The surface uniformity could be controlled within 1%. The reflectivity of the solar diffuser is higher than 90%.

2) The protection design of the diffuse reflection plate components is small.

   The solar diffuser would be organic polluted on orbit, which will directly affect its optical reference characteristics and its stability. Therefore, reliable protection is required during the diffuse plate making, storage, detection and transportation: the first layer is optical protection, the diffuse plate under the protective plate is made of PTFE with the same material and cleanliness as the diffuse plate while protecting the diffuse plate; the second layer is physical protection, the components containing the protective plate is installed in the stainless steel nitrogen box, and then tighten the sealing circumference and cover plate, and effective prevent the damage.

3) Diffuse reflection plate performance monitoring,

   The furnace test plate is installed under the diffuse plate protection and next to the product. Once the protection is removed, the furnace test plate is exposed to the same laboratory environment. Remove the furnace test plate before launch, the difference from the diffuse plate is delivered, and the BRDF correction factor of the diffuse plate for sample monitoring before launch is obtained.

The solar diffuser stability monitoring radiometer monitors the changes of the bidirectional reflection distribution function of the incident light in the outgoing direction after a surface reflection, and corrects the reflection illumination [7] (Fig. 2).

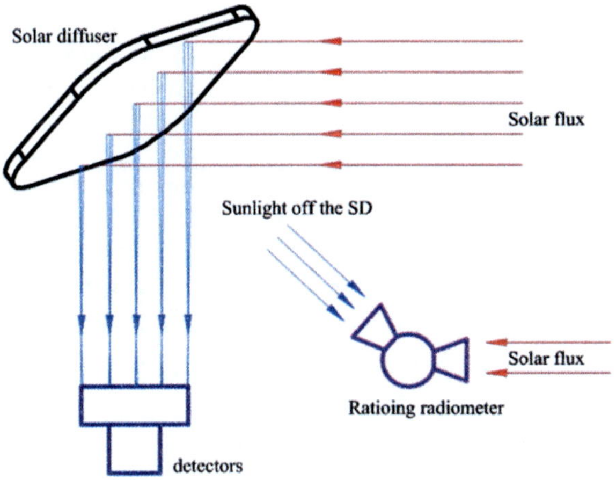

Fig. 2. Schematic diagram of solar calibration of diffuse reflective plate

2.3 Infrared Star Calibration and Data Correction Method

At present, the method of using stars for radiation calibration is mainly used in astronomical satellites and astronomical observations. The following factors need to be considered in the star calibration process:

1) The star is a typical point source target, and its solid angle is generally much smaller than the camera pixel's IFOV. The star position deviation caused by satellite attitude jitter or flutter, the image point extraction algorithm error in the point target speckle formed when the camera observed the star.
2) The influence of camera noise in the measurement process is also a part of the calibration error, so it is necessary to reduce the camera time noise as much as possible.

The camera response during star calibration could be calculated by the following formula.

$$V = k \cdot \frac{\int R(\lambda) \cdot E(\lambda) \cdot \eta d\lambda}{\Omega \cdot \cos^3(\omega) \int R(\lambda) d\lambda} \tag{1}$$

where, V: camera response; k: Radiation calibration coefficient; R(λ): Camera spectral response; E(λ): Stellar irradiance; η: Energy concentration; ω: Imaging photonics off-axis angle; Ω: instantaneous field angle of view of camera.

The calculation formula of point target receiving area source irradiance is:

$$E = \frac{\Phi}{dA^2} = I_\theta \cdot \frac{\cos\theta}{l^2} \tag{2}$$

Both infrared star calibration and onboard blackbody calibration could obtain high precision absolute radiometric calibration coefficients. However, considering the decay of onboard blackbody emissivity with time under long-term operation, the equivalent spectral radiance of onboard blackbody calculated according to operating temperature could not really represent the equivalent radiance of the phase machine entrance pupil. Therefore, the absolute radiometric calibration coefficient obtained by the black body on the star is defined as the calibration coefficient before correction, which is applied to the absolute radiometric correction after being corrected by the absolute radiometric calibration coefficient of the star [8, 9].

When the DN value of the blackbody radiometric calibration data and stellar radiometric calibration data of the pixel is the same

$$DN(n) = K'(n) \times L_e' + C'(n) \tag{3}$$

where DN is the pixel gray value; L_e' is the equivalent spectral radiance of the calibrated blackbody on the satellite K', C' are respectively the gain and intercept of the absolute calibration coefficient on the satellite.

$$K'(n) = \frac{DN_h(n) - DN_l(n)}{L_{eh} - L_{el}},\ C' = \frac{DN_l(n) \times L_{eh} - DN_h(n) \times L_{el}}{L_{eh} - L_{el}}.$$

When calibrating stars, select 2 stars from the star library, and use their known spectral radiance data to analyze and obtain:

$$DN = K(n) \times L_e + C(n) \tag{4}$$

where, is the equivalent spectral radiance of the entrance pupil Is the gain and intercept of the absolute radiometric calibration coefficient.

The onboard blackbody radiation calibration coefficient is corrected by the following formula.

$$L_e = \frac{K'(n)}{K(n)} \times L_e' + \frac{C'(n) - C(n)}{K(n)} \tag{5}$$

The correction coefficients R_k and R_c could be obtained: $R_k = \frac{K'(n)}{K(n)}, R_c = \frac{C'(n)-C(n)}{K(n)}$

For each new calibration data, new calibration coefficients could be obtained based on the calibration coefficients and correction coefficients R_k and R_c obtained from the new onboard blackbody calibration [10].

3 Radiometric Calibration Uncertainty

The calibration uncertainty of space optical remote sensor mainly depends on the stability of the calibration radiation source itself, measurement accuracy, calibration method and the stability of space optical remote sensor. From the point of view of the vacuum radiometric calibration energy transmission link, the factors that affect the vacuum radiometric calibration accuracy of the infrared camera during operation mainly include the following aspects:

1) Uncertainty of onboard blackbody temperature δ_{bb_Tacc}

The blackbody radiation source is the reference source of the camera vacuum radiation calibration, and the blackbody temperature measurement error will affect the calibration accuracy. According to the target radiation characteristics, the equivalent blackbody temperature at the entrance pupil is converted and calibrated by the metrology institute.

2) On board blackbody emissivity error δ_{bb_emi}

Through the standard transmission of emissivity, the blackbody standard could be traced back to the National Institute of Metrology, and the uncertainty of emissivity measurement mainly comes from the uncertainty of reference standards and instrument measurement error. Among them, standard traceability error is the main contribution. The emissivity test needs to ensure that the test distance and angle are as consistent as possible with the calibration. The emissivity of the calibrated blackbody is transmitted through the radiometer and compared with the standard blackbody.

3) Temperature uniformity of blackbody on satellite δ_{bb_Tuni}

The uniformity of the blackbody radiation source includes the area uniformity and angle uniformity, which are the main factors affecting the relative radiometric calibration accuracy of the camera. The influence of the temperature non-uniformity of the blackbody radiation source is calculated by comparing the radiance difference with the reference blackbody radiance. At the same time, it is necessary to consider that the angular non-uniformity within the field of view angle range.

4) Temperature stability of blackbody on satellite δ_{bb_Tsta}

The blackbody radiation source is the reference source for the camera vacuum radiation calibration. The blackbody emissivity uncertainty, temperature stability, temperature uniformity and temperature measurement error will affect the calibration accuracy.

5) Influence of camera optical and mechanical radiation fluctuation δ_{rad}

The radiation energy emitted by the camera itself will reach the camera detector together with the target/atmospheric background. The optical mechanical radiation mainly includes the radiation of the optical lens itself and the radiation of the optical mechanical structure. The optical mechanical radiation fluctuation could be calculated by using Planck's law according to the change of the optical mechanical temperature telemetry value.

6) Camera noise δ_{cam_noise}

The temporal noise of the camera is mainly composed of photon noise, detector readout noise, dark current noise and circuit noise. The temporal noise reflects the detection sensitivity of the camera, and the influence of the camera noise is shown as the random fluctuation of the output signal. The uncertainty introduced by the noise is 1.70% based on the estimation of the signal-to-noise ratio of the camera at the reference temperature.

7) Camera response nonlinearity δ_{res_non}

According to the engineering experience of the radiation calibration process in the laboratory, the corresponding nonlinearity is calculated for the values within the dynamic response range (DN value) of 5%–95%.

8) Straylight radiation $\delta_{strlight}$

The space environment, including the sun, the moon, stars and the earth's atmosphere, will affect the calibration on the satellite and generate stray radiation. In addition, the heat accumulated by the black body's long-term work will also affect the calibration accuracy of the system.

The influence amount of each factor is integrated into the total error of the laboratory radiometric calibration. The calculation method is to calculate the square root of the influence amount of each factor, which could be solved by referring to the following formula:

$$\delta = \sqrt{\delta^2_{bb\_Tacc} + \delta^2_{bb\_emi} + \delta^2_{bb\_Tuni} + \delta^2_{bb\_Tsta} + \delta^2_{rad} + \delta^2_{cam\_noise} + \delta^2_{res\_non} + \delta^2_{strlight}} \quad (6)$$

4 Conclusion

It is reasonable and feasible to calibrate the infrared space optical remote sensing camera in orbit by using blackbody+infrared star for all optical path calibration, focusing on the problems of blackbody working temperature uniformity, stability, absolute radiation performance monitoring, etc. By analyzing the factors influencing the calibration uncertainty of shortwave infrared remote sensor, the radiometric calibration accuracy could be effectively improved by improving the onboard calibration scheme, providing the blackbody temperature control accuracy of onboard calibration, and combining with the infrared star calibration method. The algorithm is verified by the ground laboratory test.

The advantage of full aperture and full optical path calibration is the low uncertainty of absolute radiometric calibration. However, for large aperture aerospace optical remote sensor, the size and weight of the onboard calibrator used to achieve full aperture and full optical path calibration are relatively large, and it requires moving parts to drive the calibrator to cut in/out the optical path, which limits its application. Therefore, it is necessary to select the onboard calibration scheme according to the specific conditions of the aerospace optical remote sensor. In the future, relevant test verification will be carried out on orbit to continuously improve the performance of infrared cameras to further reduce the uncertainty of radiometric calibration.

References

1. Chen, S.P.: Space Camera Design and Experiment, pp. 24–25, 334–360. China Astronautics Press, Beijing (2003)
2. Ma, W.P.: Aerospace Optical Remote Sensing Technology, pp. 210–211. China Science and Technology Press, Beijing (2011)
3. Slater, P.N., Biggar, S.F., Holm, R.G., et al.: Reflectance and radiance-based methods for the in-flight absolute calibration of multispectral sensors. Rem. Sens. Environ. **22**, 11–37 (1987)
4. Foukal, P.V., Kochling, H., Miller, P.: Cryogenic absolute radiometer as laboratory irradiance standards, remote sensing detectors and pyroheliometers. Appl. Opt. **29**(7), 988–993 (1990)
5. Wang, M., Zhou, S.D., He, M.Y.: Characterization and calibration method of satellite sensor radiometric calibration site. Geom. Spat. Inf. Technol. **38**(7), 24–27 (2015)

6. Zhang, B.G., Zhang, Y.F.: Analysis of relative radiometric calibration algorithm of space CCD camera. Spacecr. Recov. Remote Sens. **35**(1), 91–96 (2014)
7. Zheng, X.B., Wu, H.Y., Zhang, J.P., et al.: High accuracy radiometric calibration and standard transmission method. Chin. Sci. Bull. **45**(12), 1341–1344 (2000)
8. Liu, L., Chen, L., Xu, H.L.: On-orbit radiometric calibration in long wave infrared band of VRSS-2 satellite. Spacecr. Recov. Remote Sens. **40**(3), 94–102 (2019)
9. An, M., Liang, D.Y., Nie, Y.S.: Research on infrared detector imaging test methods for ZY-1 02E satellite. Spacecr. Recov. Remote Sens. **42**(4), 37–45 (2021)
10. Gao, H.T., Ma, Y., Liu, W., He, H.Y.: Analysis and simulation of cross calibration precision of ocean color remote sensing. Spacecr. Recov. Remote Sens. **37**(2), 116–124 (2016)

Spatial Resolution Evaluation Method Based on Edge Characteristics

Jing Zhang[1,2], Xiaohui Li[1], Jiajia Zhu[1,2](✉), Ziyang Li[1], and Xinfang Yuan[1]

[1] Key Laboratory of Quantitative Remote Sensing Information Technology, Aerospace Information Research Institute, Chinese Academy of Sciences, Beijing 100094, China
jjzhu@aoe.ac.cn

[2] College of Optoelectronics, University of Chinese Academy of Sciences, Beijing 100049, China

Abstract. The most common way to evaluate spatial resolution is visual method, but it is time-consuming, laborious, easily affected by evaluators and output environment, and resolution targets must be arranged. In practical applications, it is difficult to obtain remote sensing images containing resolution targets. Therefore, how to use images without resolution targets but with typical features (edge, linear) to develop a practical objective detection method of spatial resolution has practical significance and application requirements. To solve this problem, the basic concepts and calculation methods of several parameters: FWHM, EQW, 2σ width, ESW, $R_{0.02}$, $R_{0.05}$, etc., closely related to the spatial resolution and often used to reflect the payload's ability to reproduce details, are analyzed by using edge target. Through simulation experiments and real data verification, the relationship between these parameters and spatial resolution is analyzed and compared. It is found that these parameters are linearly related to standard deviation of the MTF curve (σ_{MTF}), and their change laws are basically consistent with the visual resolution. However, the result is far from the actual visual resolution. A new spatial resolution evaluation method based on edge characteristics is proposed in this paper. The analysis results show that there is a certain linear relationship between σ_{MTF} of the imaging system and its spatial resolution. The spatial resolution calculated by this method is consistent with the visual results, the deviation is small, and the correlation can reach 86.12%.

Keywords: Spatial Resolution · Edge Target · Standard Deviation of MTF · Visual Resolution

1 Introduction

The spatial resolution (SR) of optical remote sensing payload is simply a measure of the resolution of the payload imaging details [1], and also an indicator of the target fineness in the image. It can most intuitively reflect the payload imaging performance and image quality. However, the expression of quantitative spatial resolution is not consistent at present. The most common way to evaluate SR is visual method, that is, by laying resolution targets on the ground and interpreting the target image to obtain its SR. The

H. P. Urbach and H. Jiang (Eds.): ISSOIA 2022, SPPHY 295, pp. 298–314, 2023.
https://doi.org/10.1007/978-981-99-4098-1_27

visual resolution method is simple and intuitive, but it is actually a subjective perception process, time-consuming and laborious. The process is often affected by the emotion, experience and knowledge of the interpreter, which cannot guarantee the repeatability of the interpretation results, and is also affected by image output equipment (display, printer, etc.). In addition, the visual method must use resolution targets images, which cannot be obtained in most practical applications. Therefore, how to use images without resolution targets but with typical features (edge, linear) to develop a practical objective detection method of spatial resolution, which is consistent with the visual resolution, has practical significance and application requirements.

In many cases, parameters such as Ground Sampling Distance (GSD), Modulation Transfer Function Area (MTFA), Relative Edge Response (RER), Full Width at Half-Maximum (FWHM) and Edge Spread Width (ESW) are often used to characterize the spatial resolution of optical remote sensing payloads [2–4]. However, these parameters still have some shortcomings: either not intuitive in characterizing spatial resolution, or cannot maintain good consistency with the visual resolution.

To solve this problem, the basic concepts and calculation methods of several parameters: FWHM [5–7], Equivalent Width (EQW) [8], 2σ width ($W_{2\sigma}$) [9], ESW [10], $R_{0.02}$ [8], $R_{0.05}$ [11, 12], etc., closely related to the spatial resolution and often used to reflect the payload's ability to reproduce details, are analyzed by using edge target. Through simulation experiments, the variation rules of the above parameters and the relationship between these parameters and spatial resolution are analyzed and compared. A spatial resolution evaluation method based on the edge characteristics of ground objects was realized according to the analysis results, and finally, the evaluation method has been verified and analyzed using real data.

2 Basic Principles

Because of the diffraction of light, a point on the target passes through the optical system to form a disc called an Airy spot on the imaging surface, even if there is no aberration. The light spot is centered on the geometric image point corresponding to the original object point and takes the wavelength of light as the radius. If the imaging system aberration is taken into account, the speckle of a point on the target object will be larger and may not necessarily be disk-shaped. This imaging feature of optical systems is often represented by point spread function (PSF) in the spatial domain. The larger the sweep area and the smoother the shape of PSF, the worse the spatial resolution of the image. The modulation transfer function (MTF) of the optical system can be obtained by the Fourier transform of PSF. Therefore, the spatial resolution has a very close relationship with MTF of the imaging system.

Based on the above, several parameters related to PSF and MTF are often used to characterize spatial resolution, such as FWHM, $W_{2\sigma}$, EQW, ESW of the line Spread Function (LSF) and the corresponding spatial frequency of the relevant empirical parameters (such as MTF = 0.02 or 0.05, etc.). This section introduces the basic meanings and calculation methods of these parameters. In this article, pixel or lp/pixel is used as the resolution dimension to eliminate the impact of different GSDs.

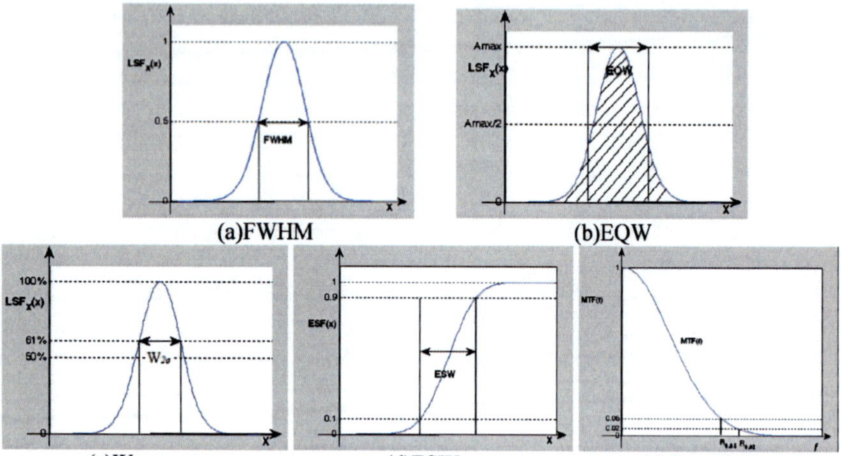

(c) $W_{2\sigma}$ (d)ESW (e) $R_{0.02}$ & $R_{0.05}$

Fig. 1. Definition of common parameters: FWHM, EQW, $W_{2\sigma}$, ESW, $R_{0.02}$ & $R_{0.05}$.

2.1 FWHM

FWHM [13] is an important parameter describing the spatial characteristics of PSF or LSF. It is defined as the abscissa difference when the value of the normalized PSF or LSF is 0.5, as shown in Fig. 1(a). In the case of the same GSD, the smaller the FWHM, the smaller the diffusion of the imaging system to the point light source or line light source, the clearer the edge of the ground object in the image, and the better the ability to reproduce and maintain high-frequency details. However, in practical application, it is usually difficult to directly measure PSF, so LSF is often used to calculate FWHM. There are also literatures that directly use FWHM to characterize the spatial resolution of the load [13, 17, 18].

Assuming that LSF is an ideal Gaussian function, after normalization,

$$LSF_X(x) = \exp\left(-(x - \mu_x)^2/2\sigma_x^2\right) \tag{1}$$

When LSF = 0.5, $x = \pm\sigma_x\sqrt{ln4} + \mu_x$. Then,

$$FWHM = x_+ - x_- = 2\sqrt{ln\,4}\sigma_x \approx 2.3548\sigma_x \tag{2}$$

The unit of FWHM is pixel.

The Gaussian model is used to represent the MTF curve of the imaging system [14], then the shape of the MTF curve depends on the standard deviation of the Gaussian model. The larger the value is, the faster the MTF curve will decline. Equation (3) gives MTF in the x direction:

$$MTF(f) \cong e^{-2\pi^2\sigma_{MTF}^2 f^2} \tag{3}$$

There is a certain linear relationship between σ_{MTF} and σ_x [9], and according to Eq. (2), the FWHM is proportional to σ_x. Therefore, there is also a certain linear relationship between σ_{MTF} and FWHM. The more severe the image degradation, the faster the MTF curve drops, the larger the MTF value, the larger the corresponding FWHM, and the lower the spatial resolution.

2.2 EQW

EQW [15] is calculated based on LSF. It represents the width of the area whose height is Amax, the maximum value of LSF, and is equal to the surrounding area, as shown in Fig. 1(b). EQW is defined as:

$$EQW = \left(\int LSF_X(x)dx \right) / A_{max} \tag{4}$$

Assuming that LSF is an ideal Gaussian function, then $A_{\max} = 1/\sqrt{2\pi}\sigma_x$ from Eq. (1), then

$$EQW = \left(\int LSF_X(x)dx \right) / A_{max} = 1/A_{max} = \sqrt{2\pi}\sigma_x \approx 2.5066\sigma_x \tag{5}$$

The unit of EQW is pixel.

Similarly, according to the reasoning of the relationship between FWHM and σ_{MTF} in Sect. 2.1, there is a similar linear relationship between EQW and σ_{MTF}. The physical significance of EQW is that it represents the concentration degree of the image energy distribution. The smaller the value is, the more concentrated the energy distribution is and the greater the corresponding spatial resolution is. So it is reasonable to use the equivalent width to represent the spatial resolution of the image.

2.3 2σ Width

Assuming that the normalized LSF presents a Gaussian distribution, the maximum slope of LSF is at the point where the intensity is 0.61, and the corresponding width of LSF at this point is called 2σ width [9], $W_{2\sigma}$, and the definition is shown in Fig. 1(c).

When LSF = 0.61, $x = \pm\sigma_x\sqrt{-ln(0.61)} + \mu_x$, then,

$$W_{2\sigma} = x_+ - x_- = 2\sqrt{-ln(0.61)}\sigma_x \approx 1.4061\sigma_x \tag{6}$$

The unit of $W_{2\sigma}$ is pixel, and the variation of $W_{2\sigma}$ is the same as that of FWHM, which will not be repeated here.

2.4 ESW

ESW [10] refers to the width of Edge Spread Function(ESF), which is the abscissa difference value of 0.1 and 0.9 after zero normalization of ESF, as shown in Fig. 1(d). Suppose that the maximum value of ESF(x) is ESF_{max}, and the minimum value is ESF_{min}.

$$\Delta E = ESF_{max} - ESF_{min} \tag{7}$$

$$ESF(x_{-}) = ESF_{min} + \Delta E \times 0.1 \tag{8}$$

$$ESF(x_{+}) = ESF_{min} + \Delta E \times 0.9 \tag{9}$$

$$ESW = x_{+} - x_{-} \tag{10}$$

2.5 $R_{0.02}$ and $R_{0.05}$

In optical practice, a single value such as 0.02 or 0.05 is often taken as the contrast threshold of human visual system [8]. This is equivalent to a straight line parallel to the frequency axis through the threshold contrast 0.02 (0.05), that is, the corresponding spatial frequency when MTF = 0.02 (0.05) is taken as the SR of the payload, namely $R_{0.02}$ ($R_{0.05}$), as shown in Fig. 1(e). Make Eq. (3) equal to the corresponding spatial frequency value when 0.02 and 0.05, then,

$$R_{0.02} = \sqrt{-ln(0.02)/2\pi^2\sigma_{MTF}^2} \approx 0.4452/\sigma_{MTF} \tag{11}$$

$$R_{0.05} = \sqrt{-ln(0.05)/2\pi^2\sigma_{MTF}^2} \approx 0.3896/\sigma_{MTF} \tag{12}$$

The units of $R_{0.02}$ and $R_{0.05}$ are lp/pixel.

2.6 Visual Resolution

In this paper, the subjective result of spatial resolution, namely visual resolution, is obtained by subjectively evaluating the radial target image using the visual method. And is represented by the minimum distinguishable target width D_{Res}. For the radial target image, D_{Res} refers to the chord length corresponding to the position that can just be resolved on the image, and the spatial resolution Res is the spatial frequency corresponding to the position, and the relationship between D_{Res} and Res is as follows.

$$\text{Res} = 1/2D_{Res} \tag{13}$$

When using the visual method to evaluate the spatial resolution, it is usually necessary to determine the location by selecting a point with the mouse. In the experiment, it is found that due to different observers, mouse misoperation and other reasons, the obtained visual limit radius will have a variation range of ±1.5–2 pixel widths, expressed in ΔR. This will result in a visual deviation of formula (14) for the width of the minimum resolvable line.

$$\Delta D = 2 \times \Delta R \times \sin(\theta/2) \tag{14}$$

where ΔD represents the deviation of line width, Δr represents the deviation of visual limit radius selection, θ represents the target strip angular width of the radial target.

3 Experiment and Analysis

The spatial resolution is largely affected by the degradation level. The higher the degradation level, the lower the spatial resolution, the wider D_{Res},the minimum resolvable line width, and the larger σ_{MTF}. According to the definition and calculation formula of the parameters in Sect. 2, FWHM, $W_{2\sigma}$, EQW and ESW are contrary to the change rule of spatial resolution, but consistent with the change rule of D_{Res}, while $R_{0.02}$ and $R_{0.05}$ are consistent with the change rule of spatial resolution. Therefore, in order to study the quantitative relationship between each parameter and spatial resolution, in the subsequent analysis and comparison, D_{Res} is used as the evaluation standard for the former type of parameters, while the spatial resolution is used as the evaluation standard for $R_{0.02}$ and $R_{0.05}$.

The design ideas of the experimental analysis are as follows:

a) Use simulation experiments to study the change rules of various parameters under different degradation levels, verify the authenticity of the above rules, and establish the qualitative relationship between various parameters and spatial resolution according to the experimental results.
b) The quantitative relationship between the parameters and the minimum resolvable line width is compared and analyzed to find a relatively close characterization parameter.

The calculation flow chart of each parameter is shown in Fig. 2.

In the simulation experiment, the evaluation results of the visual method are obtained by using the radial target results, while the characterization parameters introduced in Sect. 2 are obtained by using the edge images [16, 17]. The simulation images are shown in Fig. 3.

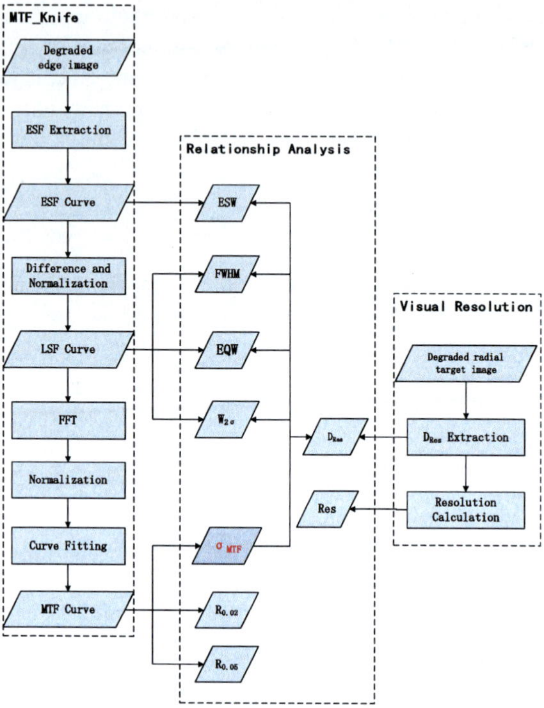

Fig. 2. Flow Chart of Experimental Analysis.

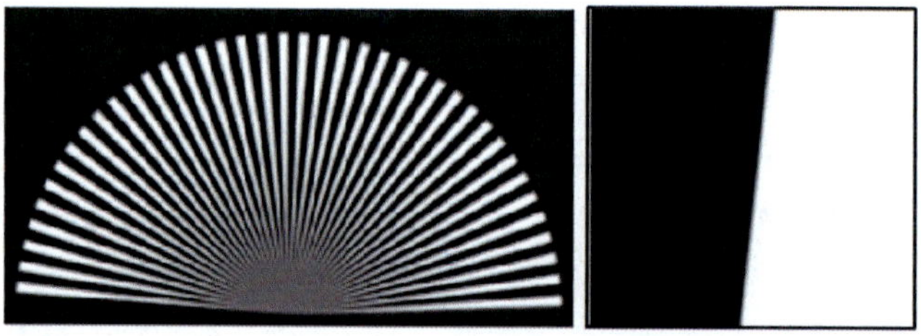

Fig. 3. Simulation Image

3.1 Analysis on the Change Rule of Each Parameter

Regardless of the influence of noise, the input simulation image is assumed to have light and dark gray values of 225 and 15 respectively, the radial target strip angle width is set to 3°, the target radius is 120 pixels, the edge target tilt angle is 6°, and the standard deviation of the degenerated Gaussian template is set from 0.38 to 0.75 in steps of 0.01. The test results of each parameter at different degradation levels are follows.

FWHM

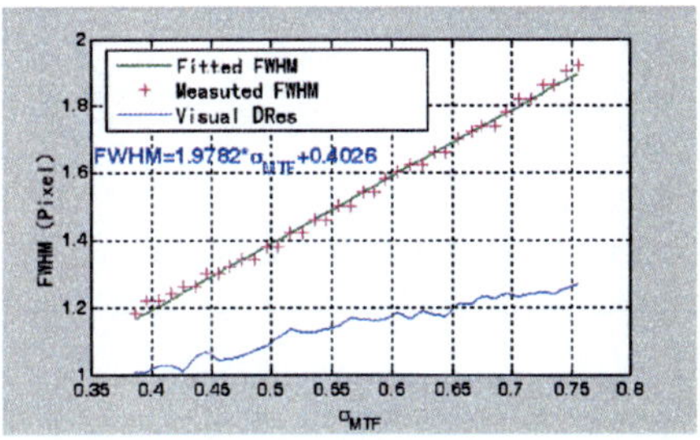

Fig. 4. FWHM change curve under different degradation levels. The horizontal axis is σ_{MTF}, the vertical axis is the FWHM, the purple cross point is the FWHM measured, and the blue line represents D_{Res} observed visually.

The results in Fig. 4 show that the FWHM measured shows an obvious linear relationship with σ_{MTF}, and FWHM gradually increases with the increase of the degradation level, which is consistent with the theoretical derivation results, and is basically consistent with the change rule of D_{Res} observed visually. The measured FWHM curve is polynomial fitted:

$$FWHM = 1.9782\sigma_{MTF} + 0.4026 \tag{15}$$

In Fig. 4, the green line indicates the change rule of FWHM with σ_{MTF} calculated by Eq. 15 under different degradation levels. It can be seen that the fitting curve is very close to the actual measured result, and the sum of residual squares is 0.0081.

EQW

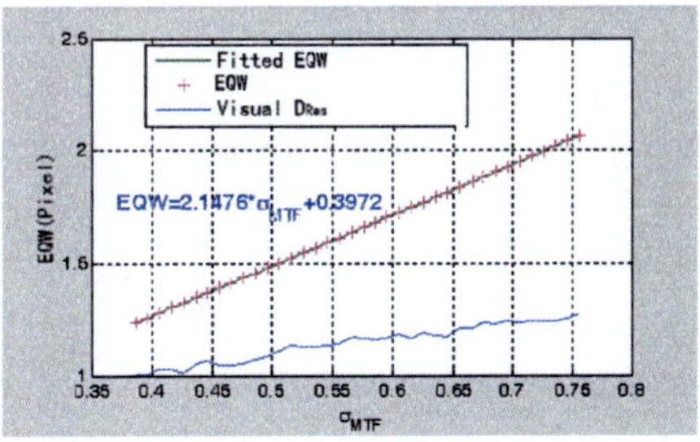

Fig. 5. EQW change curve under different degradation levels. The horizontal axis is σ_{MTF}, the vertical axis is EQW. The purple cross point is the EQW measured, and the blue line represents D_{Res} observed visually

The results in Fig. 5 show that the measured EQW shows an obvious linear relationship with σ_{MTF}, and EQW gradually increases with the increase of the degradation level, consistent with the theoretical derivation results, and is basically consistent with the change rule of D_{Res} observed visually. The measured EQW curve is polynomial fitted:

$$EQW = 2.1476\sigma_{MTF} + 0.3972 \tag{16}$$

In Fig. 5, the green line indicates the change rule of EQW with σ_{MTF} calculated by Eq. 16 under different degradation levels. It can be seen that the fitting curve is very close to the actual measured result, and the sum of residual squares is 0.0016.

$W_{2\sigma}$

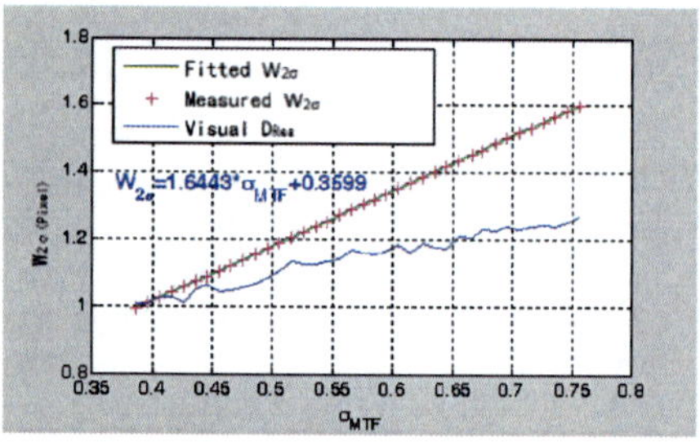

Fig. 6. $W_{2\sigma}$ change curve under different degradation levels. The horizontal axis is σ_{MTF}, the vertical axis is $W_{2\sigma}$, the purple cross point is the $W_{2\sigma}$ measured, and the blue line represents D_{Res} observed visually

The results in Fig. 6 show that the $W_{2\sigma}$ measured shows an obvious linear relationship with σ_{MTF}, and $W_{2\sigma}$ gradually increases with the increase of the degradation level, and is basically consistent with the change rule of D_{Res} observed visually. The measured W2σ curve is polynomial fitted:

$$W_{2\sigma} = 1.6443\sigma_{MTF} + 0.3599 \tag{17}$$

In Fig. 6, the green line indicates the change rule of $W_{2\sigma}$ with σ_{MTF} calculated by Eq. 17 under different degradation levels. It can be seen that the fitting curve is very close to the actual measured result, and the sum of residual squares is 0.0048.

ESW

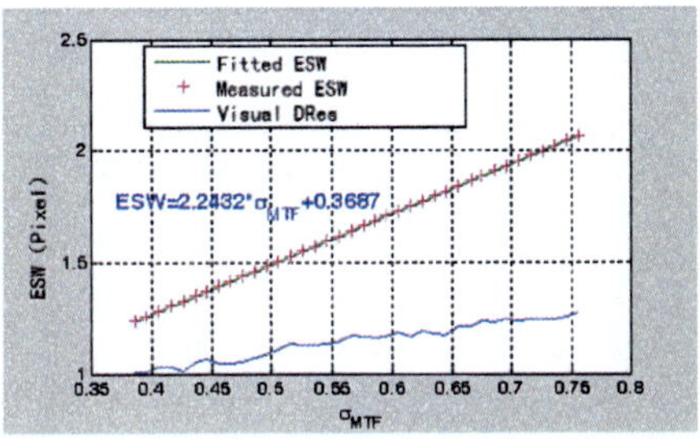

Fig. 7. ESW change curve under different degradation levels. The horizontal axis is σ_{MTF}, the vertical axis is the ESW, and the purple cross point is the ESW measured.

The results in Fig. 7 show that the ESW measured shows an obvious linear relationship with σ_{MTF}, and ESW gradually increases with the increase of the degradation level, which is consistent with the theoretical derivation results, and is basically consistent with the change rule of D_{Res} observed visually. The measured ESW curve is polynomial fitted:

$$ESW = 2.2432\sigma_{MTF} + 0.3687 \quad (18)$$

In Fig. 7, the green line indicates the change rule of ESW with σ_{MTF} calculated by Eq. 18 under different degradation levels. It can be seen that the fitting curve is very close to the actual measured result, and the sum of residual squares is 0.0058.

$R_{0.02}$ and $R_{0.05}$

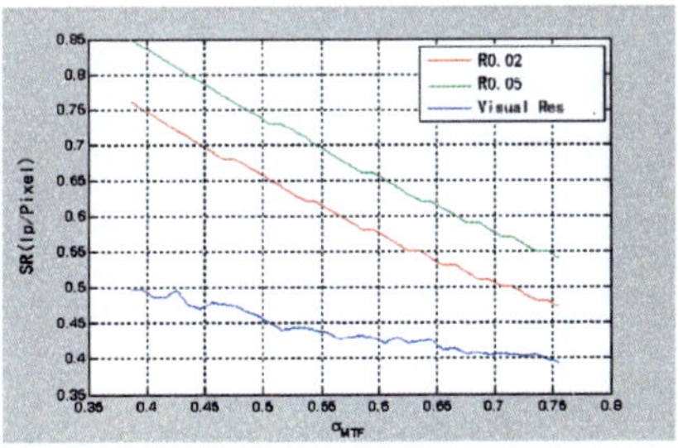

Fig. 8. $R_{0.02}$ & $R_{0.05}$ change curve under different degradation levels.

In Fig. 8, the horizontal axis is σ_{MTF}, the vertical axis is resolution, the red line shows $R_{0.02}$ measured, the green line represents $R_{0.05}$ measured, and the blue line represents D_{Res} observed visually. The results show that $R_{0.02}$ and $R_{0.05}$ differ greatly from the actual visual measurement results and are not suitable for characterizing visual spatial resolution.

D_{Res}

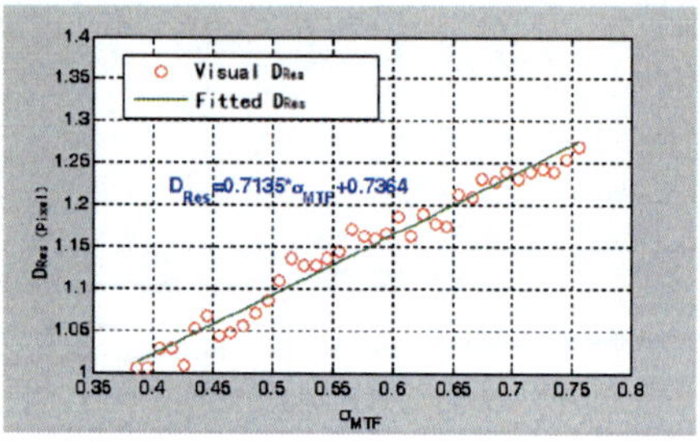

Fig. 9. D_{Res} change curve under different degradation levels.

In Fig. 9, the horizontal axis is σ_{MTF}, the vertical axis is minimum distinguishable target width, the red circle point is D_{Res} observed visually. It can be found that there is a positive linear relationship between the standard deviation of D_{Res} and σ_{MTF}..The measured D_{Res} curve is polynomial fitted:

$$D_{Res} = 0.7135\sigma_{MTF} + 0.7364 \tag{19}$$

In Fig. 9, the green line represents D_{Res} calculated by Eq. 19, and the sum of residual squares is 0.0081.

3.2 Comparison and Analysis of Each Parameters

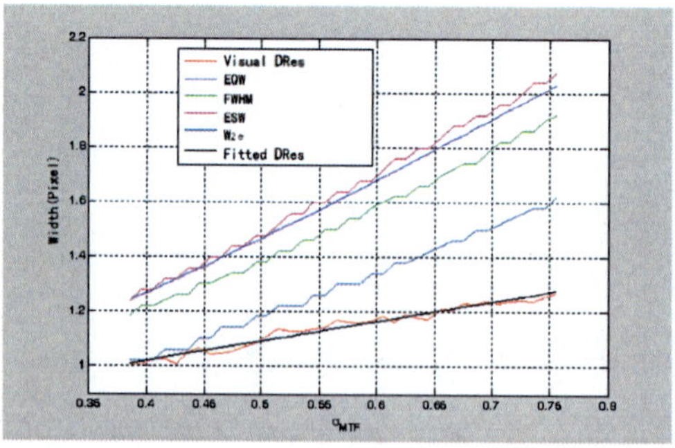

Fig. 10. Change curve of each parameter at different degradation levels

Figure 10 shows the change curve of each parameter. The red line represents the minimum visual resolvable line width. The analysis results show that all the above parameters can qualitatively characterize the spatial resolution of the load, but the use of $R_{0.02}$ and $R_{0.05}$ given in Sect. 2.5 is not applicable to the detection of the actual spatial resolution (large deviation), and these parameters will not be considered in the subsequent verification of real data. The parameters described in Sects. 2.1, 2.2, 2.3 and 2.4 has a large deviation from the visual results, and the deviation is in the order of ESW > EQW > FWHM > $W_{2\sigma}$, but whether the specific parameters adapt to the detection of real spatial resolution still needs to be further verified.

4 Real Data Validation

In this paper, the measurement and calculation of relevant parameters are carried out for the experimental data of UAV aerial photography that actually laid radial targets on the ground. Since the flight attitude of the aircraft changes more than that of the satellite, the target image obtained by the remote sensing load carried on the UAV platform will be distorted and deformed to a certain extent due to the change of the flight attitude of the aircraft, so it is better to avoid these areas when selecting the measuring edge.

4.1 Verification of the Change Rule

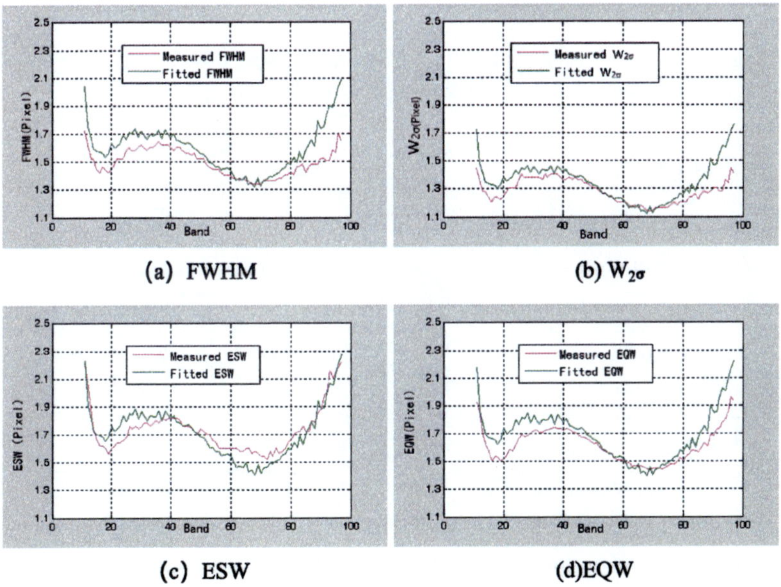

Fig. 11. Measured & fitted value of each parameter

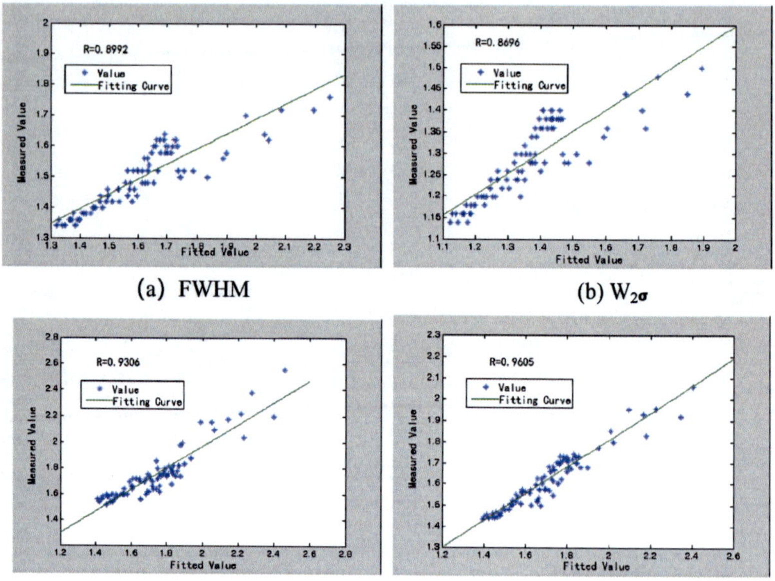

(a) FWHM (b) $W_{2\sigma}$

(c) ESW (d)EQW

Fig. 12. Consistency analysis between measured and fitted values of each parameter

From Fig. 11 & Fig. 12, FWHM, $W_{2\sigma}$, EQW, ESW correlation coefficients measured results and calculated results are in more than 85%, showed high consistency.

4.2 Verification of D_{Res}

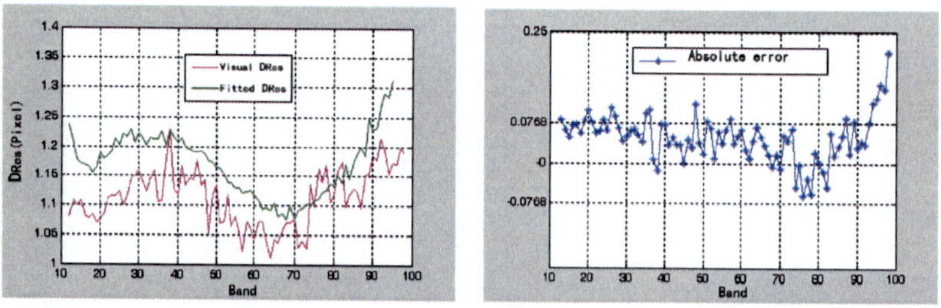

(a) Curve of visual and calculated D_{Res} (b) Absolute error between calculated and visual value

Fig. 13. The visual and fitted SR values of hyperspectral data (band 11 ~ 99)

Figure 13 shows the visual values of D_{Res} and the values calculated according to the fitting formula and their absolute errors. Elected by the formula 14 can be driven, with the target of 2.2°, the minimum perceptible article target width deviation as $\Delta D < 0.0768$. As can be seen from Fig. 13(b), in the band 30–93, the absolute error between the calculated value and the visual value is basically within this range. After the band 93, the absolute error will increase significantly, while the absolute error of band 11–30 has little change but is basically on the boundary of allowable deviation. After analyzing the image (Fig. 14), it is found that it is mainly affected by noise.

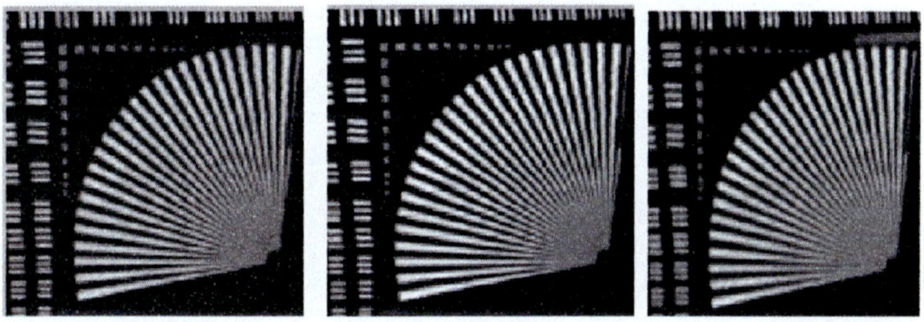

(a) Band12 (b) Band30 (c) Band95

Fig. 14. Hyperspectral radiative target image

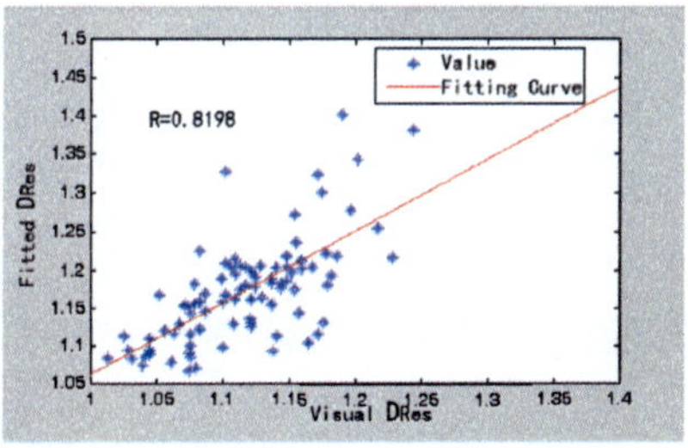

Fig. 15. Consistency analysis between the visual & fitted value of DRes

Figure 15 shows the consistency analysis curve of the visual values and fitted values of hyperspectral data in bands from 11 to 99, and Table 1 shows the comparison between the visual values and calculated values of the minimum resolvable DRes of hyperspectral data in different bands. It can be found that although the correlation coefficient obtained by using band 30–93 as the sample is higher than that obtained by using band 11–99 as the sample due to the influence of noise, the correlation coefficient obtained by the two samples is relatively high. It can be considered that the visual value of the minimum discernable line width is basically consistent with the results calculated according to the fitting formula.

Table 1. Consistency analysis between the fitted and visual values of hyperspectral data in different bands.

Band Interval	Coefficient of correlation	Maximum deviation (pixel)	Minimum deviation (pixel)	Mean deviation absolute value (pixel)
11–99	0.8198	0.2258	– 0.0614	0.0606
30–93	0.8612	0.1142	– 0.0614	0.0451

4.3 Comparison and Analysis of Each Parameters

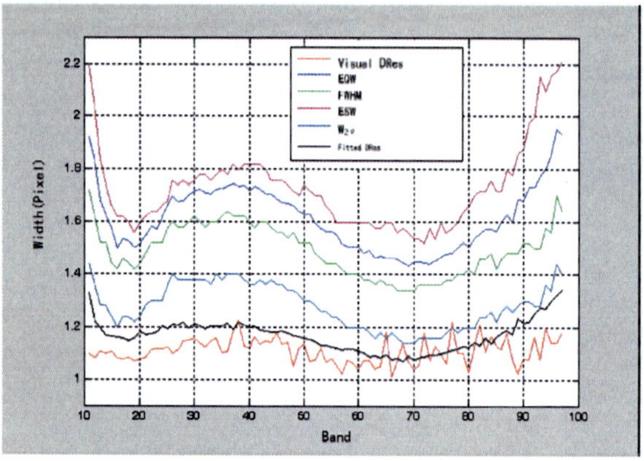

Fig. 16. Visual D_{Res} and the measured values of each characterization parameter

Figure 16 shows the comparison curve of the visual D_{Res} and the measured values of each characterization parameter, in which the red line represents the visual value of D_{Res}, and the black line represents the fitting value calculated by Eq. (19). From Fig. 16 and Table 2 the parameters mentioned in Sect. 2.1, 2.2, 2.3 and 2.4 cannot be used to directly quantitatively characterize the spatial resolution of the payload, as these parameters have a large deviation from the visual results, while the fitted D_{Res} values obtained by using the relevant formulas have a high consistency with the visual results.

Table 2. Deviation between characterization parameters and visual D_{Res} (unit: pixel)

	FWHM	$W_{2\sigma}$	ESW	EQW	Fitted D_{Res}
Maximum	0.3773	0.1981	1.4781	1.5543	0.2258
Minimum	−0.4110	−0.5510	0.2213	0.2824	−0.1125
Mean	0.3814	0.1701	0.6317	0.5077	0.0681

5 Conclusion

The basic concepts and calculation methods of several physical parameters commonly used to characterize the spatial resolution of load, such as FWHM, $W_{2\sigma}$, ESW, EQW, $R_{0.02}$ and $R_{0.05}$, are introduced. By means of simulation experiment, the variation law of these parameters under different degradation levels was studied, the linear relationship between these parameters and σ_{MTF} was obtained, and the relevant formula was obtained by polynomial fitting method. By comparing and analyzing the relationship between these parameters and spatial resolution, it is found that there are basically consistent with the variation law of spatial resolution: with the enhancement of degradation level, FWHM, $W_{2\sigma}$, ESW and EQW show a linear increase trend, while the results obtained from the empirical parameters show a decreasing trend, but these parameters are somewhat deviated from the actual visual results. Finally, the variation rule of these parameters and the relationship between them and the spatial resolution were verified with real data. The verification results showed that the calculated values obtained by using σ_{MTF} and related formulas had the smallest deviation from the visual results, and the spatial resolution R could be represented as

$$R = 1/2D_{Res} = 1/2(0.7135\sigma_{MTF} + 0.7364) \tag{20}$$

In the formula, D_{Res} represents the minimum resolvable line width (unit: pixel), and represents the standard deviation of MTF curve obtained by edge method. Here, the unit of spatial resolution is lp/pixel.

The edge method can use natural features as targets to estimate in-orbit MTF. The above conclusions can provide a basis for further research on the evaluation method of spatial resolution of images without targets.

Acknowledgements. This work was supported by the National Key Research and Development Program of China (2021YFC3000302) and Strategic Priority Research Program of the Chinese Academy of Sciences (XDA17040301).

References

1. ISO12233, Photography—Electronic still-picture cameras—Resolution measurements (2000)
2. Dial, G., Bowen, H., Gerlach, F., et al.: IKONOS satellite, imagery, and products. Remote Sens. Environ. **88**(1), 23–36 (2003)
3. Helder, D., Coan, M., Patrick, K., et al.: IKONOS geometric characterization. Remote Sens. Environ. **88**(1), 69–79 (2003)
4. Guenther, B., Xiong, X., Salomonson, V., et al.: On-orbit performance of the earth observing system moderate resolution imaging spectroradiometer; first year of data. Remote Sens. Environ. **83**(1), 16–30 (2002)
5. Ryan, R., Baldridge, B., Schowengerdt, R.A., et al.: IKONOS spatial resolution and image interpretability characterization. Remote Sens. Environ. **88**(1), 37–52 (2003)
6. Blonski, S., Ross, K., Pagnutti, M., et al.: Spatial resolution characterization for aerial digital imagery. SSTI-2220-0071, NASA Technical Report Server (2006)

7. Helder, D., Choi, J.: On-orbit modulation transfers function (MTF) measurement of QuicBird. In: Proceedings of the 2003 High Spatial Resolution Commercial Imagery Workshop. Reston, VA, USA: NASA/NIMA/USGS Joint Agency Commercial Imagery Evaluation Team (2003)
8. Li, L.: The measurement and the application of the modulation transfer function (MTF) of the photographic system. Geom. Inf. Sci. Wuhan Univ. **13**(3), 22–32 (1988)
9. Becker, S., Haala, N., Reulke, R.: Determination and improvement of spatial resolution for digital aerial images. In: Proceedings ISPRS Workshop on High Resolution Earth Imaging, Hannover, pp. 17–20 (2005)
10. Smith, S.W.: The Scientist and Engineer's Guide to Digital Signal Processing. FreeTech Books (2003)
11. Jia, J.Y.H.: The determination and application of the modulation transfer function of the digital imaging system. Geom. Inf. Sci. Wuhan Univ. **4**, 003 (1991)
12. Guo, C., Lai, G., Tong, X., Ding, L.: A Quantification method of spatial resolution improved value on remote sensing image. J. Geom. Sci. Technol. **34**(1) (2017)
13. Sitter, D.N., Jr., Goddard, J.S., Ferrell, R.K.: Method for the measurement of the modulation transfer function of sampled imaging systems from bar-target patterns. Appl. Opt. **34**(4), 746–751 (1995)
14. Wang, Q.: Principle and Technology of Electronic Color Separation. Wuhan University of Surveying and Mapping Press, Wuhan (1993)
15. Xie, Z.X., Liu, Y. H., Wang, Z. F, et al.: A nonlinearly compensatory principle and method for human vision contrast resolution. J. Chengdu Med. Coll. **3**, 003 (2009)
16. Wang, Q.: Research on comprehensive analysis and evaluation technology of optical load MTF on orbit. Academy of Opto-electronics, Chinese Academy of Sciences (2010)
17. Xu, H.: MTF assessment algorithm accuracy analyze and optimization based on knife-edge method. Academy of Opto-electronics, Chinese Academy of Sciences (2012)

P-FSM and M-FSM Piezoelectric and Magnetic Fast Steering Mirrors

F. Claeyssen[1(✉)], G. Aigouy[1], T. Maillard[1], F. Barillot[1], O. Sosnicki[1], E. Betsch[1], H. Grardel[1], J. Rebufat[1], and L. I. Fengfan[2]

[1] Cedrat Technologies, 38240 Meylan, France
frank.claeyssen@cedrat-tec.com
[2] IAE Industrial Group Co., Ltd., Room JT3365, No. 500, Xincheng Road, Jiading District, Shanghai 201822, China
ericfung@iae-oe.com

Abstract. To meet needs in air, space, defence & other embedded or demanding applications CEDRAT TECHNOLOGIES (CTEC) has developed for more than 20 years Fast Steering Mirrors (FSM). Typical applications are for example Lidar, free space optical communication (FSO) and Laser interferometry. As the specifications met in these applications can vary, CTEC has designed and patented different actuation technologies:

- FSM based on Amplified Piezo Actuators APA®, as for PHARAO, ATLID, PSYCHE missions, and for future FSO space constellations as well as other large FSM for space & instrumentation as ESO ELT.
- FSM based on Magnetic Actuation, as being developed for FSO and CO-OP optical communication toward Geosat.
- FSM based on Stepping Piezo Actuators, as for IASI-NG for METOP-SG, as well as other piezomotor-based FSM.

The presentation describes the working principle, design and performance of these different actuation concepts.

1 Introduction

Needs for Fast and/or Fine Steering Mirrors (FSM) are dramatically increasing due to new optic applications linked to laser and imaging. Those applications can be found for example in:

- Instrumentation: interferometer, Lidar, Laser Guide Stars
- Space: free space optical (FSO) communication (Fig. 1),
- Defence: Laser designator, Directed Energy Weapon,
- Optronics: image stabilization, micro scanning,
- Medical: ophthalmological laser,
- Industry: laser marking, welding, 3-D printers…

To cover the specifications met in these applications, CEDRAT TECHNOLOGIES (CTEC) has developed and patented different FSM technologies:

H. P. Urbach and H. Jiang (Eds.): ISSOIA 2022, SPPHY 295, pp. 315–331, 2023.
https://doi.org/10.1007/978-981-99-4098-1_28

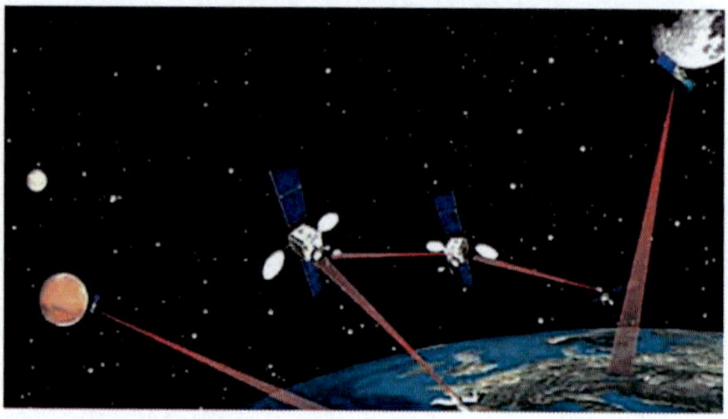

Fig. 1. Free space optical (FSO) communication

- FSM based on Amplified Piezo Actuators APA®,
- FSM based on Magnetic Actuation,
- FSM based on Stepping Piezo Motors.

These developments have been largely initiated by space application needs and supported by CNES, DGA, ESA.

Because of space launching requirements, these mechanisms should not only perform high precision functions but should also be extremely reliable while being compact, lightweight and resistant to external vibrations and shocks.

These features make these FSM interesting for other markets such as embedded, severe-environment or demanding applications, as optronics, defence, instrumentation...

Therefore, CTEC develops many standard and customised high-performance products deriving from its space FSM.

The presentation describes the general working principle, the design and the performance of these different actuation concepts. This review allows to see technologies evolutions for 20 years such as increases of strokes, mirror specs, mirror supports, and the variety of size and applications.

2 Amplified Piezo Actuators FSM

Thanks to its space heritage with ROSETTA MIDAS stage developed in 1998 [1] for ESA, CTEC developed its first FSM named MEF for ACES/PHARAO [2] for SODERN and CNES in 2002 (Fig. 2).

PHARAO is a cold-atom caesium clock using interferometry, embedded on the ICSS. The MEFs are used to balance the laser beams. It is a Tip-tilt mechanism based on two pairs of APA35XS.

The APA® are patented compact space-qualified Amplified Piezo Actuators (Fig. 3) from CTEC [3–5] based on a low voltage multilayer piezo ceramic (MLA) and a shell ensuing both MLA pre-stressed and amplification of the MLA displacement. They include strain gages (SG) for sensing the stroke. The APA® offers a stroke proportional to the voltage, in closed loop control with SG.

The APA35XS used in PHRAO MEF and DTT35XS are based on $2 \times 5 \times 10\ mm^3$ piezo MLA. Their stroke is about 40 μm, while the MLA stroke is 10 μm.

Fig. 2. PHARAO MEF Flight Model & DTT35XS

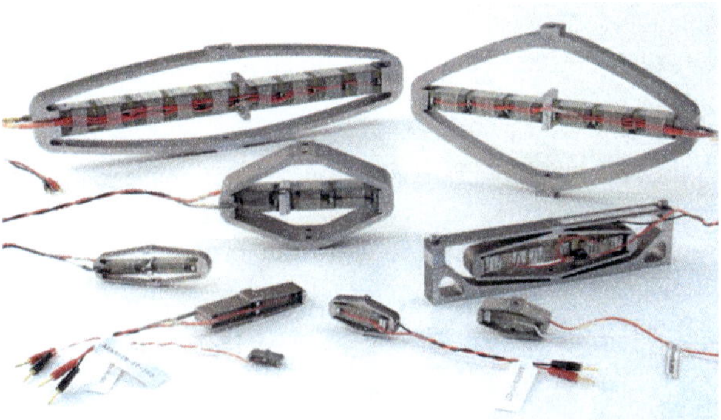

Fig. 3. Various APA® from CTEC

Driving the APA® in push pull by opposite pairs, this Tip-tilt allows deflections of a Ø 4 mm diameter glued mirror, by ±2 mrad around X and Y axis, as in Fig. 2. A series of 10 Flight Models (FM) was delivered to AIRBUS SODERN in 2004.

This FEM has inspired many different space & defence applications, and commercial products especially the Tilt platform DTT35XS-SG (Fig. 2) also based on APA35XS.

ATLID Beam Steering Assembly (BSA) (Fig. 4) was required by AIRBUS SODERN in 2014, exploiting CTEC PHARAO MEF space heritage. ATLID is an Atmosphere Lidar instrument for EarthCARE space mission. This BSA included in the emission path, aims at deviating a pulsed high energy UV laser beam to compensate the pointing misalignment between the emission and reception paths of ATLID.

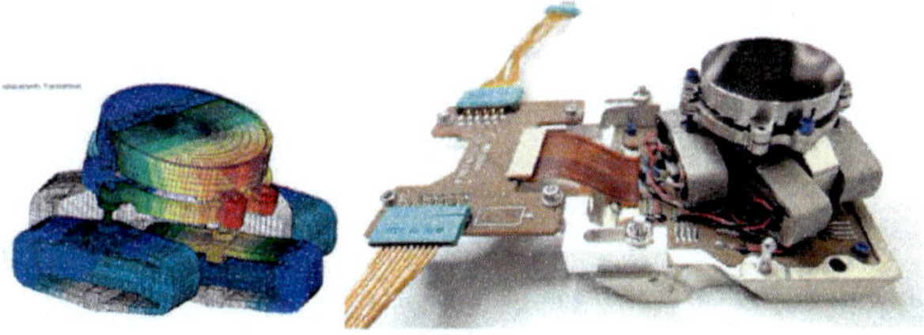

Fig. 4. ATLID BSA: FEM & Flight Model

As key new feature, the requirements were particularly severe regarding the long-term stability and the cleanliness.

Compared with PHARAO, the BSA has also to steer a larger & thicker mirror, Ø 27 mm diameter, on larger mechanical strokes: 4.3 mrad in Rx and 3 mrad in Ry.

The developed BSA [6] is a Tip-Tilt mechanism based on 4 APA60SM-SG based on a $5 \times 5 \times 20\ mm^3$ MLA. The APA60SM stroke is about 80μm, while the MLA stroke is 20 μm. The required Rx and Ry strokes being different, a non-symmetric mechanism configuration of the 4 APAs is chosen, which optimizes the mechanism size. The 4 APAs move a 3-point isostatic mirror support avoiding stress in the mirror. Figure 4 shows the tilt motion computed by FEM, accounting for push pull drive of opposite APA®.

High stiffness of APA60SM (1.3 N/μm) and additional guiding have led to the high resonance frequency of the mechanism, which is one key to pass vibration qualification tests. This high resonance frequency (>2 kHz) offers a wide range of operational frequencies but in this fine positioning application, operational range is 0 to 10 Hz, and actuators are used in quasi static modes (no inertial effect).

To obtain the desired stability and accuracy in closed loop, this mechanism is controlled by strain gauges (SG) position sensors deposited on the MLA piezo ceramics of the APA60SM. A particular development effort has been undertaken to get adequate SG within the application.

High Cleanliness for the whole mechanism was required to avoid laser-induced contamination (LIC) of ATLID instrument optics. It was achieved by careful cleaning, selection of low-outgassing materials and assembly in CTEC ISO5 assembly room.

A long qualification campaign was performed including shocks tests, vibrations tests, EMC and thermal cycle tests to establish the compliance with requirements.

Achieved experimental performances according to CTEC and AIRBUS tests are given in [7]. Targeted strokes have been got with resolution better than 0.4 μrad and long-term stability better than 200 μrad. The electric power consumption in static (DC) was about 0.6 W, with 30 mW from MLA and 0.57 W from SG.

As regard launching, qualification demonstrates a resistance to 15.5 g rms Random Vibration level, 26 g static and 100 g shock.

CTEC delivered the Flight Models to SODERN in 2015. Launching is delayed but planned in 2023.

In NASA Psyche mission, JPL is planning the assessment of a first Deep Space Optical Communication (DSOC) module. The communication distance will be about 3UA (500 million km), so the largest ever tested for a FSO link.

In this module, a Point Ahead Mechanism (PAM) aims at steering accurately the optical downlink signal towards anticipated earth position during DSOC communication.

In this context, CTEC was sub-contracted in 2018 by L3-Harris, to design, manufacture and test the performances of the PAM engineering and flight models for JPL PSYCHE DSOC. The developed PAM [8] Fig. 5 is a new tip-tilt mechanism based on 4 low voltage APA®.

The main specifications for this mechanism were to ensure an angular mechanical stroke of ±2.8 mrad throughout the full operational temperature range of the mission (−25/+50 °C full perf, −40/+65 °C reduced perf) and a mirror surface flatness under 63nm while remaining inside a very limited volume and surviving launch vibrations.

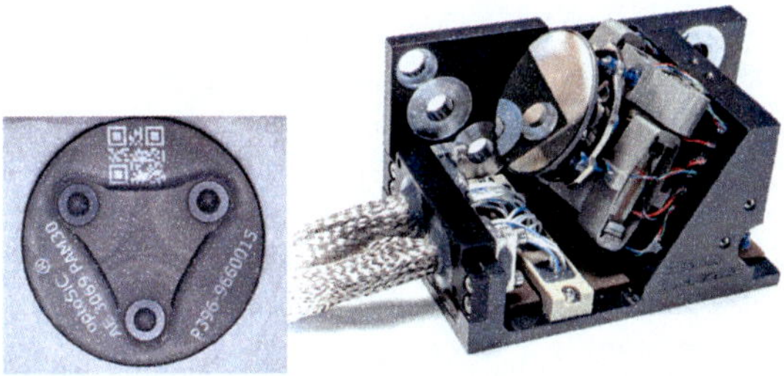

Fig. 5. PSYCHE PAM30: Mirror & Flight Model

To achieve such low mirror wave front error (WFE), CTEC has designed and integrated a specific mirror. This is a 3 feet Ø 30 mm SiC mirror manufactured by OPTOSIC. This option offers isostatic mounting. With this type of mounting, the mirror is no more glued but attached with screws to the mirror support. This provides better guaranties for vibration and shocks. The risk was to induce stress and mirror deformation, increasing WFE.

Finally, after integration, the measured reflected flatness errors (RFE) were typically less than 10 nm RMS.

The Flight models were successfully delivered in 2020. Launching is planned for August 2022.

Space constellations based on Free-Space Optical Communication (FSO) are a new challenge from many perspectives. Considering the mandatory cost efficiency, with repeatability of performances and reliability, requires an upheaval in space production and acceptance test methods, when the quantities are beyond several thousands of units.

For such future FSO constellations, new Fast Steering Mirror (FSM) named P-FSM150S (Fig. 6) have been developed by CTEC under ESA ARTES project TELCO-B.

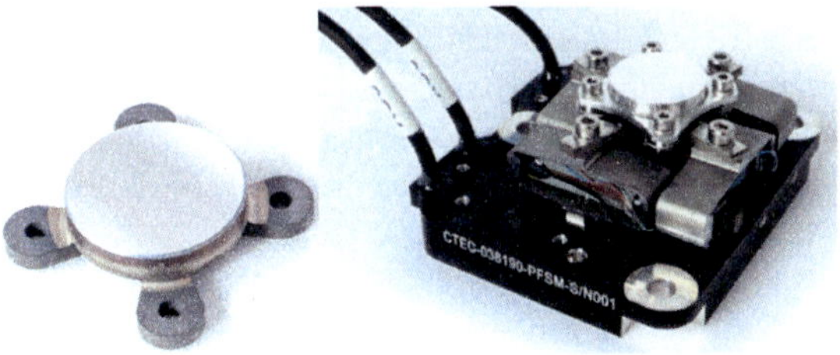

Fig. 6. P-FSM150S: Mirror & Engineering Model

The main specifications for this FSM were to ensure an angular stroke of ± 7 mrad throughout the full operational temperature range ($-10/+60$ °C) and a mirror surface flatness under 40 nm rms RWE (Reflected Wavefront Error) while remaining inside a very limited volume (especially less than 30mm height) and surviving launch vibrations.

In this need, a 4-feet flat Ø17 mm mirror concept has been designed considering 2 options for materials and suppliers: SiC from OPTOSIC and SiSiC from IDEX. Compared with choice done in PAM30 above, the advantages are a lower height, a lower rotation of mirror centre, an easier integration, and a lower cost. To avoid hyper statism and induced mirror stresses and deformation, a new flexible mirror support has been developed. Based on this choice, a P-FSM150S product has been defined, see Fig. 6.

This FSM has been fully tested (accuracy, lifetime, Vibration & shock). The measured Mirror RWE after integration is less than 20 nm rms. The final mechanical stroke is larger than ±10 mrad. Thanks to advanced electronic control, the angular position can be dynamically a bandwidth larger than 700 Hz (Fig. 7). Other details are given in [9].

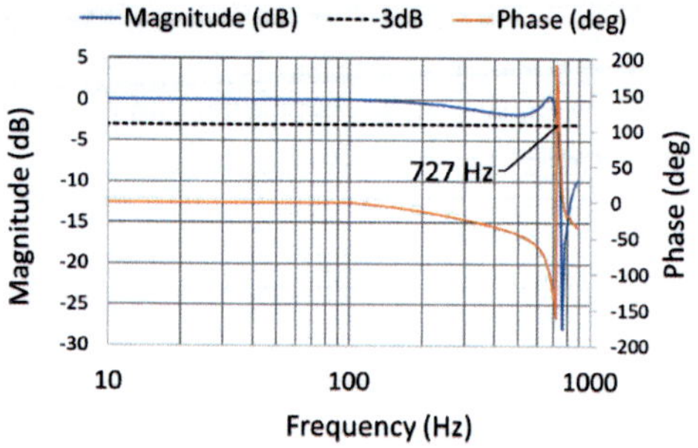

Fig. 7. P-FSM150S closed loop control at ±0.5 mrad

To get prepared for manufacturing series of FSM for new space and defence markets, CTEC has worked at industrialisation of its P-FSM150S product in MEPCOS on-going framework. In this context, CTEC has also invested in new facilities such as Zygo interferometer for measurement mirror flatness (WFE) and a large Clean room. In 2022, CTEC has started delivering P-FSM150S Flight Models.

From this large space heritage, for 10 years, CTEC has derived small FSMs for optronics & Infra-Red (IR) cameras used in defence applications. They ensure both image stabilisation and micro scanning. Micro scanning is an oversampling technique increasing the image resolution by 4 which is also performed from a long time in THALES IR cameras with XY25XS piezo micro scanner from CTEC [10], as well as in IAE solutions for cameras, as presented in this conference [11].

Typically for this market, a FSM is built on a DTT15XS-SG tip tilt mechanism driving a 10gr mirror. It is based on 4 APA15XS micro actuators. These are customised extra small (XS) APA®, deriving from the APA35XS. The DTT15XS offers angular strokes of 2 mrad and a bandwidth of up above 2 kHz.

Because of 2 functions required by the camera application, the strokes are controlled differently in two separates bandwidths. In 0–100 Hz, the desired stroke is 700 µrad, while in the 100–480 Hz the stroke is 200 µrad. This impacts both the mechanical and electronic design. Each mechanism embeds an electronic board performing the SG conditioning and including an EEPROM memory, a temperature monitoring, and a thermal compensation. With each FSM, Fig. 8, CTEC delivers a custom embedded electronic and controller board called CCBU20-PROX for driving and controlling the

FSM [11]. These FSM & CCB have achieved a high industrial maturity and is produced in series Fig. 9.

Fig. 8. CCBu20 and DTT15XS-SG for IR cameras

Fig. 9. Batch of DTT15XS-SG

Another FSM application is the line-of-sight stabilization. On a FSM50 provided by CTEC (Fig. 10), IAE has demonstrated this function implementing a compensation of «motion blurring». Figure 11 shows the same image without compensation and with compensation. The image with IAE algorithm is dramatically improved.

Large FSM are not so usual, and are also considered for space [13, 14]. For such mechanisms, as anticipated in [14], exported microvibration [15] is a major concern.

In the framework of the MEFISTO R&D project [16] funded by DGA and in collaboration with SODERN, CTEC have developed a large space mechanism which is at the opposite of previous FSM in terms of size and objectives. Its application is the satellite motion compensation for space telescopes of CNES future observation satellites. The objective of the mechanism is to move a large mirror of 730 gr with a fast saw tooth motion.

A back-and-forth stroke of typically 50 to 70 μm has to be done in less than 20 ms. Additionally, this motion should include a constant speed time called the integration

Fig. 10. CCBu20 driver & FSM50

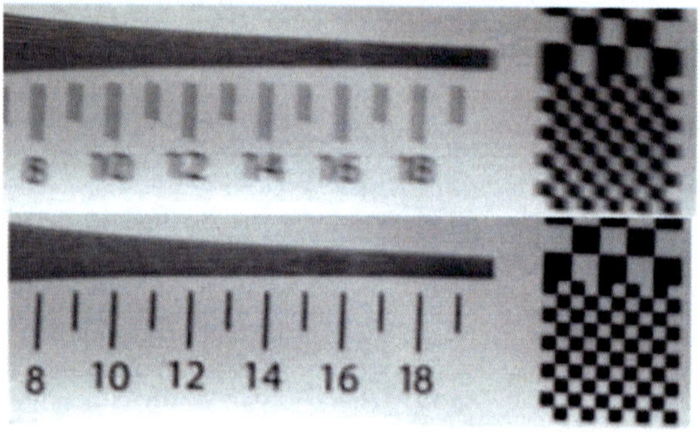

Fig. 11. Image without (top) and with «motion blurring» compensation (bottom).

time of the telescope. This is the most important performance for application: This time allows a stable telescope image acquisition in spite of the satellite motion.

At last, the mechanism should also limit the generation of micro vibrations in the satellite structure. Therefore, the symmetrised structure and the stroke shape as Fig. 12 are optimised to reduce the excitation of vibration mode of the structure, and so the exported micro-vibrations.

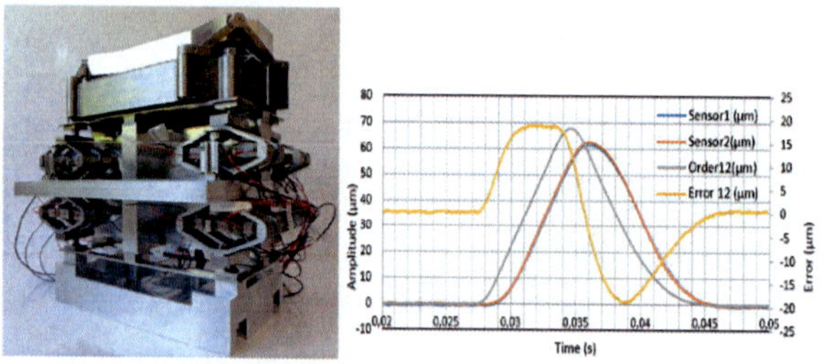

Fig. 12. MEFISTO FSM & transient stroke

The mechanism is based on 2 sets of 4 APA120ML piezo feet separated by a base plate in the plane of quasi-symmetry. Such a quasi-symmetric configuration is obtained with a counter-mass having same mass as the payload. The base plate is used for fixing the mechanism to the satellite structure. Each foot is equipped with CTEC Eddy Current Sensors ECS75. Thanks to this sensor and a close loop, the tracking error is 20 μm but this error is nearly constant in the integration time. So, the speed error is less than 1% [16]. A trigger could be added to overpass this issue to give the top to the image acquisition processing.

CTEC FSM solutions can be found in ground telescopes. A large FSM for atmospheric disturbance compensation in the laser guide star (LGS) of a ground telescope is shown in Fig. 13 [12]. The SiC mirror is elliptical, 200 × 140 × 36 mm^3. It supports high power laser. This FSM is based on 4 APA300MML-SG piezo actuators, the stroke of which is 370 μm. It offers 5.4 mrad tilt Rx Ry motion up to 100 Hz.

Fig. 13. DTT300MML-SG and LA75C driver

As a 40-m class optical, near- and mid-infrared, ground-based telescope, the ESO ELT will be the largest optical telescope ever built. It is based on 5 folding mirrors. Among these, the M5 Unit Fig. 14 stabilizes the image movements induced by Telescope mechanisms and wind shaking vibrations, by adjusting the M5 Mirror tip and tilt angles to a few tens of milli-arcsec accuracy, within a 10 Hz bandwidth [17]. The M5 mirror assembly is manufactured by SAFRAN REOSC with a 430 kg 2.7 × 2.2 m^2 SiC mirror blank provided by Mersen BOOSTEC.

The M5 cell is manufactured by SENER with 3 APA500XXL provided by CTEC, inheriting from a demonstrator [18]. Each APA500XXL includes a proper protection of the active material against environment. This 19 kg actuator is providing the force >30 kN and displacement >500 μm performance.

The ability to drive heavy payload at high frequency is achieved through high dynamic capabilities (better than 10 kN dynamic force). Special care in the design was taken to target more than 80 years lifetime and extreme reliability of 30 million cycles. M5 cell should be ready in 2023.

Fig. 14. ESO ELT M5 cell based on APA500XXL

3 Magnetic Actuation FSM

Low-power high-stroke high-bandwidth Fast Steering Mirrors (FSM) are demanded for Free-space optical communication (FSO).

Piezo FSM offer and very fine resolution and large bandwidth but strokes below 1° (17 mrad) when launching resistance is required. But as such small strokes are too constraining for some optical functions, CTEC has developed a new FSM technology based on magnetic actuation called M-FSM, targeting larger tilt stroke.

These strokes are typically from ±1 to ±5 mechanical degrees (±20 to 100 mrad). Additionally, other requests are low power, low heating and high bandwidth, from 200 Hz full stroke to 1 kHz at low level.

The M-FSM exploits a derivation and an integration of the CTEC MICA™ magnetic actuators based on variable reluctance forces [18] to drive a mobile part in Rx and Ry motions in an efficient way.

MICA actuators offer higher force density than Voice Coil [20], while being more linear than electromagnets.

MICA advantages are particularly interesting in embedded applications, such as new space compressors [21] for ZBO liquefaction and cryocoolers. These space applications benefiting of CNES and ESA support provides heritage on magnetic components as well as on ultra-long lifetime flexure bearing, useful for space-grade M-FSM.

As for MICA™, the M-FSM stator, Fig. 14, contains the coils and the magnetic poles. Each pole, located in front of the moving part, integrates a permanent magnet to bias the air gap. The moving part of the M-FSM is made of soft iron. It is guided thanks to a flexure bearing.

In a XZ cut, the permanent magnets are oriented in the same direction and generate the same bias field in the air gaps. When a current is supplied in one coil, the produced

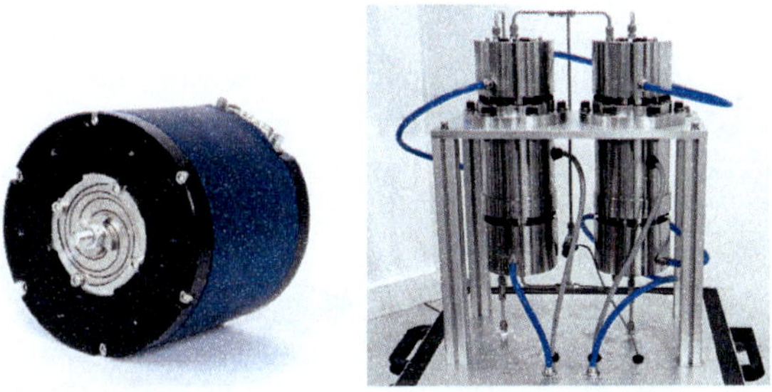

Fig. 15. MICA300CM & Space Hydrogen compressor

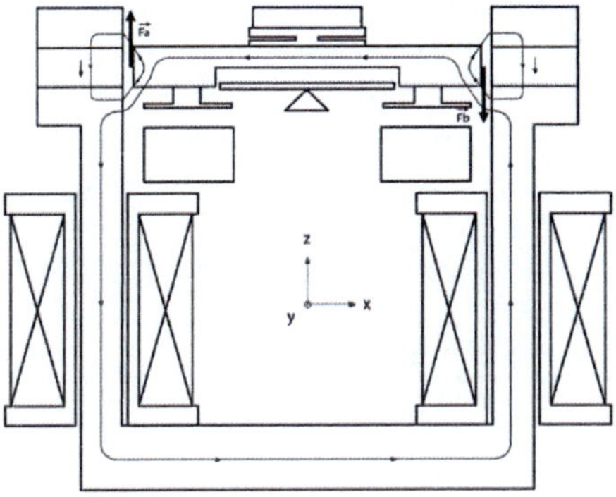

Fig. 16. M-FSM schematic

field is added to the bias field in one air gap, while it is subtracted to the bias field in the opposite air gap. This produces a pure torque on the moving plate and a rotation Ry. This torque is strictly proportional to the applied current around the equilibrium position. The same structure is used in a YZ plane, producing the same effects, giving a rotation Rx.

The M-FSM62 [22] is the first M-FSM of a series, Fig. 15. Its size is Ø62 mm × H56 mm; It drives a Sic mirror Ø31 mm.

Fig. 17. M-FSM62 & M-FSM45

Due to internal Eddy Current Sensor (ECS) acting also as end stops, the M-FSM62 offers max strokes ±35 mrad (without such internal sensors, the stroke can reach more than ±70 mrad.). For example, a measured ±35 mrad sine stroke is produced at 20 Hz with 1.56Vrms and 2.64Arms meaning a power of 3.75 W. The required power per mrad is larger than what is needed in piezo solutions. However, using a powerful amplifier such as CTEC MCSA480 switching amplifier, large strokes amplitude up to 35 mrad can be produced at frequencies up to 200 Hz. These are only limited by power requirement and thermal heating. Nevertheless, such unique performance reveals that the main interest of the M-FSM is for both large stroke and high dynamic mirror steering.

The fine control of the M-FSM in a large bandwidth implies to manage low resonant frequency of the tilt mode, which is about 100 Hz. Therefore, advanced control has been developed using the ECS and a state feedback-based controller [23]. Correlatively, to provide a high position resolution, CTEC has developed a new MCLA18 linear amplifier as an alternative to its MCSA480 switching amplifier. Test results using the MCLA18 show the closed loop control is achieved above the resonance frequency, offering a 3 dB bandwidth up to 156 Hz even at high level, Fig. 16 and 17 (Fig. 19).

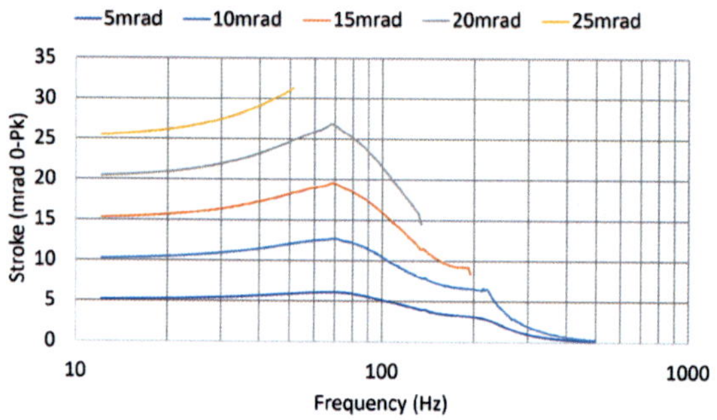

Fig. 18. Closed loop stroke at different stroke, versus frequency, accounting for MCLA18 power limits

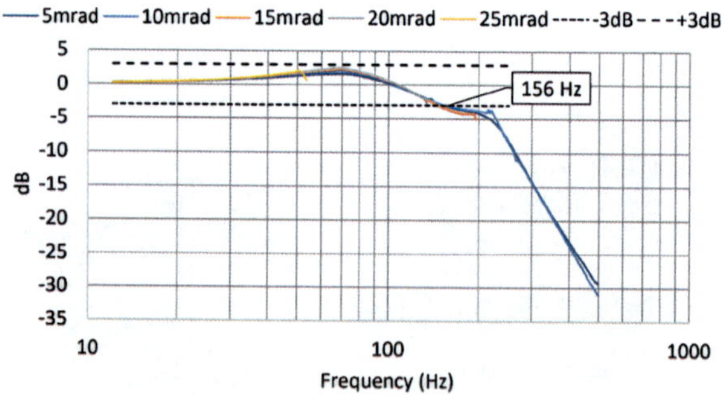

Fig. 19. Closed loop stroke bandwidth vs frequency

The M-FSM45 Fig. 15 is a miniaturization of the M-FSM62 benefiting of further improvement [24].

The M-FSM45 size is Ø45 mm × H36 mm, drives a SiC mirror Ø17 mm, similar to that of P-FSM150S. The measured Mirror RWE after integration is close to 10 nm rms, which is better than P-FSM150S. It benefits from the strong mechanical decoupling of the mobile part.

This FSM45 has been fully tested with special efforts on accuracy. Max stroke is ±26 mrad but nominal is 15mrad. The required power is 1.3W in AC quasistatic. The accuracy error of ECS position sensors was in the range of ±0,08% of the ±15 mrad measurement full scale. The −3 dB Bandwidth @ Full stroke is increased to 320 Hz.

A customization of this product is in progress for a feeder link for AIRBUS CNES CO-OP project targeting ground-GEO FSO optical links.

4 Stepping Piezo Actuators FSM

Space interferometry and instrumentation are looking for FSM and tip-tilt mechanisms for angular motion of mirrors, optical elements or samples, offering both clamping at rest and high resolution.

For these functions, piezo motors are very interesting as leading to compact and low power solutions.

A SPA-FSM developed by CTEC illustrate this possibility, Fig. 18. It is based on 3 modular stepping piezo actuator MSPA [25] driving by friction a central body carrying a mirror.

The MSPA is a driving unit relying on the stick-slip effect to produce steps. To produce one step, a slow motion drives the moving part by sticking while a reverse fast motion slips on the moving part. By repeating this operation, a stroke of several millimetres can be reached. Based on this effect, both linear and rotating motors can be developed.

More original, a direct drive FSM is also feasible. The SPA-FSM provides 3 degrees of freedom: a linear stroke Tz of 5 mm and two angular stroke Rx Ry of ±3°. The linear speed is 1 mm/s. The resolution is 0.5 μm.

This SPA-FSM concept is used in customized products for instrumentation markets [26]. It is also inspiring mechanisms for future space mission such as LISA. LISA is an early-stage project from ESA targeting a Laser Interferometer Space Antenna to be launched in 2032 (Fig. 20).

The Infrared Atmospheric Sounding Interferometer New Generation (IASI-NG) is a key payload element of the second generation of European meteorological polar-orbit satellites (METOP-SG) developed by AIRBUS under CNES contract.

This interferometer embeds a beam splitter blade. The blade tip-tilt positioning is critical with respect to IASI NG performances.

Therefore, to readjust in orbit the Beam Splitter Tip/Tilt alignment, two specific actuators called BSMA are required. Although not for a mirror, this Tip/Tilt function is close to a fine steering mirror.

At the BSMA level, the application requires very high unpowered position stability (0.15 μm over 24 h/1 K stability; 0.30 μm over 6 months; 1.4 μm long term) as well as high resolution (45 nm) over the stroke (±40 μm) with a significant force (20 N). These requirements imply piezo stepping motors offering locking at rest without backlash.

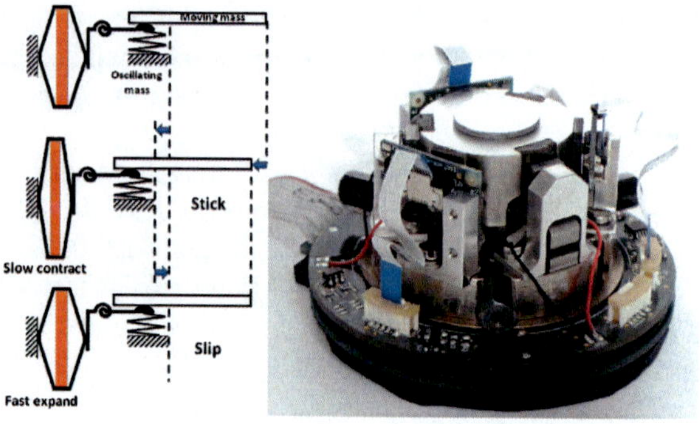

Fig. 20. MSPA principle & SPA-FSM

To design the BSMA, CTEC has derived its FSPA fine step piezo actuator [27] Fig. 13. The FSPA is a linear motor based on a screw driven by piezo rotating motor. A MPSA drives a rotating clutch, generating the linear motion of the screw. FSPA main advantages are:

- typical stepping size adjustable from 50 nm to 250 nm,
- fine mode with resolution better than 10 nm
- 5 mm stroke with up to 120 N driving force
- compact casing (Ø50 mm × 45 mm) & low mass (150 g),
- hold position while unpowered against forces that can exceed 1 kN (Fig. 21).

Fig. 21. FSPA35XS & Driving electronic

Basically, a BSMA is a linear motor based on 2 modified FSPA (one nominal and on redundant) driving 1 central active axis (Fig. 14). A main change to initial FSPA is to select a differential screw allowing very small steps. The MSPA has also been changed to operate as an inchworm.

The qualification campaign showed that repetitive step of 40nm was obtained both before and after 1 million steps (Fig. 15). The motors also pass successfully the other qualification tests [28]: stroke, force, stability, vibration…

BSMA FMs was delivered by CTEC to AIRBUS in 2021 and 2022. Launching on METOP-SG is planned on 2024 (Figs. 22 and 23).

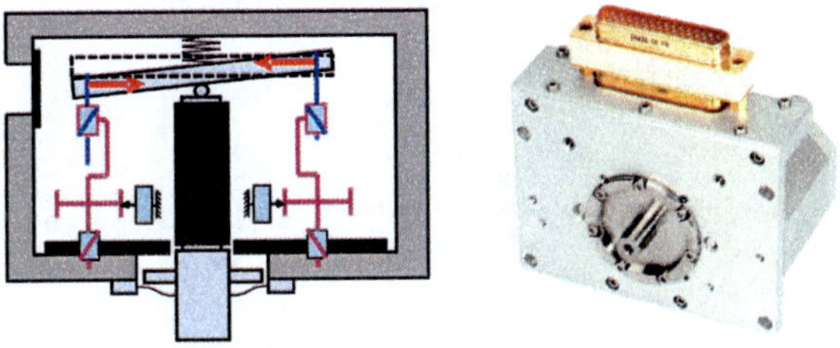

Fig. 22. IASI-NG BSMA: Schematic & Model

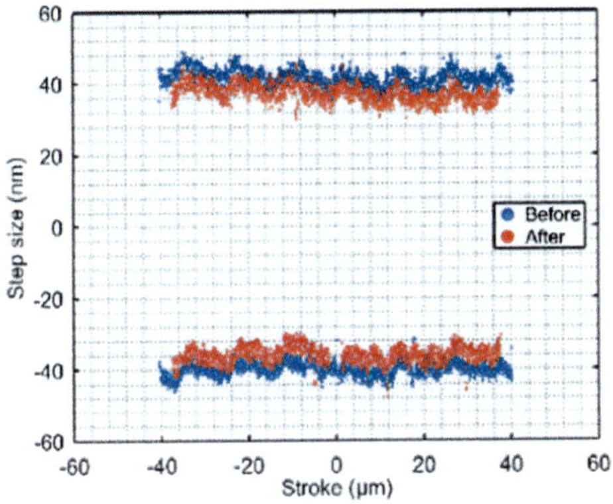

Fig. 23. IASI-NG BSMA Steps measurement

5 Conclusion

For 20 years the need for FSM and tip-tilt mechanisms has led to the development of different actuation technologies at CEDRAT TECHNOLOGIES:

- P-FSM based on Amplified Piezo Actuators APA® offer strokes up to 1°, low power in static, high bandwidth and high resolution. These are scalable and so have generated a wide range of products. They integrate innovative solutions to move mirrors without mirror deformation while supporting severe environment as met in space.
- M-FSM based on Magnetic Actuation offer strokes up to 5°, high bandwidth and high resolution. Their drawback is a power consumption at rest larger than P-FSM. Therefore, they are more adapted to fast pointing.
- SPA-FSM based on Stepping Piezo Actuators offer potentially unlimited angular strokes with nano rad resolution and clamping at rest, but they are slower than P-FSM and M-FSM. Thus, they are suited to fine static pointing.

This range of solutions provided by CEDRAT TECHNOLOGIES [29] allows for covering a wide range of functions such as fine static & dynamic pointing, active stabilization and micro scanning. Because of their space heritage and robustness, they find also markets in embedded and other demanding applications.

References

1. Le Letty R.: The scanning mechanism for ROSETTA/MIDAS from an Engineering model to the flight model. In: Proceedings of 9th ESMATS Conference (B), ESA SP-480, pp. 75–81 (2001)
2. Le Letty R.: Miniature piezo mechanisms for optical and space applications. In: Proceedings of the ACTUATOR 2004 Conference, Messe Bremen, Germany, pp. 177–180 (2004)
3. Le Letty R.: A new amplified piezoelectric actuator for precise positioning and active damping. In: Proceedings of the SPIE, Smart Systems and Materials, San Diego (1996)
4. Guay P.: Piezo qualification for space applications. In: Proceedings of the Actuator 2002, Messe Bremen, Germany, pp. 284–287 (2002)
5. APA Amplified Piezoelectric Actuators ® a CTEC registered trademark for products. http://www.cedrat-technologies.com/en/products/actuators/apa.html
6. Bourgain F.: Beam steering mechanism for EarthCARE atmospheric LIDAR instrument - an improved piezoelectric tip-tilt mechanism. In: ESMATS 2013 (2013)
7. Prevost E.: Beam steering mechanism for earthcare atmospheric Lidar instrument ATLID – an ultra-stable piezoelectric tip tilt mechanism. In: ICSO 2016, Biarritz, France (2016)
8. Guignabert A.: Point ahead mechanism for deep space optical communication: development of a new piezo based fine steering mirror. In: AMS Conference, USA (2021)
9. Aigouy G.: P-FSM150S fast steering mirror for optical space constellations. In: ACTUATOR 2022 (DE) (2022)
10. Crawford, S.: THALES long wave advanced IR QWIP cameras. In: SPIE XXXII-6206-17, Orlando (2006)
11. Li, D.: Quick demo station for microscanning super-resolution. In: ISSOAO 2022, Beijing, China (2022)
12. Claeyssen, F.: Beam steering Mirrors from space applications to optronic solutions. In: Proceedings of the OPTRO Conference, Paris (2018)
13. Cadiergues, L.: A mirror control mechanism for space telescope. In: Proceedings of the ESMATS (2003)
14. Pagès, A.: Highly dynamic and high precision self balanced optical mechanism. In: Proceedings of the ACTUATOR, Bremen (2012)
15. Smet, G.: A mechanisms perspective on microvibration - good practices and lessons learned. In: ESMATS 2018 (2018)
16. Barillot, F.: MEFISTO Design & tests of a demonstrator for filet compensation mechanism. In: ESA Mechanism Final Presentation Days (2016)
17. Vernet, E.: ESO ELT M5 unit – design and manufacturing status. In: Proceedings of the SPIE (2020)
18. Casalta, J.M.: E-ELT M5 field stabilisation unit scale 1 demonstrator design and performances evaluation. In: Proceedings of the SPIE 7736, Adaptive Optics Systems II, 77360M (2010)
19. Claeyssen, F.: Moving iron controllable actuators. In: Proceedings of the ACTUATOR, Bremen, Germany, pp. 445–448 (2008)
20. Aigouy, G.: Development of the plain bearing and flexure bearing MICA300CM actuators. In: Proceedings of the ACTUATOR, Bremen, Germany, A.5.5, p. 150 (2018)

21. Aigouy, G.: Design of a 2 stages compressor for mobility applications, using compact and efficient moving iron controllable actuators. In: Proceedings of the ISEF, Nancy (2019)
22. Claeyssen F.: Large-stroke fast steering mirror for space free-space optical communication. In: OPTRO 2020, n°0062 (2020)
23. Claeyssen, F.: Magnetically actuated fast steering mirrors. In: Proceedings of the Actuator, VDE, Bremen, Germany (2021)
24. Aigouy, G.: P-FSM150S and M-FSM45 for large scale free-space optical communication. In: Proceedings of the OPTRO, Versailles, France (2022)
25. Dubois, F.: Module stepping piezoelectric actuator - a versatile way of micro-positioning actuation. In: Proceedings of the ACTUATOR, Bremen, Germany (2016)
26. Pagès, A.: Improvement of MSPA module of stepping piezo actuator and industrial applications in piezo motors. In: Conference IWPMA 2019, Lyon, France, 1–4 October 2019, Proceedings P143 (2019)
27. Barillot, F.: Fine stepping piezoelectric actuator (FSPA) for IASI-NG. In: ACTUATOR, Bremen, Germany, Proceedings of B5.5 p. 26 (2018)
28. Barillot, F.: Nanometric positioning with IASI-NG's beam splitter mechanism actuator. In: Proceedings of the AMS (2021)
29. CEDRAT TECHNOLOGIES FSM 2022 products. https://www.cedrat-technologies.com/en/products/fine-and-fast-steering-mirrors.html

Comparison Test Method and Test Result of Star Sensor's Baffle Based on Carbon Nanotube

Shi-tong Liang(✉), Yun-cai Hao, and Cheng-wu Yu

Beijing Institute of Control Engineering, Beijing 100190, China
cast2009@163.com

Abstract. Star sensor is a kind of autonomous navigation equipment suitable for deep space exploration, which has the characteristics of small size and high accuracy. The carbon nanotube (CNT) has very high absorption, which is an ideal coating for star sensor to suppress stray light in space. In this paper, a comparison test method of star sensor based on carbon nanotube coating is proposed. A new type of baffle based on CNT paint was designed. Three types of baffles were manufactured, namely, the newly designed baffle, the traditional structure plus the traditional black paint baffle, and the traditional structure plus CNT paint baffle. The stray-light-suppression ability of the three types of baffle was compared and tested. The test results show that the performance of the new baffle with CNT paint is the best, and its stray-light-suppression ability is about three times that of the traditional black paint baffle. The second is the baffle with traditional structure and the CNT paint. The performance of the baffle using traditional black paint and traditional structure is weak. Therefore, the baffle structure design and CNT paint have an important impact on the stray-light-suppression ability of the star sensor. CNT paint can effectively improve the stray light suppression ability of star sensor.

Keywords: Spacecraft · Star sensor · Baffle · Carbon nanotubes · Stray light

1 Introduction

Star sensor is favored for its high accuracy, small size, strong autonomy and no accumulated error, and becomes one of the key equipment for spacecraft's attitude determination [1, 2]. The star sensor can obtain the attitude data of the spacecraft by imaging the stars in the celestial sphere. The energy of the stellar light is weak, while the optical environment where the spacecraft is in orbit is relatively complex and harsh, such as the sun, the earth and the light scattered by the spacecraft's surface. These stray-light sources have the characteristics of high light radiation intensity, wide radiation spectrum, etc., and the angle between the stray-light source and the star sensor changes with the spacecraft in orbit operation, which will interfere with the normal operation of the star-sensor. In severe cases, the stray light may submerge the star image, the star sensor cannot work [1]. Therefore, the stray light of star sensor must be suppressed.

Carbon nanotube (CNT) is a new type of high absorption paint that has emerged in recent years. The absorption can reach 99.9% or more, and it has a very strong optical

H. P. Urbach and H. Jiang (Eds.): ISSOIA 2022, SPPHY 295, pp. 332–338, 2023.
https://doi.org/10.1007/978-981-99-4098-1_29

absorption capacity. It is currently recognized as the blackest material in the world. If the carbon nanotube coating can be grown on the inner wall or the light shielding ring surface of the light shield, the absorption efficiency of the light shield can be greatly improved, and the design of the light shield can be simplified. With the same structure, it has a stronger ability to eliminate stray light. Therefore, it is necessary to carry out the research on the design and verification method of the baffle based on carbon nanotube coating to promote the application of carbon nanotube coating in star sensor products.

In this paper, the test method of stray light suppression of the light shield based on carbon nanotubes is analyzed, and the test results of stray-light comparison between the light shield based on CNT and the light shield based on ordinary black paint are given.

2 Proposal of CNT Baffle

The existing star sensors are basically equipped with a baffle, which is mainly used to suppress the stray light in the space environment. There are many types of baffle used for star sensor, which can be divided into absorption baffle and reflection baffle according to the extinction principle [3–5]. The absorption baffle has a long history and mature design method, which is used by most optical systems including star sensors, as shown in the Fig. 1 below.

Fig. 1. Two kinds of star sensors and their baffles

The interior of the absorption baffle is sprayed with high absorptivity black paint to suppress the stray light outside the incident field of view. Generally, several light blocking rings are set inside to effectively suppress the stray light entering the baffle. According to the form of the baffle, the absorption baffle can be divided into a primary baffle and a secondary baffle.

The secondary baffle can be considered to be composed of two primary baffles. Through the secondary baffle, stray light will be scattered at least three times before reaching the surface of the optical system, which will enhance the stray light suppression ability of the baffle, but the size and weight of the whole baffle will increase. In the case of harsh stray light environment and less stringent requirements on the overall dimensions of the baffle, the form of secondary baffle is generally selected.

Compared with the secondary baffle, stray light can reach the surface of the optical system through twice scattering after passing through the primary baffle, so the stray light suppression ability of the primary baffle is lower than that of the secondary baffle. However, the primary baffle is small in size and light in weight. When the overall dimensions of the baffle are strictly required, only the primary baffle can be selected.

The miniaturization of the star sensor has become the development trend of the star sensor, so the star sensor uses more primary baffle to reduce the volume and weight of the system.

However, because the primary baffle can only absorb the incident stray light two times, the stray light suppression ability is weak, so it is necessary to improve the stray light suppression ability of the baffle from improving the absorption rate of the internal coating. Among the known extinction materials, CNT is widely recognized as one of the coatings with ultra-high absorption rate. Beijing Control Engineering Research Institute has carried out the research on the technology of baffle based on CNT coating, developed the corresponding baffle products, and tested the stray light suppression performance of carbon nanotube light shield.

3 Comparison Test Verification Method of Hood Structure

The purpose of this test is to verify the stray light suppression ability of CNT coating in baffle products.

Three kinds of baffles were designed and manufactured for a high-precision star sensor. The first and second baffles are designed according to the drawing method of the baffle, and the structural dimensions are consistent with the internal light shielding and environmental protection. The first product is sprayed with ordinary matt black paint, and the solar absorption rate of black paint is 0.97; The second type uses CNT coating inside the baffle, and the solar absorption ratio of carbon nanotube coating is 0.999. The third type uses CNT coating, but reduces the overall size of the baffle to ensure that there is no primary scattered light inside the baffle.

The first and second baffles are shown in the following Fig. 2.

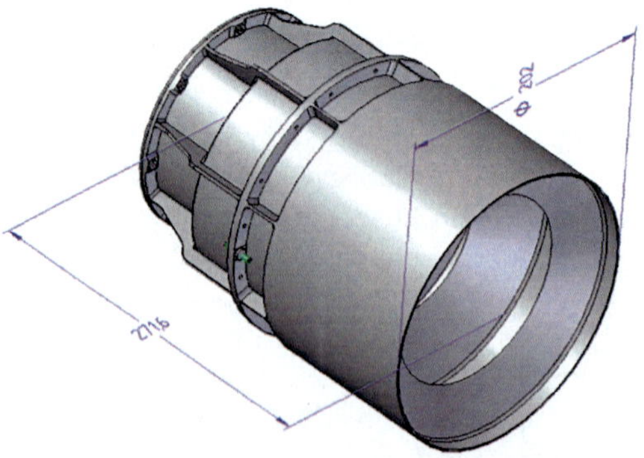

Fig. 2. Model of the first and second baffle

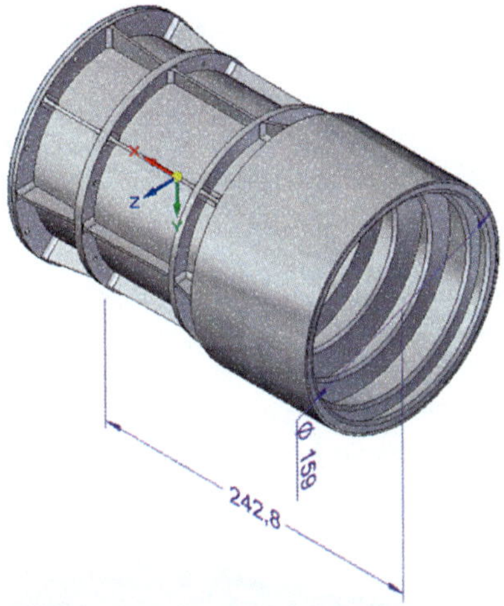

Fig. 3. Model of the CNT baffle (third baffle)

The third type of baffle is shown in the following Fig. 3.

The diameter of the outer contour of the CNT baffle is Φ 159 mm × 242.8 mm, the diameter of the outer contour of the baffle using the current conventional black paint is Φ 202 mm × 271.6 mm, and the current solar absorption rate of conventional black paint is 0.97. The volume of the CNT baffle is about 55.4% of that of the traditional baffle. With the same structural material, the weight of the CNT baffle will be reduced to about 50%. The CNT coating makes a great contribution to reducing the volume and weight of the baffle.

In the test, the star sensor is installed on the turntable, and the turntable, star sensor and extinction cone are located inside the darkroom. The interior walls and floors of the darkroom are blocked by extinction flannelette, and the absorption rate of extinction flannelette can reach 0.9. The solar simulator irradiates the baffle of the star sensor at a certain angle to form solar stray light interference. The image acquisition and processing unit is used to collect the images of the star sensor, and the specific processing algorithm is used to analyze the sunlight interference to determine the working condition of the star sensor after being interfered by the sunlight. Change the angle between the optical axis of the sun simulator and the star sensor, and test the working condition of the star sensor after the sun light irradiates the star sensor at different angles. If the star sensor works normally or meets the relevant requirements when the sun light is at a certain angle, but it cannot work normally or does not meet the relevant requirements when the angle is less than this angle, it can be determined that this angle is the sun light suppression angle of the star sensor. The criteria for whether the star sensor works normally can be whether the average value and variance of the image meet the requirements, whether there is a national bright area locally, and the image's signal to noise ratio (SNR). The criteria should be given according to the specific design of the star sensor.

4 Validation of Stray Light Suppression Performance of the Baffle

During the experiment, the CNT baffle and the traditional baffle were tested. The test angle interval is 5° in the range of 40°–90°, and 2° in the range of 26°–40°.

When the sun light enters the star sensor at an angle of not less than 90° in space, the sun light will not cause stray light interference to the star sensor, but during the ground test, due to the influence of the test environment's scatter, the environmental will generate stray light and enter the detector of the star sensor. Therefore, the stray light image when the angle between the star sensor and the solar simulator is 90° can be regarded as the ambient stray light. When processing the test data, the stray light image at each test angle should be deducted by stray light image at 90°, so the influence of the ambient stray light on the test results can be deducted.

The test site is shown in the Fig. 4 below.

Fig. 4. Test bed of the stray-light suppression ability of the baffle

After the analysis of the test data, the data analysis results of the three kinds of light shields are shown in the Fig. 5 below.

According to the test results of the solar irradiance meter, the analysis results have been converted into the gray response value of the product at 1 solar constant.

It can be seen from the above figure that the stray light energy on the image surface of the three kinds of baffles shows a downward trend with the increase of the solar incidence angle. The stray light energy of the image plane of the traditional light shield is about 1048DN at the solar incidence angle of 30°, and the stray light energy of the image plane of the carbon nanotube light shield is about 592DN and 486DN, which is significantly lower than that of the traditional light shield. At other solar incidence angles, the stray light energy of the image surface of the CNT based baffle is lower than that of the traditional baffle, indicating that the stray light suppression ability of the carbon nanotube baffle is better than that of the traditional baffle.

Among the three kinds of baffles, the CNT baffle with traditional structure has the best performance, followed by the CNT baffle with reduced size, and the light shield with traditional black paint and traditional structure has weak performance. Therefore, the structure design of the baffle and the CNT coating have influence on the stray light

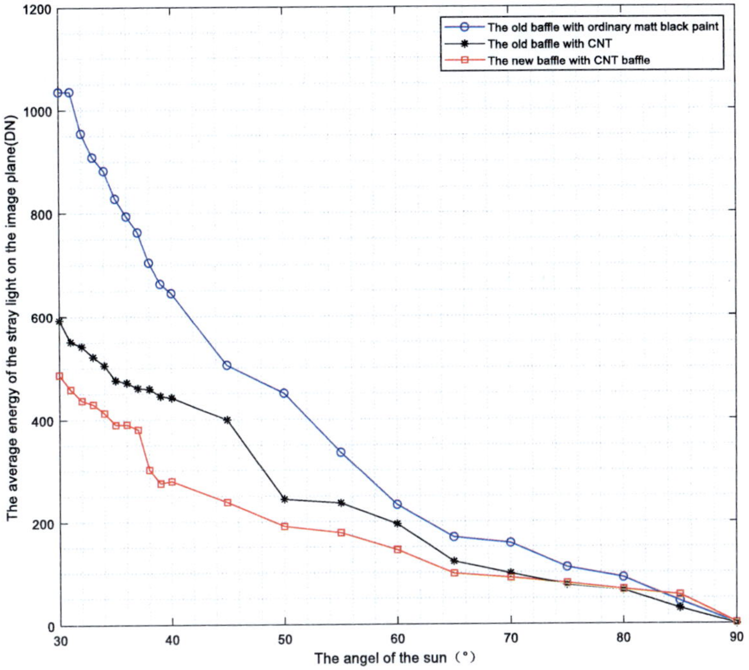

Fig. 5. Curves of the stray-light suppression ability of the baffle

suppression ability of the star sensor. Using CNT coating can effectively improve the stray light suppression ability of the star sensor shield.

5 Conclusion

The stray light test of the whole star sensor was carried out for the CNT baffle and the baffle with traditional black paint, and the test data were analyzed. Through the analysis of the output image energy of the product at different solar incidence angles, the response curves of the two kinds of shields at different solar incidence angles were obtained. It can be seen from the curve that the stray light gray of the star sensor based on the carbon nanotube shield is reduced by more than 56% than that of the conventional process shield star sensor, and the stray light suppression ability of the carbon nanotube shield is better than that of the traditional shield product using the conventional black paint. At the same time, due to the small volume and weight of the carbon nanotube shield, it can obtain a good stray light suppression effect, indicating that the carbon nanotube material has a good ability to resist stray light.

References

1. Sidi, M.J.: Spacecraft Dynamics and Control: A Practical Engineering approach. Cambridge University Press, Cambridge (1997)
2. van Bezooijen, R., Swanson, D., Boyle, P., et al.: Flight performance of the spitzer space telescope ast-301 autonomous star tracker. In: AIAA, pp. 447–466 (2005)
3. Schmidt, U.: Autonomous star tracker based on active pixel sensors (APS). ESA SP **554**(6), 355–358 (2004)
4. Zeh, T., Gal, C., Kaiser, S., et al.: MERTIS - reflective baffle design and manufacturing. In: SPIE, vol. 7808, p. 78080o-1-8 (2010)
5. Bass. Handbook of Optics, vol. 2. McGraw Hill Press (2010)

Adaptive Integral Backstepping Control Method for Space Tracking Turntable Based on NDOB

Yuchen Wang(✉), Jianbing Kang, Zhe Lin, Ziyin Chen, and Nanxing Yan

Beijing Institute of Space Mechanics and Electricity, Beijing 100094, China
542714186@qq.com

Abstract. In the low-speed tracking and pointing process of the tracking turntable for "far, dark and small" targets in the space ATP system, the friction disturbance and mechanical parameter change of the turntable cause the 'low-speed crawling' phenomenon of the tracking turntable, resulting in the tracking and pointing imaging jitter and the increase of pointing error. In order to solve this problem, an adaptive integral backstepping control method based on Nonlinear Disturbance Observer (NDO) is proposed in this paper. On the basis of establishing the vector control of permanent magnet synchronous motor on tracking turntable, an adaptive integral backstepping control method is proposed. At the same time, The NDO is designed to observe the friction disturbance in real time and serve as the disturbance term in the adaptive integral backstepping controller. The effectiveness of the proposed method is verified by numerical simulation. The simulation results show that the proposed method can effectively improve the tracking accuracy and stability of the space tracking turntable at low speed, compared with the traditional PID control method and the common backstepping control method.

Keywords: Space photoelectric tracking turntable · Pointing control · Adaptive integral backstepping · Nonlinear disturbance observer · Permanent magnet synchronous machine · Nonlinear friction disturbance

1 Introduction

Space ATP (Acquisition, Tracking, Pointing, ATP) system refers to a system deployed in a space-based motion platform to perform acquisition, tracking, and pointing operations for cooperative/non-cooperative targets. It plays an important basic supporting role in scientific projects such as space laser communication [1–3], space gravitational wave scientific detection [4], and military projects such as space laser weapons [5, 6] and space situation awareness [7–9]. As a coarse pointing mechanism in the composite axis pointing structure of the space ATP system, the space tracking turntable has an important influence on the overall performance of the space ATP system. It has important research value in improving the application efficiency of future aerospace-integrated military forces and meeting the needs of scientific research.

However, for objects that are far away in space, due to the small change of the relative geometric relationship, the tracking turntable is often work at low speed or ultra-low

H. P. Urbach and H. Jiang (Eds.): ISSOIA 2022, SPPHY 295, pp. 339–354, 2023.
https://doi.org/10.1007/978-981-99-4098-1_30

speed operation range. At this time, the friction torque of the tracking turntable is in the nonlinear section and fluctuates greatly, which leads to the phenomenon of “low-speed crawling” of a ‘fast’ and a ‘slow’ speed [10, 11] when the turntable is operating at ultra-low speed, which deteriorates the low-speed stability of the turntable. Therefore, it is important and difficult to suppress the low-speed crawling phenomenon caused by nonlinear friction to improve the stability and pointing accuracy of the turntable. At the same time, in the actual working process, the disturbance from the space environment and the satellite platform will make the motor and mechanical parameters change at any time, which will also affect the dynamic and static performance of the controlled object [12, 13].

In recent years, with the gradual development of control technology, some novel methods such as sliding mode variable structure control [14], model predictive control [15], backstepping control [16] have been gradually applied to suppress the low-speed crawling phenomenon caused by nonlinear friction of the tracking turntable. For example, reference [17] gives the sliding mode variable structure design method of speed loop and position loop based on the variable gain reaching law. Reference [18] proposes an iterative quadratic programming method to deal with the nonlinear constraints caused by input-output feedback linearization and realize nonlinear model predictive control. Simulation results show that this method can suppress the adverse effects caused by parameter changes and load disturbances. The backstepping control method is a nonlinear system control technology developed based on the backstepping control. The core idea is to select the Lyapunov function only related to the state variables of the current step and the state variables of the previous step, then to introduce virtual control variables to convert the original complex state variables. The higher-order nonlinear system is decomposed into multiple relatively simple first-order systems for processing. In reference [16], RBF neural network backstepping controller is designed for the nonlinear disturbance caused by gear backlash in the servo system of industrial robot turntable. Simulation results show that the tracking accuracy, system robustness and adaptive ability are improved, but it cannot be applied to the spaceborne platform due to the limitation of computational force. These methods can improve the control performance of the system from different angles, but the nonlinear friction disturbance in the low-speed section oscillates violently and has time-varying characteristics, which makes it difficult to accurately model. There are more dynamic uncertainties, and the computing power constraints of the spaceborne platform need to be considered. Therefore, the control effect of the above methods will also be greatly reduced.

In this paper, an adaptive integral backstepping (AIBackstepping) method is designed for the space tracking turntable, which is combined with NDO [19, 20]. Firstly, based on the vector control of the permanent magnet synchronous motor (PMSM), the backstepping control law and the adaptive law of the system are deduced, and the integral term is added to form an adaptive integral backstepping controller to suppress the change of the mechanical parameters of the motor. Secondly, a nonlinear disturbance observer is designed for the PMSM speed loop, to estimate the nonlinear friction and feedback to the adaptive integral backstepping controller. Finally, the effectiveness of the method is verified by simulation experiments.

2 Modeling and Friction Disturbance Analysis of Tracking Turntable

2.1 The Model of PMSM

The space turntable mostly uses PMSM as the actuator. In order to realize large-angle and high-reliability action, the PMSM direct drive form is often used [21]. The mathematical model of surface-mounted PMSM vector control in the rotating coordinate system *d*-*q* coordinate system (axis *d* coincides with the axis of rotor flux vector of PMSM, and the electrical Angle leading 90° along the rotation direction of the rotor is axis *q*) is shown in Eqs. (1)–(3). The block diagram of PMSM three-closed-loop vector control is shown in Fig. 1.

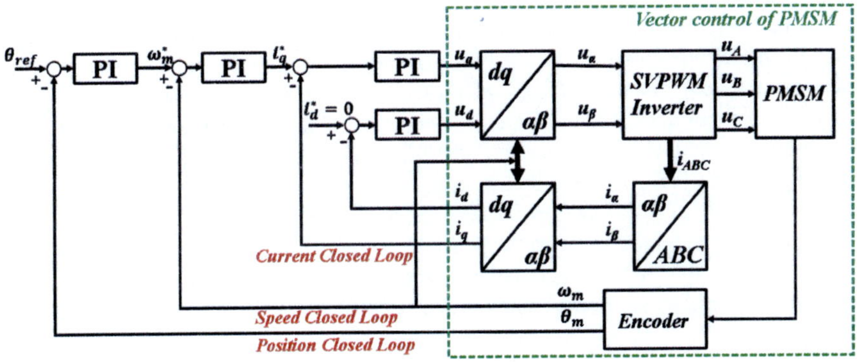

Fig. 1. PMSM three closed loop vector control block diagram

Voltage equation:

$$\begin{cases} u_d = R_s i_d + L_d \dot{i}_d - \omega_m L_q i_q \\ u_q = R_s i_q + L_q \dot{i}_q + \omega_m L_d i_d + \omega_m \psi_f \end{cases} \tag{1}$$

Electromagnetic torque equation:

$$T_e = \frac{3}{2} N_p \left[\psi_f i_q + \left(L_d - L_q \right) i_d i_q \right] \tag{2}$$

$$J \frac{d\omega_m}{dt} = T_e - T_L - T_f \tag{3}$$

where u_d and u_q are the applied voltage of *d*-axis and *q*-axis respectively; R_s is the stator winding resistance of the motor; L_d and L_q are *d*-axis and *q*-axis inductors, respectively. ω_m is the mechanical angular velocity output by the motor; i_q and i_d are the current of *d*-axis and *q*-axis, respectively; ψ_f is the rotor flux. N_p is the pole number of the motor; J is rotor moment of inertia; T_e, T_L, T_f are electromagnetic torque, load torque and disturbance torque respectively. t stands for time; θ_{ref} is the reference position input.

Let x_1, x_2, x_3, x_4 respectively represent the state variables of ω_m, i_q, i_d and the motor angle θ_m. From Eqs. (1), (2) and (3), the state space equation of PMSM vector control can be derived as shown in Eq. (4).

$$\begin{cases} \dot{x}_1 = \frac{1}{J}(-T_f - T_L) + \frac{1.5N_p}{J}(L_q - L_d)x_2x_3 + \frac{1.5N_p\psi_f}{J}x_2 \\ \dot{x}_2 = \frac{1}{L_q}\left(-R_sx_2 - L_dN_px_1x_3 - \psi_fN_px_1\right) + \frac{1}{L_q}u_q \\ \dot{x}_3 = \frac{1}{L_d}\left(-R_sx_3 - L_qN_px_1x_2\right) + \frac{1}{L_d}u_d \\ \dot{x}_4 = x_1 \end{cases} \tag{4}$$

2.2 Friction Disturbance Analysis

Based on the Coulomb model and viscous friction model, Stribeck friction model points out that the friction torque and velocity in the transition phase are continuously changing, which is of great value in engineering application. The characteristics of Stricbeck friction model are shown in Fig. 2 [22]. When the rotational speed is zero, it shows the sign switching characteristic of the maximum static friction force according to the direction trend of the speed, which is in the form of Coulomb friction (part I in Fig. 2). When the rotational speed is near zero, the friction torque is a nonlinear function of the speed (part II in Fig. 2). When the rotational speed is greater than a certain value, the friction torque is a linear function of the speed and presents a viscous friction shape (part III in Fig. 2). The nonlinear characteristics of the rotating speed between zero and zero are the main reasons for the degradation of the control performance of the turntable and the slow crawling phenomenon. The Stribeck expression is shown in Eq. (5) [23].

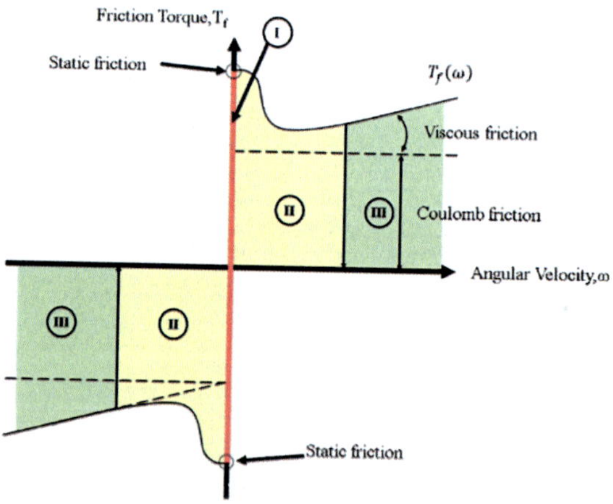

Fig. 2. Stribeck friction model characteristics

$$T_f = \begin{cases} T_\varepsilon & \omega = 0, |T_\varepsilon| \le T_s \\ T_s \operatorname{sgn}(\omega) & \omega = 0, |T_\varepsilon| > T_s \\ [T_c + (T_s - T_c)e^{-\left|\frac{\omega}{v_s}\right|^{\delta_s}} + K_v|\omega|]\operatorname{sgn}(\omega) & \omega \ne 0 \end{cases} \tag{5}$$

where, the maximum static friction torque of T_s; T_ε is the external moment; T_c is the Coulomb friction torque; δ_s is an empirical parameter; υ_s was Stribeck velocity; K_v is the coefficient of viscous friction; ω is the motor speed.

3 Adaptive Integral Backstepping Control Method Based on NDO

3.1 NDO Design

According to the general formula of NDO, the nonlinear disturbance observer designed for the speed loop of the permanent magnet synchronous motor system model is Eq. (6), which is used to observe the friction torque [19, 20].

$$\begin{cases} \dot{z} = -l(x_1)z - l(x_1)\left[-\frac{T_L}{J} - \frac{1.5N_p\psi_f}{J}i_q^* + p(x_1)\right] \\ \hat{T}_f = z + p(x_1) \end{cases} \tag{6}$$

where: $\widehat{T}_f$ is the observed value of friction disturbance, z is the internal state variable of the nonlinear state observer, $p(x)$ is the NDO function to be designed, $L(x)$ is the derivative of $p(x)$ with respect to x, called NDO gain function, as shown in Eq. (7):

$$L(x) = \frac{\partial p(x)}{\partial x} \tag{7}$$

The disturbance observation error is defined as $\tilde{T}_f = T_f - \widehat{T}_f$, and the derivative of both sides can be obtained as follows:

$$\dot{\tilde{T}}_f = \dot{T}_f - \dot{\hat{T}}_f \tag{8}$$

Substitute $\dot{\hat{T}}_f$ in Eq. (6) and $\dot{x}_1$ in Eq. (4) into Eq. (8) above and sort it out

$$\dot{\tilde{T}}_f = \dot{T}_f - \left(-l(x_1)z - l(x_1)\left[-\frac{T_L}{J} - \frac{1.5N_p\psi_f}{J}i_q^* + p(x_1)\right] + l(x_1)\cdot\dot{x}_1\right) \tag{9}$$

The following conclusions are drawn:

$$\begin{aligned} \dot{\tilde{T}}_f &= \dot{T}_f - L(x_1)\left(\frac{\dot{T}_f}{J} - \frac{\dot{\hat{T}}_f}{J}\right) \\ &= \dot{T}_f - L(x_1)\tilde{T}_f \end{aligned} \tag{10}$$

When the change rate of disturbance satisfies $|\dot{T}_f| \le \varepsilon$, that is, $\dot{T}_f$ is bounded, $\tilde{T}_f$ can be guaranteed to converge in the neighborhood with small origin by reasonable selection of NDO gain function $L(x_1)$. The bounded convergence of the disturbance error can be ensured by choosing $L(x_1) > 0$, and the functional forms of $L(x_1)$ and $p(x_1)$ are selected according to the characteristics of the actual observed objects. Considering the oscillation characteristics of friction disturbance in the low speed interval, Sigmod functional form is selected as the functional form of $p(x_1)$ in this paper. Sigmod function can smoothly map the input to between [0, 1], and is easy to take derivatives, which can be used to approximate the sign characteristics of Stribeck friction model near zero speed. The selection of $L(x_1)$ and $p(x_1)$ is shown in Eq. (11) below.

$$p(x_1) = \frac{k}{1 + e^{-ax_1}} \quad L(x_1) = \frac{ake^{-ax_1}}{\left(1 + e^{-ax_1}\right)^2} \tag{11}$$

where, a, k are the coefficients to be set.

3.2 Design of Adaptive Integral Backstepping Controller

There is a cascade relationship between ω_{m} and i_q in the PMSM vector control system. i_d and i_q, ω_{m} jointly control the motor speed, which has a dynamic coupling relationship. In this paper, the $i_d^* = 0$ control strategy is used in PMSM, and the essence of the $i_d^* = 0$ control strategy is to realize the static current decoupling of the d–q axis. The adaptive integral backstepping controller derivation design is described below.

The error variables e_1, e_2 and e_3 of the state quantity are defined as:

$$\begin{cases} e_1 = x_1^* - x_1 \\ e_2 = x_2^* - x_2 \\ e_3 = x_3^* - x_3 \end{cases} \tag{12}$$

where: x_1^*, x_2^*, x_3^* are the expected values of the corresponding state variables, and x_1^*, x_2^* are assumed to be continuously differentiable and bounded in the second order [24]. According to the $i_d = 0$ control policy, set $x_3^* = 0$.

Take the Lyapunov function as:

$$V_1 = \frac{1}{2}e_1^2 \tag{13}$$

Taking the derivative:

$$\dot{V}_1 = e_1\dot{e}_1 = e_1\left(-\frac{1.5N_{\mathrm{p}}\psi_{\mathrm{f}}}{J}x_2^* + \frac{1}{J}(T_{\mathrm{f}} + T_{\mathrm{L}})\right) \tag{14}$$

To ensure $\dot{V}_1 \leq 0$, let $\dot{e}_1 = -k_1e_1$, where $k_1 > 0$, then:

$$x_2^* = \frac{1}{1.5N_{\mathrm{p}}\psi_{\mathrm{f}}}(Jk_1e_1 + T_{\mathrm{f}} + T_{\mathrm{L}}) \tag{15}$$

Speed integral and adaptive compensation terms are added on the basis of Eq. (15). The adaptive backstepping method needs to ensure the convergence of parameter estimation errors and achieve the convergence of global tracking errors [24]. Here, the output $\hat{T}_f$ of NDO is used to replace T_f to obtain:

$$x_2^* = \frac{1}{1.5N_{\mathrm{p}}\psi_{\mathrm{f}}}\left(Jk_1e_1 + \hat{T}_{\mathrm{f}} + T_{\mathrm{L}} + J\alpha + Jk_i\theta_\omega\right) \tag{16}$$

where: α is an adaptive compensation term established to cope with the change of model parameters, and θ_ω is the rotational speed integral term, whose expression is as follows:

$$\theta_\omega = \int_0^t e_1(\tau)d\tau \tag{17}$$

Take the Lyapunov function as:

$$V_2 = \frac{1}{2}\left(e_1^2 + e_2^2 + e_3^2 + k_i\theta_\omega^2 + \frac{\alpha^2}{\lambda}\right) \tag{18}$$

Taking the derivative:

$$\dot{V}_2 = e_1\dot{e}_1 + e_2\dot{e}_2 + e_3\dot{e}_3 + k_i\theta_\omega e_1 + \frac{1}{\lambda}\alpha\dot{\alpha} \tag{19}$$

By substituting Eqs. (4), (12), (16) and (17) above, we can get:

$$\begin{aligned}\dot{V}_2 &= e_1\left(-k_1e_1 - \alpha - k_i\theta_\omega + \frac{1.5N_p\psi_f}{J}e_2\right)\\ &+ e_2\left[\frac{Jk_1(-k_1e_1 - \alpha - k_i\theta_\omega + k_1e_2) + J\dot{\alpha} + Jk_ie_1}{1.5N_p\psi_f}\right]\\ &+ e_2\left(\frac{1}{L_q}(R_sx_2 + L_dN_px_1x_3 + \psi_fN_px_1) - \frac{u_q}{L_q}\right)\\ &+ e_3\left[\frac{1}{L_d}(R_sx_3 + L_qN_px_1x_2) - \frac{u_d}{L_d}\right]\\ &+ k_i\theta_\omega e_1 + \frac{1}{\lambda}\alpha\dot{\alpha}\end{aligned} \tag{20}$$

Further arrangement can be obtained:

$$\begin{aligned}\dot{V}_2 &= -k_1e_1^2 - k_2e_2^2 - k_3e_3^2 + \alpha\left(\frac{1}{\lambda}\dot{\alpha} - e_1\right)\\ &+ e_2\left[k_2e_2 + \frac{Jk_1(-k_1e_1 - \alpha - k_i\theta_\omega + k_1e_2 + \dot{\alpha} + k_ie_1)}{1.5N_p\psi_f}\right]\\ &+ e_2\left[\left(\frac{(R_sx_2 + L_dN_px_1x_3 + \psi_fN_px_1) - u_q}{L_q}\right) + \frac{1.5N_p\psi_f}{J}e_1\right]\\ &+ e_3\left[k_3e_3 + \frac{(R_sx_3 + L_qN_px_1x_2) - u_d}{L_d}\right]\end{aligned} \tag{21}$$

The u_q, u_d and adaptive control law α are selected as follows to make $\dot{V}_2 = -k_1e_1^2 - k_2e_2^2 - k_3e_3^2$, and the parameters $k_2 > 0$ and $k_3 > 0$ are selected to ensure $\dot{V}_2 \leq 0$

$$u_q = L_q\left[\frac{1.5N_p\psi_f}{J}e_1 + k_2e_2 + \frac{Jk_1(-k_1e_1 - \alpha - k_i\theta_\omega + k_1e_2 + \dot{\alpha} + k_ie_1)}{1.5N_p\psi_f}\right]$$

$$+ \left(R_s x_2 + L_d N_p x_1 x_3 + \psi_f N_p x_1\right) \tag{22}$$

$$u_d = L_d \left[k_3 e_3 + \frac{\left(R_s x_3 + L_q N_p x_1 x_2\right)}{L_d} \right] \tag{23}$$

And let the reference adaptive control law be:

$$\dot{\alpha} = \lambda e_1 \tag{24}$$

In summary, the block diagram of the adaptive integral backstepping control system of the tracking turntable based on NDO proposed in this paper is shown in Fig. 3. The control method proposed in this paper is in the green dotted box. The overall design idea is as follows: Firstly, on the basis of PMSM triple closed-loop vector control, an adaptive integral backstepping controller is designed for the speed loop and current loop, and the position loop controller adopts PI control. Then, the nonlinear friction disturbance torque is accurately observed in real time through NDO, and the observed value is feedback to the adaptive integral backstepping controller. The motor can run stably under the action of nonlinear friction disturbance torque.

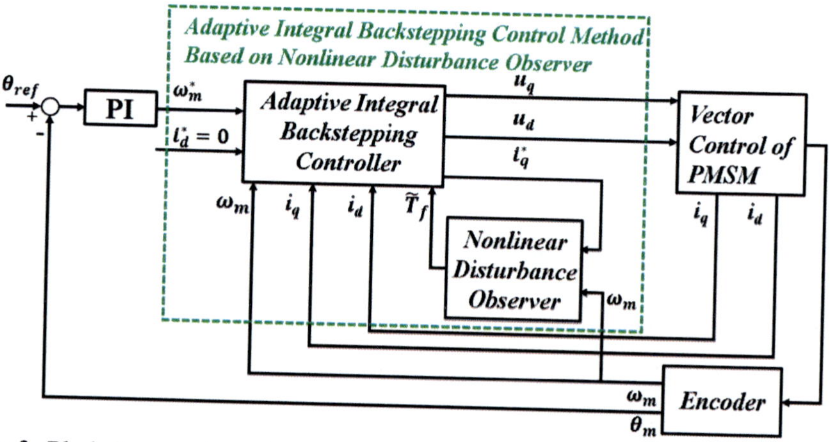

Fig. 3. Block diagram of adaptive integral backstepping control system for tracking turntable based on NDO

3.3 Proof of Stability

Through the selection of voltage input and adaptive control law in Eq. (21), it can be guaranteed that:

$$\dot{V}_2 = -k_1 e_1^2 - k_2 e_2^2 - k_3 e_3^2 \leq 0 \tag{25}$$

$\dot{V}_2 \leq 0$ gives us $V_2(t) \leq V_2(0)$, so e_1, e_2, e_3 is bounded.

Barbalats' Lemma is used to test the asymptotic stability of non-autonomous systems. According to Barbalats' lemma [25, 26], if the limit $\lim_{t\to\infty} V_2(t)$ of a positive definite function exists and is bounded, and $\dot{V}_2(t)$ is a semi-negative definite function and uniformly continuous on $t \in (0, +\infty)$, then $\lim_{t\to\infty} \dot{V}_2(t) = 0$ exists. Integrating over $\dot{V}_2(t)$:

$$\begin{aligned} V_2(t) &= V_2(0) + \int_0^t \dot{V}_2(\tau)d\tau \\ &= V_2(0) - \int_0^t \left(k_1 e_1^2(\tau) + k_2 e_2^2(\tau) + k_3 e_3^2(\tau)\right)d\tau \end{aligned} \tag{26}$$

$$\int_0^t \left(k_1 e_1^2(\tau) + k_2 e_2^2(\tau) + k_3 e_3^2(\tau)\right)d\tau = V_2(0) - V_2(t) \tag{27}$$

Since $V_2(t) \geq 0$, $\dot{V}_2 \leq 0$, we have:

$$\lim_{t\to\infty} \int_0^t \left(k_1 e_1^2(\tau) + k_2 e_2^2(\tau) + k_3 e_3^2(\tau)\right)d\tau < \infty \tag{28}$$

Next check for consistent continuity:

$$\ddot{V}_2 = -2k_1 e_1 \dot{e}_1 - 2k_2 e_2 \dot{e}_2 - 2k_3 e_3 \dot{e}_3 \tag{29}$$

Since both are bounded by e_1, e_2, e_3, then $\ddot{V}_2(t)$ is also bounded, and $\dot{V}_2(t)$ can be verified to be uniformly continuous. According to Barbalats lemma, the system is asymptotically stable.

4 Simulation Analysis

In order to verify the effectiveness of the NDO-based adaptive integral backstepping control method for the tracking turntable proposed in this paper, MATLAB simulation analysis was carried out for the operation of PMSM in the low speed interval. In order to satisfy the actual working conditions as much as possible, some parameters of Stribeck are added with random perturbation. The parameters of the simulation motor and Stribeck friction model are set in Table 1, and the parameters of the controller and NDO are set in Table 2.

In order to verify the superiority of the proposed algorithm compared with other algorithms, three-closed-loop PI control, ordinary backstepping control and backstepping control based on NDO are added in the simulation experiment as comparison. The parameter settings of the three-closed-loop PI controller are shown in Table 3. The desired command speed of the motor is set as: 0–0.24 s, $\dot{\theta}_{ref} = 0.2$(RPM), after 0.24 s, $\dot{\theta}_{ref} = 0.4$ (RPM); The load torque of the motor is $T_L = 20N \cdot m$, and when $t = 0.5$ s, it is suddenly changed to 25 N·m.It is worth noting that, at the initial stage of the system step response, the observed perturbation value feedback by the NDO to the adaptive integral backstepping controller needs to be limited to avoid the divergence of the system during

Table 1. Simulation motor and Stirbeck friction model parameter settings.

The motor parameters		Stribeck friction model parameter settings	
L_q	4.8 mH	T_s	23(±0.25) N·m
L_d	4.8 mH	T_c	2.5 N·m
J	9.8×10^{-3}kg·m^2	K_v	0.06
R_s	1.5 Ω	δ_s	2.0
N_p	4	ω_s	0.0526(±0.005)
ψ_f	0.109 Wb		

Table 2. Controller and NDO parameter settings

Adaptive integral backstepping controller		NDO		Position loop PI controller	
k_1	1×10^5	a	1.02×10^4	Proportion	850
k_2	2×10^6	k	18	Integral	5×10^5
k_3	1×10^4				
k_4	1×10^{10}				
λ	1×10^5				

Table 3. Controller and NDOB parameter settings

	Position Loop	Speed Loop	q-axis current loop	d-axis current loop
Proportion parameter	850	8500	850	850
Integral parameter	5×10^4	2×10^6	1×10^4	1×10^4

the transition stage. In this paper, the maximum value of the disturbance estimated by the NDO is limited to be twice that of the actual disturbance. The simulation results are shown in Figs. 4, 5, 6, 7, 8, 9 and 10.

Figures 4 shows the simulation curves of actual friction disturbance and NDO observed disturbance.

Figure 5 and 6 show the motor position error simulation curve and motor speed simulation curve in 0–0.24 s with different control methods. After the motor response is stable, the observation error of NDO can complete convergence in a very short time. The simulation motor with three-closed-loop PI control and backstepping control has a large fluctuation of torque due to nonlinear friction disturbance, and the speed stationarity is poor. The control method proposed in this paper benefits from the accurate observation

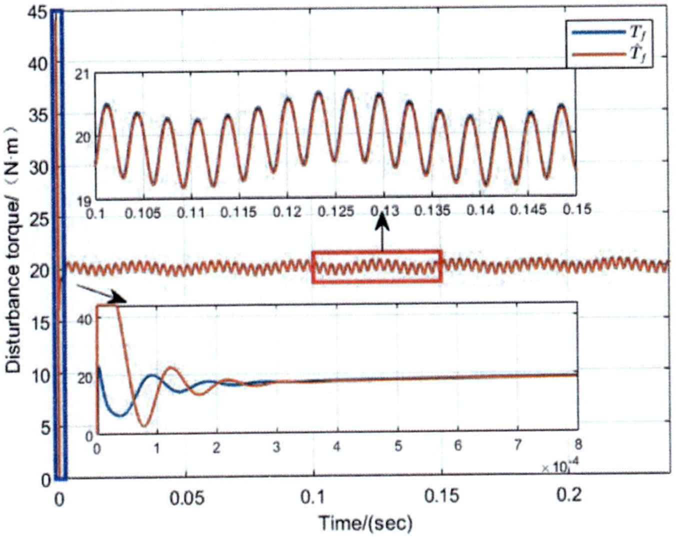

Fig. 4. Actual friction disturbance and NDO observed disturbance

of friction disturbance by NDO, which can reduce the fluctuation to a large extent and improve the speed stationarity. Therefore, the NDO is used to observe the friction disturbance and provide the backstepping controller, which can compensate for the performance degradation caused by the unknown friction disturbance torque.

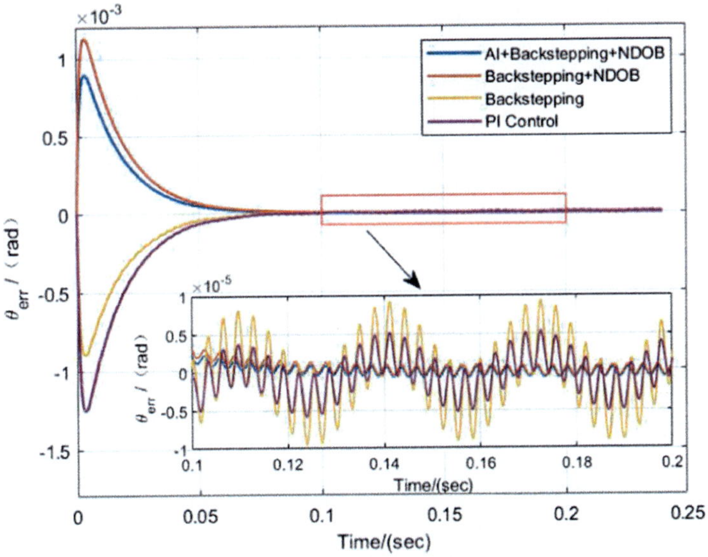

Fig. 5. 0–0.24 s motor position simulation curve

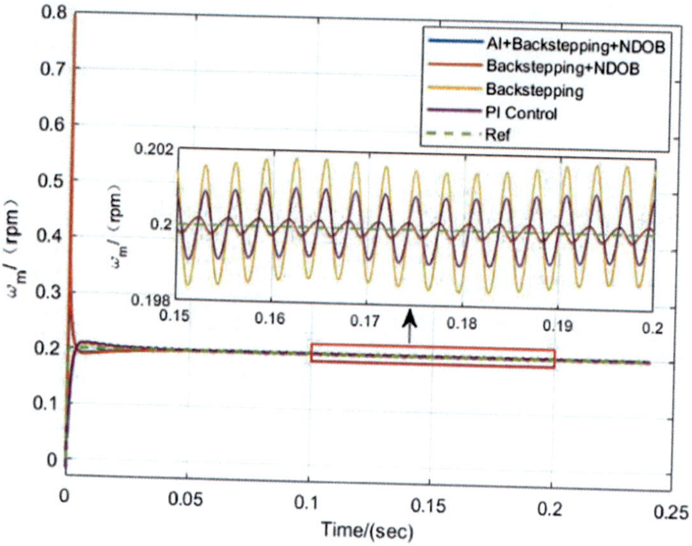

Fig. 6. 0–0.24 s motor speed simulation curve

In order to evaluate the control performance of different algorithms, the quantitative indicators of performance evaluation are introduced: the index of integral absolute error IAE $= \int|\theta_e(t)|dt$, the index of integral time absolute error ITAE $= \int t|\theta_e(t)|dt$ and steady state error. The IAE index reflects the accuracy of the control, while the ITAE can reflect both the (control accuracy) and the speed of error convergence, taking into account the control accuracy and convergence speed. Table 4 is based on the index data obtained during the 0–0.24s step simulation period.

Table 4. Algorithm performance evaluation index comparison

	Performance Quantitative Evaluation Index		
	Steady-state error/μrad	IAE($\times 10^{-7}$)	ITAE($\times 10^{-7}$)
Adaptive integral backstepping control with NDOB	1.186	0.371	0.0723
Backstepping control with NDOB	1.315	0.376	0.0733
Backstepping control	8.172	3.366	0.6566
PI control	6.126	1.937	0.3777

From the quantitative indicators shown in Table 4, it can be obtained that the adaptive integral backstepping control method based on NDOB and the backstepping control method based on NDOB are far superior to the ordinary backstepping control and PI control method in terms of steady-state error. The two indicators of ITAE are also much smaller than the backstepping control and PI control methods. After adding the adaptive term and integral term to the backstepping controller, the steady-state error can be further reduced, and the IAE and ITAE indicators can be further improved, that is, the control accuracy is further improved and the error convergence speed is accelerated.

Figure 7 and 8 show the response curves of speed and position at the step near 0.24 s with different control methods. The step response overshoot and oscillation time of the four control methods have little difference, which are mainly regulated by the parameters of position loop PI controller.

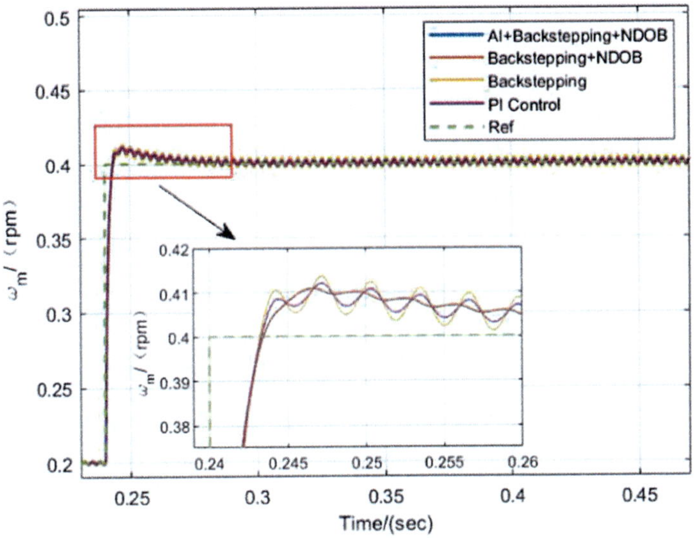

Fig. 7. Simulation curve of speed near 0.24 s

Figure 9 and 10 show the simulation curves of the speed and position of the backstepping control based on NDO and the adaptive integral backstepping control method based on NDO around 0.5 s. It can be seen that when the step disturbance is added to the load torque T_L, the designed adaptive term and integral term are compensated in response. Compared with the control method without the adaptive term and integral term, the adaptive integral backstepping control method based on NDO has less fluctuation and stronger robustness.

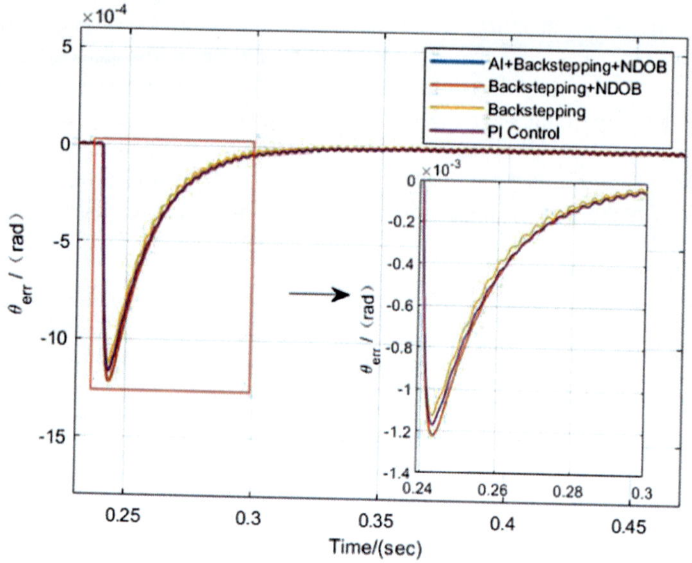

Fig. 8. Position error curve near 0.24 s

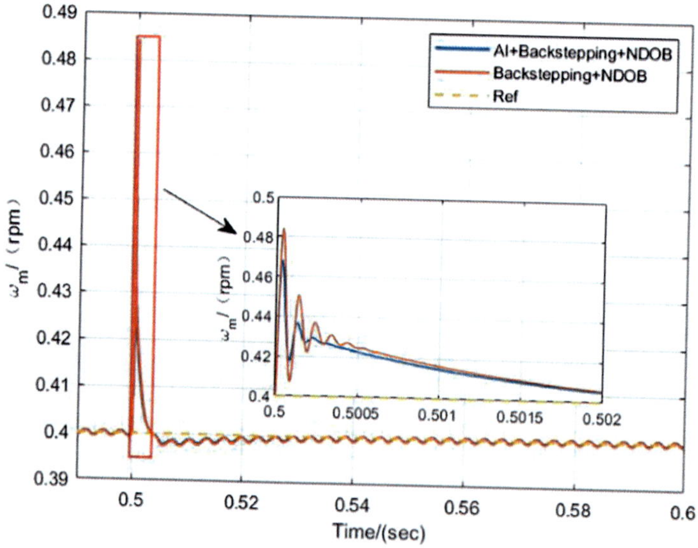

Fig. 9. Simulation curve of speed near 0.5 s

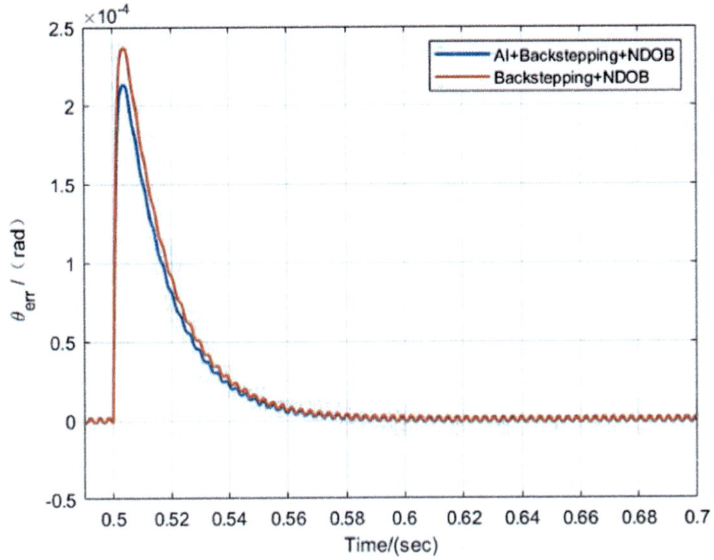

Fig. 10. Position error curve near 0.5 s

5 Conclusion

In this paper, an adaptive integral backstepping pointing control method based on nonlinear disturbance observer is proposed to solve the tracking imaging jitter and the pointing error increase due to the friction disturbance and parameter uncertainty of the tracking turntable in the low-speed tracking pointing process.

First, an adaptive integral backstepping controller is designed for the speed loop and current loop based on the PMSM three-closed-loop vector control. Then a NDOB is designed for the speed loop, the nonlinear friction disturbance torque is accurately observed in real time, and the observation value is feedback to the adaptive integral backstepping controller to achieve the stable operation of the motor under the action of the nonlinear friction disturbance torque. Finally, the proposed method is effectively verified by numerical simulation. The numerical simulation results show that compared with the PID control method, the steady-state error, IAE and ITAE indexes of this method are reduced by 80%. Therefore, the method proposed in this paper can effectively improve the low-speed tracking accuracy and stability of the space tracking turntable.

References

1. Yu, S.: Targeting Acquisition and Tracking Technology of Satellite Optical Communication. Science Press, Beijing (2016)
2. Jiang, H., An, Y., Zhang, Y.: Current situation, development trend and key technology analysis of space laser communication. J. Aircr. TT&C **34**(003), 207–217 (2015)
3. Wang, Y., Chen, P., Song, Y.: Development status and trend of foreign space laser communication technology. Flight Control Detect. **2**(01), 8–16 (2019)

4. Luo, J., Ai, L., Ai, Y: J. Sun Yat-Sen Univ. (Nat. Sci. Edn.) **60**(Z1), 1–19 (2021)
5. Wang, M., Fan, X., Huang, L.: Foreign space-based weapon system and its key technologies. Flying Missiles (08), 51–53 (2014)
6. Wu, S.: The latest development status and trend of American high energy laser weapon. Milit. Abstract (05), 40–44 (2020)
7. Wang, X.Y.: Research on the development status of foreign space target detection and recognition system. Spacecr. Eng. **27**(03), 86–94 (2018)
8. Cheng, R., Lin, Z., Zhang, A.: Research on accurate imaging and tracking method of GEO target based on relative motion. Space Return Remote Sens. **41**(01), 56–63 (2020)
9. Wang, X., Cai, Y., Chen, S.: Application and development of multi-source information fusion in space situation awareness. Space Return Remote Sens. **42**(01), 11–20 (2021)
10. Zhang, X., Zeng, H., Zhang, J.: Discussion on some performance problems of simulation turntable. Space Control **35**(05), 64–67 (2017)
11. Li, J., Zhang, A., Wang, C.: Design and simulation of current loop for space telescope scanning control system. Space Return Remote Sens. **41**(05), 72–82 (2020)
12. Li, S., Chen, X., Ni, G.: High precision satellite laser communication ground verification system. Opt. Precis. Eng. **25**(05), 1149–1158 (2017)
13. Xie, H.: Research on the control strategy of permanent magnet synchronous motor under the condition of system parameters uncertainty. University of Electronic Science and Technology of China (2021)
14. Cui, N., Zhang, Q., Wang, Q.: Sliding mode control method of photoelectric turntable based on LuGre friction model. In: Proceedings of the 2019 22nd International Conference on Electrical Machines and Systems (ICEMS) (2019)
15. Chen, Z., Qiu, J.: Parametric hyperlocal model predictive control for permanent magnet synchronous motor. Electr. Mach. Control **25**(08), 1–8 (2021)
16. Wang, S., Sun, H., Wang, J.: RBF neural network backstepping control of backlash turntable servo system. Mod. Manuf. Eng. (05), 39–46+13 (2021)
17. Liu, J., Deng, Y., Li, H.: High precision optoelectronic tracking and acquisition based on cascade sliding mode control. Opt. Precis. Eng. **28**(02), 95–107 (2020)
18. Kong, X., Liu, X.: High efficiency nonlinear model predictive control for permanent magnet synchronous motor. Acta Autom. Sinica **40**(09), 1958–1966 (2014)
19. Yang, J., Chen, W.H., Li, S.: Disturbance/uncertainty estimation and screening techniques in PMSM drives -a survey. IEEE Trans. Ind. Electron. **64**(4), 1–1 (2016)
20. Chen, W.H., Yang, J., Guo, L.: Disturbance-observer-based control and related methods - an overview. IEEE Trans. Ind. Electron. **63**(2), 1083–1095 (2016)
21. Tian, C., Hu, Y.: Research on theory and control scheme of direct torque control system for permanent magnet synchronous motor. Trans. China Electrotech. Soc. (01), 7–11 (2002)
22. Yeung, C.H., Altintas, Y., Erkorkmaz, K.: Virtual CNC system. Part I. System architecture. Int. J. Mach. Tools Manuf. **46**(10), 1107–1123 (2006)
23. Li, Y., Meng, H., Zhang, B.: Stribeck friction parameter measurement for servo turntable of photoelectric equipment. Electron. Meas. Tech. **36**(11), 92–97 (2013)
24. Yang, Y.: Research on adaptive robust control of inertial stabilization platform. University of Science and Technology of China (2017)
25. Nguyen, N.T.: Model-Reference Adaptive Control: A Primer. Springer, Berlin (2018). https://doi.org/10.1007/978-3-319-56393-0
26. Min, Y., Liu, Y.: Barbala's lemma and its application in system stability analysis. J. Shandong Univ. (Eng. Sci.) (01), 51–55+114 (2007)

Alignment of the Secondary Mirror of a Space Telescope Based on Holographic Modal Wavefront Sensor

Ming Liu and Bing Dong(✉)

School of Optics and Photonics, Beijing Institute of Technology, Beijing 100081, China
bdong@bit.edu.cn

Abstract. On-orbit wavefront sensing is essential for a space telescope to align the mirror. For a space telescope with an active second mirror (SM), we propose an automated alignment method based on the holographic modal wavefront sensor (HMWFS). Compared with the traditional alignment methods that require either an intricate instrument or a lot of iterations to optimize a metric function, only a single-shot image is captured for rapid measurement of the misalignment of the SM in five freedom of degree. The HMWFS is realized by a computer-generated hologram (CGH) into which all aberrations modes caused by the misalignment are encoded. Furthermore, the improved HMWFS applies to both point and extended objects. A kinoform CGH is adopted and the optimization is analyzed. Numerical simulation has been taken to demonstrate the feasibility of the proposed method.

Keywords: Wavefront sensing · Computer-generated hologram · Space telescope

1 Introduction

Compared with ground-based telescope, space telescope can take clearer images since there is no atmospheric interference. Large-aperture space telescope commonly adopts foldable secondary mirror (SM) to match the limited size of the launch vehicle. After deployment, the SM is expected to arrive in the correct optical position. When the SM is misaligned with respect to the primary mirror, optical wavefront error occurs. A number of approaches have been taken to align the SM, including the sensitivity table method, merit function regression, phase retrieval, and the nodal aberration method. All these alignment methods are based on wavefront sensing. Experimentally, wavefront sensing has been conventionally accomplished by using an interferometer to measure the wavefront errors directly [1–3], or using the optimization algorithm to estimate the pupil wavefront [4–6]. The interferometry-based wavefront sensing methods support precise phase measurements. However, these methods rely on the high degree of coherence of light beams to achieve interference, and the high requirement of interferometry on the mechanical stability of instrument strictly limits their applications in the space circumstance. Non-interferometric methods, including the model-based and

H. P. Urbach and H. Jiang (Eds.): ISSOIA 2022, SPPHY 295, pp. 355–363, 2023.
https://doi.org/10.1007/978-981-99-4098-1_31

model-free wavefront sensing algorithm, can provide a powerful alternative solution without using complex structure and highly coherent illumination. However, the model-free methods, such as stochastic parallel gradient descent [4], phase retrieval [5], etc., require a lot of iterations to achieve convergence, and their convergence reliability is low. The model-based methods are generally more efficient by utilizing deterministic relationships between a set of particular modes and well-defined metric functions [6]. The aberration modes can be estimated from a sequence of images which are captured in turn after introducing predetermined bias modes using an adaptive element. The model-based methods can achieve convergence within 2–3 correction cycles and avoid dropping into local optimum. However, since the bias modes are added one by one, the correction bandwidth of model-based algorithm is still seriously limited.

In this present work, we propose an automatic alignment method for the SM based on holographic modal wavefront sensor (HMWFS). The HMWFS can measure multiple aberration modes rapidly with a single-shot image. Conventional HMWFS can only work under a point beacon [7, 8]. This may be inconvenient for some remote sensing applications where there is only an extended image target. We make improvements to the HMWFS so that it can work under both cases. This is achieved by utilizing the linear relationship between the second moment of the focal intensity and the integral of modulus squared wavefront gradient. The wavefront aberration can be decomposed into multiple modes, such as the Zernike modes. In our approach, we derive a new mode basis from the wavefront aberrations caused by the misalignment of the SM. This new mode basis can be encoded into a CGH, and the misalignment of the SM in five degrees of freedom (DOFs) can be calculated simultaneously by the HMWFS. Different from the conventional HMWFS which commonly uses a binary CGH, we adopt a kinoform CGH with much higher diffraction efficiency. We show how to optimize the CGH to suppress speckle noises on the image plane. Numerical simulation results are presented to prove the effectiveness.

2 Principles

2.1 Improved HMWFS Using for Point and Extended Object

For an incoherent imaging system, it has been proved that the integral of modulus squared wavefront gradients is proportional to the second moment of the focal intensity [9].

$$\iint_{r\le R} I(u,v)r^2 dudv - \iint_{r\le R} I_0(u,v)r^2 dudv \approx c\iint_P |\nabla\Phi(x,y)|^2 dxdy \tag{1}$$

where P denotes the pupil; Φ is the wavefront aberration; ∇ is the gradient operator; I and I_0 is the focal intensity of the aberrated image and the unaberrated image; R is the integration radius; c is a constant mainly depending on R and the total energy of the focal image:

$$c = \frac{\iint\limits_{r\le R} I(u,v)dudv}{4\pi^2 \iint\limits_P P^2(x,y)dxdy} \tag{2}$$

The input aberration can be expanded by mode basis $\{E_i\}$ as

$$\Phi(x, y)=\sum_{n=1}^{N} q_n E_n(x, y) \tag{3}$$

where q_n is the modal coefficient.

If the gradients of mode E are mutually orthogonal over a unit circle:

$$\iint_P \nabla E_m \cdot \nabla E_n dxdy = \begin{cases} 1\, m = n \\ 0\, m \neq n \end{cases}. \tag{4}$$

the second moment of the focal intensity with input aberration can be written as

$$\iint\limits_{r\leq R} I(u, v)r^2 dudv - \iint_{r\leq R} I_0(u, v)r^2 dudv \approx c\sum_{n=1}^{N} q_n^2 \tag{5}$$

By introducing a positive bias mode $+b_k L_k$ into the system, we can get

$$\begin{aligned} \iint\limits_{r\leq R} I_{k+}(u, v)r^2 dudv - \iint_{r\leq R} I_0(u, v)r^2 dudv &\approx c\iint\limits_{P} |\nabla(\Phi + b_k E_k)|^2 dxdy \\ &= c\iint\limits_{P} \left\{|\nabla\Phi|^2 + b_k^2|\nabla E_k|^2 + 2b_k\nabla\Phi\cdot\nabla E_k\right\}dxdy \\ &= c\left(\sum_{n=1}^{N} q_n^2 + b_k^2 + 2q_k b_k\right) \end{aligned} \tag{6}$$

where I_{k+} is the biased focal intensity and b_k is the bias amplitude.

The kth mode's coefficient can be estimated independently by

$$q_k \approx \frac{1}{2cb_k}\iint\limits_{r\leq R} (I_{k+} - I)r^2 dudv - \frac{b_k}{2} \tag{7}$$

The above modal estimation process can be implemented in a multiplexed manner by adopting a CGH to obtain unbiased/biased images in a single shot. The desired complex-amplitude CGH is designed as

$$U_{CGH}(x, y) = \exp(j\varphi_{pos,0}) + \sum_{n=1}^{N} \exp[j(b_n E_n(x, y) + \varphi_{pos,n})]. \tag{8}$$

where $\varphi_{pos,n}$ is the wavefront tilt for the biased image and $\varphi_{pos,0}$ is the wavefront tilt for the unbiased image. After passing through the CGH, $N + 1$ spots corresponding to N biased images ($I_{1+} \sim I_{N+}$) and one unbiased image I will be arranged in order on the focal plane. Then, all aberration modes can be measured simultaneously using Eq. 7 from a single-shot.

2.2 Kinoform CGH Optimization

It is difficult to realize the complex-amplitude CGH with a single element in practice, since the currently available holographic devices such as SLMs can only provide either phase or amplitude modulation of light. The commonly used approach is to convert the complex-amplitude CGH to a quantized amplitude-only or phase-only CGH. Here we choose the kinoform CGH that takes the phase of the complex-amplitude CGH.

$$U_{KCGH} = \arg(U_{CGH}). \tag{9}$$

The reconstructed image quality of the kinoform CGH is degraded compared with the complex-amplitude CGH. The two CGHs are equivalent only if the amplitude information of the complex-amplitude CGH is negligible. The complex-amplitude U_{CGH} can be written as $U_{CGH}=|U_{CGH}|\exp(j\varphi_{CGH})$, where $|U_{CGH}|$ and φ_{CGH} is the amplitude and phase, respectively. For simplicity in analysis, the first term of Eq. 8 is cancelled here and the bias amplitude is set as $b_1 = b_2 ... = b_n = b$, then the square of the amplitude part can be expressed as

$$\begin{aligned}|U_{CGH}|^2 &= U_{CGH} \cdot U^*_{CGH} = \sum_{n=1}^{N} \exp[j(bE_n + \varphi_{pos,n})] \cdot \sum_{n=1}^{N} \exp[-j(bE_n + \varphi_{pos,n})] \\ &= N + 2\sum_{n=1}^{N}\sum_{m>n}^{N} \cos[b(E_m - E_n) + \varphi_{pos,m} - \varphi_{pos,n}].\end{aligned} \tag{10}$$

The Fourier spectrum of $|U_{CGH}|$ consists of a central zero-order (formed by the constant N) and $N(N\text{-}1)$ high-order components, as is shown in Fig. 1a. According to the scaling property of Fourier transform, these high-order components can be suppressed by increasing the bias amplitude b. Figure 1b and Fig. 1c depicts the cross section of the normalized Fourier spectrum of $|U_{CGH}|$ (denoted by the dashed line in Fig. 1a), where b is set to 5 and 40, respectively. It can be seen that the high-order components are effectively suppressed with a large bias amplitude. In this case, the kinoform CGH is a valid substitute for the complex-amplitude one.

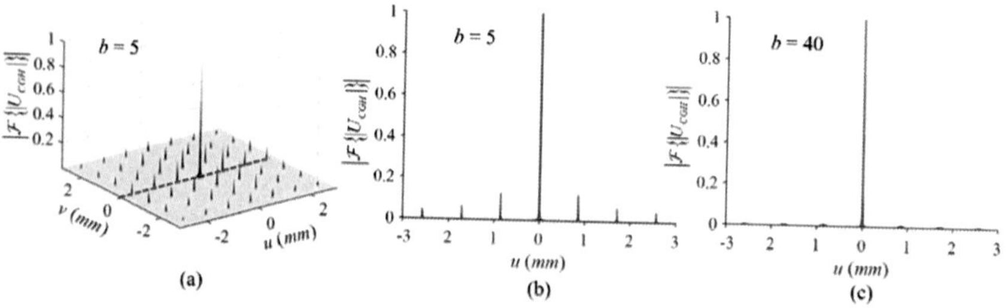

Fig. 1. The Fourier spectrum of $|U_{CGH}|$ (a) and the cross section with a bias amplitude of 5 (a) and 40 (b).

2.3 Alignment of the SM

The SM of a telescope has five independent DOFs, including the translation along X, Y, Z and the rotation about X and Y. The influence function of the SM can be defined as the pupil wavefront error induced by a unit movement of each DOF. To satisfy the orthogonality expressed by Eq. 4, a singular value decomposition can be done on the sampled influence function to export a new mode basis. Supposing **F** is a matrix whose columns are vectorized response functions of the SM, then the gradients of influence $\mathbf{F}'$ is given by

$$\mathbf{F}' = \begin{bmatrix} F_1^x & F_2^x & F_3^x & F_4^x & F_5^x \\ F_1^y & F_2^y & F_3^y & F_4^y & F_5^y \end{bmatrix} \tag{11}$$

where F_i^x and F_i^y are the numerical derivatives of the response functions in x and y dimension. $\mathbf{F}'$ can be converted into a multiplication of three matrices by singular value decomposition as

$$\mathbf{F}' = \mathbf{USV}^T \tag{12}$$

we can define a set of new modes $\{E_i\}$ whose gradients are orthogonal:

$$\mathbf{E} = \mathbf{F}\left(\mathbf{SV}^T\right)^{-1} \tag{13}$$

Since **E** is a linear transformation of **F**, modes **E** can be fully generated by the SM. The coefficient of mode **E** can be calculated using the proposed HMWFS method. The magnitude of the misalignment **d** can be calculated as

$$\mathbf{d} = \left(\mathbf{SV}^T\right)^{-1}\mathbf{e} \tag{14}$$

where **e** is the modal coefficient vector of mode **E**.

3 Numerical Simulation

Figure 2 illustrates the optical system which is a three-mirror anastigmatic telescope with a FOV of 0.64° × 0.2°. The SM is controlled with five DOFs to induce misalignment. The unit movement of each DOF is summarized in Table 1. The pupil wavefront at any field point is obtained from the optical design software (ZEMAX). The derived modes **E** along with their orthogonality are shown in Fig. 3. It can be seen that the gradients of mode **E** are mutually orthogonal and then multiplexed using a kinoform CGH.

The schematic diagram in Fig. 4 illustrates how the HMWFS works in the optical system. The telescope firstly forms an image of the object on its focal plane. Then, the kinoform CGH is used to modulate the wavefront so that the biased/unbiased images are captured with a single-shot. In the simulation, the holographic image was generated by two-dimensional discrete convolution of the object and the Fourier transform of the kinoform CGH. White Gaussian noise was added to the captured images. The noise level was set to 20dB. The modal coefficient and the misalignment of the SM are calculated

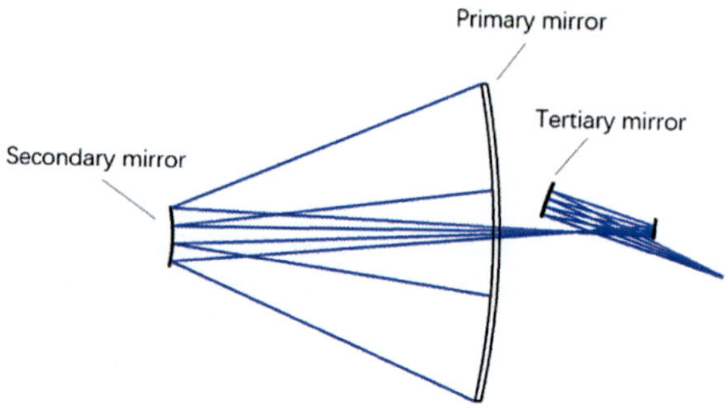

Fig. 2. Structure diagram of the three-mirror anastigmatic telescope.

Table 1. The unit movement of each DOF.

dz/um	dx/um	dy/um	tx/arcsec	ty/arcsec
1	10	10	1.44	1.44

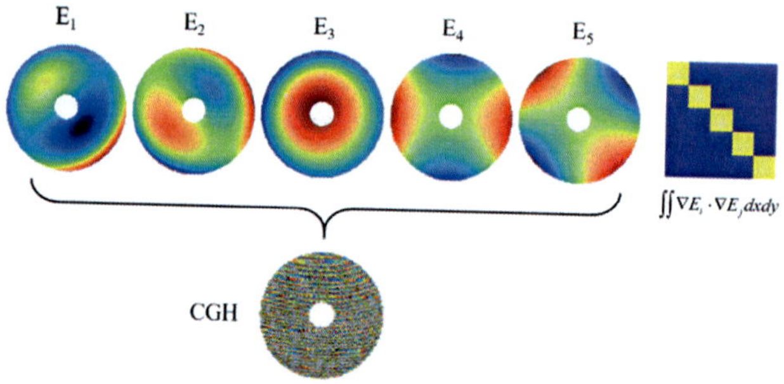

Fig. 3. The derived mode **E** whose numerical gradients are mutually orthogonal are multiplexed using a CGH.

successively according to Eq. 7 and Eq. 14. Finally, the misalignment is corrected in ZEMAX to evaluate the calculation result. Note that the constant c in Eq. 7 can be calibrated in advance by inducing a set of known aberration modes. Since c is relevant to the total energy of the image, the intensity needs to be normalized before calculating the second moment.

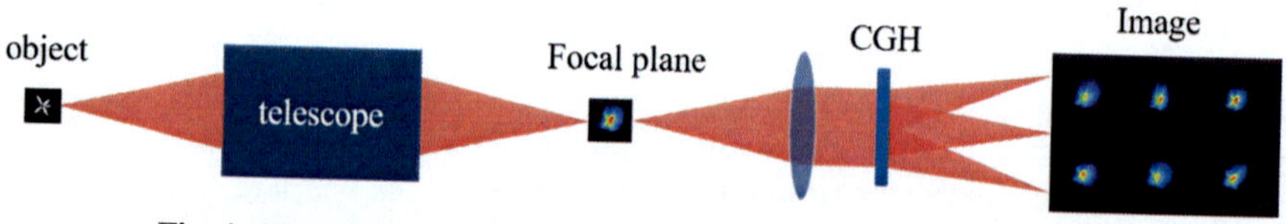

Fig. 4. The schematic diagram of the HMWFS in the telescope system.

A point and three different extended images– a harbor, USAF1951, and a speckle image are used as objects to investigate the performance of our method. When there is no misalignment error, the RMS of the initial wavefront in the field position (−0.32°, 0.36°) is 0.09λ. In this case, the HMWFS produces a base misalignment value. Since we aim to restore the system to its initial state, in the subsequent correction process, the base misalignment value should be deducted from the output of the HMWFS. When a random misalignment is induced on the SM, the wavefront RMS increases to 0.8λ. Then, the misalignment is measured and then sent back to ZEMAX for correcting. The residual wavefront RMS with respect to the correction cycle is shown in Fig. 5, where we can find the wavefront error has reduced to less than 0.1λ after the first correction, and the algorithm achieves convergence after 2 iterations, showing a quite good convergence speed. The corrected images shown in Fig. 6 verify the effectiveness of the proposed HMWFS for different objects.

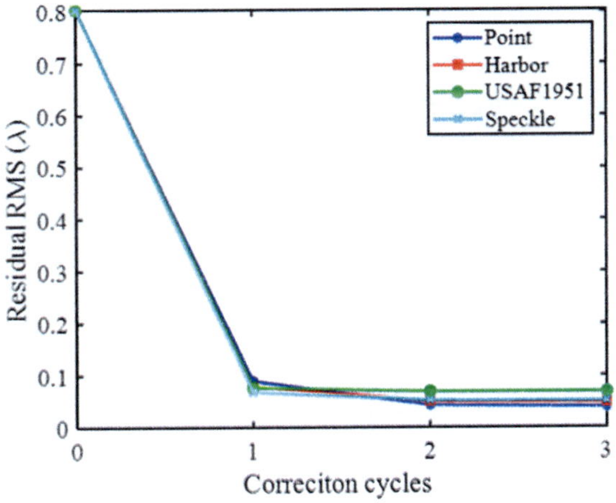

Fig. 5. The residual wavefront RMS after each correction cycles.

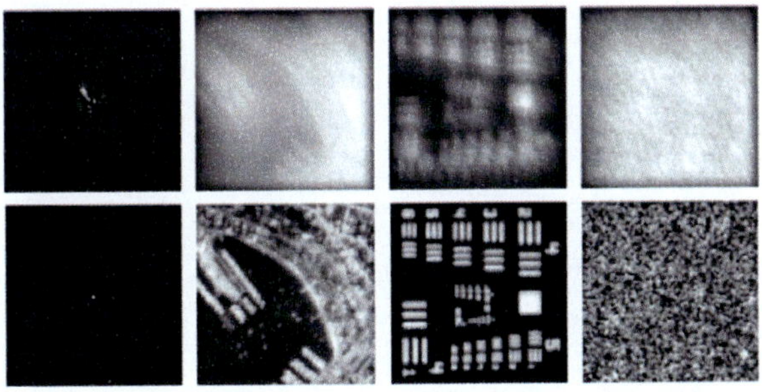

Fig. 6. The image of different objects formed by the telescope before (above) and after (below) the misalignment is corrected.

Besides the field point used for wavefront sensing, we also investigate fifty field positions across the whole field of the optical system. Figure 7a and Fig. 7b show the wavefront RMS in different field positions before and after inducing the misalignment, while Fig. 7c represents the corrected result, which is very close to the initial state.

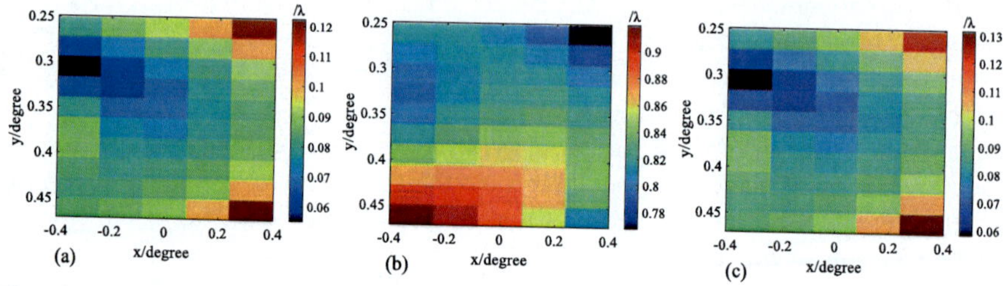

Fig. 7. The wavefront RMS in different field positions of the telescope before (a) and after (b) inducing the misalignment. (c) The wavefront RMS after the misalignment is corrected.

To further investigate the performance of the proposed method, one hundred random misalignments within the range [−20, +20] unit of each DOF are generated. Figure 8 illustrates the statistical distribution of the residual misalignments of each DOF varying with the correction cycle, where we can see that the misalignments can be effectively corrected within 2–3 iterations.

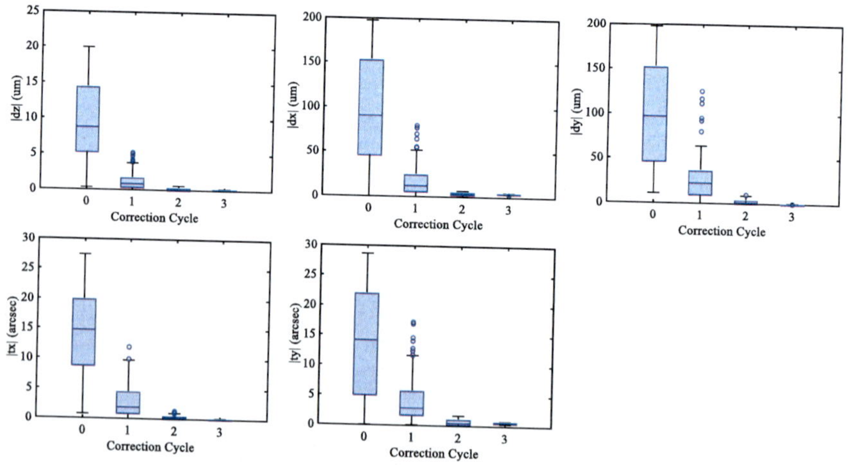

Fig. 8. The statistical distribution of the residual misalignments of each DOF varying with the correction cycle.

4 Conclusion

An improved HMWFS which is suitable for both the point target and extended target is proposed. This is achieved by leveraging the multiplexing feature of the CGH and the linear relationship between the second moment of the focal image and the integral of modulus squared wavefront gradients. We use the proposed HMWFS to measure the misalignment of the SM of a space telescope. It should be noted that the CGH operates at a specific wavelength, so a narrowband filter is needed in practical experiments. Pupil wavefront aberrations caused by the misalignment in five DOFs are used as the mode basis of the HMWFS by encoding them into a kinoform CGH. The theoretical analysis shows that the kinoform CGH can be optimized by appropriately increasing the bias amplitude. The feasibility of the proposed method is verified by numerical simulation where the misalignment of the SM of a three-mirror anastigmatic telescope is measured by the HMWFS. The simulation results show that the misalignment errors can be effectively corrected within 2–3 iterations under different kinds of objects.

References

1. Figoski, J.W., Shrode, T.E., Moore, G.F.: Computer-aided alignment of a wide-field, three-mirror, unobscured, high-resolution sensor. In: Proceedings of the SPIE 1049, Recent Trends in Optical Systems Design and Computer Lens Design Workshop II (1989)
2. Kim, S., Yang, H.-S., Lee, Y.-W., Kim, S.-W.: Merit function regression method for efficient alignment control of two-mirror optical systems. Opt. Express **15**, 5059–5068 (2007)
3. Dominguez, M.Z., et al.: Multi-field merit-function-based regression method for wide field infrared survey telescope grism system alignment. Appl. Opt. **58**, 6802–6812 (2019)
4. Li, M., Liu, X., Zhang, A., Xian, H.: Telescope alignment based on the sharpness function of under-sampled images. Chin. Opt. Lett. **17**, 061101 (2019)
5. Dean, B.H., Aronstein, D.L., Smith, J.S., Shiri, R., Acton, D.S.: Phase retrieval algorithm for JWST flight and testbed telescope. In: Proceedings of the SPIE 6265, Space Telescopes and Instrumentation I: Optical, Infrared, and Millimeter, p. 626511 (2006)
6. Ren, H., Dong, B., Li, Y.: Alignment of the active secondary mirror of a space telescope using model-based wavefront sensorless adaptive optics. Appl. Opt. **60**, 2228–2234 (2021)
7. Neil, M.A.A., Booth, M.J., Wilson, T.: New modal wave-front sensor: a theoretical analysis. J. Opt. Soc. Am. A **17**, 1098–1107 (2000)
8. Liu, M., Dong, B.: Efficient wavefront sensorless adaptive optics based on large dynamic crosstalk-free holographic modal wavefront sensing. Opt. Express **30**, 9088–9102 (2022)
9. Yang, H., Soloviev, O., Verhaegen, M.: Model-based wavefront sensorless adaptive optics system for large aberrations and extended objects. Opt. Express **23**, 24587–24601 (2015)

Static and Dynamic Error Analysis of Swing Arm Profile Measurement

Jianhua Zhang[1](✉), Mengjuan Li[2], and ChunLin Li[1,2]

[1] Beijing Institute of Space Mechanics and Electricity, Beijing 100094, China
Zhangjh2016@mails.jlu.edu.cn

[2] China Academy of Space Technology, Beijing 100080, China

Abstract. The swing arm profile measurement technology can achieve high-precision large-aperture aspheric mirror surface shape detection, and has the advantages of in-situ detection, high efficiency, and high dynamic range. However, in the process of assembly and measurement of the swing arm profiler, the static and dynamic errors caused by the spatial position and rotational motion of each component will have a great impact on the measurement effect. In order to accurately evaluate the final measurement accuracy, it is necessary to analyze the static error in the assembly process and the dynamic error in the measurement process, and realize the error budget to meet the measurement requirements. In order to solve this problem, this paper determines the object of error analysis and establishes a coordinate transformation model. By introducing deviation in coordinate transformation, mathematical modeling is carried out for different error sources, and the size of the introduced error is quantitatively calculated to analyze the sensitivity of different error sources and provide support for surface reconstruction.

Keywords: Swing Arm · Error Analysis · Aspheric surface

1 Introduction

Generally speaking, the manufacturing of large aperture aspheric mirror is a complex and long process. In order to improve the processing efficiency and shorten the processing cycle, it is necessary to improve the processing or testing process. Grinding and polishing process is the key to the manufacture of large-aperture aspheric mirrors, which often takes a long time. In recent years, silicon carbide has become the raw material of large-aperture aspheric mirrors in many projects because of its high hardness, stable chemical properties, small thermal expansion coefficient and good wear resistance, resulting in further reduction of the removal rate in the polishing stage. In order to improve the processing efficiency, it is necessary to improve the grinding accuracy and shorten the polishing time [1]. The main factor limiting the grinding accuracy is the surface shape detection accuracy in the grinding stage. Therefore, a high-precision surface shape detection method that can effectively guide the surface shape processing of large-aperture aspheric mirrors in the grinding stage is needed.

H. P. Urbach and H. Jiang (Eds.): ISSOIA 2022, SPPHY 295, pp. 364–375, 2023.
https://doi.org/10.1007/978-981-99-4098-1_32

The mirror reflection rate in the grinding stage is low, so it is not suitable for the interference measurement method. The uncertainty of the three-coordinate measurement method will increase seriously with the increase of the measurement aperture, thus greatly reducing the detection efficiency. Compared with the general coordinate measurement, the swing arm contour measurement technology can realize in-situ measurement, reduce the risk in the transportation process and save time. Using its own exquisite geometric structure, the probe of the swing arm profile measuring instrument can always be located on the closest spherical surface of the aspheric surface to be detected, so that a small-range high-precision probe can be used, thus having a high measurement accuracy. By replacing the measuring probes with different precision, changing the arm length and adjusting the structure, the swing arm profile measurement technology has the ability of large dynamic range and large error measurement in the grinding stage, and can realize the rapid detection of aspheric mirrors with different apertures and different shapes. Therefore, the swing arm profile measurement technology can effectively improve the processing efficiency of the polishing stage [2–4].

The specific geometric structure of the swing arm profile measuring instrument is shown in the Fig. 1.The main components are mirror turntable, air flotation turntable and swing arm with optical probe. The swing arm is placed tilted relative to the mirror turntable, and the arm length is L. The rotation axis of the mirror turntable and the rotation axis of the air bearing turntable intersect at the axis and intersect at C, and the optical probe at the end of the swing arm is perpendicular to the center of the aspheric mirror to be detected on the mirror turntable. The radius closest to the reference sphere is R_0. When the swing arm rotates, the trajectory of the probe tip is always located on the closest sphere of the aspheric mirror to be detected, so that the probe will always be perpendicular to the closest sphere.

The mirror table rotates $\gamma°$ for every $\alpha°$ sweep of the swing arm during operation, and the measurement is completed when the mirror table has made one revolution. The original data contains the angle of the mirror table, the angle of the swing arm and the reading of the optical probe, which is transformed into three coordinates and reconstructed to achieve a surface shape error measurement [5, 6].

2 Coordinate Transformation for Swing Arm Profiling

As can be seen from the measurement process of the swing arm profile measurement, the initial data obtained from the swing arm profile measurement is not three-coordinate information, but position information in a spherical co-ordinate system consisting of the measured value, the swing arm turntable angle and the mirror turntable angle. This requires a certain co-ordinate transformation to transform the parameters in the spherical co-ordinate system into the three-coordinate information in the Cartesian co-ordinate system.

$$(x, y, z) = T(data, \alpha, \gamma) \tag{1}$$

So after actually carrying out a round of swing arm profile measurements, the initial data needs to be processed in some way and converted to the actual spatial position distribution.

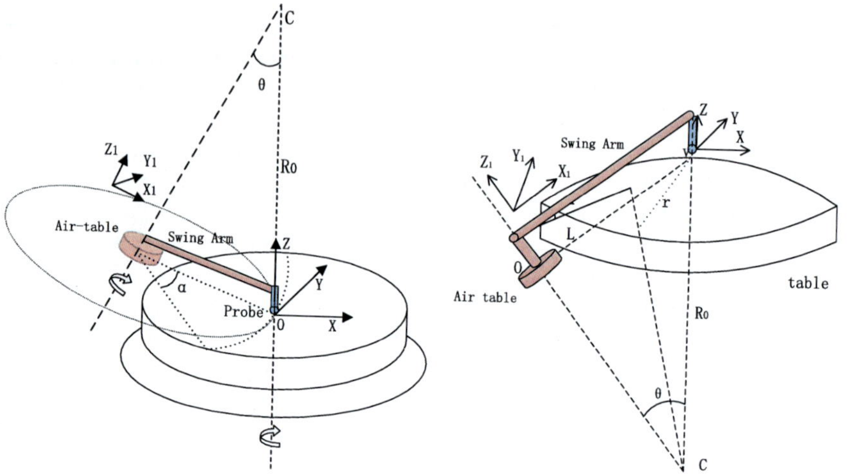

Fig. 1. Schematic diagram of concave and convex surfaces measured with the swing arm profiling technique.

For further analysis, we need to know the exact position of the probe tip, which in an ideal error-free situation would be perfectly on the closest spherical surface of the optical element to be inspected, with an arc trajectory determined by the combination of the arm length L, the swing arm rotation angle α, and the mirror turntable rotation angle γ.

The arm length L can be determined from $L = R_0 \sin\theta$, where R_0 is the closest spherical surface of the optical element to be inspected and θ is the angle between the rotation axis of the mirror turntable and the rotation axis of the pendulum arm turntable.

Therefore, taking the schematic diagram of the swing arm profile measurement principle for measuring concave surfaces in Fig. 1 as an example, the centre O of the surface of the optical element to be measured is the coordinate origin, the direction of the Z-axis is the same as the direction of the rotation axis of the mirror table, the initial state of the pendulum arm table is parallel to the X-axis, and the initial coordinates of the probe are (L, 0, 0) in the $X_1Y_1Z_1$ coordinate system, when the swing arm rotates and works, the tip of the probe will trace an arc trajectory, and the coordinates (x_1, y_1, z_1) on its trajectory can be expressed as follows:

$$\begin{bmatrix} x_1 \\ y_1 \\ z_1 \\ 1 \end{bmatrix} = \begin{bmatrix} \cos\alpha & -\sin\alpha & 0 & 0 \\ \sin\alpha & \cos\alpha & 0 & 0 \\ 0 & 0 & 1 & 0 \\ 0 & 0 & 0 & 1 \end{bmatrix} \begin{bmatrix} L \\ 0 \\ 0 \\ 1 \end{bmatrix} \tag{2}$$

By translating and rotating the $X_1Y_1Z_1$ coordinate system, the distribution of coordinates in the (X, Y, Z) coordinate system can be obtained as follows:

$$\begin{bmatrix} x \\ y \\ z \\ 1 \end{bmatrix} = \begin{bmatrix} \cos\theta & 0 & \sin\theta & 0 \\ 0 & 1 & 0 & 0 \\ -\sin\theta & 0 & \cos\theta & 0 \\ 0 & 0 & 0 & 1 \end{bmatrix} \begin{bmatrix} 1 & 0 & 0 & -L \\ 0 & 1 & 0 & 0 \\ 0 & 0 & 1 & 0 \\ 0 & 0 & 0 & 1 \end{bmatrix} \begin{bmatrix} x_1 \\ y_1 \\ z_1 \\ 1 \end{bmatrix} \tag{3}$$

Thus the coordinates of a single scan line in the (X, Y, Z) coordinate system are:

$$\begin{bmatrix} x \\ y \\ z \end{bmatrix} = \begin{bmatrix} \cos\theta(L\cos\alpha - L) \\ L\sin\alpha \\ -\sin\theta(L\cos\alpha - L) \end{bmatrix} \tag{4}$$

The coordinate points of all scan lines can be expressed as:

$$\begin{bmatrix} x \\ y \\ z \end{bmatrix} = \begin{bmatrix} \cos\gamma & \sin\gamma & 0 \\ -\sin\gamma & \cos\gamma & 0 \\ 0 & 0 & 1 \end{bmatrix} \begin{bmatrix} \cos\theta(L\cos\alpha - L) \\ L\sin\alpha \\ -\sin\theta(L\cos\alpha - L) \end{bmatrix} = \begin{bmatrix} \cos\gamma\cos\theta(L\cos\alpha - L) + L\sin\gamma\sin\alpha \\ -\sin\gamma\cos\theta(L\cos\alpha - L) + L\cos\gamma\sin\alpha \\ -\sin\theta(L\cos\alpha - L) \end{bmatrix} \tag{5}$$

For flat surfaces, the ideal position of the probe tip coordinates is calculated and the distribution of the X-Y coordinates corresponding to the measured values in the actual measurement situation is also known. Taking a 350 mm plane mirror measurement as an example, with arm length $L = 340$ mm, $\alpha = 60°$, $\Delta\gamma = 5°$, the spatial distribution of the measured values after the measurement is shown in Fig. 2.

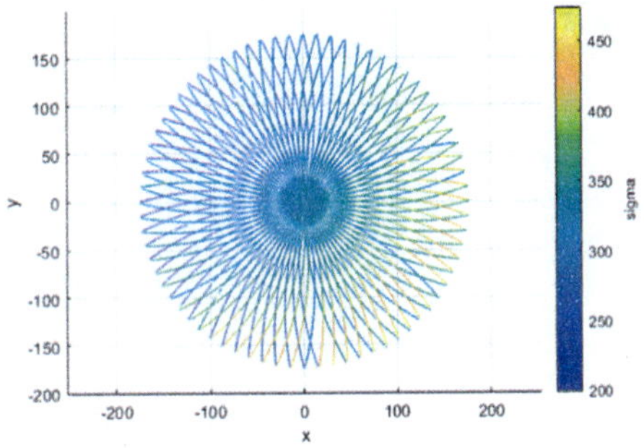

Fig. 2. Two-dimensional distribution of 350 mm measurement data.

In the case of curved surfaces, the three coordinates of the point on the aspheric surface to be measured are related to α and γ at that point, when the measurement reading h_{ij} is known, and the coordinates of the point on the surface to be measured are expressed as (u_{ij}, v_{ij}, w_{ij}), i for the i-th arc and j for the j-th point on that arc. We can invert the distribution of coordinates of the measured point at the probe tip from the ideal coordinates of the probe tip.

That is, in the $X_1Y_1Z_1$ coordinate system, the initial position of the measured value is $(L, 0, -h_0)$, and after the same coordinate transformation as above, the three-dimensional coordinates of the measured point on the aspheric surface to be detected in the XYZ coordinate system are obtained.

$$\begin{bmatrix} u_{ij} \\ v_{ij} \\ w_{ij} \end{bmatrix} = \begin{bmatrix} \cos\gamma & \sin\gamma & 0 \\ -\sin\gamma & \cos\gamma & 0 \\ 0 & 0 & 1 \end{bmatrix} \begin{bmatrix} L\cos\theta(\cos\alpha - 1) + h_{ij}\sin\theta \\ L\sin\alpha \\ -L\sin\theta(\cos\alpha - 1) + h_{ij}\cos\theta \end{bmatrix}$$
$$= \begin{bmatrix} \cos\gamma(L\cos\theta(\cos\alpha - 1) + h_{ij}\sin\theta) + L\sin\alpha\sin\gamma \\ -\sin\gamma(L\cos\theta(\cos\alpha - 1) + h_{ij}\sin\theta) + L\sin\alpha\cos\gamma \\ -L\sin\theta(\cos\alpha - 1) + h_{ij}\cos\theta \end{bmatrix}$$
$$= \begin{bmatrix} x_{ij} + h_{ij}\sin\theta\cos\gamma \\ y_{ij} - h_{ij}\sin\theta\sin\gamma \\ z_{ij} + h_{ij}\cos\theta \end{bmatrix} \quad (6)$$

Assuming that the aspherical mirror aperture D to be tested is 2000 mm, the vertex radius of curvature R_0 is 8000 mm and the aspherical coefficient k is -1, the closest comparative spherical radius calculated along the direction normal to the nearest sphere is R = 8026.7939 mm. we can calculate the position of the ideal probe tip coordinates when $\Delta\gamma = 5°$, the spatial distribution of its coordinates is shown in Fig. 3.

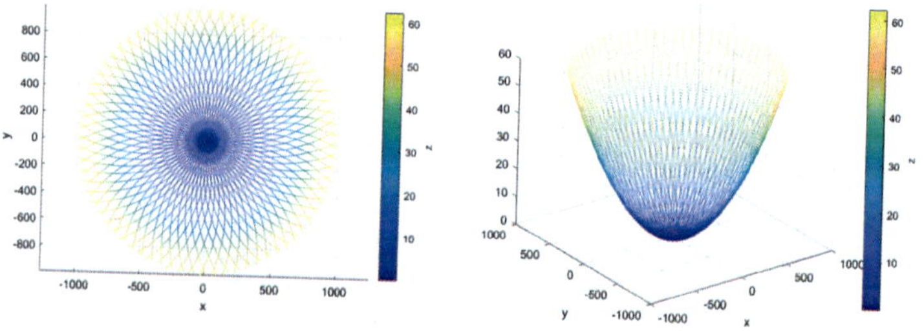

Fig. 3. Ideal coordinate distribution of probe tip.

According to the aspheric equation $x = \frac{Cy^2}{1+\sqrt{1-(k+1)C^2y^2}}$, where $C = \frac{1}{R_0}$, $k = -e^2$, the theoretical measurement along the direction closest to the comparative spherical normal can be calculated for the parameters D = 2000 mm, R_0 = 8000 mm and k = -1. This can be obtained by subtracting the two wavefronts, the aspheric and the closest comparative spherical, and we can write down the theoretical measurement as δ_{ij}, and substitute the theoretical measurement δ_{ij} as the actual measurement h_{ij} into the coordinate transformation to obtain the aspheric simulation waveform, and compare it with the ideal aspheric waveform to verify the (u_{ij}, v_{ij}, w_{ij}) accuracy of the calculation.

3 Error Analysis of the Swing Arm Profile Measurement

Considering that the swing arm profile measurement device is mainly composed of a swing arm air-bearing turntable, a mirror turntable and a rotating swing arm with a high-precision sensor for measurement, the actual reading of the probe will be affected by the assembly and calibration of the swing arm measurement device and the movement of the two turntable axes. Among them, we can call the error generated in the assembly process as static error. The main error sources are the misalignment error and angle positioning error of the two axes. The error generated during the swing arm measurement movement can be called dynamic error. The main error sources include the axial jump, radial jump and tilt generated by the swing arm turntable and the mirror turntable during the movement.

In the ideal case where there is no error, the ideal position of the probe tip during a single scan line measurement and the ideal probe reading can be represented relatively easily by simulation, as shown in Fig. 4.

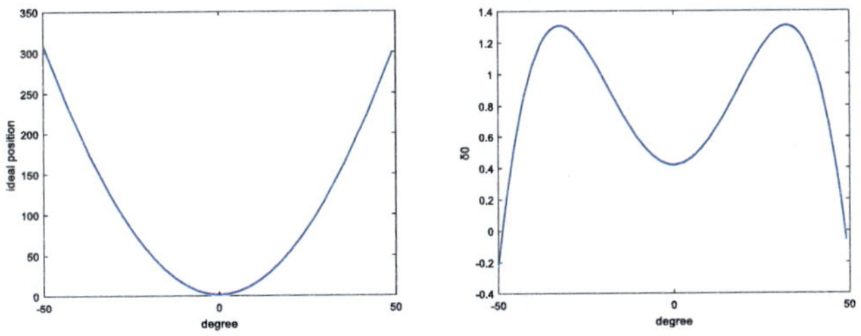

Fig. 4. Ideal position of probe tip (left) and ideal probe reading (right).

In the presence of various errors, the effect of a single error on the ideal probe tip position can be analysed to determine the effect on the probe reading, and for this reason we focus on the variation of a single scan line to assess the effect of various errors.

Static Error. The swing arm profiler is highly adaptable to different measurement requirements and its unique geometry allows for a wide range of shapes to be measured, including.

- Spherical and aspherical optics;
- Concave and convex optics;
- Optics with large surface slopes;
- Optics with large asphericities;
- Optics of different diameters;
- Optics of different radii of curvature ranging from flats to hemispheres;

Prior to the measurement of different optical elements, the swing arm needs to be mounted to suit the optical element to be measured. This requires a high degree of accuracy and the errors caused by the set-up will be an important source of error in the measurement of the swing arm profile. For static set-up errors the misalignment of the two axes is the main source of error, while arm length measurement errors and sensor alignment errors are not significant with the aid of laser tracker calibration.

In the process of mounting, the mirror turntable should be level to ensure that the mirror is level. If there is a tilted placement of the mirror itself, e.g. left or right relative to the pendulum arm, a situation as shown in Fig. 5(a) will occur, and with a certain degree of guaranteed level accuracy, the remaining error can be offset by de-tilting the mirror through 3D reconstruction.

When there is a misalignment of the two axes, for example along the X_1 direction, the measurement error situation is similar to that of the mirror itself where there is first a tilt for the front and back of the swing arm, as shown in Fig. 5(b).

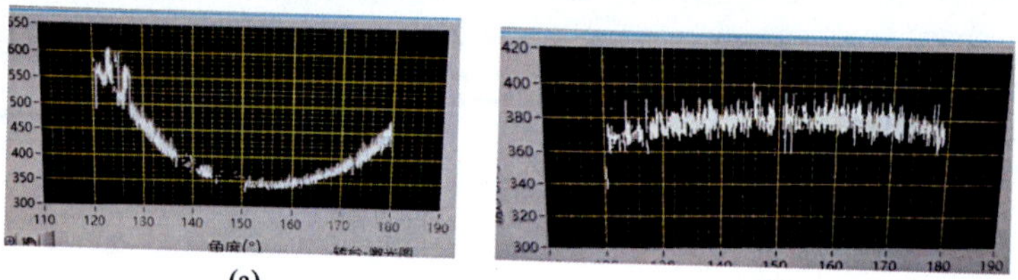

(a) (b)

Fig. 5. Effect of (a) mirror tilt (left and right) and (b) misalignment of the two axes on the measurement.

For the two-axis misalignment of the swing arm and mirror pivots, the direction of misalignment can be along either the X_1 or Y_1 coordinate axes, which are analysed according to the same principles. Here, the analysis of the case where the swing arm pivot produces a deviation in the Y_1 axis, i.e. the swing arm pivot deviates from the optical axis e in the Y_1 direction, which corresponds in essence to a rotation of the $X_1Y_1Z_1$ coordinate system, the angle of rotation being determined by the distance of the pendulum pivot to C together with the deviation e determines the angle of rotation. Multiplying the coordinate transformation in Sect. 2 before the coordinate rotation matrix T gives the new probe tip coordinate position as

$$\begin{pmatrix} x \\ y \\ z \end{pmatrix} = \begin{pmatrix} L\cos\theta(\cos\alpha - 1) \\ L\sin\theta + e \\ L\sin\theta(1 - \cos\alpha) \end{pmatrix} \tag{7}$$

When the swing arm pivot produces a deviation of e = 1mm in the Y_1 axis direction, the probe tip trajectory is obtained as shown in Fig. 6.

It is worth noting that when the swing arm pivot axis deviates in the direction of the X_1 axis, it changes the inclination angle θ and the closest comparison sphere radius R_0, which needs to be taken into account when calculating.

For the swing arm rotation angle positioning error, the probe tip remains on the nearest sphere, α = α + e. Substituting this into the coordinate transformation in sec.2 gives the measurement error at e = 1′ as shown in Fig. 7.

Dynamic Errors. Axial runout, radial runout and tilt inevitably occur during the rotational movement of the swing arm, and the error is less sensitive to radial runout and tilt perpendicular to the direction of the swing arm for the swing axis.

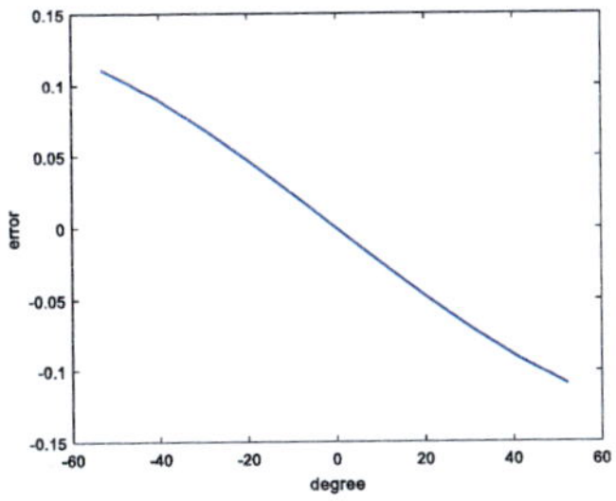

Fig. 6. Misalignment error (swing arm pivot produces deviation e = 1mm in the Y1 axis direction).

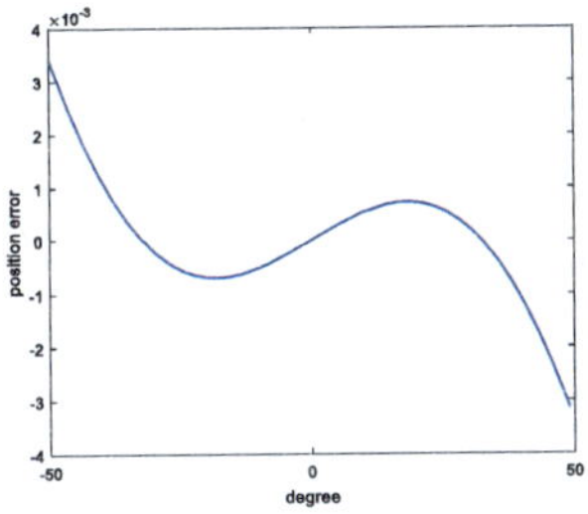

Fig. 7. Swing arm rotation angle positioning error.

For the axial runout of the swing arm, as shown in Fig. 8 (left), this mainly produces translation in the z_1 axis direction. This will result in a change in the trajectory coordinates of the probe tip. In the $X_1Y_1Z_1$ coordinate system, the initial coordinates will change from (L, 0, 0) to (L, 0, e), so that the new coordinates of the probe tip become:

$$\begin{pmatrix} x_n \\ y_n \\ z_n \end{pmatrix} = \begin{pmatrix} x_{ij} + e\sin\theta\cos\gamma \\ y_{ij} - e\sin\theta\sin\gamma \\ z_{ij} + e\cos\theta \end{pmatrix} \tag{8}$$

Also take into account that the coordinates of the probe tip are transformed without the mirror surface changing position, so that the new measured value will be determined by the new coordinates of the probe tip together with the new point corresponding to the mirror surface, as shown in Fig. 8 (right). The measured value can be calculated by using Matlab to calculate the value of the signed distance between the new coordinates of the probe tip in the direction normal to the surface of the nearest comparison sphere and the surface being measured. So the error can be expressed as $e = \delta_0 - \delta$.

If an axial runout error of e = 10 μm is produced uniformly, the error in a single scan line is shown in Fig. 9.

For the swing arm radial runout, the radial runout causes displacement of the probe up the diameter of the swing arm pivot axis. In the $X_1Y_1Z_1$ co-ordinate system, the initial value is (L+e, 0, 0), and the error produced on a single swept line when substituting the co-ordinate transformation to obtain e = 10 μm is shown in Fig. 10.

Tilting in a direction perpendicular to both the pendulum axis of rotation and the swing arm increases the measurement error in approximate proportion to the length of

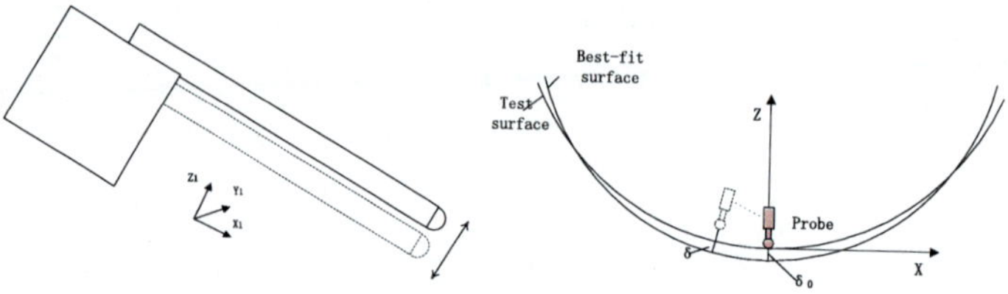

Fig. 8. Axial runout of the swing arm (left) and probe tip variation (right).

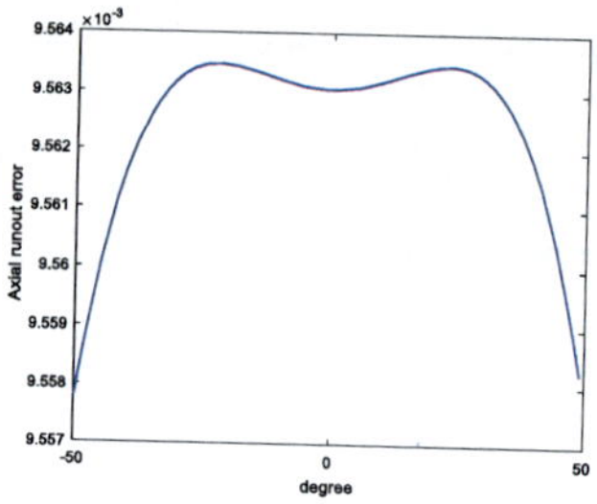

Fig. 9. Axial runout of swing arm rotary table (10 μm).

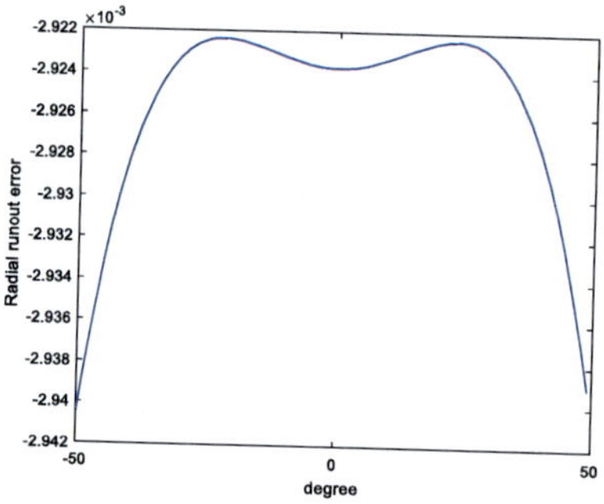

Fig. 10. Radial runout of swing arm turntable (10 μm).

the swing arm. The new coordinates of the tip of the probe are

$$\begin{bmatrix} x \\ y \\ z \end{bmatrix} = \begin{bmatrix} L\cos\theta(\cos\alpha\cos e - 1) - L\sin\theta\sin e \\ L\sin\alpha\cos e \\ L\sin\theta(1 - \cos\alpha\cos e) - L\cos\theta\sin e \end{bmatrix} \tag{9}$$

When the tilt is e = 5μrad, the tilt error on a single scan line is shown in Fig. 11.

For mirror tables, the mirror table axis and diameter jumps mainly displace the coordinate points in the Z-axis and mirror axis diameter directions, and do not change the ideal probe tip coordinate positions, but do change the initial aspheric equations, e.g. the table axis jump equation.

$$u^2 + v^2 = 2p(d + e) \tag{10}$$

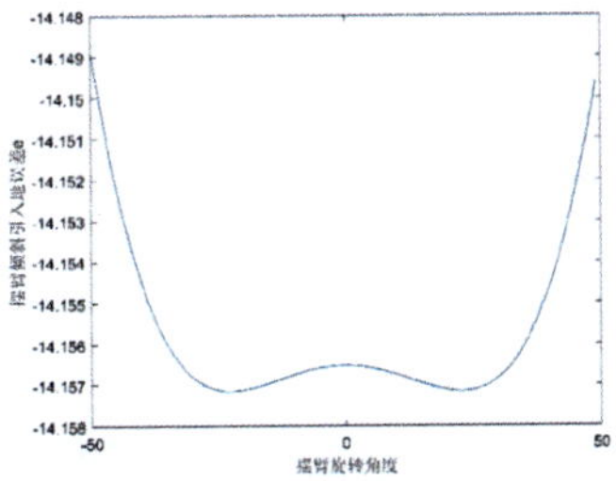

Fig. 11. Radial runout of swing arm turntable (5 μrad).

The axial runout, X-axis radial runout and Y-axis radial runout errors for a single scan line at 10μm axial and radial runout error are shown in Fig. 12.

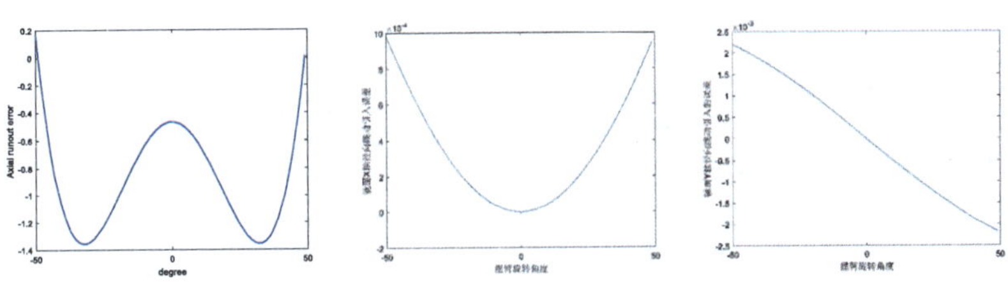

Fig. 12. Mirror axial runout error and X-axis and Y-axis radial runout error (10 μm).

Assuming that the centre of tilt is at the vertex of the surface under test, the measurement error for a single scan line with a 10μrad tilt introduced by rotation about the X and Y axes is shown in Fig. 13.

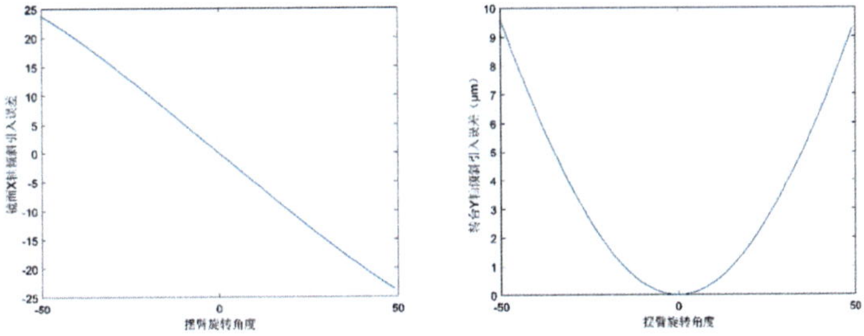

Fig. 13. X-axis and Y-axis tilt error (10 μrad).

Table 1 shows the error size after introducing the corresponding deviation after establishing the coordinate transformation model and analyzing the spatial structure influence of each error source.

The table reflects the sensitivity of the measurement error relative to the source error. Through the sensitivity, we can estimate how much measurement error will be caused

Table 1. Summary of errors introduced by individual scan lines in each section

Sources of error	Deviation size	error
Swing arm positioning error	1′	±2.88 μm
Swing arm axial runout	10 μm	9.56 μm
Radial runout of swing arm	10 μm	2.92 μm
Swing arm tilt	5 μrad	14.2 μm
Mirror turntable axial runout	10 μm	10 μm
Radial runout in the X-axis of the mirror turntable	10 μm	1.01 μm
Radial runout in the Y-axis of the mirror turntable	10 μm	±2.28 μm
Mirror turntable with X-axis tilt	10 μrad	±22.75 μm
Mirror turntable Y-axis tilt	10 μrad	10.14 μm

by a given source error, which is the basis of error budget. The error of high sensitivity is the key point to be considered when setting the adjustment tolerance. For example, in order to achieve micron-level measurement accuracy, the swing arm turntable needs to control the tilt error to sub-micro radian. For the two-axis misalignment error, the alignment tolerance can be appropriately relaxed due to the introduction of power and spherical aberration. Each error type corresponds to a motion in the direction of degree of freedom, and the real measured motion data is a combination of various error sources. The root sum square method can be used to estimate the combined uncertainty.

4 Conclusion

The swing arm profile measurement technology has become a powerful means for measuring large-aperture aspheric mirrors in the grinding stage due to its advantages of large dynamic range, large error measurement and in-situ measurement. The geometric structure of the device determines that the source of its error mainly comes from the swing arm shaft, mirror shaft and its spatial position. After establishing the swing arm coordinate transformation model, by introducing deviation in the coordinate transformation, mathematical modeling is performed for different error sources. Finally, the size of the introduced error is quantitatively calculated and the sensitivity of different error sources is analyzed to provide support for surface reconstruction.

References

1. Jing, H., Lin, Z., Ma, L., et al.: Development and experimental validation of a versatile prototype Swing Arm Profilometer for measuring optical surfaces. J. Eur. Opt. Soc. **6** (2011)
2. Anderson, D.S., Parks, R.E., Shao, L.Z.: Versatile profilometer for aspheric optics. In: Optical Fabrication and Testing. Optica Publishing Group (1990). OThA4
3. Anderson, D.S., Burge, J.H.: Swing-arm profilometry of aspherics. In: Optical Manufacturing and Testing, vol. 2536, pp. 169–179. SPIE (1995)

4. Su, P., Oh, C.J., Parks, R.E., et al.: Swing-arm optical CMM for aspherics. In: Optical Manufacturing and Testing VIII, vol. 7426, pp. 150–157. SPIE (2009)
5. Xiong, L., Luo, X., Liu, Z., et al.: Swing arm profilometer: analytical solutions of misalignment errors for testing axisymmetric optics. Opt. Eng. **55**(7), 074108 (2016)
6. Xiong, L., Qi, E., Luo, X., et al.: Stitching swing arm profilometer test for large aperture aspherics. Chin. Opt. Lett. **17**(11), 112201 (2019)

Dynamic Spectroscopic Imaging Ellipsometry for Thickness Measurement Based on a Dual-Comb System

Ruixue Zhang[1,2], Guanhao Wu[2(✉)], Qiang Cong[1], Ding Yuan[1], Fanxin Sun[1], and Jian Ding[1]

[1] Beijing Institute of Spacecraft System Engineering, Beijing, China
[2] State Key Laboratory of Precision Measurement Technology and Instruments, Department of Precision Instrument, Tsinghua University, Beijing, China
guanhaowu@mail.tsinghua.edu.cn

Abstract. Lightweight multifunctional devices, such as thin-film antennas and solar sails, are important for lightweight spacecraft development. The thickness and uniformity of thin films directly influence the performance capabilities, and need to be measured and monitored at high speed with high-precision approaches during their fabrication. Imaging ellipsometry is a standard metrology tool for evaluating thin films with the advantages of being non-contact, non-destructive, and sensitive. Conventional imaging ellipsometry using polarization modulation and phase-shifting methods is subject to trade-offs between precision, measurement speed and spectral accuracy. In this paper, we demonstrate a fast and high-precision dual-comb spectroscopic imaging ellipsometry for thin-film thickness uniformity measurement with no moving element by implementing a digital micromirror device (DMD). By utilizing our proposed dual-comb imaging ellipsometry, the spatial resolution of 3 mm × 2 mm and the thickness measurement repeatability of 5.6 nm is obtained within 5 ms, which is three orders of magnitude faster compared with the traditional ways. Our method provides a convenient and efficient solution for high-precision and fast imaging ellipsometry and practicability.

Keywords: Optical Frequency Comb · Thickness Measurement · Imaging Ellipsometry

1 Introduction

Nowadays, spacecraft are taking on more and more tasks, facilitating the development of large, lightweight and complex space structures. Lightweight multifunctional devices are an essential technical route toward lightweight spacecraft [1, 2]. For instance, the development of lightweight thin-film adaptive antennas, which provide economical cost advantage and spatial extensibility over other antennas, has the potential to enhance current radio-frequency (RF) communication capabilities [3]. Thin-film flexible photovoltaics are paving the way to low-cost electricity. Thin-film solar sails, in which electric power is generated by thin-film solar cells attached to the solar sail membrane, are a good

H. P. Urbach and H. Jiang (Eds.): ISSOIA 2022, SPPHY 295, pp. 376–385, 2023.
https://doi.org/10.1007/978-981-99-4098-1_33

solution for interstellar navigation, such as the IKAROS mission of Japan. As the thickness and uniformity of thin films directly influence the devices' performance capabilities, they have to be real-time measured and monitored with high-precision approaches during their fabrication [4].

Traditional mechanical or electrical methods have the disadvantages of contact and low measurement accuracy. As a non-contact, non-destructive, and sensitive measurement technique, ellipsometry is a standard metrology tool for evaluating thin films. The key feature of ellipsometry is that it measures the change in polarized light upon light reflection on a sample or light transmission by a sample, and the measured ellipsometric parameters (Ψ and Δ) represent the amplitude ratio Ψ and phase difference Δ between light waves known as p- and s-polarized light waves [5]. For the measurement with single-point measurement ellipsometry, sample surfaces are required to be uniform over the optical beam area. Imaging ellipsometry is a technique to characterize surfaces with micro-patterns or non-uniform distribution from a single measurement [6]. Conventional imaging ellipsometry (IE) methods include mechanical scanning based on moving stages, null imaging ellipsometry, phase-shifting IE, and coaxial spectroscopic imaging ellipsometry based on Fourier transform spectroscopy. IE based on mechanical scanning adopts polarization modulation to demodulate the ellipsometric parameters, which is time-consuming and unstable [7]. Null imaging ellipsometry needs to change the angle of the polarizer and compensator, which is not suitable for complex samples [5]. Phase-shifting IE requires a high-precision phase shift. Spectroscopic imaging ellipsometry based on Fourier transform spectroscopy bears the disadvantage of the long-range scan to improve the spectral resolution and the high linearity of the displacement stage [8]. Moreover, the first three measurement methods also need to scan wavelength if spectroscopic ellipsometry is required. In conclusion, conventional techniques are subject to trade-offs between precision, measurement speed and spectral accuracy.

In this paper, we demonstrate a fast and high-precision dual-comb spectroscopic imaging ellipsometry (DCIE) instrument for thin-film thickness uniformity measurement with no moving element by implementing a digital micromirror device (DMD). By exploiting the dual optical-frequency combs (OFC), precise spectral detection with high spectral resolution can be achieved over a wide frequency range [9]. Both the real and imaginary parts of the spectra can be obtained on a single photodetector, allowing the ellipsometric parameters to be deduced at a high update rate of 1 kHz in our system. As a proof of concept experiment, we use SiO_2 thin film as the samples to verify the effectiveness of the proposed method. The spatial resolution of 3 mm $\times$ 2 mm and the thickness measurement repeatability of 5.6 nm are obtained within several milliseconds, which is three orders of magnitude faster compared with the traditional ways.

2 Principles and System Design

2.1 Principles of Ellipsometry

Ellipsometry is an optical measurement technique that characterizes the polarization state of light reflection or transmission from samples. Based on the two ellipsometric parameters Ψ and Δ, the thickness and optical properties can be obtained by data analysis. In spectroscopic ellipsometry, Ψ and Δ spectra are measured by changing the wavelength of light. Due to its advantages of high precision, nondestructive measurement and various characterizations, ellipsometry is widely used in the areas of semiconductors, displays, optical coating, and so on.

Here, we use an air/thin film/substrate structure as an example to illustrate the optical model of ellipsometry. As shown in Fig. 1, the N_0, N_1 and N_2 denote the complex refractive indices of air, thin film, and substrate, respectively. The transmission angles (θ_1 and θ_2) can be calculated from the angle of incidence θ_0 by adopting Snell's law. From Frenel equations for p- and s-polarized waves, we obtain

$$r_{jk,p} = \frac{N_k \cos\theta_j - N_j \cos\theta_k}{N_k \cos\theta_j + N_j \cos\theta_k},\ r_{jk,s} = \frac{N_j \cos\theta_j - N_k \cos\theta_k}{N_j \cos\theta_j + N_k \cos\theta_k},$$
$$t_{jk,p} = \frac{2N_j \cos\theta_j}{N_k \cos\theta_j + N_j \cos\theta_k},\ t_{jk,s} = \frac{2N_j \cos\theta_j}{N_j \cos\theta_j + N_j \cos\theta_k}. \quad (1)$$

Here, r_{jk} and t_{jk} show the amplitude reflection and transmission coefficient at each interface. When light absorption in a thin film is small, optical interference occurs by multiple light reflections with the film. The phase variation β is expressed as

$$\beta = \frac{2\pi d}{\lambda} N_1 \cos\theta_1, \quad (2)$$

while d and λ denote the film thickness and optical wavelength, respectively. Finally, we can get the amplitude reflection coefficient as

$$r_{012,p} = \frac{r_{01,p} + r_{12,p}\exp(-j2\beta)}{1 + r_{01,p} r_{12,p}\exp(-j2\beta)},\ r_{012,s} = \frac{r_{01,s} + r_{12,s}\exp(-j2\beta)}{1 + r_{01,s} r_{12,s}\exp(-j2\beta)}. \quad (3)$$

We can see that the ellipsometric parameters Ψ and Δ are the functions of N_0, N_1, N_2, d, θ_0 and λ. As the N_0, N_1, N_2, θ_0 and λ are known, we can derive the film thickness d by the measured Ψ and Δ.

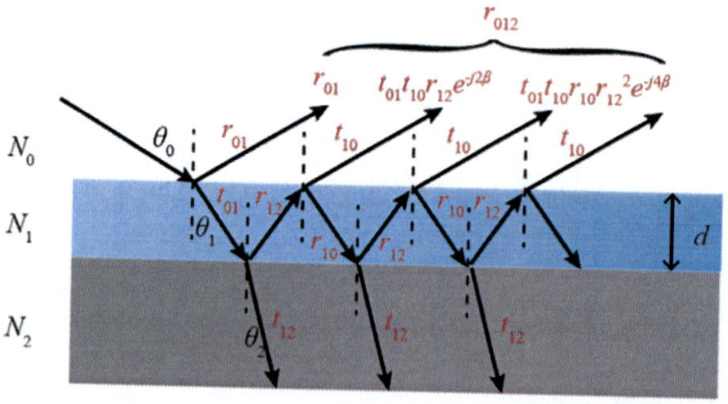

Fig. 1. The optical model of an air/thin film/substrate structure.

2.2 System Design and Principles of Dual-Comb Spectroscopic Ellipsometry

System Design of DCIE. The system of DCIE consists of a dual-comb spectroscopic ellipsometry module and the single-pixel imaging detection module. The schematic of our DCIE system is shown in Fig. 2. Two erbium-based mode-locked fiber lasers are employed as the signal (Comb1) and local comb (Comb2) oscillators with repetition rates ($f_{r1} = 56.09$ MHz, $f_{r2} = 56.091$ MHz) and offset frequencies ($f_{o1}, f_{o2} = 10.56$ MHz). The polarization states of Comb1 and Comb2 are set to linear polarizations at $-45°$ in a coordinate system in which the p-polarized direction is assumed to be X axis and the s-polarized direction is the Y axis. Then, the beam of the two combs is expanded by the two BEs. Comb1 is incident on the sample in the reflection configuration and is then spatially overlapped with Comb2 by the BS. The overlapped beam is incident on the DMD, which is composed of millions of micromirrors. Since each micromirror of DMD can be controlled independently, we can turn some micromirrors "on" or "off" separately. Here, the state of "on" indicates that the reflection light can be detected by the two PDs, while the state of "off" means the reflection light cannot be detected. After the reflection of DMD, the generated interferograms of the p- and s-polarization components (I_1 and I_2) are individually detected by a PBS and PDs.

As we use the DMD to modulate the beam in space, the dual-comb interferogram in different positions could be detected. Moreover, fast acquisition of the interference signals can be achieved for the reason that the micromirrors of the DMD can be switched to "on" or "off" quickly (nominally 105 μs).

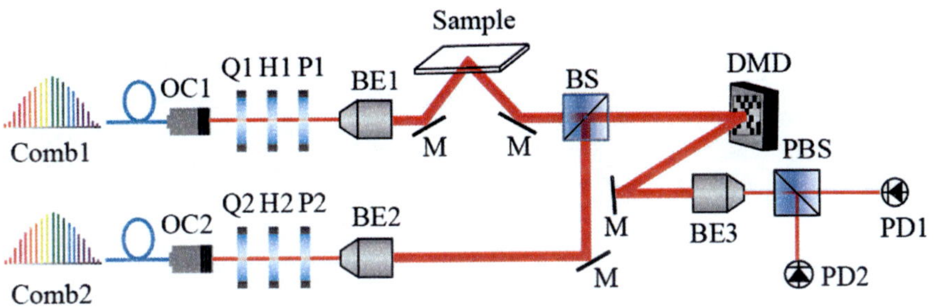

Fig. 2. The schematic setup of the dual-comb imaging ellipsometry system. OC: optical collimator, Q: quarter-wave plate, H: half-wave plate, P: polarizer, BE: beam expander, M: mirror, BS: beam splitter, PBS: polarizing beam splitter, PD: photo detector.

Principle of Dual-Comb Spectroscopic Ellipsometry. An OFC is a broadband coherent light source consisting of a series of discrete longitudinal optical modes. Each optical mode can be described in terms of repetition rate (f_r) and offset frequency (f_o) as $f(n) = nf_r + f_o$, where n is the longitudinal mode number. The dual-comb system uses a pair of combs with a slight repetition rate difference (Δf_r). This is also known as linear optical sampling in the time domain and multi-heterodyne spectroscopy in the frequency domain. As for the theory of the dual-comb ellipsometry, let's consider the optical longitudinal mode f_1 of Comb1 with number n_1 and f_2 of Comb2 with number n_2, then the two electric fields after polarizers P1 and P2 can be written as:

$$\begin{aligned}\vec{E}_1(n_1) &= \frac{\sqrt{2}}{2}\begin{bmatrix}1\\-1\end{bmatrix}A_1(n_1)\exp[j(2\pi f_1(n_1)t+\varphi_1(n_1))],\\ \vec{E}_2(n_2) &= \frac{\sqrt{2}}{2}\begin{bmatrix}1\\-1\end{bmatrix}A_2(n_2)\exp[j(2\pi f_2(n_2)t+\varphi_2(n_2))],\end{aligned} \tag{4}$$

where $f_1(n_1) = n_1 f_{r1} + f_{o1}$ and $f_2(n_2) = n_2 f_{r2} + f_{o2}$ are modes' optical frequencies. $A_{1,2}$, $\varphi_{1,2}$ and j denote the (real) amplitude, modes' phases and imaginary unit, and $[1, -1]^T$ is the Jones vector of a linear-polarized light oriented at $-45°$. Afterward, we can calculate the polarization state of light propagating through the whole system in terms of Jones calculus, and the total electric fields at PD1 and PD2 are

$$\begin{aligned}\vec{E}_{PD1}(n_1, n_2) =& \frac{\sqrt{2}}{4}A_1(n_1)\sin(\Psi)\exp[j(2\pi f_1(n_1)t+\varphi_1(n_1)-\alpha_{1-PD1}+\Delta)]\\ &+\frac{\sqrt{2}}{4}A_2(n_2)\exp[j(2\pi f_2(n_2)t+\varphi_2(n_2)-\alpha_{2-PD1})],\end{aligned} \tag{5}$$

$$\begin{aligned}\vec{E}_{PD2}(n_1, n_2) =& \frac{\sqrt{2}}{4}A_1(n_1)\cos(\Psi)\exp[j(2\pi f_1(n_1)t+\varphi_1(n_1)-\alpha_{1-PD2})]\\ &+\frac{\sqrt{2}}{4}A_2(n_2)\exp[j(2\pi f_2(n_2)t+\varphi_2(n_2)-\alpha_{2-PD2})],\end{aligned} \tag{6}$$

where $\alpha_{i\text{-PD}x}$ signifies the propagation phase from Comb*i* to PD*x* (i, x = 1,2). The photocurrent produced at PD*x* is $I_x \sim |E_{PDx}|^2$. If we neglect the detector spectral response function and use a filter to take out the ac component with $f_r/2$, the generated kth mode of the sub-comb in the rf domain ($f_{rf}(k)$) can be obtained as $I_1(f_{rf}(k))$ and $I_2(f_{rf}(k))$. Then, the ellipsometric parameters can be obtained based on the amplitude parts (||) and phase parts (arg()) of $I_1(f_{rf}(k))$ and $I_2(f_{rf}(k))$ as

$$\begin{aligned}\tan(\Psi) &= |I_1|/|I_2|,\\ \Delta &= \arg(I_1) - \arg(I_2) + \alpha_{pro},\end{aligned} \tag{7}$$

in which α_{pro} is a phase term associated with the optical path propagation ($\alpha_{pro} = \alpha_{1\text{-PD1}}\text{-}\alpha_{1\text{-PD2}}\text{-}\alpha_{2\text{-PD1}} + \alpha_{2\text{-PD2}}$) and can be calibrated by a silicon wafer (Si).

3 Results and Evaluation

3.1 Stability of Phase Difference

Before the measurement of film thickness, we first evaluate the phase difference stability ($\delta\Delta\varphi_c$) of the p- and s-polarized interferograms, as we mainly adopt the ellipsometric parameter Δ to deduce the film thickness d. Hence, the $\delta\Delta\varphi_c$ is essential to obtain high-precision thickness measurement. Figure 3 shows the normalized amplitude of the p-polarized interferogram when a Si wafer is used as a test sample. As we can see, the amplitude of the interferograms increases with the number of micromirrors "on". The amplitude of the interferogram when the micromirrors of the state "on" is 390 * 390 is much higher than that with 160 * 160. Moreover, the $\Delta\varphi_c$ is the phase difference of rf carrier between I_1 and I_2 as $\Delta\varphi_c = \varphi_{c2} - \varphi_{c1}$, and any spectral mode near the center frequency can be selected as the carrier frequency f_c in Comb1. The standard deviation (σ) of $\Delta\varphi_c$ within 1 s is adopted to evaluate the $\delta\Delta\varphi_c$. As the orthogonally polarized light shares most common path in the ellipsometry setup, the $\Delta\varphi_c$ is theoretically stable with a relatively simple independent locking scheme of the dual-comb system as we have demonstrated previously. As shown in Fig. 3, compared with the phase difference stability of 1.44 rad when the number of micromirrors "on" is 160 * 160, the $\delta\Delta\varphi_c$ decreases to 0.02 rad when the number of micromirrors "on" is 310 * 310, which indicates that we need to keep sufficient number of micromirrors "on" in order to obtain a stable phase difference. In consideration of the limited power of the light sources, we choose to use 400 * 300 micromirrors as a mask to ensure a stable phase difference, and scan a 2 * 2 area to perform the imaging ellipsometry.

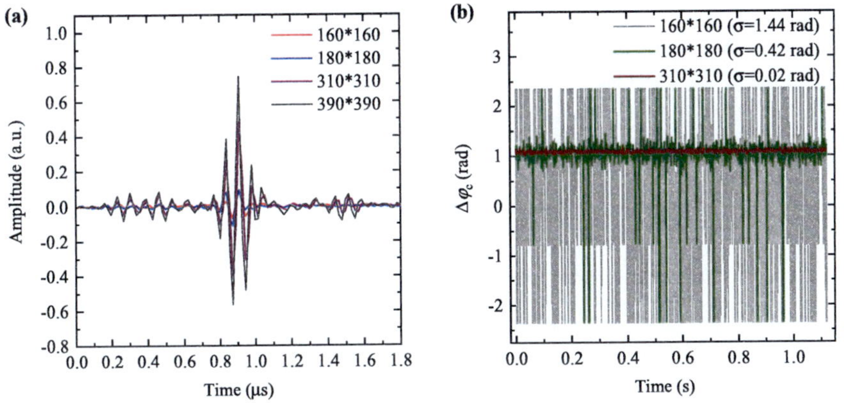

Fig. 3. The normalized amplitude of interferograms and phase difference stability ($\delta\Delta\varphi_c$). (a) The normalized amplitude of p-polarized interferograms. The legend shows the number of the micromirrors "on". (b) The phase difference between s- and p-polarized interferograms at the wavelength 1569.47 nm. Note that there are 1000 points within 1 s for the reason that the Δf_r of the dual-comb system is set as 1 kHz.

3.2 Film Thickness Evaluation

As we mentioned in Sect. 2, the ellipsometric parameters $\tan\Psi$ and Δ derived from the Fresnel formula are functions of wavelength (λ), the refractive index of the substrate, films and air, the thickness of films, and the incident angle (θ_0). As a proof of concept experiment, we use SiO_2 thin film as the samples and measure their film thickness. Although the material is different from that of the satellite thin film antenna, it should be noted that the purpose of this paper is to verify the feasibility of the imaging ellipsometry for measuring the thickness uniformity of the thin film. In our experiment, the wavelength is obtained by mapping the rf to optical frequency after transforming I_1 and I_2, and the refractive index values are known. In order to measure the film thickness, we need to calibrate the optical propagation phase α_{pro} and the incident angle beforehand. In our experiment, we use a silicon wafer to calibrate the α_{pro}, and the angle of incidence is determined by three different SiO_2 thin-film samples with the known thickness, which is determined by commercial ellipsometry (W-VASE). During the process of θ_0 calibration, the ellipsometric parameter Δ spectra of the SiO_2 thin-film samples with a thickness of 28 nm, 280 nm and 500 nm are measured while we turn on all the DMD micromirrors. After fitting the measured ellipsometric parameter with the theoretical parameters under different incident angles, the θ_0 is acquired when the root-mean-square error is minimal, and θ_0 in our system is 48.4°. Subsequently, the film thickness of 90 nm (Sample1) and 1000 nm (Sample2) SiO_2 samples is measured by the calibrated system. Note that the incident angle of each sample should remain unchanged.

We measure the film thickness using the Levenberg-Marquarelt algorithm by minimizing the RMSE of the experimental and theoretical values Δ over the spectral range of 3 nm. The number of micromirrors on a DMD mask is 400 * 300, whose actual size is 3.024 mm * 2.268 mm, considering the micromirror size (7.56 μm * 7.56 μm). During the measurement, four masks turn on alternately with the update time set to 1 ms, as depicted in Fig. 4. In addition, a blank mask is added at the end of each period as a time label.

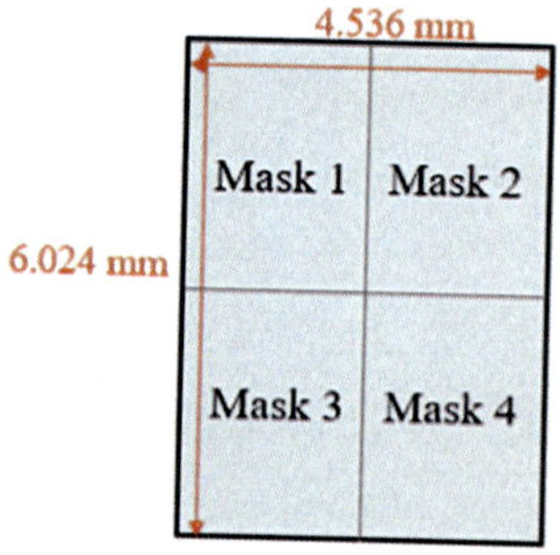

Fig. 4. The imaging area of the DMD. Masks 1–4 and a blank mask are tuned on in sequence. The total size of the area is 6.024 mm * 4.536 mm, as it is a 2 * 2 imaging area.

For each sample, we took the interference signals from both PD1 and PD2 with a time length of 1 s. A total of 200 periods of data with each of 5 ms, as four imaging masks and a label mask turned in sequence within 1 s. As can be seen from the results exhibited in Table 1 and Fig. 5, the average thickness of the two samples is 102.42 nm and 1001.22 nm, which is consistent with the nominal values 102.65 nm and 1007.19 nm obtained by the commercial ellipsometry W-VASE. For each set of data, a total of 200 measurements are performed for a single pixel. In Fig. 5, the standard deviation (*std*) of the thickness is represented as the error bar, and the average *std* of different pixels of the two samples is 5.6 nm at the measurement speed of 200 Hz.

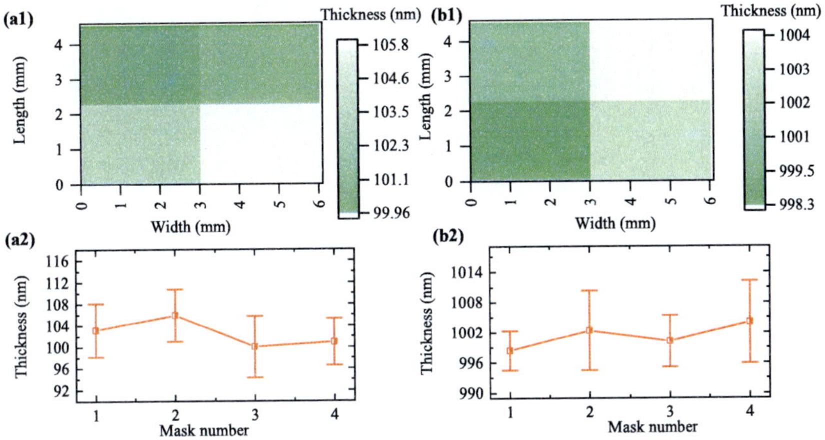

Fig. 5. Measured thickness and precision (error bar) under different masks. The average precision of different samples and masks is 5.6 nm.

Table 1. Measured film thickness of the samples in different positions ($d \pm std$).

Positions	Film thickness of Sample1	Film thickness of Sample2
Mask1	103.07 ± 4.97	998.36 ± 3.95
Mask2	105.81 ± 4.87	1002.34 ± 7.92
Mask3	99.96 ± 5.74	1000.19 ± 5.14
Mask4	100.82 ± 4.37	1004.01 ± 8.14

Furthermore, the Allan deviation for the different averaging times of the two samples is calculated to illustrate the performance of our system. As depicted in Fig. 6, the measurement precision of the film thickness can be improved to the sub-nanometer level when the averaging time is longer than 0.1 s, which means a 0.1 nm thickness measurement at the measuring speed of 10 Hz.

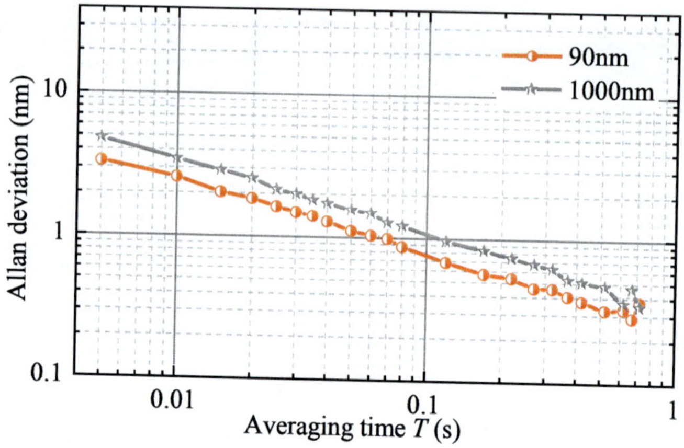

Fig. 6. The Allan deviation of the film thickness for the samples of 90 nm and 1000 nm.

4 Discussion and Conclusion

In this section, we discuss the trade-off relationship between precision, imaging resolution and measurement speed of the proposed DCIE system. In our proposed system, the measurement speed is equal to $\Delta f_r/M$, while M is the number of pixels in the imaging area. As for ellipsometry, the broader spectrum can enhance the precision, yet the available spectral range ($\Delta\nu$) is limited by the band-pass sampling theorem in a dual-comb system as $\Delta\nu < f_r^2/(2\Delta f_r)$. At the same time, the imaging resolution is proportional to M. In order to obtain the faster measurement speed, the larger Δf_r between the two combs needs to be set. Therefore, OFCs with higher repetition rates should be used, such as microresonator combs [10], to achieve higher precision at a fast measuring speed.

In summary, our proposed DCIE provides fast and high-precision imaging ellipsometry. As the repetition rates and offset frequencies of our dual-comb source utilize conventional low-bandwidth independent locking, this simplifies the locking system of the dual-comb light source, and is more suitable for actual applications. As the quantitative evaluation of the measurement performance, the high-precision thin-film thickness determination (5.6 nm) and fast measuring speed (200 Hz) verify the feasibility of the proposed system. The imaging resolution is 3 mm * 2 mm in the experiment, which can be further improved by increase the power of the laser sources. In short, our method provides a convenient and efficient solution for high-precision and fast imaging ellipsometry and practicability.

References

1. Kanimba, E., Yang, T., Huxtable, S.T., Li, L., Tian, Z.: Thermomechanical analysis of a bio-inspired lightweight multifunctional structure. Adv. Eng. Mater. **22**, 2000371 (2020)
2. Fosness, E., Guerrero, J., Qassim, K., Denoyer, S. J.: Recent advances in multi-functional structures. In: Proceedings of the IEEE Aerospace Conference, vol. 4, pp. 23–28 (2000)
3. Ivan, P., Jamie, E.R., Cory, D.C., Eric, N., Brian, J.L.: Carbon nanotube thin-film antennas. ACS Appl. Mater. Interfaces **8**(32), 20986–20992 (2016)

4. Birney, R.: Current research in thin film deposition: applications, theory, processing, and characterisation. Coatings **10**(12), 1228 (2020)
5. Hiroyuki, F.: Spectroscopic Ellipsometry: Principles and Applications. Wiley, Hoboken (2007)
6. Gang, J., Roger, J., Hans, A.: Imaging ellipsometry revisited: developments for visualization of thin transparent layers on silicon substrates. Rev. Sci. Instrum. **67**(8), 2930–2936 (1996)
7. Erman, M., Theeten, J.B.: Spatially resolved ellipsometry. J. Appl. Phys. **60**(3), 859–873 (1986)
8. Lee, S.W., Garam, C., Sin, Y.L., Yeongchan, C., Heui, J.P.: Coaxial spectroscopic imaging ellipsometry for volumetric thickness measurement. Appl. Opt. **60**(1), 67–74 (2021)
9. Ian, C., Nathan, N., William, S.: Dual-comb spectroscopy. Optica **3**(4), 414–426 (2016)
10. Boggio, J.M., Bodenmüller, D., Ahmed, S., Wabnitz, S., Modotto, D., Hansson, T.: Efficient Kerr soliton comb generation in micro-resonator with interferometric back-coupling. Nat. Commun. **13**(1), 1–11 (2022)

The Design and Implementation of High Resolution Time Interval Measurement in the Space-Borne Laser Ranging

Song Bo[1,2](✉)

[1] Beijing Institute of Space Mechanics and Electricity, Beijing 100094, China
bobosongmm@126.com
[2] China Academy of Space Technology, youyi road no. 104, Beijing, China

Abstract. The satellite was the first civilian high-resolution stereo mapping satellites in China, which set mapping and resource investigation functions. The satellite was equipped with the domestic first to ground elevation measurement space-borne laser range finder. Through studying the digital large range and high resolution time interval measurement technique based on FPGA in space-borne laser ranging, the first to achieve the space-borne laser ranging which had been reaching 1 ns level ranging resolution in the 5 ms full range. The technology for the first time using FPGA applied to satellite laser ranging, through the FPGA internal resources of DCM to achieve clock frequency, frequency doubling and shift, thus got the 1G high frequency stability clock, to achieve 1ns resolution, which including coarse counting timing and fine count measurement. At the same time the sub second accuracy was better than the 30 us index requirements, the main wave of the jump range was not greater than 500ns. By studying the working principle of time interval measurement, the FPGA hardware circuit design, the system software program design and the related index test and so on to realize the high resolution time interval measurement, the experimental results showed that the stability of the 1ns measurement resolution, the sub second time accuracy of less than 30 us and main wave jump of less than 500 ns could be obtained.

Keywords: Laser range finder · High resolution time interval measurement · 1ns resolution · sub second accuracy

1 Introduction

The laser ranging device carried by a satellite is the first satellite-borne laser rangefinder for ground elevation measurement in China. The laser is used to measure the tilt distance information of reference points along the array direction at the time of photography, and the required elevation accuracy is guaranteed by fusion with the data obtained by stereo surveying camera and GPS, so as to improve the photographic positioning ability of surveying and mapping satellites, which is of great significance for improving the surveying and mapping accuracy [1–3].

H. P. Urbach and H. Jiang (Eds.): ISSOIA 2022, SPPHY 295, pp. 386–394, 2023.
https://doi.org/10.1007/978-981-99-4098-1_34

As the first independently developed satellite-borne laser allocator system in China, the distance measurement from the photography center to the ground point is used to improve the stereo mapping ability of subsequent stars. At the same time, the technical scheme of satellite-borne laser altimeter was verified, the effective in-orbit data were obtained, the heaven-earth deduction model was improved, and the development specifications of satellite-borne laser altimeter system were established. By monitoring the in-orbit state of the rangefinder, the state difference between the satellite and the earth of the laser active remote sensing equipment was obtained, which laid the foundation for the follow-up satellite-borne laser payload, laser technology breakthrough, development and verification [4, 5].

This paper mainly studies the digital large-range and high-precision time interval measurement technology based on FPGA, achieving the time resolution of 1 ns within the full range of 5 ms, and meeting the index requirements of sub-second punctuality accuracy better than 30 us and main wave runout range less than 500 ns. It has the characteristics of high measurement resolution, better real-time performance and relatively simple equipment. It is suitable for spaceborne equipment.

2 High Resolution Time Interval Measurement Principle

Laser measurement system principle block diagram is shown in Fig. 1. The principle of laser measurement subsystem is that high energy laser pulses are generated by the laser, which are transmitted to the measured target through the optical system of the transmitting mirror, and the laser emission time signal collected by the optical fiber is used as the main wave signal. After being reflected by the measured target, the optical signal collected by the APD is used as the echo signal by the receiving optical system. Photoelectric conversion is completed on the photosensitive surface of the APD. After amplification, it enters the ranging information processing circuit, and the high resolution measurement is completed by calculating the time interval between the echo signal and the main wave signal [6–8].

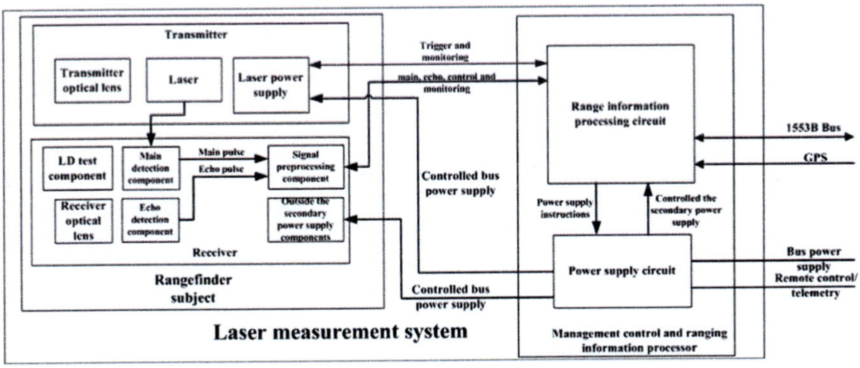

Fig. 1. Laser measurement system principle block diagram

In the time interval measurement, satellite orbit altitude biggest about $d = 650\,\text{km}$, Laser round-trip time about $t = 2 \times \frac{650\times10^3}{3\times10^8} = 0.0043\,s = 4.3\,\text{ms}$. In the time interval measurement, the balance between the counter length and the count resolution should be solved by using the time-of-flight method with a single counter. To improve the resolution, the clock frequency and counting length should be increased. In order to realize the 1ns time-measuring circuit, a 1 GHz clock is needed. Since the clock higher than 300 MHz cannot be directly entered into the FPGA chip, and it is not reliable to count only by the digital circuit, 1 GHz clock is generated by the internal phase shift, frequency division and frequency doubling of FPGA, and the coarse-fine combination method is adopted to achieve high resolution time measurement. When measuring precision test sequence as shown in Fig. 2, the measured results for $T = \Delta T_1 + nT_p - \Delta T_2$.

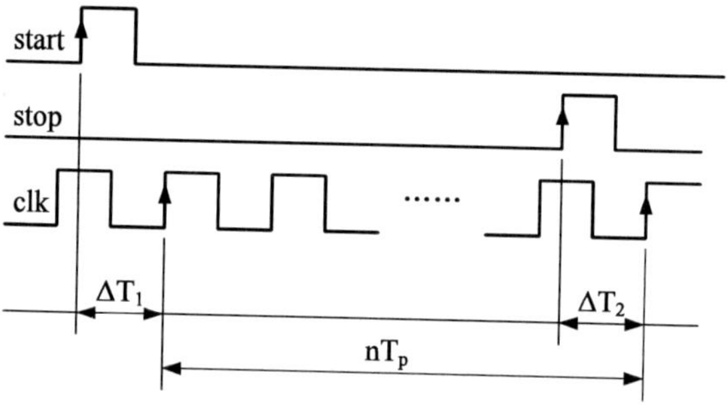

Fig. 2. Time interval measurement principle diagram

In high resolution time measurement, the precision count resolution is 1ns, realizing the main and echo with coarse count clock rise along ΔT_1 and ΔT_2 precision measurement. The coarse counter is designed as a composite counting mode of 62.5 MHz clock and 250 MHz clock. The low 2 bits of the coarse counter choose 250 MHz clock for counting, and the high 20 bits choose 62.5 MHz clock for counting. The combination counting logic is strictly synchronized with the fine counting clock to realize the whole cycle counting between the main echoes. Low frequency can effectively prevent the long range counter error jump.

3 Hardware Circuit Implementation

The satellite laser rangefinder high-resolution time interval measurement hardware circuit design based on FPGA minimum system, this is the first time in space-borne laser ranging of height measurement based on using the first-class devices.

The high resolution timing circuit is composed of FPGA minimum system, second pulse interface circuit, command and control interface circuit, laser power supply interface circuit, extravehicular pre-processing interface circuit and secondary power supply circuit, and its main functions are as follows:

- laser driver trigger work process, to achieve SPI bus data communication;

- complete pre-processing component, main signal channel and the echo signal channel of SPI bus data communication parameter settings and get the main and echo, achieve the trigger LD self-check work;
- finished space-borne laser rangefinder system control functions;
- complete time interval measurement with high precision;
- controller receives the management instruction, feedback rangefinder working parameters and condition, including the rangefinder self-checking, return status, distance, self-check state, state inspection, distance measurement, laser components work state, signal preprocessing work state, etc.;
- forward GPS second pulse signal receiving controller, punctual to complete the local operation;

Time measurement of hardware circuit diagram as shown in Fig. 3.

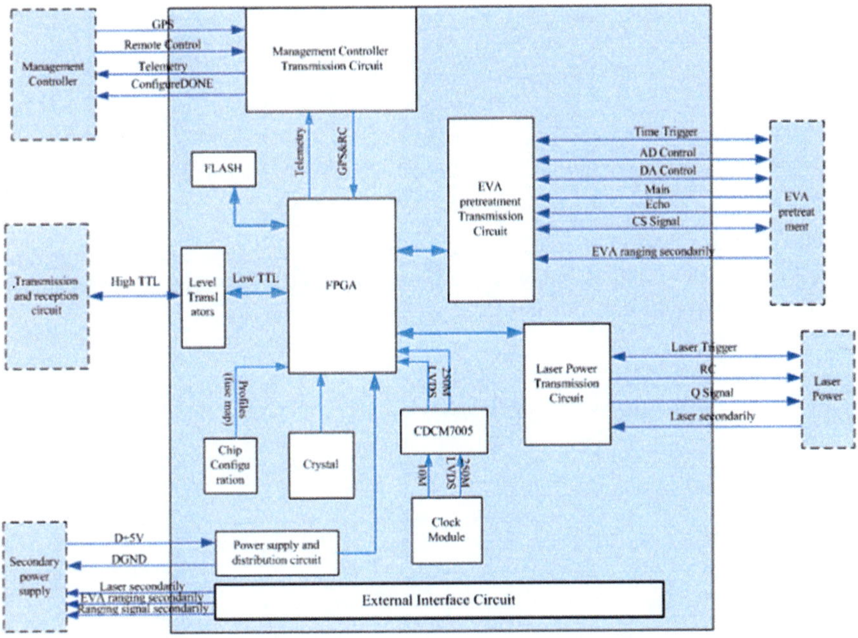

Fig. 3. Time measurement of hardware circuit function block diagram

4 System Software Implementation

High resolution time interval measurement software according to the functional requirements, mainly includes seven parts: the clock distribution, self-check, control logic and time measuring, punctuality, integration of the serial control, data storage, software module connection diagram as shown in Fig. 4.

The time measuring software uses the clock counting method to realize 1ns high resolution time measuring, 1 ns time measurement corresponding to 1 GHz clock frequency,

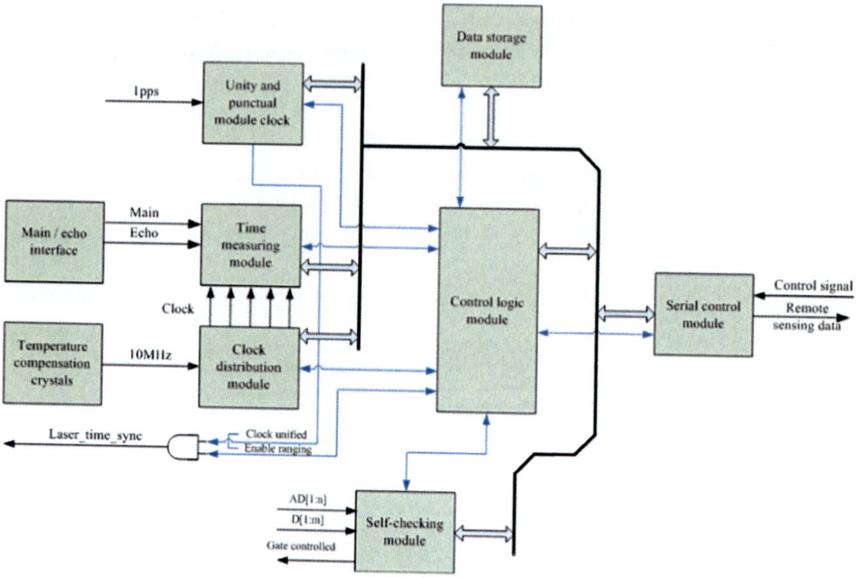

Fig. 4. High resolution time interval measurement software module connection diagram

in order to make the FPGA reliable count, reduce the burrs generated probability, the relatively low speed 250 MHz clock was used, through the four-way quadrature phase shifting to achieve the equivalent of 1 GHz clock, as shown in Fig. 5.

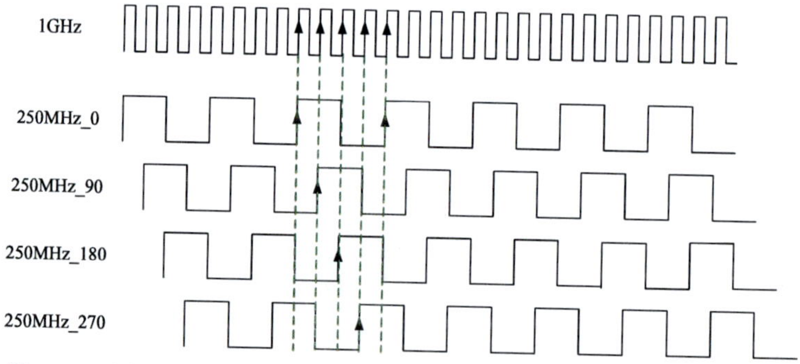

Fig. 5. Four-way 250 MHz clock quadrature phase shifting equivalent diagram of the 1 GHz clock

When multistage clock sampling is used for measuring time, the structure of temperature replenished crystal oscillator and phase-locked loop is designed. In the rough counting process, we use 250 MHz for 4 frequency division to get a 62.5 MHz low frequency clock. In the fine counting process, we use 250 MHz for 4 channel phase shift to get 250 MHz (0°), 250 MHz (90°), 250 MHz (180°) and 250 MHz (270°) as 1GHz high frequency clocks. In the process of coarse and refined multi-stage clock, we need to focus on phase alignment, which includes two aspects: First, we need to introduce the clock signal into the GCLK global clock network, where the time delay is strictly equal, and the phase alignment can always be realized at 1 GHz, 250 MHz and 62.5 MHz of the clock; Second, in the middle count of 250 MHz to the coarse count of 62.5 MHz low speed clock carry, the two modules are required to be as close as possible, with small delay, and the internal layout and wiring of FPGA should be strictly equal in length, otherwise the timing will be wrong (TDC to 250 MHz has no carry, the two are relatively independent, Long range counting from 250 MHz to 62.5 MHz requires a carry).

5 The Results of Experiments

The following test signals are extracted from the timing circuit: time series signal, main wave signal, GPS second pulse signal and echo signal. GPS second pulse signal, main wave signal and time series signal are observed on the oscilloscope as shown in Fig. 4 below (the yellow signal in channel 1 is GPS second pulse signal, the red signal is time series signal in channel 2, and the blue signal is main wave signal in channel 3) (Fig. 9, Fig. 10):

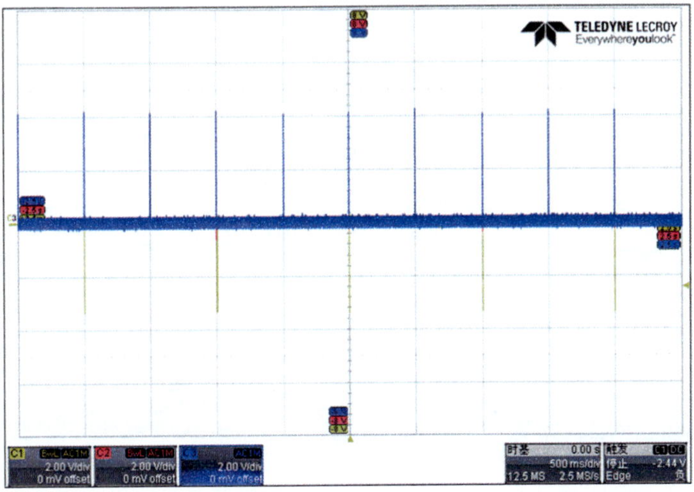

Fig. 6. GPS second pulse signal, main wave signal and time series signal location map

Through the signal source, the delay of main wave and echo is adjusted from 0ns to 33.33ms. The data jump shown in the measured distance data box is always within ± 1ns, which meets the requirement of 1ns interval measurement resolution. According to the ground inspection station, the subsecond time corresponding to the first main wave (measured whole second time) is 001d, and the corresponding decimal number is 464 us. It can be seen from Fig. 5 that the delay of the measured first main wave relative to the GPS second pulse is $\Delta X = 473.756$ us, and the difference is 9.756 us, which meets the requirements within 30 us. The sub-second time (measured whole second

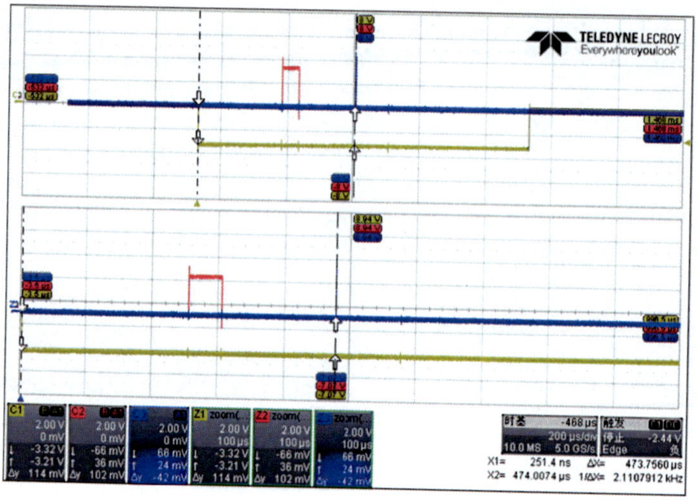

Fig. 7. GPS second pulse signal, first main wave signal and time signal relative position map

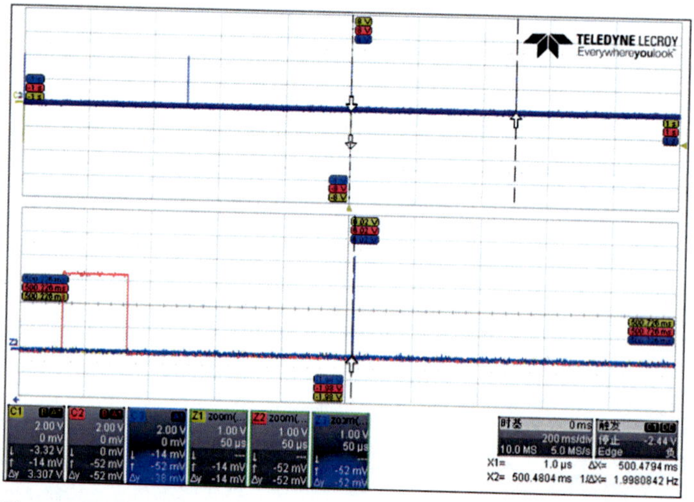

Fig. 8. GPS second pulse signal, second main wave signal and time signal relative position map

time) corresponding to the second main wave is 7a2f, and the corresponding decimal number is 500.464 ms. It can be seen from Fig. 6 that the delay of measured second main wave relative to GPS second pulse is $\Delta X = 500.4794$ ms, with a difference of 15.4 us, which meets the requirements within 30 us. Can meet the requirements of punctuality accuracy 30 us. Since the clock used to acquire the second pulse is 2 MHz, the main wave will have hopping within 500 ns. According to Fig. 7 and Fig. 8, the leading edge of the main wave at the leftmost end and the leading edge of the main wave at the rightmost

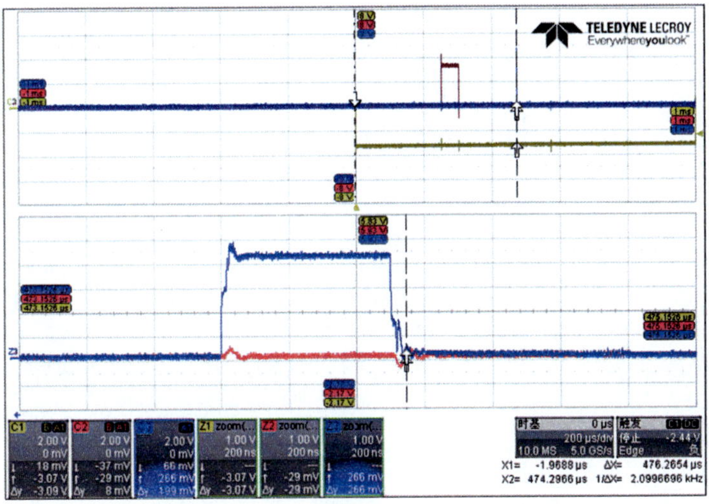

Fig. 9. Position diagram of the leftmost end of the main wave signal jump

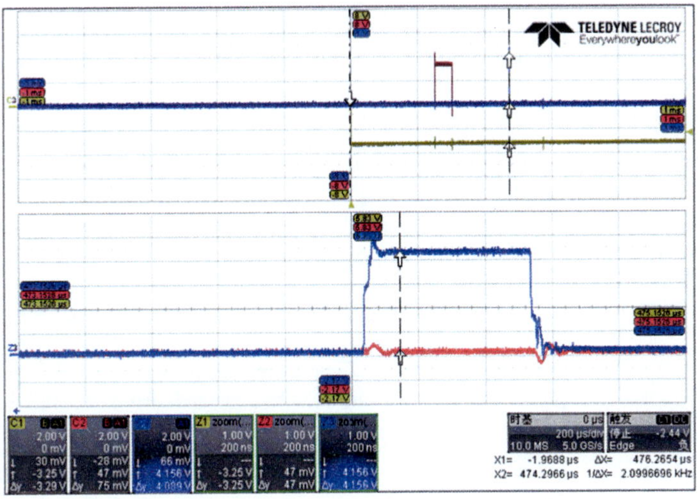

Fig. 10. Position diagram of the rightmost end of the main wave signal jump

end can be found to have hopping within 2.5 cells, and each cell can be found to be 200 ns from the figure, so the main wave hopping within 500ns meets the requirements.

6 Conclusion

In this paper, the working principle, hardware circuit design and software design of high resolution time interval measurement are described. By using the temperature compensated differential 100 MHz crystal oscillator, the time resolution, punctuality accuracy and main wave jump are studied. It can be seen that the frequency in FPGA first-level DCM (Digital Clock Managers) is partially 50 MHz by external 100 MHz temperature replenished crystal oscillator, and then the frequency is doubled to 250 MHz. In the second stage DCM, four phases of 250 MHz are shifted to obtain that 250 MHz (0°), 250 MHz (90°), 250 MHz (180°) and 250 MHz (270°) are equivalent to 1 GHz. Thus, the 1GHz clock method is feasible. The time-measuring resolution of 1ns, the punctuality accuracy of better than 30 us and the main wave hop within 500 ns can be obtained. Moreover, the FPGA and the thermal crystal oscillator have the corresponding components of high grade (aerospace class), so it can be applied to the field of spaceborne laser ranging, spaceborne laser altimeter and spaceborne three-dimensional Lidar. Follow-up research focuses on the realization of high precision time measurement data processing technology using full waveform sampling technology to obtain echo signal.

References

1. Jianhui, S.O.N.G., Feng, Y.U.A.N., Zhenliang, D.I.N.G.: High precision time interval measurement in pulsed laser ranging. Opt. Precis. Eng. **17**(5), 1046–1050 (2009)
2. Zhang Xiaopeng, X., Peng, Z.M., et al.: Study on laser imaging system experiment. Spacecraft Recovery Remote Sens. **30**(2), 62–69 (2009)
3. Wuhuang, H., Houjun, W., Peng, Y.: Research on high-speed full-waveform acquisition and high-precision positioning technology in laser remote sensing. Chin. J. Sci. Instrum. **34**(1), 73–79 (2013)
4. Huang, G., Lan, X.: High accuracy full waveform digitizing technology of spaceborne laser altimeter. Spacecraft Recovery Remote Sens. **41**(2), 39–46 (2020)
5. Yifan, W.A.N.G., Xin, X.U., Yidong, T.A.N.: Laser precision ranging technology. Spacecraft Recovery Remote Sens. **42**(6), 22–33 (2021)
6. Linjun, L.E.I., Yan, Y.A.N.G., Weibiao, C.H.E.N.: High-accuracy time-interval-unit for spaceborne laser range finder. Chin. J. Lasers **34**(10), 1422–1426 (2007)
7. Gang, W.U., Chun-Lai, L.I., Yin-Nian, L.I.U., et al.: Study on high resolution time interval measurement module in pulsed laser ranging system. J. Infrared Millimeter Waves **26**(3), 213–216 (2007)
8. Lei, Z., Geng-Hua, H., Jun-Hua, O., Rong, S., Jian-Yu, W.: Study on time interval measurement system in photon counting imaging lidar. J. Infrared Millimeter Waves **27**(6), 461–464 (2008)

Satellite Imagery Superresolution Based on Optimal Frame Accumulation

Sergey A. Stankevich[1(✉)], Mykhailo O. Popov[1], Sergiy V. Shklyar[1], Artur R. Lysenko[1], Artem A. Andreiev[1], Kun Xing[2], Shixiang Cao[2], and Rui Tao[2]

[1] Scientific Centre for Aerospace Research of the Earth, National Academy of Sciences of Ukraine, Kiev 01054, Ukraine
st@casre.kiev.ua

[2] Beijing Institute of Space Mechanics and Electricity, Beijing 100190, China

Abstract. As actual spatial resolution continues to be a primary bottleneck for satellite imagery interpretation, the aim of resolution enhancement remains very urgent. The resolution enhancement is especially important for medium- and high-resolution satellite systems targeted at small-size objects observation. One of the efficient methods for satellite imagery spatial resolution enhancement is superresolution, when several subpixel-shifted relative to each other low-resolution frames are combined into a single image that accumulates information from all input ones.

Our team has previous experience implementing this kind of superresolution in satellite imagery preprocessing. A twofold formal and 48% actual resolution enhancement has been achieved by combining two to four frames into a single image. However, the potential of the formal resolution duplication model is exhausted now.

Therefore, we propose scaling up the discretization of the joint subpixel grid over the input imageset from one-half to one-third of the pixel. Accordingly, a ninefold increase in the number of subpixels requires additional information, which is obtained from a larger number of input frames. At the moment, some remote sensing satellite systems can generate many highly-overlapping frames, such as SkySat-1/2, Gaofen-1/6, Gaofen-4/13, Jilin-1, etc. As a result, too many frames are available, which leads to the problem of selecting their optimal subset. We solve this problem by optimizing the joint subpixel grid position and achieving the maximum signal-to-noise ratio in the resulting image using the matrix of mutual subpixel shifts of the input frameset.

Keywords: resolution enhancement · superresolution · subpixel shift · frame accumulation · signal-to-noise ratio · optimal subset

1 Introduction

In a great measure, spatial resolution determines the overall informativeness of satellite images, especially under the analysis of small-size objects [1, 2]. Although its own optimal resolution is intrinsic to each remote sensing application [3], in many cases

H. P. Urbach and H. Jiang (Eds.): ISSOIA 2022, SPPHY 295, pp. 395–412, 2023.
https://doi.org/10.1007/978-981-99-4098-1_35

it is not possible to achieve it with the current particular satellite system. Technically, the resolution enhancement of the on-board imager is quite difficult. To solve one, an advanced sensor array with a huge number of photodetectors and a perfect optical lens of a multi-meter focal length is required [4]. Such systems are high-mass, oversized and very expensive [5]. Existing space instrumentation progress is often already at the forefront of modern technologies, so the further extensive improvement of their performance is usually impossible [6]. Satellite imaging hardware resolution enhancement is almost always unacceptable under modern requirements of high economic efficiency of remote sensing [7].

However, there may be another way to improve the spatial resolution of satellite imagery. It is referred to the software superresolution. A large number of varieties of superresolution techniques are known. All of them in any form involve the external additional information injection into the original image [8]. First, superresolution techniques are classified according to the presence or absence of a reference image. Reference image-based methods backed by either a higher resolution image of the same scene or, in a more complex case, some general knowledge obtained from the analysis of multiple images of similar scenes [9]. Recently, many studies have applied convolutional neural networks for superresolution [10, 11]. The top class of reference-based superresolution is the automatic extraction of information from several low resolution images of the same scene, followed by its fusion in the form of a single enhanced resolution image [12].

The key parameters of multi-image superresolution are: (i) the number of input images; (ii) a matrix of mutual subpixel shifts between ones; (iii) the formal resolution enlargement ratio; (iv) the method and chosen algorithm for superresolution and (v) the criterion used to evaluate the output image. Previously, in our superresolution approach the following parameter values were used: the number of input images is 2–4, the number of output subpixels per input pixel is 2 × 2, and the method for enhanced resolution image restoration is the Gaussian regularization within a sliding window [13]. The input images are co-registered with each other within integer-pixel and then cropped to produce a joint multi-image scene. A special software module calculates the matrix of mutual subpixel shifts of cropped images [14]. Over the set of input images, a subpixel-shifted double-resolution joint grid is determined optimally so that the minimum number of input pixels and the grid's subpixels overlap. All input images are interpolated into a joint grid and interlaced with each other. As a result, the single blurred double-resolution image is derived. The rest of the processing is performed in the frequency domain. First, statistical regularization and frequency filtering are carried out. For this purpose, the Fourier transforms of autocovariance of input images and noise are evaluated. After the inverse Fourier transform of the processed output, the superresolution image is acquired.

In our updated approach to superresolution, the following changes were made: (i) the enlargement ratio was raised to 3 × 3; (ii) the number of input images should be large: not limited, but more than 4; (iii) the procedure for selecting the optimal subset of images from all input ones according to the signal-to-noise ratio of the output superresolution image has been introduced. As before, the final evaluation of the output image is performed by the actual pixel resolution, taking into account the formal enlargement ratio.

2 Method

Since the potential possibilities of the previously used model of doubling the formal resolution have been exhausted, further scaling up of the enlargement ratio was proposed. General dataflow for superresolution of subpixel-shifted satellite imageset is shown in Fig. 1.

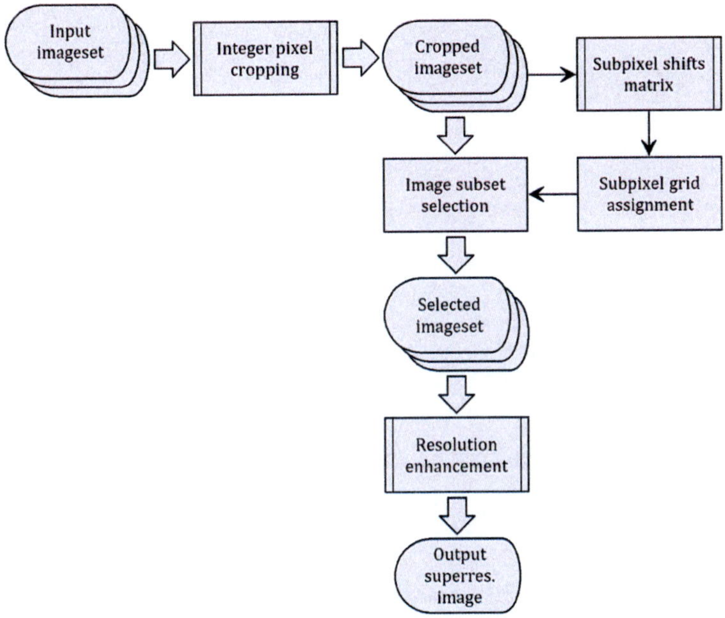

Fig. 1. General dataflow for superresolution of subpixel-shifted satellite imageset.

Each image from the input imageset must meet such requirements as: the consistency of scale, the prohibition of angular misalignment, the stationarity, and the absence of moving objects. Also, input images should be obtained in close (ideally–in the same) spectral bands, and the area of their overlap should be at least 90%.

Below, the main steps of the input imageset processing are discussed in more detail.

2.1 Integer Pixel Image Cropping

Since input images could be of different sizes and not co-registered, the procedure of integer pixel cropping is necessary. This procedure consists of three stages, namely: (i) pairwise estimation of integer pixel shift of input images and compilation of shift matrices along both X and Y axes; (ii) reference image assignment to minimize cropping area; (iii) co-registration of all input images by their cropping.

In the first stage, the fast Fourier transform (FFT) and the sliding sum algorithm is applied to calculate the mutual correlation function to determine the integer shift between pair of images. Such an approach could be used for images of different sizes [15]. To simplify the calculations, the one-dimensional case is considered below.

Let's suppose there are long and short sets named $\{ y_i\}$ and $\{ x_i\}$, respectively. Then, $\{ y_i\}$ consists of n samples, and $\{ x_i\}$ consists of m samples, where $m < n$. Thus, for the kth samples shift, the r_k correlation value will be:

$$r_k = \frac{\sum_{i=1}^{m} (x_i - \overline{x}) \cdot (y_{i+k} - \overline{y}_k)}{\sqrt{\sum_{i=1}^{m} (x_i - \overline{x})^2 \sum_{i=1}^{m} (y_{i+k} - \overline{y}_k)^2}} \quad (1)$$

where $\overline{x}$ is a $\{ x_i\}$ set mean, and $\overline{y}_k$ is a $\{ y_i\}$ set mean for m samples.

The numerator in (1) determines the covariance c_k. Since the correlation value does not depend on means, the covariance can be written as:

$$c_k = \sum_{i=1}^{m} x_i \cdot y_{i+k} - \frac{1}{m} \sum_{i=1}^{m} x_i \cdot \sum_{i=1}^{m} y_{i+k} \quad (2)$$

The sum of the $\{ x_i\}$ set in the subtracted product term from Eq. (2) can be calculated in advance. The partial sum of a $\{ y_i\}$ set is calculated quickly by the moving total algorithm. Even if the length of the short set is increased up to the length of the long one, assigning zeros to additional samples, then the Eq. (2) expression will not change; just the superior limit of the sum will be changed from m to n. The FFT can be used to calculate the $\sum_{i=1}^{m} x_i \cdot y_{i+k}$ value in Eq. (2).

The denominator in Eq. (1) is the root of the product of the variances. Both short and long sets variances can be calculated in the same way as for the case of covariance calculation.

The described method is performed N–1 times with the number of images N. After that, each image pairs shifts form the matrix D_X, which is the shift matrix along the X-axis:

$$D_X = \begin{pmatrix} \Delta x_{11} \ldots \Delta x_{1N} \\ \ldots \\ \Delta x_{N1} \ldots \Delta x_{NN} \end{pmatrix}$$

where Δx_{ij} is the shift of image i referring to image j along the X axis.

In the second stage, the virtual pixel grid minimizes the integer pixel shift between the input images and the reference image. In turn, it minimizes the cropped-out size of input images. To determine such a reference image, the sum of the minimum and maximum shifts for each row of the D_X matrix is calculated. The smallest sum assigns the reference image.

At the last stage, images are co-registered by integer pixel cropping. Then, each initial image X_i is transformed to the cropped and co-registered one X_i' by selecting start and end coordinates:

$$X_i' = X_i[\max(\Delta x_r) - \Delta x_{ri} : \eta_i - \Delta x_{ri} - \max(\Delta x_r)]$$

where r is an index number of the reference image, Δx_{ri} is the shift along the X axis of i image referred to the reference image (Δx_{ri} corresponds to element (r, i) of matrix D_X), Δx_r is a set that contains the shift values of all images referred to the reference image (Δx_r corresponds to row r of matrix D_X), n_i is a number of pixels of i image along the X axis.

Integer pixel cropping along the Y axis implies the same operations described above.

2.2 Subpixel Shift Assessment

Let's suppose that pair of images are presented as the original one and a parallel transfer from the original one. Thus, the subpixel shift value equals this parallel transfer's vector.

The following assumptions base the general algorithm for determining the subpixel shift between pair of images [14]: (i) the maximum correlation between the input images is achieved by the true shift value; (ii) to reduce the computational burden, the processing is performed in the frequency domain; (iii) the interpolation of the discrete image is performed by suppression of the high-frequency component.

The input data are two Y_1 and Y_2 images of $m \times n$ dimension at (x, y) points, $x = 0, 1, \ldots, m{-}1$, $y = 0, 1, \ldots, n{-}1$. The pixel size is selected as a linear unit.

The following equation computes discrete Fourier transform (DFT):

$$\hat{Y}_k(\theta, \vartheta) = \sum_{x=0}^{m-1}\sum_{y=0}^{n-1} Y_k(x, y)e^{-2\pi i(\theta x+\vartheta y)}$$

where $k = 1$ or $k = 2$ under $\theta = \ldots, -\frac{2}{m}, -\frac{1}{m}, 0, \frac{1}{m}, \frac{2}{m}, \ldots$ and $\vartheta = \ldots, -\frac{2}{n} -\frac{1}{n}, 0, \frac{1}{n}, \frac{2}{n}, \ldots$. $\hat{Y}_k(\theta, \vartheta)$ values will be considered for $|\theta| < \frac{1}{2}$, $|\vartheta| < \frac{1}{2}$ only.

The influence of the low-frequency image component is reduced by multiplying the Fourier transform result by the $\omega(\theta)\omega(\vartheta)$ function:

$$\hat{F}_k(\theta, \vartheta) = \hat{Y}_k(\theta, \vartheta)\omega(\theta)\omega(\vartheta)$$

where $\omega(\theta) = 1$ under $-\frac{1}{4} \le \theta \le \frac{1}{4}$; $\omega(\theta) = 2 - 4|\theta|$ under $\frac{1}{4} \le |\theta| \le \frac{1}{2}$; $\omega(\theta) = 0$ under $|\theta| \ge \frac{1}{2}$.

The corresponding transform in the spatial domain is the image interpolation by convolution with the const $\cdot \left(\cos\frac{\pi x}{2} - \cos\pi x\right)\left(\cos\frac{\pi y}{2} - \cos\pi y\right)x^{-2}y^{-2}$ function. This transform is not an exact interpolation because it changes the function values at the points where the function is defined. To ensure that the function-defined values do not change, it is recommended to take under $-\frac{1}{3} \le \theta \le \frac{1}{3}$; $\omega(\theta) = 2 - 3|\theta|$ under $\frac{1}{3} \le |\theta| \le \frac{2}{3}$; $\omega(\theta) = 0$ under $|\theta| \ge \frac{2}{3}$.

As a result of the parallel transfer of the $Y_1(x, y)$ image onto $(\Delta x, \Delta y)$, that is $Y_2(x, y) = Y_1(x + \Delta x, y + \Delta y)$, its Fourier transform changes as follows:

$$\hat{Y}_2(\theta, \vartheta) = \hat{Y}_1(\theta, \vartheta)e^{2\pi i(\theta\Delta x+\vartheta\Delta y)} \tag{3}$$

while the pixels that are on the image edge are ignored. Fourier transform, defined by Eq. (3), corresponds to a convolved image shift: the image part at the edge is transferred to the opposite edge.

To calculate the correlation between the images, the Parseval equation is used: if $\hat{Y}_k(\theta, \vartheta)$ is the Fourier transform of the $Y_k(x, y)$ function defined for $0 \le x \le m,\ 0 \le y \le n$, that is $\hat{Y}_k(\theta, \vartheta) = \int_0^m \int_0^n Y_k(x, y)e^{-2\pi i(\theta x + \vartheta y)}dxdy$, then

$$\int_0^m \int_0^m Y_1(x, y)Y_2(x, y)dxdy = \frac{1}{m \cdot n}\sum_{\theta=\ldots,\frac{-1}{m},0,\frac{1}{m},\ldots}\sum_{\vartheta=\ldots,-\frac{1}{n},0,\frac{1}{n},\ldots}\hat{Y}_1(\theta, \vartheta)\hat{Y}_2(\theta, \vartheta)$$

The Fourier transform of the second image shifted by $(\Delta x, \Delta y)$ after suppressing the high-frequency component will be

$$\hat{F}_2^{(\Delta_x,\Delta_y)}(\theta, \vartheta) = \hat{F}_2(\theta, \vartheta)e^{2\pi i(\theta\Delta x + \vartheta\Delta y)}$$

The objective function is defined as

$$\begin{aligned} Q(\Delta x, \Delta y) &= \sum_{\theta=\ldots,-\frac{1}{m},0,\frac{1}{m},\ldots}\sum_{\theta_1=\ldots,\frac{-1}{n},0,\frac{1}{n},\ldots}\hat{F}_1(\theta, \vartheta)\hat{F}_2^{(\Delta x, \Delta y)}(\theta, \vartheta) \\ &= \sum_{k=[m/2]}^{[m/2]}\sum_{l=-[n/2]}^{[n/2]}\hat{F}_1\left(\frac{k}{m}, \frac{l}{n}\right)\hat{F}_2\left(\frac{k}{m}, \frac{l}{n}\right)\exp\left[2\pi i\left(\frac{k\Delta x}{m} + \frac{l\Delta y}{n}\right)\right] \end{aligned}$$

A complete algorithm includes the following steps:

1) preparing for the $Q(\Delta x, \Delta y)$ function calculation. In this case, the $\hat{F}_1(\theta, \vartheta)\hat{F}_2(\theta, \vartheta)$ is calculated under $\theta = \frac{k}{m}$, $k = -\left\lfloor\frac{m}{2}\right\rfloor, \ldots, \left\lfloor\frac{m-1}{2}\right\rfloor$, $\vartheta = \frac{l}{n}$, $l = -\left\lfloor\frac{n}{2}\right\rfloor, \ldots, \left\lfloor\frac{n-1}{2}\right\rfloor$, and the necessary auxiliary arrays are filled;
2) finding the $Q(\Delta x, \Delta y)$ maximum by $\Delta x_1 = 0, \frac{1}{t}, \ldots, \frac{t-1}{t}$ and, $\Delta y_1 = 0, \frac{1}{t}, \ldots, \frac{1-t}{t}$, where t is a step of the iteration, and the objective function $Q(\Delta x_0 + \Delta x_1, \Delta y_0 + \Delta y_1)$ is calculated simultaneously for all $\Delta x_0 = -\left[\frac{m}{2}\right], 1 - \left[\frac{m}{2}\right], \ldots, \left[\frac{m-1}{2}\right]\Delta y_0 - \left[\frac{n}{2}\right], 1 - \left[\frac{n}{2}\right], \ldots, -\left[\frac{n-1}{2}\right]$.
3) Numerical maximization of the $Q(\Delta x, \Delta y)$ function, namely selection values of $\Delta x_0, \Delta x_1, \Delta y_0, \Delta y_1$ variables to reach the maximum in step 2.

2.3 Reference Grid Assignment

Here we align the superresolution image being constructed. We apply the method used in [13], aligning the superresolution image being constructed such that the edges of the input images are close to the edges of the superresolution image. The alignment parameter c_0 is chosen to maximize $\sum_{n=1}^{N}\cos\left(2\pi\frac{c_0 - c_n}{\Delta_{\text{fine}}}\right)$ if the superresolution factor h is an odd integer and to minimize it if h is an even integer. The optimizer is found explicitly

$$c_0 = \frac{\Delta_{\text{fine}}}{2\pi}\arg\left((-1)^{h-1}\sum_{n=1}^{N}\exp\left(\frac{2\pi i c_n}{\Delta_{\text{fine}}}\right)\right)$$

where arg(·) is the argument of a complex number.

The coordinates of the pixel centers of the input images are denoted $c_n, n = 1, \ldots, N$, while the pixel center of the grid where the superresolution images are being constructed is denoted c_0 (Fig. 2).

For 2D images, the optimal alignment parameters for each coordinate are evaluated independently.

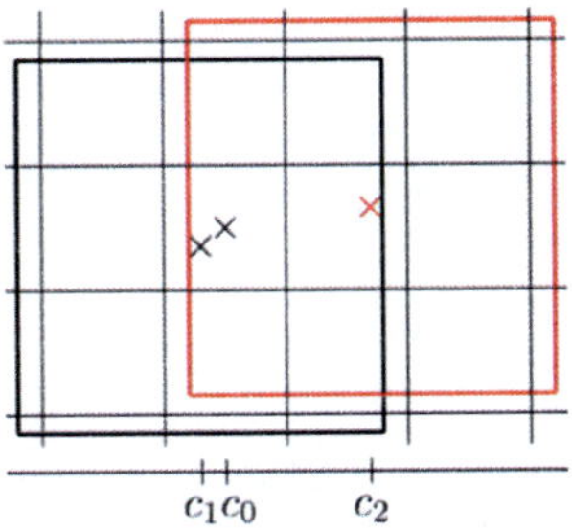

Fig. 2. Reference grid alignment

2.4 Optimal Selection of Input Images Subset

The optimal selection of the image subset aims to increase the signal energy and reduce the noise energy [16, 17]. Image energy functions are described below.

The total signal energy and noise energy that correspond to spatial frequencies $\theta, ..., \theta + \frac{h-1}{\Delta_{gr}}$ are equal to

$$PD_{Sig}(\theta) = trace(\mathrm{cov}_{Sig}(\underset{est}{\vec{X}}(\theta)))$$

$$PD_{Noise}(\theta) = trace(\mathrm{cov}_{Noise}(\underset{est}{\vec{X}}(\theta)))$$

Hence, the criterion functions are

$$P_{Sig}(\theta) = \int_0^{1/(2\Delta_{gr})} PD_{Sig}(\theta)d\theta$$

$$P_{Noise}(\theta) = \int_0^{1/(2\Delta_{gr})} PD_{Noise}(\theta)d\theta$$

Thus, the optimization criterion is considered as a ratio of the signal energy to the noise energy, namely $SNR = \frac{P_{Sig}(\theta)}{P_{Noise}(\theta)}$

For the task, the pseudo-gradient search for optimization was used. It implies the iterative discarding of one image to achieve the maximum criterion value of the subset. In detail this procedure could be described in three steps. The first one is to create *n* different subsets of *n*-1 images, taking *n* as the number of images in the current iteration. The second step engages in calculating the criterion value for each obtained subset. And the final step is to define the subset with the maximum criterion value among others. Each next iteration implies the described steps with the defined subset. As a result, the final iteration assigns the optimal input images subset.

The optimization procedure stops in two cases. The first is when the iteration reduces the criterion value concerning the previous iteration. The second case is when all possible subsets have already been considered, and no images are left to discard.

2.5 Superresolution Algorithm

The superresolution algorithm is the same as described in [13]. However, there are some differences. Namely, the aligned shift matrix *A*, which is projected onto the joint grid, is evaluated using the next formula:

$$A_{1j}^{\sin} \textstyle\sum_i \sin[(6\pi M)ij], A_{1j}^{\cos[(6\pi M)ij]}, i = \overline{1, N}, j = \overline{1, 2}$$

$$A_{1j} = \frac{\mathrm{arctg}(A_{1j}^{\sin}, A_{1j}^{\cos})}{6\pi}, j = \overline{1, 2}$$

Here M_{ij} is the mean shift matrix between all *N* source images. The superresolution image model is changed in the following way:

$$G(\Delta y, \Delta x) = \begin{pmatrix} (1-\Delta y)(1-\Delta x) & (1-\Delta y) & (1-\Delta y) & (1-\Delta y)\Delta x \\ 1-\Delta x & 1 & 1 & \Delta x \\ 1-\Delta x & 1 & 1 & \Delta x \\ \Delta y(1-\Delta x) & \Delta y & \Delta y & \Delta y\Delta x \end{pmatrix}$$

where $0 \le \Delta y \le 1$and $0 \le \Delta x \le 1$.

The frequency domain transfer function:

$T(\eta, \xi) = (1 - \Delta y + e^{2\pi i\eta/m} + e^{4\pi i\eta/m} + \Delta y e^{6\pi i\eta/m}) \times (1 - \Delta x + e^{2\pi i\xi/n} + e^{4\pi i\xi/n} + \Delta x e^{6\pi i\xi/n})$ for $0 \le \Delta y \le 1, 0 \le \Delta x \le 1$ and

$$\begin{aligned} T(\eta, \xi) &= \left(1 - \{\Delta y\} + e^{2\pi i\eta/m} + e^{4\pi i\eta/m} + \{\Delta y\}e^{6\pi i\eta/m}\right)e^{2\pi i\eta[\Delta y]/m} \\ &\times (1 - \{\Delta x\} + e^{2\pi i\xi/n} + e^{4\pi i\xi/n} + \{\Delta x\}e^{6\pi i\xi/n})e^{2\pi i\xi[\Delta x]/n} \end{aligned}$$

for general case, where $\Delta y = [\Delta y] + \{\Delta y\}$, $[\Delta y]$ is an integer and $0 \le \{\Delta y\} \le 1$, and a similar decomposition holds true for Δx.

The linear equations for the superresolution model now take the following form:

$$9\hat{Y}_k(\eta, \xi) = \sum\nolimits_{i=-1}^{1} \sum\nolimits_{j=-1}^{1} T_k(\eta + im, \xi + jn)\hat{X}(\eta + im, \xi + jn) + 9\hat{E}_k(\eta, \xi)$$

where $\hat{Y}_k$ is the Fourier transform of the *k*th input image, $\hat{E}_k$ is the Fourier transform of the errors in the *k*th input image and $\hat{X}$ is the Fourier transform of the superresolution image.

3 Implementation

The following implementation tools were used for the given approach: Python 3.10, C++ (ISO C++ 17 Standard). Python was used as a scientific and test tool, considering being an interpretive and flexible programming language with a high amount of free open-source third-party libraries. The following libraries were used in the development process: Numpy v1.23.3, Scipy v1.9.1, and Pillow v9.2.0. The development process is explained in Fig. 3.

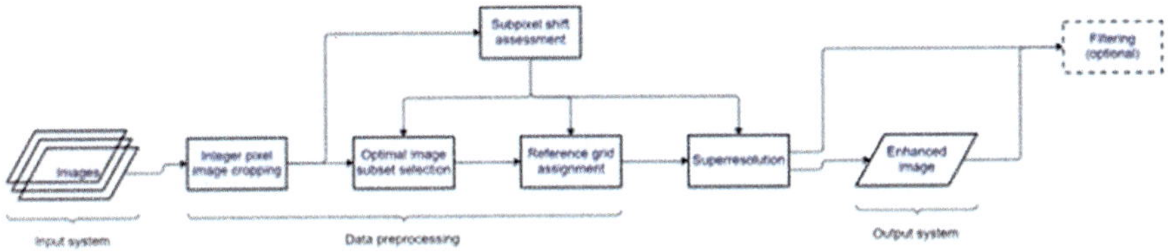

Fig. 3. Image spatial resolution enhancement flowchart.

The software product consists of the following modules: graphical user interface (GUI), source images integer pixel precision cropping module, input images optimal set selector and input satellite images spatial resolution enhancement algorithm.

A user interacts with the GUI, which contains the data input/output module. The data input module provides data transmission from the user to the data processing system, which carries out the integer pixel precision image cropping of the source images, leaving only a common imaging scene, and selecting an optimal image set of the input images. Next, the processed data is forwarded to the spatial resolution enhancement module, and its result is passed to the data output module, which saves the result in the user-selected file. The software structure is shown in Fig. 4.

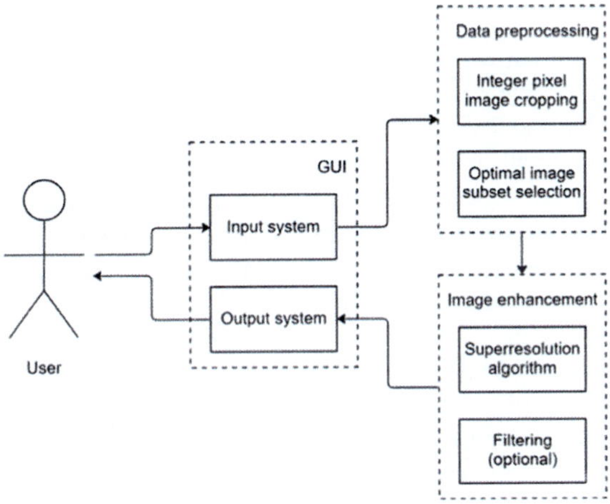

Fig. 4. Software structure.

Table 1 describes the tools used in developing the software components presented in Fig. 5.

The software modules are partially (not including input images optimal set selector module) translated to the C++ programming language with the engagement of the free open-source computer vision library OpenCV v4.5.0 resulting in a significant processing time reduction. Complete program code translation from Python to C++ is planned.

Table 1. Software modules development tools.

Module name	Development tool
Input module	Pillow
Output module	Pillow
GUI	Tkinter
Integer pixel image cropping module	Numpy, Scipy
Optimal input image set selector	Numpy
Image resolution enhancement module	Numpy, Scipy

4 Evaluation

The primary task of evaluating the developed solution is quantifying the spatial resolution enhancement achieved. Therefore, the evaluation should be performed experimentally, using test satellite imagesets. However, not every imageset is suitable for this purpose. Thus, the requirements for the input imageset must be claimed first.

4.1 Input Imageset Requirements

The proposed resolution enhancement algorithm's main component is subpixel shift estimation. Thus, specific requirements must be placed on the input satellite images to ensure valid processing. The first one is a unique approximation of the subpixel shift joint for the whole input images with some permissible accuracy. In turn, it implies substantial restrictions on the geometry of input images, namely the consistency of scale over the field of image, the prohibition of angular misalignment of input images, and minimizing the higher-order non-linear distortions over the image field, such as optical disturbances, errors of imaging rasterization, etc. [18].

The second requirement is the stationarity of input images. It implies the absence of significant changes between the images of the same scene during acquisition, for example, the presence of moving objects within the scene. In particular, images with a large area of cloud cover are not recommended to be included in the input imageset because the image's cloudiness can significantly distort the shift estimation results. Also, pairs of images obtained with a long-time interval are unsuitable for the input imageset due to significant changes in the scene, such as inter-seasonal images [19].

The third requirement is that input images should be obtained in the same or close spectral bands. Significant radiometric differences between bands may cause miscalculations or large errors in subpixel shift estimation [20].

To ensure the constancy of subpixel shift value over the field of input images, the most advanced or sophisticated co-registering techniques based on affine or non-linear spatial transforms may be necessary [21, 22].

4.2 Image Evaluation Toolkit

Spatial resolution should be considered the most important factor of satellite image informativeness [23]. Spatial resolution is a stable objective quantitative criterion for evaluating both the input and output satellite image [24]. The evaluation of the output superresolution images should be based on the achieved actual spatial resolution, taking into account the formal one. However, evaluating the actual spatial resolution of digital images is not an easy task.

A few different approaches to evaluating the spatial resolution of digital images are known. However, ground-based test objects' technical complexity and significant expensiveness for determining the satellite image resolution coerce to engage natural formations and artificial structures on the land surface for this purpose. Usually, a probabilistic assessment of the object detection in the image is usually implemented [25]. Similarly, a probabilistic approach can be applied to determine the spatial resolution of digital images [26].

In this study, the parameterization of the modulation transfer function (MTF) is applied following the results of automatic edge spread function (ESF) extraction from a digital image, as described in [27]. The actual resolution of the satellite images was estimated by the Gaussian approximation of the bidirectional MTF, which was automatically determined over the digital image using special software (Fig. 5) developed with SciLab open-source computational environment.

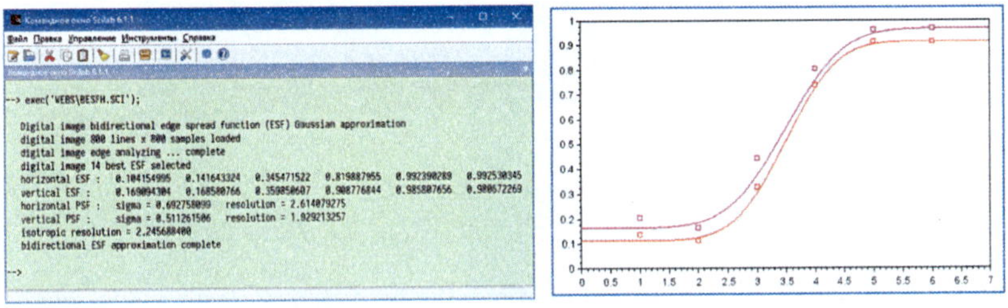

Fig. 5. Screenshot of software for estimating the digital image's actual pixel resolution

After running the software, we obtain Gaussian approximations of the bidirectional ESF (picture right part) and the isotropic pixel resolution of the image under evaluation (left one).

4.3 Test Satellite Image Superresolution Results

The GF03C image series were used for software efficiency testing purposes. The data was acquired by the Jilin-1 satellite on 17 January 2021. From 20 source images, 6 ones were selected as an optimal input image set. Input integer pixel precision optimal image set is shown in Fig. 6*a*–6*f*. Cropped image size is 463 $\times$ 358 pixels each.

Table 2 contains mutual subpixel shifts of the test imageset Fig. 6.

The result of a 3^x formal resolution enhancement of the input imageset is shown in Fig. 7.

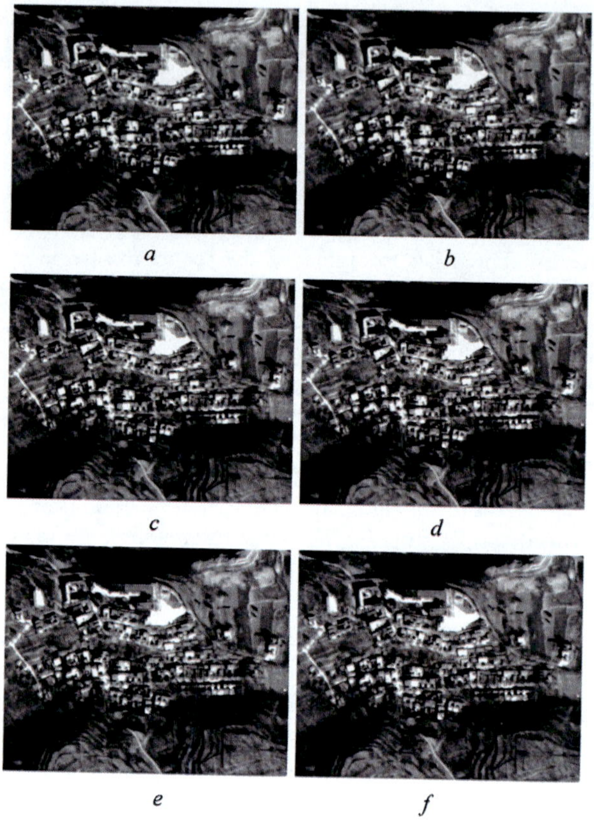

Fig. 6. Input, cropped with pixel precision satellite images:
a–JL1GF03C03_L1A_MSS_20210117100546_200037996_101_0050_001
b–JL1GF03C03_L1A_MSS_20210117100546_200037996_101_0052_001
c–JL1GF03C03_L1A_MSS_20210117100546_200037996_101_0055_001
d–JL1GF03C03_L1A_MSS_20210117100546_200037996_101_0058_001
e–JL1GF03C03_L1A_MSS_20210117100546_200037996_101_0061_001
f–JL1GF03C03_L1A_MSS_20210117100546_200037996_101_0063_001

Table 2. Subpixel shift of input images relative to the base image Fig. 6*a*.

Image	Figure 6*a*	Figure 6*b*	Figure 6*c*	Figure 6*d*	Figure 6*e*	Figure 6*f*
dx	0	–0.22	–0.25	0.03	0.2	–0.36
dy	0	0.84	0.03	0.35	0.7	0.51

The actual resolution was evaluated by a bidirectional MTF Gaussian approximation, which in turn was retrieved from the corresponding ESFs of the image. The bidirectional ESFs of the input and output images are shown in Fig. 8.

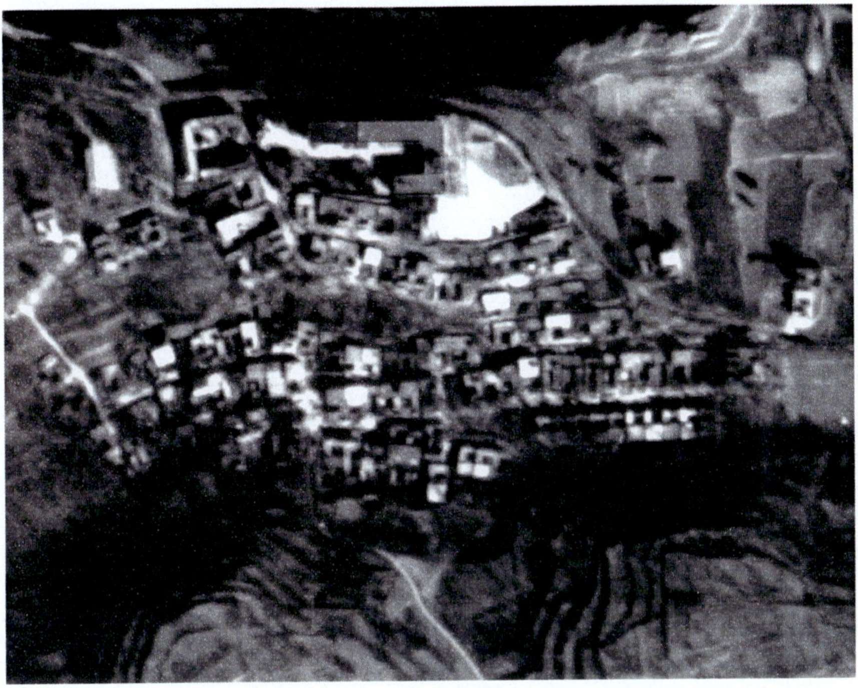

Fig. 7. Enhanced resolution 3^x image of Fig. 5 imageset

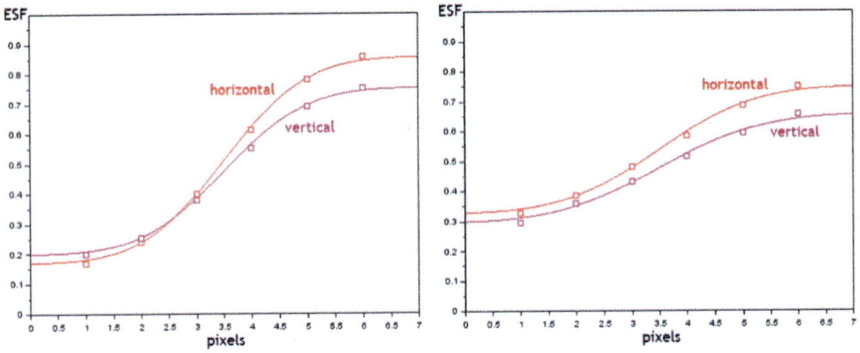

a *b*

Fig. 8. ESF of the input image (*a*) and ESF of the enhanced resolution image (*b*)

The relative enhancement δ in actual resolution is calculated as

$$\delta = \frac{d_y - \frac{d_x}{h}}{d_y} = 1 - \frac{d_x}{h \cdot d_y}$$

where d_y is the actual pixel resolution of the input image, d_x is the actual pixel resolution of the superresolution image, h is the formal multiplicity of subpixel size; in the current case, $h = 3$.

As it shown from Fig. 7, spatial resolution was enhanced by 60.2%. The testing was conducted using the personal workstation with the specifications listed in Table 3.

Table 3. Personal workstation's main specifications.

Specification	Value
CPU frequency	6 core 3.7 GHz
Operational memory	16 GB
Hard drive space	1 TB

The test example processing time was 59 s.

5 Discussion

At first glance, the obtained results are impressive: the achieved increase in the actual resolution of the satellite image practically fits the well-known paradigm of $\delta \approx \sqrt{h}$, where h is the formal resolution multiplicity [28].

At the same time, the formal resolution enhancement by three times only does not always ensure all useful information extraction from the satellite image. Currently, the de facto standard has become an increase the formal resolution of aerospace images by four times [29, 30]. Notably, almost all modern research uses various convolutional neural networks (CNN) for resolution enhancement [31–33].

Besides undoubted progress, neural network technologies in remote sensing encounter several obstacles and limitations: neural network methods provide acceptable results for formal resolution multiplicity and, accordingly, increase the size of the input image up to three times. Further magnification usually leads to the degradation of perceptual image quality [34]. The further improvement of enlarged images is achieved by the generative adversarial network (GAN) engagement, which provide a fourfold multiplicity in formal resolution with acceptable losses of visual quality [35, 36]. In addition, neural network methods for resolution enhancement require extremely large (hundreds of thousands – millions) training samples with manual expert marking of each image. The well-known approach of downsampling the original high resolution images to the size of low resolution ones to create large training samples seems poorly conditioned and does not provide the proper quality of enhanced resolution image restoration [37]. The existing datasets for resolution enhancement with neural networks include mainly small fragments of low quality amateur images [38]. And most importantly, there are no guarantees that the pre-trained neural network will not generate some artefact in the output image, which is absent in reality [39]. To eliminate the superresolution errors in real satellite images enhanced by the neural network, especial methods for controlling the output image quality are involved [40].

That is why our research uses the classic method for an output superresolution image restoration from several subpixel-shifted input ones [13]. As experiments show, the main problem of multi-image subpixel superresolution is the insufficient sharpness of the output image, which worsens its visual perception. Currently, we do not make any special efforts to improve the output image sharpness. However, many algorithms and software tools for such postprocessing are known [41], including also filtering in the frequency domain [42]. Therefore, such efforts may need to be taken in the future.

6 Conclusions

Thus, the previously described method for the spatial resolution enhancement of satellite image based on the joint processing of several input subpixel-shifted low resolution images was further developed. The presented method has undergone a severe modification: first, the formal multiplicity of the subpixel size has been increased triple; second, the optimal accumulation of frames was applied, which made it possible to process almost any number of input images; third, the signal-to-noise ratio expected in the output image was determined as a criterion for optimal selection of frames to be included into the input imageset. In addition, a significant upgrade of the resource-intensive calculation algorithms was carried out, which made it possible to improve data processing speed significantly.

The advantages of the employed approach to the superresolution of satellite images are the objectivity of the results obtained since additional information is extracted only from the original images; reliable mathematical foundation built upon classic statistical regularization; performing most operations in the frequency domain, which simplifies and speeds up heavy computing; independence from extremely large training datasets that require complicated manual preparation.

The results achieved during the tests run generally confirmed our expectations: the actual resolution enhancement reached 60% without the additional inverse filtering in the frequency domain. The developed demo prototype of software has significant reserves, improving the quality of image superresolution and processing performance.

We hope that the results obtained will be useful not only for the processing and analysis of satellite imagery but will also give impetus to the technology development of similar to the proposed one in other areas where image superresolution is required.

Acknowledgement. We sincerely thank CTTC Ltd. Company (Beijing, China) for supporting this study.

References

1. Narayanan, R.M., Desetty, M.K., Reichenbach, S.E.: Effect of spatial resolution on information content characterization in remote sensing imagery based on classification accuracy. Int. J. Remote Sens. **23**(3), 537–553 (2002). https://doi.org/10.1080/01431160010025970
2. Sheikh, H.R., Bovik, A.C.: Image information and visual quality. IEEE Trans. Image Process. **15**(2), 430–444 (2006). https://doi.org/10.1109/tip.2005.859378

3. Kononov, V.I., Stankevich, S.A.: Digital aerospace images with high and low resolution informativity comparative evaluation (in Russian). The Memoirs of Tauridic National University nam. V.I. Vernadsky, **17**(2), 88–95 (2004)
4. Somaie, A.A.: Satellite imaging system. Int. J. Image Process. **7**(3), 266–277 (2013)
5. Maglione, P.: Very high resolution optical satellites: an overview of the most commonly used. Am. J. Appl. Sci. **13**(1), 91–99 (2016). https://doi.org/10.3844/ajassp.2016.91.99
6. Murugan, P., Lakshmi, G., Gautam, V.K.: High resolution optical remote sensing satellites – challenges and techniques. Int. J. Eng. Adv. Technol. **9**(5), 495–502 (2020). https://doi.org/10.35940/ijeat.E9670.069520
7. Dubovik, O., et al.: Grand challenges in satellite remote sensing. Front. Remote Sens. **2**, 619818 (2021). https://doi.org/10.3389/frsen.2021.619818
8. Shorten, C., Khoshgoftaar, T.M.: A survey on image data augmentation for deep learning. J. Big Data **6**(1), 1–48 (2019). https://doi.org/10.1186/s40537-019-0197-0
9. Stankevich, S.A., et al.: Knowledge-based multispectral remote sensing imagery superresolution. In: van Gulijk, C., Zaitseva, E. (eds.) Reliability Engineering and Computational Intelligence. SCI, vol. 976, pp. 219–236. Springer, Cham (2021). https://doi.org/10.1007/978-3-030-74556-1_13
10. Xiao, A., Wang, Z., Wang, L., Ren, Y.: Super-resolution for "Jilin-1" satellite video imagery via a convolutional network. Sensors **18**(4), 1194 (2018). https://doi.org/10.3390/s18041194
11. Kim, S., Jun, D., Kim, B.-G., Lee, H., Rhee, E.: Single image super-resolution method using CNN-based lightweight neural networks. Appl. Sci. **11**(3), 1092 (2021). https://doi.org/10.3390/app11031092
12. Dong, L., Jin, J., Jiang, Y., Zhang, M., Xu, W.: Selection-based subpixel-shifted images super-resolution. IEEE. Access **7**, 110951–110963 (2019). https://doi.org/10.1109/ACCESS.2019.2934498
13. Stankevich, S.A, et al.: Subpixel-shifted satellite images superresolution: software implementation. WSEAS Trans. Comput. **19**(5), 31–37 (2020). https://doi.org/10.37394/23205.2020.19.5
14. Stankevich, S.A., et al.: Estimation of mutual subpixel shift between satellite images: software implementation. Ukrainian J. Remote Sens. **24**, 9–14 (2020). https://doi.org/10.36023/ujrs.2020.24.165
15. Reddy, B.S., Chatterji, B.N.: An FFT-based technique for translation, rotation, and scale-invariant image registration. IEEE Trans. Image Process. **5**(8), 1266–1271 (1996). https://doi.org/10.1109/83.506761
16. Bérubé, J., Wu, C.F.J.: Signal-to-noise ratio and related measures in parameter design optimization: an overview. Indian J. Stat. Series B **62**(3), 417–432 (2000)
17. Krbcova, Z., Kukal, J.: Relationship between entropy and SNR changes in image enhancement. EURASIP J. Image Video Process. **2017**(1), 1–8 (2017). https://doi.org/10.1186/s13640-017-0232-z
18. Dave, C.P., Joshi, R., Srivastava, S.S.: A survey on geometric correction of satellite imagery. Int. J. Comput. Appl. **116**(12), 24–27 (2015). https://doi.org/10.5120/20389-2655
19. d'Angelo, P.: Automatic orientation of large multitemporal satellite image blocks. In: Proceedings of International Symposium on Satellite Mapping Technology and Application (ISSMTA 2013), pp. 1–7. ISPRS, Nanjing (2013)
20. Ferraris, V., Dobigeon, N., Wei, Q., Chabert, M.: Detecting changes between optical images of different spatial and spectral resolutions: a fusion-based approach. IEEE Trans. Geosci. Remote Sens. **56**(3), 1566–1578 (2018). https://doi.org/10.1109/TGRS.2017.2765348
21. Stankevich, S.A.: The models for segmented discrete images automatic matching. In: Proceedings of the 2nd All-Ukrainian International Conference on Signal/Image Processing and Pattern Recognition, pp. 167–169. Institute of Cybernetics NAS of Ukraine, Kiev (1994)

22. Yakubov, G.: Methods and models of geometric rectification of high spatial resolution satellite images. Universum **11**(92), 113–116 (2021). https://doi.org/10.32743/UniTech.2021.92.11.12644
23. Artmann, U., Wueller, D.: Interaction of image noise, spatial resolution, and low contrast fine detail preservation in digital image processing. Proc. SPIE **7250**, 72500I (2009). https://doi.org/10.1117/12.805927
24. Mesner, N., Oštir, K.: Investigating the impact of spatial and spectral resolution of satellite images on segmentation quality. J. Appl. Remote Sens. **8**(1), 083696 (2014). https://doi.org/10.1117/1.JRS.8.083696
25. Hall, D., et al.: Probabilistic object detection: definition and evaluation. In: Proceedings of Winter Conference on Applications of Computer Vision (WACV 2020), pp. 1031–1040. IEEE, Snowmass (2020). https://doi.org/10.1109/WACV45572.2020.9093599
26. Sivyakov, I.N.: Calculating the resolution of optoelectronic systems. J. Opt. Technol. **65**(2), 145–147 (1998)
27. Stankevich, S.A.: Evaluation of the spatial resolution of digital aerospace image by the bidirectional point spread function parameterization. In: Shkarlet, S., Morozov, A., Palagin, A. (eds.) MODS 2020. AISC, vol. 1265, pp. 317–327. Springer, Cham (2021). https://doi.org/10.1007/978-3-030-58124-4_31
28. Zlobin, V.K., Eremeev, V.V.: Aerospace image processing. Physmathlit, Moscow (2006). (in Russian)
29. Hansen, R.S., Waldram, D.W., Thai, T.Q., Berke, R.B.: Super resolution digital image correlation (SR-DIC): an alternative to image stitching at high magnifications. Exp. Mech. **61**(9), 1351–1368 (2021). https://doi.org/10.1007/s11340-021-00729-2
30. Li, Y., Wang, Y., Li, B., Wu, S.: Super-resolution of remote sensing images for ×4 resolution without reference images. Electronics **11**(21), 3474 (2022). https://doi.org/10.3390/electronics11213474
31. Dong, C., Loy, C.C., He, K., Tang, X.: Image super-resolution using deep convolutional networks. IEEE Trans. Pattern Anal. Mach. Intell. **38**(2), 295–307 (2016). https://doi.org/10.1109/TPAMI.2015.2439281
32. Cheng, R., Wang, H., Luo, P.: Remote sensing image super-resolution using multi-scale convolutional sparse coding network. PLoS ONE **17**(10), e0276648 (2022). https://doi.org/10.1371/journal.pone.0276648
33. Wang, X., et al.: A review of image super-resolution approaches based on deep learning and applications in remote sensing. Remote Sens. **14**(21), 5423 (2022). https://doi.org/10.3390/rs14215423
34. Bashir, S.M.A., Wang, Y., Khan, M., Niu, Y.: A comprehensive review of deep learning-based single image super-resolution. PeerJ Comput. Sci. **7**, e621 (2021). https://doi.org/10.7717/peerj-cs.621
35. Singla, K., Pandey, R., Ghanekar, U.: A review on single image superresolution techniques using generative adversarial network. Optik **266**, 169607 (2021). https://doi.org/10.1016/j.ijleo.2022.169607
36. Wang, Q., Zhou, H., Li, G., Guo, J.: Single image super-resolution method based on an improved adversarial generation network. Appl. Sci. **12**(12), 6067 (2022). https://doi.org/10.3390/app12126067
37. Liu, Y., Qiao, Y., Hao, Y., Wang, F.: Single image superresolution techniques based on deep learning: status, applications and future directions. J. Image Graph. **9**(3), 74–86 (2021). https://doi.org/10.18178/joig.9.3.74-86
38. Liu, H., Qian, Y., Zhong, X., Chen, L., Yang, G.-Q.: Research on super-resolution reconstruction of remote sensing images: a comprehensive review. Opt. Eng. **60**(10), 100901 (2021). https://doi.org/10.1117/1.OE.60.10.100901

39. Liu, A., Zeng, X., Yuan, Y., Su, L., Wang, W.: Joint artifact correction and super-resolution of image slicing and mapping system via a convolutional neural network. Opt. Express **29**(5), 7247–7260 (2021). https://doi.org/10.1364/OE.413076
40. Choi, Y., Han, S., Kim, Y.: A no-reference CNN-based super-resolution method for Kompsat-3 using adaptive image quality modification. Remote Sens. **13**(16), 3301 (2021). https://doi.org/10.3390/rs13163301
41. Lin, H.-Y., Lin, C.-Y., Lin, C.-J., Yang, S.-C., Yu, C.-Y.: A study of digital image enlargement and enhancement. Math. Probl. Eng. **2014**, 825169 (2014). https://doi.org/10.1155/2014/825169
42. Dewangan, S., Sharma, A.K.: Image smoothening and sharpening using frequency domain filtering technique. Int. J. Emerg. Technol. Eng. Res. **5**(4), 169–174 (2017)

Design and Implementation of FFT Based Fast Acquiring of High Dynamic GPS Signal Based on FPGA

Yang Xiaojiang(✉), Zuo Feifei, Yu Qian, and Pei Dongbo

The 503rd Institute of the Fifth Academy of CASC, Beijing, China
yangxj721012@sina.com

Abstract. For the high dynamic GPS receiver, the system key technology is to acquire the PN signal. The high dynamic characteristic of the system refers to that the radial distance and rate between the navigation satellite and the receiver result in the GPS signal phase delay and Doppler frequency shift, which makes signal acquiring more difficult. It needs a long acquiring time when using the general slippage correlation acquisition method. In order to improve the acquiring performance, a FFT based acquiring method is analyzed, the implementation based on FPGA is introduced. By using the high dynamic GPS signal generator to conduct experiments, the availability of the method is proved. Additionally, the improving measures are presented.

Keywords: Doppler Frequency Shift · Signal Acquiring · FFT

1 Introduction

Signal acquiring is to estimate the pseudo-code phase and Doppler frequency shift of the received signal, and to initialize the track loop with such estimations. For the PN signal acquiring, slippage correlation acquisition method and matched filter method are usually adopted [1, 2]. It needs a long time to acquire signal by the slippage correlation acquisition method due to pseudo-code cycle time, resulting from the method of time domain and frequency domain serial search that related to the phase of each frequency point. To some extent, a matched filter can achieve parallel computing, not affected by pseudo-code cycle time, but it takes up a lot of hardware resources. Therefore, this system adopts the domain FFT based parallel algorithm for signal acquiring.

2 Basic Algorithm

C/A code acquiring in the slippage correlation acquisition method refers to a code related process in time domain, this process can also be carried out within the frequency domain. That is, it can be faster to adopt FFT based acquiring algorithm.

H. P. Urbach and H. Jiang (Eds.): ISSOIA 2022, SPPHY 295, pp. 413–418, 2023.
https://doi.org/10.1007/978-981-99-4098-1_36

Provided that sampling interval ($\Delta t = t_{k+1} - t_k$) is a fixed constant, the GPS after inputting signal sample can be represented as [3],

$$z[n, \widehat{\omega}] = I + jQ = \sum_{k=0}^{L-1} y_k CA_{k-n} exp\{-j[\langle\omega_{IF} - \widehat{\omega}D\rangle t_k]\} \tag{1}$$

whereas, n refers to serial number of code starting time $\widehat{t_{s(n)}} = t_0 + n\Delta t$, namely the code phase; $CA_{k-n} = CA\{[1 + \eta(k-n)\Delta t]\}$ refers to signal sampling form of pseudo code.

When local carrier frequency is fixed, that is, one Doppler frame is selected [4, 5], $z(n, \widehat{\omega}D)$ can be calculated according to the code phase n, i.e. $z(0, \widehat{\omega}D), z(1, \widehat{\omega}D), \ldots, z(L-1, \widehat{\omega}D)$. According to the convolution theorem, we can get the Formula (2):

$$Z = IFFT(Y. * C) = IFFT\begin{bmatrix} C_0 Y_0 \\ C_1 Y_1 \\ M \\ C_{L-1} Y_{L-1} \end{bmatrix} \tag{2}$$

The above formula applies the cycle characteristic of the C/A code, namely $CA_K = CA_{L+K}$ where L is the C/A code cycle [6]. The maximum value of k is the estimated value of t_s.

Therefore, frequency domain FFT based acquiring algorithm is to realize the parallel acquisition of PN code phase and Doppler frequency shift according to the property of inverse transform that the convolution in time domain is equal to the multiplication in frequency domain [7].

Figure 1 shows FFT based fastd code acquiring circuit diagram in frequency domain.

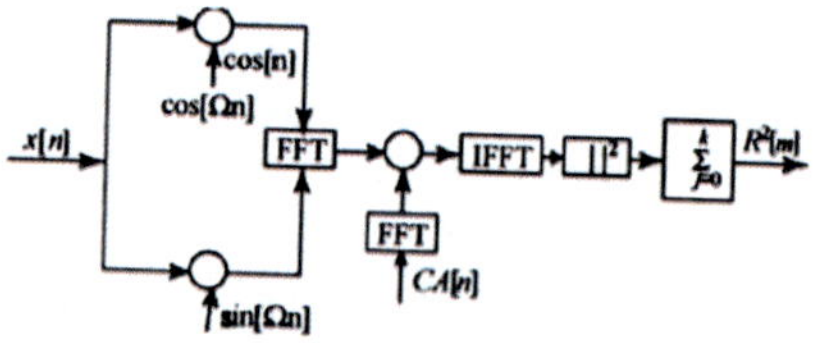

Fig. 1. FFT based fast code acquiring circuit diagram

3 Implementation

Based on the above algorithm analysis, this paper designs a more practical engineering application. See hardware function module for FFT based acquiring in Fig. 2.

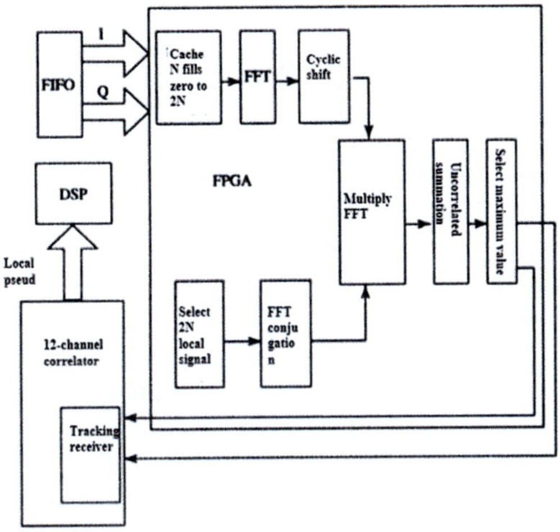

Fig. 2. Hardware function module for FFT based fast acquiring

3.1 Data Storage and FFT Arithmetic (FPGA)

Due to the larger scale of the system signal processing with high speed and high precision, system adopts FPGA chip (Vertex2XC2V3000), which can realize large-scale data parallel processing and meet the requirement. In addition, the FPGA has outstanding performance on flexibility and extensibility. FIFO chip is adopted for external storage, which avoids the short of the FPGA storage space and achieves the high-speed data transmission.

3.2 FFT Scheme Design

Based on the unit structure in Fig. 3, we can design a multi-point FFT or IFFT operation module. The Fig. 4. Shows the FFT module.

First of all, enter data (2N point) into the input data memory. These data can be divided into multiple sets of 1024 points, and each set carries out FFT arithmetic through a corresponding 1024-point FFT processing core. For the first time after the 1024-point combined FFT computation, the result is stored in the output data memory. However, only when adjust the data storage order in the memory can we get the correct result for the next 1024-point combined FFT arithmetic. Thus, the data in the output memory of FFT processing unit can be reorganized by means of the address generator and introduced to the input data memory for the next FFT arithmetic. At the last time the computing results directly stored in the output data memory for the final output.

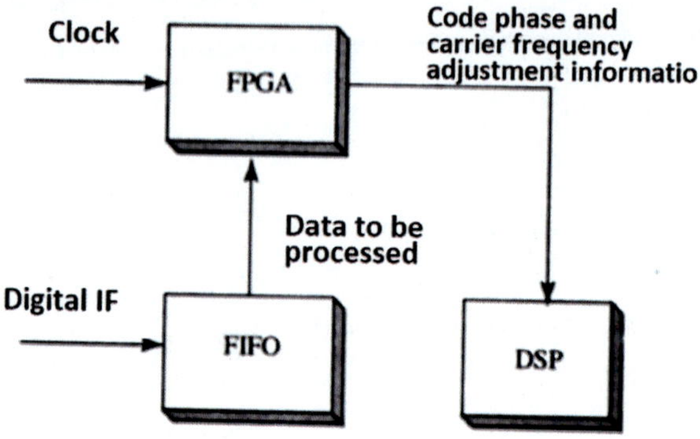

Fig. 3. Hardware structure for fast acquiring system.

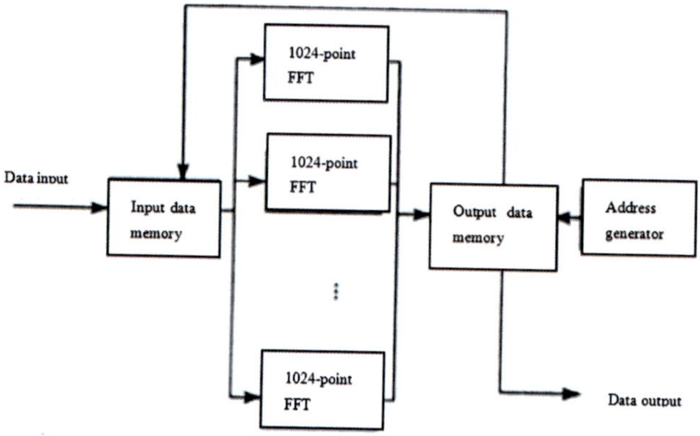

Fig. 4. FFT module internal structure in the fast acquiring system.

4 Algorithm Validation

In order to verify the validity of the algorithm, high dynamic GPS test simulator is adopted. STR4760 GPS signal simulator, which can evaluate the high positioning precision of the dynamic user receiver, can be used for the test and validation of GPS receiver.

STR4760 technical indicators: relative speed between navigation satellite and users: 120000 m/s, accelerated speed: 360 g and 500 g/s acceleration; GPS signal source controlled by ALPHA workstation, reliable in performance. Visual satellite status diagram from the simulator is as shown in Fig. 5.

The algorithm hardware platform is used to search the entire range. See search correlation matrix for No. 5 GPS satellite (visual) in Fig. 6, a significant correlation peak can be observed.

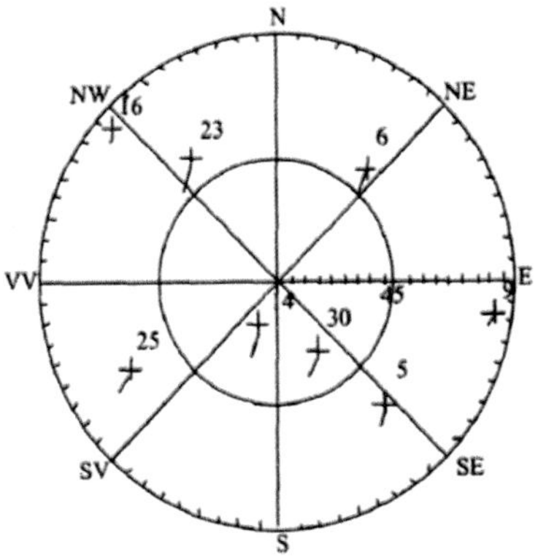

Fig. 5. Visual satellite status diagram

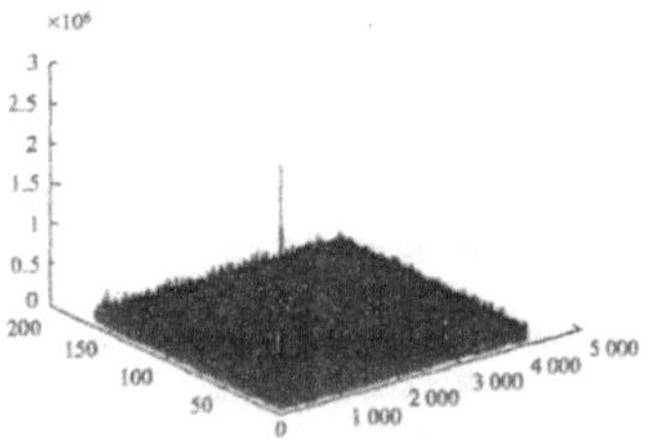

Fig. 6. Search correlation matrix for No. 5 GPS satellite

5 Conclusion

In this paper, we discuss the hardware design of fast acquiring technology for GPS receiver and the detailed industry application of circuit for each part. Experiments show that if Doppler frequency shift is between ±50 kHz and the shift length is 500 Hz, it takes 1.92 s ($0.009\,6 \times 200 = 1.92$ s) to blind acquiring a satellite at most and takes up to 42 s to collect all the ephemeris. The new technology takes up to 1 min in positioning, which has greatly improved compared with the time consuming by original satellite-borne GPS receiver.

References

1. Guangshu, H.: Digital Signal Processing: Theory, Algorithms and Implementation. Tsinghua University Press, Beijing (1997)
2. Zhihe, Q., Wanyi, W.: The Principle and Application of GPS. Electronic Industry Press, Beijing (2002)
3. Qishan, L.X.Z., Qing, C.: High dynamic GPS C/A code signal acquiring scheme and implementation. Beihang Univ. J. **28**(3), 358–361 (2002)
4. Qihua, H., Guoliang, S.: Instantaneous frequency measurement based on slippage fourier transform. Electron. Measur. Technol. **29**(3), 35–40 (2006)

5. Dingrong, X.W.S., Shujian, L.: GPS receiver pseudo random code fast acquiring technology. Beihang Univ. J. **29**(16), 489–492 (2003)
6. Xiaoyan, X., Jianshu, L., Dingrong, S., et al.: Fast Signal Acquiring in the High Dynamic Spread Spectrum Communication System. 26 (2), 22–27 2005
7. Huang Zhen, L., Jianhua, Y.S.: fast doppler frequency shift acquiring for satellite communications. J. Electron. **4**, 1053–1055 (2003)

A Dense Matching Method for Satellite Remote Sensing Images Based on Multiple Matching Primitives

Ning Yang(✉), Jian-de Zhang, Yun-sen Wang, and Qi-xing Zhu

Shandong Institute of Space Electronic Technology, Yantai 264670, People's Republic of China
yn3714@163.com

Abstract. In order to satisfy the demand of the number of matching points of satellite remote sensing images in the 3D reconstruction of geographic information space, a dense matching method for satellite remote sensing images based on multiple matching primitives is proposed in this paper, in which the matching algorithm of SIFT(Scale Invariant Feature Transform), the matching algorithm based on object space geometry constraint and the region matching point growing algorithm based on affine transformation are comprehensively used with multiple primitives. The experimental results show that the proposed method achieves the dense matching of satellite remote sensing images, which can meet the need of 3D reconstruction of geographic information space.

Keywords: Remote sensing · Image matching · Multiple primitives · Object space geometry constraint · Affine transformation

1 Introduction

Three-dimensional geographic information space has great application value in military operations, geological investigation, unmanned driving and other technical fields. As the core of 3D reconstruction of geographic information space, image matching can be divided into gray-based matching and feature-based matching according to the different matching primitives (i.e. gray matching primitive and feature matching primitive) [1, 2]. However, gray-based matching often has a large amount of computation, and when the rotation and scaling between images are very different, the gray-based matching will have a large deviation. Although feature-based matching is not affected by image geometric transformation, illumination variation and other factors, which has strong robustness, the matching results are relatively sparse, which are difficult to meet the need of 3D reconstruction. In order to overcome the shortages of single matching primitive, a dense matching method for satellite remote sensing images based on multiple matching primitives is proposed in this paper, which provides enough high-precision and dense matching points, to ensure the density and accuracy of the 3D reconstruction of geographic information space.

H. P. Urbach and H. Jiang (Eds.): ISSOIA 2022, SPPHY 295, pp. 419–426, 2023.
https://doi.org/10.1007/978-981-99-4098-1_37

2 Dense Matching of Remote Sensing Images with Multiple Matching Primitives

In this paper, the dense matching method of satellite remote sensing images uses two kinds of matching primitives: feature points (feature matching primitives) and regular grid points (gray matching primitives), and adopts multi-method comprehensive matching strategy to realize the dense matching of satellite remote sensing images. The matching of feature points adopts the feature matching algorithm based on SIFT, and the matching of grid points adopts the strategy of combining the matching algorithm based on object space geometry constraint and the region matching point growing algorithm based on affine transformation, as shown in Fig. 1.

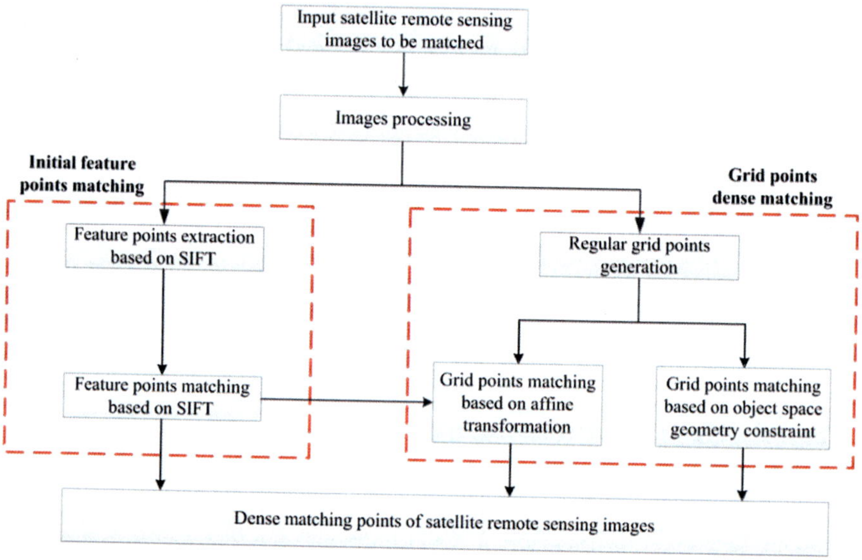

Fig. 1. Dense matching process of satellite remote sensing images based on multiple matching primitives

2.1 Feature Points Matching Based on SIFT

The initial feature points matching based on SIFT algorithm includes two parts: feature points extraction and feature points matching [3–5]. The extraction of SIFT feature point includes four parts: scale-space extrema detection, key point localization, orientation assignment, and construction of descriptor. Feature points between two images are matched by identifying their nearest neighbors. But in some cases, the second closest-match may be very near to the first. As a result, the improved k-d tree algorithm - BBF (Best Bin First) algorithm is used to find the nearest neighbor and second nearest neighbor in the matching to improve the matching accuracy.

2.2 Grid Points Matching Based on Object Space Geometry Constraint

The matching algorithm based on object space geometry constraint conducts the matching with the image-object space relationship [6–9]. In the matching process, object space geometry constraint is used to guide the matching, so as to reduce the matching time, improve the reliability and accuracy of matching, and obtain the corresponding three-dimensional information while completing the matching, as shown in Fig. 2.

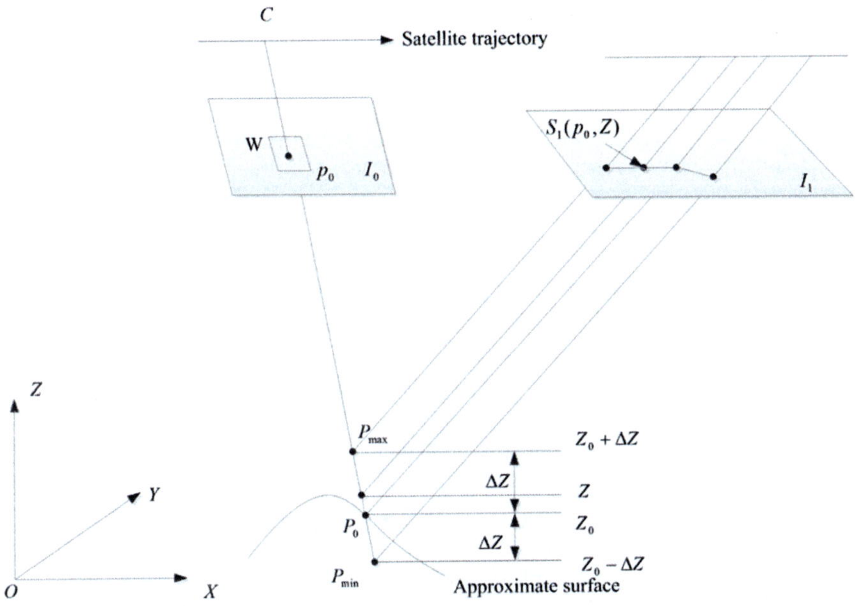

Fig. 2. Image matching based on object space geometry constraint

Assuming that two satellite images constitute the stereo image pair $I_0 - I_1$, p_0 is a given point on the reference image I_0, then the projective ray Cp_0 must pass through the corresponding point P_0 in the object space, where C is the instantaneous projective center corresponding to point p_0. Assuming that Z_0 is the estimated elevation value of p_0, the three-dimensional coordinates $P_0(X_0, Y_0, Z_0)$ of image point p_0 can be obtained by collinear equation or imaging model. Assuming that the tolerance of the estimated elevation value is ΔZ, two object space points P_{min} and P_{max} corresponding to the elevation $Z_0 - \Delta Z$ and $Z_0 + \Delta Z$ can be obtained, then P_0 must lie on the line segment $P_{min}P_{max}$. $P_{min}P_{max}$ is projected onto the search image I_1 by the exact orientation elements, and the approximate epipolar line segment of p_0 in the search image I_1 is obtained, on which the match point of p_0 must be located.

In the reference image, a matching window W centered on p_0 is selected, and then, according to the above object space geometry constraint, the corresponding matching window of p_0 in the search image is obtained. The matching similarity measure is defined as the normalized correlation coefficient NCC, as shown in Eq. (1),

$$NCC(p_0, Z) = \frac{\sum_{s \in W} \left(I_0(s) - \overline{I}_0\right) \times \left(I_1(s(Z)) - \overline{I}_1\right)}{\sqrt{\sum_{s \in W} \left(I_0(s) - \overline{I}_0\right)^2} \sqrt{\sum_{s \in W} \left(I_1(s(Z)) - \overline{I}_1\right)^2}}, \tag{1}$$

where $\overline{I}_0 = \frac{1}{m \times n} \sum_{s \in W} I_0(s), \overline{I}_1 = \frac{1}{m \times n} \sum_{s \in W} I_1(s(Z)), I_0 \cdot$ and $I_1 \cdot$ represent the gray levels of the image points in the reference image and search image respectively, m and n represent the sizes of the matching window W, s is an image point located in the matching window W, $s(Z)$ is the corresponding point of s obtained by object space geometry constraint in the search image, and the gray level of $s(Z)$ can be obtained by bilinear interpolation.

As can be seen from Eq. (1), NCC here is different from the traditional normalized correlation coefficient, which is a function of image point p_0 and its elevation Z : $Z \in [Z_0 - \Delta Z, Z_0 + \Delta Z]$. Given the image point p_0 in the reference image and its corresponding approximate elevation Z_0 and ΔZ, the elevation when NCC reaches the maximum value is exactly the correct elevation of the given point p_0, and the best matching position of p_0 in the search image is also obtained. To ensure the matching accuracy, a threshold is set for NCC. Only when the NCC is greater than the given threshold, it is considered as a correct matching.

2.3 Grid Points Matching Based on Affine Transformation

After completing the grid points matching based on object space geometry constraint, some grid points in the reference image may not find the matching points in the search image due to a series of problems, such as occlusion, fault, and so on. In order to solve this problem, the region matching point growing algorithm based on affine transformation [10] is used to re-match the unmatched grid points.

The general form of image affine transformation is

$$\begin{cases} x' = a_0 + a_1 x + a_2 y \\ y' = b_0 + b_1 x + b_2 y \end{cases}, \tag{2}$$

in which (x, y) represents the point in the reference image, and $\left(x', y'\right)$ represents the corresponding affine transformation point in the search image.

The specific steps of the algorithm are as follows.

(1) According to the unmatched grid points in the reference image, the image is divided into regions. For a certain region, the matched points in the region are used to calculate the initial affine transformation parameters.
(2) The affine transformation parameters are used to match the unmatched grid points in this region, and their initial matching points are obtained.
(3) The correlation coefficient ρ of each initial matching point in the region is calculated. If ρ is greater than the given threshold, it is considered as a correct matching point; otherwise, go to step (4).
(4) The modification values $(da_0, da_1, da_2, db_0, db_1, db_2)$ of affine transformation parameters of matched points in the region are solved with the least squares matching method [11, 12].
(5) The modified affine transformation parameters are calculated with Eq. (3), and then go to step (2) for the iteration. If the correlation coefficient ρ of each initial matching point is greater than the given threshold, the matching is considered successful and the iteration stops. If the iteration exceeds a certain number, and the correlation coefficient ρ of the initial matching point is still not greater than the given threshold, it is considered that the image point in the reference image has no corresponding matching point in the search image, and the iteration is terminated,

$$\begin{bmatrix} 1 & 0 & 0 \\ a_0^i & a_1^i & a_2^i \\ b_0^i & b_1^i & b_2^i \end{bmatrix} = \begin{bmatrix} 1 & 0 & 0 \\ da_0^i & 1 + da_1^i & da_2^i \\ db_0^i & db_1^i & 1 + db_2^i \end{bmatrix} \begin{bmatrix} 1 & 0 & 0 \\ a_0^{i-1} & a_1^{i-1} & a_2^{i-1} \\ b_0^{i-1} & b_1^{i-1} & b_2^{i-1} \end{bmatrix}. \tag{3}$$

(6) After above steps are completed, evaluate whether the number of matching points obtained meets the requirement. If not, continue to divide the regions in the reference image, and then repeat steps (1) to (5) in the sub-regions to match the unmatched grid points.

3 Experiment Results and Analysis

Two images of ISR-P5 remote sensing satellite in the same area are selected as the images to be matched, as shown in Fig. 3, in which the forward image is used as the reference image, and the backward image is used as the search image. In the matching algorithm based on object space geometry constraint, the threshold of *NCC* is set to 0.7 so as to ensure the precision of matching results. In the region matching point growing algorithm based on affine transformation, the threshold of matching point correlation coefficient ρ is set as 0.6, and the number of iterations is set as 7.

The image matching results of the proposed method in this paper are shown in Fig. 4. A total of 34,498 matching points are obtained, including 43 feature points and 34,455 grid points. Among the grid points, 17,070 are obtained by the matching algorithm based on object space geometry constraint, and 17,385 are obtained by the region matching point growing algorithm based on affine transformation. The proposed method realizes the dense matching of satellite remote sensing images and can meet the need of 3D reconstruction of geographic information space.

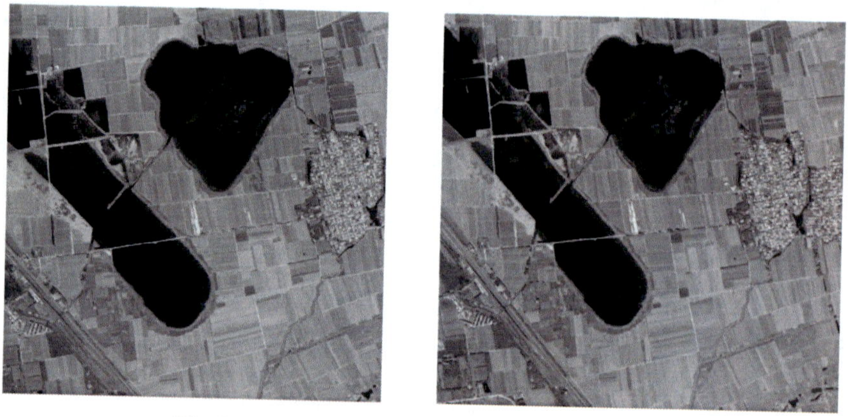

Fig. 3. Satellite remote sensing images to be matched

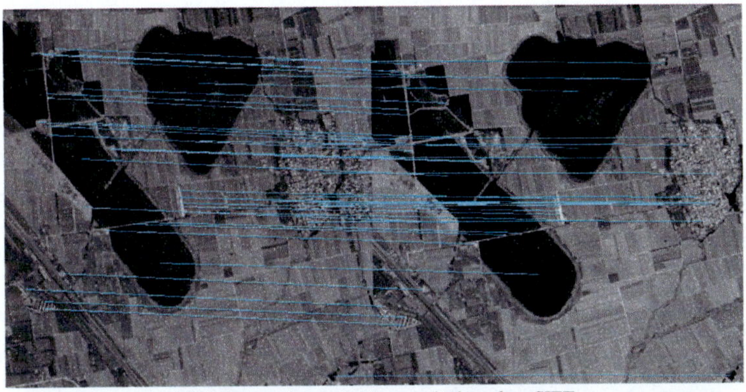

(a) Feature points matching results based on SIFT

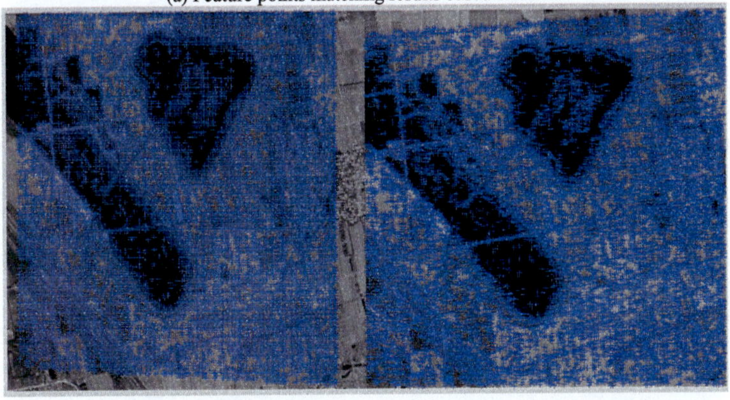

(b) Grid points matching results based on object space geometry constraint

(c) Matching results of the region matching point growing algorithm based on affine transformation

Fig. 4. Dense matching results of satellite remote sensing images based on multiple matching primitives

4 Conclusion

Satellite remote sensing images are not restricted by regions, can be acquired all-weather, and the acquisition period is short, which provides a new means for geographic information mapping. A dense matching method for satellite remote sensing images with multiple matching primitives is proposed in this paper. By combining the matching primitives and integrating multiple matching methods, the high-precision and dense matching of satellite remote sensing images is realized, which solves the shortcomings of using a single matching primitive for image matching, and can satisfy the requirement of 3D reconstruction of geographic information space with satellite remote sensing images. In the future, semi-global matching algorithm or global matching algorithm will be combined to further improve the matching accuracy of the method.

References

1. Yuan, X., Yuan, W., Xu, S., et al.: Research developments and prospects on dense image matching in photogrammetry. Acta Geod. Cartographica Sin. **48**(12), 1542–1550 (2019)
2. Wang, J., Zhu, Q., Wang, W.: A dense matching algorithm of multi-view image based on the integrated multiple matching primitives. Acta Geod. Cartographica Sin. **42**(5), 691–698 (2013)
3. Lowe, D.G.: Distinctive image features from scale-invariant key points. Int. J. Comput. Vision **60**(2), 91–110 (2004)
4. Etezadifar, P., Farsi, H.: A new sample consensus based on sparse coding for improved matching of SIFT features on remote sensing images. IEEE Trans. Geosci. Remote Sens. **58**(8), 5254–5263 (2020)
5. Zhang, X., Wang, J., Liu, S., et al.: A dense matching algorithm for remote sensing images based on reliable matched points constraint. J. Geo-inf. Sci. **23**(8), 1508–1523 (2021)
6. Zhang, L., Ai, H., Xu, B., et al.: Automatic tie-point extraction based on multiple-image matching and bundle adjustment of large block of oblique aerial images. Acta Geod. Cartographica Sin. **46**(5), 554–564 (2017)
7. Zhang, L.: Automatic Digital Surface Model (DSM) Generation from Linear Array Images. Zurich: Institute of Geodesy and Photogrammetry, ETH Zurich (2005)
8. Zhang, L., Gruen, A.: Multi-image matching for DSM generation from IKONOS imagery. ISPRS J. Photogramm. Remote. Sens. **60**(3), 195–211 (2006)
9. Zhang, K., Sheng, Y., Fu, S., et al.: Multi-view VLL matching algorithm for optical aerial images based on constraint of object space positioning consistency. Opt. Precis. Eng. **26**(7), 1784–1793 (2018)
10. Yang, N., Shao, F., Shen, J., Jia, Y.: A region dense matching algorithm for remote sensing images of satellite based on SIFT. In: Urbach, H..P., Yu, Q. (eds.) ISSOIA 2017. SPP, vol. 209, pp. 150–159. Springer, Cham (2018). https://doi.org/10.1007/978-3-319-96707-3_17
11. Dong, Y., Zhang, L., Balz, T., et al.: Radargrammetric DSM generation in mountainous areas through adaptive-window least squares matching constrained by enhanced epipolar geometry. ISPRS J. Photogramm. Remote. Sens. **137**(2), 61–72 (2018)
12. Zhang, C., Zhang, Q., Guo, B., et al.: Oblique image matching algorithm based on adaptive initial object patch. Acta Geod. Cartographica Sin. **49**(1), 108–116 (2020)

Geometric Stitching Method for Multi-detector Images of Stationary Orbit Area-Array Camera

Zhou Nan[1](✉) and Cao Jinshan[2]

[1] Beijing Institute of Space Mechanics and Electricity, Beijing 100094, China
nan_zhou2006@126.com

[2] School of Computer Science, Hubei University of Technology, Wuhan 430068, China

Abstract. Most of the stationary orbit area-array camera (SOAAC) use multiple slice detectors to stitching imaging to increase the image swath. In order to improve the efficiency of data use, users are required to provide geometrically seamless stitching products. The current stitching method is difficult to apply the situation, which no overlapping area between the sub-images, and the increased distortion after stitching. Aiming at this problem, this paper proposed a geometric stitching method for the multiple slice detectors sub-images of the SOAAC, which adopts sensor corrected method. The strict geometric calibration of the orientation of multiple slice detectors are first performed. Then, an ideal undistorted area-array image based on the calibrated orientation of each slice detector. The geometric imaging model (GIM) is built for both the virtual image and the original sub-images. Finally, based to the GIM, the coordinate mapping relationship between the image points of the virtual image and the original sub-images is built, and the gray value of the image points of the sub-images are assigned to the virtual image. The experimental results of the simulation data and the SOAAC data show that, for different detector stitching methods, the method proposed in this paper can obtain visually seamless and geometrically seamless stitched images, and the image stitching accuracy and internal geometric distortion are better than 1 pixel, which can meet the requirements of image geometric accuracy for subsequent standard product production.

Keywords: Sensor Corrected · Geometric Stitching · Area-array Camera · Stationary Orbit · Remote Sensing Products

1 Introduction

Geostationary optical remote sensing satellites have relatively fixed observation positions, high temporal resolution and wide observation range, and are an important part of earth observation imaging technology [1]. In order to meet the user's demand for large swath, most of the SOAACs use multiple slice detectors stitching imaging to increase the image swath. Different from multiple camera stitching, multiple slice detectors stitching belongs to a camera system and share a main point and focal length, and its imaging geometric characteristics are shown in Fig. 1 [2]. However, due to the problems of multiple slice detectors such as non-strict coplanarity, small or no overlapping area between

H. P. Urbach and H. Jiang (Eds.): ISSOIA 2022, SPPHY 295, pp. 427–438, 2023.
https://doi.org/10.1007/978-981-99-4098-1_38

the adjacent sub-images, and large geometric distortion after stitching, it is difficult to achieve highly precise geometric stitching through simple image-space-oriented stitching. Therefore, it is a core technology of ground preprocessing to realize the highly precise stitching of the sub-images collected by the multiple slice detectors and obtain a complete seamless image. The stitching accuracy of sub-images directly determines the geometric quality of the standard image products.

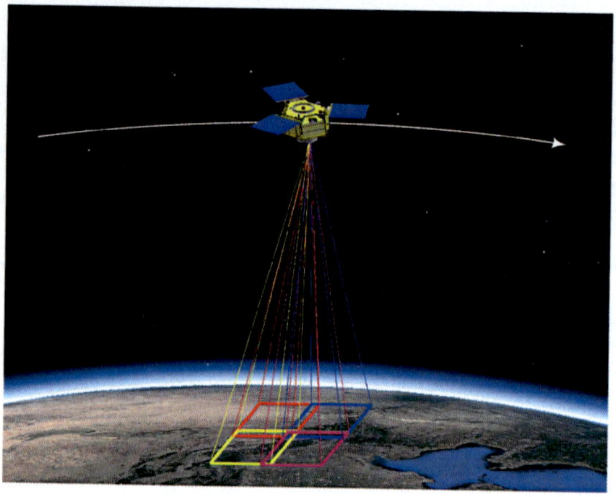

Fig. 1. The geometric characteristics of SOAAC

At present, scholars have carried out a lot of research on the geometric stitching for multiple cameras or multiple slice detectors. The geometric stitching methods include two types in general: the image transformation model method and the GIM method [3]. The image linear transformation model is used to build the geometric relationship between the sub-images for the image transformation model method. The processing steps mainly include matching homonymous points, building the transformation model, and stitching the sub-images [4, 5]. This method has a simple model, low computational complexity, and fast processing speed, and can only achieve visual seamless stitching. Due to the lack of consideration for the GIM, it is difficult to achieve sub pixel stitching accuracy. Moreover, this method cannot be applied when there is no overlapping or the overlapping area lack of texture features. Based on the rigorous imaging geometry relationship of satellite camera, the GIM stitching method adopts sensor corrected, and introduces an ideal undistorted virtual image to build the geometric mapping relationship between virtual image and sub-images, so as to achieve highly precise geometric stitching of multiple images. The GIM method has the advantages of rigorous theory and independent on the matching homonymous points between sub-images, and can achieve geometric seamless stitching, which has been successfully applied to many low orbit satellites linear-array cameras, such as the ZiYuan-1 02C, GaoFen-2, and HaiYang-1 C/D satellite [6–9]. For the image stitching of SOAAC, the verification of models and methods in the stitching processing has become an urgent need because of its imaging geometric characteristics.

In view of this, this paper proposes a geometric stitching method for multiple slice detectors sub-images of the SOAAC, which adopts sensor corrected method. In this method, the strict geometric calibration of the orientation of the multiple detectors are first performed. Then, an ideal undistorted area-array image based on the physical structure and the calibrated orientation of each detector, and the undistorted virtual image coverage is the same as that of the original sub-images. The GIM is built for both the virtual image and the original sub-images. Finally, based to the GIM, the coordinate mapping relationship between the image points of the virtual image and the original sub-images is built, and the gray value of the image points of the sub-images are assigned to the virtual image.

2 Methodology

2.1 On-Orbit Geometric Calibration of the Multiple Detectors

The premise of highly precise image stitching of the original sub-images is to accurately know the orientation of the multiple slice detectors. Before satellite launch, the orientation of the multiple slice detectors is usually accurately measured. However, due to the excessive acceleration during satellite launch and environment changes in orbit, the orientation of the multiple slice detectors will be changed. With the help of the geometric calibration, the orientation accuracy of the multiple slice detectors can be significantly improved. The geometric calibration model of the SOAAC can be expressed as follows [10, 11]:

$$\begin{pmatrix} tan(\varphi_X) \\ tan(\varphi_Y) \\ -1 \end{pmatrix} = m\left(\boldsymbol{R}_{\text{Camera}}^{Star}\right)^T \boldsymbol{R}_{J2000}^{Star}\boldsymbol{R}_{WGS-84}^{J2000}\left\{\begin{bmatrix} X \\ Y \\ Z \end{bmatrix}_{WGS84} - \begin{bmatrix} X_{GNSS} \\ Y_{GNSS} \\ Z_{GNSS} \end{bmatrix}\right\} \quad (1)$$

where (φ_X, φ_Y) are the look-angles of the detectors in the camera coordinate system; m is the scaling factor; $\boldsymbol{R}_{\text{Camera}}^{Star}$ the rotation matrix from the camera coordinate system to the star sensor coordinate system; $\boldsymbol{R}_{J2000}^{Star}$ is the rotation matrix from the J2000 celestial coordinate system to the star sensor coordinate system; $\boldsymbol{R}_{WGS-84}^{J2000}$ is the WGS 84-to-J2000 rotation matrix; $(X, Y, Z)_{WGS-84}$ 和 $(X_{\text{GNSS}}, Y_{\text{GNSS}}, Z_{\text{GNSS}})$ are the coordinates of the ground point and the satellite position, respectively, in the WGS 84 geocentric coordinate system.

If the look-angle of each detector is solved directly, it is difficult to operate in practice. A cubic polynomial look-angle model is often used to model the detector coordinates as follows [12]:

$$\begin{cases} \tan(\varphi_X) = a_0 + a_1 \cdot s + a_2 \cdot l + a_3 \cdot s \cdot l + \\ \qquad a_4 \cdot s^2 + a_5 \cdot l^2 + a_6 \cdot s^2 \cdot l + \\ \qquad a_7 \cdot s \cdot l^2 + a_8 \cdot s^3 + a_9 \cdot l^3 \\ \tan(\varphi_y) = b_0 + b_1 \cdot s + b_2 \cdot l + b_3 \cdot s \cdot l + \\ \qquad b_4 \cdot s^2 + b_5 \cdot l^2 + b_6 \cdot s^2 \cdot l + \\ \qquad b_7 \cdot s \cdot l^2 + b_8 \cdot s^3 + b_9 \cdot l^3 \end{cases} \quad (2)$$

where $(a_0, a_1, a_2, a_3, a_4, a_5, a_6, a_7, a_8, a_9, b_0, b_1, b_2, b_3, b_4, b_5, b_6, b_7, b_8, b_9)$ are the model parameters, (s, l) are column number and row number of the detector.

For the SOAAC users, satellite positions, satellite attitudes, and imaging time parameters are often unavailable, so they are unable to build the physical sensor model of the SOAAC. Instead, the rational function model (RFM) can be built, since the rational polynomial coefficients (RPCs) are often supplied by the SOAAC vendors [13], together with the corresponding image. Hence, the RFM is employed as the sensor orientation model of the SOAAC in this study [14, 15]:

$$\begin{cases} r_n = \frac{p_1(\varphi_n,\lambda_n,h_n)}{p_2(\varphi_n,\lambda_n,h_n)} \\ c_n = \frac{p_3(\varphi_n,\lambda_n,h_n)}{p_4(\varphi_n,\lambda_n,h_n)} \end{cases} \tag{3}$$

$$\begin{aligned} p_1(\varphi_n, \lambda_n, h_n) = & a_{1'} + a_{2'}\lambda_n + a_{3'}\varphi_n + a_{4'}h_n + a_{5'}\lambda_n\varphi_n + a_{6'}\lambda_n h_n + a_{7'}\varphi_n h_n \\ & + a_{8'}\lambda_n^2 + a_{9'}\varphi_n^2 + a_{10'}h_n^2 + a_{11'}\lambda_n\varphi_n h_n + a_{12'}\lambda_n^3 + a_{13'}\lambda_n\varphi_n^2 \\ & + a_{14'}\lambda_n h_n^2 + a_{15'}\lambda_n^2\varphi_n + a_{16'}\varphi_n^3 + a_{17'}\varphi_n h_n^2 + a_{18'}\lambda_n^2 h_n \\ & + a_{19'}\varphi_n^2 h_n + a_{20'}h_n^3 \end{aligned} \tag{4}$$

where (r_n, c_n) are the normalized values of the image point coordinates (r, c); and $(\varphi_n, \lambda_n, h_n)$ are the normalized values of the corresponding ground point coordinates (B, L, H), respectively; $(a_{1'}, a_{2'}, \cdots, a_{20'})$ are the coefficients of the polynomial p_1, and the coefficients of p_2, p_3, and p_4 are defined similarly. The coefficients of the polynomials p_1, p_2, p_3, and p_4 are named RPCs [16].

2.2 Geometric Modeling of the Undistorted Virtual Image

After the geometric calibration, the normalized coordinates of each detector can be obtained in the camera coordinate system. Then, an undistorted area-array image based on the physical structure of the multiple slice detectors and the geometric calibration results. As shown in Fig. 2, the virtual image coverage is the same as that of the multiple detectors sub-images. The look-angle model of the virtual image is assigned as follows:

$$\begin{cases} tan(\varphi_{X'}) = d_0 + d_1 l' \\ tan(\varphi_{y'}) = e_0 + e_1 s' \end{cases} \tag{5}$$

where $(\varphi_{X'}, \varphi_{y'})$ are the look-angles of the pixels of the virtual image in the camera coordinate system, (d_0, d_1, e_0, e_1) are the model parameters, (s', l') are column number and row number of the pixels of the virtual image. The GIM of the virtual image can be expressed as follows:

$$\begin{pmatrix} \tan(\varphi_{X'}) \\ \tan(\varphi_{y'}) \\ -1 \end{pmatrix} = m\left(\boldsymbol{R}_{Camera}^{Star}\right)^{T} R_{J2000}^{Star} R_{WGS-84}^{J2000} \left\{ \begin{bmatrix} X \\ Y \\ Z \end{bmatrix}_{WGS-84} - \begin{bmatrix} X_{GNSS} \\ Y_{GNSS} \\ Z_{GNSS} \end{bmatrix} \right\} \tag{6}$$

In order to geometric stitching, RFM for the virtual image is constructed on the basis of the GIM.

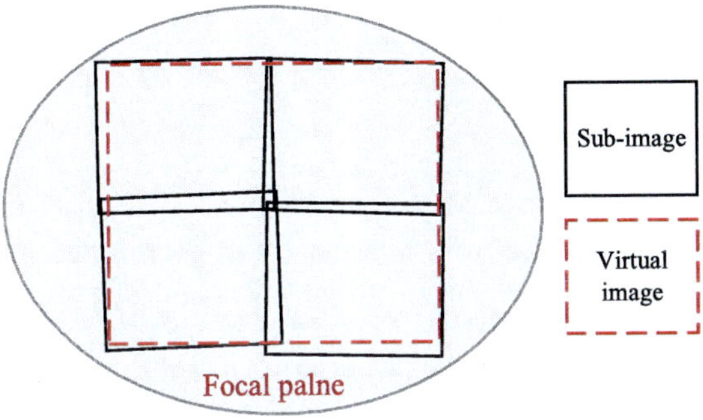

Fig. 2. The ideal undistorted area-array arrangement

2.3 Geometric Stitching of the Original Sub-images

According to the RFM for the virtual image and the original sub-images, the coordinate mapping relationship between the original sub-images and the virtual image is built. The stitching image is obtained through gray resampling. As shown in Fig. 3. The geometric stitching procedures of the sub-images are as follows:

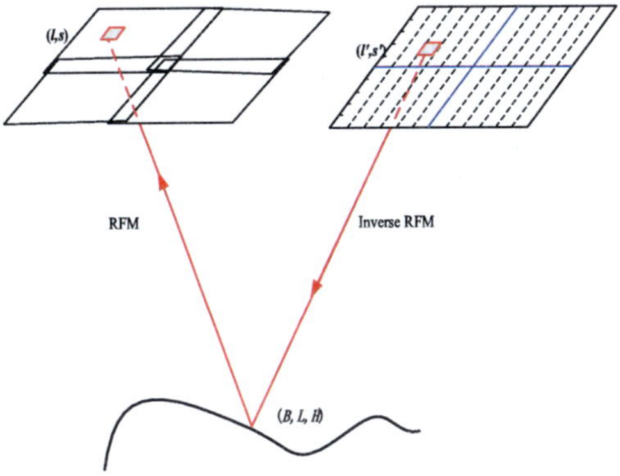

Fig. 3. The geometric stitching procedures of the sub-images

(1) According to the inverse RFM of the virtual image, each image point (s', l') is projected to the digital elevation model (DEM), and the corresponding ground point (B, L, H) is calculated.
(2) According to the RFM for the sub-images, all the calculated ground points (B, L, H) are respectively projected to the sub-images, and the corresponding image points (s, l) are calculated.
(3) The gray value of each image point (s, l) in the sub-images is obtained by bilinear interpolating, then assigning to the corresponding image point (s', l') in the virtual image.

3 Results and Discussion

3.1 Experimental Datasets

In this study, four on-orbit images collected by the SOAAC and three simulation images were tested. The general characteristics of the SOAAC are shown in Fig. 4. The SOAAC is composed of four sub-images stitching, and has no overlapping area between the sub-images, the seam size is 300 pixels between columns and 327 pixels between rows. The SOAAC images are used to verify the accuracy of the on-orbit geometric calibration and the geometric stitching. The general characteristics of the three simulation images are listed in Table 1. The simulation images are used to verify the accuracy of the geometric stitching.

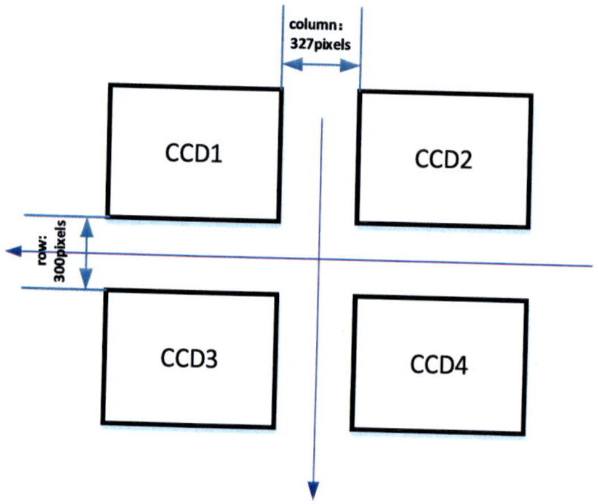

Fig. 4. The general characteristics of the SOAAC

Table 1. The general characteristics of the three simulation images

Image	Number of sub-images	Sub-image arrangement	Sub-image size (pixels)	Overlapping area
1	3	1*3	11000 × 11000	1000
2	4	2*2	11000 × 11000	1000
3	9	3*3	11000 × 11000	1000

3.2 Calibration Accuracy of the Sub-Images of the SOAAC Analysis

Due to the GSD of the SOAAC was 300 m, the 15 m resolution Landsat-8 Enhanced Thematic Mapper Plus (ETM+) digital orthophoto map (DOM) [17] and 30 m resolution Advanced Spaceborne Thermal Emission and Reflection Radiometer Global Digital Elevation Model (ASTER GDEM) [18, 19] can be used as the reference data for the geometric calibration. The plane accuracy of LandSat-8 ETM+ DOM is about 12 m, and the elevation accuracy of ASTER GDEM is about 16m, meeting the accuracy requirements of reference data for geometric calibration.

Firstly, the ground control points (GCPs) in sub-images were measured on the reference data automatically. Then, the look-angle model parameters of the multiple detectors are calculated. Finally, the RFM for the multiple slice detectors are built and all GCPs are respectively projected to the image by RFM. The errors between the image projection points coordinates and the image coordinates of the corresponding measured points are calculated as the positioning accuracy. The positioning accuracy before and after the geometric calibration are listed in Table 2. The residual errors are shown in Fig. 5.

Table 2. The positioning accuracy of the four SOAAC sub-images

Sub-image	Calibration	Number of GCPs	Max (pixel)			RMSE (pixel)		
			Column	Row	Planimetry	Column	Row	Planimetry
1	before	43481	7.547	–3.846	9.743	4.238	1.370	4.454
	after	43481	–1.330	1.281	1.828	0.531	0.512	0.738
2	before	46842	10.779	–7.788	13.210	8.420	5.307	9.953
	after	46842	1.338	1.411	1.907	0.534	0.565	0.777
3	before	19526	5.861	9.245	10.946	3.648	4.217	5.576
	after	19526	1.539	–1.367	2.045	0.614	0.547	0.823
4	before	14295	6.750	3.053	7.221	4.484	1.392	4.696
	after	14295	1.888	1.428	2.332	0.756	0.571	0.948

As shown in Table 2, the positioning accuracy of the SOAAC is 7–14 pixels before the geometric calibration. Moreover, the positioning errors of the overlapping the areas between sub-images are not consistent, resulting in the misalignment of stitching. The precise look-angle model parameters of the multiple detectors were calculated after the geometric calibration. Thus, the positioning error of the SOAAC was reduced to less than 1 pixel, and the positioning error was not shown significant systematicity. It is worth mentioning that the stitching misalignment between the sub-images after the geometric calibration, which is the premise of highly precise geometric stitching.

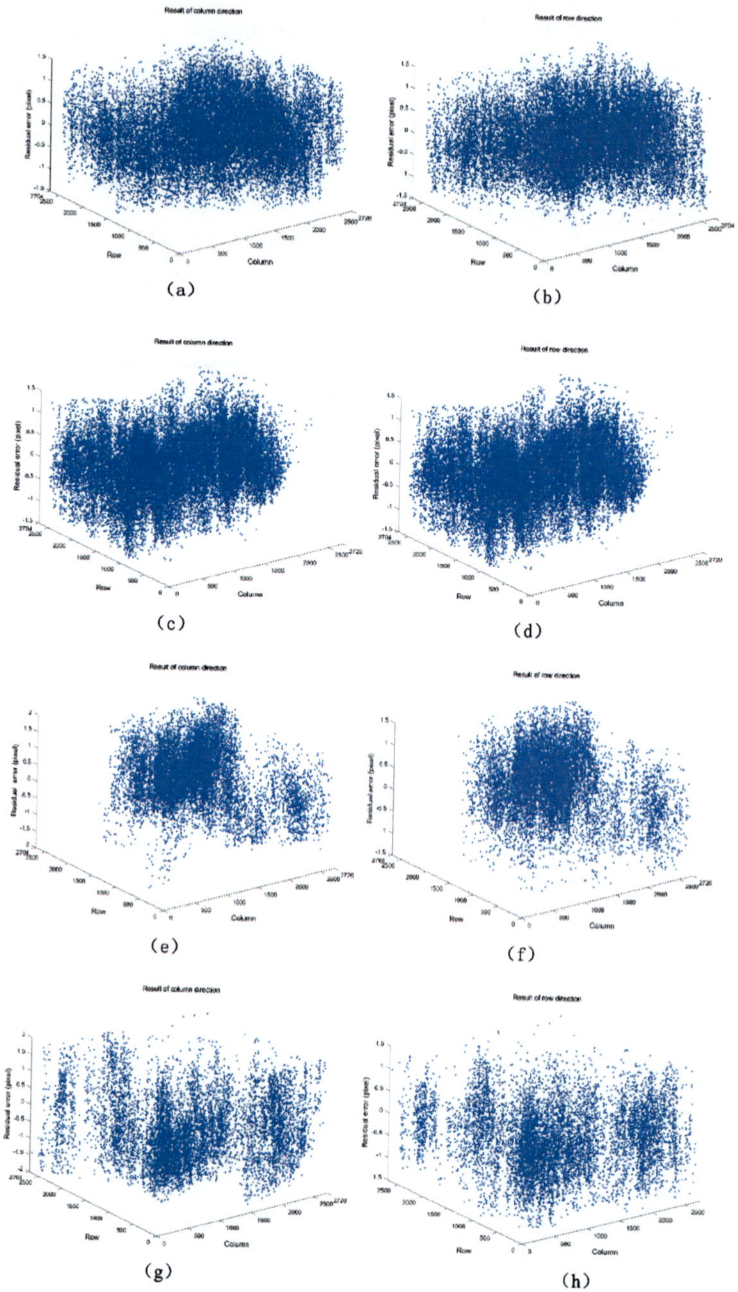

Fig. 5. The residual errors in sub-images (a, b) 1, (c, d) 2, (e, f) 3, and (g, h) 4 after the geometric calibration.

3.3 Evaluation of Geometric Stitching of the Sub-Images

On the basis of the geometric calibration, the sub-images of the SOAAC were successfully stitched together according to the method described in Sect. 2.2 and Sect. 2.3. In order to evaluate the accuracy of the proposed geometric stitching method, the same DOM and DEM described in Sect. 3.2 were also used as the reference data. The check points (CPs) were measured on the reference data automatically. Due to the lack of overlapping areas between the SOAAC sub-images, the internal geometric accuracy is only evaluated. The internal geometric accuracy of the stitched SOAAC sub-images is listed in Table 3.

Table 3. The internal geometric accuracy of the stitched SOAAC sub-images

Image	Number of CPs	Max (pixel)			RMSE (pixel)		
		Column	Row	Planimetry	Column	Row	Planimetry
1	33914	1.246	–1.201	1.681	0.499	0.479	0.691
2	46183	–1.345	–1.293	1.830	0.538	0.518	0.747
3	41965	1.337	-1.494	1.980	0.608	0.678	0.911
4	16324	1.355	–1.199	1.789	0.614	0.543	0.820

As shown in Table 3, the internal geometric error of the four stitched images was less than 1 pixel. The test results indicated that it is feasible to use sensor corrected method to perform the geometric stitching of the SOAAC. Even when there is no overlapping area between the sub-images, the highly precise geometric stitching can be achieved.

In order to verify the stitching accuracy of the method proposed in this paper for different detector stitching methods, the three sub-images of the simulation data were geometrically stitched together according to the stitching method. The CPs were measured on the reference data, and the tie points (TPs) were measured\on the adjacent sub-images. The CPs were used to calculate the positioning accuracy of the stitched SOAAC images. The TPs were used to calculate the internal geometric accuracy of the overlapping area between the adjacent sub-images.

Table 4. The positioning accuracy of the stitched simulation images

Image	Sub-image arrangement	Number of CPs	RMSE (pixel)		
			Column	Row	Planimetry
1	1*3	33914	0.499	0.479	0.691
2	2*2	46183	0.538	0.518	0.747
3	3*3	16324	0.614	0.543	0.820

Table 5. The internal geometric accuracy of the overlapping area between the sub-images

Image	Sub-image ID	Number of CPs	RMSE (pixel)		
			Column	Row	Planimetry
1	1–2	192	0.173	0.151	0.230
	2–3	234	0.176	0.164	0.240
2	1–2	245	0.167	0.169	0.237
	1–3	232	0.195	0.159	0.252
	2–4	234	0.189	0.182	0.262
	3–4	239	0.165	0.158	0.229
3	1–2	255	0.165	0.168	0.236
	2–3	245	0.162	0.134	0.210
	4–5	242	0.186	0.160	0.246
	5–6	216	0.173	0.173	0.245
	7–8	232	0.167	0.145	0.221
	8–9	217	0.198	0.204	0.284
	1–4	252	0.197	0.186	0.271
	2–5	244	0.184	0.173	0.253
	3–6	232	0.190	0.170	0.255
	4–7	236	0.168	0.166	0.236
	5–8	221	0.226	0.176	0.286
	6–9	217	0.216	0.183	0.283

As shown in Table 4 and Table 5, the positioning error of the three stitched simulation images and the internal geometric error of the overlapping area between the adjacent sub-images were less than 1 pixel and 0.3 pixel respectively. The test results indicated that it is feasible to use sensor corrected to perform the geometric stitching for different detector stitching methods.

The example of the partial enlarged image from image 1 are shown in Fig. 6. We can see from the results in Fig. 6 that the misalignment between the sub-images were corrected effectively and stitched very well.

(a) (b)

Fig. 6. The partial image from image 1 (a) before and (b) after geometric stitching.

4 Conclusion

Because the SOAAC use multiple slice detectors to stitching imaging to increase the image swath, the traditional stitching method is difficult to apply due to the small or even no overlapping area between the sub-images and the increased distortion after stitching. This paper proposed a geometric stitching method for the multiple slice detectors sub-images of the SOAAC, which adopts sensor corrected method. The strict geometric calibration of the orientation of multiple slice detectors are first performed. Then, an ideal undistorted area-array image based on the calibrated orientation of each slice detector. The geometric imaging model (GIM) is built for both the virtual image and the original sub-images. Finally, based to the GIM, the coordinate mapping relationship between the image points of the virtual image and the original sub-images is built, and the gray value of the image points of the sub-images are assigned to the virtual image. The experimental results show that the method proposed in this paper can obtain visually seamless and geometrically seamless stitched images, and the image stitching accuracy and the internal geometric accuracy are better than 1 pixel.

References

1. Yu, L.J., LIU, Y.: Development analysis on GEO civil optical earth observation satellites with mid-high resolution. Spacecraft Eng.**22**(1), 106–112 (2013)
2. Chang, X.: Research on key technologies of geometric processing for high-resolution area-array camera of stationary orbit satellite. Ph.D. Dissertation, Wuhan University, Wuhan (2010)
3. Hu, F.: Research on Inner FOV Stitching Theories and Algorithms for Sub-images of Three Non-collinear TDI CCD Chips. Ph.D. Dissertation, Wuhan University, Wuhan (2010)
4. Jacobsen, K., Hannover, U.: Geometric and information potential of IRS-1C PAN-Images. In: International Geoscience and Remote Sensing Symposium, vol. 1, pp. 428–430 (1999)

5. Hu, L., Sun, T., Zhang, T., et al.: Application of DIROEF algorithm for noncollinear multiple CCD array stitching of the Chinese mapping satellite 1–02. IEEE Trans. Geosci. Remote Sens. **14**(4), 519–524 (2017)
6. Cheng, Y., Jin, S., Wang, M., et al.: Image mosaicking approach for a double-camera system in the GaoFen-2 optical remote sensing satellite based on the big virtual camera. Sensors **17**(6), 1441 (2017)
7. Pan, J., Hu, F., Wang, M., et al.: Inner FOV stitching of ZY-1 02C HR camera based on virtual CCD line. Geomatics Inf. Sci. Wuhan Univ. **40**(4), 436–443 (2015)
8. Cao, J., Zhang, Z., Jin, S., et al.: Geometric stitching of HaiYang-1C ultra violet imager with a distorted virtual camera. Opt. Express **28**(9), 14109–14126 (2020)
9. Yang, W., Zhou, N., Cao, J.: Double-camera geometric stitching of HY-1D coastal zone imager. Spacecraft Recovery Remote Sens. **42**(2), 99–107 (2021)
10. Tang, X., Zhou, P., Zhang, G., et al.: Research on a production method of sensor corrected products for ZY-3 satellite. Geomatics Inf. Sci. Wuhan Univ. **39**(3), 288–294 (2014)
11. Wang, X., Zhang, G., Shen, X., et al.: Satellite video stabilization with geometric distortion. Acta Geodaetica et Cartographica Sinica **45**(2), 194–198 (2016)
12. Wang, M., Cheng, Y., Chang, X., et al.: High accuracy on-orbit geometric calibration of geostationary satellite GF4. Acta Geodaetica et Cartographica Sinica **46**(1), 53–61 (2017)
13. Cao, J., Yuan, X., Gong, J.: In-orbit geometric calibration and validation of ZY-3 three-line cameras based on CCD-detector Look Angles. Photogram. Rec. **30**(150), 211–226 (2015)
14. Zhang, L., Zhang, J., Chen, X., et al.: Block adjustment with SPOT5 imagery and sparse GCPs based RFM. Acta Geodaetica et Cartographica Sinica **38**(4), 302–310 (2009)
15. Wang, T., Guo, Z., Li, D., et al.: Block ortho-rectification for Satellite Images. Geomatics Inf. Sci. Wuhan Univ. **39**(7), 838–842 (2014)
16. Cao, J., Yuan, X.: Refinement of RPCs Based on Systematic Error Compensation for Virtual Grid. Geomatics Inf. Sci. Wuhan Univ. **36**(2), 185–189 (2011)
17. Liang, J., Gong, J., Li, W.: Applications and impacts of Google earth: a decadal review (2006–2016). ISPRS J. Photogram. Remote Sens. **146**, 91–107 (2018)
18. Tang, X., Li, S., Li, T., et al.: Review on global digital elevation products. Natl Remote Sens. Bull. **25**(1), 167–181 (2021)
19. CGIAR. SRTM 90m DEM Digital Elevation Database. https://srtm.csi.cgiar.org/

Real-Time Image Super Resolution System Based on Micro-scanning Technology

Dan Li[1](✉), Yanan Wu[1], Dandan Yang[1], Sen Meng[1], Fengfan Li[1], F. Maillard[2], and F. Claeyssen[2]

[1] IAE Industrial Group Co., Ltd., Shanghai, China
lidan@iae-oe.com
[2] CEDRAT TECHNOLOGIES, 38240 Meylan, France

Abstract. Different from the recent popular super resolution system based on AI technology which needs normally massive training datasets, the micro-scanning super resolution system by integrating the high-precision mechanism and the image processing system can overpass the training datasets limit to enhance the image resolution and quality considerably in real time. Such a real-time image super resolution system named Quick Demo Station (QDS) system is presented. This system is jointly developed by Shanghai IAE, China, in cooperation with Cedrat technologies, France. The system is made of two modules, one imaging module (MicroScan_VIS_Module), the other image processing module (a portable image processing workstation). The image processing module performs imaging module's control, as well as image acquisition, image registration, image super-resolution reconstruction, image contrast enhancement. Here the algorithm used to restore the low-resolution (LR) images is the iterative reconstruction method, which could achieve high quality super-resolution (SR) results. The output is a video stream whose frame rate is higher than 25 frames per second. The system is being tested in different applications and show the excellent super-resolution results, including low light level conditions and outdoor variable lighting scenarios. Furthermore, the algorithm in image processing module can be used not only on the workstations, but also in embedded processing system.

Keywords: Real-time super-resolution · Image restoration · Micro-scanning · Piezo stage

1 Introduction

To conquer the under-sampling defects of imaging system, recently the super resolution system on basis of machine learning or deep neural network technology got the huge development thanks to AI industry's boost. Super-resolution algorithms based on deep neural network can be divided into supervised and unsupervised methods [1]. Supervised learning methods require a large number of training datasets corresponding to true scenario. The datasets include pair of high-resolution (HR) and low-resolution (LR) images. However, it is difficult to obtain pairs of images in practical conditions.

H. P. Urbach and H. Jiang (Eds.): ISSOIA 2022, SPPHY 295, pp. 439–454, 2023.
https://doi.org/10.1007/978-981-99-4098-1_39

Therefore, LR images are often obtained through interpolation methods such as bilinear with images captured by camera. This down-sampling process is not consistent with the actual optical imaging process, and the trained neural network may not be able to recover to reliable high-resolution images from the actual LR images captured by the camera. Unsupervised learning method does not require a large number of training datasets. It can generate image details through generative adversarial networks (GAN) [2] and other types of deep neural networks. However, these image details are likely to be fake, which is not in line with the reality.

By consequence machine learning technology has impassable shortcomings for the application that has limited training datasets or for the application pursuing the high authenticity of the super resolution. In order to face the challenging application mentioned above, the super resolution system based on the physic model is continuing be driven in image processing domain. Among the physic modeling super resolution technologies, Micro-scanning resolution enhancement technology occupies an important position, thanks to its robustness, no requirement of datasets and in particular the authenticity of its reconstruction.

Many researchers have discussed the theory and feasibility of improving image resolution technology through micro-scanning [3] and super resolution process that generate a sequence of sub-pixels-shift frames, then to reconstruct a super-resolved image using oversampled phase information. However, there is no mature COTS software and hardware in the market combined system to real-time process LR images to achieve super-resolution images through modeling and iterative reconstruction method. As we all know, iterative method could achieve better SR image quality but heavy computation requirements. It is difficult to realize real-time image super-resolution.

Therefore, the QDS system based on micro-scanning super-resolution technology is developed to process the images in real time and evaluate the actual benefits of the current optical system. At the same time, it can also be used to verify the benefits of different micro-scanning modes (1 × 1 2 × 2 3 × 3 4 × 4) (see Fig. 1). Finally, the QDS system can help users to quickly have an intuitive understanding of the micro-scanning super-resolution technology, and evaluate how much improvement this technology can bring for a given optical system.

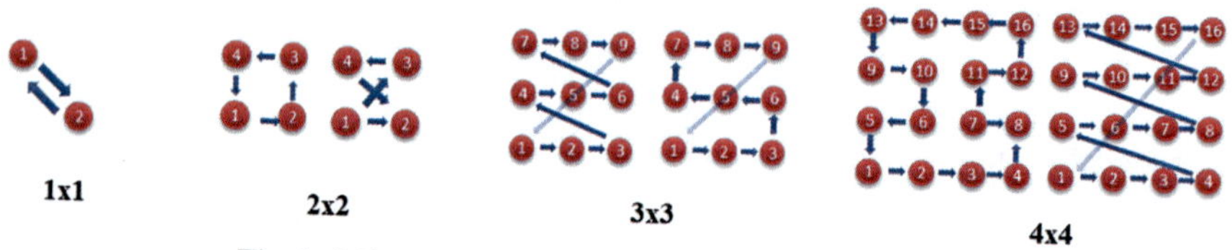

Fig. 1. Micro-scanning modes (1 × 1 2 × 2 3 × 3 4 × 4)

2 QDS System Architecture

The system architecture is shown in Fig. 2. As the main processor of QDS system, image processing module is responsible for image processing, including image acquisition, image registration, super-resolution reconstruction and so on, and complete SR results output and display. As the main controller of QDS system, the image processing module is responsible for the data communication with micro-scanning submodule, detector, and timing control between micro-scanning submodule's motion and detector exposure.

The image processing module sets the micro-scanning motion vector, and then gives instructions to the controller. The controller drives the micro-scanning stage, and then the detector mounted on the micro-scanning stage moves synchronously. Once the detector is shifted in sub-pixel position and stays stable, the image processing module sends instructions to trigger the detector's exposure. Then the image taken at the current position is acquired and transmitted to the image processing module. The cycles mentioned above repeat to realize the real-time video stream super-resolution.

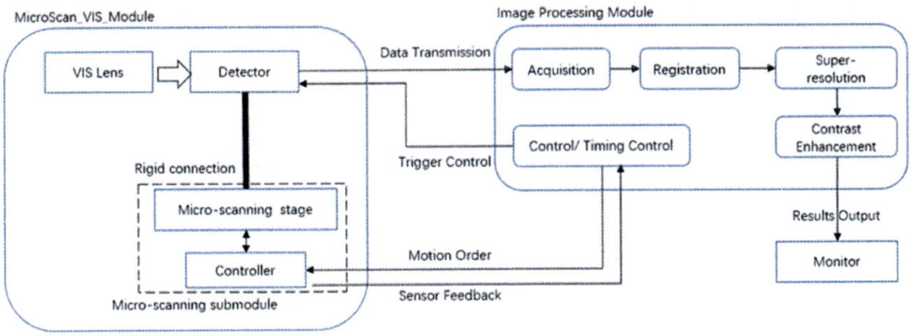

Fig. 2. QDS system architecture

3 Modules

The hardware module of QDS system consists of MicroScan_VIS_Module and a portable image processing workstation (see Fig. 3).

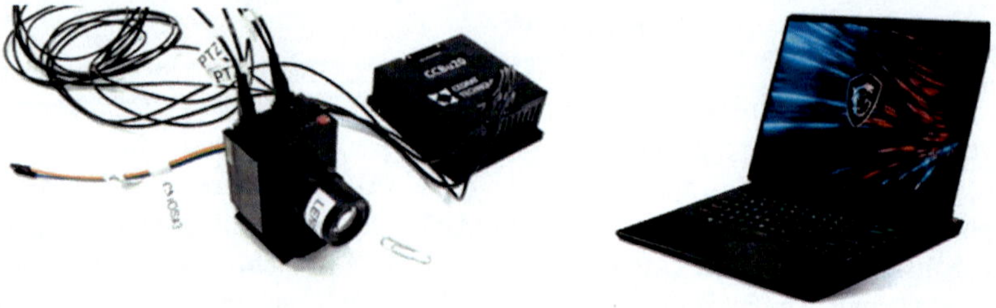

Fig. 3. MicroScan_VIS_Module (left) and portable image processing workstation (right)

3.1 MicroScan_VIS_Module

Aptina MT9V034 CMOS chip has been chosen for QDS. The pixel size of the detector is 6um and the maximum frame rate reaches 60Hz. So, the micro-scanning submodule needs to satisfy sub-micrometer accuracy and millisecond response. In high precision positioning applications, the piezoelectric ceramic mechanisms are commonly used. The piezoelectric ceramic mechanisms have the advantages of no lubrication, compact structure, strong load capacity, high resolution, and fast response. The choice of micro-scanning submodule is crucial.

Micro-scanning Submodule Selection. For micro-scanning super resolution system, the precision (0.05 pixel precision), response time (1 ms) and stability (0.02 pixel stability) of module are extremely important to assure imaging systems keep outputting the LR images that possess stable phase shift in image sequence. The performance of micro-scanning submodule is the primary condition and crucial of a good super resolution process. Cedrat Technologies (CTEC) company has extensive experience in the field of piezoelectric micro-scanning mechanisms and controllers, reputed for its compactness, robustness, dynamic, which are widely used in the industrial, medical, astronautic, and aerospace field.

Using the space heritage of its XY stage for ROSETTA MIDAS mission [4], CTEC has developed the XY200M, [5] a mechanical symmetric design. It is made of a monolithic frame including the payload support surrounded by the 2-axis made of a pair of APA200M piezo actuator each (see Fig. 4). The payload motion is driven on each axis thanks to push-pull electronic channel. This symmetric mechanical design of XY piezo stage brings significant improvements in thermomechanical behavior, compactness, integration process and dynamic performances compared to other scanning mechanisms for optics [3].

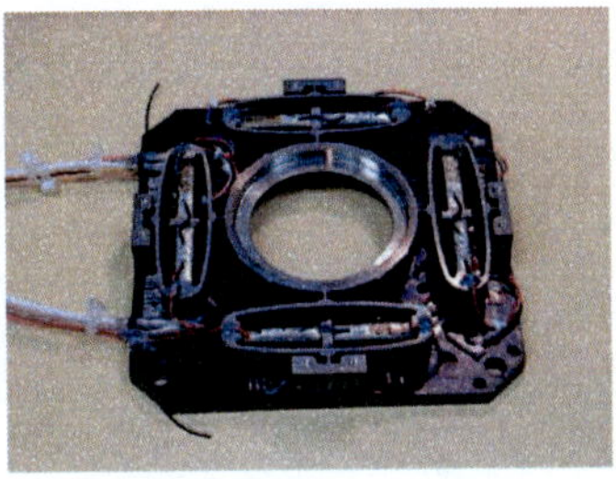

Fig. 4. View of the XY200M moving payload frame

Scaling down this symmetric XY stage design to piezo actuators APA35XS (+12 times smaller and lighter than APA200M), led to a new customized XY25XS-SV stage developed for micro-scanning function and subsequent image resolution enhancement of THALES Catherine® IR Camera in 2006 [6]. In this application, the stage moves a lens in front of the detector, and not the detector itself.

CTEC released a Commercially Off The Shelves (COTS) model of this XY25XS stage (see Fig. 5) typically used for engineering model and breadboard phases of image resolution enhancement development project. This XY25XS COTS can be equipped with Strain gages positioning sensor to allow a fine positioning control.

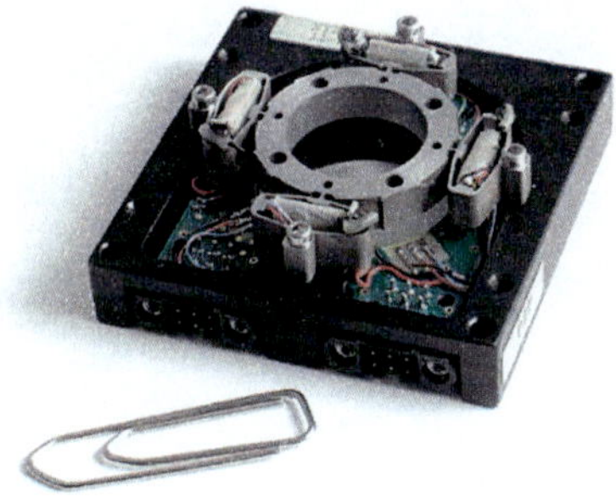

Fig. 5. XY25XS COTS model

CTEC has been constantly developed several customized XY piezo stages [7] under customer requirements for many different functions (scanning, tracking, pointing, etc....) & applications (resolution enhancement, Line of sight stabilization, anti-blurring, etc....) embedded in various electro-optic systems (IR Camera, Gyro stabilized platform, space telescope, Lidar, etc....) to name but a few like XY400M stage (see Fig. 6) for PICSAT CubeSat [8] & [9].

Fig. 6. XY400M Stage for PICSAT CubeSat

The CCBu20 is a 2 channels push-pull controller used to drive in both open and closed loop up to 2 axes piezo mechanisms like XY stages & Fast Steering Mirrors developed by CEDRAT TECHNOLOGIES (CTEC).

The CCBu40 has the same functions as CCBu20, except that it can provide more driving ability (Fig. 7).

(a)(b)

Fig. 7. CCBu20 controller (a) and CCBu40 controller (b)

The specifications of the stages and controllers are listed in Table 1 and Table 2. According to our application, the XY25XS stage and CCBu20 is our best choice.

Table 1. Characteristics of piezoelectric stages

Parameter	Unit	XY25XS	XY200M	XY400M
Stroke	μm	25	200	400
Unloaded resonance frequency	Hz	3000	580	300
Resolution	nm	3	20	40

Table 2. Characteristics of controllers

Parameter	Unit	XY25XS	XY200M
Number of axis		25	200
Power supply	V	24–28	
Typ.static power consumption	W	6	
Max.dynamic power consumption	W	28	100
Peak output current	A	0.2	0.5
Max.RMS output current	mArms	35	350

Implementation. As the Aptina MT9V034 CMOS chip chosen is light enough, it is decided to move detector instead of moving lens in QDS for facilitating the system design and economizing the system cost. In fact, both moving detector and lens can achieve the same effect of shifting the optical axis on the FPA, so no image processing algorithm need to be adapted.

MicroScan_VIS_Module adopts Aptina MT9V034 CMOS chip, which is fixed to a two-dimensional high-precision piezoelectric micro-scanning stage XY25XS from CTEC through mechanical interface. A compact controller CCBU20 from CTEC is used to control XY25XS for high precision and stable sub-pixel displacement. Shows the parts drawing and perspective drawing of MicroScan_VIS_Module (Fig. 8).

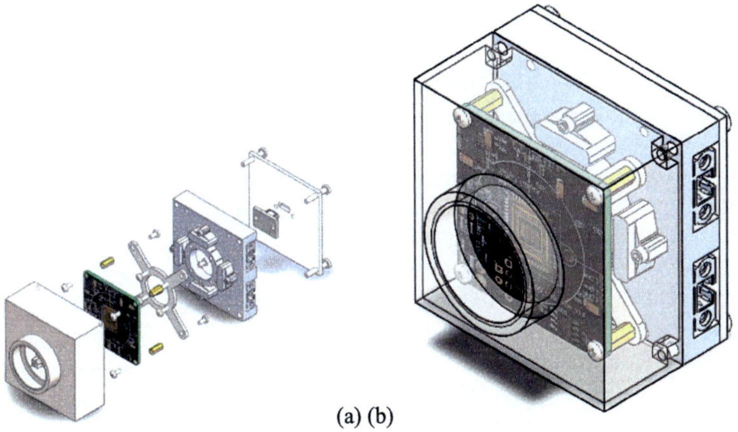

(a) (b)

Fig. 8. Parts drawing (a) and perspective drawing (b) of MicroScan_VIS_Module

The QDS system specifications are listed in Table 3.

Table 3. Characteristics of MicroScan_VIS_Module

Specification		Value
Lens	Wavelength	400 nm–1000 nm
	Focal length	25 mm
	F#	1.2
CMOS	Sensor	MT9V034
	Exposure Type	Global
	Pixel size	6 μm
	Resolution	640 × 480
	Frame rate	50 Hz
Micro-scanning submodule	Settling time	<1 ms
	Stroke	±14 μm
	Operation bandwidth	>600 Hz

3.2 QDS Portable Image Processing Workstation

QDS portable image processing workstation (CPU, Intel(R) Core (TM) i7-9750H @ 2.60 GHz; RAM, 16 GB; GPU, NVIDIA GeForce GTX1660) is not only used as a host to control MicroScan_VIS_Module, but also as an image processing center for super resolution, as well as image and video output. The image processing module in QDS is based on a normal PC, but can be integrated in embedded processing boards with some extra efforts.

Super-Resolution Algorithm Introduction. Micro-scanning super-resolution technology refers to the technique of reconstructing one high-resolution image by processing multiple LR images with sub-pixel shift complementary information. It can be roughly divided into frequency based and spatial domain algorithms.

The method based on frequency domain started earlier, using the reconstruction method of eliminating frequency aliasing. It utilizes the relationship between discrete Fourier transform and continuous Fourier transform to describe the aliasing. The frequency domain algorithm has the advantages of simple operation, low operation complexity and easy parallel processing. However, the observation model of this method does not consider the point spread function of optical systems and assumes that there is no motion blur and observation noise, so the reconstruction quality is limited. Frequency domain reconstruction methods are normally only applicable to the global motion and linear space invariant quality reduction model.

The phase-shift spatial domain reconstruction methods based on probability theory considering the real imaging process. Different motion blurs, transformations (global, local motion, optical blur, etc.), observation noise, down-sampling and all the other factors that affect the reconstruction performance will be considered in the model. Besides, it can combine the prior information of the images, which is able to achieve better results.

However, such method needs to be solved iteratively, which is difficult to realize in real time.

In order to meet the requirements of real-time image processing and reconstruction quality, the super-resolution software in QDS is based on the method of probability theory to model the imaging system, and obtain the optimal solution by iteration.

Generative Model. Super-resolution is the process of inferring high-resolution images from known LR images with different sub-pixel shifted information. Super-resolution process is essentially an ill-posed problem, in order to obtain the optimal solution as accurately as possible the mathematical modeling process should be subjected to the given optical system. Several factors affecting optical system imaging are considered, such as PSF blur, camera movement, image noise, etc. Different factors have different mathematical modeling. For example, PSF blur mostly conforms to Gaussian distribution, but with aberrations or impacted by other factors, PSF may not in line with Gaussian distribution. So other types of PSF should be taken into account and selected according to the actual situation. Camera movement is also classified into different types, such as translation, rotation, affine, which need to be considered in the model. Ill-posed problems are sensitive to noise and even a very small perturbation will have a great effect on the results. Noise modeling should be as consistent as possible with imaging system. The generative model in QDS software is as follows

$$Y_k{=}D_kH_kX{+}N_k \tag{1}$$

where Y_k is the kth LR image. D_k is the combined matrix of the kth image, including down-sampling and optical blurring of the imaging system. H_k is the geometric registration matrix related to the reference LR image. X denotes HR image. N_k is the noise of the kth image.

Assuming the noise N_k conforms to Gaussian distribution $N_k\ (0,\ \beta^{-1}\ I)$, the probability density function of the Y_k image with respect to HR image X, geometric registration H_k and combined matrix D_k may be expressed as follows

$$r^k = \|D_kH_kX - Y_k\|_2^2 = \frac{1}{2}(D_kH_kX - Y_k)^T(D_kh_kX - Y_k) \tag{2}$$

Residual error is defined as

$$r^k = \|D_kH_kX - Y_k\|_2^2 = \frac{1}{2}(D_kH_kX - Y_k) \tag{3}$$

The derivative of r^k with respect to X is denoted by

$$\frac{\partial r^k}{\partial X} = (D_kH_kX)^T(D_kH_kX - Y_k) \tag{4}$$

The derivative of r^k should be zero for minimizing residual error r^k. The high-resolution image X is obtained by

$$X = \left((D_kH_k)^T(D_kH_k)\right)^{-1}(D_kH_k)^TY_k \tag{5}$$

Thus, X is solved directly if D_k H_k and Y_k are known. It is computationally expensive to solve this equation directly. If the matrix $(DH)^T(DH)$ is singular, the direct inversion is problematic. This ill-posed problem leads to infinite solutions, so iterative based method is used to solve this equation. However, the super-resolution reconstruction algorithm contains a huge amount of data calculation. As the input image size and enlarging scale increases, the computation cost also increases, which poses great challenges to the PC capability.

Huge work has been done to accelerate calculation speed and reduce memory usage for the software, so that even a normal PC mentioned above can easily enhance a LR 640 × 480 video frames to 1280 × 960 even up to 2560 × 1920 in real time.

3.3 Software (MSRE QDS) Implementation

The QDS software developed by IAE can perform super resolution processing on the images captured by the MicroScan_VIS_Module in real time.

By executing different modes (2 × 2 3 × 3 4 × 4), QDS brings different super-resolution benefits of 2 times more resolved, 3 times more resolved or 4 times more resolved compared with input LR images. In order to compare the differences between the input LR images and output SR images with more convenience, a function in QDS software has been developed to enlarge the LR image patch and the SR image patch in the same area at the same time. The users will easily observe the huge detail improvement by micro-scanning super-resolution. The dual-window linkage function of software GUI is shown in Fig. 9.

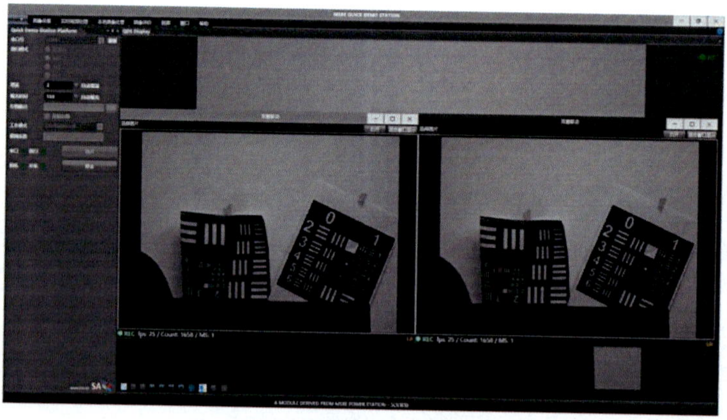

Fig. 9. MSRE QDS software GUI

4 Experiments

4.1 Real Image Sequences

Indoor Condition. First, indoor conditions were tested and the test conditions are shown in Fig. 10. The target is 3.5 m away from the lens (Fig. 11, Fig. 12).

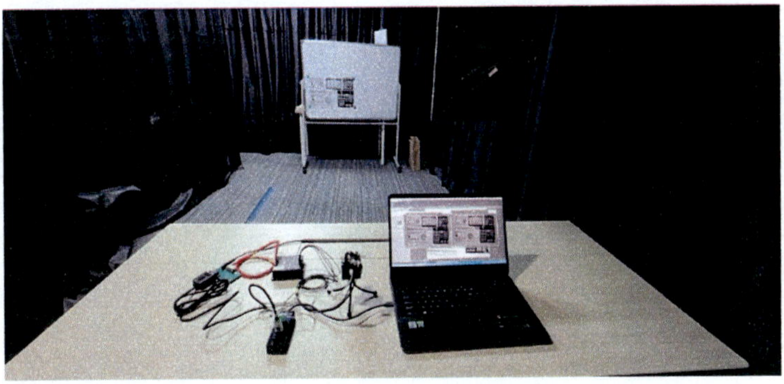

Fig. 10. Indoor Test condition

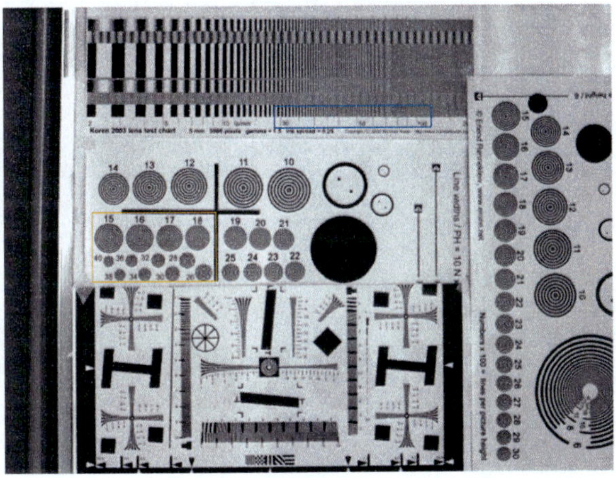

Fig. 11. Indoor LR image

It is obvious that the resolution of SR images has been significantly improved. The higher the micro-scanning factor, the greater the benefit of SR image quality improvement (Fig. 13).

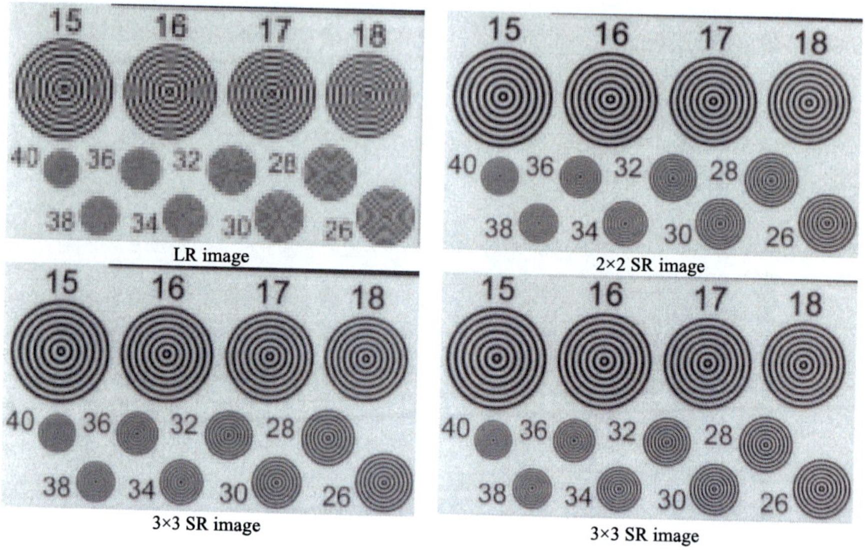

Fig. 12. Enlarged view of LR yellow part

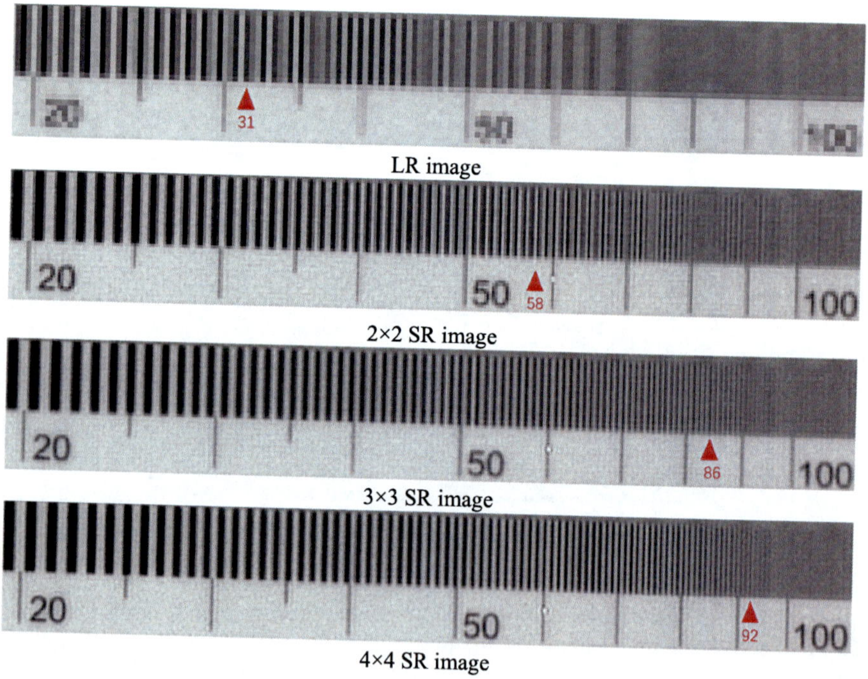

Fig. 13. Enlarged view of LR blue part

We can only distinguish 31st line pairs from the LR image, while after SR reconstruction we can distinguish 58-line pairs with 2 × 2 SR image, 86th line pairs with 3 × 2 SR image and 92nd line pairs with 4 × 4 SR image. At the same time, the artifacts of input LR image have been largely removed.

In order to quantify the micro-scanning super resolution technology benefits, by calculating the contrast at a series of spatial frequencies and normalizing with LR contrast value as a reference, the relative MTF curves under different micro-scanning modes are obtained. See Fig. 14, the figure shows the theoretical MTF and the measured MTF value.

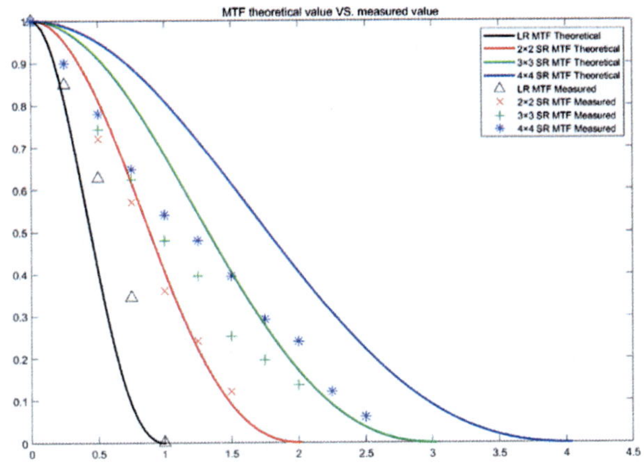

Fig. 14. MTF Theoretical value VS. Measured value

Outdoor Condition. Second, outdoor conditions were tested and the test conditions are the same as indoors. The SR results are shown in Fig. 16 and Fig. 17 (Fig. 15) .

Fig. 15. Outdoor LR image

Fig. 16. Enlarged view of LR red part

During outdoor test, micro-scanning super-resolution technology effectively removes the blur of the imaging system, makes the images clearer and restore much more detailed information. More details of the car can be clearly remarked, the words on the billboard can be easily identified.

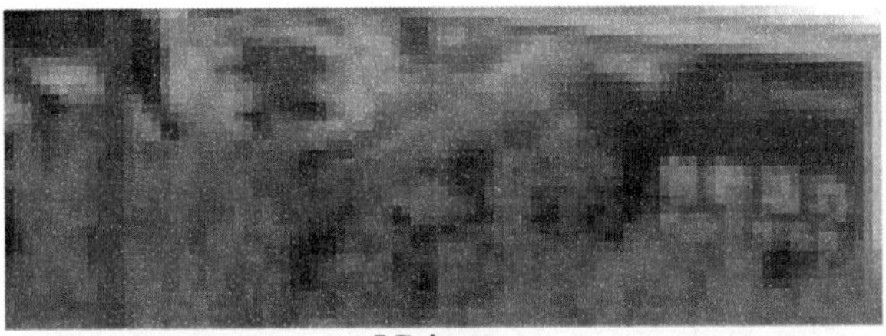

LR image

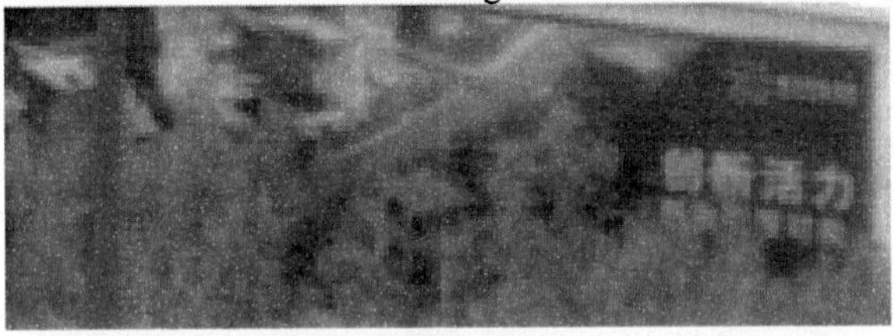

2×2 SR image

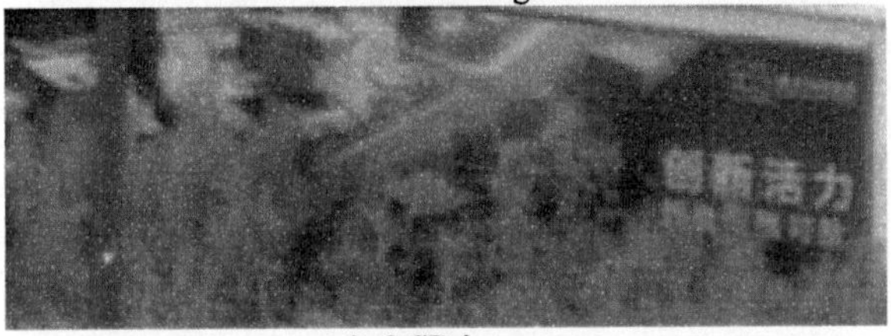

3×3 SR image

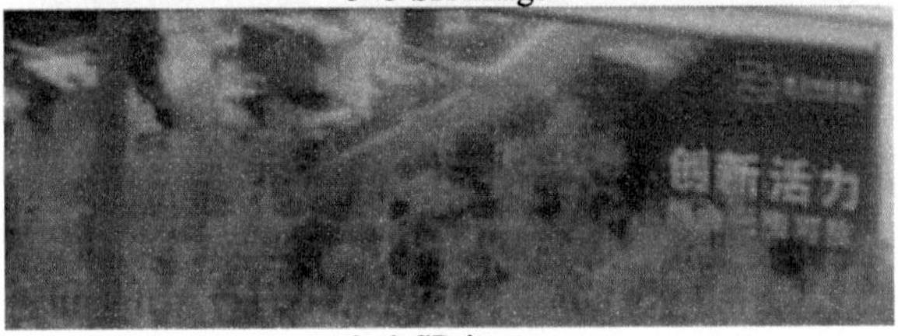

3×3 SR image

Fig. 17. Enlarged view of LR yellow part

5 Conclusion

The development of image resolution enhancement (super resolution) of any electro-optic or imaging system is exciting but challenging. It requires the combination & synchronization of hardware (piezo stage & detector) & software (Image treatment Algorithm & GUI) operating in both spatial & time domain. On top of that, each hardware & software parts to be embedded in user imaging system shall demonstrate robustness to field environmental conditions and efficiency during operations. CTEC XY piezo stages have been qualified and used for more than 20 years for micro-scanning functions in Space and Defense applications like image resolution enhancement to name but a few.

With the QDS, IAE helps any development engineer to check the efficiency and robustness of its real time Algorithm for SR image treatment before implementation in its imaging system and field test.

References

1. Shi, W.: Real-time single image and video super-resolution using an efficient sub-pixel convolutional neural network. In: CVPR (2016)
2. Zhu, J.Y., Park, T., Isola, P., Efros, A.A.: Unpaired image-to-image translation using cycle-consistent adversarial networks. arXiv preprint arXiv:1703.10593 (2017)
3. Wexler, H.J.: Effect of Electronic Microscaning on Infrared Image Aliasing and spatial resolution, MIT MsC thesis (1995)
4. Le Letty, R.: The scanning mechanism for ROSETTA/MIDAS from an engineering model to the flight model. In: Proceedings of9th ESMATS conference Lewen (B), Sept 2001, ESA SP-480, pp 75–81 (2001)
5. Guay, P.: Piezo Qualification for Space Applications. In: Proceedings of Actuator 2002, Ed. Messe Bremen (G), pp 284–287 (2002)
6. Crawford, S.: THALES Long Wave Advanced IR QWIP Cameras, SPIE XXXII-6206–17 Orlando (2006)
7. CTEC Brochure for XY stages heritage. https://www.cedrat-technologies.com/fileadmin/user_upload/CTEC/Mechatronic_products/Stages_Heritage.pdf
8. Nowak, M., Lacour, S., Lapeyrère, V., et al.: A compact and lightweight fibered photometer for the PicSat mission (2017)
9. Menegaldo, C.G., Fialho, F.D.O., Janot-Pacheco, E., et al.: PicSat's enduring legacy. Probing the flight of a small astronomical satellite. Publ. Astron. Soc. Pac. **134**(1033), 034501 (2022)

Research on Manufacturing Technology of Large-Size Baffle of Remote Sensing Camera

Meng Hongtao(✉), Liu Yangtong, Fang Haijun, Sun Donghua, Lv Chengjie, and Tian Feng

Beijing Institute of Space Mechanics and Electricity, 104# Youyi Rd., Haidian District, China
menght163@163.com

Abstract. The large-size baffle of remote sensing camera is used to control the field of view angle of the camera and inhibit it to eliminate stray light. With its large appearance size, complex structure form, high structural interface accuracy and high large heat strain in the complex space environment, great challenge is brought to the manufacturing. According to the product structure characteristics and technical requirements, the key and difficult points of manufacturing process are detailed analysis. Then decide the process scheme: overall molding the baffle, splice the flange, honeycomb and inserts to the baffle in room temperature later. This paper proposed extinction honeycomb-mold, no vacuum bag uniform compressing method for large arc honeycomb innovatively. The product developed successfully because all of these. The product passed the sinusoidal vibration test, the fundamental frequency meets the design requirements, and the single machine assembled smoothly. At present, the baffle successfully entered orbit with the satellite and works in normal. The final results show that the baffle manufacturing technology is feasible and reliable, the product can meet the requirements of eliminating stray light function, and its structural mechanical properties meet the design requirements.

Keywords: Remote sensing satellite · Baffle · Honeycomb sandwich

1 Forward

Advances in optical technology, solid-state technology and computer technology have promoted the profound transformation of space remote sensing technology. The space remote sensor system has developed in a wider spectrum range, higher spectral resolution, higher spatial resolution, higher radiation resolution, as well as multi-polarization, multi-angle, multi-direction [1, 2]. The development of space remote sensors also puts forward higher requirements for camera baffles. The shape of the hood is larger, the structural form is more complex, and the space environment is worse, which puts forward higher requirements for the manufacturing technology of the baffle.

The development of space remote sensor optical system has undergone the evolution from refraction system, refraction system to total anti-reflex system. The baffle in different optical systems is different, and the baffle in different positions in the same optical

H. P. Urbach and H. Jiang (Eds.): ISSOIA 2022, SPPHY 295, pp. 455–468, 2023.
https://doi.org/10.1007/978-981-99-4098-1_40

system is also different [3]. Although the baffle has different forms,its main purpose is to suppress stray light, which will reduce the camera imaging signal-to-noise ratio, interfere with the extraction and identification of imaging signals, and even cause damage to the detector[4, 5].Because the composite has the advantages of light quality, high specific strength and specific stiffness, and fatigue resistance, this material has become the first choice for remote sensing camera baffle. In terms of the structural form of the composite baffle, there are mainly honeycomb panel, built-in aperture, thin shell type and so on [6]. Different types of baffle manufacturing processes are also different.

This paper introduces the manufacturing technology of a large-size remote sensing camera baffle.

2 Structural Design and Technical Requirements

2.1 Structural Design

Due to the space limitation of the fairing, the baffle of remote sensing camera is designed as a four-calibration cylindrical structure, with an outer contour size of 1383 mm (larger side length) × 1038 mm (large side width) × 1847 mm (height).The wall thickness of the baffle is 30 mm, with 5 mm on the inner side and 25 mm on the outside of the honeycomb sandwich structure. The effective field of view of the baffle' cavity should meet the angle requirements to suppress sunlight. There are two windows on the +X side of the baffle, which are designed as carbon fiber pieces structure for light transmission and heat dissipation. A light port with size 106 mm × 186 mm is provided at the root of +X side of the baffle. A flange is set in the middle and lower part of the baffle to connect the baffle and the camera body. The shape of the baffle is shown in Fig. 1.

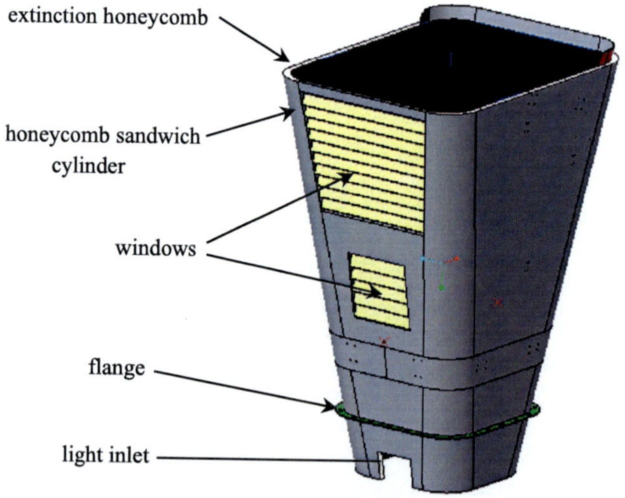

Fig. 1. The model of baffle

2.2 Technical Requirements

(1) The baffle is made of composite materials with higher stiffness, and the materials used should be light-opaque and resistant to particles and ultraviolet radiation in the space environment.
(2) Weight≤30 kg.
(3) Rail temperature of the baffle:–120°C–+120°C.
(4) Stiffness requirements: the first order fundamental frequency for X, Y side ≥ 45 Hz, the first order fundamental frequency for Z side ≥ 70 Hz.
(5) The maximum stress of the baffle meets the strength requirements, and the safety margin is 0.5.
(6) The baffle can experience the mechanical environment test without any damage.

2.3 Material Selection

According to the characteristics of the large-size baffle, honeycomb sandwich cylinder use carbon skin/aluminum honeycomb sandwich structure as its material. In order to ensure sufficient stiffness of the product, the skin is made of high-mode carbon fiber/cyanate material, and the honeycomb is made of porous durable aluminum honeycomb. The flange is made of aluminum alloy. The windows are carbon fiber/cyanate composite. All the materials used have been verified on-orbit, and the anti-irradiation properties can meet the design requirements.

3 Manufacturing Process Plan

3.1 Manufacturing Process Design

The baffle is mainly composed of four parts: honeycomb sandwich cylinder, extinction honeycomb, windows and flange. The baffle will take the "split manufacturing, then assembly and connection" plan. The reason why this plan is adopted is that the structural forms of these four parts are different, so the split manufacturing of each part is better, and the quality is better too. On the other hand, the extinction honeycomb can be prepared separately and divided into 4 petals to spray black paint, so the painting is more convenient, and the uniformity of the paint surface is better. In the post-bonding link, each part can be positioned by tools, so that the position of each part is more accurate.

3.2 Mold Design and Optimization

3.2.1 Mold form Selection

When the baffle is molding, there are two mold forms used, male mold and female mold [7]. The two different forms of molding methods have their own characteristics.

(1) Male mold method: The honeycomb sandwich cylinder structure is molded in the form of a male mold. Traditionally, three kinds of mold are needed: 1) Extinction honeycomb mold: used to prepare extinction honeycomb shaping. 2) outer skin mold: used for outer carbon fiber skin molding.3) assembly molding mold: used

for internal skin and honeycomb sandwich structure assembly molding. When the honeycomb sandwich structure is assembled, it operates directly on the outer surface of the male mold.

(2) Female mold method: When the honeycomb sandwich cylinder structure is molded, it is in the form of a female mold. The following molds are required. 1) Extinction honeycomb shaping mold: used to prepare extinction honeycomb shaping. 2) Outer skin mold: for outer skin molding. 3) inner skin mold: for inner skin molding. 4) assembly molding mold: used for honeycomb sandwich structure assembly molding. When the honeycomb sandwich structure is assembled, it needs to be operated on the inner surface of the mold.

By comparing the two mold forms, it can be found that there are fewer molds used in the male mold method, and because of the shape of the baffle, if the female mold form is adopted, it will make it difficult to operate when forming honeycomb sandwich structure. Therefore, the product chooses the positive mold method finally.

3.2.2 Mold Optimization

According to above analysis, the traditional male mold method requires three molds. This mold form has some disadvantages. On the one hand, because the baffle is large in size, and the mold needs to go through the process of raising human room temperature to high temperature when the product is cured. After the thermal expansion of the mold, the appearance of the product will increase. Extinction honeycomb made from extinction honeycomb mold and honeycomb sandwich cylinder made by assembly mold will poorly matched. On the other hand, the larger the mold size, the more quantity, the higher the cost.

According to the above mold disadvantages, the assembly mold has been cancelled. Firstly, the extinction honeycomb mold is used to molding the extinction honeycomb. Then, extinction honeycomb will be a part of the mold to form a extinction honeycomb-mold. And then use the combination mold to make the inner skin and the honeycomb sandwich cylinder structure. This optimization can effectively ensure the matching of the inner skin and the extinction honeycomb, and also save costs.

3.3 Manufacturing Process Program

The manufacturing scheme of the baffle is to first prepare the extinction honeycomb, outer skin, inner skin and metal parts, and then assemble and glue it to form a honeycomb sandwich structure. Then remove the extinction honeycomb and spray it with black paint, and finally connect the windows, post-inserts, flange and extinction honeycomb to honeycomb sandwich cylinder by glue (Fig. 2).

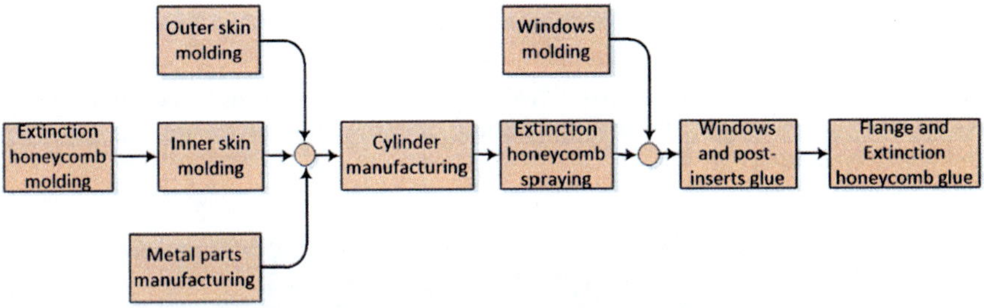

Fig. 2. Manufacturing flowchart of baffle

3.3.1 Extinction Honeycomb Molding

Due to the limitations of painting equipment, the extinction honeycomb needs to be divided into four petals for spray painting. The extinction honeycomb is first made with carbon fiber cloth on the mold, and the fixed skin is divided into flaps at each rounded corner to form 4 pieces. After the fixed skin is made, remove it and place the extinction honeycomb on the mold, then encapsulate it with vacuum bag for some hours. And then the extinction honeycomb is glued with fixed skin with the room temperature structural 420 glue (Fig. 3).

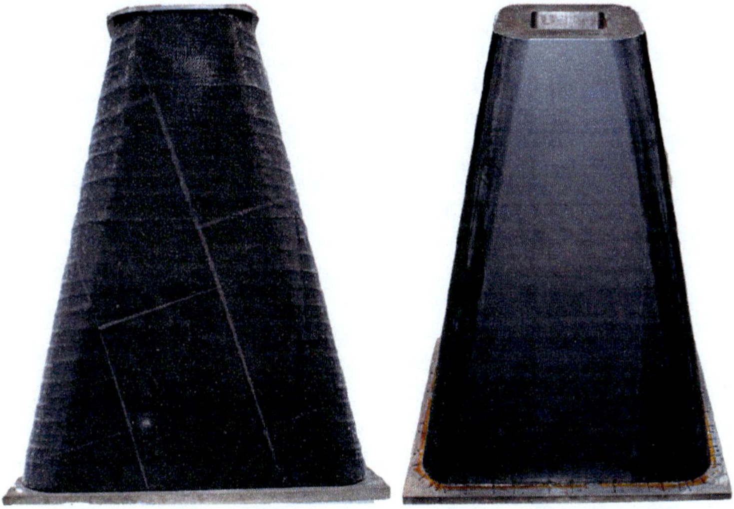

Fig. 3. Manufacturing picture of Extinction honeycomb (Left, Extinction honeycomb. Right, fixed skin)

Pressure should be applied evenly on the outside of the skin to ensure the glue quality of honeycomb and fixed skin, At the beginning, the vacuum bag pressurization method was used to pressurize the skin with vacuum pressure. However, after the glue is cured, it is found that there are many bubbles on the adhesive surface between the honeycomb and the skin. The bubble can be punctured, and there is a risk of later dropping to form excess. In order to avoid this phenomenon, this paper optimizes the process scheme and proposes "no vacuum bag uniform compression method". According to the optimization scheme, the adhesive layer of final preparation of extinction honeycomb and skin is flat and with no bubble (Fig. 4).

Fig. 4. Contrast of Extinction honeycomb (Left, with vacuum bag. Right, no vacuum bag)

3.3.2 Outer Skin and Inner Skin Molding

Apply the lay up on the extinction honeycomb-mold in the ply order: [+45/−45/0/−45/+45].The inner skin is cured as a whole piece, and each layer is staggered with each other. In order to prevent excessive distortion in the direction of 45°, the inner skin is laid in an A direction and B direction, and the layer is extended to the direction C and D, and each layer of stitching seams are staggered with each other. For the 0°layer, take the direction A(C) plane as the benchmark, along the direction of the centerline of the plane and extend it to both sides. When assembling, the width of the C-directional and D-directional layers and the laminated joints is about 30mm.When plying, you need to lay the slope in the lap area in advance. Details are shown in Fig. 5.

Apply the lay up on the outer skin mold in the ply order:[+45/−45/0/−45/+45].The outer skin is divided into two seas, which is an A-directional and B-directional overall layer, and the layer is extended to the direction C and D, and each layer of stitching seams are staggered with each other. For the 0°layer, take the direction A(C) plane as the benchmark, along the direction of the centerline of the plane and extend it to both sides. When assembling, the width of the C-directional and D-directional layers and the laminated joints is about 30 mm.When plying, you need to lay the slope in the lap area in advance. Details are shown in Fig. 6.

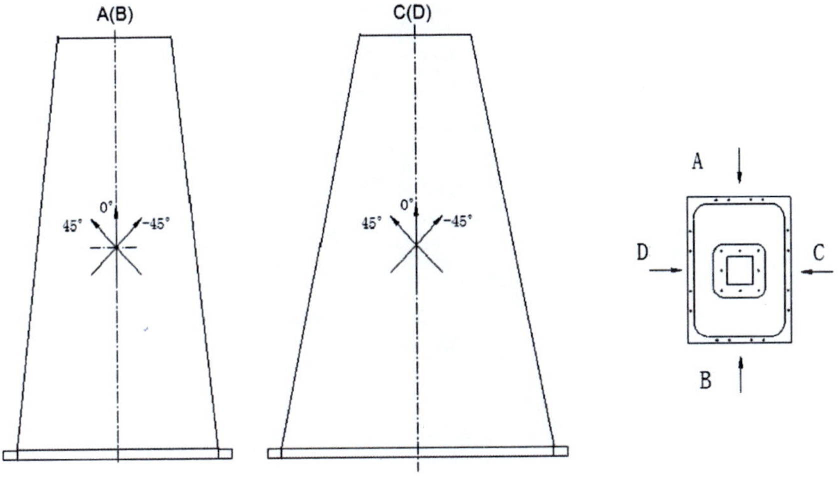

Fig. 5. Direction of inner skin ply

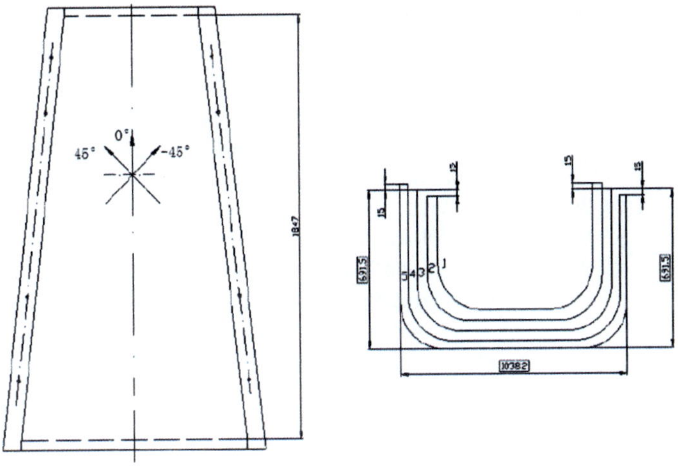

Fig. 6. Direction of outer skin ply

3.3.3 Cylinder Manufacturing

Apply J-47 glue film to the prepared inner and outer skin, and then place the honeycomb on the inner skin to ensure that is completely fitted to the inner skin. A honeycomb grid needs to be reserved for compression at the stitching joints. Then place the outer skin and the pressure plate, package it, the put it into the autoclave for curing. The cured product is shown in Fig. 7.

Fig. 7. Honeycomb sandwich cylinder after curing

3.3.4 Flange and Extinction Honeycomb Glue

As the external double-click of the baffle, the flatness of the flange is high. When it's glued, stress-free must be carried out to avoid the deformation of its product. When glued, test-assemble the flange with the baffle to check the gap and coordination. Then apply 420 glue to the adhesive surface of the flange and baffle evenly, and fill the assembly gap with carbon sheets according to the test-assembly results.

The flange material is aluminum alloy, it is inconsistent with the thermal expansion coefficient of the carbon fiber skin, and the orbital temperature difference of the baffle more than 240°C, there is a risk of deboning between the flange and the baffle cylinder. In order to avoid risk, two measures have been taken to strengthen it: First, connect the flange with baffle by gluing the carbon cloth, the one-sided width is 15mm, almost 2 layers. Second, strengthen it by mechanical connection. Install the screw hole under the flange to bury the post-insets, and use the screw to block the single connection of the reinforcement sheet. As shown in Fig. 8.

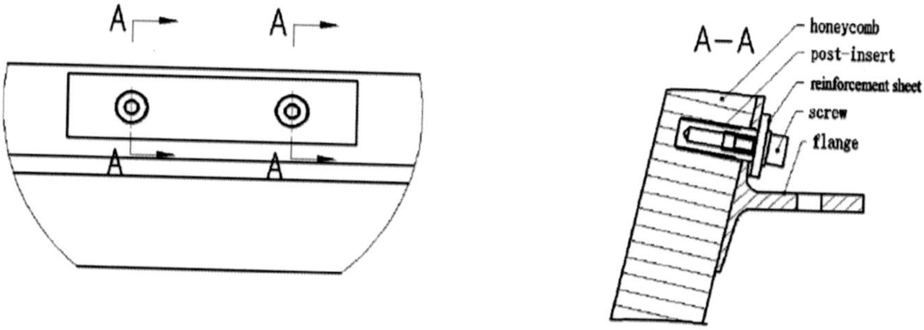

Fig. 8. Reinforcement of the flange

4 Product Results and Test Status

4.1 Product Results

The baffle was finally successfully manufactured according to the above process plan. The appearance of the product is good, the glue quality has passed the detection, and the size meets the requirements of the design index. The final product is shown in Fig. 9. The product's mass is 27.49 kg$<$30 kg, it meets the design requirements. The measured data is shown in Table 1.

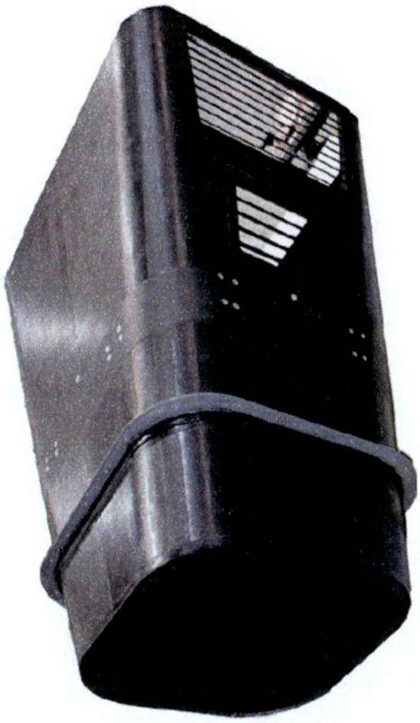

Fig. 9. The picture of product

4.2 Product Results

4.2.1 Fundamental Frequency Test

The baffle passes the characteristic sine sweeping test, and the results meet the requirements of the design index. The data is shown in Table 2.

Table 1. Measured data of the product

Inspection item	Design parameters	Measured data	Result
Envelop size/mm	$(1906 \pm 3) \times (1382 \pm 1.2) \times (1038 \pm 1.2)$	$1850.0 \times 1382.6 \times 1038.4$	Qualified
Flatness of flange/mm	≤ 0.2	0.19	Qualified
Height of flange/mm	278 ± 0.5	278.3	Qualified
Position of hole on flange/mm	φ0.1	φ0.10	Qualified
Position of post-inset/mm	φ0.3	φ0.29	Qualified
Resistance value/Ω	≤ 200	5.3 ~ 190.4	Qualified

Table 2. Data of the fundamental frequency test

Direction	Design parameters	Measured data
Z	≥70 Hz	144.3 Hz
X	≥45 Hz	68.5 Hz
Y	≥45 Hz	81.6 Hz

4.2.2 Sine Vibration Test

The qualified sinusoidal vibration test was performed, and the experimental conditions are shown in Table 3.The acceleration response of the baffle was tested in the test. The extraction point was located at key positions such as the top corner point and the midpoint of the outside surface. (Fig. 10)The acceleration response is shown in Table 4–5.The results show that some parts in the sinusoidal vibration input condition have large response. So the intensity of baffle is checked in the sinusoidal vibration condition. After analysis, the minimum safety margin of the honeycomb sandwich cylinder is 2.55>0.5, appearing in y-direction sinusoidal condition. The maximum Von-Mises stress is 44.1 Mpa of the flange and the minimum safety margin is 4.9>0.5, appearing in y-direction sinusoidal condition. In conclusion, the strength under the sinusoidal vibration condition meets the design requirements. On the other hand, the baffle is good,with no stratification, debonding, crack and other defects during the whole vibration test process, which proves that the baffle's manufacturing technology is correct and effective (Table 5, Table 6, Fig. 11, Fig. 12).

Table 3. The sinusoidal vibration test conditions

Direction	Frequency range Hz)	Acceleration	Scan frequency
X	5 ~ 10	10.43 mm(O-P)	2 cot/min
	10 ~ 25	4.2 g	
	25 ~ 40	To 12 g	
	40 ~ 70	12 g	
	70 ~ 100	To 7.5 g	
Y	5 ~ 10	10.43 mm(O-P)	2 oct/min
	10 ~ 25	4.2 g	
	25 ~ 35	To 10.5 g	
	35 ~ 80	10.5 g	
	80 ~ 100	To 6g	
Z	5 ~ 12	7.75mm(O-P)	2 oct/min
	12 ~ 30	4.5 g	
	30 ~ 40	To 9 g	
	40 ~ 80	9 g	
	80 ~ 100	To 15 g	

Table 4. Acceleration response on X-direction

Position	X-response (g/Hz)	Y-response (g/Hz)	Z-response (g/Hz)
Top four corners	36.1/81.0	18.8/81.0	9.9/81.0
Midpoint of the X side	21.2/51.0	0.1/82.0	5.8/82.0
Midpoint of the ± Yside	11.7/79.0	23.5/81.0	3.4/81.0

Table 5. Acceleration response on Y-direction

Position	X-response (g/Hz)	Y-response (g/Hz)	Z-response (g/Hz)
Top four corners	32.5/69.0	59.4/68.0	15.8/68.0
Midpoint of the X side	0.4/81.0	17.4/67.0	0.2/80.0
Midpoint of the ± Yside	8.4/68.0	44.8/68.0	10.1/69.0

Table 6. Acceleration response on Z-direction

Position	X-response (g/Hz)	Y-response (g/Hz)	Z-response (g/Hz)
Top four corners	8.3/82.0	4.8/82.0	11.4/100.0
Midpoint of the X side	4.3/52.0	0.1/83.0	10.6/100.0
Midpoint of the ± Yside	1.6/82.0	6.0/81.0	10.6/100.0

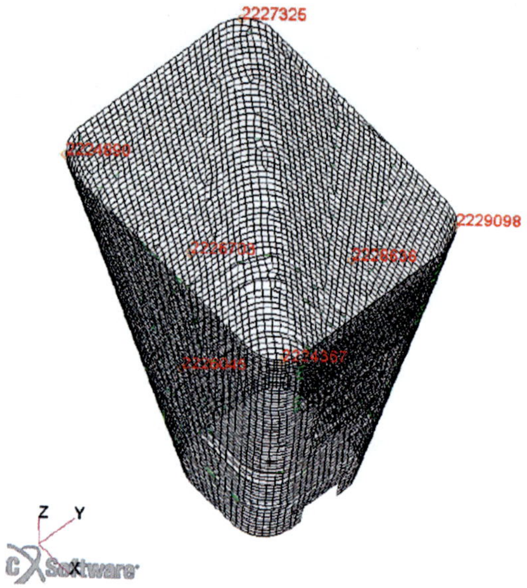

Fig. 10. The point of acceleration response

4.3 On-Orbit Applications

This type of satellite has been successfully launched recently and is operating normally in orbit. The image quality is excellent after the camera is turned on, and the imaging quality meets the expected design requirements. The successful application of this type of remote sensing camera in orbit proves that the structural mechanical properties of the baffle meet the conditions of satellite launch and on–orbit loads. At the same time, it also proves that the baffle has successfully withstands a temperature difference more than 240°C, and various performances meet the design requirements.

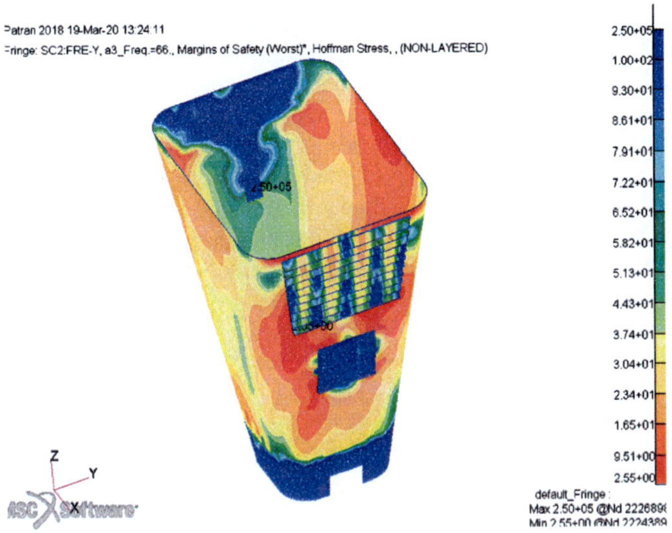

Fig. 11. Safety margin of honeycomb sandwich cylinder (sine on Y-direction, 55 Hz)

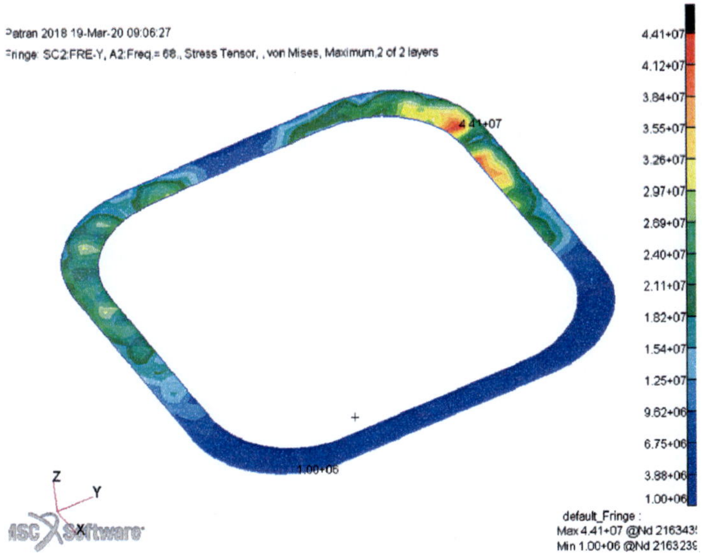

Fig. 12. The flange Von-Mises stress (sine on Y-direction, 57 Hz)

5 Concluded Remarks

With the development of space remote sensing detection, large-sized, lightweight and complex composite baffle have increasingly become an important part of remote sensing cameras [4]. Among the structural types of baffle, double-layer honeycomb sandwich structure is more and more widely used in large-size baffle. In the process of engineering application, how to take into account the special requirements of optical remote sensors and the characteristics of composites is an issue that we need to pay attention to. Combined with the design index of the large-size remote sensing camera baffle and manufacturing process characteristics, this paper formulates a manufacturing process. And by optimizing the mold form to achieve both product quality and cost results, a no vacuum bag uniform compression method is proposed to solve the problem of adhesive foaming next to the extinction honeycomb. Finally, the product has been successfully tracked with the remote sensor after a series of mechanical tests. The results prove that the manufacturing technology of large-size remote sensing cameras introduced in this paper is reliable and feasible.

References

1. Shiping, C.: Development of space remote sensing science and technology. Spacecraft Eng. **18**(2), 1–7 (2009)
2. Linghui, Z., Ping, C., Qijie, W., et al.: Composite materials and technology of baffle for space remote sensor. J. Astronaut. **35**(6), 726–733 (2014)
3. Ping, C., Linghui, Z., Shikui, L.: Development technology of shutter-type baffle in space remote sensor. Aerosp. Manuf. Technol. **12**(6), 46–49 (2012)
4. Yuncai, H.A.O., Shuqin, X.I.A.O., Lixia, W.A.N.G.: Current situation and development of stray light elimination techniques in space-borne optical remote sensors. Chinese Space Sci. Technol. **15**(3), 40–50 (1995). (in Chinese)
5. Sheng, L.I.A.O.: Study on Stray Light Elimination of Optical System. University of Electronic Science and Technology of China, Chengdu (2003). (in Chinese)
6. Ping, C., Linghui, Z., Shikui, L., et al.: Forming technology of complex curved surface baffle. Spacecraft Recovery Remote Sens. **34**(6), 81–87 (2013)
7. Dongqi, W., Donghua, S., Fang, T.: Development of special shaped bames with large opening and double layered honeycomb sandwich structure. Aerospace Manuf. Technol. **6**(3), 33–37 (2016). (in Chinese)

Check for updates

Estimation of Sensor Noise Parameters from Remote Sensing Data Using Convolutional Neural Networks and Mixed Synthetic and Real Data

Mikhail Uss[1](✉), Benoit Vozel[2], Sergey Abramov[1], and Vladimir Lukin[1]

[1] National Aerospace University, Chkalova Str.17, Kharkov 61070, Ukraine
uss@xai.edu.ua, {s.abramov,v.lukin}@khai.edu

[2] University of Rennes 1, IETR UMR CNRS 6164, 22305 Lannion Cedex, France
benoit.vozel@univ-rennes1.fr

Abstract. Characterization of remote sensing (RS) sensors noise is important for optimization and analysis of potential accuracy of subsequent image processing stages (such as filtering or compression, image registration accuracy or objects classification). Blind noise parameters estimation (BNPE) allows analyzing sensor noise directly from noisy images. BNPE is complementary to direct sensor calibration methods and is able to provide additional noise parameters measurements when calibration procedure is not accurate or cannot be performed. Recent BNPE methods show improved performance by utilizing learning-based approach. This paper aims at solving the problem of lack of training data with precise ground truth noise parameters needed for training BNPE models. We focus on learning-based approach using calibrated real data, mixture of synthetic and real data and novel semi-supervised training with uncalibrated real data. The results show that BNPE method training in semi-supervised manner performs better than model-based approach and comparable to supervised training. These results open up possibility of easy extension of training datasets for BNPE methods.

Keywords: blind noise parameter estimation · isotropic/anisotropic fractional Brownian motion · multivariate noise variance model · noise signal-dependency · convolutional neural network · deep regression with uncertainty · hard mining

1 Introduction

Multiple sources of random noise are always encountered for any digital image acquisition system. This is reflected in the images acquired by the presence of different components of noise, the strength of which can vary depending on signal content and level and wavelength. Thus, accurate knowledge of the sensor's noise model and its characteristics is a prerequisite for efficient image post-processing, including image restoration, lossy compression, registration, classification and/or fusion [1]. If a deviation occurs with respect to a prior calibration of the captured noise or, at worst, if such recent calibration data are not directly available, blind noise parameter estimation (BNPE) is a feasible

H. P. Urbach and H. Jiang (Eds.): ISSOIA 2022, SPPHY 295, pp. 469–485, 2023.
https://doi.org/10.1007/978-981-99-4098-1_41

solution. The intrinsic essence of the BNPE-based methods is to recover the image noise parameters directly from the noisy data themselves. Many BNPE methods are already available in the literature. Some of them deal with the simplistic pure additive noise, others are relevant for more realistic and also more complex situations of signal-dependent [2], spatially correlated [3], multi-parametric [4] and non-stationary [5] models of the noise.

Estimating accurate noise statistics directly from the images is a rather complex task due to the spatial and spectral variability of their contents. This is the reason for which most BNPE techniques exploit in some way a set of patches or local blocks of images detected as sufficiently suitable for this purpose. Indeed, noise parameter estimates are most likely to be accurate in homogeneous areas (with very small variations of the signal level) and conversely inaccurate in non-uniform areas (containing mainly edges or textures) [6]. It is then essential to detect inappropriate image patches and eliminate outlying estimates from any further consideration as they can only contribute unfavorably to the final estimation accuracy. This can be achieved by robustly processing a collection of noise estimates (standard deviation, SD) obtained from many image patches or on a per-patch basis by evaluating their estimation confidence individually and rejecting those with a too large uncertainty [7–9]. Both ways to do can be considered separately or together. Now the confidence of a patch-based estimation can be derived either following a model-based approach [10] by regression with accounting for uncertainty or a learning-based one [11] with a specific structure of convolutional neural network (CNN).

The model-based approach initially assumes an appropriate model to describe the local texture of a patch. Isotropic (fBm) and anisotropic (afBm) fractal Brownian motion models have been considered in our previous works [10, 12] and [13]. Once the correlation function of either field is derived, the estimation of the parameters of interest can be obtained in a second step following a maximum likelihood (ML) approach. Provided that an image patch complies with the assumed texture model (either fBm or afBm), its confidence for noise SD estimation can be characterized by Cramér–Rao lower bound (CRLB) on noise SD estimation. Note however that the lower bound calculated so only approximates a real lower bound for all deviations of the patch structure from the assumed fBm model (including edge-like patches, patches with complex structures like urban areas).

Learning is another way to handle the complexity and diversity of local textures in patches. Learning-based methods have thus recently appeared in the literature with the special feature of not having to resort to parametric modeling but at the cost of the designing of a specific network structure and performing an appropriate learning (as well as the availability of a sufficiently large learning database). However, the challenge remains the same for learning-based methods since the ultimate goal is not only to estimate the noise parameters but also its uncertainty on a per-batch basis of patches. In [11], an original CNN-based network has been proposed. It is structured around two parallel branches each having a specific output but working towards a shared objective that is to predict the noise SD for one branch and the uncertainty of the previous branch output for the other branch. Of course, the training of the proposed network with a special loss function should ensure that the uncertainty branch learns to predict the SD of noise SD estimation error. A strong constraint to be relaxed in such a process is whether knowledge of the ground truth (GT) about the noise SD is necessary for the learning process. This would make the proposed method advantageously evolving to a less supervised scheme and possibly independent of the noise SD value.

A substantial problem of training BNPE CNN is availability of GT sensor noise information. This information can be obtained by calibrating a sensor using multiple images captured with this sensor. This approach has two drawbacks. First, a sensor calibration requires significant efforts. For example, noise parameters of a DLSR (Digital Single-Lens Reflex) camera depends on ISO value and colour channel. Each combination should be calibrated independently by collecting images of a calibration pattern [14]. Second, accuracy of calibration procedure is limited. A bias in calibration data will propagate to the trained CNN output. As a result, noise parameters predicted by this CNN with be biased in a hardly predictable way. Finally, BNPE methods are designed to be applied to target sensors for which calibration procedure is too complex to perform (otherwise, BNPE approach is not needed). Data acquired by the target sensor could have specific features that are not present in the training data. In this case, CNN predictions could be affected by the domain shift [15].

This paper aims at addressing the problem of limited amount of training data for BNPE methods by proposing "calibrate-while-training" approach. According to this approach both synthetic data with known noise parameters (we use afBm data for this purpose) and real data with unknown parameters are used for training. Noise parameters for the real data are estimated in the course of the training process by special grouping of patches with the same (but unknown) noise SD.

The contributions of this work are summarized as follows:

- Analysis of data sources available for the learning-based approach including synthetic data and calibrated real data
- A new approach for training BNPE CNNs in semi-supervised mode using synthetic data and uncalibrated real data

The reminder of this paper is organized as follows. Section 2 gives a brief survey on blind noise parameter estimation approaches. Section 3 recalls the BNPE method developed using the fBm model. Subsequently, Sect. 4 describes the learning-based method (the NoiseNet CNN) and presents new results on its semi-supervised training. For all described methods, comparative experimental results are provided. Section 5 concludes this work with several remarks and perspectives.

2 Review of Blind Noise Parameters Estimation Methods

The goal of BNPE is to estimate parameters of a sensor Noise Level Function (NLF) using only noisy images. Let us define sensor noise variance as σ_n^2. In the cases of practical interest, σ_n^2 is a polynomial function of image intensity $\sigma_n^2 = \sigma_n^2(I)$:

- $\sigma_n^2(I) = \sigma_{n.a}^2$ for pure additive noise (e.g. thermal sensor noise);
- $\sigma_n^2(I) = \sigma_{n.SD}^2 I$ for pure Poisson-like noise (typical for optical sensors);
- $\sigma_n^2(I) = \sigma_\mu^2 I^2$ for pure multiplicative noise (typical for radar sensors);
- $\sigma_n^2(I) = \sigma_{n.a}^2 + \sigma_{n.SD}^2 I$ for mixture of additive and Poisson-like noise;

- $\sigma_n^2(I) = \sigma_{n.a}^2 + \sigma_\mu^2 I^2$ for mixture of additive and multiplicative noise.

In the more general case, $\sigma_n^2(I)$ can be approximated by a polynomial of power n as $\sigma_n^2(I) = \sigma_n^2(I, \mathbf{c}) = c_0 + c_1 I + c_2 I^2 + ...c_n I^n$, where $\mathbf{c} = (c_0, c_1, c_2, ...c_n)$ is the vector of coefficients.

Methods for BNPE have a long development history. Currently there exists a whole spectrum of methods that cover different noise and sensor types: pure additive [16, 17], Poisson [2, 18], multiplicative [19], general signal-dependent [20, 21] noise NLF; spatially uncorrelated and correlated noise [3, 22]; univariate and multivariate noise [4]; single-channel [10], multi-channel [23] and hyperspectral sensors [24–26]. Apart from the main goal of sensor noise calibration, BNPE methods could be used for Digital Elevation Model (DEM) products quality characterization [4] or even image forensics [27].

The main idea behind BNPE methods is to use the different properties of the true image texture and the sensor noise. A large group of methods utilized differences in spatial and/or spectral domains. This could be done either heuristically or with a model-based approach. Following recent tendency, learning-based BNPE approaches are gaining more interest because of their ability to better adapt to the complex structure of real images. Let us briefly recall both groups of methods.

2.1 Methods Operating in Spatial, Spectral or Spatial-Spectral Domains

BNPE methods operating in the spatial domain utilize homogeneous areas (HA) in the noisy image [28]. Noise variance is then estimated from these areas simply by calculating sample variance. Although manual selection of HAs can be used in some applications [29], automatic selection is preferable. The idea consists in splitting an image into small fragments, calculating for them local means and local SDs [6]. For pure additive noise, noise SD can be estimated as an histogram mode of the local SD estimates [30]. For signal-dependent noise, the scatterplot approach employs a robust curve fit to local variance estimates in mean-variance plane [31]. Liu et al. [21] incorporated a priori information on NLF to improve robust curve fit accuracy in the digital camera applications. As hypothesis on HAs could be quite restrictive due to a limited area of HAs, spatial methods may result in biased noise SD estimates or fail providing a practically desired accuracy [32].

To improve the performance of spatial methods, methods operating in the spectral domain utilize the fact that an image texture is usually smoother than noise. As a result, after applying suitable decorrelation transform, image texture is concentrated in low frequency coefficients while noise (spatially uncorrelated one) spreads uniformly among low and high transform coefficients.

Several methods have been proposed for the spectral separation of image texture and noise. High-pass filtering is used to suppress smooth texture variations in data-masking method [33]. In [34], a data-masking approach has been adapted to handle the signal-dependent noise. Among the orthogonal transforms, wavelet [35], DCT [36] and 3D DCT [37] transforms have been successfully applied for noise SD estimation. For multi- or hyperspectral data, one more information source should be used in addition to the spatial/spectral dimensions, namely the inter-band correlation of image data [26, 38].

2.2 Model-Based Approach

Model-based approaches use the same spatial and spectral information, but are formalized using an assumed image model. The model-based approach presumes that small image patches are consistent with an assumed parametric model of the local image texture. Both isotropic and anisotropic fractal Brownian motion models, with their related correlation functions have been considered in our previous works [10] and [12]. Then, a maximum likelihood estimation of the whole set of parameters, including those of the assumed field (either fBm or afBm) plus the noise SD has been naturally preferred, as it allows to characterize the estimation accuracy of the involved parameters quite straightforwardly with a CRLB of parameters' estimation variance [39]. In addition to the noise SD (main parameter of interest here), the set of field parameters includes the Hurst exponent and the texture amplitude in case of a presupposed fBm field augmented with the texture rotation angle and the texture amplitude along and across rotation direction in case of the more complex but of more general application afBm field.

The model-based approach assumes that the set of image texture parameters and noise SD should be estimated for each small image patch. However, these two groups of parameters are conflicting with each other. Noise SD can be estimated from patches with relatively high level of noise and low level of image texture (ideally a pure homogeneous patch composed of only sensor noise). For such patches, image texture parameters cannot be accurately estimated. On the contrary, image texture parameters can be accurately estimated for noise-free patches. For these noise-free patches, the noise SD cannot be accurately estimated. We called the first type of patches Noise Informative (NI) and the second type – Texture Informative (TI). The value of CRLB $\sigma_{\mathrm{n.LB}}$ can be used as a criterion for detecting NI patches.

The noise SD estimation scheme called NI+fBm [10] uses both NI and TI areas by iteratively refining image texture parameters using TI patches and noise SD using NI patches. A more detailed description of the whole procedure is given in [39] and [10]. Note that the model-based approach and the underlying idea of splitting NI and TI patches can be applied to pure additive or signal dependent noise, spatially uncorrelated or correlated noise, univariate or multivariate NLF, single-, multi- or hyperspectral remote sensing data.

The model-based approach requires that image patches follow an assumed model, in our case either the fBm or afBm models. For patches that deviate from these models (patches with edges, small-sized objects, regular textures), results could become biased. To cope with this problem, we proposed in [39] a robust version of NI+fBm called NI+DCT. In this estimator, the final estimation stage of the noise SD is implemented in DCT domain. In NI+DCT, for each NI patch of size 7 by 7 pixels, DCT transform is applied. We consider only the six highest frequency coefficients with indices (7, 7), (6, 7), (7, 6), (6, 6), (5, 7), and (7, 5). These DCT coefficients are collected from all NI patches and the final noise SD estimate is obtained as sample SD of the array thus formed of these DCT coefficients. In what follows, we use the result of NI+DCT approach as a baseline for comparison with learning-based methods.

2.3 Learning-Based BNPE Approaches

The advantage of the learning-based approach is in its ability to better adapt to complex structure of image texture during training process on a large database of real examples. J. Yang et al. [20] proposed to estimate NLF of signal-dependent noise using multi-column CNN (MC-Net) with an end-to-end training architecture. The multi-column architecture allows using kernels of different size for NLF analysis providing information at different scales or spatial frequencies. Output of three columns with input kernels of size 5×5, 7×7 and 9×9 are fused together using SPP (Spatial Pyramid Pooling) CNN [40], fed to fully-connected layers to provide NLF estimation.

A deep convolutional neural network named "deep residual noise estimator" (DRNE) is proposed in [41] to estimate pixel-wise noise variance values. In this approach, the assumption that the noise variance is the same for patches with the same intensity is dropped. However, it is not clear what kind of sensor could produce this type of noise.

The main challenge with an end-to-end learning approach that directly predicts a NLF function is the need in large image databases that are hard to collect. S. Bahnemiri et al. designed in [5] a deep convolutional autoencoder (DCAE) for estimation of a non-stationary noise variance. In contrast to [40] and [41], DCAE is trained on local image patches and not on large scale images. In this manner, variability of the training dataset could be assured using a smaller amount of real images.

In the next section, we will discuss the NoiseNet CNN that operates on image patches similar to DCAE. The advantage of the NoiseNet is in its ability to characterize uncertainty of noise variance estimation in local patches.

3 Learning-Based BNPE Using Uncalibrated Data

In this section, we address the problem of training dataset collection by developing a semi-supervised training scheme where samples with the true noise variance are synthesized and real data can also be used for training without knowing the true noise parameters.

3.1 Confidence of Noise SD Estimation

Whichever the BNPE method is followed to estimate the parameters of the noise inherently present in observed images, either model-based or learning-based, the confidence or uncertainty of these estimates has to be quantified to make these outputs reliably useful in the subsequent processing tasks. In the estimation setting, a lower bound on the SD of the estimates (noise SD here) is commonly considered as a meaningful confidence. We refer to this confidence as $\sigma_{\mathrm{n.LB}}$ where LB stands for Lower Bound and "n" relates to noise. In the remainder of this paper, this confidence is by default normalized by the standard deviation of the noise for ease of convenience of future analysis ($\sigma_{\mathrm{n.LB}} = 0.1$ means that the estimation SD of noise SD equals to $0.1 \cdot \sigma_{\mathrm{n}}$, where σ_{n} is the noise SD). We denote the potential SD corresponding to the estimation error of the noise SD using afBm model as $\sigma_{\mathrm{n.LB.afBm}}$.

Once this confidence has been calculated (depending on the BNPE approach followed), a thresholding rule $\sigma_{\mathrm{n.LB}} < \sigma_{\sigma_t}$ on its value can be set appropriately to separate reliable estimates from too approximate ones, keeping the former and discarding the latter with the goal to improve the accuracy of the final estimate. Indeed, the distribution of noise SD estimates often tends to be non-Gaussian from a certain value of this threshold $\sigma_{n.LB} > \sigma_{\sigma_t}$ [39].

3.2 Training of CNN for Noise Parameters Estimation

According to the learning-based approach, the estimation of the noise SD is solved indirectly in the form of a regression with uncertainty. For each patch, we have proposed a neural network, denoted NoiseNet, to predict an estimate of the noise SD as well as its associated uncertainty. The underlying idea behind this learning-based approach is that it may compensate for the failure of the model-based approach for particular structures in the image not adapted to the isotropic or anisotropic fractal Brownian field (especially the isotropic one).

The general structure of our learning-based BNPE approach (Fig. 1) consists of a basic branch typically in charge of feature extraction, which then splits into two parallel specialized sub-branches, one is for estimating the noise SD via regression and the other one for quantifying the uncertainty observed at the output of the regression sub-branch.

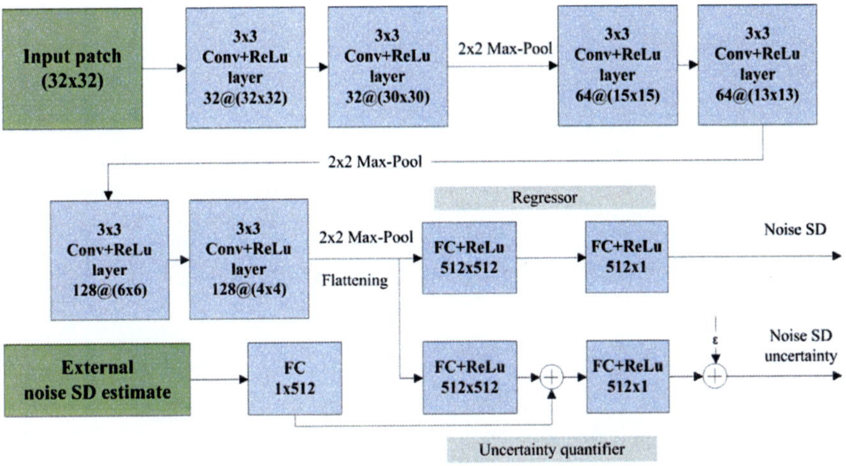

Fig. 1. The NoiseNet CNN structure. "Conv" denotes 2D convolutional layer, "FC" – fully connected layer, "ReLu" -rectified linear unit, "SD" – Standard Deviation.

3.3 Isotropic and Anisotropic Fractal Brownian Motion Model

Isotropic fBm model $B_H(t, s)$ is a non-stationary Gaussian process. It is zero at the origin, $B_H(0, 0) = 0$, and has correlation function that can be presented as follows

$$\langle B_H(t,s) \cdot B_H(p,q) \rangle = 0.5\sigma_x^2 \left(\sqrt{t^2+s^2}^{2H} + \sqrt{p^2+q^2}^{2H} - \sqrt{(t-p)^2+(s-q)^2}^{2H} \right), \quad (1)$$

where $H \in (0, 1)$ denotes the Hurst exponent that describes texture roughness ($H \to 0$ for a very rough texture, $H \to 1$ for a very smooth one), σ_x^2 denotes the variance of increment of fBm process on unit distance.

Anisotropic fBm can be obtained by applying a linear spatial transform to the isotropic fBm [42]. Let the fBm field to be compressed for axes t and s in such a manner that the parameter σ_x for them is equal to σ_{x1} and σ_{x2}, respectively. For simplicity, we assume that $\sigma_{x1} < \sigma_{x2}$. The new field is anisotropic and its structural function is expressed as

$$D_x(t, s, p, q) = \left\langle (x(t, s) - x(p, q)^2 \right\rangle = \left(\sigma_{x1}^{\frac{2}{H}} \cdot (t-p)^2 + \sigma_{x2}^{\frac{2}{H}} \cdot (s-q)^2 \right)^H. \quad (2)$$

Introducing the change of variables $t - p = \rho \cdot \cos(\varphi)$, $s - q = \rho \cdot \sin(\varphi)$, where $\rho = \sqrt{(t-p)^2 + (s-q)^2}$, $tg(\varphi) = (s-q)/(t-p)$, we obtain

$$D_x(t, s, p, q) = \left(\sigma_{x1}^{\frac{2}{H}} \cdot \cos(\varphi)^2 + \sigma_{x2}^{\frac{2}{H}} \cdot \sin(\varphi)^2 \right)^H \rho^{2H}. \quad (3)$$

If the field is additionally rotated by an angle α, then

$$D_x(t, s, p, q) = \left(\sigma_{x1}^{\frac{2}{H}} \cdot \cos(\varphi - \alpha)^2 + \sigma_{x2}^{\frac{2}{H}} \cdot \sin(\varphi - \alpha)^2 \right)^H \cdot \rho^{2H} = \sigma_x^2(\varphi, \alpha) \cdot \rho^{2H}. \quad (4)$$

The parameter $\sigma_x^2 = \sigma_x^2(\varphi, \alpha)$ of the obtained field depends on the direction. If one draws a straight line from the origin at an angle φ, then the values of the anisotropic field lying on this straight line represent a one-dimensional Brownian motion with the parameters H and $\sigma_x^2(\varphi, \alpha)$.

The correlation function of the resulting field is determined by the following expression

$$R_x(t, s, p, q) = 0,5 \cdot \left(\sigma_x^2(\varphi_1, \alpha) \cdot \rho_1^{2H} + \sigma_x^2(\varphi_2, \alpha) \cdot \rho_2^{2H} - \sigma_x^2(\varphi_3, \alpha) \cdot \rho_3^{2H} \right), \quad (5)$$

where $\rho_1 = \sqrt{t^2 + s^2}$, $\rho_2 = \sqrt{p^2 + q^2}$, $\rho_3 = \sqrt{(t-p)^2 + (s-q)^2}$, $tg(\varphi_1) = s/t$, $tg(\varphi_2) = q/p$, $tg(\varphi_3) = (s-q)/(t-p)$.

For the afBm field, we denote the vector of parameters as $\boldsymbol{\theta}_{afBm} = (H, \sigma_{x1}, \sigma_{x2}, \alpha)$. Examples of afBm realizations are depicted in Fig. 2 (in these examples $\boldsymbol{\theta}_{afBm}$ depends on spatial coordinates).

3.4 Training of CNN for Noise Parameters Estimation Using Mixture of Simulated and Real Data

In the original version of the NoiseNet, for jointly learning noise SD and its variance we have used a special form of loss function [43] defined as

$$L_{regression} = \left(\frac{\hat{\sigma}_n - \sigma_{n.0}}{\sigma_{uncert}\, \sigma_{n.0}} \right)^2 + 2\,ln(\sigma_{uncert}\, \sigma_{n.0}). \quad (6)$$

Fig. 2. Examples of simulated afBm field realizations with spatially varying parameters.

where $\hat{\sigma}_n$ is predicted value of noise SD, $\sigma_{n.0}$ is the true value, $\sigma_{uncert} = \sigma_{\mathrm{n.LB.NNet}}$ is output of uncertainty branch that corresponds to the $\sigma_{\mathrm{n.LB}}$ value predicted by the NoiseNet.

The estimation uncertainty of the noise SD predicted by the NoiseNet is utilized in two ways jointly: to weight appropriately the noise SD estimates and to reject the unreliable estimates. Therefore, the NoiseNet solves a mixed regression-classification problem. Using only real data and loss (6), the classification term is not present in the loss. With synthetic afBm data, this becomes possible.

We split the range of $\sigma_{\mathrm{n.LB}}$ variation into two ranges: less than 0.3 and greater than 0.3. In the first range, we require $\sigma_{\mathrm{n.LB.NNet}}$ to be as close to $\sigma_{\mathrm{n.LB}}$ as possible. In the second range, we only require that $\sigma_{\mathrm{n.LB.NNet}} > 0.3$. The latter requirement is sufficient to discard unreliable patches; $\sigma_{\mathrm{n.LB.NNet}}$ should not regress $\sigma_{\mathrm{n.LB}}$ in this upper range. We proposed the following loss function for synthetic afBm data [13]:

$$L = \left(L_{regression} + L_{confidence}\right)(1 - h(\sigma_{\mathrm{n.LB}} - 0.3)) + L_{classification}h(\sigma_{\mathrm{n.LB}} - 0.3), \quad (7)$$

where $h(x) = 1$ if $x > 0$ and 0 otherwise denotes the step function, $L_{regression}$, $L_{confidence}$, and $L_{classification}$ are the loss terms introduced for carrying out noise SD regression, confidence approximation and discrimination between reliable and non-reliable patches:

$$L_{confidence} = \beta_1 \left(\frac{\sigma_{uncert}}{\sigma_{\mathrm{n.LB}}} + \frac{\sigma_{\mathrm{n.LB}}}{\sigma_{uncert}} - 2 \right)^2 \quad (8)$$

$$L_{classification} = \beta_2 \mathrm{ReLu}(0.3 - \sigma_{uncert}), \quad (9)$$

The term $L_{regression}$ is described by. For real data, only the term $L_{regression}$ is applied. Based on extensive experiments, we selected $\beta_1 = 10$, $\beta_2 = 100$.

For training, samples corresponding to 5120 afBm parameter vectors were generated. Each vector was sampled randomly according to the following rule: the Hurst exponent is sampled randomly in the interval 0.1 to 0.95, $r_\sigma = \sigma_{x1}/\sigma_{x2}$ - in interval from 0.1 to 1 ($\sigma_{x1} = \sqrt{r_\sigma}\sigma_{x2} = 1/\sqrt{r_\sigma}$), α in the interval $-\pi/2 \ldots \pi/2$ and the noise SD – in the interval 0.1…10. Corresponding value of $\sigma_{\mathrm{n.LB}}$ has been calculated for each parameter vector. For each afBm parameter vector, 25 samples of size 64 by 64 pixels were generated. In total, 128000 afBm samples of the size 64 by 64 pixels were generated. To improve the NoiseNet performance, in [13] we proposed to generate more versatile and complex set of afBm samples using hard mining approach. A sample was selected for being included in the training batch with higher probability proportional for those afBm samples for which NoiseNet results in high bias of noise SD estimates or a significant deviation of the predicted uncertainty of the noise SD estimate from the lower bound $\sigma_{n.LB}$ occurs.

3.5 Semi-supervised Training of CNN for Noise Parameters Estimation

Addition of synthetic data to the NoiseNet training is able to improve the model performance (see results in Table 1 in the experimental section). However, synthetic data need to be used together with real data that may have no noise calibration parameters. To overcome the problem of using data with unknown noise parameters, we propose semi-supervised learning by using both synthetic data with available GT noise parameters and GT-free real data directly from the target sensor. Our approach can be described as "calibrate-while-training" (see Fig. 3).

To inject real data without GT into the NoiseNet training, we rely on the following assumptions: (1) the sensor noise parameters are unknown but depend on known predictors; (2) the noise parameters can be transformed in a predictable way by specific data transformations or by adding artificial noise to real data. Let us discuss these assumptions more in details.

For optical and radar sensors, noise depends on image intensity in a functional way $\sigma_n^2 = f(I, \boldsymbol{\lambda})$, where $\boldsymbol{\lambda}$ denotes the vector of parameters of the NLF. In turn, $\boldsymbol{\lambda}$ can depend on colour channel or spectral band, polarization, radar operation mode, DLSR camera ISO or other settings. It is possible to fix sensor parameters that influence the vector $\boldsymbol{\lambda}$ and collect a representable set of images that cover both homogeneous and heterogeneous areas, different textures and structures. Noise SD for these images will be an unknown function of image intensity. We propose to group small image patches by intensity. In each intensity group, noise SD will take the same (unknown) value. Therefore, for a target sensor it is possible to collect groups of patches with the same but unknown noise SD. Taking into account that modern sensors tend to have high productivity, practically unlimited amount of data could be collected for each group.

Unknown noise SD could be transformed in a predictable way. First of all, assuming that noise distribution is Gaussian, simulated Gaussian noise with SD $\sigma_{n.add}$ could be added to an image patch. Resulting noise SD will become $\sqrt{\sigma_n^2 + \sigma_{n.add}^2}$. Second, sequence of image filtering with averaging filter with support size L_{filter} and subsampling will result in noise SD given by σ_n / L_{filter}. These two transforms are non-trivial in the sense they change signal-to-noise ratio and effectively generate new training sample (again with unknown noise SD). This is in contrast to trivial transforms – data intensity bias or scaling – that do not generate a new sample.

For training NoiseNet using GT-free real data, we compose each batch with 50% of simulated afBm data and 50% of real data. The structure of the synthetic part of the batch and the loss function term applied to it are the same as in the previous section. Real patches for the batch are sampled from a random group of patches with the same noise SD. To each *ith* patch, random Gaussian noise with variance $\sigma_{n.add.i}^2$ is added yielding patch noise variance $\sigma_{n.group.i}^2 = \sigma_{n.group}^2 + \sigma_{n.add.i}^2$.

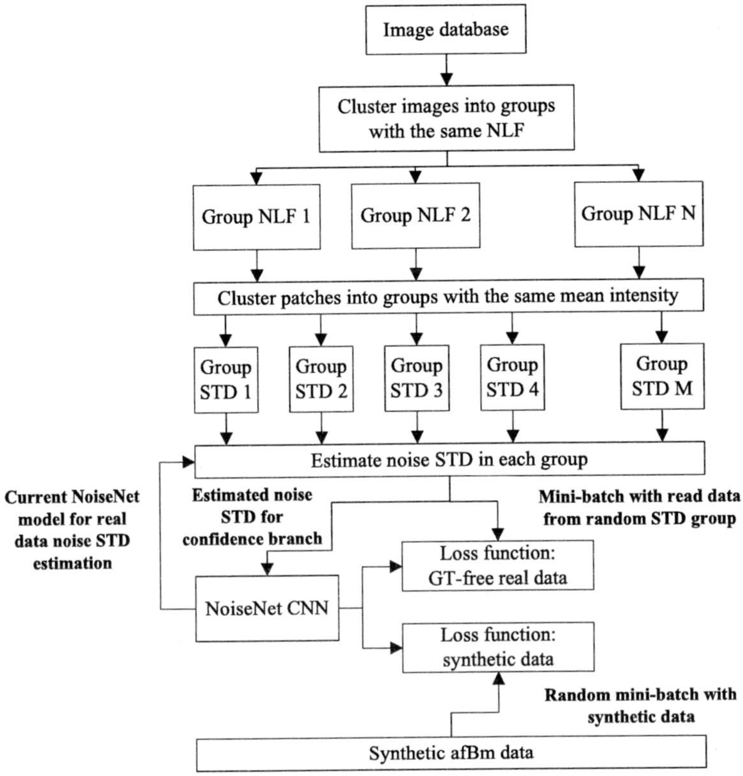

Fig. 3. Scheme of NoiseNet training with mixture of simulated and GT-free real data.

Training of the NoiseNet CNN depends on GT noise SD in two ways: as GT data for noise SD regression branch and as input value for SD uncertainty branch. Let us discuss these two ways of use. The value of noise SD for the uncertainty branch is used only for uncertainty calculation and does not affect noise SD regression branch. In this case, deviations from GT values are acceptable. Therefore, we propose to replace GT noise SD value in each group by the estimate obtained with the NoiseNet itself. More in detail, noise SD in each group is initialized with average patch intensity SD value. After each 30 epochs, NoiseNet CNN is applied to each group to update noise SD estimate. To make this procedure more robust, it is possible to apply polynomial NLF model to a set of groups for which NLF parameter vector $\boldsymbol{\lambda}$ is known to be the same. Noise SD in each group is then calculated as $\sigma^2_{n.group} = f(I_{group}, \hat{\lambda})$, where I_{group} is the mean intensity in the current group, $\hat{\lambda}$ is the NLF parameter vector estimate for the set of groups. This procedure is effectively the target sensor calibration using BNPE CNN. Both calibration parameters and CNN predictions are interdependent and their quality improves in the training process. Because of this dependence, we call the training procedure “calibrate-while-training”.

The value of $\sigma^2_{n.group}$ is used in the uncertainty branch. However, it cannot be used in the noise SD regression branch because it will lead to biased noise SD prediction values. The unknown value $\sigma^2_{n.batch}$ can be estimated for each batch using the current CNN predictions $(\sigma_{n.i}, \sigma_{\sigma_{n.i}})$ as $\hat{\sigma}^2_{n.batch} = \sum_i \frac{\sigma^2_{n.i} - \sigma^2_{n.add.i}}{\sigma^2_{\eta.i}} / \sum_i \frac{1}{\sigma^2_{\eta.i}}$, where $\sigma^2_{\eta.i} = 2\sigma^2_{\sigma_{n.i}} \left(\sigma^2_{n.group} + \sigma^2_{n.add.i} \right)$.

For GT-free real data, we apply the same loss function (6) as for real data with GT but replace GT noise SD value with batch estimate $\sigma^2_{n.GT.i} = \hat{\sigma}^2_{n.batch} + \sigma^2_{n.add.i}$.

4 Experimental Part

4.1 NED2012 Database

For testing the different BNPE methods, we created NED2012 database of 12-bit raw images from the Nikon D80 DSLR camera with a 10.2-Mpx CCD sensor (Fig. 4). This database represents various real-life textures, both homogeneous and heterogeneous areas. True sensor noise NLF for this database was obtained via a calibration procedure.

Fig. 4. Images of the NED2012 database.

4.2 NoiseNet Training on NED2012 Database in "CALIbrate-While-Training" Mode

Training of the NoiseNet in semi-supervised mode has the same settings as for the original model except for the target sensor NLF update. This new update is performed after each 30 epochs. Results of the trained model performance for NED2012 database is illustrated in Fig. 5 where calibrated NLF and estimated during training are compared.

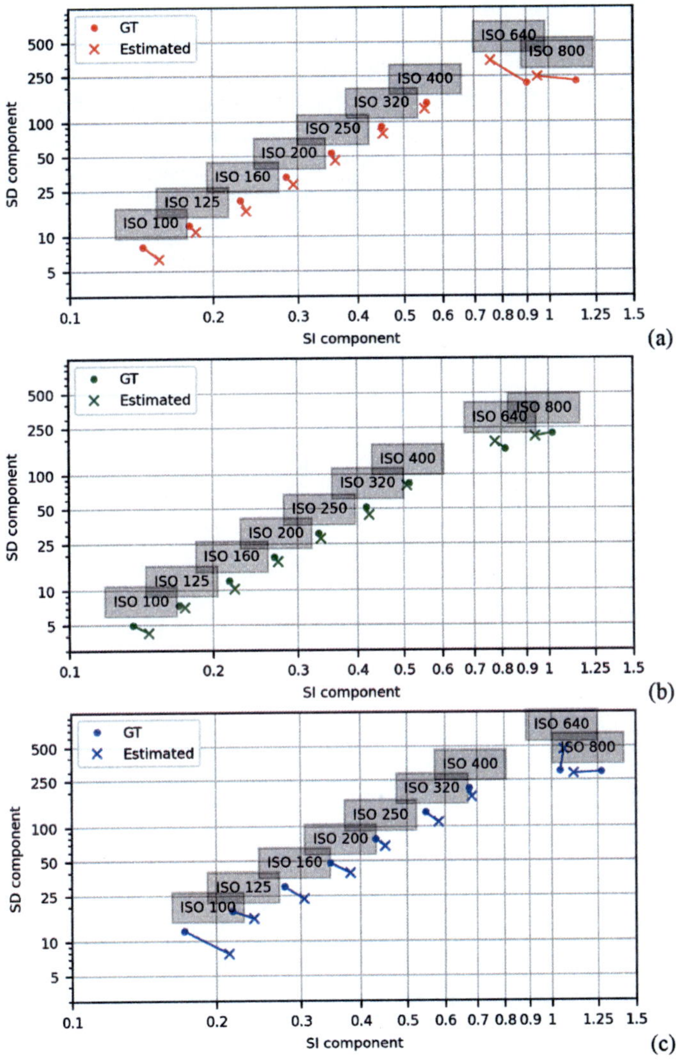

Fig. 5. Comparison of GT sensor noise NLF parameters for NED2012 database and estimated during training in GT-free mode: (a) red channel; (b) green channel; (c) blue channel. ISO values are shown in gray boxes.

The accuracy of the original NoiseNet model, the accuracy of the same model trained with mixed synthetic and real data and finally the accuracy of the model trained without GT noise parameters are shown in Table 1.

The use of synthetic data in training has improved accuracy of both signal-independent and signal-dependent noise components estimation as compared to the original model. The main gain of about 40% is seen for signal-independent component. As compared to the model-based NI+DCT approach, learning-based BNPE improves noise SD estimation accuracy by a factor of 2…2.5. This significant improvement can be attributed to the high efficiency of CNN models in adapting to the image content and complex textures. Training NoiseNet without GT noise parameters lead to efficiency comparable to that of the original model and slightly worse than the one of the training with both synthetic and GT noise data. This result is promising as it opens up the possibility of easily expanding the amount of the training data without compromising the accuracy of the training model. Even more importantly, data from targeted sensors for which calibration procedure is impossible can also be used for training purposes.

Table 1. Performance of the original and modified versions of the NoiseNet model on NED2012 database

Algorithm	Signal Dependent component	Signal Independent component
	relative SD * 100%	
NI + DCT	7.39	44.85
Original NoiseNet*	3.62	29.52
NoiseNet: synthetic and real data	3.54	18.25
NoiseNet: synthetic and GT-free real data	4.13	24.01

* Results for the original model slightly deviate from the paper [11]. We attribute this to changes in the testing environment (e.g. different versions of TensorFlow framework). Testing environment is the same for the original and modified models.

5 Conclusions

This paper considers the problem of blind estimation of noise parameters using the model-based or the learning-based approaches. The model-based approach is based on the isotropic or anisotropic fBm models. Its advantage lies in its ability to characterize the confidence of noise SD estimates in each image patch. However, the model-based noise SD estimates may become biased when the image structure deviates from the assumed model (for example, regular structures and neighbourhood of edges).

The learning-based approach (implemented as CNN) outperforms the model-based approach thanks to its ability to adapt to the data. But it requires data with calibrated noise parameters. For some sensors, these data are hard or even impossible to collect (for example, BNPE may be considered as an alternative method for calibration data verification; in this case, calibration data cannot be used for BNPE CNN training). We got around this difficulty in two ways: by addition of synthetic afBm data to the training process and through training BNPE CNN on real data without GT-noise parameters.

The latter approach is the most promising as it implements the "calibrate-while-training" approach and allows easily extending training dataset using data from uncalibrated sensors. The only assumption is that a sensor used could provide a large amount of data with the same NLF (dependence of noise SD on image intensity). This could be easily done when a sensor operates in the same mode (same ISO value for DLSR cameras, same polarization for radar sensors). We provide experimental evidence that NoiseNet CNN training on GT-free mode has accuracy close to the model training using GT noise parameters. Such a model with this new way of training was proven to correctly estimate the calibration noise parameters for the NED2012 database.

References

1. Lukin, V. V., et al.: Methods and automatic procedures for processing images based on blind evaluation of noise type and characteristics. APPRES. 5:053502–053502–053526 (2011)
2. Azzariand, L., Foi, A.: Gaussian-cauchy mixture modeling for robust signal-dependent noise estimation, acoustics, speech and signal processing (ICASSP). In: 2014 IEEE International Conference on2014, pp. 5357–5361 (2014)
3. Colom, M., Lebrun, M., Buades, A., Morel, J.M.: Nonparametric multiscale blind estimation of intensity-frequency-dependent noise. IEEE Trans. Image Process. **24**(10), 3162–3175 (2015)
4. Uss, M.L., Vozel, B., Lukin, V.V., Chehdi, K.: Estimation of variance and spatial correlation width for fine-scale measurement error in digital elevation model. IEEE Trans Geosci Remote Sens. **58**, 1941–1956 (2020)
5. S.G. Bahncmiri, M. Ponomarenko, and K. Egiazarian. Deep Convolutional Autoencoder for Estimation of Nonstationary Noise in Images, *2019 8th European Workshop on Visual Information Processing (EUVIP)*2019, pp. 238–243
6. Gao, B.-C.: An operational method for estimating signal to noise ratios from data acquired with imaging spectrometers. Remote Sens. Environ. **43**, 23–33 (1993)
7. Vozel, B., et al.: Blind determination of noise type for spaceborne and airborne remote sensing. In: Collet, C., Chanussot, J., Chehdi, K. (eds.), Multivariate image processing, Wiley-ISTE2010, p. 464
8. Dong, L., Zhou, J., Tang, Y.Y.: Effective and fast estimation for image sensor noise via constrained weighted least squares. IEEE Trans. Image Process. **27**, 2715–2730 (2018)
9. Azzari, L., Foi, A.: Indirect estimation of signal-dependent noise with nonadaptive heterogeneous samples. IEEE Trans. Image Process. **23**(8), 3459–3467 (2014)
10. Uss, M.L., Vozel, B., Lukin, V.V., Chehdi, K.: Image informative maps for component-wise estimating parameters of signal-dependent noise. J Electron Imaging **22**, 013019 (2013)
11. Uss, M., Vozel, B., Lukin, V., Chehdi, K.: NoiseNet: signal-dependent noise variance estimation with convolutional neural network. In: Blanc-Talon, J., Helbert, D., Philips, W., Popescu, D., Scheunders, P. (eds.) ACIVS 2018. LNCS, vol. 11182, pp. 414–425. Springer, Cham (2018). https://doi.org/10.1007/978-3-030-01449-0_35
12. Uss, M.L., Lukin, V.V., Vozel, B., Chehdi, K.: Accuracy of model-based and learning-based approaches for image noise variance estimation. In: 2020 IEEE Ukrainian Microwave Week (UkrMW), pp. 438–442 (2020)

13. Uss, M., Vozel, B., Lukin, V., Chehdi, K.: Comparison of learning-based and maximum-likelihood estimators of image noise variance for real-life and synthetic anisotropic textures. In: SPIE (2020)
14. Ortizandm, A., Oliver, G.: Radiometric calibration of CCD sensors: dark current and fixed pattern noise estimation. In: IEEE International Conference on Robotics and Automation, 2004 Proceedings ICRA 2004 2004, vol. 52004, pp. 4730–4735 Vol.4735
15. Zhao, S., Li, B., Xu, P., Keutzer, K.: Multi-source domain adaptation in the deep learning era: A systematic survey. arXiv preprint arXiv:200212169 (2020)
16. Ponomarenko, M., Gapon, N., Voronin, V., Egiazarian, K.: Blind estimation of white Gaussian noise variance in highly textured images. Electronic Imaging. 2018:382–381–382–385 (2018)
17. Zoranand, D., Weiss, Y.: Scale invariance and noise in natural images. In: IEEE 12th International Conference on Computer Vision, 2009, pp. 2209–2216 (2009)
18. Foi, A., Trimeche, M., Katkovnik, V., Egiazarian, K.: Practical Poissonian-gaussian noise modeling and fitting for single-image raw-data. IEEE Trans. Image Process. **17**, 1737–1754 (2008)
19. Aja-Fernández, S., Vegas-Sánchez-Ferrero, G., Martín-Fernández, M., Alberola-López, C.: Automatic noise estimation in images using local statistics. Additive and multiplicative cases. Image Vis. Comput. **27**(6), 756–770 (2009)
20. Yang, J., Liu, X., Song, X., Li, K.: Estimation of signal-dependent noise level function using multi-column convolutional neural network. In: 2017 IEEE International Conference on Image Processing (ICIP), pp. 2418–2422 (2017)
21. Liu, C., Szeliski, R., Kang, S.B., Zitnick, C.L., Freeman, W.T.: automatic estimation and removal of noise from a single image. IEEE Trans Pattern Anal Mach Intell. **30**, 299–314 (2008)
22. Uss, M.L., Vozel, B., Lukin, V.V., Chehdi, K.: Maximum likelihood estimation of spatially correlated signal-dependent noise in hyperspectral images. Opt. Eng. **51**(11), 111712–111712 (2012)
23. Uss, M., Vozel, B., Lukin, V., Chehdi, K.: Analysis of signal-dependent sensor noise on JPEG 2000-compressed Sentinel-2 multi-spectral images. In: SPIE Remote Sensing, vol. 10427, p. 12. SPIE (2017)
24. Uss, M.L., Vozel, B., Lukin, V.V., Chehdi, K.: Local signal-dependent noise variance estimation from hyperspectral textural images. IEEE J Sel Topics Signal Process. **5**, 469–486 (2011)
25. Uss, M., Vozel, B., Lukin, V., Chehdi, K.: Blind noise parameters estimation for multichannel images using deep convolutional neural networks. In: SPIE (2019)
26. Mahmood, A., Robin, A., Sears, M.: Estimation of the noise spectral covariance matrix in hyperspectral images. IEEE J. Sel. Top. Appl. Earth Observations Remote Sens. **11**, 3853–3862 (2018)
27. Gardella, M., Musé, P., Morel, J.-M., Colom, M.: Noisesniffer: a fully automatic image forgery detector based on noise analysis. In: 9th IEEE International Workshop on Biometrics and Forensics (IWBF) 2021, Rome (virtual), Italy (2021)
28. Van der Meerand, F.D., De Jong, S.M.: Imaging spectrometry: basic principles and prospective applications, Kluwer Academic, Dordrecht, The Netherlands (2011)
29. Xu, B., Gong, P.: Noise estimation in a noise-adjusted principal component transformation and hyperspectral image restoration. Can. J. Remote Sens. **34**(3), 271–286 (2008)
30. Lukin, V., Abramov, S., Zelensky, A., Astola, J., Vozel, B., Chehdi, K.: Improved minimal inter-quantile distance method for blind estimation of noise variance in images. In: SPIE (2007)

31. B. Aiazzi, C. Istituto di Fisica Applicata «Nello Carrara» , Firenze, Italy, L. Alparone, U.d.S.d.F. Dipartimento di Elettronica e Telecomunicazioni, Italy, A. Barducci, C. Istituto di Fisica Applicata «Nello Carrara» , Firenze, Italy, S. Baronti, C. Istituto di Fisica Applicata «Nello Carrara» , Firenze, Italy, P. Marcoionni, C. Istituto di Fisica Applicata «Nello Carrara» , Firenze, Italy, I. Pippi, C. Istituto di Fisica Applicata «Nello Carrara» , Firenze, Italy, M. Selva, and C. Istituto di Fisica Applicata «Nello Carrara» , Firenze, Italy. Noise modelling and estimation of hyperspectral data from airborne imaging spectrometers2006
32. Lukin, V.: Testing of methods for blind estimation of noise variance on large image database. Book chapter in Theoretical and practical aspects of digital signal processing in information-telecommunication systems, Shakhty, Russia2009, pp. 43–70
33. Corner, B.R., Narayanan, R.M., Reichenbach, S.E.: Noise estimation in remote sensing imagery using data masking. Int. J. Remote Sens. **24**, 689–702 (2003)
34. Hensel, M., Pralow, T., Grigat, R.-R.: Modeling and real-time estimation of signal-dependent noise in quantum-limited imaging. In: Proceedings of the 6th WSEAS international conference on signal processing, robotics and automation Corfu Island, Greece, Citeseer2007, pp. 183–191
35. Stefano, A.D., White, P.R., Collis, W.B.: Training methods for image noise level estimation on wavelet components. EURASIP J. Appl. Signal Process. **2004**, 2400–2407 (2004)
36. Ponomarenko, N.N., Lukin, V.V., Zriakhov, M.S., Kaarna, A., Astola, J.: An automatic approach to lossy compression of AVIRIS images. In: 2007 IEEE International Geoscience and Remote Sensing Symposium2007, pp. 472-475
37. Danielyanand, A., Foi, A.: Noise variance estimation in nonlocal transform domain. In: 2009 International Workshop on Local and Non-Local Approximation in Image Processing2009, pp. 41–45
38. Sun, L.: Signal-dependent noise parameter estimation of hyperspectral remote sensing images. Spectrosc. Lett. **48**, 717–725 (2015)
39. Uss, M., Vozel, B., Lukin, V., Abramov, S., Baryshev, I., Chehdi, K.: Image informative maps for estimating noise standard deviation and texture parameters. EURASIP J. Adv. Signal Process. **2011**, 806516 (2011)
40. He, K., Zhang, X., Ren, S., Sun, J.: Spatial Pyramid Pooling in Deep Convolutional Networks for Visual Recognition, pp. 346–361. Springer, Cham (2014)
41. Tan, H., Xiao, H., Lai, S., Liu, Y., Zhang, M.: Pixelwise Estimation of signal-dependent image noise using deep residual learning. Comput. Intell. Neurosci. **2019**, 12 (2019)
42. Pesquet-Popescu, B., Véhel, J.L.: Stochastic fractal models for image processing. IEEE Signal Process. Mag. **19**(5), 48–62 (2002)
43. Gurevichand, P., Stuke, H.: Learning uncertainty in regression tasks by deep neural networks. arXiv preprint arXiv:170707287(2017)

Study on System-Level Test Verification Technology for Optical Remote Sensing Satellite Payload

Haichao Wu(✉), Jianfeng Dai, Chuan Zhao, Zhi Yang, and Shaonan Liu

DFH Satellite Co., Ltd., Beijing 100094, China
151524320@qq.com

Abstract. Space optical remote sensing satellite payload is one of the most technically demanding payloads for test verification technology in all kinds of satellites. In the process of the ground system-level test, the important guarantee of ensuring the reliable and stable operation and executing missions successfully of remote sensing satellites in orbit is to verify adequately compliance with the requirements of spacecraft system-level technical specification that includes the interface correctness between optical payload and each system of the platform and combinations of working conditions under specified conditions to realize the comprehensive testing of the system functions and performance for Space optical remote sensing payload. The main domestic and international research results on system-level test for optical remote sensing payload are introduced briefly. The current situation of system-level test verification technology for optical remote sensing payload is summarized systematically. The advantages and disadvantages of system-level test verification compared with subsystem-level or unit-level test verification are analyzed. The technical problems that exist are discussed. Technology development current situation and developing trends of the Space optical remote sensing satellite payload are analyzed. The developing trends of the system-level test verification technology for optical remote sensing satellite payload are pointed out.

Keywords: Optical Remote Sensor · Payload · System-level test

1 Introduction

With the rapid development of small satellites in batches and clusters, flexible and efficient small satellite development demands promote the development of space optical remote sensing payloads to light miniaturization and high integration. More and more space optical remote sensing payloads adopt integrated electronic devices for centralized control, and adopt open architecture and plug-and-play standards to meet the internal control, communication and component management of payloads. At the same time, integrated electronic software can be customized to meet the different needs of different small satellites, so as to achieve local modular design and subsystem software definition.

H. P. Urbach and H. Jiang (Eds.): ISSOIA 2022, SPPHY 295, pp. 486–495, 2023.
https://doi.org/10.1007/978-981-99-4098-1_42

In this paper, the main research achievements of optical remote sensing payload at home and abroad are fully investigated, the present situation of system-level testing and verification technology of optical remote sensing payload is expounded, the advantages and disadvantages of system-level testing and verification compared with sub-system or single-machine testing and verification are analyzed, and the future development trend of system-level testing and verification technology of optical remote sensing payload is prospected, which can provide some reference for system-level testing and verification of space optical remote sensing payload.

2 Development Status of Optical Remote Sensing Payload at Home and Abroad

2.1 Development Status of Optical Remote Sensing Payload Abroad

In 2017, Maxar Technologies acquired Digital Earth and began to develop the remote sensing constellation of "World Observation Corps". The "Legion" satellite maintains the high agility of the "World Watch" series satellites, and it takes only 10 s to sweep the ground 200 km sideways. This capability is roughly the same as that of previous "World Watch" satellites, but because the orbital altitude of the Legion satellite is nearly 200 km lower than that of other "World Watch" satellites, the agility capability of the Legion satellite is actually higher. The company comprehensively considers the technical indicators such as camera caliber, agility, field of view, detector type, image acquisition rate and data transmission rate of the satellite, and comprehensively optimizes the cost. For example, the large aperture optical system has many advantages, such as providing higher resolution, or enabling the satellite to be deployed in a higher orbit with a specific resolution, achieving a wider field of view, and reducing the need for platform agility. "Legion" constellation provides a reference for the development of high-end commercial optical imaging satellite technology in the next 10 years. The multi-satellite networking, multi-orbit coordination, large El-diameter optical system, platform load integration design, agile satellite platform, long-line array push-and-scan detector and other technologies adopted by it represent the development direction of high-end commercial optical imaging satellite technology in the future [1].

Foreign space companies are actively developing intelligent optical remote sensing satellites, including autonomous satellite operation, data acquisition, data processing, data application and satellite transportation and management. The National Defense Advanced Research Projects Agency (DARPA) put forward the "Reconfigurable Imaging" project in 2016. The reconfigurable imaging technology developed by this project will improve the traditional imaging mode of the camera. After the introduction of artificial intelligence technology, the camera can automatically optimize the imaging mode according to the scene and obtain more valuable information. American Earth Observation-1 (EO-1) and other satellites adopt on-board image processing technology, which has the functions of emergency monitoring, etc. Using channel threshold algorithm, the on-board automatic cloud judgment is initially realized. In September, 2020, the European Space Agency (ESA) launched PhiSat-1 satellite, which was equipped with Intel's video processing unit (VPU) chip for cloud judgment. When 70% of the area of the image taken by the development of artificial intelligence algorithm was blocked by the cloud, the image was automatically deleted, thus avoiding wasting on-board storage capacity and downloading useless images. The "Dual Spectral Infrared Detection" (BIRD) satellite launched by Germany in 2001 uses the on-board neural network chip to analyze the multi-spectral infrared images, and realizes fire identification. NI1000 neural network chip on the satellite generates fire theme map to reduce the data rate. On the one hand, the ground processing verifies the results of on-board processing. On the other hand, it optimizes the algorithm, and injects the new algorithm into the satellite, so as to continuously optimize the on-board processing capability iteratively. Tacsat-3, a military tactical satellite launched by the United States in 2009, is based on hyperspectral image processing technology, and has the ability of automatic target recognition. Tacsat-3 is equipped with the payload of Advanced Rapid Response Tactical Military Imaging Spectrometer (ARTEMIS), and the core of ARTEMIS is hyperspectral imaging processor (HSIP). HSIP has the capabilities of real-time hyperspectral image processing, image storage, and payload control and power conversion adjustment. HSIP will compare and analyze the spectral data of the target area according to the satellite-borne spectral characteristic database, so as to find the target and proofread the geographic coordinates of the target on the panchromatic image. After the above-mentioned on-board data processing, ARTEMIS can provide full-color images [2–5] marked with target location information to combat users.

NASA attaches great importance to the research of on-board autonomous mission planning to improve the efficiency of mission execution and information transmission. NASA launched the Autonomous Spacecraft Experiment (ASE) project on the EO-1 satellite in 2003. With ASE on-board continuous planning, behavior decision based on tasks and objectives, and on-board pattern recognition capabilities, it can quickly and independently select communication links to quickly send data back to the ground. IPEX technology test satellite is loaded with CASPER planner. Space-borne CASPE planner is used to manage spacecraft resources, and CASPER can adjust the operation of IPEX according to constraints (such as power constraints, time constraints, task constraints, etc.) [6].

2.2 Domestic Development Status of Optical Remote Sensing Payload Technology

In recent years, many research institutes and universities in China have explored the intelligent processing technology of remote sensing images in orbit. The related literature of Wuhan University points out that the optical remote sensing satellite system has good measurement and data acquisition capabilities, but lacks intelligent observation and perception capabilities. It is necessary to focus on breaking through the key technologies of on-board data processing, establishing a task-driven satellite-to-ground collaborative processing mechanism of remote sensing data with the user's needs as the core, and giving the development direction of the subsequent intelligent improvement of remote sensing satellites. On-orbit image intelligent processing is the key technology to realize the intelligence of remote sensing satellite, and it is a hot spot in the development and engineering application of remote sensing satellite technology. Our national commercial satellites, such as Tsinghua-1, have the functions of cloud detection [7].

After the successful launch of a domestic commercial remote sensing satellite in 2021, according to the mission plan of the whole satellite, the special test and key technology verification of on-orbit intelligent processing were carried out, and the on-orbit functional performance of intelligent processing was tested and evaluated. The results of several rounds of on-orbit tests fully verify and demonstrate the innovation and practicality of many key technologies such as on-board deep learning acceleration, real-time processing of cloud detection, multi-level reconstruction, etc. On-orbit intelligent processing system meets the users' requirements for intelligent observation and efficient information acquisition, and effectively improves the intelligent level of on-board processing. Based on the design idea of software-defined function, the intelligent processor realizes the sharing of software and hardware resources between intelligent processing and traditional data processing, and the multi-level reconfiguration design of on-board system also provides strong support for subsequent optimization and upgrading [8].

3 Present Situations of System-Level Test and Verification Technology of Optical Remote Sensing Payload

According to the composition and principle of optical remote sensing payload, the system-level test of payload can be divided into four test items: camera imaging and control function test, camera temperature control function test, data transmission subsystem test, and payload data transmission combined working mode test. The main purpose of system-level test is to check whether the electrical performance and function of the camera and data transmission subsystem meet the requirements, and qualitatively check whether the imaging function of the camera meets the requirements in the whole satellite state, including verifying the correctness of various working modes of the camera subsystem, whether the electrical performance and function of the interface between the camera subsystem and the power supply, OBDH and data transmission subsystem meet the requirements, and the reliability, accuracy and stability of information channel transmission (Fig. 1).

The camera imaging and control function test mainly includes power-on initialization check, camera secondary power supply check, focal plane locking and unlocking function test, series and gain setting, image data format check, imaging and focusing test in active and standby state, instruction check and its corresponding function and telemetry parameter check, and upper injection function test. Under the scheduling of the OBDH subsystem, the camera subsystem should be able to collect telemetry data, send telemetry parameters to the OBDH through the lower computer, and receive remote control instructions and broadcast data from the OBDH at the same time. It can realize the backup switching of the lower computer, receive the second pulse and the whole second time data of the OBDH time reference, generate high-precision time scales, and have the ability to operate the camera on and off and set imaging parameters in batches. The camera temperature control function test is to check the active temperature control capability of the camera subsystem to ensure the normal imaging of the camera. It should have the functions of setting the thermal control temperature threshold and telemetering the temperature.

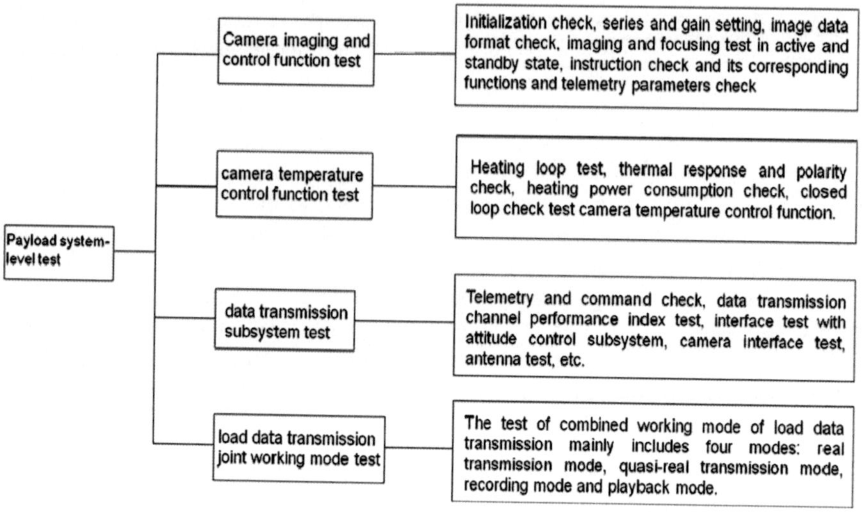

Fig. 1. Composition diagram of payload system-level test

Data transmission subsystem test mainly includes telemetry and command check, channel performance index test, interface test with attitude control subsystem, camera interface test, antenna test, etc. The test of data transmission combined working mode mainly includes four modes: real transmission mode, quasi-real transmission mode, recording mode and playback mode. Each mode should be subdivided into several sub-modes to cover the main parts and backups of data processor, memory interface and servo controller of data transmission subsystem, as well as the main parts and backups of management controller and video processor of camera subsystem. Each mode should also cover the commonly used imaging parameters (series, gain, etc.)

The detailed performance of the payload subsystem needs to be tested under special laboratory conditions. In the whole satellite state, due to the limitation of the test site and environmental conditions, the technical indexes of the imaging quality of the camera subsystem can no longer be tested in detail and accurately, and the qualitative test of the function is mainly carried out. The test should cover the main functions and performances of optical system, imaging circuit, recording system and data transmission subsystem.

The camera subsystem function test includes telemetry check, remote control instruction check, controller main and backup switching function test, integration time circuit main and backup switching function test, parameter adjustment function test, algorithm correction function test, auxiliary data arrangement function test and safety shutdown function test. These function tests judge whether the function is executed correctly according to the relevant telemetry response after sending the remote control instruction. Working mode test includes focusing mode test and imaging mode test. In the system-level comprehensive test, the surface light source and the scene simulator are used to simulate the target. The surface light source and its control equipment are special ground test equipment for cameras, which can emit five kinds of light, full color, red, green, blue and near infrared. The five kinds of light are independently provided by five groups of LEDs.

The camera temperature control function is tested by downloading function test, heating loop test, thermal response and polarity check, heating power consumption check and closed loop check. In the open-loop test, according to the corresponding relationship between thermistor and temperature, the heater of a certain temperature control loop is turned on or off, and the change of instantaneous load current and the change of temperature telemetry after a period of time are observed to detect whether the power consumption and polarity of the heating loop are correct. During the closed-loop test, observe the current temperature telemetry value of a temperature control loop, modify the temperature control threshold by injecting data blocks, make the lower limit of the temperature control threshold higher than the current temperature value, observe whether the circuit will open soon, and record the instantaneous load current change and the temperature telemetry change after a period of time to check whether the heating power consumption and polarity are correct.

The digital transmission subsystem tests the corresponding functions according to the design requirements and criteria for instruction checking and telemetry interpretation; the channel performance index tests are measured and counted according to the professional test method of communication performance index with the help of measuring equipment such as spectrometer, power meter and frequency meter; the antenna tests are conducted by using antenna simulator or installing electric antenna, and the telemetry data of antenna angle or the actual rotation direction and angle of antenna are observed.

The joint working mode of camera subsystem and digital subsystem mainly includes real-time transmission mode, quasi-real-time transmission mode, recording mode and playback mode, in which quasi-real-time transmission is the working mode of recording and playback, and the test step is the combination of recording mode and playback mode. The test contents and methods of real-time transmission mode, recording mode and playback mode are as follows:

In the ground test, the surface light source or the ground scene simulator is used as the image source, and the light emitted by the image source enters the camera optical lens through the hood and is imaged on the focal plane. Through the remote control and telemetry functions, the working states of relevant parts are controlled and monitored, and the video data stream output by the image and video system is recorded in real time by the on-board fixed memory according to the mode setting during testing, and the working states of the camera subsystem and the fixed memory are monitored and checked during the recording process. Check the format of the demodulated bit data stream received by the ground equipment, check the correctness of the synchronization signal and the camera auxiliary service data, and qualitatively observe the received image quality and quantitatively calculate the number of error frames through the quick-view image monitor. During the test, it is necessary to monitor the key telemetry of the camera and the digital transmission and distribution system, and observe whether the image displayed by the quick view is normal or not, and interpret the image data recorded by the quick view afterwards to ensure that the image data output by the satellite is consistent with the design.

4 Advantages and Disadvantages of System Level Test Verification Compared with Subsystem Test Verification

The traditional optical remote sensing satellite focuses on the function inspection and performance evaluation of the payload system itself, while the new intelligent remote sensing satellite has more complex mission mode and operation control mode. With the development cycle of the satellite greatly shortened, it is necessary to shift the function and performance test of the payload system to the subsystem test and verification. In the system-level test and verification, more emphasis is placed on the correctness, integrity and consistency of the payload system's completion of the whole satellite planning task, and the payload system's completion of intelligent image preprocessing, intelligent image processing and intelligent image processing. The advantages and disadvantages of system-level test verification compared with sub-system or single-machine test verification are mainly reflected in:

1) In the system-level test verification, the platform and the payload subsystem cooperate with each other, which is more similar to the on-orbit technical state and the environment of the imaging task. Through the test verification of independent task planning, the ability of the optical load system to perform the planned task can be more fully verified. If it is tested in a subsystem or a single machine, it is difficult to effectively check the ability and timing matching with other subsystems.
2) Test and verify the intelligent image processing of optical remote sensing load. In the system-level test environment, you can access the digital transmission subsystem through the ground image simulation source, and verify different image preprocessing or processing modes by setting a variety of image test cases. However, the existing test methods only examine the intelligent processing capability of the digital transmission subsystem, and assume that the camera load can output ideal standardized image data. The subsequent camera load and data transmission subsystem should input the original image through the scene simulation equipment in a unified way, so as to carry out more systematic and comprehensive intelligent image processing verification, which is difficult to fully carry out and simulate at the subsystem or stand-alone level.
3) At present, optical remote sensing satellites need to improve the on-board effective data rate and solve the bottleneck problem of data transmission. At present, the bandwidth of satellite download can't meet the requirement of real-time data download, and even after compression, it is still difficult to download data in real time. On-orbit intelligent compression and object of interest extraction will effectively reduce the data rate. In the whole satellite system level test, it is necessary to fully verify the data consistency and transmission rate of the original image or compressed image, so as to ensure the matching and stability of the data transmission rate between the payload system and the data transmission subsystem. However, this kind of verification is also difficult to fully develop and realize at the subsystem or single machine level.
4) In the test and verification of remote sensing load subsystem's imaging performance, transfer function, radiation calibration and other load performance, the system-level test environment is complex, which is restricted by the safety and convenience of multi-system use, and it is difficult to be as efficient as subsystem or single-machine

test. Therefore, it is necessary to shift the basic function and performance of optical load system to subsystem-level test to complete the verification.

5 Future Development Trends of Optical Remote Sensing Payload System-Level Test and Verification Technology

With the gradual deepening of the application of optical remote sensing, all kinds of users have higher and higher requirements for image data information extraction, which requires higher resolution image quality and faster and more flexible image generation mode. At the same time, the realization of high spatial resolution and high temporal resolution (large width) has become the development trend of high-resolution remote sensing satellites, and it is required to improve imaging efficiency and reduce satellite cost. High-resolution agile imaging remote sensing satellite, which has the ability of large-angle quick attitude maneuver, high-precision space-time information measurement and mission planning for on-board resources, has become the leading role of space optical remote sensing satellite [9]. At present and in the future, agile remote sensing satellites are developing towards the goal of intelligent remote sensing satellites driven by user tasks, with emphasis on the task planning technology and intelligent image processing technology for on-board resources.

The working mode of traditional optical remote sensing satellite is relatively simple, and the imaging mode is mainly sub-satellite point imaging or simple side swing imaging, so the implementation of planning and scheduling instructions in task operation management is relatively simple. For the high-resolution optical remote sensing satellite, because the agile imaging mode will face a large number of complex control instruction sequences, and the on-board actions are strongly interrelated, the failure of single-step implementation will affect the subsequent tasks and actions, and the real-time performance of the agile imaging control actions is required to be high. If the traditional method of injecting program-controlled instructions on the ground is adopted to carry out operation control, there are many instructions and it is easy to make mistakes, especially in the case of staggered tasks. Therefore, it is necessary to adopt on-board resource-oriented mission planning technology and drive the operation of on-board autonomous mission planning software to evaluate the overall application capability of the satellite system. Under the condition of limited satellite resources (remote sensor capability, observation time, observation time length, attitude angle requirements, image type requirements, etc.), by solving the task planning problem, the best task control instruction sequence can be found, the best imaging effect and the lowest engineering cost can be achieved, the on-board energy utilization can be optimized, and the on-orbit fault handling capability can be improved. The autonomous task planning technology also requires the optical remote sensing payload to have stronger autonomous control and processing capability. The optical remote sensing payload should analyze the task planning instructions (task planning data blocks) sent by the satellite host in real time, and generate an instruction sequence with time marks, so as to reduce the instruction generation workload of the task planning system, and cooperate with the task planning system to complete the remote sensing payload imaging task of complex scenes [10].

A large number of high-resolution optical remote sensing satellites and constellations have been launched one after another. The remote sensing satellites are characterized by a new era of "large number, high quality and high requirements". At present, they are faced with new challenges: First, the number of optical remote sensing satellites has increased greatly, which brings a huge burden to transportation and control. Second, the amount of remote sensing data is expanding rapidly, which causes great pressure on data transmission and data processing, and there are many invalid data on the satellite, which wastes the storage resources and data transmission channel resources on the satellite. Third, the demand for rapid response from natural disasters and special needs is increasing day by day. Traditional satellite data processing takes a long time, which makes it difficult to meet the timeliness requirements of products. In order to solve these problems, space companies at home and abroad are actively developing the research and application of intelligent remote sensing satellites, including autonomous satellite operation, data acquisition, data processing, data application and satellite transportation and management. For optical remote sensing payload, it also puts forward the task-driven autonomous task planning requirements and the intellectualization of image data processing, including intelligent data acquisition, intelligent routine preprocessing (radiation correction, geometric correction, atmospheric correction, etc.), intelligent image processing (cloud judgment, target identification, target indication, etc.), intelligent distribution, intelligent transportation and management, intelligent orbit control, intelligent planning and scheduling, intelligent health management of satellite and intelligent planning and operation of constellation, etc. However, there are many differences between these technologies and traditional optical load testing in terms of function and performance. For example, for the test verification of autonomous mission planning, it is necessary to fully verify the correctness of the information flow and communication protocol of the optical payload and mission planning output system (onboard data handling system) (OBDH), fully verify the capability of the optical payload subsystem to parse the payload instruction sequence according to the mission block of the whole satellite planning, and examine the capability of the optical remote sensing payload to cooperate with other subsystems to complete the system-level imaging tasks and modes. For example, for testing and verification of intelligent image processing, it is necessary to fully verify the ability of optical remote sensing payload to complete the preprocessing, intelligent processing and management of original image data in real time, and fully verify the correctness of intelligent processing mode and data stream transmission of optical payload subsystem. These new testing technologies are matched with new ground equipment such as image simulation source and real-time intelligent interpretation software, which makes the verification of these new technologies faster, smarter and more reliable.

6 Concluding Remarks

System-level testing is one of the most important means to ensure the successful development of small satellites. It is necessary to conduct comprehensive and sufficient system-level testing for payloads. However, in actual projects, it is often faced with the reality of tight time and heavy tasks. Fully grasping the task requirements of optical remote sensing satellites for payloads and formulating targeted testing items, contents, methods and steps can accomplish the testing verification with half the effort and effectively improve the testing coverage.

References

1. Li, N., Liu, T.: American "world observation corps" satellite launch prospect. Int. Space (10), 22–26 (2021). https://doi.org/10.3969/J.ISSN.1009-2366.2021.10.005
2. DARPA. Broad Agency Announcement Reconfigurable Imaging (ReImagine) [EB/OL] (2021). https://research-authority.tau.ac.il/sites/resauth.tau.ac.il/files/DARPA-BAA-16-56_ReImagine_2016.pdf
3. Artificial intelligence for Earth obser vation [EB/OL] (2021). https://www.esa.int/Applications/Observing_the_Earth/Ph-sat
4. Dan, M.: Validation of on-board cloud cover assessment using EO-1, NASA (2004)
5. Wang, X., Chen, X., Su, Y.: Frontier technology of contemporary high-resolution optical payload. Int. Space (5), 13–18 (2013)
6. Li, N., Liu, T.: Research on the development of intelligent earth observation satellites abroad. Int. Space (11), 31–37 (2021). https://doi.org/10.3969/J.ISSN.1009-2366.2021.11.007
7. Wang, M., Yang, F.: Intelligent remote sensing satellite and remote sensing image real-time service. J. Surv. Mapp. **48**(12), 1586–1594 (2019)
8. Xiao, H., Wang, P., Zhang, J., et al.: Design and verification of intelligent processing of remote sensing satellite images in orbit. Spacecr. Eng. **37**(2), 43–48 (2022). https://doi.org/10.3969/j.issn.1673-8748.2022.02
9. Ge, Y., Zhao, J., Yang, F.: Overview of high-resolution optical remote sensing satellite platform technology. Int. Space (5), 2–8 (2013)
10. Tian, Z., Li, X., Yang, Y., et al.: Analysis of constraints on the availability of remote sensing satellites in orbit and countermeasures. Spacecr. Eng. **27**(1), 37–44 (2018). https://doi.org/10.3969/J.ISSN.1673-8748.2018.01

Evaluation of Target Detectability Performance Under Specific Image Complexity

Boyuan Bian(✉) and Feng Zhou

Beijing Institute of Space Mechanics and Electricity, 100094 Beijing, China
boyuan_b@163.com

Abstract. Infrared image complexity measurement is an important component of target detection performance evaluation. By analyzing the current studies on image complexity and measurement methods at home and abroad, it is found that most of them only calculate the complexity of a single frame image, and there are few studies on the complexity measurement methods of image sequences. Based on the generated image sequences, the target detectability under specific image complexity is evaluated. Two generally recognized indexes are used for evaluation of target feature and local background feature, which are the confusion degree of target and the occultation degree of target. The complexity of infrared image sequence is measured. The calculation result show that the occultation degree has a higher impact on the complexity of image sequence than the confusion degree. The target detectability in different image sequences is analyzed when the false alarm probability is 10% and the gray threshold is 10. It is found that the target detectability decreases with the increase of image sequence complexity. Therefore, the threshold of target detectability should be set to match the image complexity. The method proposed in this paper can provide reference for the study of relationship between the complexity of image sequence and the target detection algorithms.

Keywords: Complexity of infrared image sequence · Confusion degree of target · Occultation degree of target · Target detectability

1 Introduction

Infrared image clutter quantization is an important component of target detection performance evaluation. The target detection algorithm can only meet the requirements in a specific scene. The mathematical method is used to quantify the clutter of infrared image and make reasonable classification of infrared image, which can properly simplify the problem of target detection. Therefore, in order to improve the efficiency of target detection algorithm, it is urgent to classify infrared image complexity, and it is meaningful to evaluate target detectability under different image complexity. By analyzing the current researches on image complexity and measurement methods at home and abroad, it is found that most of them only calculate the complexity of a single frame image [1–5], and there are few researches on the complexity measurement methods of image sequences. In this paper, the complexity of image sequence is used as the clutter quantization factor to evaluate the target detectability of infrared image sequence.

H. P. Urbach and H. Jiang (Eds.): ISSOIA 2022, SPPHY 295, pp. 496–503, 2023.
https://doi.org/10.1007/978-981-99-4098-1_43

2 Complexity Measurement of Infrared Image Sequence

In terms of image complexity measurement, two generally recognized indicators are used to evaluate target features and local background features [6]. These two indexes correspond to the confusion degree of target and the occultation degree of target respectively. The confusion degree of target represents the similarity between the background clutter and the target itself. The greater the similarity degree, the smaller the difference between the target and the background clutter, the higher the false alarm, and the more difficult the target can be detected. The occlusion degree of target represents the occlusion performance of the local background to the target. The larger the occlusion degree, the more difficult the feature extraction of the target and the difficulty of the target detection. The image features of the point target are mainly divided into two categories, which are the gray level of the target and the interframe features of the target motion information. The local background feature of the target represents the difference between the location of the target and the adjacent location of the background, which can be divided into two categories, namely, the contrast and the statistical difference of the grayscale distribution between the target and local background.

The complexity measurement of infrared image sequence is closely related to the target tracking algorithm. Neural network Sigmoid function is introduced to improve the weight of high-complexity infrared image and reduce the weight of low-complexity infrared image. The complexity of infrared image sequence F can be expressed as:

$$F = \left(\sum_{1}^{N} f_c \cdot k(f_c) \right) / N \tag{1}$$

where, N is the frame number of infrared image sequence and f_c is the complexity of a single frame of infrared image. The subscript c represents complexity.

$$f_c = \frac{2 \cdot F_{img} \cdot S_{img}}{F_{img} + S_{img}} \tag{2}$$

$$k(f_c) = \frac{2 \cdot F_{img} \cdot S_{img}}{1 + e^{-5(f_c - 0.5)}} \tag{3}$$

where, F_{img} is the confusion degree of target and S_{img} is the occultation degree of target. The subscript img represents a single frame of infrared image.

Target confusion is a measure of the similarity between clutter and target region of the whole background image. It is necessary to comprehensively calculate the similarity and number between the real target and the suspected target. This index represents the complexity of the global infrared image. Since the moving target can be regarded as a moving point target on the geostationary orbit image plane, the confusion degree of the point target in the whole background is more difficult to extract from a single frame image, but the background feature law changes little between different frames. In order to improve the target detection performance, based on the imaging characteristics of the geostationary orbit infrared imaging system, that is, the background imaging features have little change between different frames, the confusion degree calculation method for the moving point target is as follows: Firstly, each two frames of images are processed by difference, and the area occupied by the target is eliminated in the whole infrared image. Meanwhile, the infrared image of the eliminated target is scanned with the target template. The size of the scanning window is consistent with the size of the target template. Calculate the normalized cross relation G_i between the scan window and the target template. i represents the scan serial number. When $G_i \geq 0.8$, the scan window number is m, as the suspected target number. Then the grayscale feature similarity f_s between the suspected target and the target template is calculated:

$$f_s(t, o_j) = \frac{min(F(t), F(o_j))}{max(F(t), F(o_j))} \tag{4}$$

where, $F(t)$ is the gray feature of the target template t, and $F(o_j)$ is the gray feature of the suspected target o_j. The subscript s represents similarity. At the same time, based on the gray feature and position feature of the target template extracted from each frame, the change rate of the target gray information f_{gi} and the change rate of the target motion information f_{mi} between two frames are calculated respectively. The subscript gi and g represent gray information, the subscript mi and m represent movement information.

$$f_{gi} = \frac{f_g(t_{i+1}) - f_g(t_i)}{f_g(t_i)} \tag{5}$$

$$f_{mi} = \frac{f_m(t_{i+1}) - f_m(t_i)}{f_m(t_i)} \tag{6}$$

where, f_g is the gray information of the target, f_m is the movement information of the target, that is, the change rate of the moving distance between two frames of the target, i is the number of frame. Then, the similarity between the suspected target and the target template in a single frame of infrared image can be expressed as follows:

$$f_{img} = c[f_s(t, o_j) + f_{gi} + f_{mi}] \tag{7}$$

where, $c = 1/3$ represents the equal weight of each feature. The confusion degree of target in a single frame of infrared image is calculated by normalization of the above equation:

$$F_{img} = f_{img} / (f_{img} + 1) \tag{8}$$

With the target as the center, the background is taken as the local background with the area twice of the pixel occupied by the target as the local background. The occlusion degree of target calculation method is as follows:

$$S_{img} = \frac{d(E_T, \mathrm{E}_K)}{d(\mathrm{E}_0, \mathrm{E}_K)} \tag{9}$$

where, E_T is the feature vector of the target local background in the infrared image to be evaluated, E_K is the feature vector of the target local background without masking, that is, let the gray value of the local background be 0, E_0 is the zero vector, and $d(x, y)$ is the Euclidian distance between x and y vectors. According to the above formula, when the feature vector E_T of the infrared image to be evaluated is consistent with the feature vector E_T of the target without masking, namely $d(E_T, E_K) = 0$, then the local background has no masking effect on the target; otherwise, the local background has masking effect on the target. The calculation method of the three characteristic components of the obscuration is as follows:

$$SBR = 10\left|\lg \frac{p_{max}(x, y)}{p_{min}(x, y)}\right| \tag{10}$$

$$H = \sum_{i=0}^{n} P_i \log P_i \tag{11}$$

$$corr = \frac{\sum_i \sum_j (T_{ij} - \overline{T}) \cdot (B_{ij} - \overline{B})}{\sqrt{\left(\sum_i \sum_j (T_{ij} - \overline{T})^2\right)\left(\sum_i \sum_j (B_{ij} - \overline{B})^2\right)}} \tag{12}$$

where, *SBR* is the contrast feature, *H* is the entropy difference feature and *corr* is the gray correlation feature, p_{max} and p_{min} represent the maximum and minimum gray values of the target and local background image respectively, p_i is the proportion of pixels whose gray value is i in the image, T_{ij} is the gray value of the target, B_{ij} is the gray value of the local background, and $\overline{T}$ is the mean gray value of the target. $\overline{B}$ is the mean value of local background gray.

3 Complexity Measurement Experiment

Based on the above analysis, the complexity of generated aircraft target and missile target infrared image sequences in different spectral bands is measured respectively, and the result is shown in Table 1. The tracks of aircraft target and missile target in different infrared image sequences are shown in Fig. 1.

It can be seen from the calculation result that the occultation degree has a higher influence on the sequence complexity of infrared images than the confusion degree. The main reason is that the targets are subpixel targets in high-orbit imaging systems and are easily affected by background clutter. By comparing the complexity of four sequences, it is found that the complexity of missile target infrared image sequence is lower than that of aircraft target infrared image sequence because the imaging feature of missile target is more obvious.

4 Target Detectability Performance Evaluation

Detection probability is an important index to measure the detection capability of infrared detection system. For the infrared image sequence of N frame, the image sequence is processed according to the image processing algorithm, and the total frame number of each sequence is N_1 and the frame number of target detection is N_2, then the detection probability P_d of the target can be written as:

$$P_d = \frac{N_2}{N_1} \tag{13}$$

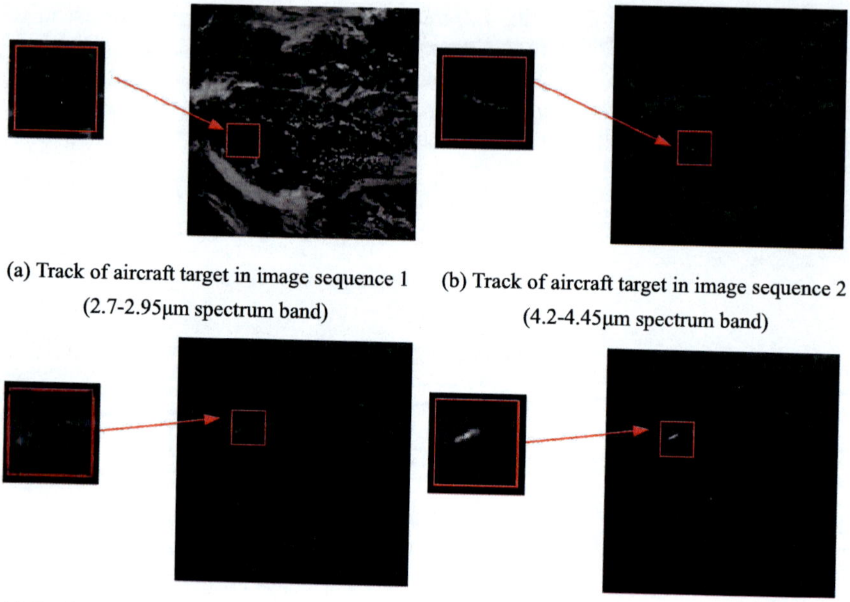

(a) Track of aircraft target in image sequence 1 (2.7-2.95μm spectrum band)
(b) Track of aircraft target in image sequence 2 (4.2-4.45μm spectrum band)
(c) Track of missile target in image sequence 3 (2.7-2.95μm spectrum band)
(d) Track of missile target in image sequence 4 (4.2-4.45μm spectrum band)

Fig. 1. Tracks of aircraft target and missile target in infrared image sequences

where, the subscript d represents detection.

It is generally expected that the detection probability of false alarm is low and the detection probability is high when detecting the target. However, the two interact and restrict each other. In order to reduce the false alarm probability, it is necessary to increase the threshold value. Since the threshold of the detection algorithm determines the detection performance of the system, the detection probability will decrease accordingly. According to different image detection algorithms and resolution levels, the image signal-to-clutter ratio threshold corresponding to the detection target can be determined, and the detection performance can be determined according to the relationship between signal-to-clutter ratio and detection performance. Assume that the image signal-to-clutter ratio threshold TNR of the detection algorithm is known, that is, the false alarm probability P_{FA} of the detection system is known, which can be expressed as:

$$P_{FA} = \frac{1}{2}\left[1 - erf\left(\frac{TNR}{\sqrt{2}}\right)\right] \tag{14}$$

where, the subscript FA represents false alarm.

Under a certain false alarm probability, the target detectability probability P'_d in a single frame image can be denoted as:

$$P'_d = \frac{1}{2}\left[1 + erf\left(\frac{SNR - TNR}{\sqrt{2}}\right)\right] \tag{15}$$

Table 1. Complexity measurement of infrared image sequence

Serial number of each image frame	Aircraft target infrared image sequence 1 (2.7–2.95 μm spectrum band)			Aircraft target infrared image sequence 2 (4.2–4.45 μm spectrum band)		
	Confusion degree	Occultation degree	Complexity	Confusion degree	Occultation degree	Complexity
1	0.795	0.991	0.882	0.478	0.962	0.871
2	0.378	0.991	0.547	0.738	0.961	0.542
3	0.334	0.988	0.499	0.752	0.961	0.496
4	0.364	0.992	0.533	0.521	0.960	0.528
5	0.752	0.992	0.855	0.762	0.961	0.844
6	0.788	0.992	0.878	0.844	0.963	0.867
7	0.584	0.992	0.735	0.833	0.962	0.727
Complexity of sequence	0.678			0.644		
Serial number of each image frame	Missile target infrared image sequence 3 (2.7–2.95 μm spectrum band)			Missile target infrared image sequence 4 (4.2–4.45 μm spectrum band)		
	Confusion degree	Occultation degree	Complexity	Confusion degree	Occultation degree	Complexity
1	0.612	0.989	0.756	0.698	0.986	0.755
2	0.433	0.989	0.602	0.695	0.986	0.602
3	0.517	0.988	0.679	0.392	0.986	0.679
4	0.359	0.985	0.526	0.643	0.958	0.526
5	0.433	0.977	0.599	0.493	0.984	0.601
6	0.406	0.973	0.573	0.648	0.984	0.575
7	0.638	0.978	0.773	0.785	0.983	0.774
Complexity of sequence	0.440			0.421		

where:

$$erf(x) = \frac{2}{\sqrt{\pi}} \int_{0}^{x} e^{-t^2} dt \tag{16}$$

$$TNR = \frac{Th}{\sigma_{total}} \tag{17}$$

$$SNR = \frac{G_t - G_b}{\sigma_{total}} \tag{18}$$

where, *Th* is the detecting algorithm of gray threshold, G_t represents the mean gray level of the target, G_b represents the mean gray level of the background, σ_{total} represents the image clutter level.

In order to analyze whether the target in an image frame can be detected, under a certain false alarm probability, it is considered that the target can be detected if the target detection probability in a single frame is greater than or equal to 90%. If the target can be detected, it is denoted as 1; otherwise, it is denoted as 0. The detection situation of the target in each frame of the image sequence is counted to calculate the target detectability of the image sequence. Figure 2 shows the target detectability in different infrared image sequences when the false alarm probability is 10% and the gray threshold is 10.

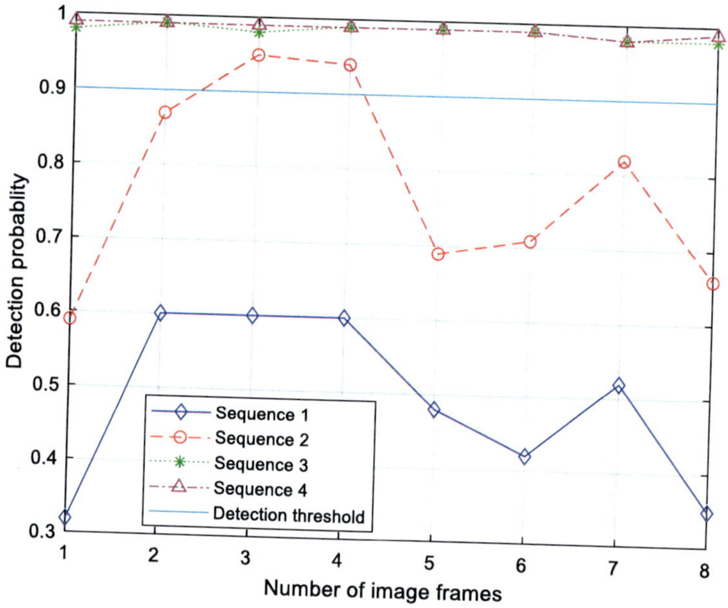

Fig. 2. Point target detection probability in different infrared image sequences

As can be seen from Fig. 2, when the detectability threshold is set to that the detection probability of a single frame is 90%, each frame of the image sequence 3 and 4 meets the requirements, so the target detectability probability of the image sequence 3 and 4 is 100%. For image sequence 2, the frames meeting the above detection requirements are the third frame and the fourth frame, so the target detectability probability is 25%. The main reason is that the difference between the target and the background in the image sequence 1 and 2 is small, the complexity of the image sequences is high, and the target detectability is low due to the unreasonable setting of the target detectability threshold. By comparing Table 1 and Fig. 2, it can be seen that the target detectability decreases with the increase of image sequence complexity. Therefore, the target detectability threshold should be set to match the image complexity.

The work in this paper provides research basis for the applicable relationship between complexity of infrared image sequence and target detection algorithm.

References

1. Harper, S., Jay, C., Michailidou, E., et al.: Analysing the visual complexity of web pages using document structure. Behav. Inf. Technol. **32**(5), 491–502 (2013)
2. Corchs, S.E., Ciocca, G., Bricolo, E., et al.: Predicting complexity perception of real world images. PLoS ONE **11**(6), e0157986 (2016)
3. Tebano, R., Ciocca, G., Corchs, S., et al.: Does color influence image complexity perception. In: Trémeau, A., Schettini, R., Tominaga, S. (eds.) CCIW 2015. LNCS, vol. 9016, pp. 139–148. Springer, Cham (2015). https://doi.org/10.1007/978-3-319-15979-9_14
4. Zhou, B., Xu, S., Yang, X.X.: Computing the color complexity of images. In: International Conference on Fuzzy Systems & Knowledge Discovery, pp. 1942–1946 (2016)
5. Mao, X., Diao, W.H.: Criterion to evaluate the quality of infrared small target images. J. Infrared Millimeter Terahertz Waves **30**(1), 56–64 (2009)
6. Zheng, X., Peng, Z., Dai, J.: Criterion to evaluate the quality of infrared target images based on scene features. Elektron. Elektrotech. **20**(10), 44–50 (2014)

Automatic Decision Undertaking on Expedience of Image Denoising Based on Filter Efficiency Prediction

Andrii Rubel(✉), Oleksii Rubel, Rostyslav Tsekhmystro, Volodymyr Rebrov, and Vladimir Lukin

National Aerospace University, Chkalov Str. 17, Kharkiv 61070, Ukraine
{a.rubel,o.rubel,v.lukin}@khai.edu

Abstract. Image denoising (filtering) is one of the most important procedures in image pre-processing chain. The aim of applying denoising is to reduce the noise while preserving image details and features. However, in practice, there are quite many cases when image denoising can lead to degradation of image quality which affects further thematic (high-level) image processing steps. Thus, the benefit of applying denoising is not always observed. In these cases, it is better to skip image denoising. This paper discusses decision-making on image denoising expedience based on denoising efficiency prediction for several modern filters. Analysis of visual quality for denoised images using subjective evaluation is carried out. Performance evaluation of existing image quality measures in the context of their application to assess the quality of denoised images is conducted. Finally, an algorithm for decision-making on image denoising expedience is proposed. It is shown that reliable decision on image denoising expedience can be undertaken in advance without carrying out denoising itself.

Keywords: Denoising expedience · image denoising · filtering efficiency · performance prediction · image quality measures · subjective image quality assessment

1 Introduction

Over the last decades, there has been a significant increase of using remote sensing data for different applications [1]. With the growing volume of data received by various remote sensing imaging systems, development of automatic remote sensing image processing methods has become an active research topic. Acquired remote sensing images can suffer from various distortions like noise caused by imperfect characteristics of sensors, image acquisition method, etc. [2]. The presence of noise greatly reduces effectiveness of subsequent image processing stages, including compression, segmentation, classification, and so on [1, 2]. Image denoising (filtering) methods are usually used at the preliminary processing stage to reduce noise for acquired remote sensing images [1, 3].

H. P. Urbach and H. Jiang (Eds.): ISSOIA 2022, SPPHY 295, pp. 504–524, 2023.
https://doi.org/10.1007/978-981-99-4098-1_44

It is usually assumed that image denoising methods enhance image quality after processing and hence improve the overall performance of high-level image processing methods. Meanwhile, existing image denoising methods along with noise suppression can introduce different annoying effects or artifacts [4, 5]. This can be caused by different reasons, including unwanted oversmoothing which is especially critical for highly textured and low contrast images or images that contain a lot of edges and small details [5–7]. Thus, visual quality of denoised image significantly depends on image properties [6, 7] or the so-called image complexity (presence of complex textures), type of noise and its intensity, methods of processing (i.e. filters) and their parameters.

In order to understand in which cases it is better to apply denoising, it is necessary to clarify its impact on visual quality for denoised images. There are two general approaches to evaluate image quality: subjective and objective [8]. Subjective methods based on aggregating and averaging opinions from human observers have been attracted to perform image quality assessment. Such methods are considered the most adequate and reliable. One of the main drawbacks of subjective assessment is that it is time-consuming [9, 10]. Objective methods (represented by image quality measures) are used to automatically assess image quality with accordance to subjective (human) scores [10]. Objective methods are often used to optimize algorithms and parameters of image processing systems as well as to evaluate efficiency of image processing methods. According to the availability of a reference (distortion-free) image, objective image quality assessment (IQA) measures are classified into full-reference (FR IQA) and blind or no-reference measures [8]. Full-reference IQA are designed to measure similarity between reference and distorted images. No-reference IQA are used when reference image is absent. Development of no-reference image quality measures is desired for real-life applications and it is still a challenging problem. FR IQA measures demonstrate more stable and reliable efficiency across different test images datasets, while effectiveness for no-reference measures are limited by training datasets [11, 12].

Image denoising efficiency is commonly evaluated by improvement of image quality measures (difference between the image quality measure values after and before applying denoising) [5, 7, 13]. Generally, a positive value of such improvement is considered as image quality enhancement after applying the image denoising. At the same time, such a positive value does not guarantee that visual quality of denoised image has been improved compared to original noisy image [4, 5, 7]. In the works [6, 10, 12], it has been demonstrated that existing image quality measures can be not adequate enough. According to this, even for a high denoising efficiency in terms of improvement for a given image quality measure, denoised image may look oversmoothed. One can expect that there are images or image fragments for which applying the denoising can be useless since it requires computational time and resources but does not result in sufficient image quality improvement. Therefore, it is highly recommended to detect such cases before starting to apply the denoising and decide whether it is worth applying the denoising or not. Note that some image pre-processing (including pre-filtering) can be performed on board of a satellite or other carrier of a remote sensing system [14]. If denoising is performed on board, decision on its expedience should be made in a fully automatic mode due to limited energy and computational resources.

This paper concentrates on decision-making on image denoising expedience and aims to propose an algorithm to undertake a reliable decision on expedience of image denoising in an automatic mode. The remainder part of this paper is organized as follows. Section 2 describes analysis of subjective evaluation for denoised images. In Sect. 3, a comparative analysis of performance for existing full-reference image quality measures for image denoising is carried out. In Sect. 4, an algorithm for decision-making on image denoising expedience is proposed. Finally, the conclusions are given.

2 Subjective Evaluation of Visual Quality for Denoised Images

Since the positive effect of denoising can be negligible or the negative effects are greater than the positive ones in terms of image visual quality, we have conducted experiments to obtain subjective scores.

Now we briefly describe the conducted subjective evaluation for denoised images. The purpose of this subjective evaluation is to understand when denoising is expedient

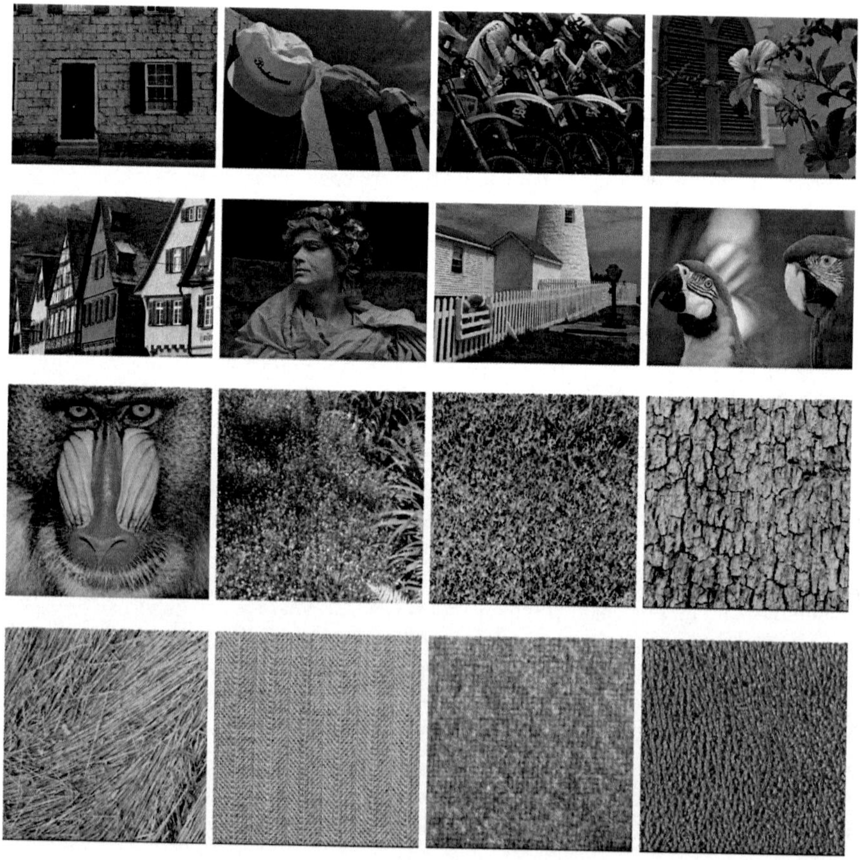

Fig. 1. Test images from the created dataset.

and worth applying. To this end, sixteen grayscale images of different complexity have been chosen (see Fig. 1).

The case of additive white Gaussian noise (AWGN) with the zero mean and seven values of standard deviations (STD), namely 3, 5, 10, 15, 20, 25 and 30 has been considered. Two filters, namely the DCT filter [15] and the BM3D filter [16] have been analyzed. Each reference image has been corrupted by AWGN with seven noise levels (STD). Then, each distorted image has been denoised by both the DCT and BM3D filters. The standard DCT filter belongs to a family of transform based filters. This filter produces good results in terms of widely used image quality measures and works very fast [15]. The BM3D filter is one of the most effective image denoising methods and it belongs to the modern family of nonlocal filters based on similar blocks search and their collaborative processing [16].

Thus, the created dataset consists of 16 reference images, 112 corrupted by AWGN images and 224 denoised images.

Subjective evaluation has been carried out by pairwise comparisons between two images - noisy and denoised. Each participant had to choose an image with better visual quality. Methodology of conducted subjective evaluation is shown in Fig. 2.

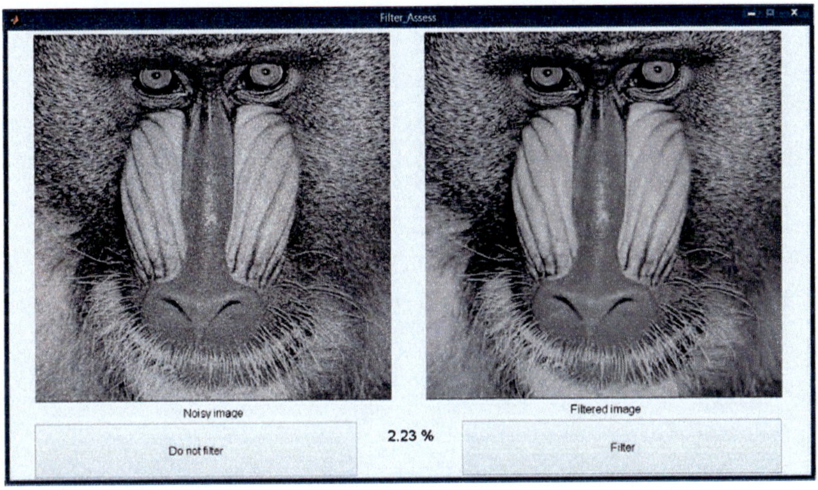

Fig. 2. Subjective evaluation (noisy image left and denoised right).

Subjective evaluations have resulted in obtaining scores represented by probabilities of voting for denoising P_{vote} for each test image, for each value of noise STD and for both filters. Such probability means that the denoised image has a better visual quality than the corresponding noisy one. If probability P_{vote} is close to unity, then denoising is obviously expedient. Probabilities P_{vote} greater than 0.5 mean that the use of image denoising is, probably, expedient. Probability of voting for denoising lower than 0.5 means that it is not worth applying the denoising for a given image and noise intensity.

The total number of participants who have been attracted to perform subjective evaluations for the denoised images is 145.

Scatterplots of the obtained probabilities P_{vote} of voting for denoising and noise STDs for both filters are shown in Fig. 3.

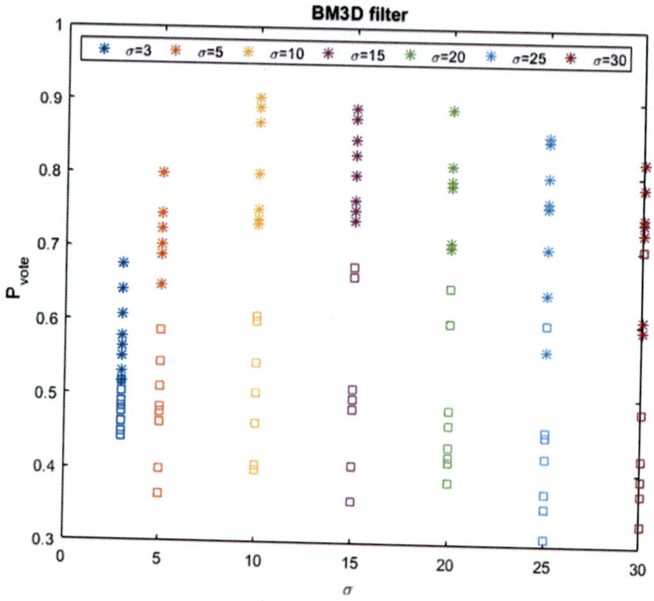

a

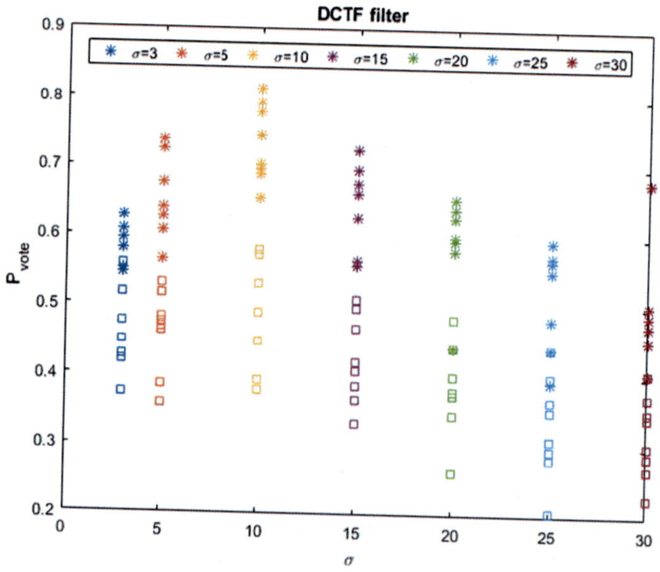

b

Fig. 3. Scatterplots of the obtained probabilities of voting for denoising P_{vote} vs noise STDs for the BM3D filter (a) and the DCT filter (b).

In Fig. 3, asterisk denotes the first eight images from the created dataset with a rather middle complexity and squares denote highly textured images (the second part of the used test images). It is well seen that for a given noise intensity there is a large diversity of values of probabilities of voting for denoising (it varies in wide limits - from 0.2 to 0.9). It is also possible to see that the use of denoising is mostly expedient for images having rather simple structures (low complexity, large homogeneous areas) and moderate noise intensity (STDs equal to 10 and 15) for both filters. Another interesting observation is that the BM3D filter performs better in terms of subjective visual quality scores than the DCT filter. This happens due to better edge preservation provided by the nonlocal BM3D filter [5, 16]. Meanwhile, practically for all highly textured images both filters do not provide sufficient visual quality enhancement.

Figure 4 shows examples of images when the use of denoising is obviously expedient. The obtained probability of voting for denoising P_{vote} are 0.8879 and 0.903, respectively, in the case of using the BM3D filter and for AWGN noise STD equal to 10. It is well seen that the BM3D filter demonstrates good noise suppression along with edge preservation. Figure 5 shows an example of image fragment when it is worth applying denoising in case of using the DCT filter and noise STD equal to 10 (obtained probability of voting for denoising is about 0.79). Here, the denoised output looks very similar to the corresponding reference image.

Fig. 4. Example of images for which applying the denoising is expedient (reference image – left, noisy image with the noise STD equal to 10 – middle, denoised image - right).

Figure 6 and 7 show an example of images for which denoising is useless – for both filters the obtained probabilities do not exceed 0.35 and 0.4, respectively. All images look very similar. These images are highly textured. In these cases, it is unclear which image has the best visual quality. There is no noticeable difference between the denoised images and noisy one. It demonstrates that improvement of visual quality for textured images is very problematic.

Fig. 5. Example of image for which applying the denoising is expedient (the reference image fragment – left, the noisy fragment with noise STD equal to 10 – middle, the denoised fragment - right).

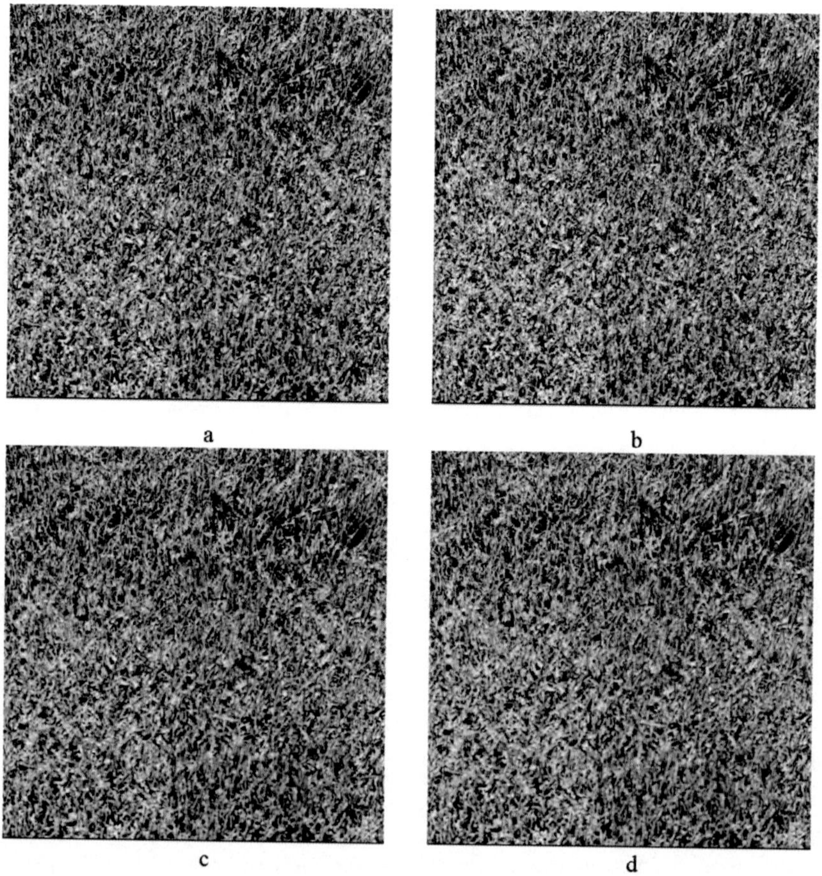

Fig. 6. Example of the image for which applying the denoising is useless (the reference image (a), the noisy image with the noise STD equal to 15 (b), the DCT filter output (c), the BM3D filter output (d)).

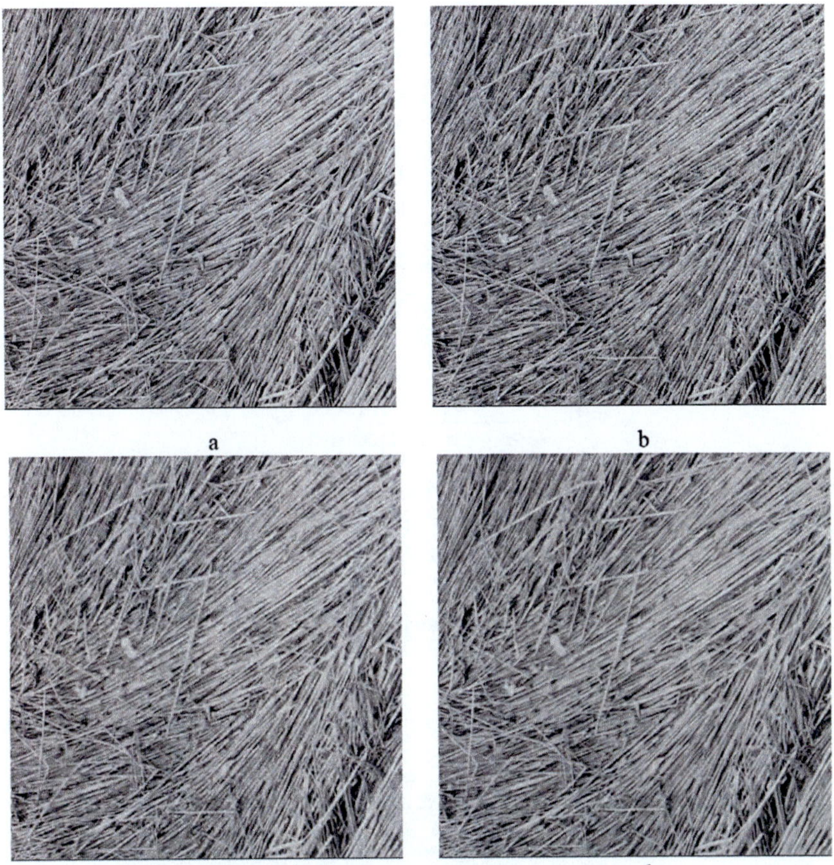

a b c d

Fig. 7. Example of the image for which applying the denoising is useless (the reference image (a), the noisy image with the noise STD equal to 10 (b), the DCT filter output (c), the BM3D filter output (d)).

3 Analysis and Performance Evaluation of Full-Reference IQA Measures for Denoised Images

Since the subjective quality assessment of denoised images requires significant resources and cannot be performed in real time applications [8, 10, 12], there is a need to use objective image quality measures. More than one hundred image quality measures have been proposed during the last decades [8–11]. However, existing image quality measures have not been studied enough for image denoising [6, 10]. In this study, we focus on full-reference IQA measures. It is still unknown which of them can most adequately characterize visual quality for denoised images. First, we evaluate the performance of full-reference image quality assessment measures using the created database.

We have selected 46 full-reference image quality measures. Evaluation of each image quality measure has been carried out by comparing the calculated values of image quality measures with subjective scores obtained from human observers (in our case, scores are represented by probabilities P_{vote} of voting for denoising) using correlation methods [10, 17]. Spearman rank-order correlation coefficient (SROCC) is used for performance comparison of FR IQA measures. Note that SROCC is calculated between the values of improvement for a given image quality measure and probabilities P_{vote} of voting for denoising.

The values of the correlation coefficients lie in the range from −1 to 1. Values close to 0 indicates that there is no correlation between analysed data. If the values of the correlation coefficients are close to 1 or −1, then there is a strict dependence.

SROCC is defined as

$$\rho = 1 - \frac{6}{n(n-1)(n+1)} \sum\nolimits_{i=1}^{n} (R_i - S_i)^2, \tag{1}$$

where R and S are the ranks of the i-th image in the sorted arrays of the considered measure and subjective score values, n is the number of images and their estimates in the arrays of measures and subjective scores.

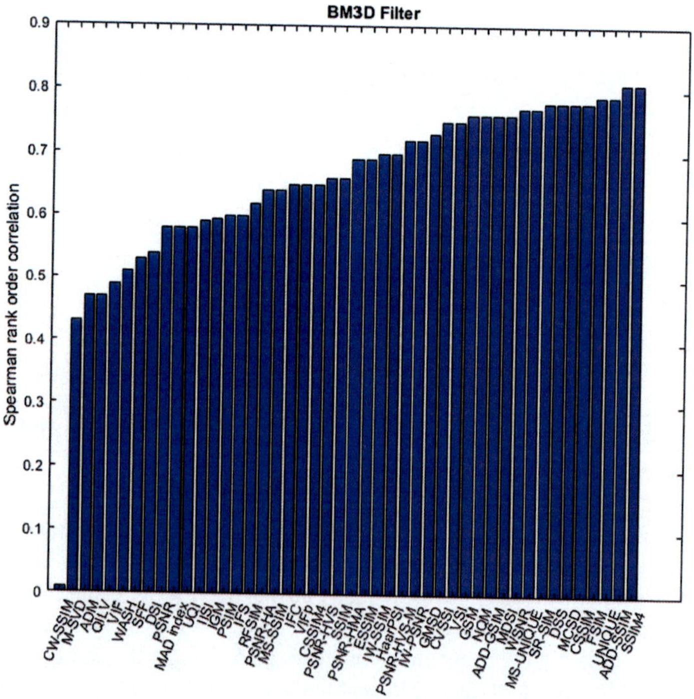

Fig. 8. SROCC values obtained for the BM3D filter.

The obtained results of SROCC for the BM3D and the DCT filters are shown in Figs. 8 and 9, respectively.

In case of analysis of the values obtained for the BM3D filter, it is well seen that the best SROCC value is 0.81. Among the best measures are the well-known FSIM [18] and the recently proposed FR IQA measures UNIQUE [19], ADD-SSIM [20] and SSIM4 [21]. Most of the well-known and widely used measures like SSIM [22] and PSNR demonstrate only moderate correspondence to human subjective scores (SROCC values are about 0.6).

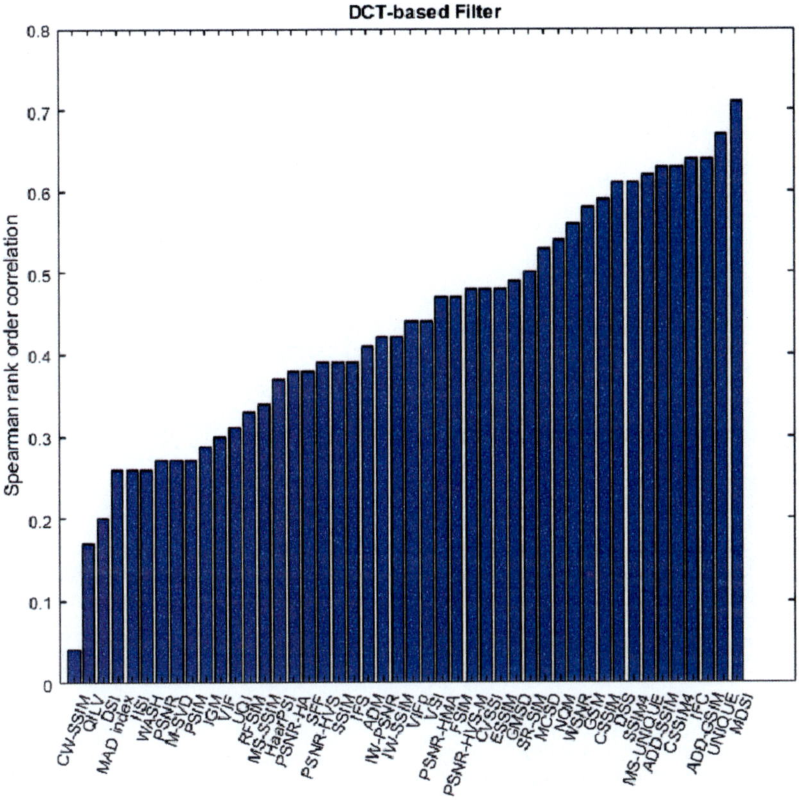

Fig. 9. SROCC values obtained for the DCT filter.

SROCC values obtained for images denoised by the DCT filter are slightly different (see Fig. 9). Here the best SROCC value is 0.71 for the MDSI image quality measure [23]. However, the measures UNIQUE, ADD-SSIM and SSIM4 are among the best for this case too (SROCC values are larger than 0.61). Although the DCT filter provides rather high improvement for a given image quality measures, it has a tendency to oversmooth texture and small details which is noticeable by visual inspection. Low values of SROCC show that image quality measures do not always adequately characterize the denoising efficiency.

Let us analyse the best image quality measures more in detail. Scatterplots of values of improvement for image quality measures FSIM, UNIQUE, ADD-SSIM and SSIM4 (denoted as I-FSIM, I- UNIQUE, I- ADD-SSIM, and I-SSIM4, respectively) in case of using the BM3D filter are shown in Fig. 10.

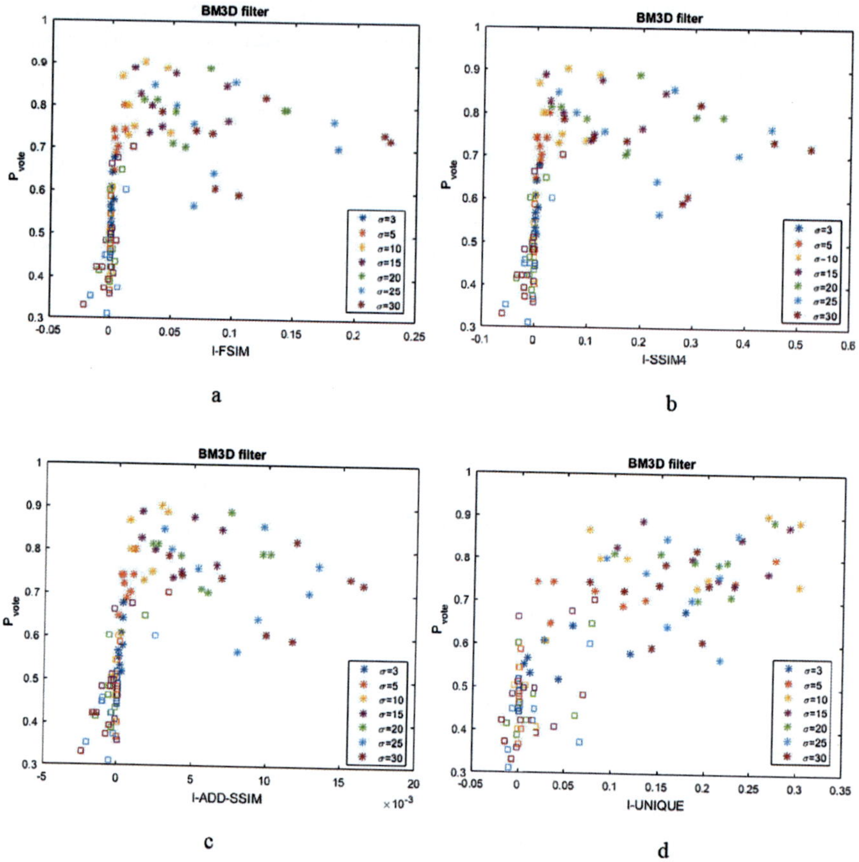

Fig. 10. Scatterplots of probabilities P_{vote} and improvements of I-FSIM (a), I-SSIM4(b), I-ADD-SSIM (c) and I-UNIQUE (d) for the BM3D filter.

From the results presented in Fig. 10, it is seen that the obtained scatterplots are practically the same especially for I-FSIM, I-SSIM4 and I-ADD-SSIM. Note that the range of possible values for these measures is from 0 to 1. The larger the value, the better the denoising efficiency. Since FSIM measure is widely used and is well-studied, we will further analyse the scatterplot and values for this IQA measure.

From the analysis of the scatterplot of values of improvement for FSIM measure and probabilities of voting for denoising P_{vote}, , it is possible to see that negative values of improvement I-FSIM (values for the measure FSIM are larger for noisy images than for the corresponding denoised ones) indicate that image denoising is useless since the values of P_{vote} are lower than 0.4. Second, if the value of I-FSIM is greater than 0.01, it is possible to conclude that applying the denoising is expedient. But there are quite many images for which the values of improvement lie between 0 and 0.01 while probabilities vary in a wide range from 0.4 to 0.7. This happens in the case of low intensity noise (noise STD is equal to 3) and highly textured images due to ability of texture to mask the noise [6, 24]. It would be also helpful to analyze the scatterplots for measures with other properties such as traditional PSNR and PSNR-HVS-M [25]. Recall that image quality measures PSNR and PSNR-HVS-M are expressed in dB, where larger values are supposed to correspond to better image visual quality. The scatterplots of the probabilities P_{vote} vs improvements I-PSNR (a) and I-PSNR-HVS-M (b) are shown in Fig. 11.

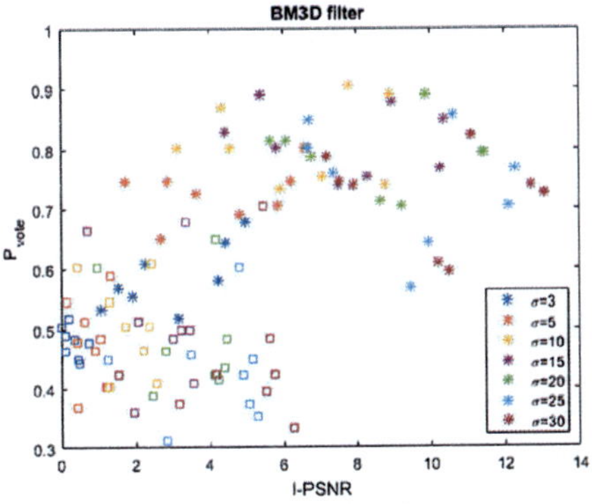

a

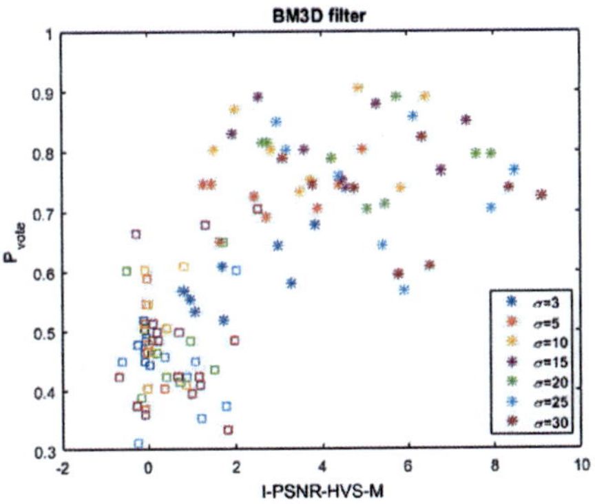

b

Fig. 11. Scatterplots of the probabilities P_{vote} vs improvements I-PSNR (a) and I-PSNR-HVS-M (b) for the BM3D filter.

Analysis of the obtained scatterplots shows that the values of improvement for I-PSNR-HVS-M measure larger than 2 dB practically guarantee that applying denoising is expedient, while for I-PSNR measure it should be larger than 6 dB. It is still challenging to undertake a decision for highly textured images for which it depends on texture characteristics (its regularities, contrast etc.).

Next, we conduct the same analysis for the DCT filter. The scatterplots of values of improvements for image quality measures I-MDSI, I-UNIQUE, I-ADD-SSIM, I-SSIM4 and for the measures I-PSNR and I-PSNR-HVS-M in case of using the DCT filter are shown in Fig. 12 and Fig. 13. It is clear that the scatterplots are very similar to those obtained for the BM3D filter. The only difference is that probabilities of voting for denoising P_{vote} for the DCT filter are usually smaller than for the BM3D filter. Note that, for the MDSI measure, small values correspond to better visual quality. The recommendations described above are also valid for this case.

Moreover, we can also analyze dependences between the values of improvement for a given measures and noise intensity (STDs). Figure 14 shows the scatterplots of values of improvements for the measures FSIM and PSNR-HVS-M and noise STDs for both filters. From the scatterplots for the I-FSIM measure (Fig. 14a, b), it is well seen that for STDs larger than 10 for highly textured images there is no sense to apply denoising.

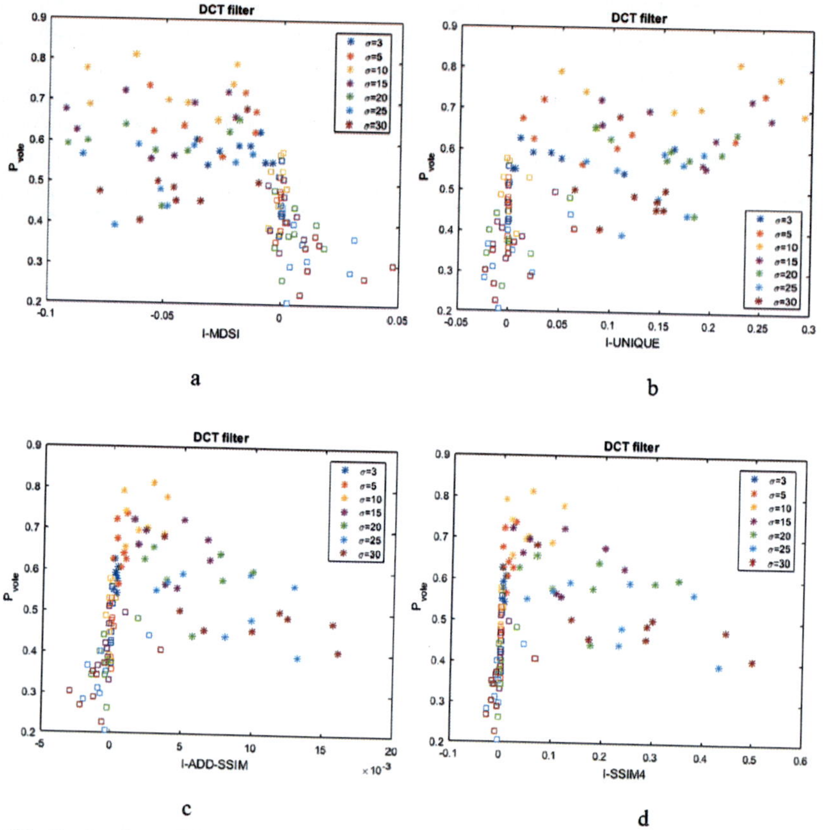

Fig. 12. Scatterplots of the probabilities P_{vote} vs improvements of I-MDSI (a), I-UNIQUE (b), I-ADD-SSIM (c) and I-SSIM4 (d) for the DCT filter.

The scatterplots of the values of improvements for the measure PSNR-HVS-M is especially interesting. In this case, one can clearly identify images for which denoising is useless. The values of I-PSNR-HVS-M do not exceed 2 dB and the probabilities of voting for denoising mostly less than 0.5. Thus, it is possible to establish such a rule: if I-PSNR-HVS-M $> 0.08 \cdot \sigma_n$, the applying denoising is expedient, otherwise it is not worth to apply denoising.

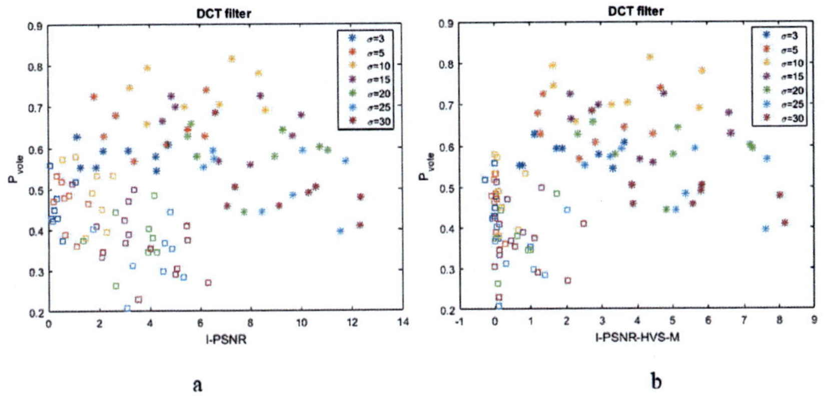

Fig. 13. Scatterplots of the probabilities P_{vote} vs improvements I-PSNR (a) and I-PSNR-HVS-M (b) measures for the DCT filter.

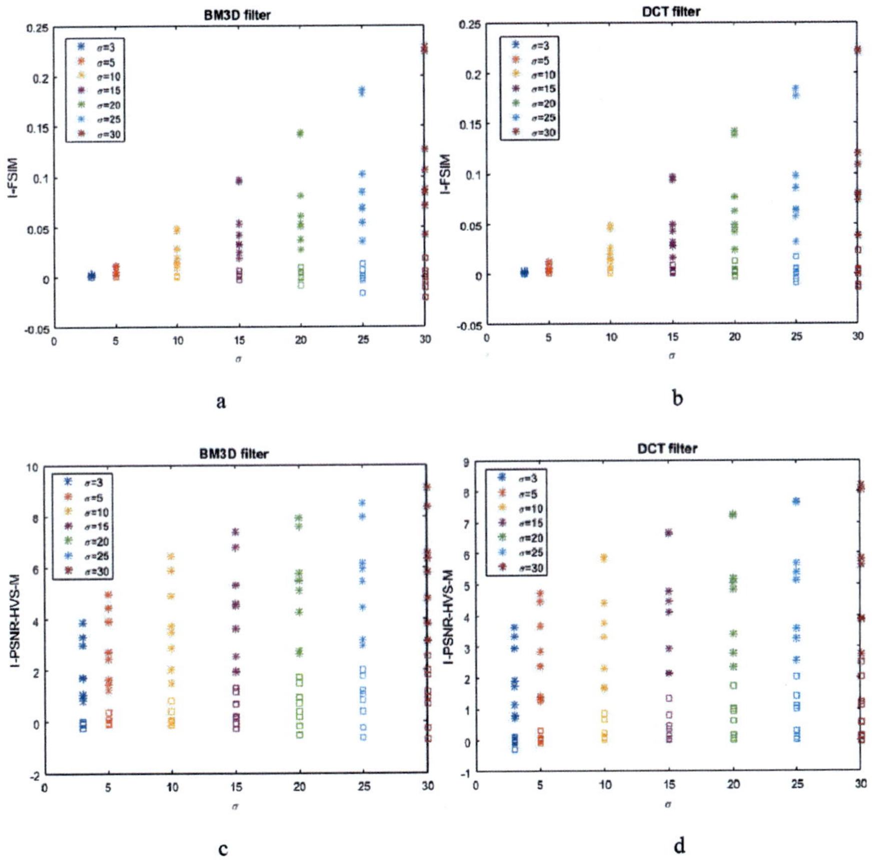

Fig. 14. Scatterplots of improvements vs noise STD for the measures I-FSIM (a, b) and I-PSNR-HVS-M (c, d) for both filters.

4 Decision-Making on Image Denoising Expedience

As it has been shown above, there are situations when image denoising provides only a small improvement for a given image quality measure, while subjective scores P_{vote} from human observers vary in wide limits. To get rid of this issue, it is also necessary to analyze the quality for the original noisy image. Sometimes noise in images may not be noticeable (for example, masked by textures) [24, 25] and, therefore, it makes denoising useless since it will take computational resources and time but will not provide quality enhancement compared to the corresponding original noisy image.

In this regard, we need to analyze the values of FR IQA measures for noisy images corrupted by AWGN from our database. To do this, we use FSIM and PSNR-HVS-M measures. In studies [6, 10, 26], it has been shown that these measures have high correspondence to human perception for noisy images corrupted by AWGN. The scatterplots

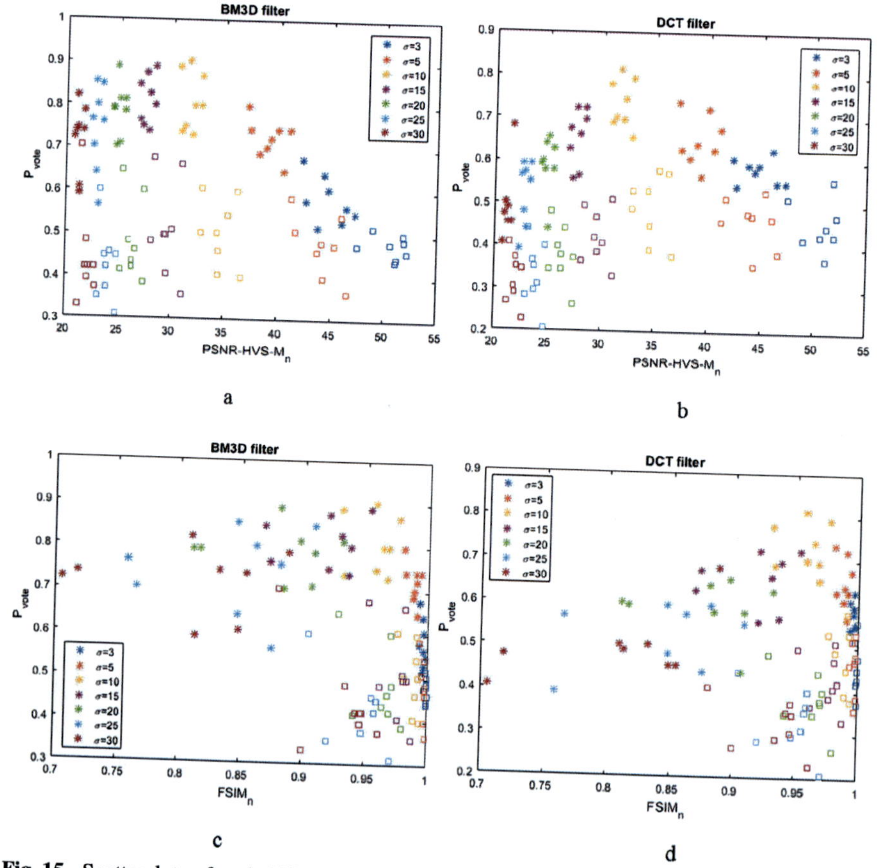

Fig. 15. Scatterplots of probabilities of voting for denoising P_{vote} for both filters and values of PSNR-HVS-M (a, b) and FSIM (c, d) measures calculated for noisy images.

of probabilities of voting for denoising P_{vote} and values of FSIM and PSNR-HVS-M measures for both filters are shown in Fig. 15.

From the obtained scatters, one can state the following. First, for noisy images for which the values of FSIM measure are about 0.97 and more, the denoising is useless. Second, it is worth applying the denoising for noisy images with FSIM values less than 0.95 and moderate complexity (without highly structured textures). Third, from the presented scatterplots for PSNR-HVS-M measure, it is seen that for noisy images with PSNR-HVS-M greater than 45dB, the use of denoising is not expedient.

Summarizing the above analysis and observations, we propose an algorithm for automatic decision-making on image denoising expedience. The proposed algorithm is shown in Fig. 16.

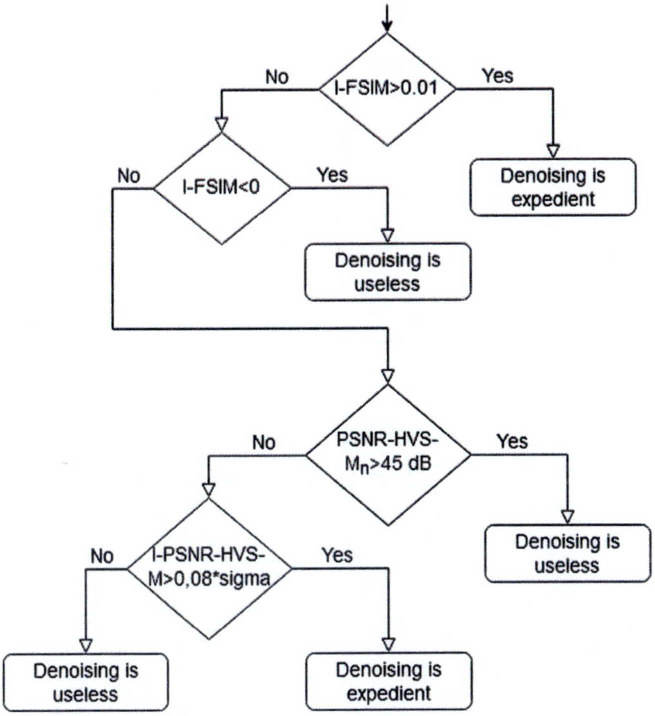

Fig. 16. Algorithm for decision-making on image denoising expedience.

Here we verify the proposed algorithm by comparing the obtained results with the probabilities of voting for denoising P_{vote} from the created database. Note that we will consider the use of denoising expedient if this probability is larger than 0.6.

Thus, the correctness of the decision-making according to the proposed algorithm for images denoised by the BM3D filter is 92.86%, while for the DCT filter it is only 71.43%. The main problem in decision-making on expedience of applying the DCT filter is that despite rather large improvements for the analyzed measures for intensive noise

(STD values equal to 20, 25 and 30), the probabilities P_{vote} indicate that denoising is useless (P_{vote} are less than 0.6). The DCT filter is very effective in noise suppression in homogeneous areas but it introduces distortions (oversmoothing) for textures and small-sized details.

Figure 17 presents an example of image fragment for which applying the denoising is obviously expedient and the proposed algorithm confirms it. Here the test image is corrupted by additive white Gaussian with noise STD equal to 10. It is easy to see that the noise is visible, especially in image homogeneous regions (see Fig. 17b). The use

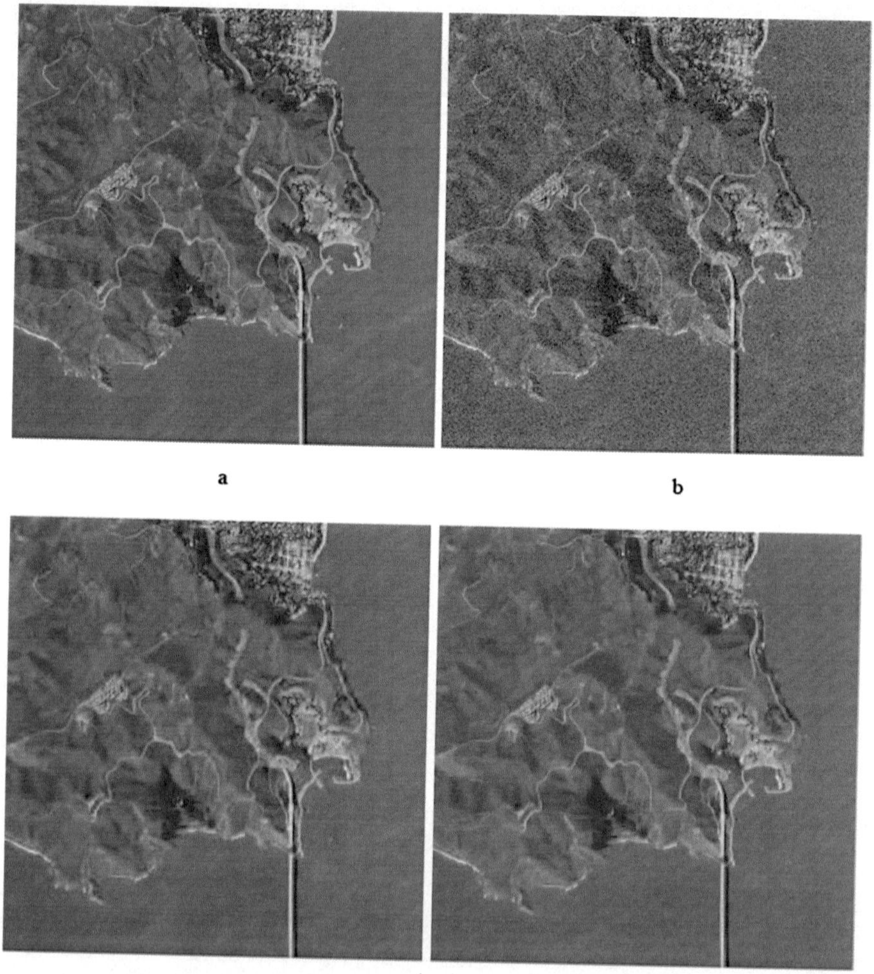

a b c d

Fig. 17. Example of the RS image fragment for which applying the denoising is expedient: the reference image (a), the noisy image with the noise STD equal to 10 (b), the DCT filter output (c), the BM3D filter output (d).

of both filters results in noticeable image quality enhancement (see Fig. 17c, d). The values of improvement of I-FSIM measure for the DCT filter and the BM3D filter are 0.0277 and 0.0288, respectively; according to the proposed algorithm indicates that the denoising is expedient (the I-FSIM values are larger than 0.01).

Figure 18 shows an example of enlarged image fragment for which there is no sense to apply denoising for both filters. There is no much visual difference between the images (noise is masked by textures). This result has been obtained for the noise STD equal to 5. The proposed algorithm also indicates the useless of applying denoising for both filters.

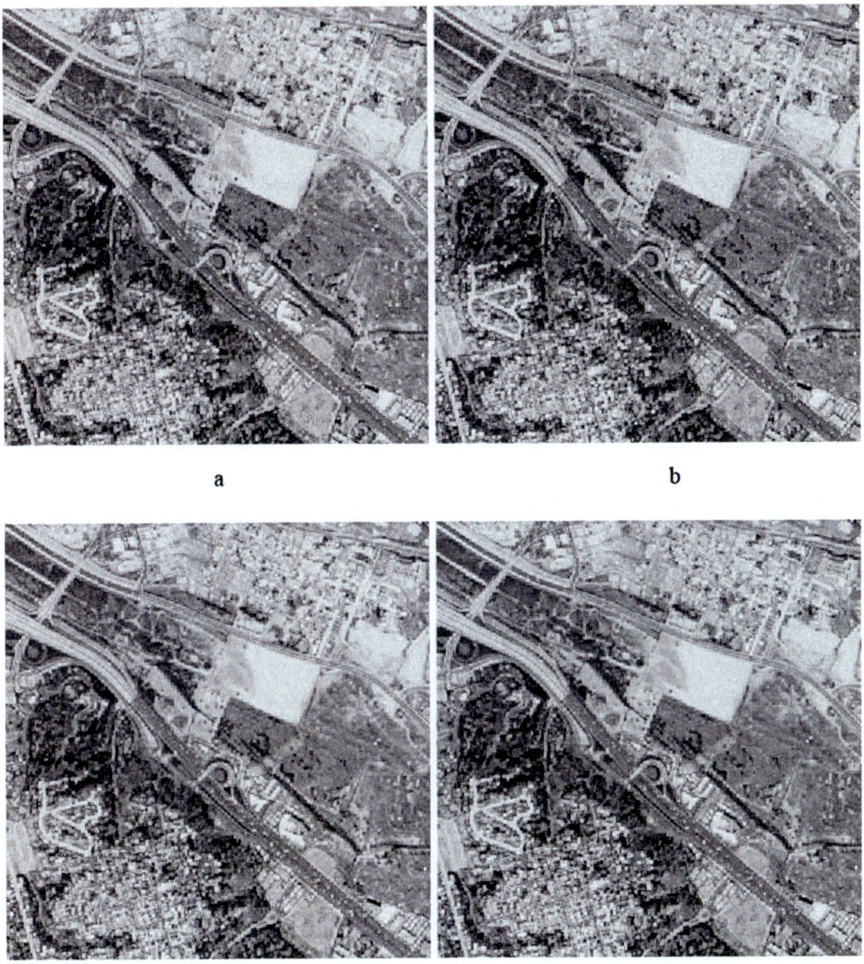

a b

c d

Fig. 18. Example of enlarged image fragment for which applying the denoising is useless: the reference image (a), the noisy image with noise STD equal to 5 (b), the DCT filter output (c), the BM3D filter output (d).

For both filters, the values of improvement of I-FSIM measure are close to zero, while for the I-PSNR-HVS-M measure they are about 0.2–0.28 dB. Since in accordance with the proposed algorithm, the values of improvement of PSNR-HVS-M measure are less than $0.08 \cdot \sigma_n$ ($\sigma_n = 5$, , i.e. I-PSNR-HVS-M < 0.4 dB), then the applying of denoising is useless. Visual inspection allows to agree with this decision.

One can see that the proposed algorithm requires preliminary calculation of full-reference IQA measures, while they cannot be used in real-life applications where reference noise-free image is not available. A way to overcome this problem is to predict the values of the considered full-reference image quality measures. It is worth noting that such methods to predict improvements due to denoising (filtering efficiency) as well as values of measures for original noisy images have been proposed recently [27–29]. These methods employ trained multilayer neural networks to accurately predict improvements for different image quality measures with high prediction accuracy.

5 Conclusions

In this paper, analysis of visual quality for the denoised images has been carried out. Two denoising methods, namely the standard sliding window DCT filter and BM3D filter have been considered. Subjective evaluation of visual quality for the denoised images using the test image dataset has been conducted. It has been shown that the denoising is expedient for images having quite simple structure and for moderate noise intensity. Based on the conducted analysis, the algorithm for automatic decision-making on image denoising expedience for images corrupted by AWGN in the form of a sequence of metric comparisons to thresholds has been proposed. It has been shown that it is possible to make rather reliable decision about image denoising expedience especially for the BM3D filter in advance without carrying out denoising itself.

References

1. Schowengerdt, R.: Remote Sensing: Models and Methods for Image Processing, 3rd edn., p. 560 (2007)
2. Lee, J.-S., Pottier, E.: Polarimetric Radar Imaging: From Basics to Applications, p. 422 (2009)
3. Oliver, C., Quegan, S.: Understanding Synthetic Aperture Radar Images, p. 486 (2004)
4. Ponomarenko, N., Lukin, V., Egiazarian, K.: HVS-metric-based performance analysis of image denoising algorithms. In: 3rd European Workshop on Visual Information Processing (EUVIP), pp. 156–161 (2011). https://doi.org/10.1109/euvip.2011.6045554
5. Rubel, O., Ponomarenko, N., Lukin, V., Astola, J., Egiazarian, K.: HVS-based local analysis of denoising efficiency for DCT-based filters. In: 2015 Second International Scientific-Practical Conference Problems of Infocommunications Science and Technology (PIC S&T), pp. 189–192 (2015). https://doi.org/10.1109/infocommst.2015.7357309
6. Egiazarian, K., Ponomarenko, M., Lukin, V., Ieremeiev, O.: Statistical evaluation of visual quality metrics for image denoising. In: 2018 IEEE International Conference on Acoustics, Speech and Signal Processing (ICASSP), pp. 6752–6756 (2018). https://doi.org/10.1109/icassp.2018.8462294
7. Rubel, A., Lukin, V., Uss, M., Vozel, B., Pogrebnyak, O., Egiazarian, K.: Efficiency of texture image enhancement by DCT-based filtering. Neurocomputing **175**, 948–965 (2016). https://doi.org/10.1016/j.neucom.2015.04.119

8. Wang, Z., Bovik, A.: Modern image quality assessment. Synthesis Lect. Image Video Multimed. Process. **2**, 156 (2006). https://doi.org/10.2200/s00010ed1v01y200508ivm003
9. Lin, W., Jay Kuo, C.: Perceptual visual quality metrics: a survey. J. Vis. Commun. Image Represent. **22**, 297–312 (2011). https://doi.org/10.1016/j.jvcir.2011.01.005
10. Ponomarenko, N., et al.: Image database TID2013: peculiarities, results and perspectives. Signal Process.: Image Commun. **30**, 57–77 (2015). https://doi.org/10.1016/j.image.2014.10.009
11. Zhang, W., Ma, K., Yan, J., Deng, D., Wang, Z.: Blind image quality assessment using a deep bilinear convolutional neural network. IEEE Trans. Circuits Syst. Video Technol. **30**, 36–47 (2020). https://doi.org/10.1109/tcsvt.2018.2886771
12. Ma, K., et al.: Waterloo exploration database: new challenges for image quality assessment models. IEEE Trans. Image Process. **26**, 1004–1016 (2017). https://doi.org/10.1109/tip.2016.2631888
13. Lukin, V., Abramov, S., Ponomarenko, N., Egiazarian, K., Astola, J.: Image filtering: potential efficiency and current problems. In: 2011 IEEE International Conference on Acoustics, Speech and Signal Processing (ICASSP), pp. 1433–1436 (2011). https://doi.org/10.1109/icassp.2011.5946683
14. van Zyl Marais, I., Steyn, W., du Preez, J.: Construction of an image quality assessment model for use on board an Leo satellite. In: IGARSS 2008 - 2008 IEEE International Geoscience and Remote Sensing Symposium, pp. II-1068–II-1071 (2008). https://doi.org/10.1109/igarss.2008.4779183
15. Lukin, V., Oktem, R., Ponomarenko, N., Egiazarian, K.: Image filtering based on discrete cosine transform. Telecommun. Radio Eng. **66**, 1685–1701 (2007). https://doi.org/10.1615/telecomradeng.v66.i18.70
16. Dabov, K., Foi, A., Katkovnik, V., Egiazarian, K.: Image denoising by sparse 3-D transform-domain collaborative filtering. IEEE Trans. Image Process. **16**, 2080–2095 (2007). https://doi.org/10.1109/tip.2007.901238
17. Kendall, M., Stuart, A., Ord, J.: The Advanced Theory of Statistics, p. 457 (1994)
18. Zhang, L., Zhang, L., Mou, X., Zhang, D.: FSIM: a feature similarity index for image quality assessment. IEEE Trans. Image Process. **20**, 2378–2386 (2011). https://doi.org/10.1109/tip.2011.2109730
19. Temel, D., Prabhushankar, M., AlRegib, G.: UNIQUE: unsupervised image quality estimation. IEEE Signal Process. Lett. **23**, 1414–1418 (2016). https://doi.org/10.1109/lsp.2016.2601119
20. Gu, K., Wang, S., Zhai, G., Lin, W., Yang, X., Zhang, W.: Analysis of distortion distribution for pooling in image quality prediction. IEEE Trans. Broadcast. **62**, 446–456 (2016). https://doi.org/10.1109/tbc.2015.2511624
21. Ponomarenko, M., Egiazarian, K., Lukin, V., Abramova, V.: Structural similarity index with predictability of image blocks. In: 2018 IEEE 17th International Conference on Mathematical Methods in Electromagnetic Theory (MMET), pp. 115–118 (2018). https://doi.org/10.1109/mmet.2018.8460285
22. Wang, Z., Bovik, A., Sheikh, H., Simoncelli, E.: Image quality assessment: from error visibility to structural similarity. IEEE Trans. Image Process. **13**, 600–612 (2004). https://doi.org/10.1109/tip.2003.819861
23. Nafchi, H., Shahkolaei, A., Hedjam, R., Cheriet, M.: Mean deviation similarity index: efficient and reliable full-reference image quality evaluator. IEEE Access **4**, 5579–5590 (2016). https://doi.org/10.1109/access.2016.2604042
24. Ieremeiev, O., Ponomarenko, N., Lukin, V., Astola, J., Egiazarian, K.: Masking effect of non-predictable energy of image regions. Telecommun. Radio Eng. **76**, 685–708 (2017). https://doi.org/10.1615/telecomradeng.v76.i8.30

25. Ponomarenko, N., Silvestri, F., Egiazarian, K., Carli, M., Astola, J., Lukin, V.: On between-coefficient contrast masking of DCT basis functions. In: 3rd International Workshop Video Processing Quality Metrics Consumer Electron, pp. 1–4 (2007)
26. Ponomarenko, N., Lukin, V., Astola, J., Egiazarian, K.: Analysis of HVS-metrics' properties using color image database TID2013. In: Battiato, S., Blanc-Talon, J., Gallo, G., Philips, W., Popescu, D., Scheunders, P. (eds.) ACIVS 2015. LNCS, vol. 9386, pp. 613–624. Springer, Cham (2015). https://doi.org/10.1007/978-3-319-25903-1_53
27. Rubel, O., Lukin, V., Rubel, A., Egiazarian, K.: NN-based prediction of sentinel-1 SAR image filtering efficiency. Geosciences **9**, 22 (2019). https://doi.org/10.3390/geosciences9070290
28. Rubel, O., Rubel, A., Lukin, V., Egiazarian, K.: Blind DCT-based prediction of image denoising efficiency using neural networks. In: 2018 7th European Workshop on Visual Information Processing (EUVIP), pp. 1–6 (2018). https://doi.org/10.1109/euvip.2018.8611710
29. Rubel, A., Rubel, O., Lukin, V.: Neural network-based prediction of visual quality for noisy images. In: 2019 IEEE 15th International Conference on the Experience of Designing and Application of CAD Systems (CADSM), pp. 1–5 (2019). https://doi.org/10.1109/cadsm.2019.8779325

Design and Spectral Analysis of Medium and Long Wave Infrared Filter for HRD

Yong Chen(✉), Xin Chen, Yifan Zhang, Weidong Lv, Libing Jin, Haibo Hu, and Ji Zhou

Beijing Institute of Space Mechanics and Electricity, Beijing 100094, China
475028635@qq.com

Abstract. Medium and long wave infrared (MDIR) filters are one of the key devices on aerospace optical remote sensing cameras. The spectral response of the high-resolution detector is determined by the spectral characteristics of the filter, of which the medium and long wave infrared accounts for a large proportion. The medium and long wave infrared filter prepared in the traditional way has a gap with the theoretical value, which makes the medium and long wave infrared filter suffer from spectral angular drift or temperature drift, resulting in a mixed superposition of high-frequency and low-frequency spectra in the high-resolution detector, leading to distortion of the recovery spectrum. This paper presents the design and spectral characteristic of a medium and long wave IR filter using for high-resolution detectors in the operating band 3.7 μm–6.2 μm. In order to achieve the wavelength-specific filtering characteristics of the medium and long wave IR filter (99% or more pass-through in the 3.7 μm–6.2 μm band and cut-off in the 2.7 μm–3.65 μm and 6.25 μm–7.2 μm bands), a structural iteration was carried out using Si material as the filter substrate and a band-pass filtered high and low reflectivity film stack structure. The structure of the film is designed with TiO_2 for the high refractive index and SiO_2 for the low refractive index. The number of layers is greatly reduced, and the medium and long wave infrared filter achieves a dual-band front-to-back cut-off and high transmittance through temperature drift tests, angular drift tests, spectral characteristic and surface shape tests. Environmental tests have shown that the medium and long wave infrared filters are well matched to the substrate material and have high stability, making them suitable for the harsh temperature variations and high-energy particle irradiation in space.

Keywords: optical thin films · high resolution detectors · medium and long wave infrared light · filters

1 Introduction

1.1 A Subsection Sample

Filters play a very important role in aerospace optical remote sensing detectors as a device for filtering light and selecting spectral lines [1, 2]. The detection of medium and long wave infrared light is of great importance for remote sensing in the field of defence and

H. P. Urbach and H. Jiang (Eds.): ISSOIA 2022, SPPHY 295, pp. 525–534, 2023.
https://doi.org/10.1007/978-981-99-4098-1_45

security [3]. The design parameters of the filter directly affect the image quality, and the environmental durability and reliability of the filter determines the working condition and lifetime of the detector in space [4]. Most of the current mainstream filters are multispectral filters with a large number of layers and complex film systems. The spectral response of a high-resolution detector is determined by the spectral characteristics of the medium and long wave infrared filters. The structural design, optical performance, environmental tolerance and stability of the medium and long wave infrared filters are crucial for the entire analysis system [5, 6]. Medium and long wave infrared filters will be increasingly used in future infrared remote sensing cameras.

The level of preparation of medium and long wave infrared filters is different from the theoretical value, allowing phenomena such as spectral angular drift or temperature drift to occur, resulting in a mixed superposition of high and low frequency spectra in high-resolution detectors leading to distortion of the recovered spectrum [7]. In the 1990s, Dr. Cushing et al. designed a bilayer system structure with an adjustable intermediate layer that could be superimposed to achieve depolarisation for the preparation of depolarisation filters [8]. In the 21st century, Gu Peifu et al. systematically studied the central wavelength of filters and conducted numerous experiments on the distribution of film materials, finally overcoming the influence of temperature on the central wavelength, so that temperature is no longer an influencing factor on the wavelength [9]. The processing cycle of filters is long and costly, especially for medium and long wavelength infrared filters, where the yield is very low and the time cost is very high according to the current production process [10]. Therefore, it is important to consider the optical system, structure, processing technology and alignment bonding factors at the early stage of filter design to ensure smooth processing and installation. In order to better meet the needs of high-resolution detectors for medium and long wave infrared detection, this paper designs a medium and long wave infrared filter film system structure based on the theory of thin-film optics and the characteristics of the medium and long wave infrared band. $_2$and SiO_2 for the low refractive index material to achieve a dual-band cut-off before and after the passband..

2 Membrane System Design

2.1 A Subsection Sample

The concept of potential transmittance was first introduced by the Americans Bering and Turner [11]. The most common optical film materials are TO_2, Ta_2O_5, Nb_2O_5, ZrO, HfO, SiO_2, MgF_2 [12, 13]. SiO_2 is physicochemically stable at room temperature and has the lowest refractive index (~1.45–1.47) among oxide film materials, and is far more transparent in the 150 nm to 8.5 μm wavelength range [14]. The refractive index versus photon energy curve for silicon dioxide thin film materials is shown in Fig. 1 shown at [15].

TiO_2 thin film materials have a relatively high refractive index (around 2.2–2.5) compared to other oxide thin film materials, good densities, chemical stability, strong adhesion to the substrate material, high strength, and high transmittance in the infrared spectral range seen at [16–18]. Therefore it is often combined with SiO_2 as a high

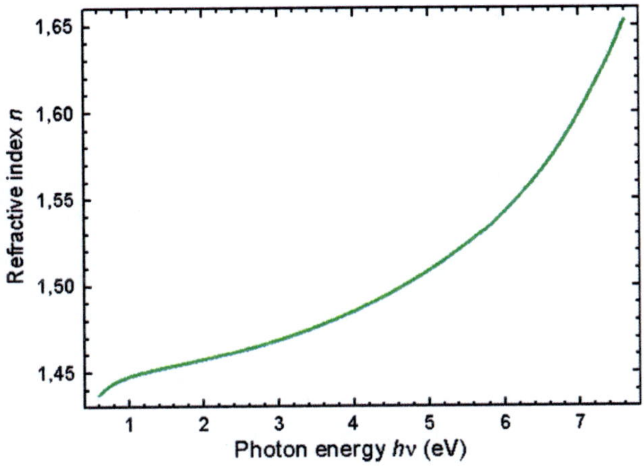

Fig. 1. Photon energy versus refractive index of SiO_2

refractive index material. The relationship between photon energy and refractive index of TiO_2 thin film materials is shown in Fig. 2 as shown.

In order to reduce the number of film layers and minimise the effect of angle on the filtering effect when preparing narrow-band filters, we chose TiO_2 for the high refractive index material and SiO_2 for the low refractive index material in order to achieve a dual-band cut-off. Therefore, the working band designed in this paper is 3.7 μm–6.2 μm and the base film system is designed as S|HLH2LHLH|Air, where S is the substrate, (HL) is the high and low reflectivity film stack and Air is the surface. Based on the spectral design requirements, the design process was optimised with Macleod software and the optimised film system structure was S|1.34H 1.75L 1.28H 1.31L 0.22H 0.31L 0.85H|Air.

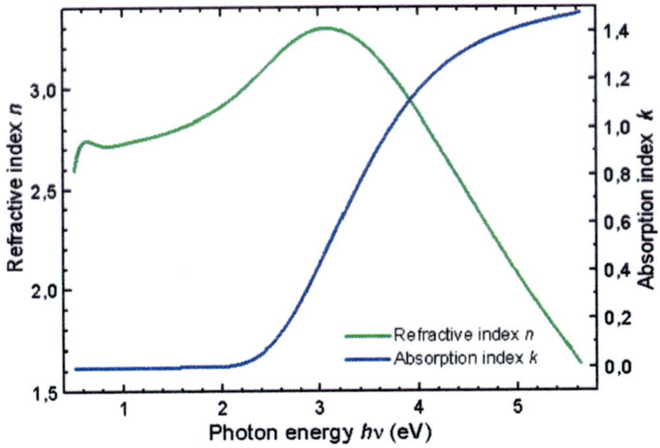

Fig. 2. Photon energy versus refractive index of TiO_2 thin film material

3 Experimental Procedure

3.1 A Subsection Sample

The high-throughput electron beam (90 A/cm^2 °C) vapour deposition method is used to heat a high-resistance, high-melting point tungsten metal ring. The kinetic energy of the surface electrons (4.5 eV) is greater than their binding energy and escapes, accelerating in a uniform electric field (10 kV) and starting to bombard the target material when the velocity reaches 6×10^4 km/s, enabling the material to evaporate and move and cool down to form a thin film structure near the substrate surface. In order to avoid contamination of the electron source due to free diffusion during the evaporation of the film material, the layout of the electron source is improved by setting the electron source at a position below the bottom of the crucible containing the film material, and when the electrons escape from the electron source and reach the top of the crucible, a strong magnetic field is used to turn the direction of the electron beam and hit the film material for evaporation [7, 19–21]. The principle of operation of the electron gun is shown in Fig. 3. The electron gun works as shown. The optical properties of the film structure are highly dependent on the thickness of the film layer and therefore the requirements for the thickness accuracy of the film layer are extremely stringent, so the monitoring of the film layer becomes a key item in the preparation of the film layer. The vacuum level for this experiment is 1×10^{-4} Pa, the substrate maintenance temperature is 100 °C ± 0.5 °C and the deposition rate of the film material is 5 nm/min.

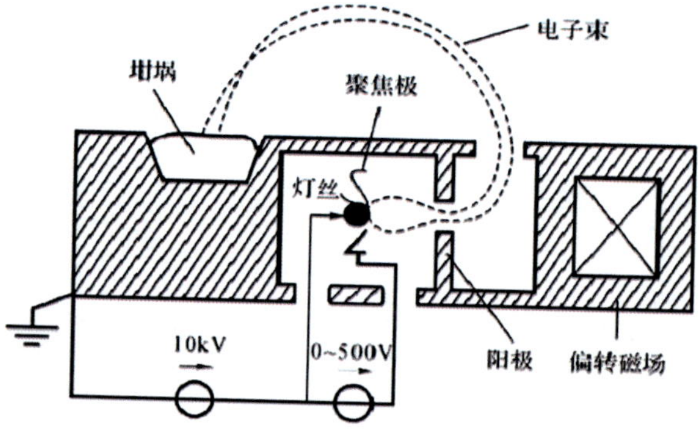

Fig. 3. Principle of evaporation by electron gun for evaporation

4 Experimental Results and Analysis

4.1 Analysis of Temperature Drift Results

As Fig. 4 shown, we took the range of 3.0 μm to 7.2 μm and tested at 80 K and 293 K. As the temperature increases, the spectral curve shifts towards the long-wave direction. A comparison of the curves shows that the effect of temperature on the spectral properties of the filter such as peak transmittance, spectral bandwidth and rectangularity of the graph is within 0.6%. It can also be analysed from the graph that λ_1–λ_2 is always in the transmission spectral range and its transmittance is above 90%. The spectral curve maintains a rectangular shape before and after, with no significant changes. Table 1 the difference in angular drift of 0.016 μm at 80 K compared to 293 K incident λ_1 and 0.013 μm compared to 293 K incident λ_2 is listed in the figure, with a maximum deviation of less than 0.6%.

Table 1. Temperature drift test results for medium and long wave IR filters before and after cut-off wavelength.

Test temperature/K	Front cut-off λ_1/μm	Post cut-off λ_2/μm
293	3.711	6.196
80	3.751	6.244

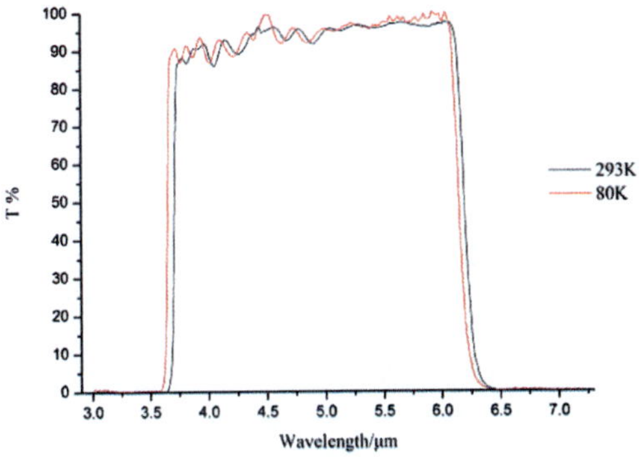

Fig. 4. Medium and long wave IR filter temperature drift test results.

4.2 Analysis of Angular Drift Results

As Fig. 5 as shown, we took the range of 2.7 μm to 7.2 μm and tested the angles of incidence of 0°, 5° and 10.7°, with increasing angles, the spectral curve shifted towards the medium and long wave direction. A comparison of the curves shows that the effect of the angle change on the spectral characteristics such as peak transmittance of the filter, spectral bandwidth and rectangularity of the graph are all within 0.5%. The graph also shows that λ_1–λ_2 is within the transmission band and the transmittance is above 90%. The spectral curve maintains a rectangular shape before and after, with no significant changes. Table 2 the maximum deviation is less than 0.01% with an angular drift of 0.005 μm compared to 0° incident λ_1 at 5°, 0.006 μm compared to 0° incident λ_1 at 10.7° and 0.003 μm compared to 0° incident λ_2 at 10.7°.

Table 2. Results of cut-off wavelength angle drift tests before and after medium and long wave infrared filters.

Test temperature/K	Front cut-off $\lambda_1/\mu m$	Post cut-off $\lambda_2/\mu m$
0°	3.712	6.201
5°	3.711	6.191
10.7°	3.706	6.184

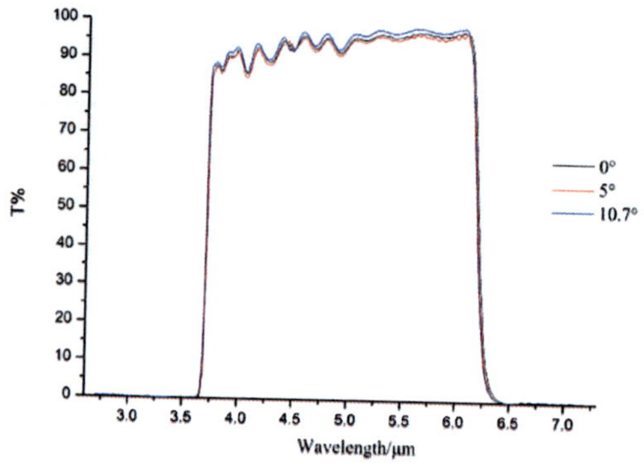

Fig. 5. Angular drift test results for medium and long wave infrared filters.

4.3 Analysis of Angular Drift Results

As Fig. 6 the spectra of the medium and long wave IR filters are shown in Fig. 6. From the test curves at 293 K and 80 K, it can be seen that the average transmittance of the IR medium and long wave filters prepared on Si substrates is above 90% at 3.7 μm–6.2 μm, where 80 K is lower than 293 K at 92.4%, and the transmittance outside the cut-off region is $T < 2\%$, and the spectral curves of the spectra was significantly better. The slope of the spectrum is steep in both the rising and falling transmittance regions (Table 3).

Table 3. Results of medium and long wave IR filter spectroscopy.

Test temperature/K	2.7 μm to 3.7 μm average transmittance/%	3.7 μm to 6.2 μm average transmittance/%
293	0.59	92.0
80	0.96	92.4

4.4 Facial Analysis

The comparison of the surface shape accuracy of the medium and long wave IR filters A/B shows that the PV and RMS values remain within a high accuracy range and the

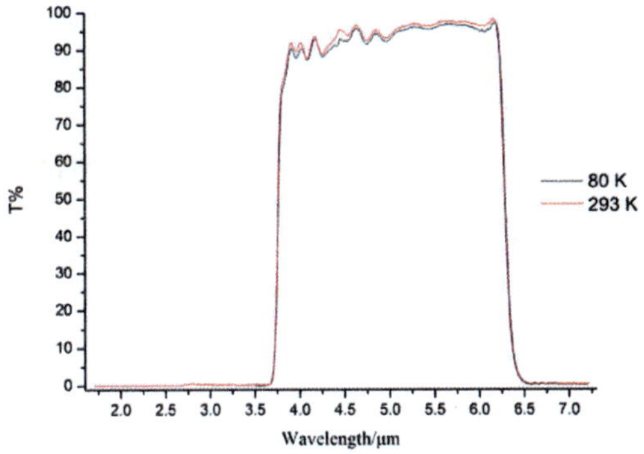

Fig. 6. Spectral profile of medium and long wave infrared filters.

overall performance of the filter A/B surface layer is relatively good. As Fig. 7 shown here (Table 4).

Table 4. Medium and long wave infrared filter face pattern results.

Face	PV/μm	RMS/μm
Positive	0.362	0.043
against	0.430	0.065

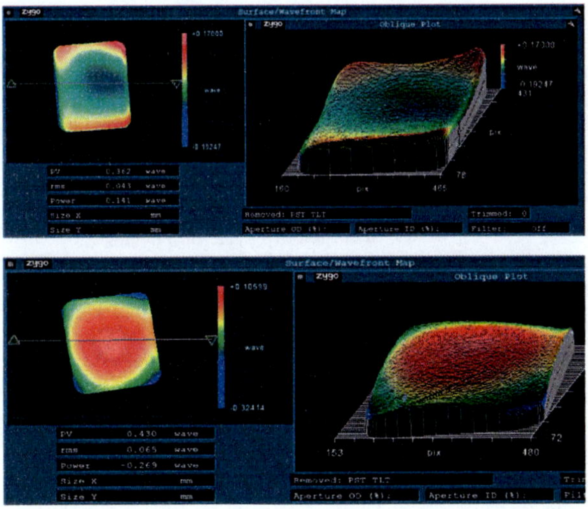

Fig. 7. Medium and long wave infrared filter A/B surface shape.

4.5 Space Environment Test Analysis and Validation of Medium and Long Wave Infrared Filters

The medium and long wave infrared filter space environment test was carried out in accordance with the aerospace industry standard QJ1697-89 "General Technical Conditions for Infrared Interference Filters", and the assessment conditions and service life of the high-resolution detector were made more stringent for the unique assessment conditions. The assessment conditions and test results adopted for the test are shown in Table 5 below.

Table 5. Medium and long wave infrared filter face pattern results.

Pilot projects	Test conditions	Test results
Soak test	Soak in pure water at 45 °C for 8 h	Experiments passed
Adhesion test	The film adhesion strength test is carried out with polyester tape by applying the adhesive side of a standard polyester tape to the film surface to avoid air bubbles and testing the peel area of the film by pulling up on one end of the tape	Experiments passed
Humidity test	In a constant temperature of 45 °C and a relative humidity of 95% or more for 72 h in a constant humidity chamber	Experiments passed
Temperature alternation test	The temperature is maintained at 45 °C for 30 min, then cooled down at a rate of 3 °C/h to −10 °C, then stopped, maintained for 30 min, and then ramped up to 45 °C at a rate of 3 °C/h. The cycle is completed by a total of 5 cycles	Experiments passed
Low temperature shock test	The filters were placed in a container filled with liquid nitrogen at room temperature and left to stand for 5 min, then warmed up naturally and kept at room temperature for 15 min to complete a cycle of 5 cycles	Experiments passed
UV irradiation test	The infrared filter is located at the entrance to the optical path, where the UV irradiation comes from direct solar irradiation and solar irradiation after reflection from the earth's surface and is attenuated by several optical elements and analysed by stray light to obtain a dose of 5200 ESH	Experiments passed
Co_{60} gamma irradiation test	Total radiation dose of 40 krad(Si) based on a 700 km track with an 8 year lifetime and a guaranteed design margin	Experiments passed

5 Conclusion

This paper presents the design and spectral analysis of a medium and long wave infrared filter for high-resolution detectors in the operating band 3.7 μm–6.2 μm, whose spectral profile basically meets the desired objectives. In order to achieve a filter film on a Si substrate with dual-band cut-off colour separation (front cut-off wavelength range of 2.7 μm–3.65 μm, back cut-off wavelength range of 6.25 μm–7.2 μm, passband wavelength of 3.7 μm–6.2 μm), transmittance above 98% and maximum deviation of angular drift less than 0.6%, the initial design was carried out in an F-P type film system structure. By means of electron beam heating and evaporation with ion-assisted precipitation technology, Si material is used as the substrate, TiO_2 is used for the high refractive index layer and SiO_2 is used for the low refractive index material to finally achieve the effect of double band cut-off before and after the passband. The medium and long wave infrared filter meets the design objectives with dual-band cut-off and high transmittance. Through environmental testing, the medium and long wave IR filter has demonstrated excellent structural stability, indicating a suitable match between the film layers. The medium and long wave IR filter is well suited for use in extreme space conditions.

References

1. Liu, Q., Zhou, J.S., Nie, Y.F., et al.: Study on the spectral spectroscopic properties and detection methods of linear gradient filters. Spectrosc. Spectral Anal. **35**(04), 1142–1145 (2015)
2. Hu, J.N., Wang, S.Y., Ruan, N.J., et al.: Design of optical systems for submicron image components in space applications. Space Return Remote Sens. **40**(1), 50–58 (2019)
3. Liu, D., Luo, Y., Fu, X., et al.: Scattering characteristics of low-loss 1064 nm bandpass filter multilayers. China Laser **48**(9), 0903004 (2021)
4. Wang, K., Wang, B., Tian, C., et al.: Design of a dual resistive band infrared frequency selective surface. Infrared Laser Eng. **47**(07), 0704003 (2018)
5. Chen, G., Wang, K., Gao, L., et al.: Spectral modulation of visible and near-infrared dual bandpass thin-film filters. J. Infrared Millimeter Waves **39**(04), 791–795 (2020)
6. Wang, F., Su, J., Wang, W., et al.: Parasitic oscillation in high power laser facility with multi-pass-cavity amplification. High Power Laser Particle Beams **21**(8), 1183–1186 (2009)
7. Liu, Y.L., Zhao, H.B., Zhong, X.M., et al.: Research on space-based composite computational spectral detection and identification methods. Space Return Remote Sens. **42**(6), 74–81 (2021)
8. Luo, H., Shi, H.-L., Li, Z.-W., et al.: Study on the spectral characteristics of narrow-band filters in spatial external difference spectrometry. Spectrosc. Spectral Anal. **35**(04), 1116–1119 (2015)
9. Wang, D., Li, M., Lu, R., et al.: Principle, preparation and application of multilayer structured colour filters. J. Artif. Cryst. **50**(07), 1187–1306 (2021)
10. Liu, R., Wang, J., Dong, L.: Study on the method of improving the energy resolution of CCD-type soft X-ray detectors. Space Return Remote Sens. **41**(1), 102–112 (2020)
11. Namgoong, J.W., Kim, S.H., Chung, S.W., et al.: Aryloxy and chloro substituted zinc(II) phthalocyanine dyes: synthesis, characterization and application for reducing the thickness of color filters. Dyes Pigm. **154**(01), 128–136 (2018)
12. Holsteen, A.L., Cihan, A.F., Brongersma, M.L.: Temporal color mixing and dynamic beam shaping with silicon metasurfaces. Science **365**(6450), 257–260 (2019)

13. Zou, Y., Zhang, L., Han, C., et al.: Research on high-precision spectral calibration of Fourier spectrometer. Spectrosc. Spectral Anal. **38**(04), 1268–1275 (2018)
14. Duan, M., Li, D., Yu, D., et al.: Design and development of an advanced subspectral integrated filter for hyperspectral imaging systems. Infrared Millimeter Waves **35**(4), 430–434 (2016)
15. Liu, H., Yang, X., Liu, D., et al.: Physical modeling of the optical constants of SiO_2 thin films. Infrared Laser Eng. **46**(09), 0921003 (2017)
16. Joo, W.J., Kyoung, J., Esfandyarpour, M., et al.: Metasurface driven OLED displays beyond 10, 000 pixels per inch. Science **370**(6515), 459–463 (2020)
17. Lee, Y., Park, M.-K., Kim, S., et al.: Electrical broad tuning of plasmonic color filter employing an asymmetric-lattice nanohole array of metasurface controlled by polarization rotator. ACS Photonics **4**(8), 1954–1966 (2017)
18. Sun, S., Yang, W., Zhang, C., et al.: Real-time tunable colors from microfluidic reconfigurable all-dielectric metasurfaces. ACS Nano **12**(3), 2151–2159 (2018)
19. Sun, S., Zhou, Z., Zhang, C., et al.: All-dielectric full-color printing with TiO_2 metasurfaces. ACS Nano **11**(5), 4445–4452 (2017)
20. Li, Z.Y., Butun, S., Aydin, K.: Large area, lithography free super absorbers and color filters at visible frequencies using ultrathin metallic films. ACS Photonics **2**(2), 183–188 (2015)
21. Yang, B., Liu, W.W., Li, Z.C., et al.: Polarization-sensitive structural colors with hue-and-saturation tuning based on all-dielectric nanopixels. Adv. Opt. Mater. **6**(4), 1701009 (2018)

Micro-vibration Modeling and Suppression Technology and Prospects for High Resolution Remote Sensing Satellite

Zhenwei Feng, Jing Chang, Yongqiang Hu, Zhijia Liu(✉), and Yufu Cui

Dong Fang Hong Satellite Co., Ltd., Chengdu 100094, China
jiankang_53@163.com

Abstract. With the development of remote sensing satellite technologies, and the pressing demand for high resolution, high agility and low costs for remote sensing satellites, integrated design of platform and payload has been a growing tendency, which could remarkably reduce cost, volume and mass. However, compared to traditional designs, remote sensors in the integrated design are much closer to vibration sources, implying that the micro-vibration environment of remote sensors become even worse. Although the micro-vibration is so small that its influence to structure is negligible, it could result in considerable reduction of image quality of high resolution remote sensors. Micro-vibration has become a critical factor for high resolution image quality especially in the integrated satellites. In this paper, the micro-vibration modeling technology including disturbance source modeling, coupling of structure and control system and comprehensive evaluation technology will be discussed. In order to verify the correctness of the micro-vibration integrated modeling and comprehensive evaluation technology and evaluate the effectiveness of the micro-vibration control system, ground test and in orbit measurement have been done, but the differences between the ground and the earth need further study. For the effects of micro-vibration on image quality, given the distribution of micro-vibration amplitude and frequency, the performance indicators of payload such as MTF and image motion can be given, but for the inverse problem, that is, given the desired performance of payload, the conditions that the vibration distribution should meet cannot be clearly given. This is very important for the selection of micro-vibration suppression methods and related parameters. To reduce the effects of micro-vibration on image quality, many vibration isolation methods have been adopted in high resolution satellites. According to the different vibration isolation modes, vibration isolation can be divided into four types: passive, active, active-passive integrated and semi-active. Some advanced nonlinear vibration isolation technologies, such as nonlinear stiffness and nonlinear damping technology, have achieved good results in low-frequency and broadband vibration isolation, and have high reliability, which may be a trend to solve high-precision spacecraft vibration isolation in the future. Finally, some suggestions on reducing the influence of micro-vibration on the imaging quality are proposed. Firstly, it is necessary to control the generation of disturbance vibration from the source, reducing the manufacturing and assembly errors of disturbance vibration sources such as CMG, momentum wheel and refrigerator. Secondly, we should keep on developing system modeling and measurement technology, analyzing the transmission

H. P. Urbach and H. Jiang (Eds.): ISSOIA 2022, SPPHY 295, pp. 535–546, 2023.
https://doi.org/10.1007/978-981-99-4098-1_46

characteristics of micro-vibration to find out the key links of micro-vibration generation and transmission. Thirdly, micro-vibration suppression technology should be vigorously studied to make the parameters of high-frequency vibration and low-frequency vibration meet the requirements of payload and create a "quiet" mechanical environment for payload. Fourthly, we should improve the "tolerance" ability of the payload itself, and study the corresponding vibration compensation measures and image processing methods.

Keywords: Micro-vibration · Integrated modeling · Micro-vibration suppression · Configuration and layout optimization · Optimization of vibration isolator parameters

1 Introduction

Micro-vibration on spacecraft is a special kind of vibration which has tiny amplitude and high frequency, whose influence to structure is negligible. However, as to high resolution satellites, micro-vibration has become a critical factor for image quality [1]. From the in orbit measurement data, it can be seen that the micro-vibration has the characteristics of small amplitude, wide frequency range, and difficult to detect. Its frequency can reach 2000 Hz, and its energy is mainly concentrated in 0.1 Hz–300 Hz [2]. CMG, momentum wheel, refrigerator are the main sources of disturbance. High-precision spacecraft, such as sub-meter remote sensing satellites, are increasingly used in earth observation, laser communication and deep space exploration. Micro-vibration seriously affects the in orbit application of these high-precision spacecraft, and greatly reduces their important performance indicators such as image quality. Therefore, the influence of micro-vibration must be considered seriously in the design of high-precision spacecraft.

In this paper, the micro-vibration modeling technology including disturbance source modeling, coupling of structure and control system and comprehensive evaluation technology will be discussed. In order to verify the correctness of the micro-vibration integrated modeling and comprehensive evaluation technology and evaluate the effectiveness of the micro-vibration control system, ground test and in orbit measurement have been done, but the differences between the ground and the earth need further study. For the effects of micro-vibration on image quality, given the distribution of micro-vibration amplitude and frequency, the performance indicators of payload such as MTF and image motion can be given, but for the inverse problem, that is, given the desired performance of payload, the conditions that the vibration distribution should meet cannot be clearly given. This is very important for the selection of micro-vibration suppression methods and related parameters. To reduce the effects of micro-vibration on image quality, many vibration isolation methods have been adopted in high resolution satellites. According to the different vibration isolation modes, vibration isolation can be divided into four types: passive, active, active-passive integrated and semi-active. Some advanced nonlinear vibration isolation technologies, such as nonlinear stiffness and nonlinear damping technology, have achieved good results in low-frequency and broadband vibration isolation, and have high reliability, which may be a trend to solve high-precision spacecraft vibration isolation in the future. Finally, some suggestions on reducing the influence of micro-vibration on the imaging quality are proposed. Firstly, it is necessary to control the generation of disturbance vibration from the source, reducing the manufacturing and assembly errors of disturbance vibration sources such as CMG, momentum wheel and refrigerator. Secondly, we should keep on developing system modeling and measurement technology, analyzing the transmission characteristics of micro-vibration to find out the key links of micro-vibration generation and transmission. Thirdly, micro-vibration suppression technology should be vigorously studied to make the parameters of high-frequency vibration and low-frequency vibration meet the requirements of payload and create a "quiet" mechanical environment for payload. Fourthly, we should improve the "tolerance" ability of the payload itself, and study the corresponding vibration compensation measures and image processing methods.

2 Influence and Compensation of Micro-vibration on Image Quality

2.1 Effect of Micro-vibration on Image Quality

Due to the difference of frequency characteristics, the influence of micro-vibration on the imaging quality is shown in two different ways. Low-frequency micro-vibration makes the optical camera and other payloads move as a whole, and its optical axis is shown as approximately linear sliding in the integration time, which is shown as imaging distortion from the image. The high frequency micro-vibration may be coupled with the internal structure of the payload, resulting in the optical components in the payload such as the optical camera deviating from their ideal alignment positions, causing relative motion, resulting in defocus and aberration of the optical system, reducing the MTF of the optical system, and blurring the image. The division of low frequency and high frequency is relative to the integration time of the optical camera. The frequency far greater than the corresponding frequency of the integration time of the camera is called high frequency, and the frequency far less than the corresponding frequency of the integration time is called low frequency. High frequency micro-vibration leads to image blurring, and low frequency micro-vibration leads to image distortion. There is also an intermediate frequency between the low frequency and the high frequency, in which the motion characteristics of payloads such as optical cameras are between the low frequency and the high frequency, and its impact on the imaging quality is manifested by the coupling of blur and distortion. It should be noted that due to the broadband characteristics of micro-vibration, micro-vibration has an impact on low, medium and high frequency bands, and the impact on image quality is also blurred and distorted coupling. At present, most of the research on the influence of low-frequency micro-vibration on image quality regards the camera as a rigid body, that is, the influence of low-frequency micro-vibration is mainly considered, and the research on the influence of high-frequency micro-vibration is not sufficient.

The relationship between micro-vibration and image quality can be divided into positive and negative aspects. At present, the research on the positive problem tends to be mature, that is, given the distribution of micro-vibration amplitude and frequency, the performance indicators of payload such as MTF and image motion can be given, but for the inverse problem, that is, given the performance of payload, the conditions that the vibration distribution should meet cannot be clearly given. This is very important for the selection of micro-vibration suppression methods and related parameters.

2.2 In Orbit Image Quality Prediction

Due to the influence of boundary conditions, test state, gravity and other factors, there is a certain gap between the ground test and the in orbit state. At present, two methods are mainly used to predict the image quality of the satellite in orbit operation: one is to correct the finite element model of the clamped boundary by using the test data of the ground test state, and then change the clamped boundary into a free boundary. The finite element model of the in orbit flight state is obtained by unfolding the solar array and antenna and unloading the gravity, and the in orbit micro-vibration environment of the satellite is analyzed. The second is to compare and analyze the test data of the model under research with the test data of the launched model, and predict the image quality of the model under research according to the image data of the launched model. These methods are based on certain assumptions and approximations. The ideal method is to analyze the ground test data and in orbit measurement data to get the difference between the two States of micro-vibration, which provides a basis for accurate evaluation of in orbit image quality.

2.3 Compensation for the Impact of Micro-vibration

It is an economical and effective method to compensate the impact of micro-vibration. However, at present, the impact of image quality caused by vibration is mainly measured by MTF, and there is little research on the image motion caused by vibration, which is of great significance for the high-precision image motion measurement and compensation of high-resolution space camera. Moreover, the actual in orbit data is disturbed by multiple factors. The influence of platform vibration cannot be considered separately; In addition, it is difficult to evaluate the compensation effect without the ideal ground object information as the compensation standard.

3 Micro-vibration Modeling and Measurement Technology

The micro-vibration effects on high-precision spacecraft payloads are complex, involving structural, control and optical systems. The integrated modeling technology is an effective method at present. In order to verify the correctness of the integrated modeling and comprehensive evaluation technology and evaluate the effectiveness of the micro-vibration suppression system, many research institutions have designed ground measurement systems, and carried out a large number of in orbit micro-vibration measurements.

3.1 Integrated Modeling Technology

Integrated modeling technology refers to the process of establishing a system-level input and output mathematical model by using different disciplinary modeling tools. The model contains various models related to system performance and their interactions, including disturbance source model, structure model, optical model, control model and so on, which can provide a method of system-level comprehensive performance evaluation and error analysis, and guide satellite design from the system level. Among the integrated modeling technologies, the research on structural modeling, control system modeling, optical system modeling and other subsystem modeling has been relatively mature, but there are still some problems to be further explored, such as the modeling of disturbance source, the coupling of structure and control system, and the comprehensive evaluation technology.

(1) **Modeling of disturbance source**

There are many disturbance sources, and their own movement is more complex. The impact on the satellite is also different. Taking the flywheel which has great influence on the payload as an example, there are mainly two kinds of disturbance force models of flywheel at present: empirical model and analytical model. The empirical model assumes that the disturbance force generated by the flywheel is composed of discrete harmonics related to the rotational speed, and the amplitude of each order component is proportional to the rotational speed. The static and dynamic unbalance of the flywheel and the elasticity of the flywheel installation structure are considered in the analysis model, and the disturbance force prediction is given by establishing the dynamic equation. However, both the ground test [3] and the in orbit measurement [4] show that, for a given flywheel speed, the measured micro-vibration generally appears as a broadband multi-peak random vibration, as shown in Fig. 1. There is a certain difference between the above model and the test data. Due to the different transmission characteristics of vibration in different frequency bands and the different effects on the payload, the validity of the integrated modeling analysis results obtained by this vibration source model cannot be guaranteed, so the vibration source modeling technology still needs further research.

(2) **Coupling of structure and control system**

During in orbit operation, the spacecraft is subject to uncertain disturbances such as environmental moments and internal disturbances, as well as control moments. These factors are easy to excite the vibration of the flexible structure of the spacecraft. The coupling of structural vibration and spacecraft control will disturb the attitude of the spacecraft, affect the stability and control accuracy of the attitude, and even make the spacecraft unstable. Especially for modern complex spacecraft, some low-frequency modes may be within the bandwidth of the control system, and the coupling resonance will occur between the structural vibration and the control effect. The coupling effect should also be considered in the integrated modeling, so as to eliminate the rigid body attitude "drift" phenomenon in the open-loop dynamic simulation, and make the structural response characteristics closer to the actual situation of the satellite in orbit.

(3) **Comprehensive evaluation technology**

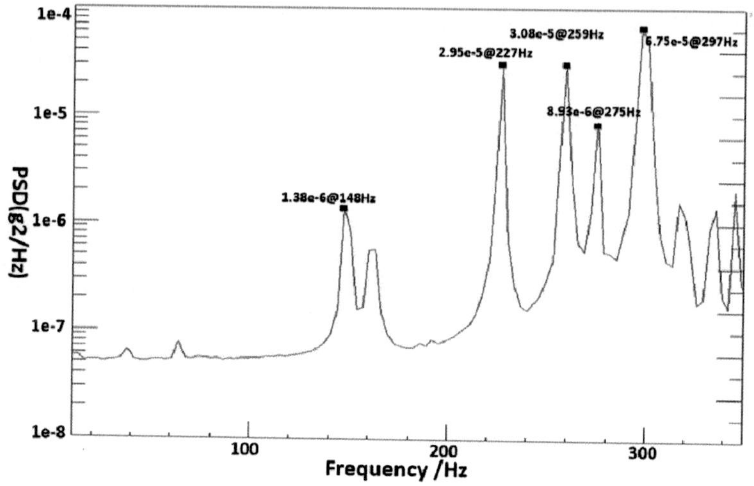

Fig. 1. Acceleration power spectrum of momentum wheel mounting point measured by a satellite

Comprehensive evaluation is mainly to evaluate whether the important performance of the system under the influence of micro-vibration dynamics environment can meet the index requirements on the basis of the integrated model, and to improve the system design by comprehensively considering the influence of various factors from the system level. Related technologies mainly include high-precision spacecraft micro-vibration disturbance performance evaluation, parameter sensitivity analysis, performance uncertainty analysis and other performance analysis technologies, key parameter optimization design technology, disturbance performance error allocation technology and so on. There are many researches on the theory of comprehensive evaluation technology, but because of the complexity of the theory itself, the coupling influence of satellite systems and the fact that many engineering experiences cannot be integrated into the analysis theory after being expressed in mathematical language, the application of comprehensive evaluation technology in the development of engineering models is rare, and further research is still needed.

3.2 Ground Test and Measurement

In order to verify the correctness of the micro-vibration integrated modeling and comprehensive evaluation technology and study the effectiveness of the micro-vibration control system, various research institutions have designed a large number of experimental equipment. Satellite control experiment platform of Honeywell, high-precision interference experiment platform of JPL and origin experiment platform of SSL have been relatively mature. Domestically, Dong Fang Hong Satellite Company, Beijing University of Aeronautics and Astronautics, Tsinghua University, Beijing Institute of Control Engineering and other research institutes have also set up micro-vibration test platforms. Among them, the ultra-high precision dual-frequency laser interferometer and supporting lifting table developed by Dong Fang Hong Satellite Company are used for high-precision micro-vibration displacement/angular displacement measurement of satellite remote sensors, and also for high-precision measurement of satellite thermal deformation. Dual-frequency laser interferometer system is the highest precision equipment for multi-axis displacement measurement at present, which can measure the displacement, yaw angle and pitch angle of the target position at the same time. The method can be expanded to nine axes at most, and the displacement, yaw angle and pitch angle of the target are measured from three directions at the same time. The method has been applied to the micro-vibration measurement of multiple small satellites (Fig. 2).

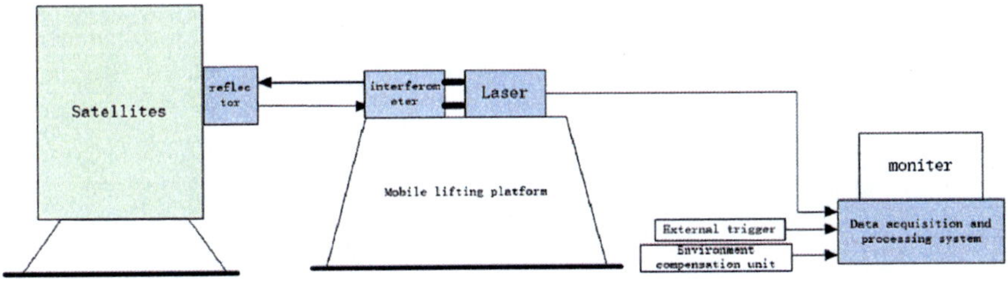

Fig. 2. Micro-vibration measurement system of Dong Fang Hong Satellite Company

3.3 In Orbit Micro-vibration Measurement

In the ground micro-vibration test, it is difficult to simulate the micro-gravity environment in orbit, and the theoretical research and digital simulation model of satellite micro-vibration in orbit are lack of measured data to modify, so it is difficult to verify the research and calculation results. It is necessary to measure the micro-vibration in orbit. Relatively mature measurement systems have been developed abroad, such as the Space Acceleration Measurement System (SAMS) and the Microgravity Acceleration Measurement System (MAMS) developed by the Glenn Research Center of the National Aeronautics and Space Administration (NASA), which are used to measure the micro-vibration environment of the International Space Station (ISS). The Acceleration Instrumentation (PAX) system, developed by the European Space Agency (ESA), measures the micro-vibration of communications satellites to obtain design data for high-precision pointing of optical payloads; Japan's Advanced Land Observing Satellite (ALOS) uses an angular displacement sensor (ADS) for micro-vibration measurements. China Academy of Space Technology (CASC) has carried out micro-vibration measurements on some remote sensing satellites in orbit, and measured the micro-vibration of the main moving parts on the satellite and the influence of the moving parts on the payload on the satellite [5].

The in orbit micro-vibration measurement system must meet the requirements of low power consumption, high precision, small size, real-time measurement, and can adapt to high impact, high temperature difference, high radiation and other space environments, and need to consume energy and occupy telemetry channels. Some researchers try to measure micro-vibration without using special measurement equipment on the satellite. Zhao Hongzhi proposed a method for measuring the image shift caused by micro-vibration on high-resolution optical remote sensing satellites by means of the time-sharing imaging characteristics of multi-color CCD devices used in high-resolution optical remote sensing satellites, and verified the feasibility of the method through research examples [6]. Shi Junxia realized the detection of vibration parameters by using fast area array CCD. As shown in Fig. 3, the fast area array CCD is installed between two TDICCDs in the same line. During the exposure time of TDICCD, the image is blurred due to the vibration of the satellite, while the fast area array CCD can be considered to have no relative motion during the exposure time due to its very short exposure time, so the image sequence obtained by the fast area array CCD is clear. These clear image sequences are used to calculate the relative displacement between them, so that the displacement versus time curve can be obtained, and so the vibration parameter of the satellite can be obtained [7].

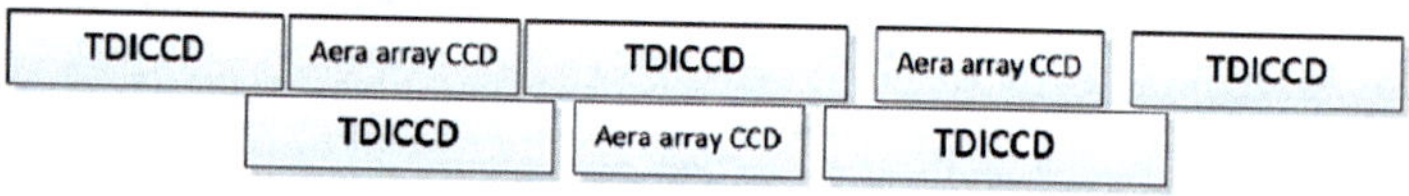

Fig. 3. Satellite Vibration Detection Device

Generally speaking, the ground test cannot fully simulate the in orbit free boundary condition, and there are differences between the test state and the in orbit state. At the same time, considering the gravity unloading, how to modify the ground micro-vibration test data and finite element model needs to be studied urgently. The influence of the above differences on the micro-vibration measurement can only be concluded by comparing the ground micro-vibration measurement data with the in orbit micro-vibration measurement data, but at present, there are few satellites in China that measure the micro-vibration on the ground and in orbit at the same time, and the test data obtained cannot meet the needs to study the differences between the ground and the earth.

4 Micro-vibration Suppression Technology

With the continuous improvement of spacecraft precision, micro-vibration has become a key factor affecting the performance of spacecraft. Many foreign high-precision spacecraft have taken vibration isolation and reduction measures for vibration sources or payloads. The Hubble Space Telescope (HST) launched in 1990 used a viscous liquid damper isolator, and the EOS-AMI satellite launched in 1999 added a viscoelastic material to its mounting seat and uses a piezoelectric actuator for active vibration isolation of the payload. The TacSat-2 Earth observation satellite launched in 2006 also adopted the active-passive hybrid vibration isolation scheme to isolate the payload. In order to obtain high-quality images, the WorldView-2 satellite launched in 2009 and the GFDM-1 satellite, GF-9(05) satellite launched in 2020 took vibration isolation measures for the control moment gyroscope and the camera. The Webb Telescope launched in 2022, used a dual passive vibration isolation system to reduce the effects of flywheel vibration. Some space camera and optical camera manufacturers have provided vibration isolators as standard accessories for satellites.

There is no essential difference between micro-vibration and traditional vibration in theory. The main difference is that the extremely low amplitude of micro-vibration and the harsh space environment put forward new requirements for the engineering realization of vibration isolation device. According to the different vibration isolation modes, vibration isolation can be divided into four types: passive, active, active-passive integrated and semi-active (Fig. 4).

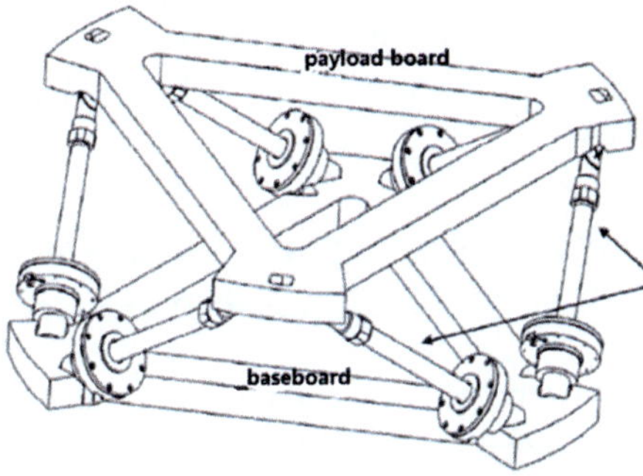

Fig. 4. Six-axis Stewart active vibration isolation platform

Passive vibration isolation is widely used in engineering because of its high reliability, simple structure and no energy consumption, but its low-frequency vibration isolation performance is not good. Active vibration isolation can achieve better low-frequency vibration isolation effect, but the general structure is complex, the reliability is low, and it needs to consume a lot of energy. Active-passive hybrid vibration isolation has the advantages of both active and passive vibration isolation, and has good effect on narrowband and broadband vibration isolation, but it also has the disadvantages of complex structure, low reliability and high energy consumption of active vibration isolation [8]. Semi-active vibration isolation can also achieve better isolation effect than pure active vibration isolation or pure passive vibration isolation, and has the advantages of shorter response time and less energy consumption compared with active and passive hybrid vibration isolation, but its weight is larger, which is unacceptable for spacecraft with strict weight control, and the use of some controlled smart materials alleviates this problem [9]. However, its weight and low-frequency vibration isolation performance still need to be further improved. Some advanced nonlinear vibration isolation technologies, such as nonlinear stiffness and nonlinear damping technology, have achieved good results in low-frequency and broadband vibration isolation, and have high reliability, which is a trend to solve high-precision spacecraft vibration isolation in the future.

In the selection of the installation position of the vibration isolation system, in some spacecraft, it is installed at the mounting plate of the payload. In this situation, the problem of ectopic control caused by this needs to be considered, and the interaction between the flexible mode of the structure and the control system is very easy to form unstable interaction, which leads to the instability of the control system. In other spacecraft, it is installed at the disturbance source such as momentum wheel. At present, there is no spacecraft installed on other vibration transmission paths, and there are few related theoretical studies. On the one hand, because of the complexity of vibration transmission, it is difficult to establish an accurate mathematical model. Even if a mathematical model is established, how to determine where to install the vibration isolation system is also a very difficult problem. On the other hand, it may have an additional impact on other systems and even affect the performance of the whole satellite. However, the research on the transfer path can not only optimize the installation position of the vibration isolation system to achieve the best vibration isolation effect, but also provide a reference for the layout of satellite equipment, optimize the structural design, and maximize the attenuation effect of satellite structure on micro-vibration, so the vibration transfer characteristic is also an important research direction.

5 Some Suggestions on Reducing the Influence of Micro-vibration on the Imaging Quality of high Resolution Satellite

With the continuous improvement of spacecraft precision, the requirements of payloads such as cameras for micro-vibration environment are becoming more and more stringent, and the requirements for micro-vibration control are also getting higher and higher. The control of micro-vibration is a systematic project. Firstly, it is necessary to control the generation of disturbance vibration from the source, reducing the manufacturing and assembly errors of disturbance vibration sources such as CMG, momentum wheel and refrigerator. Secondly, we should keep on developing system modeling and measurement technology, analyzing the transmission characteristics of micro-vibration to find out the key links of micro-vibration generation and transmission. Thirdly, micro-vibration suppression technology should be vigorously studied to make the parameters of high-frequency vibration and low-frequency vibration meet the requirements of payload and create a "quiet" mechanical environment for payload. Fourthly, we should improve the "tolerance" ability of the payload itself, and study the corresponding vibration compensation measures and image processing methods. However, under the current level of design, manufacturing and testing, it is expensive to reduce the micro-vibration load generated by the disturbance source, even beyond the scope of current technical capabilities. Reducing the sensitivity of sensitive loads to micro-vibration is often contrary to the resolution and other indicators, and vibration compensation measures are difficult to apply in orbit, so it is a feasible scheme to change the transmission characteristics of micro-vibration in the satellite structure by optimizing the configuration and layout of the satellite.

5.1 Optimize the Layout of Disturbance Vibration Source and Sensitive Load

Although it is difficult to reduce the micro-vibration of the disturbance vibration source itself with the current technology and process level, the micro-vibration transmitted to the sensitive load can be reduced by analyzing the transmission characteristics of the micro-vibration and optimizing the layout of the disturbance vibration source and the sensitive load. For example, the micro-vibration load transmitted to the sensitive load can be reduced by placing the momentum wheel with large interference on the path with small transmission ratio and placing the momentum wheel with small micro-vibration amplitude on the path with large transmission ratio, which has been confirmed in the micro-vibration test of a certain type of satellite [10]. When designing the layout of sensitive load, the relative relationship between the sensitive load and the disturbance vibration source should be considered under the condition that other constraints are met: one is the length of the transmission path from the disturbance vibration source to the sensitive load, and the other is the number of discontinuous interfaces on the transmission path from the disturbance excitation source to the sensitive load.

5.2 Optimize the Main Structure Material and Size

The main structure of the satellite mainly refers to the structure that bears the load on the satellite and provides the installation position for the on-board equipment. For the satellite with box-plate load-bearing structure, the main structure of the satellite mainly refers to the structural plates such as the platform cabin bottom plate, platform cabin partition plate, load cabin bottom plate, load cabin partition plate and side plate. The material and size of the structure are different, and the local stiffness and damping will be different, which will affect the transmission characteristics of micro-vibration.

5.3 Optimizing Frequency Allocation

The current frequency allocation mainly considers the fundamental frequency of the whole satellite and the frequency of large accessories such as solar wings and antennas, but does not consider the natural frequency of the internal structure of the equipment, which causes the coupling resonance between the micro-vibration and the structure when it is transmitted inside the equipment, resulting in the amplification of the micro-vibration on the transmission path. The natural frequencies of critical structures within sensitive equipment shall also be considered. In addition, for the disturbance sources such as the refrigerator with larger amplitude and less disturbance frequency, the frequency design requirements can be put forward to avoid the sensitive frequency of the sensitive load as far as possible. For the vibration sources such as momentum wheel with dense vibration frequency distribution, the vibration frequency with large amplitude should be avoided in the design of sensitive load. For the coupling structure, the additional constrained damping structure can be used to suppress the micro-vibration.

5.4 Optimization of Isolator Parameters

At present, the design and development of vibration isolators are lack of systematic theoretical guidance, and the design and development of vibration isolators mostly adopt the method of experience and test, which has great blindness in design and test, low efficiency, waste a lot of manpower and material resources, and even affect the progress of satellite development. Moreover, the selection of vibration isolator is concentrated on the level of single machine, which is coupled with the satellite structure after being installed on the satellite. Therefore, it is of great significance to study micro-vibration suppression and optimize the parameters of vibration isolators from the system level, which can shorten the time of model development, save manpower and financial resources, and improve the quality of imaging.

6 Conclusion

This paper systematically introduces the research status of micro-vibration modeling analysis and test, micro-vibration impact on image quality and compensation, micro-vibration suppression and other technologies. Some urgent research problems in modeling and test technology combined with engineering practice have been pointed out and some suggestions are proposed to reduce the micro-vibration impact on high resolution satellite imaging quality. It can provide a reference for the future study of micro-vibration.

References

1. Feng, Z., Cui, Y., Yang, X., Qin, J.: Micro-vibration issues in integrated design of high resolution optical remote sensing satellites. In: Urbach, H.P., Zhang, G. (eds.) 3rd International Symposium of Space Optical Instruments and Applications. SPP, vol. 192, pp. 459–469. Springer, Cham (2017). https://doi.org/10.1007/978-3-319-49184-4_45
2. Kwon, S.-C., Jeon, S.-H., Oh, H.-U.: Performance evaluation of spaceborne cryocooler micro-vibration isolation system employing pseudoelastic SMA mesh washer. Cryogenics **67**, 19–27 (2014)
3. Dong, Y., Zhou, X., Shen, J., et al.: Study on micro-vibration suppression technology of FY4 satellite. Aerosp. Shanghai **34**(4), 20–27 (2017)
4. Zhou, D., Cao, R., Zhao, H., et al.: Micro-vibration measurement and analysis of remote sensing satellites in orbit. Spacecraft Environ. Eng. **30**(6), 627–630 (2013)
5. Wang, G., Zhou, D., Zhao, Y.: Analysis of in orbit micro-vibration measurement data of remote sensing satellite. Acta Astronaut. **36**(3), 261–267 (2015)
6. Zhao, H., Huang, J., Cao, Q., Han, X.: A method for measuring image shift caused by micro-vibration on satellite. Spacecraft Eng. **23**(5), 35–40 (2014)
7. Shi, J., Xue, X., Guo, Y.: Influence of satellite vibration on TDICCD imaging quality and its compensation method. Optoelectronic Eng. **37**(12), 11–16 (2010)
8. Thayer, D., Campbell, M., Vagners, J., von Plotow, A.: Six-axis vibration isolation system using soft actuators and multiple sensors. Spacecr. Rockets **39**, 206–212 (2002)
9. Onoda, J., Makihara, K., Minesugi, K.: Energy-recycling semi-active method for vibration suppression with piezoelectric transducers. AIAA **41**, 711–719 (2003)
10. Yang, X., Bai, Z., Yang, D., Li, Y.: Research on satellite ground micro-vibration characteristics induced by momentum wheel and in orbit simulation analysis. Equipment Environ. Eng. **12**(3), 15–21 (2015)

Design of Large Array Infrared Detector Imaging Electronics System for Space Application

Xiu-rong Yan(✉), Bo Song, Shuai Cai, and Cui-cui Qi

Beijing Institute of Space Mechanics and Electricity, Beijing 100094, China
yanxr1978@163.com

Abstract. For space application imaging electronics system, it needs to face many problems such as many detector modules, low noise, small size, light weight, long life and high reliability. In this paper, the imaging electronics system is optimized based on 2 * 2 spliced large array infrared detector. In the system design, the previous backup redundancy design scheme is abandoned, and the idea of dividing the imaging electronics into two half fields of view for electronics system design is proposed, which can reduce the volume and weight by up to 50% . By adopting the architecture scheme of focal plane circuit and signal processing circuit separation and the scheme of full differential analog signal long distance transmission, the problem of extremely small focal plane position space layout is solved. The data reservation transmission mode is designed to directly transmit target data to satellite data transmission, greatly saving the demand for satellite storage resources and transmission resources. By selecting the appropriate high-precision operational amplifier and high-speed ADC, the highly integrated and low noise imaging circuit design is realized. With FPGA as the main processing platform, the real-time nonuniformity correction and double integration time imaging exposure functions are developed, which effectively improve the imaging quality and dynamic range of the infrared imaging system. The experimental results show that the imaging system has excellent image quality, and the real-time nonuniformity residual error reaches 1.33%, meeting the current engineering application requirements.

Keywords: infrared detector · high dynamic range · Imaging circuit

1 Introduction

As one of the mainstream development directions of infrared detectors in the future, large array infrared detectors have developed very rapidly in recent years [1]. With the application of civil disaster reduction, land resources, environment and meteorology, the demand for large scene width, large dynamic, high temporal resolution and high spatial resolution of satellite remote sensing dada is becoming more and more urgent [2]. Large scene width, high temporal resolution and high spatial resolution require high performance large array infrared detector and high quality imaging electronics system to

H. P. Urbach and H. Jiang (Eds.): ISSOIA 2022, SPPHY 295, pp. 547–561, 2023.
https://doi.org/10.1007/978-981-99-4098-1_47

achieve imaging requirements. At present, the array format of the array infrared detectors is small, which cannot meet the requirement of single scene width. The system requirement is met by splicing detector modules. For space application imaging electronics, it needs to face many problems such as many detector modules, low noise, small size, light weight, long life and high reliability. In this paper, the imaging electronics system is designed based on 2 * 2 spliced large array infrared detector, and the overall layout of the electronics system, the working mode, circuit performance and reliability are optimized. In the system design, the performance of imaging circuit is verified to ensure high performance and operate stably and reliably.

2 Design of Detector Component

The infrared detector is mainly used to realize the conversion of target infrared signal and electrical signal. A domestic infrared is composed of HgCdTe infrared detector chip and readout circuit. HgCdTe infrared detector chip and readout circuit realize electrode interconnection through flip chip interconnection to form a hybrid infrared focal plane detector. The high performance snapshot mode readout circuit is used, and the readout circuit completes the integration, storage, row gate, cell gate and signal voltage drive output of the signal of photovoltaic diode. The output channels have 4 or 16 optional outputs. It can operate in ITR mode or TWR mode. The pixel rate is 10 MHz.

The output internal resistance of the infrared detector is large, and the load is required to meet the requirements for parallel load of $R \geq 100$ kΩ and $C \leq 15$ pF, as shown in Fig. 1. Under this external load, the output signal level at maximum flux is 0.9 V(ITR mode)/1.4 V(IWR mode), and the output signal level at 0 flux is 3.0 V, as shown in Fig. 2. The dynamic range of detector is 2.1 V(ITR mode)/1.6 V(IWR mode). The signal output cycle is the master clock cycle. The output signal of each pixel maintains a level value after establishment. To master the typical electrical interface characteristics of the detector is the premise for designing high-quality imaging circuit.

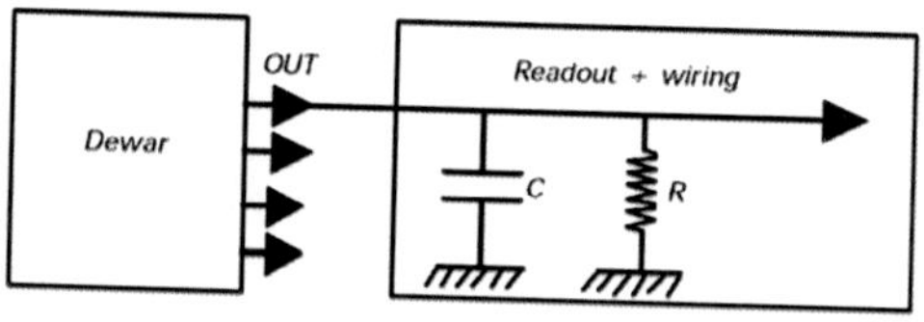

Fig. 1. Detector signal characteristics

The 2 * 2 splicing component is mainly composed of single detector module and splicing substrate. The splicing substrate is integrated with electrical, thermal and mechanical interfaces. Four single module signals are independent and physically isolated, which greatly reduce the circuit conductive coupling between channels [3].

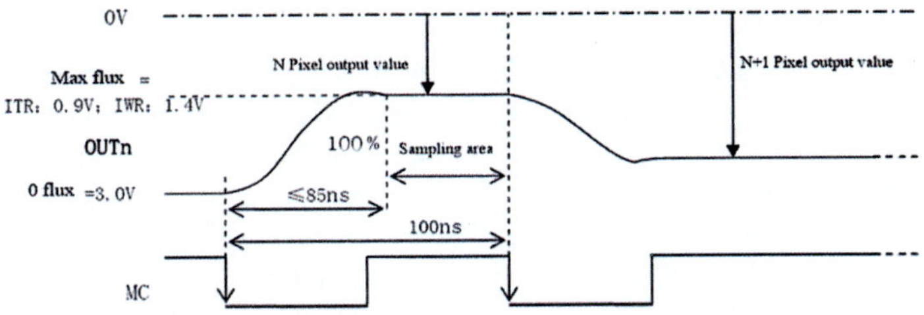

Fig. 2. Timing diagram between MC and output signal

3 Scheme Design of Imaging Electronics System

3.1 Architecture Design of Imaging Electronics System

The infrared target radiation information is converged to the focal plane infrared detector through the optical system. The target radiation energy is converted into electrical signals by the detector. The imaging circuit receives the electrical signals, performs AD conversion and sends them to the satellite data transmission. The imaging electronics system block diagram is shown in Fig. 3.

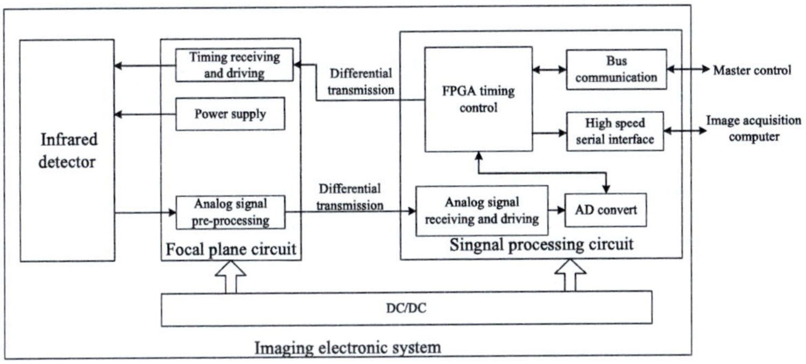

Fig. 3. Block diagram of imaging electronics system

The imaging circuit is mainly composed of focal plane circuit, signal processing circuit and DC/DC power supply. The focal plane circuit mainly generates the bias voltage of the detector, completes the timing reception of the detector and drives the detector to work normally. After the analog signal is preprocessed, the focal plane circuit outputs the differential analog signal. The signal processing circuit generates the timing signal of the detector and data acquisition timing control signal of AD conversion, performs AD conversion on the analog signal, and sends the data to the satellite data transmission system through the high speed serial interface. The bus communication completes the bus communication with the master control system, receives control commands and parameter settings, and returns telemetry parameters. The DC/DC power supply filters the satellite bus power and provides voltages for the imaging circuit.

Considering the factors such as the size, volume and weight of the imaging circuit, and combining the system application requirements, the single module detector is designed to use 4 outputs. 4 pieces of single module detectors have a total of 16 outputs. Due to the strict restriction of the space layout near the focal plane detector, the system abandons the previous integrated design scheme of focal plane circuit and signal processing circuit, and adopts a separate design scheme. The focal plane circuit is installed near the focal plane position. The signal processing circuit is installed at other position. A full differential signal transmission scheme is adopted, which can suppresses the common mode noise in the signal transmission process, reduces the signal drift, and greatly improves the anti-interference capability of the system [4]. Coaxial cable differential transmission is adopted to ensure long distance and high-quality signal transmission.

In order to meet the specified life and reliability requirements, the backup design structure is generally adopted, and the backup design will lead to an increase in product cost, volume and weight [5]. In this design, the idea of dividing the large array infrared detector into two half fields of view for electronic system design is adopted instead of the previous backup design. The four detector modules are divided into two half fields, and two detector modules correspond to one focal plane circuit module. One signal processing circuit completes the acquisition and processing of all channels analog signals. The topological relation of circuits is shown in Fig. 4. The design of focal plane circuit module 1 and focal plane circuit module 2 is completely the same, and the failure of any focal plane circuit module in physical isolation will not lead to complete failure of the system. This redundancy scheme can reduce the volume and weight of the circuit board by up to 50%. Through the highly reliable design scheme without backup redundancy, the circuit is miniaturized, lightweight and highly reliable design.

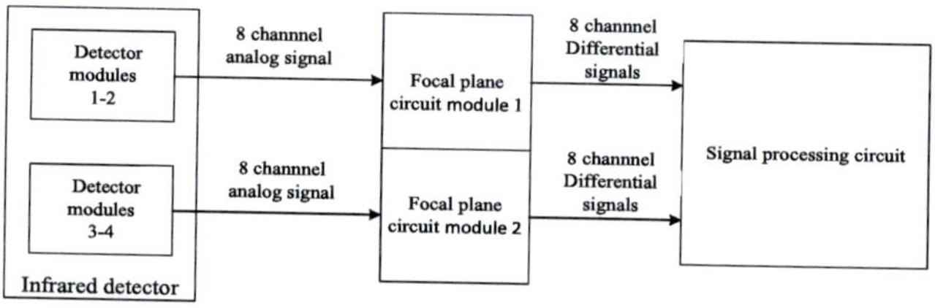

Fig. 4. Topological relation of circuits

3.2 Optimization Design of Working Mode and Imaging Mode

Large array detector modules have high frame rate, and imaging system code rate is up to hundreds of megabytes per second. The satellite data transmission capacity is limited. Based on the conventional calibration mode and imaging mode, a data reservation transmission mode is designed to determine whether to output image dada to satellite data transmission through the command identification. The design can directly transmit target data to satellite data transmission system, avoiding transmission of a large number of invalid data, and greatly saving the demands for satellite storage resources and transmission resources.

The imaging circuit is also designed with two imaging mode, the autonomous imaging mode and the external synchronous imaging mode. The imaging circuit receives the external synchronous imaging control signal and triggers the detector imaging. The frame rate of the external synchronous imaging depends on the trigger rate of the external synchronous signal. In order to prevent the external synchronous signal false trigger imaging caused by the burr pulse signal, the imaging circuit is designed to detect the high level width of a single external synchronous pulse which is $\geq$10 μs, so as to ensure that the external synchronous signal effectively triggers imaging. If the imaging circuit does not detect the external synchronous signal, it will enter the autonomous imaging mode. During the autonomous imaging, the imaging circuit is not controlled by the external synchronous signal, and the integrated imaging and signal readout are independently. The timing relation between the external synchronous signal and the integrated signal of the detector is shown in Fig. 5. External synchronous signal pulse width T1 is 1 ms $\pm$ 1 μs. The detector integral signal time delay with external synchronous signal is 10 μs $\pm$ 1 μs.

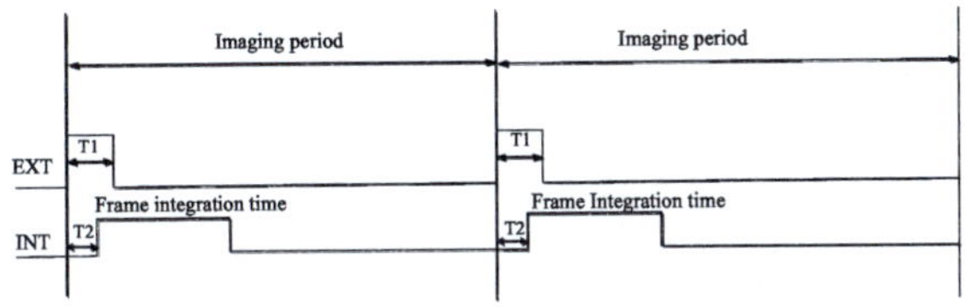

Fig. 5. Timing relation between external synchronous signal and detector INT signal

4 Hardware Design of Imaging Circuit

4.1 Design of Detector Drive Circuit

Infrared detector works at 80 K. In addition, it is necessary to provide a certain voltage and pulse signal to the detector, which performs the photoelectric conversion to form photocurrent, integrates the photocurrent by CMOS readout circuit to form voltage signal and then outputs the voltage signal [6]. Therefore, the detector drive circuit is mainly used to provide the bias voltage and timing signal to the detector. The power supply and bias voltage noise of the detector are required to be high, the voltage fluctuation range shall reach mV level, and the root mean square of the noise shall reach μV level. The Gpol voltage RMS noise is required to be less than 0.1 mV, which directly affects the performance of the detector [7].

According to the above requirements, the drive voltage power supply circuit is designed. The design idea is: first, a stable voltage is generated through the voltage reference. Then the appropriate voltage value is obtained by precise resistance voltage division, and the voltage obtained by voltage division is buffered or amplified by the operational amplifier with low noise, low drift and high common mode rejection ratio to obtain the required drive voltage. The output terminal is connected with small resistance in series and shunt capacitance to ground to filter the noise. The schematic diagram of the drive voltage circuit is shown in Fig. 6. According to the noise transfer theory, since the voltage reference is the first stage of the circuit, the better the smaller noise figure is. Therefore the selected voltage reference chip has the characteristics of ultra-low noise, high precision and low temperature drift to ensure the voltage stability and noise [8, 9]. Considering the number of modules, the number of bias power required is also large. In order to save the circuit board volume, four channel precision operational amplifier is selected.

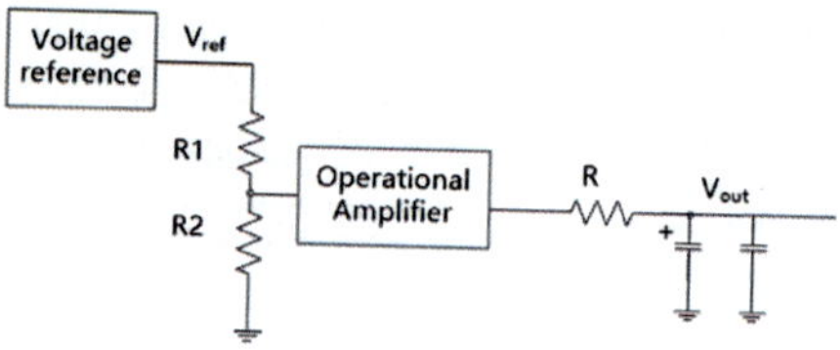

Fig. 6. Bias power supply circuit

The relation between the output voltage of operational amplifier and the reference voltage is shown in Formula 1.

$$V_{OUT} = \frac{R2}{R1 + R2} \times V_{REF} \times G_{OP} \quad (1)$$

In the Formula, V_{REF} is generated by the voltage reference, and G_{OP} is the operational amplifier gain.

Precision voltage reference can reduce the influence of input power changes on output voltage to ensure a stable output voltage. The operational amplifier adopts the design method of voltage series negative feedback. Considering that voltage fluctuates with the resistance temperature drift, the high common mode rejection ratio operational amplifier can suppress the influence of the voltage fluctuation on the output voltage. In addition, the negative feedback network of the operational amplifier can stabilize the output signal and amplifier gain.

4.2 Design of Analog Signal Processing Circuit

The analog signal processing circuit mainly includes impedance conversion circuit module and differential transmission circuit module. The diagram of analog signal processing circuit is shown in Fig. 7.

The impedance converting circuit uses high input impedance and low output impedance operational amplifier design a voltage following circuit to buffer the output signal. The output impendence of the detector is large. The impendence converting circuit is used to meet the requirements of the infrared detector for load resistance and load capacitance. Then, the single-ended to differential signal processing is carried out. Because the impedance converting circuit is located at the front of the entire processing link, high-precision operational amplifier with high input impedance, low output

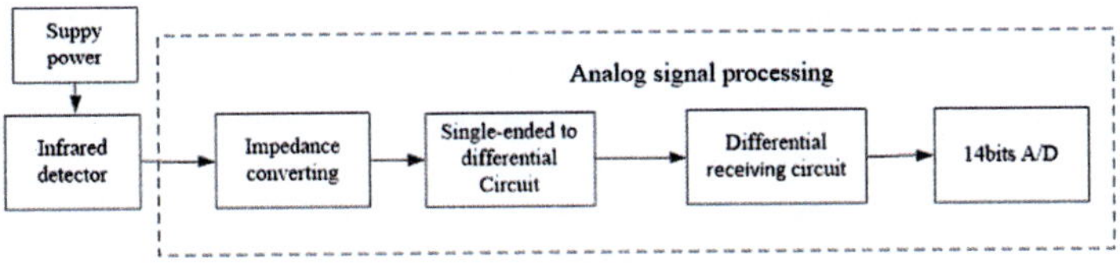

Fig. 7. Analog signal processing circuit diagram

impedance, low current noise, low temperature drift coefficient can reduce the influence of noise and signal drift on the detector output signal [10]. In order to reduce the size of the circuit board, the design selects a dual channel operational amplifier, which has the characteristics of high input resistance ≥100 kΩ and low input capacitance ≤15 pF. The impendence converting circuit is shown in Fig. 8.

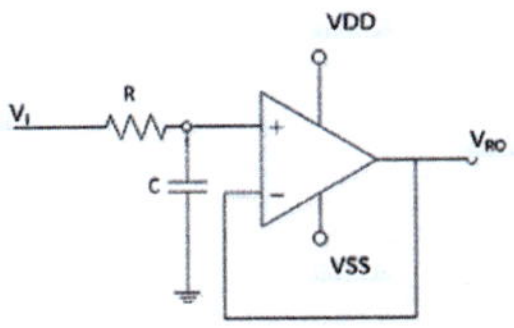

Fig. 8. Impedance converting circuit

DC coupling design is adopted in the circuit, and then low-pass filter circuit is composed of RC. The capacitance value of C+C_{in} should be less than 15 pF to meet the detector load requirements. C_{in} is the input capacitance of operational amplifier. The cut-off frequency of the low-pass filter is designed to be 3–5 times the output frequency of the detector.

The differential transmission circuit mainly realizes two functions: 1) complete the conversion from single-ended to difference, so that the subsequent signal can be transmitted in differential mode, and at the same time the signal is adjusted to make the dynamic range of the detector analog signal match the input dynamic range of AD converter [5]. The input dynamic range of AD is ±0.84 V. 2) the differential analog signal transmitted by the coaxial cable is received through the differential circuit and gets into the AD after the signal is adjusted once again. The differential transmission circuit is shown in Fig. 9.

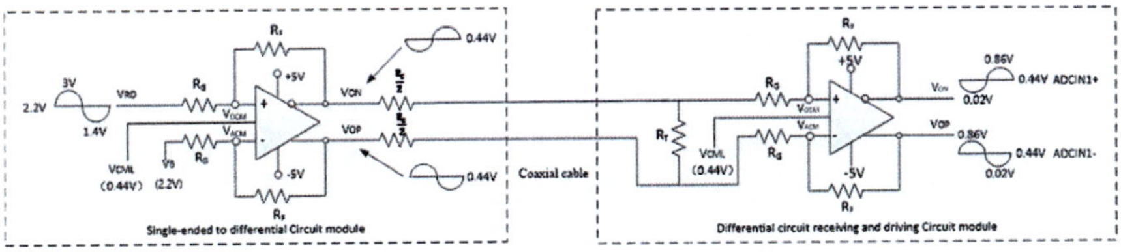

Fig. 9. Differential transmission circuit

The single-ended to differential circuit is realized by fully differential operational amplifier. The fully differential operational amplifier is an operational amplifier with differential input and differential output. It can also realize single-ended to differential conversion and differential to differential processing by configuring peripheral resistors. The selected differential operational amplifier has the characteristics of low noise, high SFDR and wide bandwidth, and can be used to drive the AD converter. The common mode voltage of the output differential signal can be set through the V_{OCM} pin. The designed signal amplification is realized by configuring the values of the R_F and R_G. To meet the voltage range of AD input in differential mode, the common mode voltage V_{OCM} is connected to the V_{CML} of AD. The voltage of V_{OCM} is 0.44 V. The output of single-ended to differential circuit is matched with $R_s/2$, and then gets into the subsequent circuit via coaxial cable.

The differential mode input voltage V_{INDM} and differential mode output voltage V_{OUTDM} of the single-ended to differential circuit are shown in Formula 2 and Formula 3.

$$V_{INDM} = V_{RO} - V_B \tag{2}$$

$$V_{OUTDM} = V_{OP} - V_{ON} \tag{3}$$

In the Formula, V_{OP} is positive output voltage, V_{ON} is negative output voltage, V_{RO} is positive input voltage and V_B is negative input voltage. In order to make the circuit amplify only the effective dynamic signal for the detector, the bias voltage V_B is 2.2 V which can subtract the DC level in the detector output signal.

The differential mode gain of differential operational is shown in Formula 4.

$$\frac{V_{OUTDM}}{V_{INDM}} = \frac{R_F}{R_G} \tag{4}$$

In the Formula, R_F and R_G are determined by magnification.

In order to match the impedance of the source terminal and the coaxial cable, the differential receiving circuit is designed with R_T as the matching termination resistance to ensure the quality of the analog signal. In addition, the amplifier gain is designed to be 1 ($R_F = R_G$). The differential output voltage range is −0.84 V~+0.84 V, completely matching with AD input voltage range. It should be noted that because the detector output signal and the detector optical input signal are inversely proportional, the signal should be connected inversely to AD converter, so as to ensure that the AD quantized image data DN and the detector optical input signal are positively proportional.

4.3 Design of Data Acquisition and Processing Circuit

The AD of acquisition and processing circuit uses a high precision AD chip. The quantization bits of the chip are 14 bits, and the maximum sampling rate is ≥30 Msps. The AD has the integrated sample and hold function and reference function. AD input voltage range is ±0.84 V.

Data processing is mainly realized by FPGA. FPGA is the control center of imaging electronics system [11], which is used to realize the driving timing of detector, generate the driving timing of AD, receive and cache high speed image data after AD conversion, output high-speed serial data, bus communication and other functions. This design uses Virtex 5 series FPGA chips to achieve timing control and data processing functions. The system is externally connected with 4 pieces of DDR2 SDRAMS to realize image data storage, which can meet the data storage capacity at different frame rates. 16 channels of image data are processed and integrated by FPGA and send to 2711 chip. 2711 is used to convert parallel digital signals into high speed serial signals to complete high speed data transmission. There are 2 high speed serial interfaces in the circuit, which is a hot backup interface each other, and every interface transmission data rate is designed as 2.0 Gbps. The coding scheme is the 8b/10b, which can keep the same number of "0" and "1", ensuring the accuracy of the high speed signal transmission [12].

5 Key Software Modules Design of Imaging Circuit

Limited by the material detects and manufacturing process of the infrared detector, the infrared image has defects such as large noise, many bad spots, large nonuniformity and low contrast [6]. To overcame these defects, image pre-processing and high dynamic range imaging function are designed to effectively improve the imaging performance and application effect of the infrared imaging system.

The FPGA processing circuit provides the detector drive timing, obtains 16 outputs signals to form data stream, and performs processing functions such as blind pixel replacement and nonuniformity correction. In addition, the software has also designed to a double integration time exposure mode. The two frames image after double integration time exposure is fused into a frame image, so as to improve the dynamic range of the image.

5.1 Design of Nonuniformity Correction Module

On the satellite, computing resources are limited compared with the ground. The two-point nonuniformity correction algorithm can meet the residual error requirement of real-time nonuniformity according to the statistical analysis of residual error of different algorithms on the ground.

In the calibration mode, the software first collects 32 frames low temperature black-body image data. Then the blackbody warms up, and then the software collects 32 frames high temperature blackbody image data. The nonuniformity correction coefficients are calculated from the two-point calibration image data, and the nonuniformity correction coefficients are stored.

The nonuniformity correction coefficients are calculated by Formula 5 and Formula 6.

$$k_i = \frac{\overline{DN}(\varphi_H) - \overline{DN}(\varphi_L)}{DN_i(\varphi_H) - DN_i(\varphi_L)} \tag{5}$$

$$c_i = \frac{\overline{DN}(\varphi_L) \times \mathrm{DN}_i(\varphi_H) - \overline{\mathrm{DN}}(\varphi_H) \times \mathrm{DN}_i(\varphi_L)}{\mathrm{DN}_i(\varphi_H) - \mathrm{DN}_i(\varphi_L)} \tag{6}$$

In the Formula, $\overline{\mathrm{DN}}(\varphi_H)$ is average value of all index of high temperature blackbody image data, $\overline{\mathrm{DN}}(\varphi_L)$ is average value of all index of low temperature blackbody image data, $\mathrm{DN}_i(\varphi_H)$ is value of i index of high temperature blackbody image data, and $\mathrm{DN}_i(\varphi_L)$ is value of i index of low temperature blackbody image data.

For i index, the DNG_i after nonuniformity correction is calculated by Formula 7.

$$DNG_i = K_i \times DN_i + C_i \tag{7}$$

Software real-time nonuniformity correction flow is shown in Fig. 10.

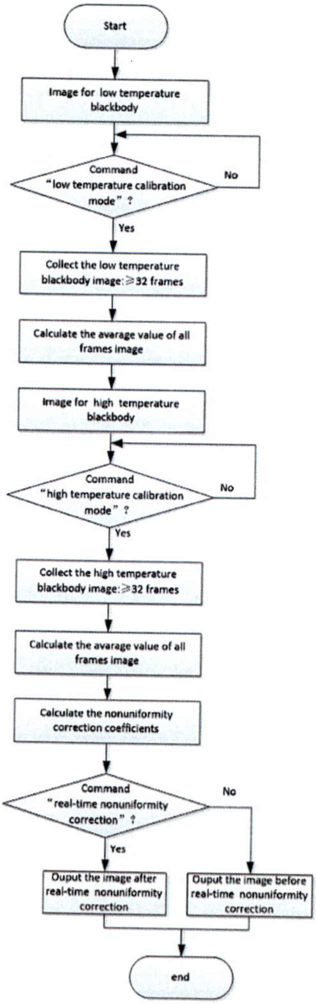

Fig. 10. Flow chart of Real-time nonuniformity correction

5.2 Design of Double Integration Time Exposure Module

In the frame period, the detector integrates with t1 and t2 successively, and then the circuit collects image data. When processing data, the image data of bright targets are extracted from short integration time image data, and the image data of dark targets are extracted from long integration time image data. Finally, they are fused into a frame image data to obtain a target image with high dynamic range and strong contrast.

Double integration time exposure module flow is shown in Fig. 11. In double integration time exposure, the integration time and the number of imaging frames can be adjusted by command. In addition, dual integration time exposure and single integration time exposure can be switched by command to meet different application requirements.

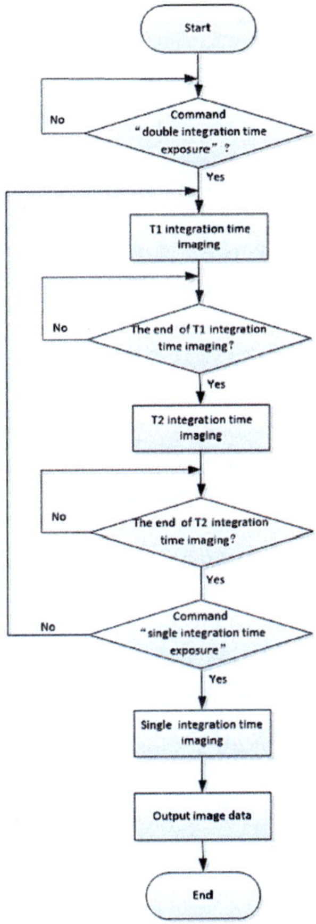

Fig. 11. Flow chart of double integration time exposure

6 Test Results

6.1 Noise Test

The noise characteristic is a key index to measure the infrared imaging system [13], which plays a key role in the sensitivity of the imaging system. It is very important to test the noise whether the noise can meet the requirements of engineering applications.

Noise Test of Imaging Circuit. In order to verify the noise performance of the imaging circuit, the noise of the circuit is tested without the detector load. The test method is as follows: connect the input of the circuit to a fixed level, and collect the image data. The standard deviation of the image data is calculated as the noise of the circuit. The noise of different detector modules is shown in Table 1. The calculation results show that the noise of circuit is 2.17 LSB, and the converted voltage is 0.223 mV.

Table 1. Circuit noise results of different detector modules

Different detector modules	D1	D2	D3	D4
Average (DN)	12658	12762	12618	12565
Noise (DN)	2.07	2.11	2.17	2.06
Noise (mv)	0.212	0.216	0.223	0.211

Noise Test of Imaging System. The imaging circuit with an infrared detector load and an optical system is evaluated the noise of the infrared imaging system. In order to ensure that the integration time is the same as the actual application. The imaging system noise is tested in a vacuum tank use a large area blackbody.

During test verification, the blackbody aperture completely covers the field of view of the imaging system, without the background radiation outside the field of view [14]. By adjusting the blackbody temperature the blackbody images are obtained, and the image data are processed to obtain the imaging system noise. The calculation results show that the noise of the imaging system is 4.54 LSB, and the converted voltage value is 0.466 mv (Table 2).

Table 2. Infrared imaging system noise results of different detector modules

Different detector modules	D1	D2	D3	D4
Average (DN)	5416	5328	5229	5333
Noise (DN)	4.41	4.33	4.21	4.54
Noise (mv)	0.452	0.444	0.432	0.466

6.2 Test of Nonuniformity Correction

By adjusting the blackbody temperature to 293 K and 308 K respectively, and the imaging circuit collects image data and calculates the nonuniformity correction coefficient in real time. The image data in 300 K blackbody temperature is real-time nonuniformity corrected. The images before nonuniformity correction and after nonuniformity correction are shown in Fig. 12.

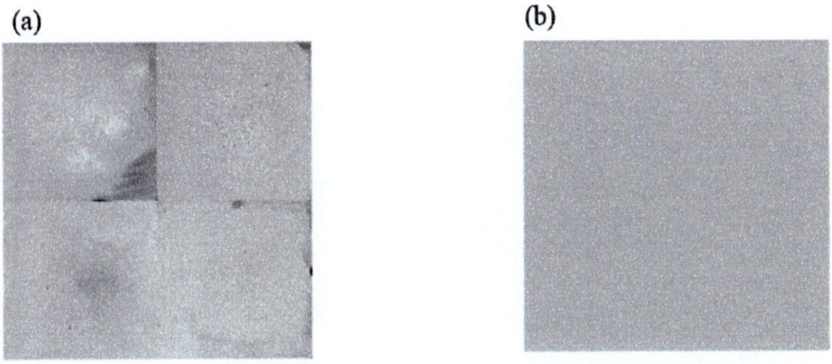

Fig. 12. (a) Image before nonuniformity correction (b) Image after nonuniformity correction

For array imaging system, the ratio of the standard deviation of all effective pixels after nonuniformity correction to average DN value of all effective pixels is the residual error of nonuniformity correction. The residual error of nonuniformity correction is 1.33%, which meets the requirement of 3% for imaging system.

6.3 Test of Double Integration Time Exposure

The infrared imaging system works in double integration time exposure mode, acquires the image data with different integration times, and fuses the two frames image data to achieve high dynamic range target image. The image before double integration time fusion and after double integration time fusion is shown in Fig. 13.

Fig. 13. (a) Short integration time image (b) Short integration time image (c) Double integration time fusion image

7 Conclusion

In this paper, the imaging electronics system is optimized based on 2 * 2 spliced large array infrared detector. The volume and weight are reduced by more than 50%. The data reservation transmission mode is designed to directly transmit target data to satellite data transmission, which greatly saves the demand for satellite storage resources and transmission resources. High integration and low noise imaging circuit design is realized by selecting appropriate high-precision operational amplifiers and high-speed AD converter. Real-time nonuniformity correction and double integration time exposure are developed on the processing platform of FPGA, which effectively improve the image quality and dynamic range. The test results show that the image quality is excellent and meets the current engineering application requirements.

References

1. Wang, X., Zhou, L., Tan, Z.: Developments of cooled large-format infrared detectors. Infrared **40**(12), 1–9 (2019)
2. Liu, T., Zhang, Y., Li, L., et al.: FPGA design of "GF-4" satellite large-array infrared camera video processing circuit. Spacecraft Recovery Remote Sens. **38**(3), 109–115 (2017)
3. Zhou, P., Wang, X., Dong, J., et al.: Imaging noise analyzing and suppressing for Tianwen-1 high-resolution camera. Opt. Precis. Eng. **30**(2), 218–226 (2022)
4. Ma, T.: Reliability optimization design of large-scale imaging electronic system. Chin. J. Sci. Instrument **35**(6), 82–86 (2014)
5. Wang, H., Wei, Z., Zhang, W., et al.: Design of 480 $\times$ 6 infrared focal plane array signal processing circuit. Infrared Technol. **31**(9), 504–512 (2009)

6. Zhang, B., Li, Z., Wu, J., et al.: Design of mid-wave 1 280 × 1 024 infrared imaging components (Invited). Infrared Laser Eng. **50**(4), 20211023 (2021)
7. Yang, X., Shi, M., Ling, L.: Design of the key driving and signal processing circuit for cooled infrared detector. Infrared Technol. **38**(7), 556–560 (2016)
8. Liu, W., Dong, F.: Design of image acquisition system for 320 × 256 type-II superlattice infrared detector. Mod. Electron. Tech. **42**(6), 18–21 (2019)
9. Chen, X., He, X.: Design of key driving circuits for 288 × 4 long-wavelength infrared detector. Laser Infrared **41**(5), 562–564 (2011)
10. Bo, S., Wang, H., Li, T.: Design of 320 × 256 long-wave area array infrared detector focal plane circuit. Electron. Measur. Technol. **39**(9), 20–25 (2016)
11. Zhang, H., Ma, Q., Wang, S.: High speed CMOS imaging electronics system for ultraviolet remote sensing instrument. Opt. Precis. Eng. **26**(2), 471–479 (2018)
12. Sun, Z., Liu, D., Fang, W., et al.: Design of high density modularity TDI CCD imaging system. Infrared Laser Eng. **47**(6), 06180001 (2018)
13. Ma, N., Liu, S., Li, J.: Analysis on spatial noise and temporal noise of IR thermal imager. Laser Infrared **47**(6), 717–721 (2017)
14. An, M., Liang, D., Nie, Y., et al.: Research on infrared detector imaging test methods for ZY-1 02E satellite. Spacecraft Recovery Remote Sens. **42**(4), 37–45 (2021)

Peculiarities of Thematic Processing of Data of Earth Monitoring from Space

Olga Butenko[1(✉)], Inesa Krasovska[1], Stanislav Horelyk[1], Artem Nechausov[1], Maksim Mariushko[1], and Ruslan Pashchenko[2]

[1] National Aerospace University "Kharkiv Aviation Institute", 17 Chkalovf str., Kharkiv 61070, Ukraine
{o.butenko,s.horelik,a.nechausov}@khai.edu, ines75ma@ukr.net

[2] Usikov Institute for Radiophysics and Electronics of the National Academy of Sciences of Ukraine, Academic Proskury str. 12, Kharkiv 61085, Ukraine
r.paschenko@i.ua

Abstract. In this study, main stages of thematic processing of Earth remote sensing data are presented: space imagery segmentation, space imagery data interpretation, prediction of phenomena development using thematic processing data. Main space imagery segmentation methods are examined, and a fractal analysis method for thematic processing is described. Additionally, we provide a description of fuzzy set theory elements for assessing the consequences of dangerous phenomena. Further, we present a detailed discussion of the results of Typhoon Hagibis space imagery segmentation using fractal dimensional fields and assessing the dynamics of its development with fuzzy logic. Results reveal that space imagery fractal analysis makes it possible to study typhoon structure characteristics, as well as to distinguish zones with different energy characteristics. Assessment of Typhoon Hagibis developmental dynamics in conditions of informational uncertainty using fuzzy logic allows predicting its consequences and timely generation of danger warnings.

Keywords: Thematic Processing · Segmentation · Fuzzy Logic · Fractal Dimensional Field · Remote Sensing · Typhoon

1 Introduction

Earth remote sensing (ERS) is aimed at ensuring the needs of land use, cartography, geology, environmental monitoring, obtaining information about the situation in emergency areas, as well as assessing the scale of emergencies, continuous observation of large-scale atmospheric processes [1, 2].

Meteorological and geodetic satellites are used to observe cloud coverage changes, development of cyclones, hurricanes and typhoons. Space imagery obtained from satellites is also used to study sea currents and monitor inhomogeneities arising on the sea surface associated with anthropogenic human activity. Data obtained by means of remote sensing constitutes an important part of glaciology research for determining the parameters of sea and freshwater ice, glaciers, etc. Space imagery is effectively utilized in

H. P. Urbach and H. Jiang (Eds.): ISSOIA 2022, SPPHY 295, pp. 562–585, 2023.
https://doi.org/10.1007/978-981-99-4098-1_48

agriculture and forestry, due to the possibility of seasonal monitoring of vegetation and soil cover characteristics. Remote sensing data is the main information source when building land development maps and topographic maps.

At present, when solving ERS problems, a wide range of technical observation means is used, each operating in different wavelength ranges (visible range of 0.4–0.9 μm, which are subdivided into multispectral and panchromatic, as well as infrared range of 3 to 1000 μm). Depending on a task, satellite imagery of low (more than 1 km) and ultra-high (less than 1 m) spatial resolution is utilized. Information obtained with the help of satellites is presented in the form of separate or chronologically sequential digital images, which are then processed and analyzed [3]. A large volume of digital space imagery is processed, which leads to a low analysis rate for remote sensing data. Usage of automated digital imagery processing methods can increase the speed and analysis efficiency.

Various segmentation methods are used in Earth remote sensing for thematic space imagery processing [3, 4]. General task of remote sensing thematic data processing reduces to identifying and localizing various natural and anthropogenic objects on the Earth's surface, as well as assessing their properties by satellite imagery parameters, contact sensing methods data as well as auxiliary information (results of previously performed decryption, cartographic materials, statistical data analysis results, etc.). Thematic processing procedures are divided into sequential stages, the main ones being data segmentation and interpretation. Based on the results of thematic processing, an assessment of change dynamics in the analyzed phenomena or forecasting of their negative consequences can be carried out. At the same time, the main component analysis is carried out for a time-differentiated series of images, then presented in a single coordinate system. However, decision making perturbing to the tendency of probable changes often occurs in conditions of information uncertainty associated with an insufficient amount of a priori data. In such conditions, it is advisable to use interval estimates and fuzzy logic methods.

This chapter is organized as follows. First, we provide fundamentals of data segmentation and interpretation as well as fuzzy logic theories. Second, we discuss the results of Typhoon Hagibis space imagery segmentation applying the fractal dimensions field. Additionally, an assessment of development dynamic for consequences prediction using fuzzy logic is provided.

2 Fundamentals

During thematic processing, various operations can be carried out, the main ones being: color transformations and analysis of imagery in pseudo colors; formation and analysis of index images; spectral, correlational, texture-based and fractal analysis; imagery segmentation and ERS object classification.

Image segmentation is generally defined as the process of dividing an image into homogeneous regions so that each region is homogeneous, and no union of two adjacent regions is homogeneous. Such areas are usually called segments. An important role in the segmentation process is determined by the choice of a suitable homogeneity measure, which makes it possible to distinguish the analyzed objects from each other. Spectral and textural characteristics are mainly used in that process [5]. During segmentation, other features can also be used (shape, size, context, etc.). Along with traditional methods of spatial imagery analysis (visual [1], spectral, correlation [2] and contour-texture [5]), fractal analysis methods are gaining more and more adoption when solving remote sensing tasks [6–15]. In addition, a large amount of information received requires the creation of methods that allow automation of data interpretation process for space imagery.

Space imagery thematic processing results can be combined with contact measurement results and statistical data of the phenomena being assessed to gauge their development dynamics, as well as to provide a forecast.

2.1 Remote Sensing Data Thematic Processing Segmentation Methods

There are various methods for space imagery segmentation [3, 16], which allow the selection of either homogeneous areas or the object boundaries in the provided imagery. When analyzing natural phenomena, for example, studying the structure of a hurricane, these methods allow determining only their outer boundaries and do not allow the separation of their structural elements. Cluster analysis use for segmentation [17, 18] makes it possible to effectively identify objects with pronounced features, i.e. characteristics that differ significantly from the background. However, cluster analysis does not allow to determine the structure of objects close in brightness, which are often present in hurricane structural elements.

Fractal method of digital imagery analysis allows estimating the geometric complexity of the analyzed structure (space imagery) using fractal dimension, which, as a rule, is a non-negative non-integer number. Different complexity of underlying surfaces and observation objects in the image leads to different fractal dimension values. This fact allows the use of fractal dimension for the segmentation of space imagery [19].

For fractal segmentation with the use of publicly available (open) geoinformation portal resources, original space imagery is obtained, which is then written into the permanent computer memory. Then, the digital image is presented in three-dimensional form (in three coordinates): x and y are the dimensions of the original image horizontally and vertically, z are the brightness values of the image pixel. The original image is then transformed onto another analysis plane – the fractal dimension field (FDF) is constructed. Then, the original image is processed using a "sliding window" or "jumping window", starting from its upper left corner. The user sets the dimensions of the "window", which are smaller than the size of the image, and the step "window" movement. If the step of movement is equal to one, then the "window" is called "sliding", with a different step value - "jumping". The fractal (fractional) dimension is calculated for each position of the "window". When processing images, the coverage method (calculation of box-dimension) is most often used to determine the fractal dimension [20].

$$D = \lim_{\varepsilon \to 0} \frac{\log N(\varepsilon)}{\log(1/\varepsilon)},$$

where ε is the size of the covering cube edge, $N(\varepsilon)$ are the number of cubes covering the image in the "window".

In the practical implementation of this method, covering cube edge value ε_1 is set. The three-dimensional representation of the image is filled with cubes of a given size and the number of cubes that cover all z points in the "window" is counted. One value is generated - $N(\varepsilon_1)$. Then, calculations are carried out, $N(\varepsilon)$ for different edge length of the cube covering ε (for example, $\varepsilon_2 = \varepsilon_1/2$, $\varepsilon_3 = \varepsilon_1/4$). Based on the data obtained, a dependency is obtained: $\log N(\varepsilon)$ on $\log 1/\varepsilon$. The obtained points are approximated, for example, by the method of ordinary least squares (OLS), by a straight line of the form $\log N(\varepsilon) = -b \cdot \log \varepsilon + C$. As an estimate of the fractal dimension, $D = -b$ is chosen, tangent is the angle of inclination of a straight line (OLS).

The obtained values of fractal dimensions in all "windows" are written into a separate matrix, which we will call the fractal dimensions field. The values of fractal dimensions depend on the complexity (irregularity) of the changes in the brightness of the image in the "window". With similar image structures in neighboring "windows", fractal dimensions have similar values (differing starting at fourth decimal). "Windows" with close values of fractal dimension can be combined into separate segments, which provide image segmentation.

After the construction of the FDF, its visualization is carried out, i.e. the entire range of fractal dimension variations is divided into several levels (segmentation thresholds). Each level is displayed in a corresponding color (segments with specified fractal dimensions are highlighted), which allows separating parts of the image with different structures.

For automated separation (segmentation) of various structures on the image, a histogram of the calculated PFR can be built and selective images can be created (images on which elements corresponding to a given range of fractal dimensions are displayed). When constructing selective images, the correspondence between the position of the "window" on the original image and the value of the fractal dimension in this "window" is preserved. After specifying the range of variation of the fractal dimension, the original image displays "windows" in which the fractal dimensions are included in the specified range. The constructed selective images make it possible to analyze parts of the initial image that have differing complexity.

2.2 Comprehensive Assessment of the Development Dynamics of Processes and Phenomena According to Earth Monitoring Data Using Fuzzy Logic

Revealing the dynamics (tendencies) of possible changes in the object under monitoring is associated with a large number of variables on which further forecast is dependent on. First of all, additional variables are introduced at the stage of modeling various changes in the state of an object under the influence of various disturbing factors and analyzing the cause-and-effect relationships of their occurrence. Unaccounted factors that have an insignificant effect on the object, but significantly affect the forecast when additionally considering the additivity property (for factors that have this property) and calculating the integral indicator of their combinatorial effect on the object under study, can have great consequences. When forecasting, special attention should be paid to the analysis of the initial data and the compilation of the vector of input parameters and the corresponding informative features.

Peculiarities of performing these tasks in real time often lead to a lack of information about the initial conditions. The presence of uncertainty in the decision-making process does not allow to accurately assess the impact of control actions on the objective function [28]. Since the uncertainties that exist in observations are presented as stochastic processes, then stochastic control methods are applicable to such problems. However, this approach is not always effective. This is explained by the fact that the set of standard probabilistic concepts and methods turns out to be inadequate for describing the situations under consideration [29], since there is difficulty in obtaining the necessary statistical characteristics of the parameters, the absence of ergodicity of the processes as a whole, and their significant non-stationarity. In the case when the source of uncertainty is not random and partially or completely deterministic, the problem is simplified. Despite all the indicated difficulties and uncertainty of information, the requirements for the accuracy of the obtained solution are increasing. Unlike statistical methods, in which all factors are assumed to be random or uncertain, the forecast model is the total impact of the cyclical component, deterministic factors and random components. When forecasting, it is advisable to use fuzzy logic methods. Indirect methods for determining the membership function values to obtain interval estimates at each stage of modeling are used when there are no elementary measurable properties through which the fuzzy set of interest is determined. This pertains to the maximin ratio, which allows establishing a relationship between parameters that are not directly related to each other:

$$\mu_{R_1 \circ R_2}(x, z) = \bigcup_y \left[\mu_{R_1}(x, y) \cap \mu_{R_2}(y, z)\right] = \max_y\left[\min(\mu_{R_1}(x, y), \mu_{R_2}(y, z)\right],$$

where $\mu_{R_1}(x, y)$ is the matrix of direct interactions (mutual influence) of factors $\{x\}$ on $\{y\}$; $\mu_{R_2}(y, z)$ is the matrix of direct interactions (mutual influence) of factors $\{y\}$ on $\{z\}$; $\mu_{R_1 \circ R_2}(x, z)$ denotes fuzzy relationships that determine the degree of indirect interaction between the factors $\{x\}$ and $\{z\}$.

When working with Earth space monitoring data, namely with decryption signs, at the stage of identifying the dynamics of changes according to the set of images for the analyzed time series, it is advisable to use the top-down inferences method:

$$R_1 = \begin{bmatrix} \mu(x_1,y_1)\ldots\ldots\ldots\mu(x_1y_j) \\ \mu(x_2,y_1)\ldots\ldots\ldots\mu(x_2y_j) \\ \ldots\ldots\ldots\ldots\ldots\ldots\ldots\ldots\ldots\ldots\ldots \\ \mu(x_i,y_1)\ldots\ldots\ldots\mu(x_i,y_j) \end{bmatrix},$$

where R_1 is the reference matrix of correspondence of classification characteristics $(x_1, \ldots, x_i)$ and admissible (predicted) changes in the images $(y_1, \ldots, y_i)$, where the degree of compliance is determined by the membership function $\mu(x, y)$;

$$R_2 = \begin{bmatrix} \mu(y_1) \\ \mu(y_2) \\ \ldots\ldots \\ \mu(y_j) \end{bmatrix},$$

where R_2 denotes the assessment of real changes in images $(y_1, \ldots, y_i)$, where the degree of compliance with the classification criteria is determined by the membership function $\mu(x, y)$.

As a result, the interaction by the method of top-down inference is reduced to the system of interval equations

$$\begin{bmatrix} \mu(y_1) \\ \mu(y_2) \\ \cdots\cdots \\ \mu(y_j) \end{bmatrix} = \begin{bmatrix} \mu(x_{1,}y_1)\ldots\ldots\ldots\mu(x_1y_j) \\ \mu(x_2, y_1)\ldots\ldots\ldots\mu(x_2y_j) \\ \cdots\cdots\cdots\cdots\cdots\cdots\cdots\cdots\cdots \\ \mu(x_{i,}y_1)\ldots\ldots\ldots\mu(x_{i,}y_j) \end{bmatrix} \circ \begin{bmatrix} (x1) \\ (x_2) \\ \cdots\cdots \\ (x_i) \end{bmatrix}.$$

Solution to these equations is carried out as follows

$$\mu(y_1) = max\{min[\mu(x_{1,}y_1); x_1]; min[\mu(x_1, y_2); x_2]; \ldots; min[\mu(x_1; y_j); x_i]\};$$

$$\mu(y_j) = max\{min[\mu(x_{i,}y_1); x_1]; min[\mu(x_i, y_2); x_2]; \ldots; min[\mu(x_i; y_j); x_i]\}.$$

To establish the relationship and highlight a set of features inherent in a particular situation in modeling with a full set of a priori data, direct methods can be used. This approach allows obtaining an indicator in a single scale at the final stage of building a forecast that determines the trend of further changes in the observed object or phenomenon.

3 Typhoon (Hurricane) Spatial Imagery Thematic Processing Characteristics

3.1 Typhoon (Hurricane) Structure Description

Remote sensing satellite data can be used to monitor the change in cloudiness and the cyclone development, such as hurricanes and typhoons (tropical cyclones are called typhoons in the Pacific Ocean, and hurricanes in the Atlantic Ocean). Tropical cyclones arise over the warm sea surface, they are characterized by low pressure and are accompanied by powerful thunderstorms, heavy rainfall and storm winds of more than 117 km/h (or 30 m/s). On satellite imagery, hurricanes (typhoons) are characterized in the form of an extensive bright white spot with short spiral branches extending from it. The more intense the hurricane, the more extensive (but not more than 500–700 km in diameter) its cloud system and the better the spiral structure is expressed [21]. The densest and most powerful mass of clouds concentrate around the center of the cyclone, forming a zone of continuous clouds. A visual analysis of space images of hurricanes makes it possible to generally assess the distribution of clouds in a hurricane, as well as to study the inhomogeneities of the cloud cover. However, using visual analysis, it is difficult to identify and classify the structural elements of a hurricane. It is possible only to highlight the boundaries of the hurricane and estimate the size of its peripheral (tail) regions.

Existing descriptions of hurricane (typhoons) structure [22] distinguish on their surface annular zones with different temperatures, pressures and humidity. The central part of the hurricane, in which the air descends downward, is called the eye. The eye, as a rule, has a regular round shape, 30–60 km in diameter. The wind speed within the eye is not high, the air is warm, the atmospheric pressure is very low, and the sky near strong hurricanes is clear. The eye of a weak hurricane may be partially or completely covered with clouds. Behind the eye, an area is situated (the middle part of the hurricane), which is characterized by the highest wind speeds and precipitation near the water / land surface. It is called the wall of the eye. The diameter of the eye wall is not constant and can change several times during the existence of a hurricane [21, 22]. At the same time, the old wall of the eye narrows to 10–25 km, and the new one has a larger diameter. The outer part of the hurricane (with a radial width of 30–50 km) is organized into rain strips that slowly move towards the center of the hurricane. The wind speed also increases evenly as it approaches the center. Around the outer edge of a cyclonic vortex, cloud cover is usually not evenly distributed around the entire periphery of the hurricane.

Let us consider the possibility of analyzing hurricane structures using fractal segmentation of a digital image. For example, we receive a satellite image of Typhoon Hagibis from the Aqua satellite in channels 1, 4 and 3 from October 09, 2019, which is freely available on the NASA Rapid Response website [23]. Figure 1a shows a synthesized satellite image of Typhoon Hagibis, which raged off the coast of Japan in the Pacific Ocean from 07 to 13 of October, 2019. The temperature distribution in the typhoon and the surrounding space is shown in the image (Fig. 1, b), which can also be found on the site [23] (low temperatures are indicated in dark blue, and high temperatures are in beige).

a b

Fig. 1. Satellite image of Typhoon Hagibis from the Aqua satellite (09/10/2019) (a) and temperature distribution in the typhoon (b)

The typhoon Hagibis had a wind force of 160 miles per hour (259 km/h), which corresponded to a fifth category typhoon (category 5 - wind speed exceeding 157 miles per hour), the pressure in the center of the vortex was about 908 hPa, the radius of storm winds was 370 km east of the center. The typhoon Hagibis had one feature that is characteristic of some active hurricanes and typhoons, which is called the "rental eye" [24]. This happens when the typhoon's eye is much smaller than normal, making it more powerful as the typhoon revolves around it much faster. In this case, the small eye, which rotates in the center, is covered around by the large eye, as if the small eye was stuck inside the large one [24].

A visual analysis of the Typhoon Hagibis space image (Fig. 1, a) shows that some features of its structure can be highlighted. The entire hurricane is a compact solid cloudy array in the shape of a circle (the disk of the main typhoon area is visible in the picture). Its eye is located in the center of the typhoon (in the picture, it is represented as a light spot with a dark dot in the center). Behind it, there is a dense cloud cover, which is called the wall of the eye (displayed in the picture as a denser belt of white around the eye). However, it is almost impossible to visually determine its boundaries. Further, there are swirling stripes extending from the main massif of the typhoon, called precipitation strips (the image shows spiral-shaped stripes at the periphery of the typhoon). However, these spiral structures are clearly distinguished only in the outer region of the typhoon; in the middle part, they are not distinguished.

As noted above, the intensity of the wind and the temperature of the clouds in different areas of the typhoon (hurricane) are different. In the center, the wind speed is low, in the middle part, the speed reaches a high destructive force, and at the periphery, the speed is somewhat lower. Looking at Fig. 1b, it can be seen that in the center of the typhoon (eye) the temperature is higher (beige) than in the middle part, in which the temperature is the lowest (dark blue in the image). On the periphery of the typhoon, the temperature rises (the color changes from blue to beige). Additional information on the temperature distribution in typhoon makes it possible to more confidently identify its main structural parts.

3.2 Segmentation of the Typhoon (Hurricane) Structure Using the Fractal Dimensions Field

Remote sensing satellite data can be used to monitor the change in cloudiness and the cyclone development, such as hurricanes and typhoons (tropical cyclones are called typhoons in the Pacific Ocean, and hurricanes in the Atlantic Ocean). Tropical cyclones arise over the warm sea surface, they are characterized by low pressure and are accompanied by powerful thunderstorms, heavy rainfall and storm winds of more than 117 km/h (or 30 m/s). On satellite imagery, hurricanes (typhoons) are characterized in the form of an extensive bright white spot with short spiral branches extending from it. The more intense the hurricane, the more extensive (but not more than 500–700 km in diameter) its cloud system and the better the spiral structure is expressed [21]. The densest and most powerful mass of clouds concentrate around the center of the cyclone, forming a zone of continuous clouds. A visual analysis of space images of hurricanes makes it possible to generally assess the distribution of clouds in a hurricane, as well as to study the inhomogeneities of the cloud cover. However, using visual analysis, it is difficult to identify and classify the structural elements of a hurricane. It is possible only to highlight the boundaries of the hurricane and estimate the size of its peripheral (tail) regions.

Existing descriptions of hurricane (typhoons) structure [22] distinguish on their surface annular zones with different temperatures, pressures and humidity. The central part of the hurricane, in which the air descends downward, is called the eye. The eye, as a rule, has a regular round shape, 30–60 km in diameter.

To study the features of the typhoon structure, an automated image segmentation is carried out. The fractal dimension value is used [25]. For this, a fractal dimension field is constructed and visualized. Figure 2, a and b show the visualization of the FDF, when processing the Typhoon Hagibis space image, shown in Fig. 1, a.

Fractal dimensions field was constructed using a "jumping window" with a size of 16 × 16 pixels using the MathCAD computer program (the value of the "jump" was equal to half the size of the "window"). Afterwards, the range of changes in fractal dimensions was divided into three (Fig. 2, a) and five (Fig. 2, b) levels. Each level was displayed in the corresponding color automatically by the MathCAD program.

As can be seen from the analysis of Fig. 2, the use of fractal dimension as a sign of segmentation makes it possible to distinguish various structural parts of the typhoon. Thus, when the FDF is divided into three levels, the eye and the outer part of the typhoon (red in the center and at the periphery), as well as its middle part (green with elements of blue and red) are distinguished. An increase in the number of levels of division of fractal dimensions up to five makes it possible to highlight the features of the main parts of the typhoon. As such, in the outer part, spiral stripes are clearly distinguished (red on a yellow background). Additionally, in the middle, a blue area is highlighted.

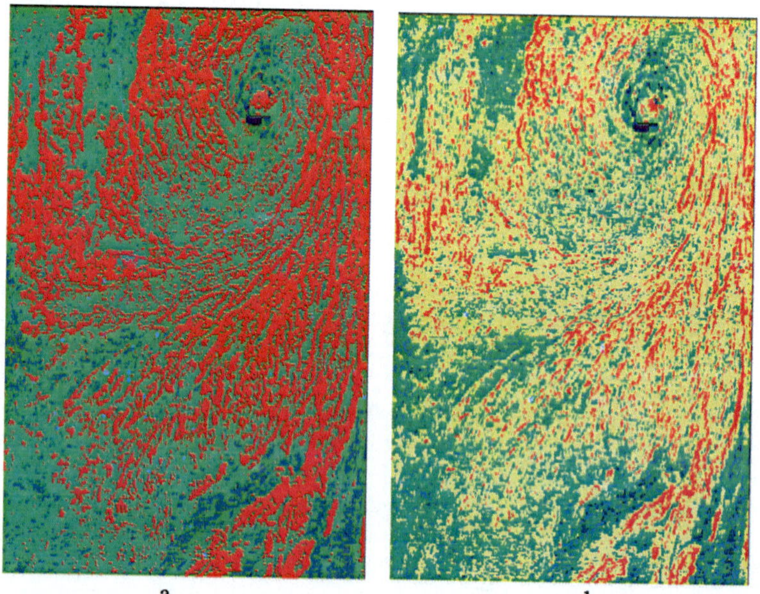

a b

Fig. 2. Fractal dimensions field visualization when dividing the range of changes in fractal dimensions into three (a) and five (b) levels

Comparison of the constructed FDF (Fig. 2) and the temperature distribution in the typhoon (Fig. 1, b) shows that there is a correlation between the magnitude of the fractal dimension and the magnitude of the temperature in different parts of the typhoon. Green and yellow colors on the FDF correspond to lower temperatures, and red and blue colors in the FDF correspond to lower temperatures. Thus, the color visualization of the FDF makes it possible to represent the general structure of the typhoon, however, the analysis of its features is carried out visually, and the identification of its parts is carried out qualitatively.

Further analysis of the typhoon structure will be carried out using a histogram of the distribution of fractal dimensions on the FDF and construction of selective images. Figure 3 shows a histogram of the FDF values of the Typhoon Hagibis image, on which the fractal dimension values D are plotted along the abscissa, and the ordinate is the relative frequency of the fractal dimension f_D on the analyzed FDF.

Analysis of the FDF histogram shows that there is a long "tail" of fractal dimensions with a low frequency of their appearance in the FDF (the left side of the histogram, $D < 2.9$). Let us construct selective images corresponding to the characteristic ranges of variation of the fractal dimension ΔD.

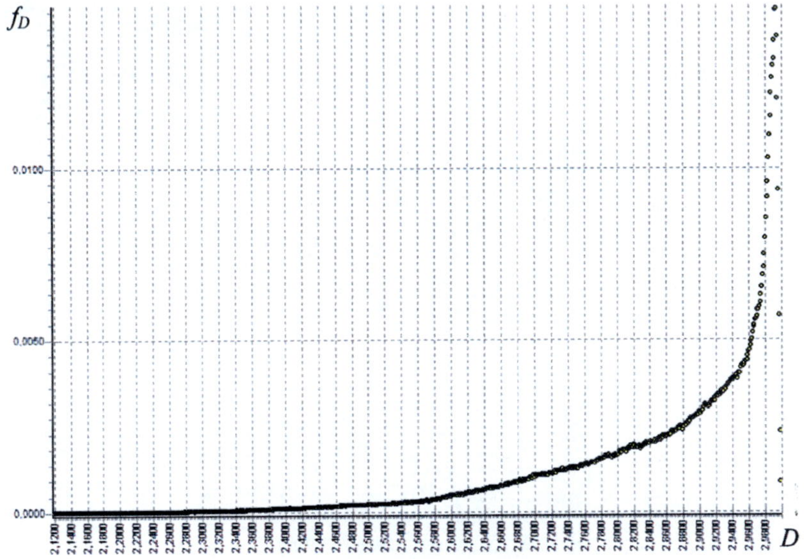

Fig. 3. Typhoon image FDF values histogram

Figure 4 shows selective images for different ranges of fractal dimension at the first level of subdivision $\Delta D_1 = 2.0 - 2.9$ (a), $\Delta D_2 = 2.9 - 2.99$ (b) and $\Delta D_3 = 2.99 - 3.0$ (c). The first level of partitioning of fractal dimensions corresponds to two decimal places. In the figure, "windows" on the original image are marked in red, corresponding to the given ranges of fractal dimension.

As seen in Fig. 4a to the first range (left part of the histogram, $\Delta D_1 = 2.0 - 2.9$) corresponds mainly to the water surface of the ocean (background). A cloud mass in the

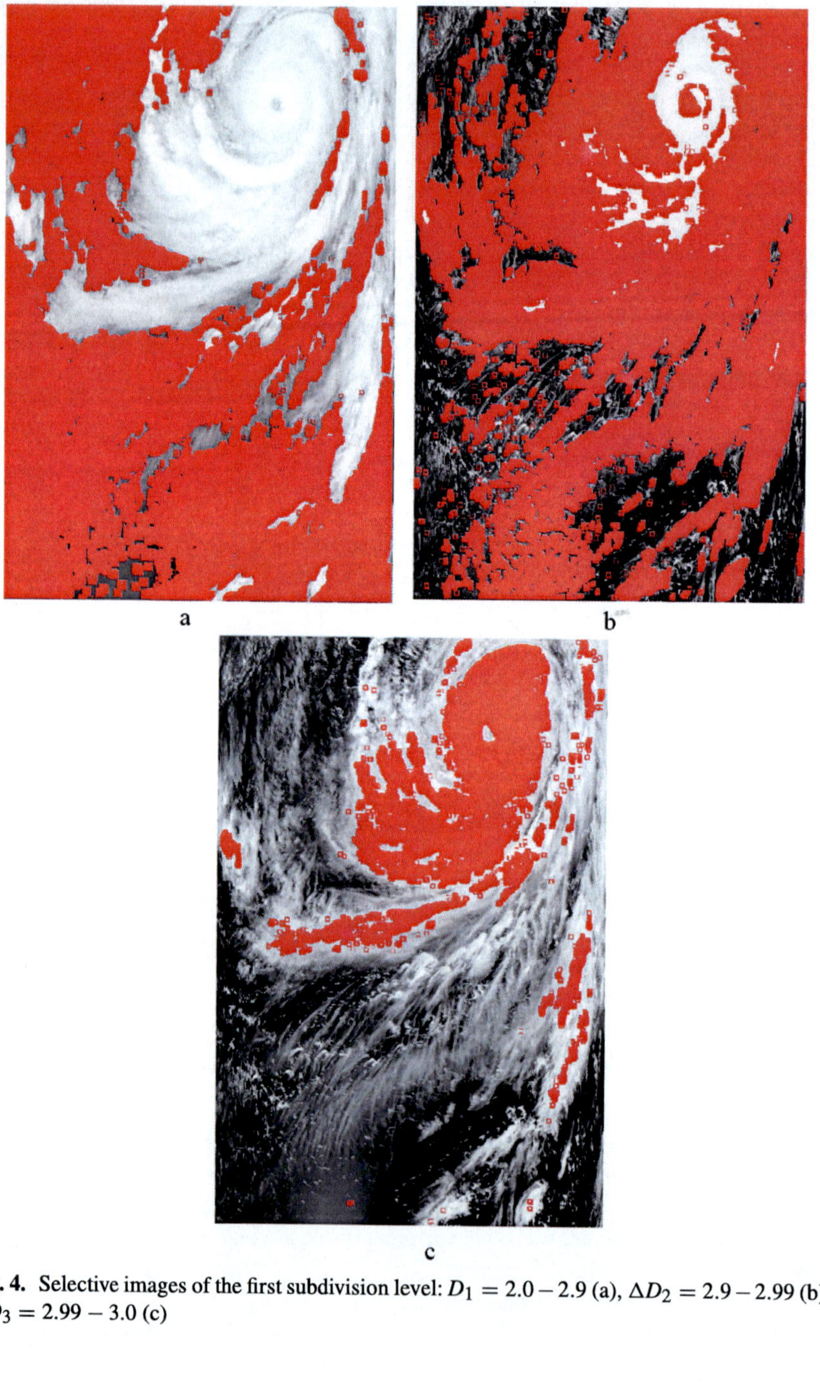

a b

c

Fig. 4. Selective images of the first subdivision level: $D_1 = 2.0 - 2.9$ (a), $\Delta D_2 = 2.9 - 2.99$ (b), $\Delta D_3 = 2.99 - 3.0$ (c) form of a rotating disk, from which separate spiral stripes of clouds depart, remained unselected. At the periphery of the typhoon cloud system (bottom right), separate parts of the spiral also did not stand out. In Fig. 4b, it can be seen that the second range (a sharp increase in the histogram, $\Delta D_2 = 2.9 - 2.99$) corresponds to the center of the typhoon (the eye stood out), as well as to the bands of precipitation on its periphery. However, on the selective image of the second range of fractal dimensions, it was not possible to assess the features of the typhoon's eye. When the third range is selected (histogram decay, $\Delta D_3 = 2.99 - 3.0$) the wall of the typhoon's eye and the main (most powerful) cloud arrays stand out well, while the edge looks sharp (Fig. 4, c).

As noted above, the Typhoon Hagibis eye consists of two parts, the inner (the small eye that rotates in the center) and the outer (the large eye around the small one). To analyze the peculiarities of the Typhoon Hagibis eye, let us analyze the narrower ranges of fractal dimensions included in the range ΔD_2, a selective image of which is shown in Fig. 4, b.

Figure 5 shows selective images where the small Typhoon Hagibis eye is highlighted, $\Delta D_{25} = 2.95 - 2.96$ (a) and the big eye $\Delta D_{27} = 2.97 - 2.98$ (b).

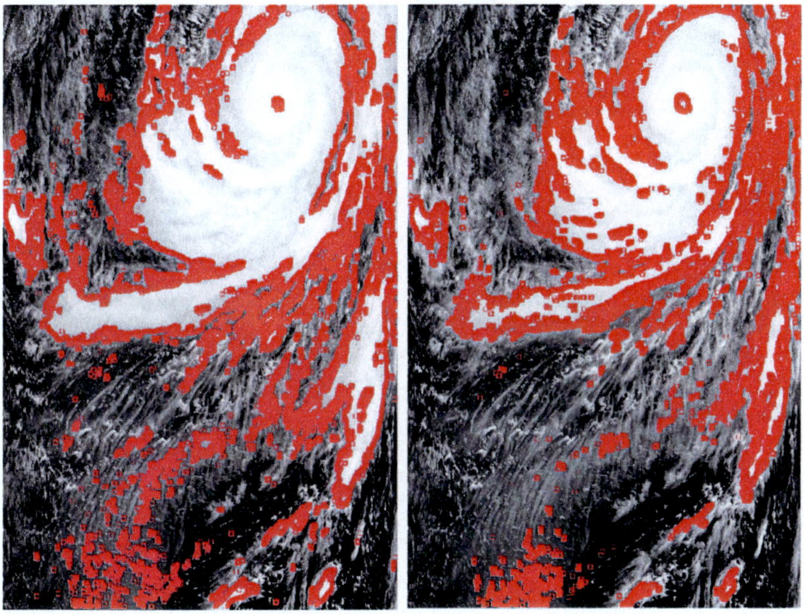

a b

Fig. 5. Selective images of Typhoon Hagibis eye features: $D_{25} = 2.95 - 2.96$ (a), $\Delta D_{27} = 2.97 - 2.98$ (b)

As seen in Fig. 5a, the small Typhoon Hagibis eye stands out, which is located in its center, and in Fig. 5b, a large eye stands out, which covers a small eye. Figure 5 also clearly shows that in these ranges of variation of fractal dimensions, the boundaries of precipitation bands at the periphery of the typhoon are more clearly distinguished.

Thus, using the FDF histogram and the constructed selective images of the first level of fractal dimension partitioning (two decimal places), it is possible to distinguish the main parts of the typhoon: the eye, the wall of the eye and the outer part of the typhoon from which the cloud stripes diverge, and its boundaries are also well displayed.... For further investigation of the structure of typhoon, we will carry out the second partition of the third range of fractal dimensions (three decimal places) and construct its selective images.

Figure 6 shows selective images that detail the third range of fractal dimension variation (Fig. 4, c), which correspond to the second subdivision level $\Delta D_{31} = 2.991 - 2.992$ (a), $\Delta D_{37} = 2.997 - 2.998$ (b). At the second subdivision level, the ranges of fractal dimensions that differ by 0.001 are analyzed.

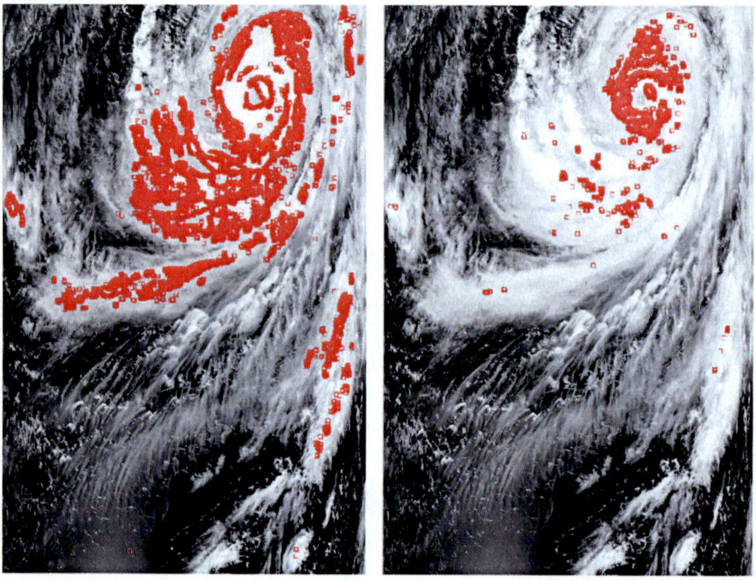

a b

Fig. 6. Selective images of the second subdivision level: $\Delta D_{31} = 2.991 - 2.992$ (a), $\Delta D_{37} = 2.997 - 2.998$ (b)

Data analysis in Fig. 6 shows that further fractal dimensions analysis makes it possible to identify the typhoon structural features more precisely. Figure 6, a ($\Delta D_{31} = 2.991 - 2.992$) spiraling cloud stripes are observed at the periphery of the typhoon and at the border of the eye wall. If we compare the obtained results with the temperature distribution in the typhoon (Fig. 1, b), it can be seen that the highlighted areas correspond to the lowest temperatures. In Fig. 6, b ($\Delta D_{37} = 2.997 - 2.998$) it can be observed that in the fractal dimension variation range, the middle part of the typhoon eye's wall is clearly distinguished, where the temperatures are lower.

3.3 Hurricane Radar Image Features Analysis

Let us consider the possibility of analyzing the structure of a hurricane using fractal segmentation of a digital radar image [9]. Figure 7 shows a radar image of the tropical hurricane Diana, which arose during the evolution of the extratropical low-pressure region north of the Bahamas. In the image, the dark area in the center represents the eye of the hurricane. Here the speed of the near-surface wind is significantly weakened. The bright ring surrounding the dark core region is associated with significant wind speeds. If cloudiness is observed within the light ring, then with a small water content, precipitation falls of low intensity. Heavy cloudiness and precipitation are concentrated further from the center, mainly in the area of the eye wall, which is visible in the radar image as a dark ring. Variations in brightness outside the outer wall of the eye are mainly caused by variability in the surface wind, and, to a lesser extent, variability in cloudiness with precipitation attenuation characteristics [26].

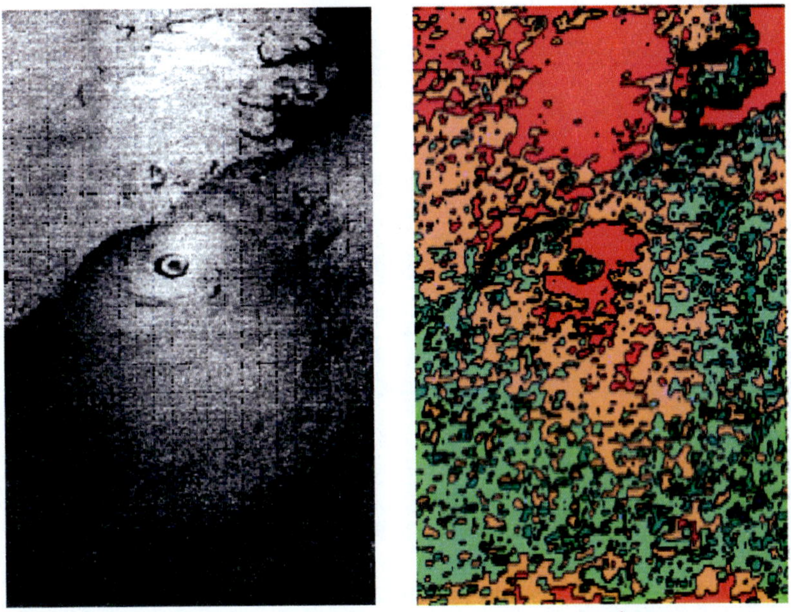

a b

Fig. 7. Radar image (a) of the tropical Hurricane Diana and its FDF (b)

It is possible to separate the different areas of the hurricane using the FDF, which is shown in Fig. 7, b. Fractal dimensions field was built using the coverage method, using a "sliding window" of 8x8 pixels and five levels of range division of changes in the fractal dimension using the MathCAD computer program.

FDF analysis in Fig. 7, b shows that the use of fractal dimension as a segmentation characteristic of the Hurricane Diana radar image allows to distinguish various structural hurricane parts, just like in the segmentation of the Typhoon Hagibis synthesized image

(Fig. 1, a). This is how the eye and the outer part of the typhoon stand out (green in the center and at the periphery), as well as its middle part (red with orange elements).

Additionally, just like in the typhoon Hagibis synthetized image segmentation, we will analyze the Hurricane Diana structure using a histogram of the fractal dimensions distribution on the FDF and construct selective images. Figure 8 shows Hurricane Diana radar image FDF values histogram. After analyzing the histogram of the fractal dimension field (Fig. 8), several characteristic ranges can be distinguished: first – $\Delta D_1 = 2.6 - 2.8$, second – $\Delta D_2 = 2.8 - 2.9$ and third – $\Delta D_3 = 2.9 - 2.96$.

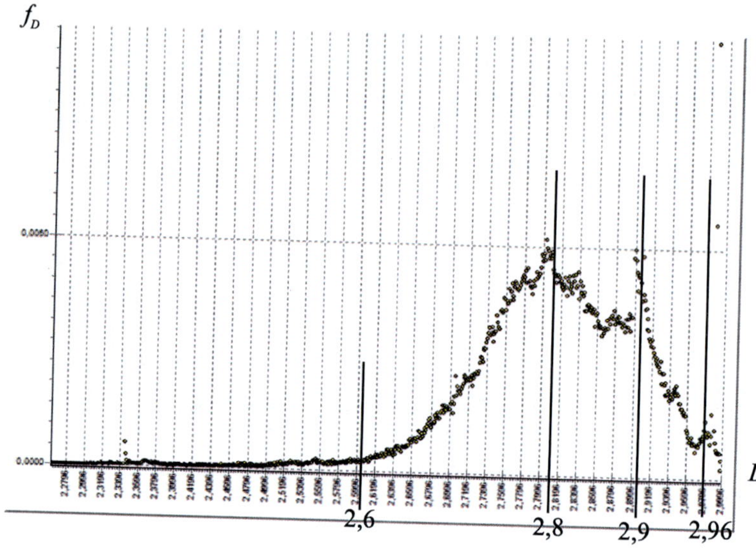

Fig. 8. FDF histogram of the original Hurricane Diana image

Figure 9 shows selective images for different ranges of fractal dimensions at the first subdivision level $\Delta D_1 = 2.6 - 2.8$ (a), $\Delta D_2 = 2.8 - 2.9$ (b) and $\Delta D_3 = 2.9 - 2.96$ (c). In the figures, "windows" on the original image are marked in white, corresponding to the given ranges of fractal dimension.

According to grades provided by authors [26], the wind speed in the hurricane section was in the range of 10 to 50 m/s. Construction of selective images at the first subdivision level made it possible to identify zones with wind speeds of 10–20, 20–30 and 30–50 m/s, which are presented in Fig. 9, a–c, respectively.

Thus, fractal segmentation of both synthesized and radar images of typhoons (hurricanes) makes it possible to identify their structural features and well as zones with different energy characteristics.

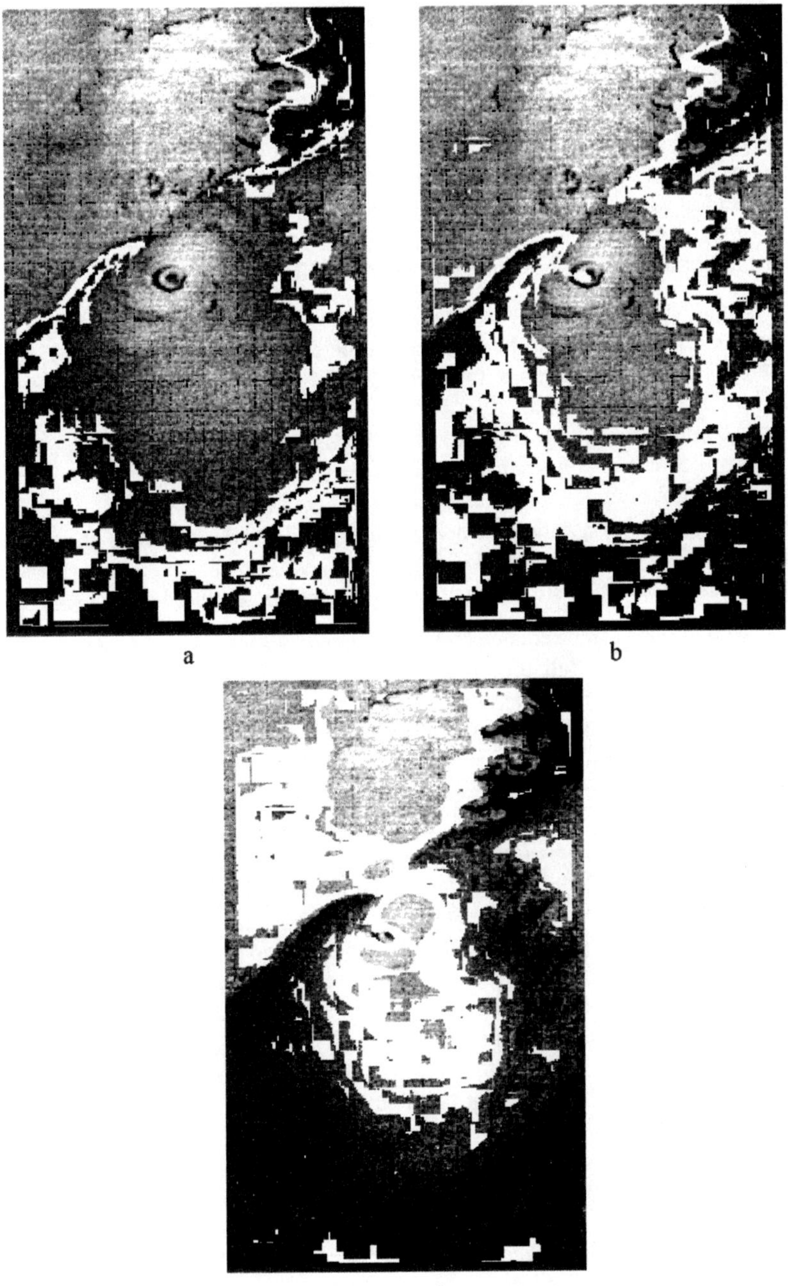

a b

c

Fig. 9. Selective Hurricane Diana images: $\Delta D_1 = 2.6 - 2.8$ (a), $\Delta D_2 = 2.8 - 2.9$ (b), $\Delta D_3 = 2.9 - 2.96$ (c)

4 Typhoon (Hurricane) Spatial Imagery Thematic Processing Characteristics

Remote sensing satellite data can be used to monitor the change in cloudiness and the cyclone development, such as hurricanes and typhoons (tropical cyclones are called typhoons in the Pacific Ocean, and hurricanes in the Atlantic Ocean). Tropical cyclones arise over the warm sea surface, they are characterized.

Formation of typhoons (hurricanes) is still not fully understood. However, the main factors necessary for the formation can be identified:

- the temperature of the near-surface layer of ocean water is more than 26.5 °C at a depth of more than 50 m. This is the minimum sufficient temperature of near-surface water, which can cause instability in the atmosphere above it and maintain the existence of a thunderstorm system [28–30];
- rapid cooling of air with altitude, which makes it possible to release condensation energy, the main energy source for a typhoon (hurricane);
- high air humidity in the lower and middle layers of the troposphere;
- low vertical wind gradient, since a high wind gradient breaks the typhoon's (hurricane's) circulation system;
- existing area of low atmospheric pressure or unstable weather over the years.

Such conditions can be created by low-level and low-latitude flares, which are associated with the Madden-Julian oscillation [28]. Studies of Earth space monitoring data make it possible to observe and predict the consequences of typhoons (hurricanes). However, this does not allow for even rough estimates of expected negative consequences scale in order to prevent them in a timely manner. It requires an integrated approach and the joint use of space monitoring data, statistical data, meteorological data and the results of contact-based research. It is also imperative to consider the specifics of the locality and the typhoon type features.

In order to integrate the interpretation results with statistical and meteorological data when deciding on the possible scale of destruction after a typhoon, there is often no complete set of direct signs that would make it possible to draw a conclusion about the probable degree of destruction. Therefore, the use of various combinations of indirect factors affecting the typhoon makes it possible to draw conclusions under conditions of uncertainty caused by an insufficient amount of a priori information. For this, it is proposed to utilize fuzzy logic methods, which make it possible to consider all possible combinations between the factors of typhoon occurrence, terrain features and those a priori data that are available. One of the components of the forecast is to determine the indicator of the typhoon development dynamics to assess its possible consequences.

It is known that typhoons are ranked according to one of five intensity scales according to their maximum wind resistance and in which basin they are located. There also exist alternative scales, such as the accumulated energy of the cyclone scale, the dispersion index, the integrated kinetic energy index and the hurricane severity index [31].

Let us consider the possibility of predicting the consequences of typhoons (hurricanes) under conditions of a priori uncertainty using fuzzy logic, with the typhoon Hagibis as the example.

Figure 10 shows graphs of changes in the average wind speed (a) and average pressure (b) of the typhoon Hagibis for the time interval from 5 to 12 of October, 2019.

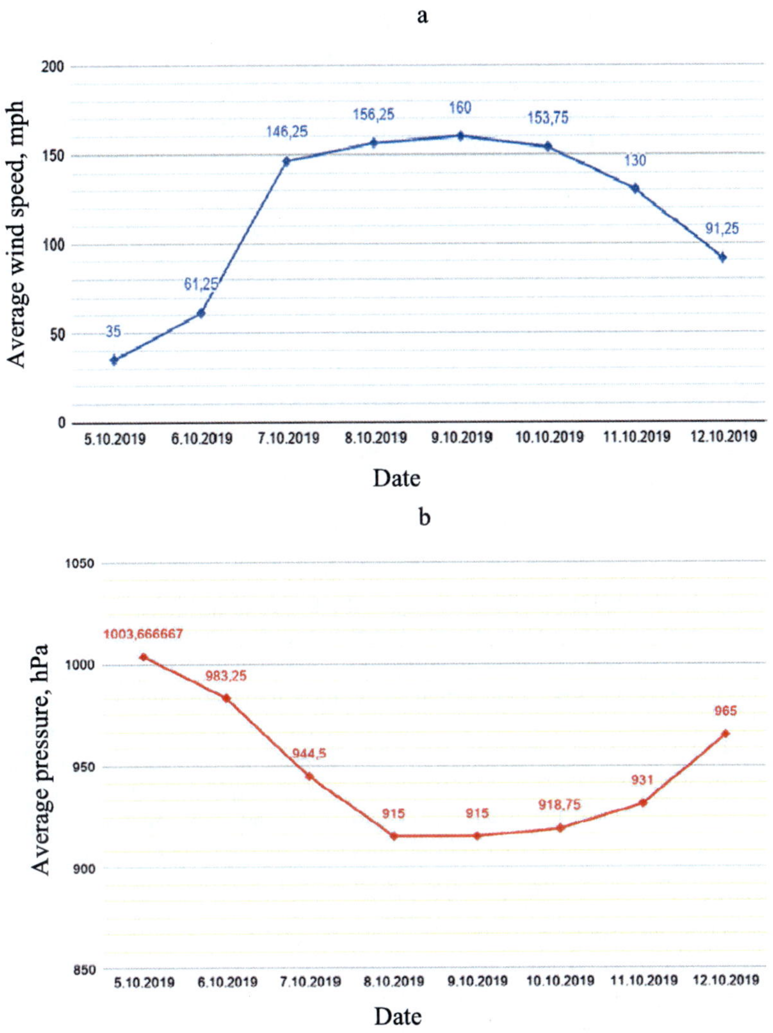

Fig. 10. Changes in average wind speed (a) and average pressure (b) of the typhoon Hagibis between 5 and 12 of October, 2019

The graphs show that at the maximum typhoon intensity, the pressure drops, and the wind speed increases. The obtained data make it possible to determine the possible scale of negative typhoon consequences, considering the properties of their type (typhoon category), wind intensity, pressure and precipitation intensity.

Since the decision on the possible typhoon consequences is made in conditions of informational uncertainty, when it is not known what strength the typhoon will be and what category it will belong to, it becomes necessary to establish a relationship between the minimum set of indirect factors influencing the intensity of the typhoon with all five possible categories (according to the Saffir-Simpson scale) and an assessment of the degree of their combinatorial sets, obtained on the basis of statistical data and expert assessments. The Saffir-Simpson scale [27–31] was developed to measure the potential damage from a hurricane and is based on wind speed, it includes an assessment of storm waves in each of the five categories (Table 1).

Table 1. Saffir-Simpson Scale.

Category	Wind speed, km/h	Storm waves, m
1 – Very dangerous winds will produce some damage	119–153	12–15
2 – Extremely dangerous winds will cause extensive damage	154–176	18–23
3 – Devastating damage will occur	179–209	27–36
4 – Catastrophic damage will occur	210–249	39–55
5 – Catastrophic damage will occur	>250	>55

The typhoon Hagibis has been raging in the Pacific Ocean since October 6, 2019, when the tropical cyclone was assigned category 1 with a wind speed of V = 75 mph at a pressure of P = 970 hPa. A feature of the typhoon Hagibis was its sharp (sudden) increase. As such, on October 7, the wind speed (by day) increased to 160 mph (night - 115 mph, morning - 150 mph, day and evening - 160 mph), and the pressure dropped to 922 hPa (night - 964 hPa, morning - 952 hPa, day - 940 hPa, evening - 922 hPa), thus, the typhoon reached the 5th category by day. On October 8 and 9, the typhoon was assigned category 5 with V = 160 mph and P = 915 hPa. On October 10 and 11, the typhoon weakened to category 4 (V = 130 mph, P = 933 hPa). On October 12, at night (V = 105 mph, P = 944 hPa) and in the morning (V = 100 mph, P = 961 hPa), the typhoon was assigned category 2, by evening it was assigned category 1 (V = 750 mph, P = 980 hPa). Figure 11 shows the trajectory of Typhoon Hagibis and the change in its category (http://ftp.emc.ncep.noaa.gov/wd20vxt/hwrf-init/decks/bwp202019.dat), as well as space imagery of the typhoon obtained at one time on the NASA Rapid Response website [23]. As seen in Fig. 11, along with changes in the characteristics of the typhoon (wind speed, pressure, category), the typhoon space imagery changes. As such, at typhoon category 4 and 5 its eyes, the wall of the eye and the stripes of precipitation stand out well.

Thus, the typhoon Hagibis passed through the north of the Mariana Islands to the shores of Japan and had a mesoscale size, which significantly complicated its forecast in terms of its dynamics, both trajectory and wind strength-wise.

According to the results of statistical indicator analysis for different types of typhoons and the results of the consequence scale analysis from each of them based on analogies, a correspondence matrix is built. That matrix contains factors influencing typhoon effects.

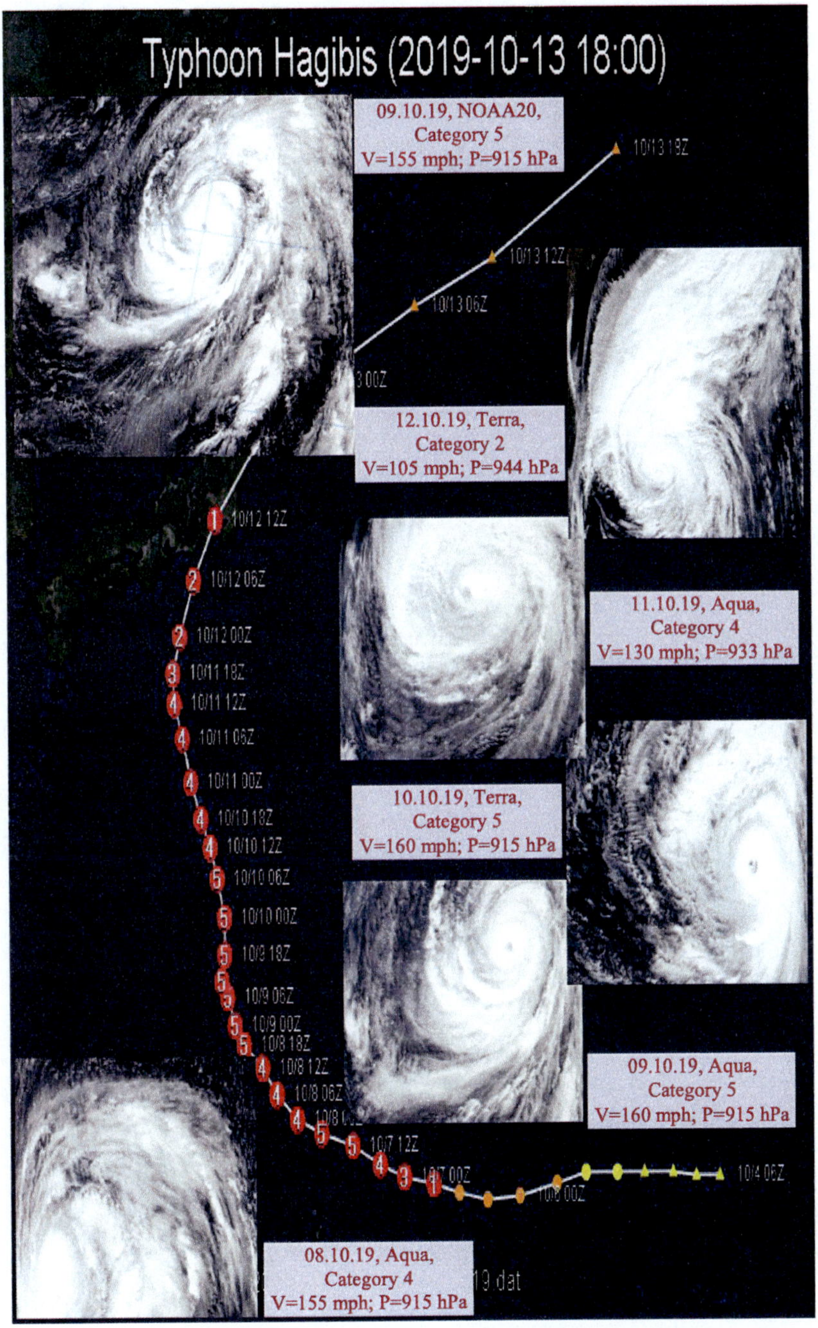

Fig. 11. Changes in Typhoon Hagibis category and trajectory

Analysis of statistical data on the characteristics of a typhoon makes it possible to evaluate the minimum and maximum values of each parameter in accordance with the category of typhoon.

Taking into account that each of the analyzed parameters, namely x_1 is the pressure, x_2 is the wind speed, x_3 denotes the precipitation intensity and typhoon categories y_1, y_2, y_3, y_4, y_5, are considered in various measurement systems, it is proposed to introduce a unified grading scale from 0 to 1 (0 corresponds to the minimum possible value of the parameter, 1 - the maximum possible).

Evaluation results of the analyzed parameters for each of the typhoon categories in a single scale are presented in Table 2.

Table 2. Unified grading scale.

$R1$	$y1$	y_2	y_3	y_4	y_5
$x1$	0.9	0.6	0.5	0.3	0.1
x_2	0.1	0.3	0.4	0.8	0.9
x_3	0.1	0.2	0.6	0.8	0.9

Also, according to statistical data, we will estimate the likelihood of large-scale negative consequences typical for different categories of typhoons, as shown in Table 3.

Table 3. Likelihood of large-scale negative consequences.

$R2$	$z1$	z_2
$y1$	0.2	0.6
y_2	0.2	0.7
y_3	0.3	0.8
y_4	0.4	0.8
y_5	0.5	0.9

In Table 3, z_1 corresponds to the indicator of the likelihood of large-scale negative consequences, and z_2 – to the indicator of the likelihood of occurrence of minor negative consequences.

In order to establish a correlation between the change in the values of the analyzed parameters with an estimate of the likelihood of large-scale negative consequences, it is advisable to use the maximin composition (convolution).

$$\mu_{R_1 \circ R_2}(x, z) = \bigcup_y \left[\mu_{R_1}(x, y) \cap \mu_{R_2}(y, z)\right] = \max_y\left[\min(\mu_{R_1}(x, y), \mu_{R_2}(y, z)\right],$$

where $\mu_{R_1}(x, y)$ is the matrix of direct interactions (mutual influence) of parameters x_1 is the pressure, x_2 denotes the wind speed, x_3 is the precipitation intensity for

y_1, y_2, y_3, y_4, y_5 – typhoon categories; $\mu_{R_2}(y, z)$ is the matrix of direct interactions (mutual influence) y_1, y_2, y_3, y_4, y_5 – typhoon categories, where z denotes the destruction indicator corresponding (according to statistical studies) to each of the possible categories; $\mu_{R_1 \circ R_2}(x, z)$ - fuzzy relations that determine the degree of indirect (indirect) interaction between the parameters x_1 (pressure), x_2 (wind speed), x_3 (precipitation intensity) and z – an indicator of typhoon consequences magnitude (z_1 corresponds to the indicator of the likelihood of large-scale negative consequences, and z_2 – to the indicator of the likelihood of minor negative consequences).

Expected scale of typhoon consequences, without knowing a priori its category and in the presence of only such parameters as pressure, wind speed and precipitation intensity, are presented in Table 4.

Table 4. Expected typhoon impact magnitude.

$\mu \times R3$	$z1$	z_2
$x1$	0.3	0.6
x_2	0.5	0.9
x_3	0.5	0.9

As can be seen in Table 4, the largest scale of the typhoon Hagibis impact will be as a result of simultaneous changes in wind speed and precipitation intensity. In the event of a sudden drop in pressure and an increase in wind speed, it can be concluded that the typhoon is assigned to the largest category and, as a consequence, will pertain greatest expected negative consequences. The largest scale of destruction is expected with the expected value of wind speed and / or an increase in precipitation intensity of 0.9. That value is obtained in the introduced scale of weighting coefficients and can be simply recalculated into the system of corresponding measurements for each of the estimated parameters. With regard to these results, measures should be taken to prevent possible significant negative consequences in the case when the wind speed reaches 112 mph and the pressure reaches 936 hPa.

Comparison of calculation results with real data on the consequences of the typhoon Hagibis in Japan showed the adequacy of this approach. However, these results are only an indicator of the degree of possible typhoon consequences in conditions of limited a priori data. The forecast is possible only with the joint heterogeneous time-differentiated data analysis, which includes archived data.

5 Conclusions

In this study, space imagery fractal analysis was applied to highlight the typhoon (hurricane) structural features. Construction of fractal dimension field and its colorized visualization makes it possible to represent the general typhoon (hurricane) structure and to carry out a qualitative analysis of its features. To automate the selection of a given typhoon (hurricane) part, it is proposed to construct selective images, which display the peculiarities of the typhoon (hurricane) structure. Multilevel subdivision of the fractal dimension variation range makes it possible to outline all parts of the typhoon. Choice of appropriate fractal dimension range allows to investigate feature of typhoon (hurricane) structure, as well as to identify zones with different energy characteristics. Practical implementation of the fuzzy logic method for determining the possible consequences scale of the typhoon Hagibis showed that certain management decisions should be carried out to prevent negative consequences of a typhoon where the pressure suddenly drops and the wind speed increases. The threshold wind speed value is 112 mph, and the threshold pressure value is 936 hPa.

References

1. Krasovsky, G.Ya., Petrosov, V.A.: Introduction to environment space monitoring methods. State Aerospace University "Kharkiv Aviation Institute", Kharkiv (1999). (in Russian)
2. Yanutsh, D.A.: Deciphering aerospace imagery. Nedra, Moscow (1992). (in Russian)
3. Pal, N.R., Pal, S.K.: A review on image segmentation techniques. Pattern Recogn. **26**(9), 1277–1294 (1993)
4. Richards, J.A.: Sensing Digital Image Analysis: An Introduction John A. Richards. Springer, New York (2012)
5. Potapov, O.A., Andreev, N.A., Murashova, I.B., Pomosov, A.V., Kozhanov, V.N., Petrova, T.A.: Dependence of the processing characteristics of a copper powder and its dendrite structure on electrolysis conditions at a constant overvoltage. Soviet Powder Metallurgy Metal Ceramics **29**(2), 91–96 (1990). https://doi.org/10.1007/BF00794528. (in Russian)
6. Pashchenko, R.E., Shapovalov, A.V.: Image segmentation by method of fractal dimensions field ranking. Collection of scientific works. Inf. Process. Syst. **8**(36), 103–107 (2004). (in Russian)
7. Ivanov, V.K., Pashchenko, R.E., Stadnik, A.M., et al.: Forest imagery fractal analysis. Adv. Mod. Radio Electron. **12**, 55–62 (2005). (in Russian)
8. Ivanov, V.K., Pashchenko, R., Yatsevich, S.Ye., et al.: Radar remote sensing images segmentation using fractal dimension field. In: Proceedings of the 3rd European Radar Conference (EuRAD) as Part of European Microwave Week (EuMV2006), Manchester, UK, pp. 217–220 (2006)
9. Dolya, G.N., Ivanov, V.K., Pashchenko, R.E., et al.: Fractal analysis of processes, structures and signals. In: Pashchenko, R.E. (ed.) XOOO NEO "EcoPerspectiva", Kharkiv (2006). (in Russian)
10. Ivanov, V.K., Pashchenko, R.E., Yatsevich, S.E., et al.: Assessment of snow cover characteristics from radar imagery using fractals. Collection of scientific works. Inf. Process. Syst. **2**(51), 30–36 (2006). (in Russian)
11. Grushenko, M.V., Pashchenko, R.E., Shapovalov, A.V.: Sea surface monitoring using the "k-tree" method of the fractal dimension fields. Eastern Eur. J. Adv. Technol. **1/2**(19), 100–104 (2006). (in Russian)
12. Grushenko, M.V., Pashchenko, R.E., Petrushenko, N.N., et al.: Processing of aerospace imagery using fractal dimensions field. Collection of scientific works. Inf. Process. Syst. **1**(50), 35–41 (2006). (in Russian)
13. Ivanov, V.K., Pashchenko, R., Yatsevich, S.Ye., et al.: Fractal analysis of sea ice images. In: Proceedings of the Sixth International Kharkov Symposium on Physics and Engineering of Microwaves, Millimeter and Submillimeter Waves and Workshop on Terahertz Technologies (MSMW 2007), Ukraine, Kharkov, vol. 2, pp. 989–991 (2007)

14. Butenko, O., Pashchenko, R., Mariushko, M., Topchiy, A.: Analysis of the earth's surface type in remote sensing using fractal dimension. In: The 6th Microwaves, Radar and Remote Sensing Symposium (MRRS-2020), Ukraine, Kharkiv (2020). https://doi.org/10.1109/UkrMW49653.2020.9252811
15. Maryushko, M.V., Paschenko, R.E.: Fractal analysis of SENTINEL-2 space imagery for agricultural culture monitoring. Radioelectron. Comput. Syst. **4**(96), 34–47 (2020). https://doi.org/10.32620/reks.2020.4.03. (in Russian)
16. Bo, P., Lei, Z., David, Z.: A survey of graph theoretical approaches to image segmentation. Pattern Recogn. **46**, 1020–1038 (2013)
17. Thilagamani, S., Shanthi, N.: A survey on image segmentation through clustering. Int. J. Res. Rev. Inf. Sci. **1**(1), 129–137 (2011)
18. Kosarevich, R.Ya., Rusyn, B.P., Korniy, V.V., et al.: Image segmentation based on estimating the tendency to cluster formation by image elements using point field characteristics. Cybern. Syst. Anal. **51**(5), 45–55 (2015). (in Russian)
19. Feder, J.: Fractals. Plenum Press, New York (1988)
20. Falconer, K.J.: Fractal Geometry. Mathematical Foundations and Applications. Wiley, Chichester (2003)
21. Chan Johny, C.L., Kepert Jeffrey, D.: Global Perspectives on Tropical Cyclones: From Science to Mitigation (World Scientific Series on Asia-pacific Weather and Climate). World Scientific Publishing, Singapore (2010)
22. James, P., Terry, P.: Tropical Cyclones: Climatology and Impacts in the South Pacific. Springer, New York (2010)
23. NASA Rapid Response (2018). https://lance2.modaps.eosdis.nasa.gov/realtime
24. Typhoon Hagibis: characteristics, formation and key features (2021). https://www.meteorologiaenred.com/ru/tifon-hagibis.html. (in Russian)
25. Kobliuk, N., Pashchenko, R., Mariushko, M.: Information technology of aerospace images fractal analysis. In: 14th International Scientific and Technical Conference on Computer Sciences and Information Technologies (CSIT-2019), Ukraine, Lviv, 17–20 September, vol. 2, pp. 102–105. IEEE (2019). https://ieeexplore.ieee.org/xpl/conhome/8910522/proceeding?isnumber=8929738&searchWithin=Information%20technology%20of%20aerospace%20images%20fractal%20analysis
26. Nazirov, M.V., Pichugin, A.P., Spiridonov, Yu.G., et al.: Earth's surface radiolocation from space. In: Mitnik, L.M., Viktorov, S.V. (eds.) Gidrometeoizdat, Moscow (1990). (in Russian)
27. Romashova, T.V.: Geography in figures and facts. In: Maloletko, A.M. (eds.) Toipkro, Tomsk (2008)
28. Kachinsky, A.B.: Safety, contaminate and risk: scientific concepts and mathematical methods, p. 472 (2003). (in Russian)
29. Mochalov, I.A., Petrunin, N.G., Redkin, A.S., et al.: Fuzzy probabilistic-statistical methods. Inf. Technol. **4**, 24 (2003). (in Russian)
30. Atlantic Oceanographic and Meteorological Laboratory, Hurricane Research Division. Frequently Asked Questions: What regions around the globe have tropical cyclones and who is responsible for forecasting there
31. Integrated Kinetic Energy. Atlantic Oceanographic and Meteorological Laboratory. National Oceanic and Atmospheric Administration

High Precision Preloading of Angular Contact Bearing in Movable Mechanism of Space Optical Remote Sensor

Hao Pan[1,2](✉), Jiang Chang[1,2], and Qihui Tu[1,2]

[1] Beijing Institute of Space Mechanics and Electricity, Beijing 100190, China
phyj12004827@163.com
[2] Beijing Engineering Technology Research Center of Aerial Intelligent Remote Sensing Equipments, Beijing 100094, China

Abstract. The bearing in the movable mechanism of the optical remote sensor is one of the important components that affect the function realization, motion accuracy and working life. The precise control of bearing preload is the key link. A test system of bearing loading is built to measure the load-displacement relationship and friction torque accurately. And the data obtained by experiment are compared with the results of numerical calculation and simulation analysis. The results indicate that, with the increase of the load, there is an obvious "inflection point" in the bearing stiffness value, the friction torque of bearing rises first, than falls and then rises steadily; under the same load, the displacement and friction torque of the double-row bearing are two times of that of the single-row bearing. Finally in the precision preloading experiment, we control the friction torque error within 7.9% through eliminating the structural deviation, determining the "zero point" of the preload reasonably, and increasing the compensation preload, which effectively improves the reliability of the mechanism products.

Keywords: Contact bearing · Friction torque · Preload · Error compensation

1 Preface

The movable mechanism of optical remote sensor is characterized by high precision, long service life and high reliability. A typical mechanism is shown in Fig. 1. It is very important to the function realization, movement accuracy and working life of the bearing movable mechanism. Bearing preload is the most critical technical index in the process of bearing assembly, and its precise control is the key link of bearing assembly. For the angular contact bearing pre-tightened by positioning method, there are mainly the following problems during assembly: 1) In the process of bearing pre-tightening, the changing law and mechanism between pre-tightening load and "bearing displacement and friction torque" have not been studied very clearly; 2) In the process of bearing pre-tightening, the influence of the rigid deformation of the system on the pre-tightening amount needs to be removed; 3) The contact timing between the inner and outer rings

H. P. Urbach and H. Jiang (Eds.): ISSOIA 2022, SPPHY 295, pp. 586–598, 2023.
https://doi.org/10.1007/978-981-99-4098-1_49

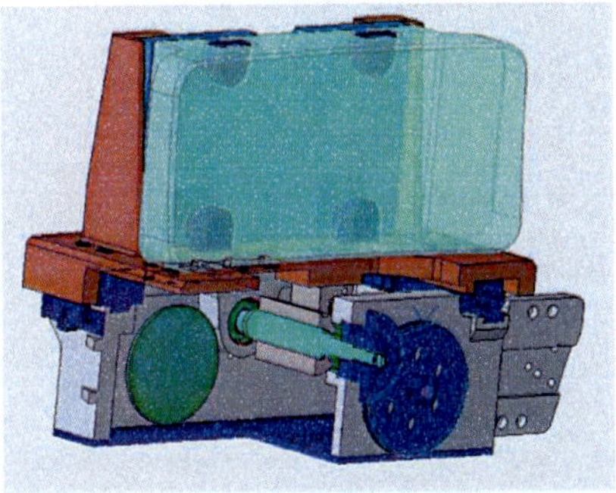

Fig. 1. A focusing mechanism of the space remote sensor.

of the bearing and the balls cannot be accurately determined, that is, the preload "zero point" of the bearing cannot be determined; In this study, the relationship between preload of diagonal contact bearing and displacement and friction torque was measured accurately, and the correctness of the measurement results was verified by theoretical calculation and simulation analysis. Finally, according to the measurement results and rules, this study improves the pre-tightening accuracy of bearings from three aspects: eliminating the influence of structural rigidity error, determining the zero point of bearing pre-tightening and increasing pre-tightening compensation, so as to ensure the motion accuracy and reliability of mechanism products.

2 Measurement of Load-Displacement Relationship

2.1 Bearing Preload Measuring System

In this paper, an integrated measurement system for studying the relationship between bearing preload, displacement and friction torque was established. The moment was measured by balance method. The outer bearing ring was fixed on the positioning mandrel of the base. And the guide loading device applies axial load to bearings, structural parts and shaft parts through force sensors. The displacement sensor measured the displacement of the inner bearing ring relative to the outer bearing ring. Torque sensor measured the friction torque of the assembled bearing (Fig. 2).

When this system measured axial displacement, the displacement of bearing inner and outer rings was measured by length sensor. A length sensor was respectively installed at the left and right ends of the loading ring of the inner and outer rings. During loading, the final displacement of the inner and outer rings of the bearing was determined by analyzing the readings of the two length sensors. When the friction torque was measured, the motor was directly connected in series with the torque sensor to control the motor to rotate at a stable low speed. The torque measured by the torque sensor was the sum of the friction torque of the supporting shafting and the friction torque of the measured bearing, and the friction torque of the supporting shafting can be measured separately. The difference between the torque measured by the torque sensor and the friction torque of the supporting shafting is the bearing friction torque.

Fig. 2. Bearing preload measuring system.

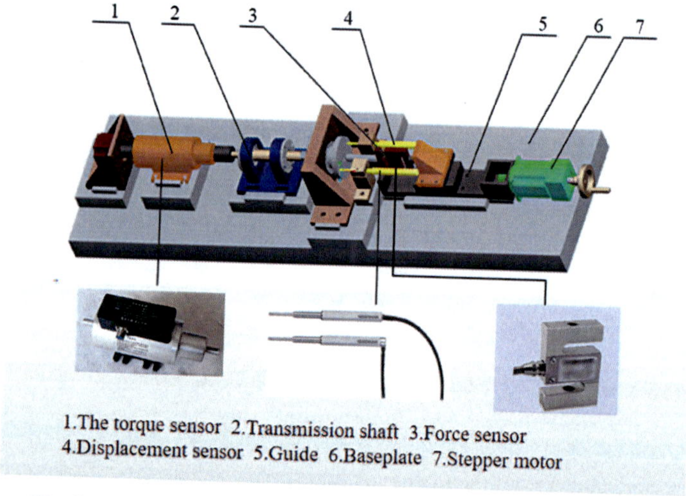

Fig. 3. Schematic diagram of system composition and sensor layout.

See Table 1 for the types, measuring ranges and precision indexes of sensors involved in the measurement system. See Fig. 3 for the measurement system and sensor arrangement.

2.2 Influence of System Rigidity on Measurement Results

When measuring, the influence of the rigidity of the bearing itself and the rigidity of the measuring system itself needs to be considered. When using tooling to complete the test, the rigidity of the system itself needs to be tested separately. When the rigidity of the test system is equal to that of the bearing, it will have a great impact on the test results, so it is necessary to remove the rigidity of the system from the test results. The results

Table 1. Sensor specifications in the measurement system.

Sensor name	model	Range	Precision Requirement
Torque sensor	CKY-805	0.01–0.5 N.m	0.02–0.05%F·S
Force sensor	CKY-1S	1-500N	0.02–0.05%F·S
Displacement sensor	ST300	0–5 mm	±1 um

in Fig. 4 show that the test system will produce displacement of about 8 um under the load of 200 N, which can't be ignored for the bearing preload and must be deleted.

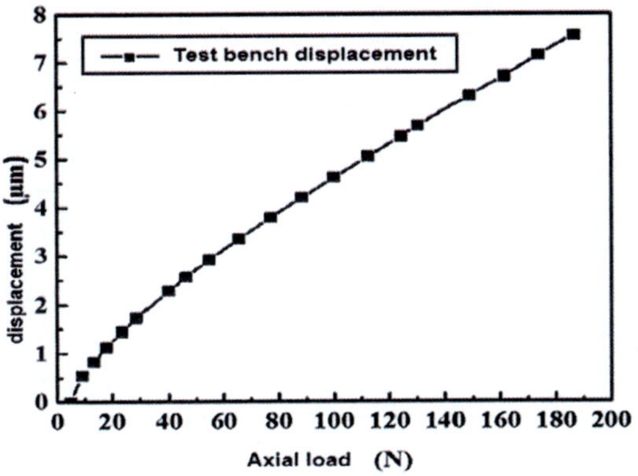

Fig. 4. Relation between load and displacement of test system.

2.3 Determine the Bearing Preload Zero Point

Bearing preload is the most critical technical index in the process of bearing preload. For the angular contact bearing pre-tightened by positioning method, the pre-tightening amount refers to the deformation that continues to be loaded after all the balls in the bearing just contact the inner and outer rings. This deformation amount can be precisely realized by grinding the boss or gasket of the bearing cover. How to determine the proper contact between the bearing inner ring, outer ring and balls, that is, to determine the "zero point" of bearing preload, is the key issue of bearing preload.

Bearing loading must go through three stages: partial contact, just complete contact (i.e. pre-tightening zero point) and ball deformation. When the "zero point" of preload is uncertain, firstly, a certain preload is applied to the bearing. The load should ensure that the preload displacement of the bearing has crossed the zero point; Then, the bearing is unloaded. Theoretically, when the load is reduced to zero, the bearing should be at the pre-tightening zero point. Considering the actual error, when the load is reduced to 1N, it can be considered that the bearing is in the "zero point" state.

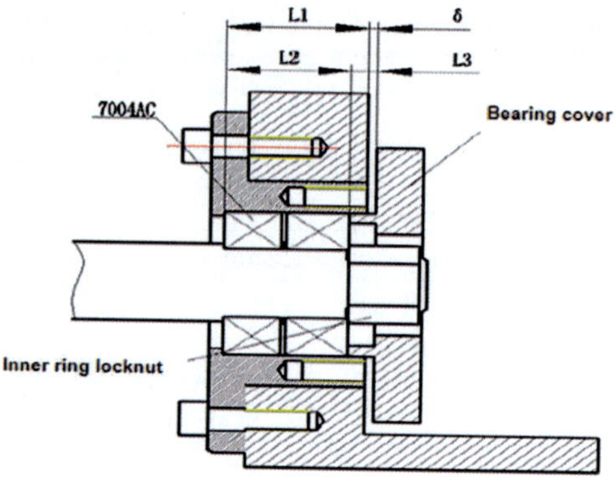

Fig. 5. Structural assembly drawing of angular contact bearing.

In the actual assembly process, the pre-tightening δ is determined by measuring the relevant dimensions of structural members, bearings and bearing cover. According to the dimension chain in Fig. 5, $\delta = L2 + L3 - L1$, where L2 is the distance between the two outer rings when the bearing is at the preload "zero" point; L1 is the depth of the bearing mounting hole on the structural member; L3 is the height of the boss on the bearing cover. L3 and L1 in the formula can be measured accurately, and L2 should be measured at the "zero point" of pretension. In the actual assembly process, when measuring L2, the bearing is often not in the "zero point" state, which is also the main reason for the pretightening error, and error compensation is needed.

2.4 Determination of Preloading Error

In this paper, the bearing was tested with and without preload. No-load test is to directly control the pressing force of the bearing cover to 1 N through the force sensor, establish the initial contact of the bearing, and take this state as the starting point to continuously measure the displacement values corresponding to different loads. Pre-load test is to apply a load of 10 N to the bearing first, then reverse unload it to 1 N, and then take this state as the starting point to continuously measure the displacement values corresponding to different loads.

The results of Fig. 6 show that the displacement with preload is smaller than that without preload, and the difference is fixed after a certain load value. After the single-row bearing is loaded to 20 N, the difference is 4 um; When the double-row bearing is loaded to 40 N, the difference is about 8 um. When the load is small, because there is no preload, the contact state in the bearing is unstable, and the internal clearance of the bearing needs to be eliminated, which will bring additional displacement. This displacement should be compensated. When the load reaches a certain level, the bearings of the two loading modes are in stable contact, and the difference of their displacement is fixed. The difference of the displacement is the compensation Δ of the pre-tightening error.

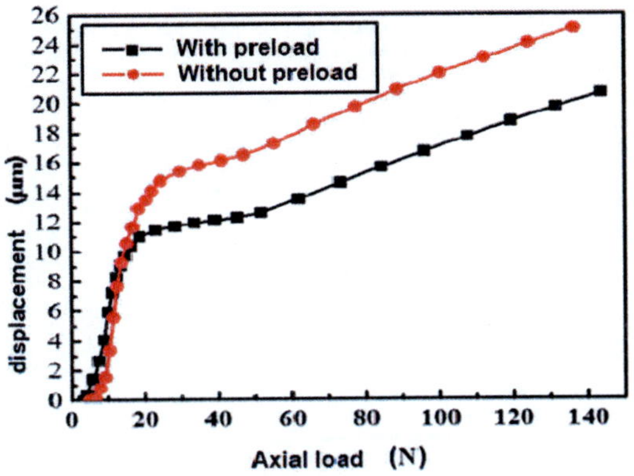

(a) 7004AC Single row preload displacement

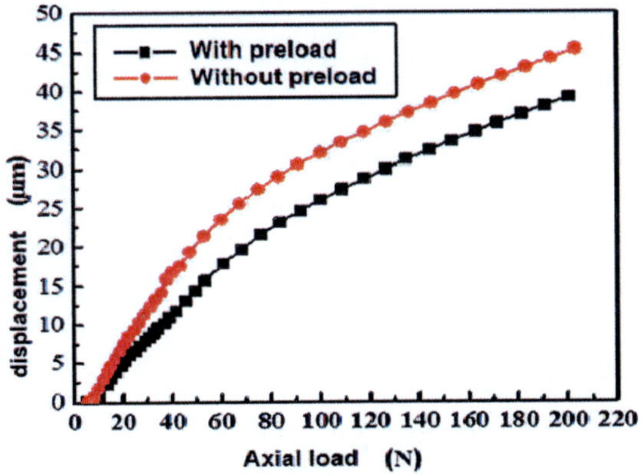

(b) 7004AC Double row preload displacement

Fig. 6. Displacement curve under bearing preload.

2.5 Measurement Result

The relationship between load and displacement of 7004AC bearing was measured by preloading, and the system rigidity error was deleted. The results in Fig. 7 show that: 1) there is an obvious "inflection point" in bearing stiffness, that is, the slope of load-displacement curve changes obviously before and after the inflection point; 2) Under the same load condition, the displacement of double-row bearing and single-row bearing is 2 times linear.

According to GB/T 32562-2016 "Measurement Method of Friction Torque of Rolling Bearings", the rotational friction torque of 7004AC and 7300 bearings under single-row loading and double-row loading were measured respectively. The results in Fig. 8 show that: 1) During the loading process, the friction torque first rises, then falls and then rises. At the initial stage of loading, the contact state between the ball and the raceway in the bearing is unstable, and the local contact stress is too large, resulting in a large friction torque. With the increase of load, the internal contact state of the bearing is stable, the

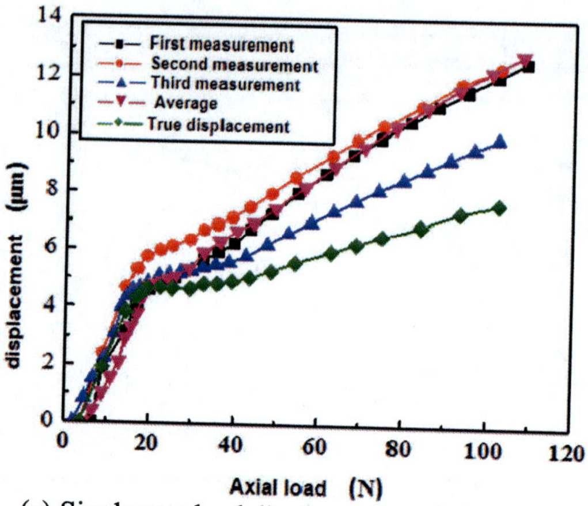

(a) Single row load displacement relation curve

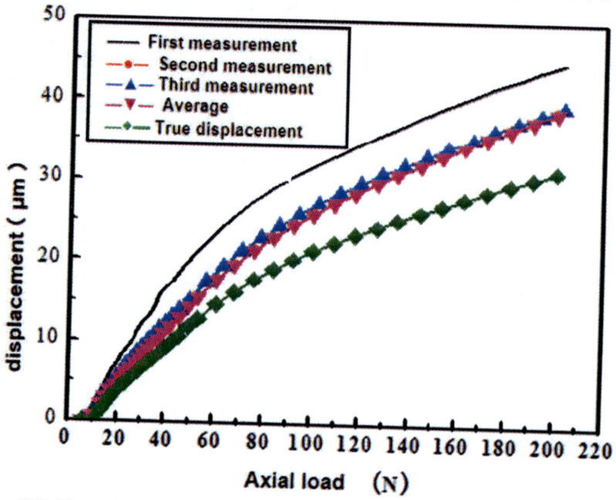

(b) Double row load displacement relation curve

Fig. 7. Load—Displacement relation curve.

raceway and the balls contact evenly, and the friction torque will decrease relatively, and then the friction torque will increase steadily with the increase of load. 2) Under the same load condition, the friction torque value of double-row bearing and single-row bearing is 2 times of linear relationship.

3 Theoretical Calculation and Simulation

3.1 Calculation and Simulation of Load-Displacement Relationship

According to "Analysis, Calculation and Application of Rolling Bearings", under the action of pure axial load Fa, and assuming that the contact angle after being stressed is still the initial contact angle, the relationship between the load and displacement of the

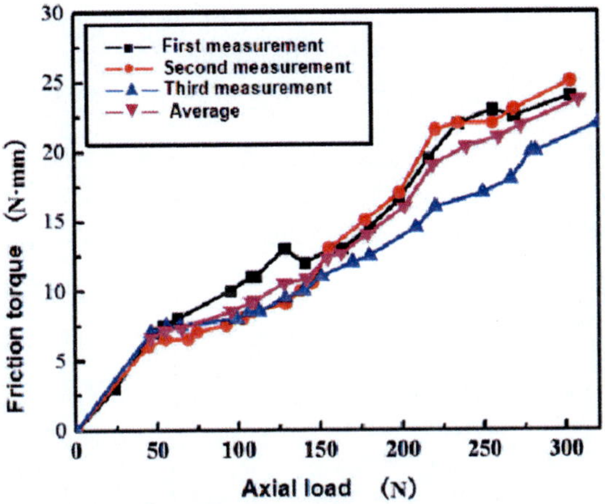

(a) 7004AC Single row friction torque (n=1.9rpm)

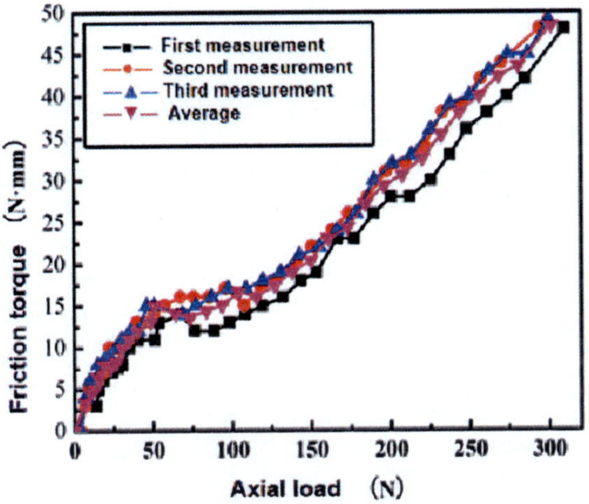

(b) 7004AC Double row friction torque (n=1.9rpm)

Fig. 8. Load—Friction torque relation curve.

bearing is:

$$F(a) = K_n Z(\sin \alpha_0)^{2.5} \delta_a^{1.5} \quad (1)$$

K_n is the bearing stiffness coefficient; Z is the number of balls; α0 is the initial contact angle; δa is the axial displacement. For steel bearings:

$$K_n = 2.1343 \times 10^5 \left(\delta_i^* \Sigma^{1/3} \rho_i + \delta_o^* \Sigma^{1/3} \rho_0 \right)^{-1.5} \quad (2)$$

δ_i^* and δ_o^* are the contact areas of steel balls in the inner and outer rings of the bearing, which are related to the elliptical eccentricity parameters; $\sum \rho_i$ and $\sum \rho_o$ are the curvature sum of the inner ring and the outer ring of the bearing, which are related to the geometric parameters of the specific bearing. Taking 7004AC single-row bearing as an example, the relationship between axial displacement and load is as follows:

$$\delta_a = \frac{1}{4104} F(a)^{2/3} \tag{3}$$

The specific geometric parameters of 7004AC contact bearing are shown in Table 2.

Table 2. The specific geometric parameters of 7004AC contact bearing.

Internal diameter D_i(mm)	External diameter D_o(mm)	Radius of curvature of inner raceway r_i(mm)	Radius of curvature of outer raceway r_o(mm)	Diameter of bearing section circle d_m(mm)	Diameter of steel ball D(mm)	Number of steel balls Z	Ring width B/mm	Contact angle α_0/(°)
20	42	3.26	3.32	30.95	6.35	9	12	25

In this study, abaqus software was used for modeling and simulation. According to the bearing test conditions, the steel ball was set as a rigid body, and the contact properties among the inner ring, the outer ring and the rigid body were set as universal contact; A linear load of 200 N was applied between the inner and outer rings of the bearing. The simulation results were as follows: at 200 N, the displacement of the inner ring was 0.0184 mm, and the output displacement curve was shown in the figure. The contact stress between the steel ball and the raceway was 36.9 Mpa, and the friction moment was 15.2 N·mm (Fig. 9).

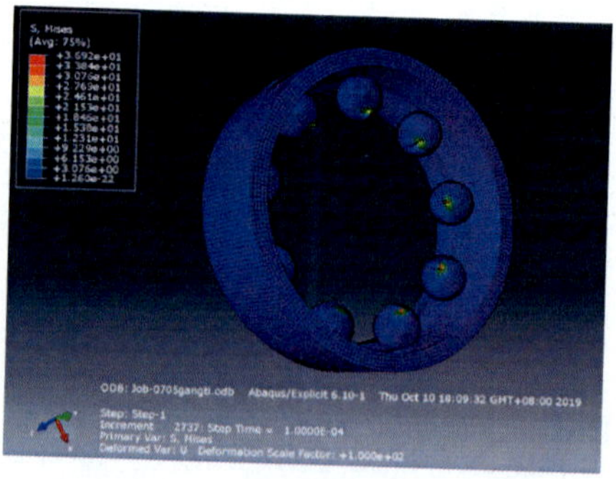

Fig. 9. Simulation results of bearing preloading process.

3.2 Friction Moment Calculation

According to the NSK rolling bearing technical manual, the bearing friction moment M is calculated as follows:

$$M = M_s Z \sin\alpha \quad (4)$$

M_s is based on contact angle α The shaft of is the rotary friction center, Z is the number of steel balls, and α is the contact angle (°)

$$M_s = \frac{3}{8}\mu_s Q a E(k) \quad (5)$$

μ_s is the sliding friction coefficient of the contact surface, Q is the rolling body contact load, a is the major axis radius of the ellipse, and $E(k)$ is the second type of ellipse integral. The relationship between the friction torque and the load of 7004AC single row bearing is calculated as follows:

$$M = 0.021 F_a^{\frac{4}{3}} \quad (6)$$

3.3 Result Analysis

The results of the load-displacement relationship in Fig. 10 show that the ABAQUS finite element analysis results are close to the curve results of the measured values, and the error is about 10%. Both of them can see the obvious "inflection point" of the bearing stiffness, that is, when the loading force is 20 N, the bearing stiffness increases obviously. Compared with the measured and simulated values, the calculated values have a large error of about 30–40%. In the initial stage of loading (within 20 N), the error is close to 50%, and the axial stiffness of the bearing has no "inflection point" phenomenon.

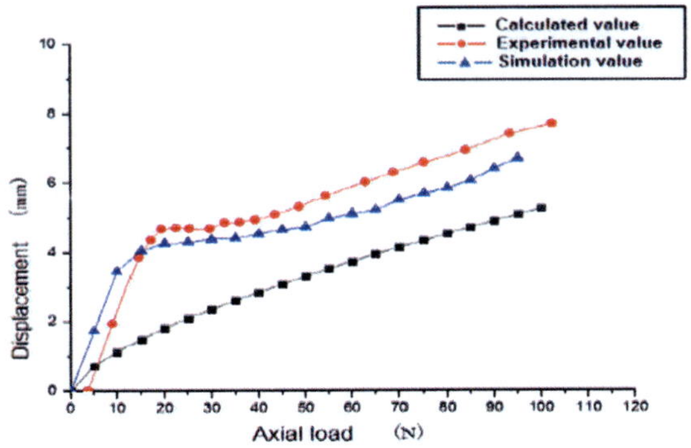

Fig. 10. Comparison of analysis results of the relationship between load and displacement.

Cause analysis: 1) The calculated value is based on Hertz contact theory and bearing geometry theory. This value is mainly related to the curvature radius and contact angle of the bearing inner and outer rings. However, with the increase of the load, the radial component of the load on the bearing inner and outer rings also increases, which leads to the change of the size and contact angle of the bearing inner and outer rings. This change is not considered in Hertz analytical solution in this paper, which is the main reason for the error generation. In addition, the theoretical calculation value does not consider the actual contact conditions (friction and sliding) between the bearing steel ball and the inner and outer rings, so that the stiffness "inflection point" will not appear.

2) At the initial stage of loading, the contact between the steel ball and the inner and outer rings has not been established stably, and its contact stiffness will increase sharply. When the steel ball is in stable contact with the inner and outer rings, its stiffness curve will become relatively flat, which is the cause of the bearing stiffness "inflection point".

3) Finite element simulation can overcome the above shortcomings. By setting the contact attributes and analyzing the friction and sliding between the bearing contact parts, the results of the solved bearing contact problems are close to the actual measured values. Considering the deviation between the simulated parameter setting and the actual measurement, the error of about 10% is acceptable. However, the actual measured value curve lags behind the simulated calculation curve. This is mainly because in the initial stage of loading, due to the manufacturing errors of bearing raceway and steel balls, the eccentric loading of loading force and other reasons, the nine steel balls only partially contact with the inner and outer rings, and gradually all contact with the increase of loading force. In actual measurement, a certain preload will be applied to make the bearing contact stably, so there will be a certain "lag" on the actual measurement curve. Therefore, the simulation calculation results in errors because the actual situation is not considered.

The result of load-friction torque in Fig. 11 shows that the measured value of friction torque is close to the calculated value of simulation, and the error is about 10%, which is within an acceptable range. The error of the calculated value is about 20%–30%, and the calculated and simulated values of friction torque are steadily rising, while the measured values show a trend of "first rising, then falling and then rising".

The reason for the measured value "rising first, then falling and then rising" is as follows: during the loading process, due to the deviation of the loading force, the contact between the inner ball and the raceway is uneven, the local contact stress is too large, and at the same time, the friction torque is large; With the increase of loading force, all the balls are in uniform contact with the raceway, and the local contact stress disappears, which reduces the friction torque. The further increase of the subsequent loading force makes the friction torque rise steadily. Both calculated and simulated values are ideal conditions of bearing loading, that is, the bearing balls are always in uniform contact with the raceway, so there will be no phenomenon of "first rising, then falling and then rising". As for the reason for the larger error of the calculated value, it is consistent with the relationship between displacement and load analyzed above, so it will not be repeated here.

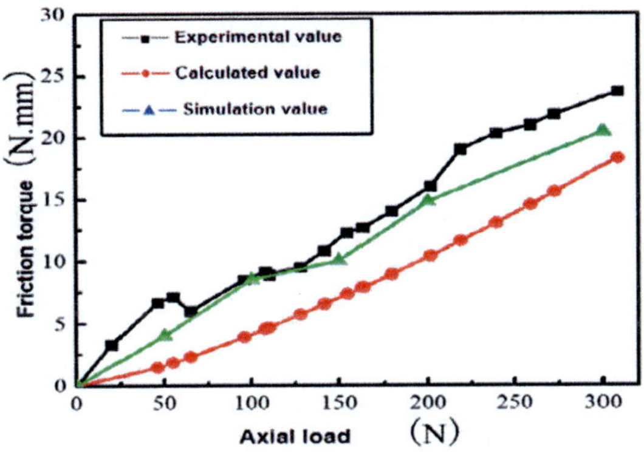

Fig. 11. Result analysis and comparison.

4 Implementation and Verification of Precision Preloading Process

Take the preload of 7004AC bearing as an example, load the preload of 0.01 mm, 0.02 mm, 0.03 mm and 0.04 mm respectively. According to the content in Sect. 3.3, the compensation amount Δ under the corresponding pre-tightening amount was determined respectively, and the height value of the boss of the bearing cover was machined respectively, and then the bearing was pre-tightened. Measure the rotational friction torque when the rotating speed should be 2 r/min, and compared it with the friction torque obtained by loading the test platform. The results were consistent, with the maximum error of 7.9%. The results were as follows (Table 3).

Table 3. Comparative analysis of bearing preload and friction torque.

Number	Preload (mm)	Compensation Δ(mm)	Friction moment loaded on the base plate (N.mm)	Friction moment loaded on bearing cover (N.mm)	Error rate (%)
1	0.01	0.005	6.4	6.1	4.6
2	0.02	0.007	10.1	10.9	7.9
3	0.03	0.007	14.3	15.1	5.6
4	0.04	0.007	28.3	26.5	6.3

5 Conclusion

(1) During the loading process, there is an obvious "inflection point" in the stiffness of the bearing, and the inclination of the load-displacement curve first increases and then decreases; Under the same load condition, the displacement of double-row bearing and single-row bearing is 2 times linear.

(2) During the loading process, the value of bearing friction torque first rises, then falls, and then rises steadily; Under the same load condition, the friction torque value of double-row bearing and single-row bearing is 2 times linear.
(3) By eliminating the structural rigidity error, reasonably determining the "zero point" of pre-tightening and increasing the compensation pre-tightening, the precision pre-tightening test of the bearing is carried out. The control error of the bearing friction torque is 7.9%, which effectively improves the reliability of the mechanism products.

References

1. Jia, D., Zhang, M., Zhang, Y.: Determination of precompact capacity of rolling bearing. Mech. Res. Appl. **14**(4), 12–13 (2001)
2. Ozturk, E., Kumar, U., Turner, S., et al.: Investigation of spindle bearing preload on dynamics and stability limit in milling. CIRP Ann. Manuf. Technol. **61**(1), 343–346 (2012)
3. Gunduz, A., Dreyer, J.T., Singh, R.: Effect of bearing preloads on the modal characteristics of a shaft-bearing assembly: experiments on double row angular contact ball bearings. Mech. Syst. Signal Process. **31**, 176–195 (2012)
4. Lewis, S.D., Humphries, M.E.: An introduction to bearing active preload systems-Technology and performance benefits. In: Esa Sp, pp. 261–269 (2005)
5. Xu, T., Xu, G., Zhang, Q., et al.: Experimental study on bearing preload optimum of machine tool spindle. J. Phys: Conf. Ser. **364**, 012–030 (2012)
6. Zhang, X., Han, Q., Peng, Z., et al.: A new nonlinear dynamic model of the rotor-bearing system considering preload and varying contact angle of the bearing. Commun. Nonlinear Sci. Numer. Simul. **22**(1–3), 821–841 (2015)
7. Li, C., Duan, H., Wang, Y., et al.: Simulation and analysis on radial stiffness of angular contact ball bearing. Torpedo Technol. **25**(5), 351–358 (2017)
8. Si, S.: Study on bearing preload of space joint and its dynamic characteristic, pp. 1–4, 32–56. Harbin Institute of Technology, Harbin (2010)
9. Deng, S., Li, X., Wang, Y., et al.: Analysis on the friction torque fluctuation of angular contact ball bearings. J. Mech. Eng. **47**(23), 104–112 (2011)
10. Ning, F., Chen, R., Yao, Y., et al.: Influences of preload and assembly deviation on friction torque of aerospace bearing. Acta Armamentarii (7), 1436–1442 (2018)
11. Hu, H., Pan, B., Sun, J.: Friction torque modeling of spacecraft driving mechanism shafting. J. Mech. Eng. **53**(11), 75–80 (2017)
12. Xie, N., Hao, Z.: Refrigeration control system of infrared detector based on FPGA. Spacecr. Recov. Remote Sens. **40**(4), 48–56 (2019)

Infrared Small Target Detection Algorithm Based on Full Feature Measure and Tensor Recovery

Jinhao Kuang, Bo Zhou(✉), Chunbo Chen, Xiaohang Hong, and Shutong Xu

School of Electronic Information and Communications, Huazhong University of Science and Technology (HUST), Wuhan, China
zhoubo@hust.edu.cn

Abstract. Infrared small target detection is the crucial technology of infrared search and track systems. The targets' characteristics of small size, few features, and low signal-to-noise ratio result in considerable difficulties with detection. At present, many infrared small target detection algorithms based on low-rank of backgrounds and sparse representation of targets have good detection performance. But their abilities of scene generalization and resistance to sparse interference still need to be improved.

To solve this problem, based on the infrared patch tensor model, an infrared and small target detection algorithm based on tensor recovery is proposed in this paper: a regularization term that combines the full feature measure and the non-convex rank relaxation method is used as the constraint low-rank property of the background tensor. The full feature measure method based on N-tubal rank makes better use of the feature information between different modes of the background tensor. The non-convex rank relaxation method based on the Log function can better approximate the rank function of the tensor. In addition, an efficient algorithm based on the alternating direction method of multipliers is developed to solve the proposed model to obtain the target tensor.

The experimental results show that the scene adaptability and resistance to sparse interference of the proposed algorithm are greatly improved. In terms of target enhancement, background suppression, and detection accuracy, the proposed algorithm is better than the comparison algorithms.

Keywords: Infrared small target detection · Low-rank and sparse · Full feature measure · Non-convex rank relaxation

1 Introduction

As defined [1] by the International Society for Optical Engineering, "small" means that the size of the target generally does not exceed 9 by 9 pixels and the area does not exceed 0.15% of a 256 by 256 image. Infrared small target detection plays an important role in the military field. Through the lack of target features and the complex background scene, Infrared small target detection becomes challenging research [2]. Recently, due to the

H. P. Urbach and H. Jiang (Eds.): ISSOIA 2022, SPPHY 295, pp. 599–611, 2023.
https://doi.org/10.1007/978-981-99-4098-1_50

development of tensor recovery algorithms, the infrared small target detection algorithm [3] based on tensor representation-based low-rank sparse recovery using the infrared patch-tensor model (IPT) [4] has gradually become a research hotspot. However, these algorithms generally have the following problems: on the one hand, the local information of the sparse target is not fully utilized, which leads to the inaccurate approximation of the target regularization term and miss detection; on the other hand, the global information of the background is not fully utilized, which leads to the incomplete extraction of the background regularization term for the edge part and corner point, resulting in the mixed clutter in the target and generating false alarms. In this paper, IPT based infrared small target detection algorithm is studied from two aspects of robustness and accuracy.

2 The Proposed Method

The infrared small target detection algorithm based on the IPT model mainly uses the non-local prior characteristics of the target and background for detection. The detection approaches are roughly included the following four steps. First, the infrared block tensor of the original infrared image is constructed; Second, two regular terms that can accurately describe the sparsity of the target and the low rank of the background are introduced in our model; Third, the target tensor is obtained by solving the target function through an optimization algorithm; Fourth, The target images are reconstructed from the target tensor, and the final target detection result is output through the adaptive threshold segmentation algorithm. The flowchart of our proposed method is shown in Fig. 1.

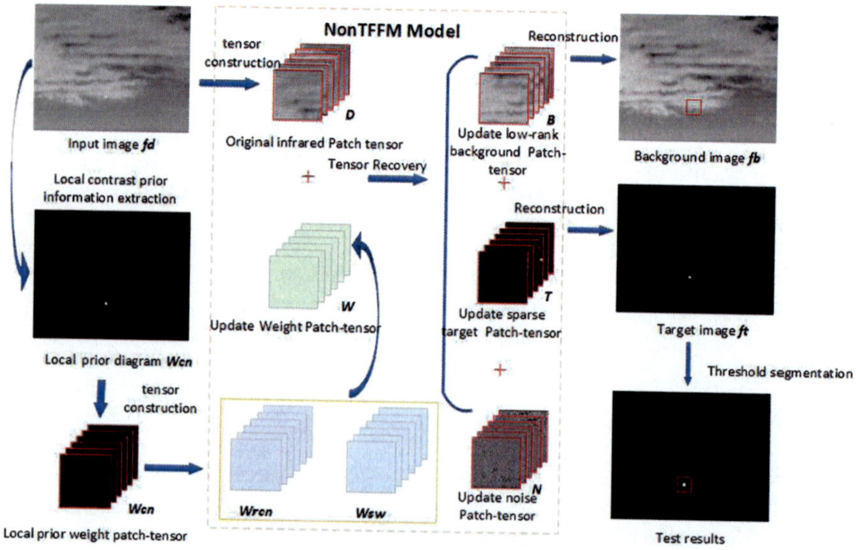

Fig. 1. The flowchart of our proposed method.

2.1 Background Prior Analysis

Considering the strong autocorrelation among the local blocks of the IR image background, it can be assumed that low rank is an intrinsic property of the background patch-tensor. Tensor rank [5] is the most direct way to measure this low-rank property, but the tensor decomposition is not uniform, which leads to the definition of tensor rank is also not unique. In the IPT detection model, we need to find a suitable and easy-to-compute tensor rank measure. Zhang et al. proposed the Tensor Full Feature Measure (TFFM) method [6], which can be simply the regularization of different features of each tensor dimension. This measure not only contains information about the relevant features between any dimensions and the features of each dimension itself but also connects the Tucker rank and the pipeline rank, further facilitating the connection between different ranks. It is defined as follows.

$$\|\mathcal{X}\|_{TFFM} = \sum\nolimits_{1 \le k_1 \le k_2 \le N} \beta_{k_1 k_2} rank\left(\mathcal{X}_{(k_1 k_2)}\right) \tag{1}$$

where $\beta_{k_1 k_2} \ge 0(1 \le k_1 \le k_2 \le N, k_1, k_2 \in \mathbb{Z})$, and $\sum\nolimits_{1 \le k_1 \le k_2 \le N} \beta_{k_1 k_2} = 1$, $\mathcal{X}_{(k_1 k_2)} = unfold_{k_1}(k_1 = k_2)$, $\mathcal{X}_{(k_1 k_2)} = t - unfold(\mathcal{X}, k_1, k_2)(k_1 \ne k_2)$.

2.2 Proposed Model Based on IPT

The Surrogate of Tensor Rank. Since the Log function can better approximate the norm of the rank, the tensor rank can be approximated by the sum of the Log functions of the singular values [7–9]. Directly using TNN as its relaxation approximation cause many problems. Therefore, formula (2) is further integrated into the tensor kernel norm based on the full feature metric by using the Log function. The proposed method is called Nonconvex tensor rank relaxation of the Log function-based TFFM (NonTFFM). This method not only effectively compensates for the solution deviation caused by the kernel norm, but also simplifies the solution process of the objective function. The definition of NonTFFM is as follows:

$$\|\mathcal{X}\|_{NonTFFM} = \sum\nolimits_{1 \le k_1 \le k_2 \le N} \beta_{k_1 k_2} Log\left(\mathcal{X}_{(k_1 k_2)}\right) \tag{2}$$

$$Log\left(\mathcal{X}_{(k_1 k_2)}\right) = \sum\nolimits_{m} \left(log\left(\sigma_m\left(\overline{\mathcal{X}}_{(k_1)}\right) + \varepsilon\right)\right), k_1 = k_2 \tag{3}$$

$$Log\left(\mathcal{X}_{(k_1 k_2)}\right) = \sum\nolimits_{i=1}^{J_{k_1 k_2}} \sum\nolimits_{m} \left(log\left(\sigma_m\left(\overline{\mathcal{X}}^{(i)}_{(k_1 k_2)}\right) + \varepsilon\right)\right), k_1 \ne k_2, J_{k_1 k_2} = \prod\nolimits_{s \ne k_1, k_2} I_s \tag{4}$$

Among them, Eqs. (3) and (4) are in the form of Log sum functions, $\sigma_m\left(\overline{\mathcal{X}}_{(k_1)}\right)$ and $\sigma_m\left(\overline{\mathcal{X}}^{(i)}_{(k_1 k_2)}\right)$ are the m singular value of $\overline{\mathcal{X}}_{(k_1)}$ and $\overline{\mathcal{X}}^{(i)}_{(k_1 k_2)}$, respectively.

NonTFFM Model Construction. Considering that NonTFFM can better approximate the tensor rank, it can be used as the constraint regularization term of the infrared low-rank background tensor, which can further improve the separation effect of the target tensor and the background tensor in the IPT model. Then, the objective function represented can be further rewritten as:

$$\min_{\mathcal{B},\mathcal{T}} \|\mathcal{B}\|_{NonTFFM} + \lambda\|\mathcal{W} \odot \mathcal{T}\|_1$$

$$s.t.\mathcal{D} = \mathcal{B} + \mathcal{T} \tag{5}$$

In addition, since the IPT model is an ideal model, it does not take into account the existence of noise, which is called the NonTFFM model, and is expressed as follows:

$$\min_{\mathcal{B},\mathcal{T},\mathcal{N}} \|\mathcal{B}\|_{NonTFFM} + \lambda\|\mathcal{W} \odot \mathcal{T}\|_1 + \alpha\|\mathcal{N}\|_F^2$$

$$s.t.\mathcal{D} = \mathcal{B} + \mathcal{T} + \mathcal{N} \tag{6}$$

Here $\alpha \geq 0$, is the regularization parameter for the noise term.

Solution of NonTFFM Model. ADMM algorithm [10] is currently the most widely used in solving constrained optimization problems. In this paper, the ADMM algorithm is used to solve the NonTFFM model.

First, introduce auxiliary variables to Eq. (6) and rewrite it as the following equivalence constraint problem.

$$\min_{\mathcal{B},\mathcal{T},\mathcal{N}} \sum_{1\leq k_1\leq k_2\leq N} \beta_{k_1k_2} Log\left(\left(\mathcal{Z}_{k_1k_2}\right)_{(k_1k_2)}\right) + \lambda\|\mathcal{W} \odot \mathcal{T}\|_1 + \alpha\|\mathcal{N}\|_F^2$$

$$s.t.\mathcal{D} = \mathcal{B} + \mathcal{T} + \mathcal{N}$$

$$\mathcal{B} = \mathcal{Z}_{k_1k_2} \tag{7}$$

Equation (7) can be further expressed as an augmented Lagrangian function as follows.

$$\begin{aligned} & L_{\mu_{k_1k_2},\rho}\left(\mathcal{B}, \mathcal{Z}_{k_1k_2}, \mathcal{T}, \mathcal{N}, \mathcal{Q}_{k_1k_2}, \mathcal{P}\right) \\ & = \sum_{1\leq k_1\leq k_2\leq N} \beta_{k_1k_2} Log\left(\left(\mathcal{Z}_{k_1k_2}\right)_{(k_1k_2)}\right) + \lambda\|\mathcal{W} \odot \mathcal{T}\|_1 \\ & + \alpha\|\mathcal{N}\|_F^2 + \langle \mathcal{Q}_{k_1k_2}, \mathcal{B} - \mathcal{Z}_{k_1k_2}\rangle + \frac{\mu_{k_1k_2}}{2}\|\mathcal{B} - \mathcal{Z}_{k_1k_2}\|_F^2 \\ & + \langle \mathcal{P}, \mathcal{D} - \mathcal{B} - \mathcal{T} - \mathcal{N}\rangle + \frac{\rho}{2}\|\mathcal{D} - \mathcal{B} - \mathcal{T} - \mathcal{N}\|_F^2 \end{aligned} \tag{8}$$

where $\mu_{k_1k_2}$ and ρ are penalty factors greater than 0, $\mathcal{Q}_{k_1k_2}(1 \leq k_1 \leq k_2 \leq N)$ and $\mathcal{P}$ are the Lagrange multiplier.$L_{\mu_{k_1k_2},\rho}\left(\mathcal{B}, \mathcal{Z}_{k_1k_2}, \mathcal{T}, \mathcal{N}, \mathcal{Q}_{k_1k_2}, \mathcal{P}\right)$ is decomposed a series of subproblems by fixing a certain variable to iteratively and alternatively update the other variables. At the $l+$ 1st iteration, the variables are solved in the following specific way:

$$\mathcal{B}^{l+1} = \underset{\mathcal{B}}{argmin} \frac{\mu_{k_1k_2}^l}{2}\left\|\mathcal{B} - \mathcal{Z}_{k_1k_2}^{l+1} + \frac{\mathcal{Q}_{k_1k_2}^l}{\mu_{k_1k_2}^l}\right\|_F^2 + \frac{\rho^l}{2}\left\|\mathcal{D} - \mathcal{B} - \mathcal{T}^l - \mathcal{N}^l + \frac{\mathcal{P}^l}{\rho^l}\right\|_F^2 \tag{9}$$

$$\mathcal{Z}_{k_1k_2}^{l+1} = \underset{\mathcal{Z}_{k_1k_2}}{argmin}\sum\nolimits_{1\le k_1\le k_2\le N}\beta_{k_1k_2}Log\left(\left(\mathcal{Z}_{k_1k_2}\right)_{(k_1k_2)}\right) + \frac{\mu_{k_1k_2}}{2}\left\|\mathcal{B}^{l+1} - \mathcal{Z}_{k_1k_2} + \frac{\mathcal{Q}_{k_1k_2}^{l}}{\mu_{k_1k_2}^{l}}\right\|_F^2 \tag{10}$$

$$\mathcal{T}^{l} = \underset{\mathcal{T}}{argmin}\lambda\|\mathcal{W}^{l}\odot\mathcal{T}\|_1 + \frac{\rho^{l}}{2}\left\|\mathcal{D} - \mathcal{B}^{l+1} - \mathcal{T} - \mathcal{N}^{l} + \frac{\mathcal{P}^{l}}{\rho^{l}}\right\|_F^2 \tag{11}$$

$$\mathcal{N}^{l+1} = \alpha\|\mathcal{N}\|_F^2 + \frac{\rho^{l}}{2}\left\|\mathcal{D} - \mathcal{B}^{l+1} - \mathcal{T}^{l+1} - \mathcal{N} + \frac{\mathcal{P}^{l}}{\rho^{l}}\right\|_F^2 \tag{12}$$

The Eq. (10) and (11) can be solved by Singular value decomposition and the Eq. (9) and (12) can be solved by derivation.

$$\mathcal{Z}_{k_1k_2}^{l+1} = \begin{cases} fold_{k_1}\left(U\Sigma_{\frac{\beta_{k_1k_2}^{l}}{\mu_{k_1k_2}^{l}}}V^{T}\right), & if k_1 = k_2; \\ t-fold\left(\mathcal{U} * \mathcal{S}_1 * \mathcal{V}^{H}, k_1, k_2\right), if k_1 \neq k_2 \end{cases} \tag{13}$$

$$\mathcal{B}^{l+1} = \frac{\sum_{1\le k_1\le k_2\le N}\mu_{k_1k_2}^{l}\left(\mathcal{Z}_{k_1k_2}^{l+1} - \frac{\mathcal{Q}_{k_1k_2}^{l}}{\mu_{k_1k_2}^{l}}\right) + \rho^{l}\left(\mathcal{D} - \mathcal{T}^{l} - \mathcal{N}^{l} + \frac{\mathcal{P}^{l}}{\rho^{l}}\right)}{\sum_{1\le k_1\le k_2\le N}\mu_{k_1k_2}^{l} + \rho^{l}} \tag{14}$$

$$\mathcal{T}^{l+1} = S_{\frac{\lambda\mathcal{W}^{l}}{\rho^{l}}}\left(\mathcal{D} - \mathcal{B}^{l+1} - \mathcal{N}^{l} + \frac{\mathcal{P}^{l}}{\rho^{l}}\right) \tag{15}$$

$$\mathcal{N}^{l+1} = \frac{\rho^{l}\left(\mathcal{D} - \mathcal{B}^{l+1} - \mathcal{T}^{l+1}\right) + \mathcal{P}^{l}}{2\alpha + \rho^{l}} \tag{16}$$

$$\mathcal{W}^{l+1} = \mathcal{W}_{rcn}\odot\mathcal{W}_{sw}^{l} \tag{17}$$

$$\mathcal{P}^{l+1} = \mathcal{P}^{l} + \rho^{l}\left(\mathcal{D} - \mathcal{B}^{l+1} - \mathcal{T}^{l+1} - \mathcal{N}^{l+1}\right)$$

$$\mathcal{Q}^{l+1} = \mathcal{Q}^{l} + \mu^{l}\left(\mathcal{B}^{l+1} - \mathcal{Z}^{l+1}\right) \tag{18}$$

$$\mu^{l+1} = min\left(\gamma\mu^{l}, \mu_{max}\right)\rho^{l+1} = min\left(\gamma\rho^{l}, \rho_{max}\right) \tag{19}$$

The above is the specific solution process for each variable in the NonTFFM model, and we can summarize the process of the algorithm:

Table 1. NonTFFM Model Solving Algorithm

Input: $\mathcal{D}$, $\mathcal{W}_{rcn}$, $\boldsymbol{\beta}$, $\boldsymbol{\lambda}$,$\boldsymbol{\alpha}$
Output: $\mathcal{B}$, $\mathcal{T}$, $\mathcal{N}$
1. Initialize: $\mathcal{B}^0 = \mathcal{T}^0 = \mathcal{N}^0 = 0$, $\mathcal{P}^0 = 0$, $\mathcal{Z}^0_{k_1k_2} = 0$, $\mathcal{Q}^0_{k_1k_2} = 0$, $\mathcal{W}^0_{sw} = 1$, $\mathcal{W}^0 = \mathcal{W}_{rcn} \odot \mathcal{W}^0_{sw}$, $\boldsymbol{\beta_{max}} = \boldsymbol{u_{max}} = 1 \times 10^{10}$, $\boldsymbol{\gamma} = 1.2$, $\boldsymbol{tol} = 1 \times 10^{-4}$, $\boldsymbol{l_{max}} = 200$
2. While not converged and $\boldsymbol{l} < \boldsymbol{l_{max}}$ do
3. Via(13) update $\mathcal{Z}^{l+1}$
4. Via(14) update $\mathcal{B}^{l+1}$
5. Via(15) update $\mathcal{T}^{l+1}$
6. Via(16) update $\mathcal{N}^{l+1}$
7. Via(17) update $\mathcal{W}^{l+1}$
8. Via(18) and(19) update $\mathcal{Q}^{l+1}$、$\mathcal{P}^{l+1}$ and μ^{l+1}、$\boldsymbol{\rho^{l}}$
9. Inspect the stop conditions: $\left(\frac{\Vert\mathcal{B}^{l+1}-\mathcal{B}^{l}\Vert_F}{\Vert\mathcal{B}^{l}\Vert}\ _{\boldsymbol{F}}\right) < \boldsymbol{tol}$ or $\Vert\mathcal{T}^{l+1}\Vert_0 = \Vert\mathcal{T}^{l}\Vert_0$
10. Update $\boldsymbol{l}$: $\boldsymbol{l} = \boldsymbol{l} + 1$
11. End while
12. $\mathcal{B} = \mathcal{B}^{l}$, $\mathcal{T} = \mathcal{T}^{l}$,$\mathcal{N} = \mathcal{N}^{l}$

For the target canonical term, we used an enhanced local contrast algorithm (Enhanced LCM, ELCM) based on a three-level window structure for calculating the local structure weights. Finally, combining the two background constraint terms and the target constraint term, we designed the NonTFFM-ELCM model for experiments.

3 Experiment and Analysis

To further demonstrate the effectiveness of the proposed algorithms, eight classic and current mainstream IR small target detection algorithms, including Top-hat filter [11], LCM [12], MPCM [13], IPI [14]. NRAM [15], RIPT [4], PSTNN [16], and SRWS [17], are selected for comparison. The algorithm simulation software for the experiments in this chapter is MATLAB R2018a. PC configured as Core i5–7500 CPU @ 3.40GHz, 8.00GB RAM. Note that we are using the NonTFFM-ELCM algorithm for the experimental comparison.

3.1 Dataset

This chapter will conduct experiments on multiple existing public datasets, mainly including the SIRST dataset provided by Dai et al. [18]. SIRST is a dataset specially constructed for single-frame infrared small target detection, in which the images are selected from hundreds of infrared sequences for different scenarios. Twenty single-frame and five-sequence infrared images of typical scenes are selected here. These images have been used in algorithm experiments many times, which are representative and challenging, and can reflect the comprehensive performance of the algorithm in multiple types of scenes. As shown in Fig. 2, (a) through (t) are twenty single-frame IR images, and (u) through (y) are representative frames of five sequential IR images. For visualization purposes, the size of all images is standardized in this chapter, where the targets and their zooms are marked with red borders. In addition, Table 2 provides a detailed description of the five sequences.

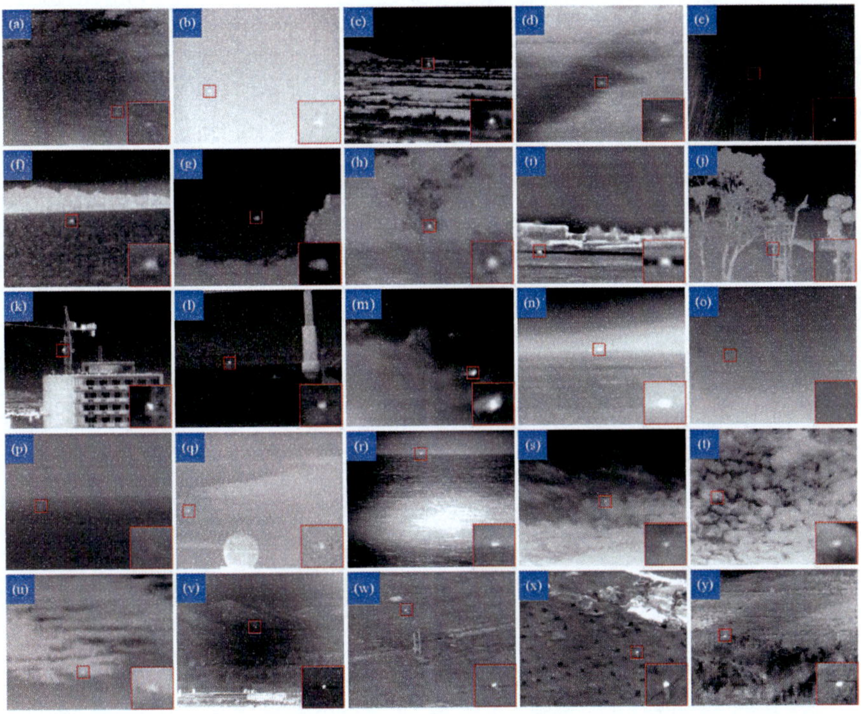

Fig. 2. Twenty-five single-frame and sequence infrared images under different scenes

Table 2. Description of five sequences.

Sequence	Frames	Size	Description of target	Description of background
Seq.1(Fig. 2(u))	30	256 × 200	Single target, regular shape, moving along the cloud layer	Banded clouds, high noise
Seq.2(Fig. 2(v))	80	256 × 256	Single target, slow maneuvering, small size	Scenes with strong artificial building interference
Seq.3(Fig. 2(w))	100	256 × 256	Single target, fast maneuver, size change	Ground clutter interference
Seq.4(Fig. 2(x))	100	256 × 256	Single target, fast maneuver, size change	Non-target radiation highlights such as snow
Seq.5(Fig. 2(y))	100	256 × 256	Single target, fast maneuver, regular shape	Trees, grass shading

3.2 Subjective Analysis

This chapter further conducts experimental tests on five infrared image sequences. This chapter uses sequence 1 to sequence 5 to verify the scene robustness of this algorithm and the other eight comparison algorithms. For more direct observation, we extract a frame from each sequence to reflect the detection results, as shown in Fig. 3. From the three-dimensional grayscale distribution of the original infrared image, it can be seen that the background components of sequences 1 to 5 are relatively complex, the saliency of the target is weak, and the detection difficulty is relatively high. Experiments show that the algorithm in this paper has excellent scene robustness.

3.3 Target Enhancement Analysis

Object enhancement is equivalent to enlarging the contrast between the object and its neighbors within a local area, thereby reducing the output map. The segmentation difficulty of the object in the image. Based on the infrared image sequence, the target enhancement ability of the algorithm is quantitatively analyzed by the average value of the signal-to-noise ratio gain ($\overline{\text{SCRG}}$). Table 3 shows the detection results of all algorithms on five infrared image sequences. The best detection results are marked in bold and underlined, respectively. The NonTFFM algorithm has obtained the optimal value on most of the test sequences, and only obtained the sub-optimal value on sequence 2. This is because its ability to maintain the integrity of the target is slightly weaker than that of the PSTNN algorithm, and the detected target grayscale value is low. In general, the algorithm combining local prior and non-local prior will obtain higher values $\overline{\text{SCRG}}$ than the algorithm based only on a single prior and have better target enhancement ability, and the NonTFFM-ELCM algorithm proposed in this paper can enhance the target The performance is the best and more stable.

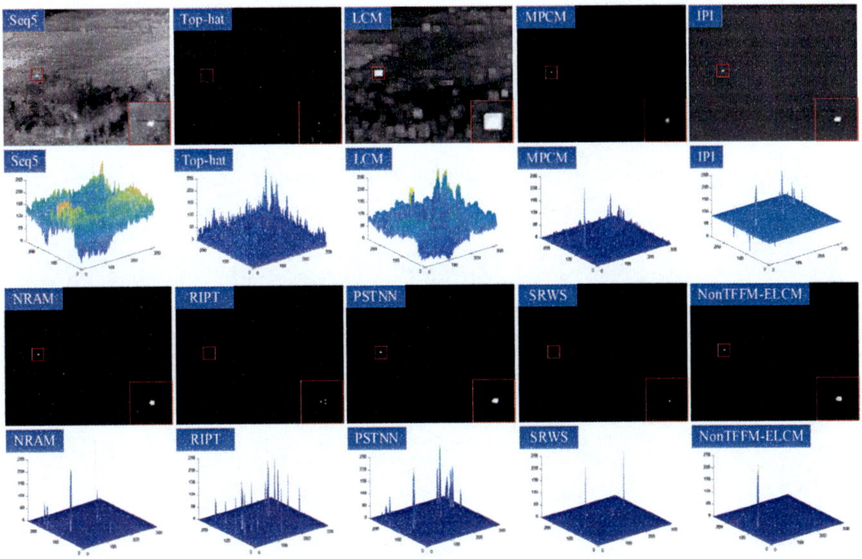

Fig.3. The detection results of some algorithms on five infrared image sequences and the corresponding three-dimensional grayscale distribution map

Table 3. $\overline{\text{SCRG}}$ in nine algorithms under five test sequences

algorithms	Seq1	Seq2	Seq3	Seq4	Seq5
Top-hat	2.50	6.58	2.45	1.67	0.45
LCM	1.83	2.25	1.14	1.14	2.55
MPCM	6.29	97.65	7.71	5.64	15.62
IPI	200.39	53.28	3.42	395.51	28.94
NRAM	3723.23	2042.43	227.96	967.00	1060.07
RIPT	3854.10	467.58	285.36	423.09	11.46
PSTNN	4653.69	2176.60	208.68	1074.40	1512.90
SRWS	2652.03	1735.92	63.37	697.97	3402.61
NonTFFM- ELCM	5751.44	2108.41	1936.73	1361.44	4774.79

3.4 Background Suppression Analysis

Background suppression is equivalent to attenuating background pixels as much as possible to zero within a given area, leaving only target pixels. Based on the infrared image sequence, the background suppression ability of the algorithm is quantitatively analyzed by using the average value of the background suppression factor ($\overline{\text{BSF}}$). Table 4 shows the detection results of all algorithms on five infrared image sequences. Combining the

$\overline{SCRG}$ indicators, it is not difficult to find that NonTFFM-ELCM is particularly outstanding in both target enhancement ability and background suppression ability, and the overall performance is better.

Table 4. $\overline{BSF}$ in nine algorithms under five test sequences

algorithms	Seq1	Seq2	Seq3	Seq4	Seq5
Top-hat	1.89	4.72	1.20	3.14	2.27
LCM	0.82	1.76	0.57	1.14	1.37
MPCM	3.92	13.90	10.02	10.33	13.74
IPI	19.18	17.58	3.90	15.77	21.48
NRAM	42.44	29.73	7.09	34.80	24.70
RIPT	49.00	22.29	5.83	14.24	12.28
PSTNN	118.46	10.00	4.77	6.35	15.99
SRWS	547.80	76.21	42.40	469.77	66.89
NonTFFM- ELCM	650.44	123.96	107.24	532.62	154.19

3.5 Detection Performance Analysis

The detection performance of the infrared small target detection algorithm mainly includes one aspect: detection ability. The detection ability can be analyzed by the ROC curve in Fig. 4.

The calculation method of the ROC curve is to segment the target image by setting different thresholds and then calculate the detection rate and false alarm rate of the segmented image respectively. Since the number of target pixels in the infrared image is much lower than the number of background pixels, the false alarm rate is relatively low. The ROC curves of all algorithms in this chapter in five infrared image sequences are shown in Fig. 4. NonTFFM-ELCM achieves the highest detection rate and lowest false alarm rate on all sequences, with the best overall detection performance. Experiments show that it is difficult for the comparison algorithm to achieve a high detection rate under all sequences, but NonTFFM-ELCM still shows a relatively stable detection ability.

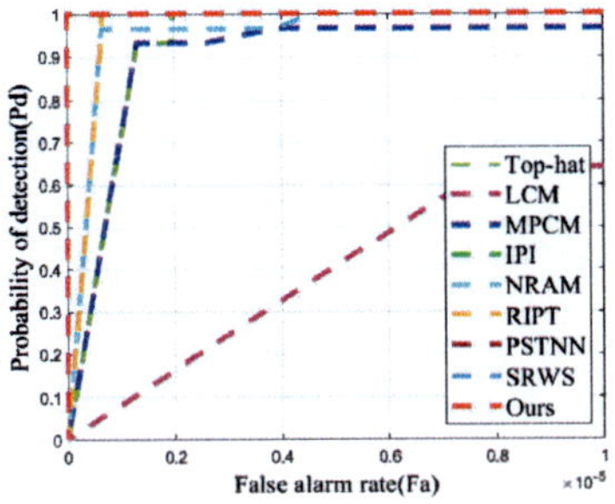

(a)

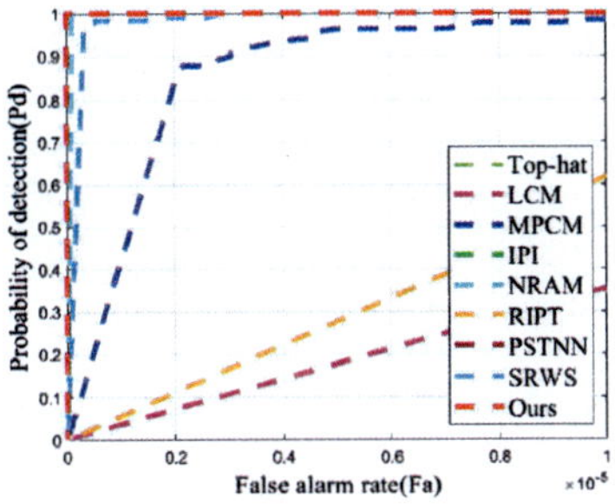

(b)

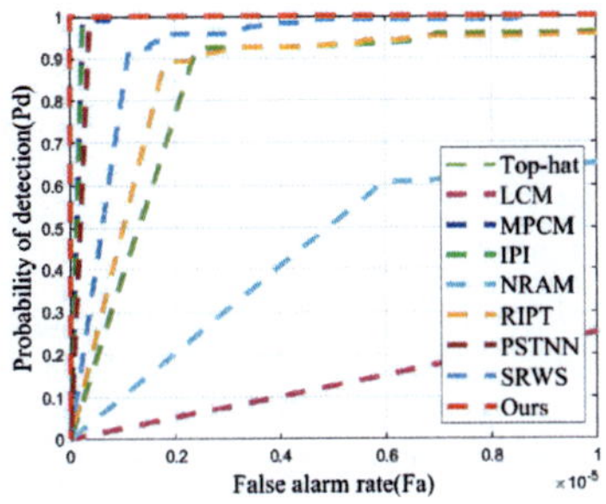

(c)

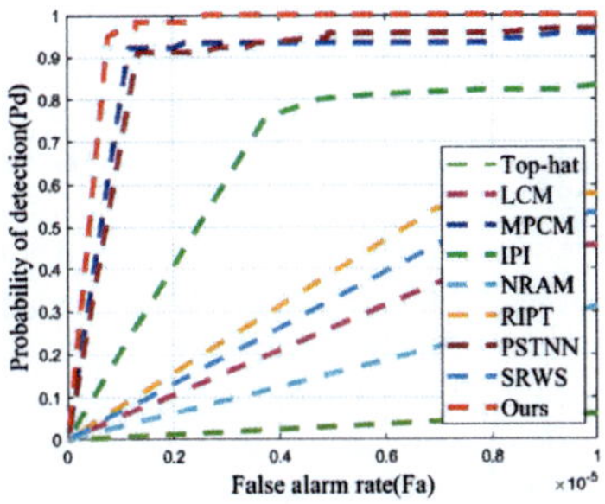

(d)

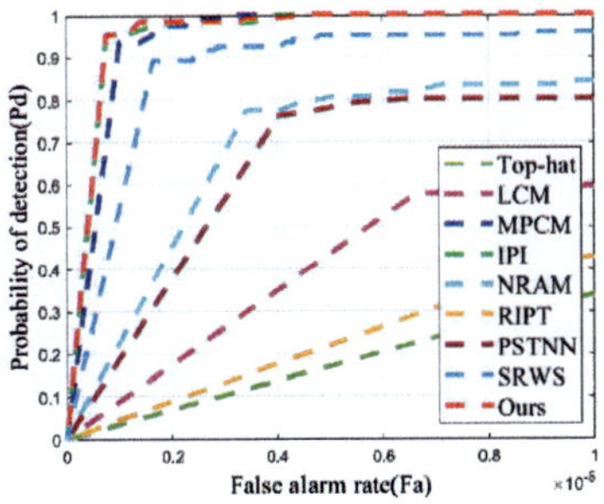

(e)

Fig.4. ROC curves in nine algorithms on five infrared image sequences: (a)-(e) Sequences 1–5 respectively

4 Conclusion

To deal with the scene generalization and resistance to sparse interference. We proposed the NonTFFM model that combined N-tubal rank and Log function. The experimental results show that our method has a good advantage in enhancing the target suppressing the background and detection ability. Its performance in evaluation indicators is mostly better than other baseline algorithms, which also shows the effectiveness of the algorithm improvement.

Funding. This research is supported by the CAST Innovation Fund.

References

1. DsaZhang, W., Cong, M., Wang, L.: Algorithms for optical weak small targets detection and tracking. In: International Conference on Neural Networks and Signal Processing, 2003. Proceedings of the 2003, vol. 1, 643–647. IEEE (2003)
2. Lee, A., Kim, S.Y., Hong, S.J., et al.: Phenotypic analysis of fruit crops water stress using infrared thermal imaging. J. Biosyst. Eng. **44**(2), 87–94 (2019)
3. Zhao, M., Li, W., Li, L., et al.: Single-frame infrared small-target detection: a survey. IEEE Geosci. Remote Sens. Magaz. **10**, 87–119 (2022)
4. Dai, Y., Wu, Y.: Reweighted infrared patch-tensor model with both nonlocal and local priors for single-frame small target detection. IEEE J. Select. Top. Appl. Earth Obs. Remote Sens. **10**(8), 3752–3767 (2017)
5. De Lathauwer, L., De Moor, B., Vandewalle, J.: A multilinear singular value decomposition. SIAM J. Matrix Anal. Appl. **21**(4), 1253–1278 (2000)
6. Zhang, H., Liu, X., Fan, H., et al.: Tensor full feature measure and its nonconvex relaxation applications to tensor recovery. arXiv preprint arXiv:2109.12257 (2021)
7. Gu, S., Xie, Q., Meng, D., et al.: Weighted nuclear norm minimization and its applications to low level vision. Int. J. Comput. Vision **121**(2), 183–208 (2017)
8. Zheng, Y.B., Huang, T.Z., Zhao, X.L., et al.: Mixed noise removal in hyperspectral image via low-fibered-rank regularization. IEEE Trans. Geosci. Remote Sens. **58**(1), 734–749 (2019)
9. Xue, J., Zhao, Y., Liao, W., et al.: Nonconvex tensor rank minimization and its applications to tensor recovery. Inf. Sci. **503**, 109–128 (2019)
10. Boyd, S., Parikh, N., Chu, E., et al.: Distributed optimization and statistical learning via the alternating direction method of multipliers. Found. Trends® Mach. Learn. **3**(1), 1–122 (2011)
11. Tom, V.T., Peli, T., Leung, M., et al.: Morphology-based algorithm for point target detection in infrared backgrounds. In: Signal and Data Processing of Small Targets 1993. SPIE, vol. 1954, pp. 2–11 (1993)
12. Chen, C.L.P., Li, H., Wei, Y., et al.: A local contrast method for small infrared target detection. IEEE Trans. Geosci. Remote Sens. **52**(1), 574–581 (2013)
13. Wei, Y., You, X., Li, H.: Multiscale patch-based contrast measure for small infrared target detection. Pattern Recogn. **58**, 216–226 (2016)
14. Gao, C., Meng, D., Yang, Y., et al.: Infrared patch-image model for small target detection in a single image. IEEE Trans. Image Process. **22**(12), 4996–5009 (2013)
15. Zhang, L., Peng, L., Zhang, T., et al.: Infrared small target detection via non-convex rank approximation minimization joint l 2, 1 norm. Remote Sens. **10**(11), 1821 (2018)
16. Zhang, L., Peng, Z.: Infrared small target detection based on partial sum of the tensor nuclear norm. Remote Sens. **11**(4), 382 (2019)

17. Zhang, T., Peng, Z., Wu, H., et al.: Infrared small target detection via self-regularized weighted sparse model. Neurocomputing **420**, 124–148 (2021)
18. Dai, Y., Wu, Y., Zhou, F., et al.: Asymmetric contextual modulation for infrared small target detection. In: Proceedings of the IEEE/CVF Winter Conference on Applications of Computer Vision, pp. 950–959 (2021)

Design and Research of Light Space Polarization Camera Based on Hyperlens Integration Technology

Gong Li[1,2](✉), Guolong Liu[1], Jing Wang[1], and Yongli Hu[1]

[1] Beijing Institute of Mechanics and Electricity, Beijing, China
13651058529@139.com
[2] Space Engineering University, Beijing, China

Abstract. Polarization optical imaging technology is a new earth observation method based on the detection of the polarization information of the reflected light of the target. Compared with the traditional intensity imaging technology, it is of great significance in improving the ability of target detection. Hyperlens is a planar optical device whose metasurface has great application potential in the field of imaging. It has promising development prospects in the light weight and mass manufacturing of devices. Combining the advantages of the two, from the perspective of small, simple and easy integration design of optical multi-dimensional information acquisition device, the integration of traditional hyperlens array and image sensor has monolithic integration and hybrid integration. According to the practical application requirements, the implementation of structural integration is studied, and system scheme of high spatial resolution polarization imaging is proposed, and the engineering of the system is realized. In order to verify the rationality of the scheme, an engineering prototype of the polarization camera is designed and developed. The location imaging experiment of the camera is carried out, and the results are consistent with the expectation. The development process of the engineering prototype is summarized, and the problems that need to be considered are put forward, which lays a foundation for the subsequent application of polarization camera in the model.

Keywords: Hyperlens integration · Spatially polarized camera · High resolution · engineering

1 Introduction

A metalens is a planar optical device composed of metasurfaces, which can precisely adjust multiple parameters such as the phase, amplitude, and polarization direction of light. It has the advantages of flexible design, small mass, planarization, and compact structure [1, 2]. It can be manufactured by existing microelectronic processing technology, and can realize the integration and planarization of traditional large and complex optical systems, and has great application prospects. In addition to the traditional imaging information of the target, the polarization imaging detection technology can also

H. P. Urbach and H. Jiang (Eds.): ISSOIA 2022, SPPHY 295, pp. 612–626, 2023.
https://doi.org/10.1007/978-981-99-4098-1_51

obtain the polarization information of the target. It has a strong ability to identify stealth, camouflage, and false targets. Has a strong detection ability. At present, polarization detection equipment mainly uses mechanical adjustment or transformation of optical path to extract target polarization information. There are deficiencies in this aspect, which restrict the large-scale application of polarization imaging devices [3].

According to the characteristics of the metalens, which can flexibly adjust the polarization direction and phase of light, it realizes polarization recognition and focusing imaging at the same time [4]. Applying the metalens to the space optical camera, the ability to obtain the polarization spectrum information and visible light spectrum information of the ground objects on orbit, and verify the feasibility of the polarization imaging technology of the metalens for ground monitoring, will be of great significance to the development of remote sensing technology and the application of space-based technology in China. The demand for developing aerospace technology innovation is of great significance.

2 Feasibility and Process Design of Detector and Metalens Integration

2.1 Research on Integration of Metalens and CCD

According to the characteristics of metalens that can flexibly adjust the polarization direction and phase of light, a variety of different polarization-sensitive structures are used to form macrostructures [5], and a silicon-based full-Stokes polarization imaging metalens (wavelength of 800 nm) is designed and fabricated. Identify and focus imaging functions. By coupling with the detector, one pixel corresponds to one metalens unit, and the polarization detection function is realized by the detector by adopting a pixel division method. The schematic diagram is as follows (Fig. 1):

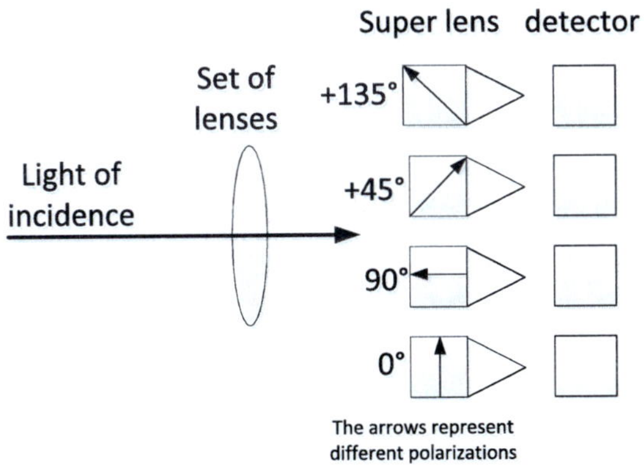

Fig. 1. Schematic diagram of the metalens' working principle

In the four directions of 0°, 90°, +45°, and +135°, the four superstructure units form a complete group in the line array direction, and the resolution in the line array direction is 4 times that of panchromatic. The resolution in the push-broom direction is the same as panchromatic.

2.2 Integration Solution Selection

Generally, there are three methods for the integration of metalenses and image sensors, namely monolithic integration, hybrid integration, and structural integration. Monolithic integration, for visible light CCD, is to make a flat layer on the front of the sensor chip, and then make a super lens on the flat layer, as shown in Fig. 2; hybrid integration is to make a super lens on silicon or quartz glass, and then make a super lens The lens and the sensor chip are packaged together by UV-curable optical glue. The main processes include photolithography, reactive ion etching, and chip/super-lens alignment bonding process, as shown in Fig. 3; the structural integration is through the structure on the photosensitive surface of the sensor. Attach the metalens, and then package the metalens and the sensor chip together, as shown in Fig. 4.

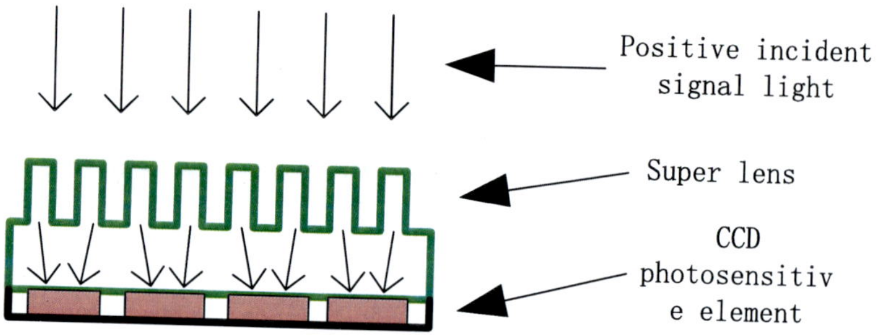

Fig. 2. Monolithic integration of visible light CCD and metalens

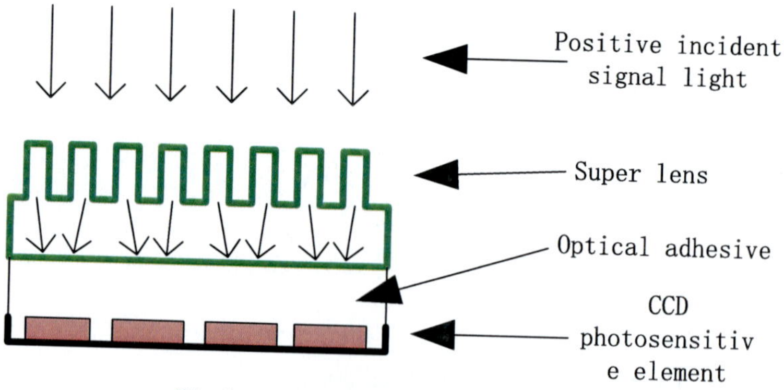

Fig. 3. Hybrid integration structure diagram

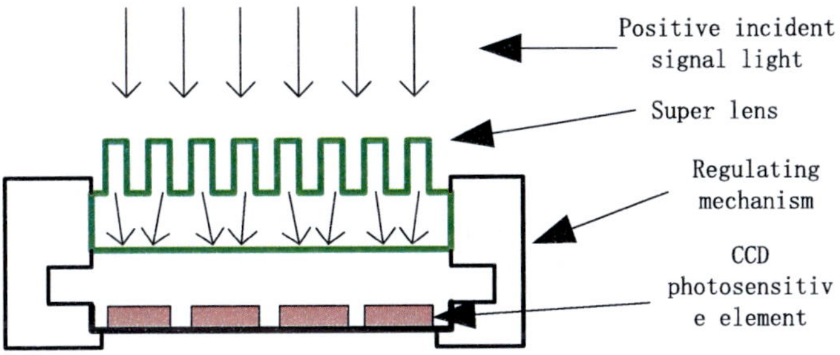

Fig. 4. Structural integration structure diagram

Table 1. Comparison of three different integration methods

Heading level	advantages	disadvantages
Monolithic integration	Integrated compact structure, simplified process, high reliability	There is damage to the chip during the manufacturing process. For visible light CCDs, since the lens material is a polymer, the chemical stability is low
Hybrid integration	The lens material is silicon or quartz, which has good chemical stability; because the metalens manufacturing process is relatively independent, the integration process has almost no damage to the CCD	Since the integrated structure is encapsulated by optical glue, the chemical stability and transmittance of the optical glue have a great influence on the performance of the device, and bubbles are easily generated inside the optical glue during the bonding process
Structural integration	Since the fabrication process of the metalens is relatively independent, the integration process hardly damages the CCD and does not affect the light transmittance	The structural reliability is slightly poor; the integration process is complicated; the external structure design and the metal lens design require high precision

The advantages and disadvantages of monolithic integration, hybrid integration and structural integration are listed in Table 1.

Combining the advantages and disadvantages of monolithic integration, hybrid integration and structural integration, as well as various tests for the three integrations in the early stage, in order to avoid damage to the chip and ensure the minimum impact of light transmittance on device performance, structural integration is chosen to complete the task.

2.3 Detector and Metalens Integrated Process Design

Difficulties in integrating detectors and metalens:

1) Solve the problem of short focal depth of metalens and poor detector flatness;
2) Solve the problem of precise alignment between the metalens unit and the detector pixel plane in the X and Y directions;
3) Solve the Z-direction precise focusing problem between the super lens unit and the detector pixel.

The research on the focal plane integration process of metalens polarization imaging involves four aspects: detector flatness measurement, structural adjustment design, alignment integration process, sealing and leak detection, and functional performance testing. The surface integration process meets the requirements for the use of metalens and various optical, mechanical, force, and thermal requirements in the background of aerospace remote sensing.

Detector Integrated Structure Design

a) Coupling requirements analysis

From the theory of metalens design, we know that the focal depth range of metalens is very small, which is on the order of microns. Therefore, the coupling accuracy of the entire detector photosensitive surface in the Z direction of the focus direction is required, and the accuracy of the focus Z direction reaches 1 μm order of magnitude.

To achieve this goal, the following issues need to be addressed:

Question 1: Through actual measurement, the flatness of the photosensitive surface of the detector is not ideal, showing an undulating wave surface, and the peak-to-peak value is much larger than the focal depth range of the metalens, which is on the order of 10 μm;

Question 2: The undulating wave surface within the range of the photosensitive surface of the detector is affected by the structural installation, the free state and the fixed and tightened state are different, and the magnitude of the change is about 5 μm;

Question 3: The detector is a positive-illuminated detector, which has a metal layer on the surface and has a certain thickness, and the penetration depth of 800 nm light in silicon material is 10 μm, how to accurately focus within the range of 1 μm.

Requirements for integrated geometric accuracy: the integrated accuracy requirements in the X and Y directions are less than 1 μm, and the integrated accuracy requirements in the Z direction are less than 1 μm.

b) Coupling Design

Through actual measurement, the flatness of the detector is not ideal, showing an undulating wavefront, and the peak-to-peak value is much larger than the focal depth range of the metalens. In order to ensure that the metalens can focus accurately, the mounting surface of the metalens is repaired to match the wavefront of the detector. Undulations are used to adjust and ensure the Z-direction distance and equidistant relationship between the photosensitive surface and the metalens.

From this adopt the scheme:

The surface undulation in the whole detector range is tested by infrared interferometer;

Solve the problem of the impact of screw tightening on the detector plane during fixation. The method is: when testing the flatness of the full detector range, the detector in the tightened state is not used in the free state detector, and the tightening torque is quantitatively controlled to solve the problem 2;

After testing the surface undulations within the entire range of the detector, it is planned to repair the structural surface of the metalens installation to form an undulating wavefront to match the wavefront undulations of the detector, which is used to adjust and ensure the Z-direction distance between the photosensitive surface and the metalens and equidistant relationship, solve problem 1;

In order to realize the adjustment during integration, the metalens support structure needs to design an adjustment link, and provide precise adjustment capability and alignment adjustment margin in the X/Y plane.

Design a detector box structure and a metalens support structure that meet the requirements of light, mechanics, force, and heat for aerospace applications. This structure not only meets the needs of ground assembly and detection, but also has good adaptability to the space environment: on-orbit The thermal deformation that affects the performance of the optical system does not occur within the temperature fluctuation control requirements, and the position and dimension accuracy of the device should be maintained; in the orbital irradiation environment, ensure that the device is not damaged by the total radiation dose; in the vacuum environment at the altitude of the satellite orbit Under these conditions, cold welding of moving parts will not occur. Considering that this structure is applied near the detector and belongs to the stray light sensitive area, it is also necessary to consider the requirements of stray light elimination during design to avoid the introduction of stray light, which will affect the imaging.

The detector box structure is used to fix the detector on the circuit board or other structures. The box structure and the detector shell are glued size.

The metalens support structure is used for surface shape preservation and integrated adjustment of the metalens. The structure has a certain rigidity and strength, and at the same time, there is a structural reference, and the surface on which the metalens is installed can be precisely repaired (Fig. 5).

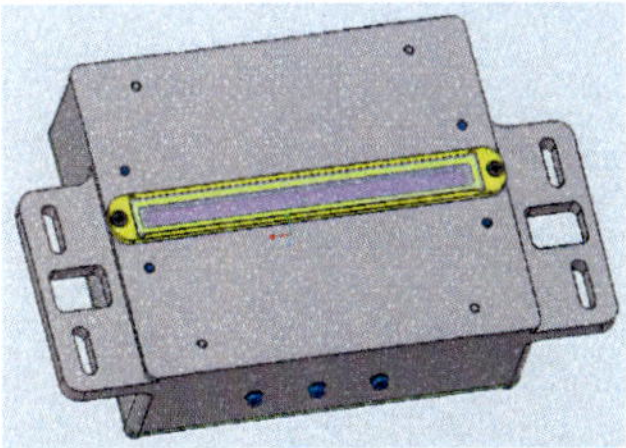

Fig. 5. Detector box structure

Detector and Metalens Integration

The polarization integration of the detector and the metalens needs to be carried out on the splicer using a multi-dimensional precision adjustment stage.

The high-precision splicing instrument is mainly composed of the following parts: optical system, high-precision guide rail (X, Y, Z), microscopic camera system, image acquisition and processing system, motion control system, etc. The light source of the splicing instrument illuminates the CCD pixel of the splicing component through the semi-reflective semi-lens and the optical system. The reflected light of the CCD pixel is received by the microscopic camera system through the optical system of the long line array splicing instrument, and then passes through the image acquisition system to make the CCD image The pixel is imaged on the display system, and the focal plane position of the pixel can be calculated by the image processing system. The working spectral range of the stitching instrument is 0.4 μm–0.9 μm, covering the spectral range of the metalens device at 800 nm ± 40 nm. The total magnification of the imaging system of the splicing instrument is 3600 times, the indication accuracy of the X and Y directions is 0.1 μm, and the indication accuracy of the Z direction is 1 μm, which can meet the integration accuracy requirements (Figs. 6 and 7).

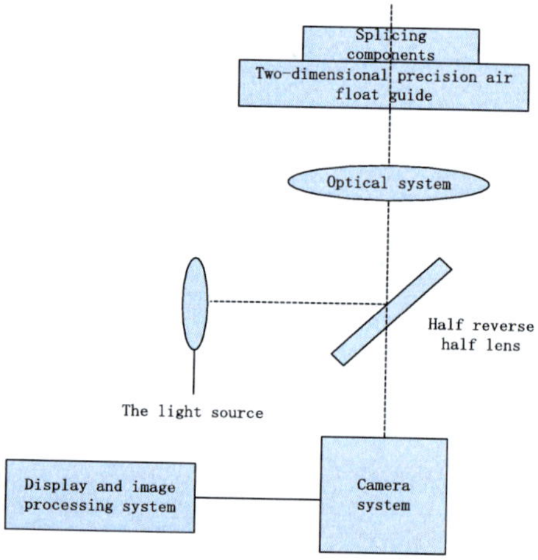

Fig. 6. Schematic diagram of the optical path of the splicing instrument

Fig. 7. High-precision splicing instrument in kind

The multi-dimensional precision adjustment table is an outsourced part, which can realize six-dimensional high-precision adjustment of the metalens bracket. The adjustment accuracy is better than 1 μm.

The adjustment structure adopts a designed and modified multi-dimensional precision adjustment table. First, fix the detector box structure, adjust the support arm to clamp the super lens bracket, manually rotate the differential head, and align the super lens and the cross reticle on the photosensitive surface under the focal plane stitching instrument. or contour lines.

Detector Metalens Polarized Z-Direction Integration
Solve problem 3 through polarized photoelectric imaging, design the adjustment gasket, and adjust the thickness of the gasket by grinding to achieve the best focusing effect. The grinding accuracy requires 1 μm.

The imaging method is used to test the effect of integrated focusing, and the functional performance is assessed by the pixel response of the metalens to linearly polarized light. The polarization modulations of 4 adjacent clusters (corresponding to 4 pixels) are 0°, 45°, 90° and 135° respectively. Rotate the polarizer of the test light source to form linearly polarized light at different angles, and use the linearly polarized light to calculate the DN value. Response test. Acquire the integrated detector response image via snapshot. Change the polarization angle of linear polarized light, and test and evaluate it many times. The entire image surface will also be imaged in a regular "line" style.

3 System Solution for Spatially High-Resolution Polarization Imaging

There are two main considerations in the development of the polarization imaging camera: one is the optical design of the polarization imaging camera, and the other is the realization of the polarization information detection technology using the split focal plane.

3.1 Optical Design Method

The camera of this project is a system with medium focal length, medium field of view, and large F number. Both the coaxial reflective system and the off-axis reflective system can meet the requirements of the transmission index. Although the off-axis reflective system is unobstructed, it is heavy and bulky, and it is not suitable for miniaturized cameras.

For the coaxial three-mirror system, if it only starts from meeting the requirements of the transmission, the conventional coaxial three-mirror partial field of view system can fully meet the requirements of the transmission, and has a compact structure, good system symmetry, and relatively light weight. However, since the previous main optical system needs to be used in conjunction with the metalens optical elements, the incidence angles of the different fields of view of the front main optical system on the metalens change with the farther the central field of view and the larger the angle of incidence. If the incident angle is too large, the polarization effect achieved by the metalens will decrease.

Through simulation analysis, in order to achieve a good polarization effect, when the metalens is used, the incident angle cannot be greater than 15° design constraints.

The conventional coaxial three-mirror anti-polarization field of view usually has a large incident angle on the image plane, and the incident angle of the designed three-mirror coaxial system will reach 20° at the edge, as shown in the figure below, which exceeds the use requirements of the super lens and cannot meet the requirements of this project. Therefore, design improvements need to be made on this basis (Fig. 8).

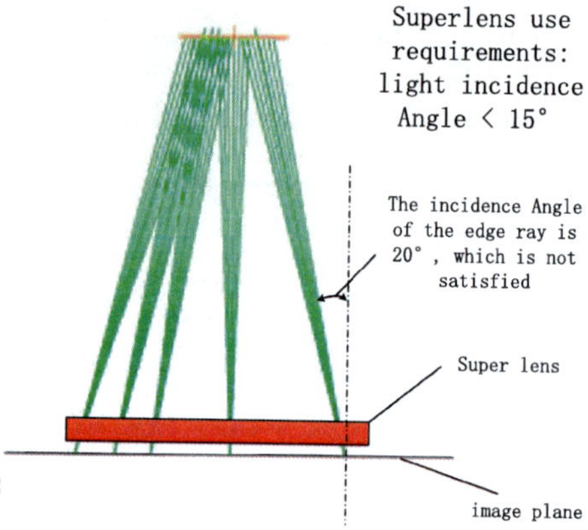

Fig. 8. Schematic diagram of light rays incident on the metalens

3.2 Split Focal Plane Scheme

The focal plane polarization imaging form has the characteristics of small size, good compatibility, simultaneous imaging, and being applicable to dynamic scenes. The focal plane polarization information detection technology is to integrate the polarization element into the focal plane. Each pixel on the detector corresponds to a metalens unit. Every four pixels form a group, and four polarization directions are marked respectively. The polarization parameter can be calculated by using the light intensity information of the four pixels. The super-lens unit is to carve silicon nano-columns on the material with SiO2 as the substrate. Figure 9 is the result of the new circular polarization carvement, which can realize full-polarization imaging.

The imaging method of the split-focus plane polarization imaging system is shown in Fig. 10. In addition to the selection of the micro-polarization array substrate and the coupling process, it is necessary to focus on controlling the alignment error and distance between the micro-polarization array and the focal plane.

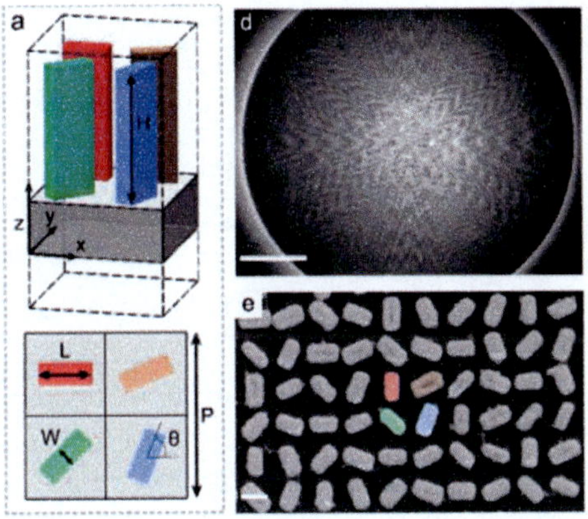

Fig. 9. Schematic diagram of silicon-based full-Stokes polarizing metalens

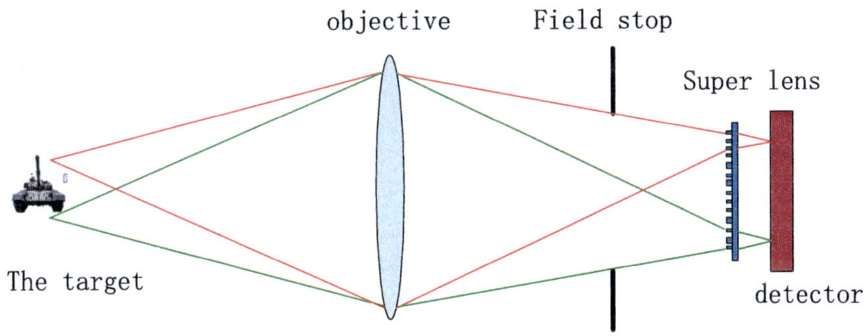

Fig. 10. Schematic diagram of metalens polarization imaging

4 Development of Polarization Camera Engineering Prototype

The polarized optical camera adopts the integrated design of the camera light-mechanical body and the circuit, and the whole camera is composed of two devices. The camera optical system is a three-mirror off-axis system. The camera body is composed of a primary mirror, a secondary mirror, a triple mirror, a quad mirror and a main frame. The field of view is covered, and the two detectors are designed to work independently and independently, and can image at the same time. The working principle of the camera is shown in the Fig. 11.

The camera subsystem is composed of camera body and camera integrated electronics. The camera body is composed of optical lens, panchromatic focal plane assembly, polarization focal plane, thermal control assembly, and hood assembly. The optical lens is composed of the main frame assembly, the primary mirror assembly, the secondary mirror assembly, the third mirror assembly, the folding mirror assembly, the beam splitter assembly, etc. The composition of the camera is shown in the Fig. 12. Camera thermal control is controlled by integrated electronics.

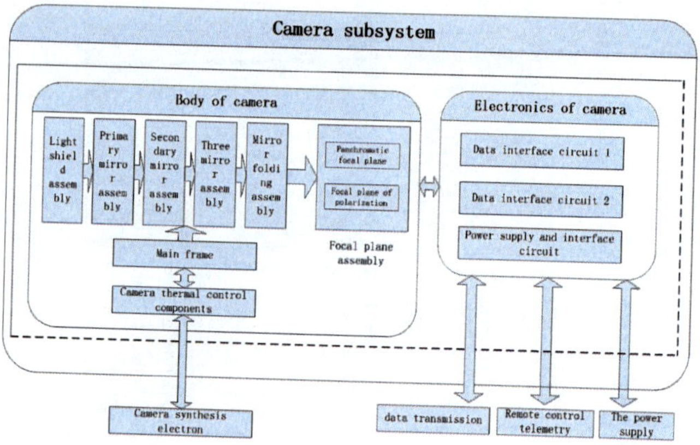

Fig. 11. Schematic block diagram of space metalens polarization camera

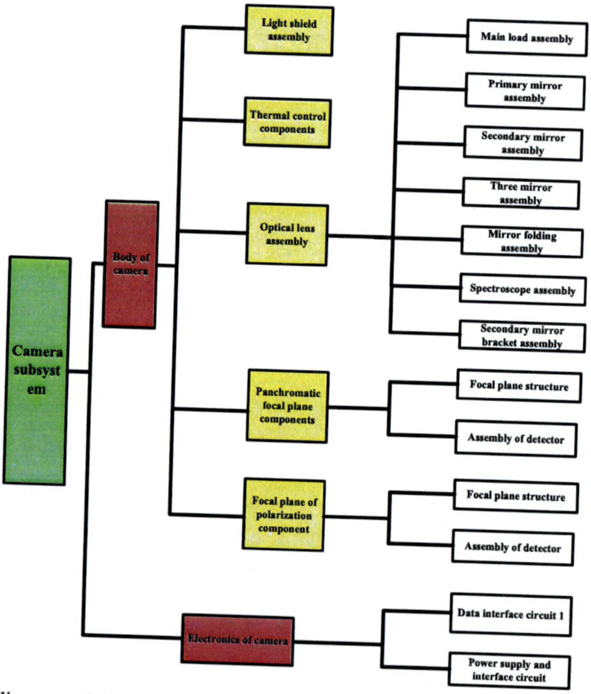

Fig. 12. Block diagram of the component components of the space metalens polarization camera

4.1 Optical System Design

In order to control the incident angle of light on the image plane, the off-axis design based on the coaxial reflective system is carried out, and the asymmetrical aberration caused by the reduction of the incident angle of the image plane is corrected through the eccentric tilt of the secondary mirror and the third mirror. Through the reasonable allocation of optical power, the incident angle of marginal light is controlled, and finally the requirement that the incident angle of light in the full field of view is $<15°$ is met. The specific configuration is shown in the Fig. 13.

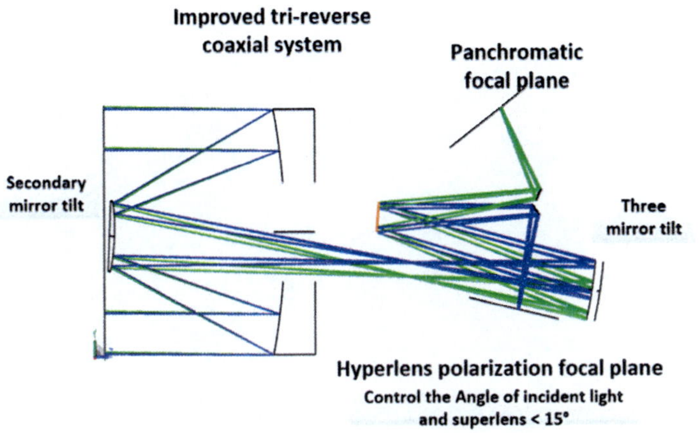

Fig. 13. Improved coaxial three-anti scheme

The optical system of the space optics camera adopts the form of a three-mirror coaxial system, and the optical path diagram is shown in Fig. 14. The color channel and polarization channel realize ground imaging through the form of field of view light splitting. The primary mirror is an ellipsoid, the secondary mirror is a hyperboloid, and the third mirror is a ten-degree aspheric surface. The other reflectors are flat mirrors, which play the role of refracting the light path.

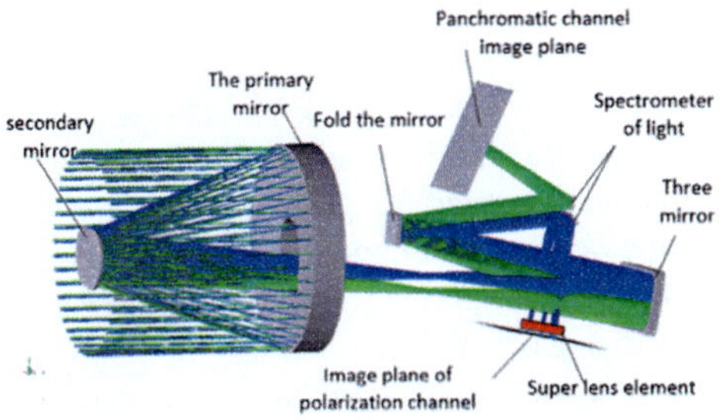

Fig. 14. Optical path diagram of space hyperlens polarized optical camera

4.2 Camera Body Design

The camera subsystem is composed of camera body and camera integrated electronics. The camera body is composed of optical lens assembly, panchromatic focal plane assembly, polarization focal plane assembly, thermal control assembly, hood assembly and main body direct parts. The optical lens is composed of the main bearing assembly, the primary mirror assembly, the secondary mirror assembly, the third mirror assembly, the folding mirror assembly, the beam splitter assembly, the secondary mirror support assembly, the stray light suppression assembly and the lens direct parts. Camera thermal control is controlled by integrated electronics shown in Fig. 15.

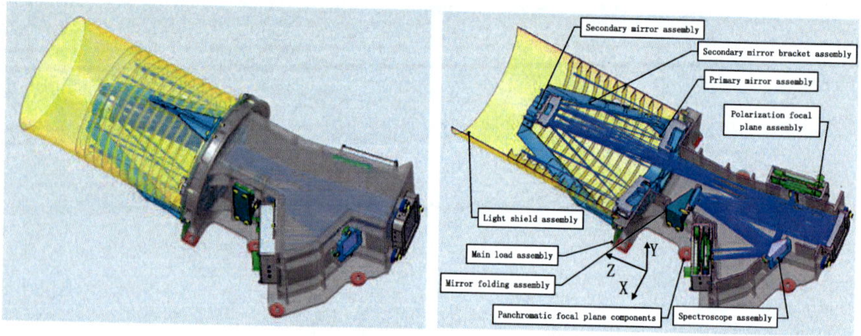

Fig. 15. Main structure diagram of metalens polarization camera

4.3 Electronics Design

The camera electronics are designed to meet the requirements and the main principles are rapid development. The system is composed of a camera integrated electronic stand-alone and a focal plane detector. The camera detector adopts the industrial TDICMOS linear array detector with two-way integration function as the imaging core. The focal plane detector is installed on the camera body, which greatly improves the system integration. The focal plane detector is used to realize the imaging function, and the integrated electronic stand-alone completes the bus communication and data transmission interface.

Among them, the camera integrated electronic stand-alone needs to complete the power management function, direct remote control telemetry control, CAN bus communication function, time management function, detector control function, image data receiving function, auxiliary data generation, arrangement and transmission function.

The camera electronics are composed of 2 sets of industrial detectors. The camera integrates the electronics to receive the original image output by the detector, packs the corresponding auxiliary data, and outputs it to the satellite data transmission subsystem through the hot backup CoaXPress standard interface, and configures the main/standby hot backup 2-way data transmission output interface.

4.4 Field Imaging Experiment

The engineering prototype of the hyperlens polarized optical camera is undergoing imaging experiments. In the field imaging experiment, the image of the haze scene is selected, and the image is simultaneously imaged through the field of view of the camera's panchromatic and polarization channels, and the image is output. The image post-stage of the polarization channel is based on the physical model of atmospheric degradation, and the image is restored by inverting the degradation process of the real scene light to achieve the dehazing effect. Figure 16 shows the output image of the polarization imaging camera panchromatic channel under foggy weather conditions. Figure 17 shows the image of the haze scene output by the polarization channel. From the comparison between the defogged effect image and the original image, it can be seen that the scene detail resolution ability in the defogged effect image is stronger, the image contrast is higher, and the imaging quality is better.

Fig. 16. The image of the haze scene output by the panchromatic channel of the polarization imaging camera

Fig. 17. The image of the haze scene output by the polarization channel of the polarization imaging camera

5 Conclusion

In this paper, the characteristics of metalens and polarization imaging are analyzed. On this basis, from the perspective of small, simple and easy-to-integrate design of optical multi-dimensional information acquisition devices, the advantages of both are combined, and the realization of structural integration is studied, proposed a system scheme for high spatial resolution polarization imaging, designed and developed a polarization camera engineering prototype, verified the rationality of the scheme, the imaging experiment results were consistent with expectations, and verified the polarization ability of the metalens Apply the foundation. However, the extinction ratio index of the polarization imaging detector needs to be further improved, and the polarization characteristics of the target and the background and the polarization transmission characteristics need to be further deepened, etc., which are the direction of future efforts.

References

1. Shrestha, S., Overvig, A.C., Lu, M., et al.: Broadband achromatic dielectric metalenses. Light Sci. Appl. **7**, 85 (2018)
2. Tseng, M.L., Hsiao, H.H., Chu, C.H., et al.: Metalenses: advances and applications. Adv. Opt. Mater. **6**(18), 1800554 (2018)
3. Luo, H., Zhang, J., Gai, X., Liu, Y.: Development status and prospects of polarization imaging technology. Infrared Laser Eng. **51**(1), 20210987-1–20210987-10 (2022)
4. Zhao, F., Chen, X.N., Wang, D.C.: Metalens design and simulation for simultaneous focusing of orthogonal circularly polarized light. **40**(10), 1024001-1–1024001-6 (2020)
5. Lishujun, Jianghuilin, Zhujingpin, et al.: Design and simulation of a hyperlens for simultaneous focusing of orthogonal circularly polarized light. Chin. Opt. **6**(6), 803–809 (2013). (in Chinese)

Current Status and Progress of Cooperative Target Technology for Spatial Pose Measurement

Jing Liu[1](✉), Loulou Deng[1], Xingtan Liu[1], and Liqing Tao[2]

[1] Beijing Institute of Control Engineering, Beijing 100190, China
bigmirror@sina.com

[2] Hangzhou Center of China Academy of Space Technology, Hangzhou 310024, China

Abstract. Because of the advantages of the high accuracy and reliability, the spatial pose measurement technology based on the cooperative target has been widely applied in the various fields such as aerospace, industrial intelligent manufacturing and medical intelligent diagnosis and treatment. As the key reference for the measurement, the cooperative target is an important part of the spatial pose measurement system, which design is one of the important factors affecting the measurement accuracy. In this paper, the technology status of the cooperative target for the spatial pose measurement is presented, focusing on the analysis and review of the structure design and the characteristics of the common components such as coating, cone corner prism and glass beads, etc. For more clearly description, many typical applications of the cooperative target for spatial pose measure are listed in this paper, and the prospect of the field is described.

Keywords: Spatial Pose Measurement · Cooperative Target · Cone Corner Prism · Glass Beads

1 Introduction

With the development of photoelectric detection and information processing technology, spatial pose measurement technology is widely used in various fields such as aerospace, industrial intelligent manufacturing, medical intelligent diagnosis and treatment. The cooperative target-based spatial pose measurement technology has the advantages of high accuracy, high reliability and high timeliness, and can be applied to space rendezvous and docking, workpiece grasping, large equipment pose measurement and other complex scenarios [1].

How to design the cooperative target is the primary problem to be solved in the study of spatial pose measurement system. The general cooperative target is a pattern marker of known size, shape and structure [2]. The cooperative target is the key reference for spatial pose measurement, and its selection is determined by the measurement conditions, installation environment and other factors, while its characteristics determine the spatial pose measurement algorithm design [3]. The reasonable selection and design of the cooperative target is one of the important factors affecting the accuracy index of spatial pose measurement.

H. P. Urbach and H. Jiang (Eds.): ISSOIA 2022, SPPHY 295, pp. 627–636, 2023.
https://doi.org/10.1007/978-981-99-4098-1_52

2 Key Technologies of Cooperative Target

2.1 Design Principles

The design of the shape and pattern of the cooperative target should contain the exact information that can resolve the measured object's positional parameters, meet the requirements of high measurement accuracy in complex operating environments. To ensure the robustness of the measurement system, the design should have enough information and the ability to resist light changes, pollution and occlusion interference [4]. The design of cooperative targets should accord with the following design principles:

(1) The cooperative target should contain enough feature information to ensure that the detection system can acquire the information required for spatial pose measurement.
(2) The feature pattern and color of the cooperative target should be significantly different from the surrounding environment in order to facilitate the detection system to identify the target.
(3) The cooperative target should be simple and easy to follow for visual recognition processing in order to achieve the goal of real-time.
(4) The structure of the cooperative target should be easy to install on the platform of the object to be measured, with good stability and redundant design of information that is resistant to pollution and obscuring.
(5) It should be easy to produce and the mechanical accuracy can meet the engineering requirements [5].

2.2 Design Methods

As shown in Fig. 1, The design of the cooperative target includes the design of shape, size, color and material.

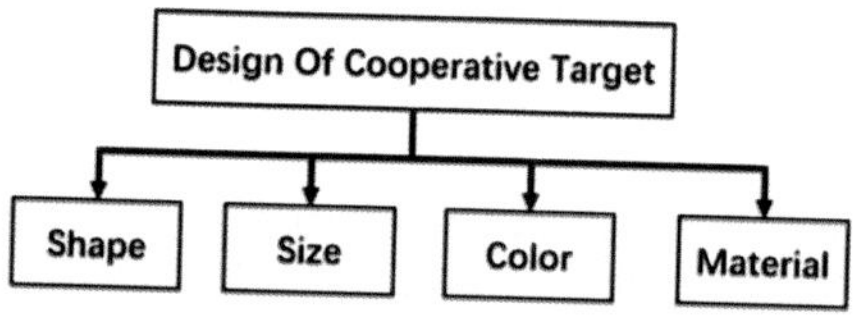

Fig. 1. The Design contents of cooperative target

Firstly, according to the principle that the cooperative target should be simple and have enough feature information, the type of feature information of the cooperative target, such as point features, line features or curve features, is determined in combination with the application, the positional measurement algorithm, and the directivity. Cooperative targets are usually designed using shapes such as "H-shaped", "T-shaped", "circular" and "square". The shape is related to the type, number and distribution of features. Secondly, the size of the cooperative target is designed according to the mutual constraints of the camera parameters and the cooperative target, including the location and size of the features. Then the color and material of the cooperative target are designed according to the application environment. Finally, the validation for the rationality of the cooperative target design is carried out through simulation and experimentation.

The measurement accuracy is directly affected by the size and the mechanical accuracy of the cooperative target. The geometric size and the mechanical accuracy of the cooperative target should be designed according to the actual external measurement environment and the systemic accuracy requirements [6].

3 Status and Progress of Cooperation Target Technology

3.1 Types of Cooperation Target

Cooperative targets for spacial pose measurement consist of active targets and passive targets according to the light-emitting mechanism. Generally active targets consist of a number of actively emitting target lights, and each target light is driven by a constant-current power supply to emit light autonomously. Passive targets cannot emit light autonomously and do not need to be driven by a power supply, but need to be irradiated by an external light source.

Cooperative targets can be divided into flat targets and three-dimensional targets according to the geometric configuration. Flat targets are the flat pattern targets or several targets distributed in a plane. Three-dimensional targets are several targets distributed in a three-dimensional space, with a certain height difference between targets. The three-dimensional target configuration can improve the axial accuracy of spatial post measurement and is mostly used in monocular vision measurement systems.

3.2 Features and Applications

(1) **Active Targets**

Active targets can be used in the absence of light and have a wider range of applications than passive targets. However, active targets require a power supply to drive them, and their size is generally larger than that of passive targets. Therefore, there is a certain amount of power consumption in active targets. Active targets generally use light-emitting diodes, lasers as active target light source [2], the design mainly includes luminous intensity, field of view, uniformity, power and other indicators.

Light source can be used for the design of the target background. In the microsatellite rendezvous and docking pose measurement applications of the large line-of-sight and no light conditions, Zhang Liu et al. [2] designed the active target, and chose the light-emitting diodes as the background light source for the entire target. The whole target can be seen as two different targets at long range and close range, and the background light sources of the target are several diodes to ensure that the light intensity is evenly distributed over the whole target surface. The remote target and the close range target are nested and designed independently, using different light sources and can be controlled independently. The overall layout of the target is shown in Fig. 2. The background color of the target is white and the color of the feature points is black, taking into account the fact that the white feature points scatter more strongly than the black feature points, which affects the image edges

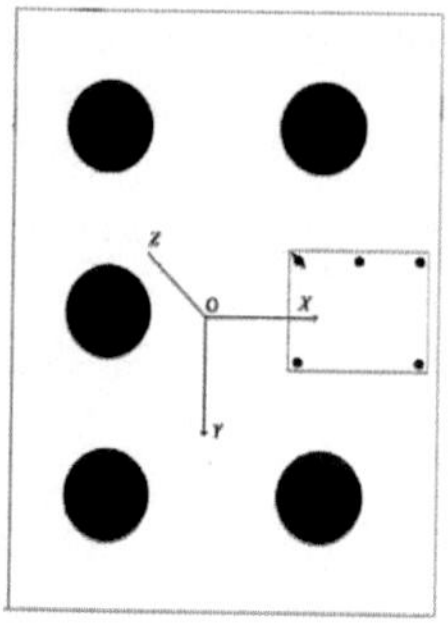

Fig. 2. Diagram of cooperative target for microsatellite rendezvous and docking pose measurement

of the feature points during imaging and is detrimental to the subsequent image processing work.

The light source can also be used for the design of feature points. In the Shenzhou 8 spacecraft and Tiangong 1 space station rendezvous and docking with the cooperative target using target marker lamp autonomous light. The camera takes direct imaging measurements of them. Its advantages are active target marker image spot uniform, stable, high measurement accuracy, high reliability of target identification [7]. From the perspective of easy image point extraction, easy identification and mass centre calculation, the shape of the marker lamp was chosen as circular, and the engineering realisability of the target lamp was also considered in the design process.

(2) **Passive Targets**

Passive targets are generally composed of materials such as coatings, cone corner prisms and glass beads, and the choice of material is determined by the application scenario, light source characteristics, etc.

1) Coatings

Coated pattern targets are generally used in well-light scenes, which have diffuse reflective properties and can be used in visible or infrared light detection systems, depending on the coating material.

Target patterns with a high-contrast composition of light and dark colors (mainly black and white) and with internal and external contours are more widely used. Corner feature colors should stand out relative to the background color for easy identification, usually with black or white coatings or colored coatings filling the target shape, which are easy to segment relative to the background color. Zhang Bin et al. [8] designed a room-type black-and-white target pattern, as shown in Fig. 3, and obtained pose data of the cargo platform module by detecting the pattern target to make a smooth landing of airborne equipment dropped from a long distance.

Coated pattern targets are widely used in UAV autonomous landing applications. In the design of UAV visual localisation and navigation in a strong denial environment, Gao Hailing et al. [5] designed a "concave" target, as shown in Fig. 4. The "concave" interior of the cooperative target, except for the circular

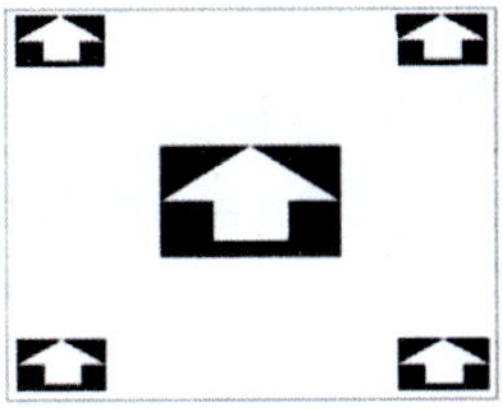

Fig. 3. Diagram of cooperative targets for airdrop systems

area, is designed with black infrared paint, which makes the cooperative target significantly different from the surrounding environment and can assist the infrared sensor in the visual navigation system to achieve autonomous landing of the UAV in poor lighting conditions. The design of the circle inside the cooperative target as blue enables the extraction of the target based on color threshold processing method only when extracting the circle inside the cooperative target at close range.

Fig. 4. Diagram of cooperative target for UAV visual localisation and navigation

Ni Lixue et al. [9] used an infrared thermal imaging camera as an unmanned airborne vision system and designed a long cross-shaped cooperative target to guide the UAV landing using the long side in line with the landing direction. A black infrared coating with an emissivity of $\varepsilon > 0.92$ and corresponding to a wavelength band of 8–14 μm prepared from a high emissivity powder was chosen to be sprayed on the surface of the cooperative target to improve the thermal radiation efficiency (Fig. 5).

The above coated targets are filled with color paint or infrared paint inside the pattern, making the target significantly different from the environmental background and facilitating target extraction. At the same time the design of the coated target needs to take into account the need for water resistance and subsequent cleaning and maintenance. The coated target design is widely used in the field of autonomous landing position measurement for UAVs, as well as in civil applications such as industrial mating and positioning systems and medical navigation and positioning systems.

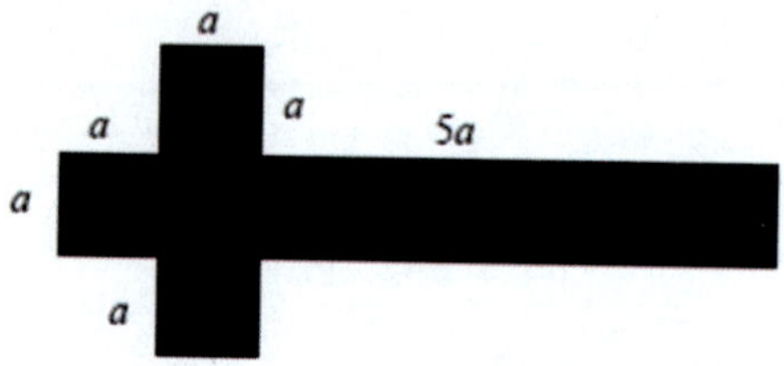

Fig. 5. Diagram of a drone infra-red locating a cooperative target

2) Cone corner prism

The cone corner prism has high reflectivity and good directionality of the reflected beam in the original direction, and is widely used in laser ranging, lidar and other laser detection technologies.

As shown in Fig. 6, the cone corner prism is a tetrahedron consisting of three mutually perpendicular reflection planes (commonly known as the small plane) and one incident plane (commonly known as the large plane). The cone corner prism allows the incident light to return in its original direction, with the incident and exit rays symmetrical according to the centre of the cone point [10]. The cone corner prism cannot be applied directly as an optical element. Usually the cone corner prism is assembled with a mechanical structure to form an angular reflector.

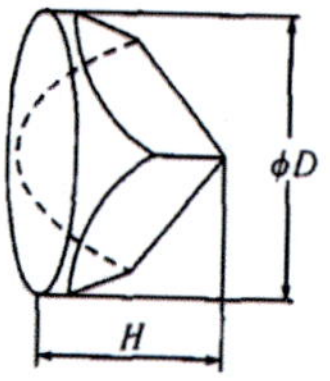

Fig. 6. Schematic diagram of an cone corner prism

The measurement accuracy of angular cone prisms decreases as the angle of incidence increases, with a sudden change at the maximum permissible angle of incidence [11]. This is why tracking measurements with angular reflectors take into account the effect that changes in the angle of incidence have on the measurement accuracy. The relationship between the effective reflective area of an angular reflector and the incident angle of the beam is usually analysed to obtain the relationship between the measurement accuracy of an angular reflector and the angle of incidence.

With the characteristics of directional reflection of the angular reflector, the laser angular reflector can be used on satellites and launch vehicles to achieve accurate ranging and effective tracking and orbiting of high-speed moving satellites and other targets, detected by the laser detection equipment [12].

In the rendezvous and docking of Shenzhou XI and Tiangong II space stations, angular cone prisms are used as passive targets. The cooperative targets include far-field, near-field and super-near-field targets, of which the near-field and super-near-field targets are mounted to the target vehicle after forming a three-dimensional target assembly through a common mount. The far-field target marker also contains six angular reflectors, which are mounted directly to the target vehicle according to certain geometric position relationships [13].

Foreign rendezvous and docking missions also mostly use angular reflector arrays, such as the US MSFC in the Orbital Express autonomous rendezvous and docking mission using the Advanced Video Guidance Sensor, whose cooperation target is shown in Fig. 7a. The French company Sodern is using a Video Guidance Sensor (Videometer) for the Orbital Express autonomous rendezvous and docking mission, with the collaboration objectives shown in Fig. 7b.

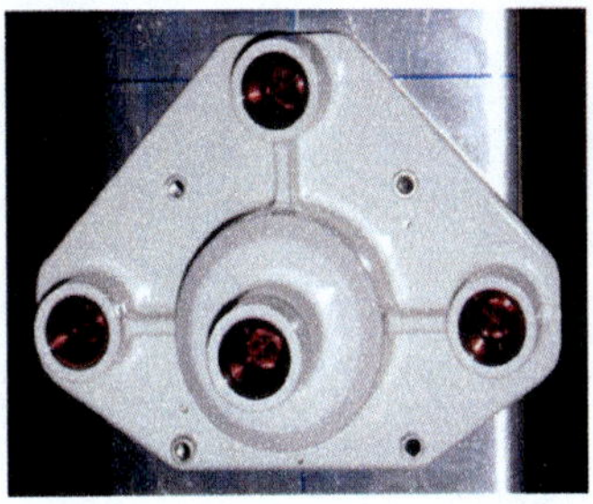

a Diagram of AVGS' cooperative target

b Diagram of VGS' cooperative target

Fig. 7. Diagram of cooperative target for the rendezvous and docking

Corner reflectors on laser ranging satellites are usually distributed in spherical, hemispherical or planar shapes.Spherical and hemispherical shapes mostly used on low-orbiting satellites and planar shapes mostly used on high-orbiting satellites. For example, the Russian low-orbit high-precision laser ranging satellite WESTPAC-1 carries 60 angular reflectors in spherical form, the low-orbit scientific satellite ADEOS Earth observation satellite carries 9 angular reflectors in hemispherical form, and the US high-orbit GPS satellite carries 32 angular reflectors in planar form, as detailed in Fig. 8 [14].

3) Glass beads

Glass beads are glass spheres with diameters ranging from a few microns to a few hundred microns and have primary reflective properties. Glass beads with high reflectivity need to have a refractive index of not less than 1.93, an ellipsoidal index of not more than 5%, no precipitation in the glass beads, and no impurities mixed with them [15]. At present, the domestic bead forming process is realized by using negative pressure system and gasification chamber, which can meet the above index requirements.

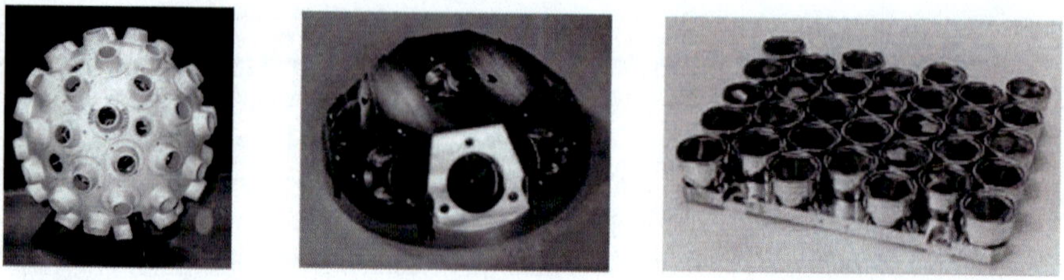

a Diagram of cooperative target for WESTPAC-1

b Diagram of cooperative target for ADEOS

c Diagram of cooperative target for High-orbiting GPS satellites

Fig. 8. Diagram of cooperative target for Laser Ranging Satellite

The bead glass original directional reflective screen is composed of glass beads coated on pressure-sensitive adhesive film, and then pasted onto the flat plate. The structure is illustrated in Fig. 10. If the subsequent surface is coated with aluminium or silver film, the reflective brightness can be greatly improved. The effective angle of the glass beads can reach ±70°, reflectivity of them is about 20%, the main reflection area reflection dispersion angle of about 1°, the maximum dispersion angle can reach 3–5°. Compared with cone corner prisms, glass beads have the advantages of low cost, light weight, easy installation and no change in the shape of the equipment, but the disadvantage is that their reflected beam dispersion angle is larger, and the reflectivity is lower [10] (Fig. 9).

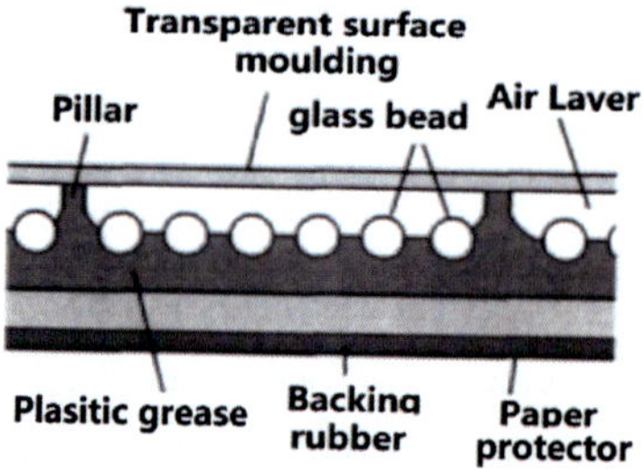

Fig. 9. The structure of glass beads original directional reflection screen

The glass bead primary reflective structure is available in flat and spherical shapes, as shown in Fig. 10. The planar shape has circular and coded patterns, and can be fixed by pasting, which does not take up dimensional space and is convenient and simple. The spherical shape requires mechanical fixation and takes up dimensional space, but the measurement field of view is large and the measurement accuracy is relatively high compared to the planar shape. The glass bead material is widely used in industrial, medical and aerospace fields.

Fig. 10. Diagram of the primary reflection structure of glass beads

3.3 Summary

The above describes the current state of cooperative target technology from the perspective of the types of cooperative targets, and clarifies that the selection and design of cooperative targets need to be determined according to indicators such as illumination mechanism, application scenario, accuracy requirements and cost. Coatings, cone corner prisms and glass beads are the main application materials of passive cooperative targets at present, among which cone corner prisms and glass beads have in-situ reflection characteristics and are widely used.

4 Conclusion

Cooperative targets are an important part of the spacial pose measurement system. This paper analyses the principles, methods and the technology state of the cooperative target, which is a reference value for the design of cooperative targets. Different applications determine the shape, size, color and material of the cooperative target. Miniaturisation, light weight and low cost are common requirements for all applications. In the aerospace sector, miniaturisation and lightweighting have been the focus of cooperative target technology due to resource constraints. In special applications, such as medical and steel mills, where the cooperative target is consumables, low cost design is the focus of collaborative target design. This shows that miniaturisation, lightweighting and low-cost design will become increasingly important in cooperative target technology, while meeting the technical level of simplicity, high machining accuracy and high stability.

References

1. Wu, Y.: Study of cooperative target-based UAV positioning with monocular vision problem. Nanjing University of Science and Technology, Nanjing (2017)
2. Zhang, L.: Cooperative design method in microsatellite rendezvous and docking pose measurement. Chin. J. Liq. Cryst. Displays **33**(3), 415–424 (2022)
3. Lu, Y.: Research on mono-vision pose measurement for space cooperative target. University of Chinese Academy of Sciences, Beijing (2018)
4. Chen, Y.: Novel landmark design of autonomous landing in unmanned aerial vehicle. J. Univ. Electron. Sci. Technol. China **45**(6), 934–938 (2016)
5. Gao, H.: Vision aided localization and navigation for unmanned aerial vehicle in GPS-denied environment. Xidian University, Xian (2018)

6. Qu, Y.: Graphics design of cooperative targets on monocular vision high precision measurement. Acta Opt. Sin. **40**(13), 1315001–1–1315001–8 (2020)
7. Zhao, C.: A vision guidance sensor for Shenzhou-8 spacecraft autonomous. Aerosp. Control Appl. **37**(6), 6–14 (2011)
8. Zhang, B.: A vision-based method for solving heading attitude of airdrop cargo platform. Tianjin University of Technology, Tianjin (2022)
9. Ni, L.: Research and design of a sort of cooperative target on the ground. Opto-Electron. Eng. **34**(7), 8–13 (2007)
10. Zhang, D.: A laser cooperative target design with adjustable divergence angle of reflected beam. Technol. Innov. Appl. **13**, 99–101 (2018)
11. Liu, W.: Effect of incident laser beam angle varying on cube corner retro-reflector measurement accuracty. Opt. Precis. Eng. **17**(2), 286–291 (2009)
12. Zhou, H.: Design of satellite laser retro-reflector. Opto-Electron. Eng. **32**(11), 25–29 (2005)
13. Liu, Q.: New generation camera-type rendezvous and docking sensor. Aerosp. Control Appl. **44**(2), 56–61 (2018)
14. Wan, Q.: Current status and progress of laser cooperative targets technology for SLR. Adv. Lasers Optoelectron. **42**(5), 20–29 (2005)
15. Li, H.: Study on divergence characteristics of glass microbeads from retroreflective screen. Transducer Microsyst. Technol. **38**(3), 14–16 (2019)

An Efficient Hyperspectral Band Selection Method Based on TLBO and Improved Quantum Particle Swarm Optimization

Yang Li(✉), ShuLi Dong, Ze Li, LiNan Yuan, BingJie Liu, and RunXin Zhang

Beijing Institute of Space Mechanics and Electricity, 104 Youyi Road, Beijing, People's Republic of China
2531987249@qq.com

Abstract. Hyperspectral sensor can obtain images with hundreds of very narrow contiguous bands, which provides high spectral resolution and abundant ground information. This makes the research of hyperspectral data is of great significance in the field of target recognition and classification. However, the problems in terms of the high-dimensional non-linearity, the high inter-band correlation, information redundancy and large volume of the hyperspectral image data have posed a huge challenge to hyperspectral data processing. This might lead to the "Hughes" phenomenon which must be taken into account. Therefore, how to select the band subset with rich information and better classification ability from numerous bands has become an inevitable problem to be solved. This is the prerequisite for precisely target recognition and classification. In this paper, an effective band selection method based on teaching-learning-based optimization and quantum particle swarm optimization (TLBO_QPSO) is proposed. In this method, teaching-learning based optimization (TLBO) is incorporated into quantum particle swarm optimization (QPSO) to quickly select the optimal band set, which can improve the global optimization performance of the algorithm. By this way, a satisfied balance between diversity and convergence can be maintained. Different from the traditional particle swarm optimization algorithm, the idea of teaching and learning is exploited to particle state evolution equation, which can increase the diversity of particles as well as the convergence performance of QPSO. Specifically, the main content of the achievement of the method is as follows. First of all, the correlation coefficient matrices of the entire range of bands are used to divide the spectrum space of hyperspectral data, obtaining band subsets with smaller relevance. After that, TLBO-improved quantum particle swarm method is performed by means of Optimum Index Factor (OIF) and Jeffries Matusita (JM) distance weighted summation to iteratively search optimal bands in those subsets. This allows comprising an optimal band set. Finally, the algorithm mentioned in this paper together with classical algorithms such as SPSO, SAPSO, QPSO and DE-QPSO are applied to simulative experiments respectively using Indian Pines and Pavia datasets. The selected band sets by these algorithms are then employed to classification experiments using SVM classifier. The experimental results show that the algorithm proposed in this research can select the optimal band set which meets the requirement of ground classification.

H. P. Urbach and H. Jiang (Eds.): ISSOIA 2022, SPPHY 295, pp. 637–653, 2023.
https://doi.org/10.1007/978-981-99-4098-1_53

Keywords: hyperspectral image · band selection · teaching-learning-based optimization · quantum particle swarm optimization

1 Introduction

Hyperspectral remote sensing uses spectral imaging technology to image the spatial spectral features of features, and the image elements are able to obtain continuous coverage of hundreds of bands, which can obtain a nanometer-level spectral resolution. The spectral spatial resolution of hyperspectral data is extremely high. Therefore, how to more fully extract the feature information contained in the data and how to more effectively analyze and process the hyperspectral data has important research significance in the field of target identification and classification technology. The high spectral dimension of hyperspectral images, high band correlation, and large data volume pose great difficulties for storing, transmitting, and processing data [1]. Therefore, it is necessary to perform dimensionality reduction on hyperspectral data. The dimensionality reduction is divided into feature extraction and feature selection. Band selection reduces the computational effort by directly selecting part of the band, and can effectively avoid the "Hughes" phenomenon [2]. Feature extraction is to obtain a new feature space by linear or nonlinear transformation of the original feature space, and recombine and optimize each spectral band [3]. The feature extraction can cause the correlation between bands to change as well, which destroys the physical properties of the spectrum and leads to the loss of spectral information. In order to retain the original spectral information, this paper performs dimensionality reduction by means of feature selection. Article [4] selected the best band of vegetation index by adaptive band selection method. In recent years, with the development of intelligent optimization algorithms, many search algorithms have also been used for dimensionality reduction, with representative ones such as genetic algorithm, ant colony algorithm, particle swarm algorithm, etc. [5, 6]. Particle swarm algorithm is commonly used for hyperspectral data dimensionality reduction as an optimal intelligent optimization algorithm. Particle Swarm Optimization (PSO) was first proposed by Eberhart and Kennedy in 1995 and originated from Complex Adaptive Systems (CAS) [7–9]. The PSO algorithm is simpler in operation, and also features low operational complexity and fewer parameters. However, PSO still has many shortcomings, such as theoretically not globally optimal, easy to fall into local optimality, etc. PSO is not a globally convergent algorithm. Inspired by quantum mechanism and PSO trajectory theory research [10], Sun et al. found that the quantum system is extremely stochastic due to the superposition of states, so the way of particle motion in quantum space is also highly stochastic and more similar to the human population intelligent behavior. Therefore, Sun et al. introduced the quantum mechanism into the particle swarm algorithm and proposed a probability-based particle swarm algorithm combined with the 8-potential well model [11], and thereafter introduced the mean optimal position and proposed a quantum-behaved particle swarm optimization (QPSO) algorithm based on this [12]. Quantum particle swarm inherits the characteristics of simple operation, low operation complexity and few parameters of particle swarm algorithm, and introduces quantum uncertainty on this basis, which is a good solution to the drawback that particle swarm is easy to fall into local optimum.

The particles in the quantum particle swarm algorithm (QPSO) are in the bound state, and it can obtain the global optimal solution by searching the whole solution space, but the algorithm will reduce the diversity of the population in the later stage of the search. Therefore, in order to improve the performance of the quantum particle swarm algorithm, this paper introduces the TLBO algorithm into the quantum particle swarm algorithm for the last band subset search of hyperspectral images, and proves the effectiveness of the method in this paper through simulation experiments.

2 Quantum Particle Swarm Algorithm and Improvement Algorithm

2.1 Quantum Particle Swarm Algorithm

In particle swarm algorithms, the states of particles are described by position and velocity vectors, while in quantum states, the states of particles are described by wave functions. Compared with the PSO algorithm, QPSO has no velocity vector and is much easier to operate. By taking advantage of the quantum uncertainty, the QPSO algorithm can effectively avoid the "premature" phenomenon of the particle swarm and guarantee the theoretical optimality of all of them to converge to its local attractor. Clerc and Kennedy analyzed the characteristics of the trajectory of the particles in the PSO algorithm [10] and pointed out that each particle in the PSO algorithm is able to converge to its local attractor if given a suitable value. Each particle in the PSO algorithm is able to converge to its local attractor. The local attractor is defined as follows:

$$p_{ij} = \varphi * pbest_{ij} + (1 - \varphi) * gbest_j, (i = 1, 2, ..., M, j = 1, 2, ..., \mathrm{D}) \quad (1)$$

where, $\varphi = c_1 r_1/(c_1 r_1 + c_2 r_2)$, $\varphi \in U(0, 1)$, M is the particle population size, D is the particle dimension, c1 c2 is the learning factor, and usually set $c_1 = c_2$, then φ is a random number between (0, 1). $pbest_i$, $gbest$ are the individual optimal position of the particle and the global optimal position of the population, respectively.

From the above equation, the local attraction point p_i is located in the hyper-rectangle formed by the vertices of the individual optimal position $pbest_i$ and the global optimal position $gbest$ of the particle, and follows the $pbest_i$ and $gbest$, which means that the particle will move in the direction of the attraction point and the final particle velocity is zero. From the mechanical point of view, each particle must be confined to the range centered on the local attraction point, and the particles gradually decelerate and approach p_i, and finally the particles in PSO are under bound states. The QPSO algorithm describes the change of particle states from the quantum space, in which the particles are not moving in the form of trajectories with strong uncertainty. The particles can be in searching for the optimal solution in the whole space of feasible solutions. Sun et al. proposed that the QPSO in which the attraction potential is established at the local attractor using a potential well δ, so that the particle is in a quantum bound state, and the particle state is described using a wave function.

The particles are updated iteratively in a potential well δ centered on a local attractor, and eventually converge. Assuming that the particle population is in a D-dimensional quantum space, the i_{th} particle in the n_{th} iteration moves in a potential well centered on a local attractor p_i. The probability density function Q and the distribution function F of the particles are:

$$Q(x_{ij}(n+1)) = \frac{1}{L_{ij}(n)} e^{-2|p_{ij}(n) - x_{ij}(n+1)|/L_{ij}(n)} \quad (2)$$

$$F(x_{ij}(n+1)) = e^{-2|p_{ij}(n) - x_{ij}(n+1)|/L_{ij}(n)} \quad (3)$$

where $L_{ij}(n)$ is the standard deviation of the double exponential distribution, describes the characteristic length of the wave function, and characterizes the particle search range size. Finally, applying the Monte Carlo method, the particle update formula after the nth+1 iteration is obtained as

$$x_{ij}(n+1) = p_{ij}(n) \pm \frac{L_{ij}(n)}{2} \ln(1/u), \; u = rand(0, 1) \quad (4)$$

where u is the random number uniformly distributed on the (0, 1) interval.

For calculation $L_{ij}(n)$, the global average optimal position $Mbest(n)$ is introduced to characterize the average centroid of all particles' historical optimal positions, which is calculated as follows:

$$Mbest(n) = \frac{1}{M}\left(\sum_{i=1}^{M} pbest_{i1}(n), \sum_{i=1}^{M} pbest_{i2}(n), \ldots\ldots, \sum_{i=1}^{M} pbest_{iD}(n)\right) \quad (5)$$

where M is the population size, $L_{ij}(n)$ it is calculated as follows.

$$L_{ij}(n) = 2\alpha_n |Mbest_j(n) - x_{ij}(n)| \quad (6)$$

Combining (5) and (6), the particle state update equation in QPSO can be obtained as

$$x_{ij}(n+1) = p_{ij}(n) - \alpha |Mbest_j(n) - X_{ij}(n)| * \ln(1/u), \; u = rand(0, 1) \quad (7)$$

where the parameter α is the shrinkage expansion factor, which is used to balance the local and global searchability of particles in the iterative process and control the convergence of the population. After the particle state update, the particle individual optimal position $pbest_i$ and the global optimal position $gbest$ need to be updated in the following way:

$$pbest_i(n+1) = \begin{cases} x_i(n+1) & \text{if } f(x_i(n+1)) < f(pbest_i(n)) \\ pbest_i(n) & \text{if } f(x_i(n+1)) \geq f(pbest_i(n)) \end{cases} . \quad (8)$$

$$gbest_n = pbest_{g,n}, \; g = \arg\min(f(pbest_{i,n})) \quad (9)$$

The block diagram of QPSO algorithm is shown in Fig. 1.

Let M be the population size, the i_{th} particle $\{X_i\}_{i=0}^{M}$, and the particle dimension is D. The corresponding fitness value of each particle is then calculated, and the individual optimal position $pbest_i$ and the global optimal position $gbest$ are updated. Set the current iteration number as n.

Initial population: random X_i and set $pbest_i = X_i$

While

Do

$n = n+1$;

According to equation (5), the mean optimal position $Mbest$ is calculated

Initialize α

for $i = 1:M$

for $j = 1:D$

$\varphi = rand(0,1)$;

$p_{ij} = \varphi * pbest_{ij} + (1-\varphi) * gbest_j$;

$u = rand(0,1)$;

if $rand(0,1) > 0.5$

$X_{ij} = p_{ij} - \alpha * abs(Mbest_j - X_{ij}) * \ln(1/u)$;

else

$X_{ij} = p_{ij} + \alpha * abs(Mbest_j - X_{ij}) * \ln(1/u)$

end if

end for j

if $f(X_i) < f(pbest_i)$ then $pbest_i = X_i$

if $f(pbest_i) < f(gbest)$ then $gbest = pbest_i$

end for i

end Do

End Until termination criterion is met.

Fig. 1. QPSO algorithm flow chart

2.2 Improved Quantum Particle Swarm Algorithm

TLBO Algorithm

The teaching-learning-based optimization (TLBO) algorithm is a new swarm intelligence optimization algorithm proposed by Rao et al. in 2011 [13]. The TLBO algorithm is mainly divided into initialization, teaching and learning phases, where the teacher teaches the participants in the teaching phase and the participants learn from each other in the learning phase to select the better The TLBO algorithm simulates the teaching and learning process and has the advantages of fewer parameters, simple and easy to understand algorithm, fast convergence speed, high accuracy and strong convergence ability.

Optimization problem: $z = \min_{X \in S} f(X)$, The search space $S = \{X | x_i^L \leq x_i \leq x_i^U, i = 1, 2, \ldots, d\}$ contains NP individuals, and the individual participants X_j are denoted as $X^j = (x_1^j, x_2^j, \ldots, x_d^j)$, where d is the individual vector dimension (the number of decision quantities), x_i^L is the upper bound and x_i^U is the lower bound, $f(X)$ is the fitness function. The TLBO algorithm steps are as follows.

The initialized population is randomly generated in the initial data space S, and the number of iterations is set to G.

$$x_i^j = x_i^L + rand(0, 1) * (x_i^U - x_i^L)$$
$$j = 1, 2, \ldots, NP; i = 1, 2, \ldots, d \tag{10}$$

Calculate the average value of individuals $Xmean = (x_1^m, x_2^m, \ldots, x_d^m)$, calculate the fitness function value of all individuals, and select the individual with the best fitness value as the teacher $X_{teacher}$.

$$X_{teacher} = \arg\min f(X^j)(j = 1, 2, \ldots, NP) \tag{11}$$

Each individual learns to "teach" according to the difference between the average and the teacher, and the teaching process is realized according to Eq. (12)

$$X_{new}^i = X_{old}^i + r_i * (X_{teacher} - TF_i * X_{mean}) \tag{12}$$

where the learning factor $r_i = rand(0, 1)$ and the teaching factor $TF_i = round[1 + rand(0, 1)]$. Each individual compares the adaptation values before and after learning, and updates X^i them.

For individuals $r_i = rand(0, 1)$, learning individuals $X^j (j = 1, 2, \ldots, NP, j \neq i)$ are randomly selected for mutual learning among individuals, and their respective fitness function values are compared as follows

$$X_{new}^i = \begin{cases} X_{old}^i + r_i * (X^j - X^i), if (f(X^i) > f(X^j)) \\ X_{old}^i + r_i * (X^i - X^j), if (f(X^i) < f(X^j)) \end{cases} \tag{13}$$

where, the learning factor $r_i = rand(0, 1)$, $f(X^i)$ and $f(X^j)$ are the fitness function value of individual X^i and X^j. Compare the fitness values X^j, $f(X_{new})$, $f(X_{old})$, and update $f(X_{old})$.

Improved QPSO Algorithm Based on TLBO

With the introduction of quantum particle swarm (QPSO) algorithm, some improvement algorithms of QPSO have appeared one after another. The literature [14, 15] makes improvements from the direction of parameter adjustment, probability distribution function, etc., and Coelho L D et al. proposed the chaotic quantum particle swarm algorithm [16]. In the process of iterative update of QPSO, the particles will gradually aggregate and converge to the optimal solution. Especially in the later search process, the particles are aggregated while the population diversity is lost, and the aggregated particles lose the ability to expand the search space, which leads to fall into local optimal solutions.

In this paper, from the perspective of particle diversity, the TLBO algorithm is introduced into QPSO to increase the diversity of particles and make up for the lack of particle diversity in the later search process, and the advantages of both are combined to propose the TLBO-based quantum particle swarm (TLBO_QPSO) algorithm. Specifically, TLBO_QPSO incorporates the "learning" phase of TLBO into the particle state update equation, which can effectively maintain the diversity of individuals and ensure the global search capability of the algorithm in the entire search space.

The combination of Eq. (13) with Eq. (7) of the particle state update equation is improved as follows.

$$x_{ij}(\mathrm{n}+1) = \begin{cases} p_{ij}(\mathrm{n}) - \alpha|Mbest_j(\mathrm{n}) - X_{ij}(\mathrm{n})| * \ln(^1/_u) + \beta(x_{kj}(\mathrm{n}) - x_{ij}(\mathrm{n})), \text{ if } f(x_k(\mathrm{n})) < f(x_i(\mathrm{n})) \\ p_{ij}(\mathrm{n}) - \alpha|Mbest_j(\mathrm{n}) - X_{ij}(\mathrm{n})| * \ln(^1/_u) + \beta(x_{ij}(\mathrm{n}) - x_{kj}(\mathrm{n})), \text{ if } f(x_i(\mathrm{n})) < f(x_k(\mathrm{n})) \end{cases} \tag{14}$$

where, $u = \mathrm{rand}(0, 1)$, $\beta = \mathrm{rand}(0, 1)$, $x_k(\mathrm{n})$ is the randomly selected particles.

3 Hyperspectral Band Selection Method Based on TLBO and Improved Quantum Particle Swarm Optimization

The basis of band selection lies in two important physical characteristics of hyperspectral data: hyperspectral data contain a large amount of redundant information, and the correlation between neighboring bands is generally higher than that between bands that are farther apart. According to the feature of strong correlation between neighboring bands of hyperspectral images, it is necessary to divide all bands into subspaces in advance to obtain several subspaces with weak correlation, then search for the optimal bands in each subspace to form a subset of bands, and finally select the most suitable subset of bands by using TLBO_QPSO.

3.1 Subspace Division

The main basis of subspace division is the correlation coefficient matrix image chunking characteristics between images of different bands in hyperspectral data and the nearest neighbor transferable correlation, and subspaces of different dimensions can be obtained by subspace division, and the band correlation within each subspace is larger. The specific steps [17] are as follows:

1) Convert the two-dimensional band hyperspectral image into a one-dimensional band vector.
2) Calculate the correlation coefficients between the band vectors to obtain the correlation matrix R of the hyperspectral data
3) Obtain the nearest neighbor transferable correlation vector from the correlation matrix $r^{NTR} = [r^{1,2}, r^{2,3}, \ldots, r^{i,i+1}, \ldots, r^{j-2,j-1}, r^{j-1,j}]^T$, and calculate D − 1 minimal values for r^{NTR}.
4) Divide the hyperspectral data into D band subspaces based on the D − 1 minima.

As an example, the correlation coefficient matrix and the correlation coefficient vector are calculated for the Indiana agroforestry data (200-band) collected by AVIRIS, as shown in Fig. 2. Figure 2(a) shows the correlation matrix as a grayscale image, and the bright areas visible in the figure are mainly concentrated in the main diagonal, indicating

(a) correlation matrix

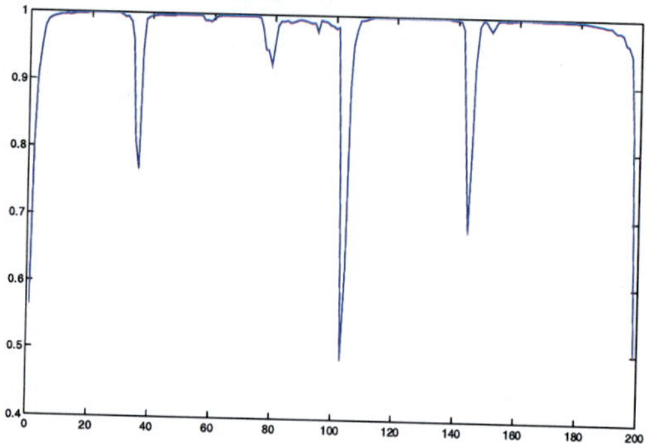

(b) correlation coefficient

Fig. 2. Correlation matrix and correlation coefficient

a stronger correlation between adjacent bands. Figure 2(b) shows the correlation coefficient curve, the horizontal coordinate is the number of bands and the vertical coordinate is the correlation coefficient, which can visualize the range of each band with stronger correlation. According to the band characteristics of hyperspectral data, these bands with larger correlations can be divided in order to accelerate the subsequent processing.

3.2 Evaluation Function of Band Selection

The band selection method is generally based on the information quantity of each band and the correlation degree between bands to select the band with high information quantity and low correlation [18]. Optimum Index Factor (OIF) integrates the two factors of information quantity and correlation between bands, which is suitable as the selection criterion for hyperspectral image band selection. The larger the value of OIF, the more information the band contains, which is calculated as shown in (15).

$$O_{OIF} = \frac{\sum_{i=1}^{N} S_i}{\sum_{i=1}^{N} \sum_{i=1}^{N} |R_{ij}|} \tag{15}$$

where, S_i is the standard deviation of the $\mathrm{i_{th}}$ band, N is the number of bands, and R_{ij} represents the correlation coefficient value of the $\mathrm{i_{th}}$ band and the $\mathrm{j_{th}}$ band.

Meanwhile, to make the subsequent classification better, the criterion measuring the class separability is added as a selection criterion together with OIF. The main purpose of the inter-category separability is to calculate the classification distance between different categories, and to find those bands that make the classification distance the largest, so that the categories of features can be distinguished more easily. The JM (Jeffries Matusita) distance is based on the B distance and takes into account the first-order and second-order statistics of the data, and the second-order statistics are very important for the improvement of classification accuracy in high-dimensional data space. In this paper, the JM distance is chosen as the criterion of interclass separability, and the expressions of B distance and J-M distance are shown in (16) and (17):

The B distance equation is

$$B_{ij} = \frac{1}{8}(\mu_i - \mu_j)^T \left(\frac{C_i + C_j}{2}\right)^{-1} (\mu_i - \mu_j) + \frac{1}{2} \ln \frac{\left|\frac{1}{2}(C_i + C_j)\right|}{\sqrt{|C_i|}\sqrt{|C_j|}} \tag{16}$$

where, μ_i and μ_j are the mean vector of the spectra of class i and j features, and C_i, C_j represents the covariance matrix of class i and j features in different combinations of bands. B_{ij} represents the B distance between features of class i and features of class j at different combinations of bands. The larger the value, the closer the obtained solution is to the optimal solution.

The J-M distance equation is

$$JM_{ij} = [2 \times (1 - e^{-B_{ij}})]^{\frac{1}{2}} \tag{17}$$

$$JM = \sum\nolimits_{i=1}^{C-1} \sum\nolimits_{j=i+1}^{C} JM_{ij} \tag{18}$$

where S is the feature type of hyperspectral data and JM_{ij} represents the JM distance between features of class i and features of class j in different combinations of wavebands. Considering that the hyperspectral flutter data contains many kinds of features, the fitness function should consider the separability between each class of features and other features, using the sum of interclass distances JM between different features.

In order to obtain the interclass separability and informative waveform at the same time, the combination of both interclass separability distance JM and optimal index (OIF) is selected as the fitness function in this paper. Its weighting formula is derived

$$fit = \frac{1}{x \times JM + y \times OIF} \tag{19}$$

where x and y are the weighting coefficients, and the fitness values of different subsets of bands are calculated according to Eq. (19).

3.3 Flow of Hyperspectral Image Band Selection Algorithm Based on TLBO_QPSO

The process steps of TLBO_QPSO based hyperspectral image band selection algorithm are as follows.

Step 1. Hyperspectral image data preprocessing: convert the hyperspectral data cube into a two-dimensional band matrix, divide the hyperspectral data into two parts: the training set is used for training to select bands, and the test set verifies the classification of the subset of bands.

Step 2. Initialize the population: set the maximum number of cycles t_{max}. According to the divided D subspaces, one band is randomly selected in each subspace to generate a subset of bands containing D bands. Assuming M is the population size, the i_{th} particle is $\{X_i\}_{i=0}^{M}$, with particle dimension D. Initialize the individual optimal position $pbest_i = X_i$ of the particle, and calculate the corresponding fitness value $f(X_i)$ of each particle by Eq. (19); select the minimum value from it, and initialize the global optimal position *gbest*.

Step 3. Calculate the mean optimal position *Mbest* according to Eq. (5).

Step 4. Calculate the local attraction point p_{ij} according to Eq. (1).

Step 5. Update the particle positions according to Eq. (14).

Step 6. Calculate the value of the fitness function of the updated particles according to Eq. (19).

Step 7. Update the individual optimal position $pbest_i$ and the global optimal position *gbest* according to the principle of minimum fitness function value based on the fitness function value obtained from step 6;

Step 8. Add 1 to the number of iterations and determine whether the loop termination condition is satisfied, if not, go back to step 3, and if the condition is satisfied, terminate the loop and get the global optimal position *gbest* as the selected wave subset.

4 Band Selection Experiment and Result Analysis

To verify the effectiveness and feasibility of the algorithm in hyperspectral band selection in this paper, band selection experiments are conducted on two real hyperspectral images. According to the selected subset of bands, this experiment uses SVM as the base classifier to classify a variety of features. The SVM classification results are used as the index of feature separability, and combined with the band correlation and the amount of band combination information as the evaluation criteria to verify the merits of the selected subset of bands. Four groups are set up for the comparison experiments of this experiment, namely, the standard particle swarm algorithm SPSO, simulated annealing particle swarm algorithm SAPSO, quantum particle swarm algorithm QPSO, and DE-based quantum particle swarm algorithm DE-QPSO.

4.1 Datasets Introduction

Indiana Agroforestry Hyperspectral Dataset and Pavia College of Engineering Dataset, the feature distribution of both datasets are shown in Fig. 3 and Fig. 4. Separate test samples were selected for the two images, and eight of the categories were chosen to participate in the experiment. Among them, the eight main categories of Indiana AVIRIS hyperspectral images include Corn-no till, Corn-min, Grass/Pasture, Grass/Trees, hay-windrowed, Soybeans-no till, Wheat, Woods, as shown in Table 1. The number of bands involved in the experiment is 200. The eight main categories of Pavia Engineering Institute include Asphalt, Meadows, Trees, Metal Sheets, Soil, Bitumen, Bricks, Shadows, as shown in Table 2. The regions involved in the experiment are 144 × 144, the original data set contains 125 bands, and after removing the noise, the number of bands involved in the experiment is 103.

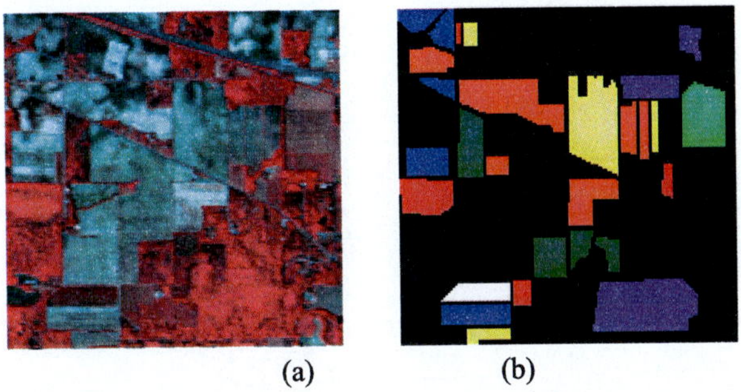

Fig. 3. Indian Pine dataset. (a) Three-band composition (bands 17,27and 50 for RGB) (b) Reference land cover map.

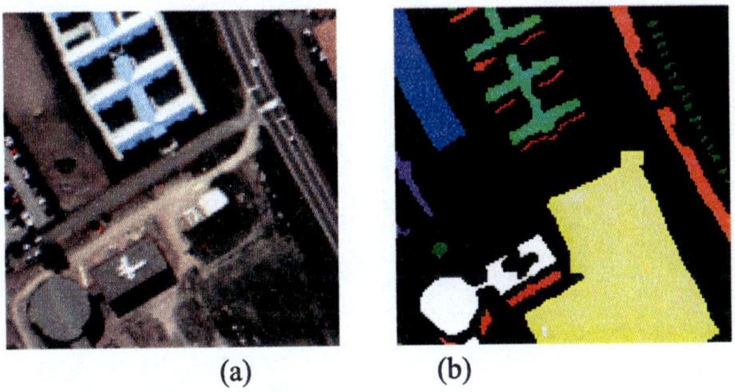

Fig. 4. University of Pavia dataset. (a) Three-band composition (bands 17,27and 50 for RGB) (b) Reference land cover map.

4.2 Parameter Settings and Evaluation Function

The base classifier is chosen as standard SVM, and the kernel function is chosen as radial basis kernel function. Multi-classification method: one-against-rest approach. The SVM classifier penalty factor C, kernel parameters σ are selected in [10 103] and [10^{-2} 102] using grid search method.

Inter-Band correlation

$$r = \frac{1}{C^2} \sum_{i=1}^{N} \sum_{j=i+1}^{N} R_{ij} \tag{20}$$

Table 1. Hyperspectral image of 8 types of features schematic of Indian Pine dataset

Number	Category	Number of samples
1	Corn-notill	1434
2	Corn-min	834
3	Grass/Pasture	497
4	Grass/Trees	747
5	Hay-windrowed	489
6	Soybeans-notill	968
7	Wheat	212
8	Woods	1294

Table 2. Hyperspectral image of 8 types of features schematic of University of Pavia dataset

Number	Category	Number of samples
1	Asphalt	559
2	Meadows	785
3	Trees	285
4	Metal Sheets	154
5	Soil	721
6	Bitumen	4021
7	Bricks	785
8	Shadows	129

where, C is the number of categories, N is the number of bands, and R represents the correlation coefficient values of the i_{th} and j_{th} bands.

$$R_{ij} = \frac{\sum_{k=1}^{C} (x_{ik} - \overline{x_i})(x_{jk} - \overline{x_j})}{\sqrt{\sum_{k=1}^{C} (x_{ik} - \overline{x_i})(x_{jk} - \overline{x_j})}} \tag{21}$$

where x_i is the i_{th} band, and $\overline{x_i}$ is the mean value of the i_{th} band. The overall classification accuracy OA is calculated as follows.

$$OA = \sum_{i=1}^{c} m_i \Big/ C \tag{22}$$

where C is the number of categories, c is the total number of samples tested in category C, and m, represents the number of samples correctly classified into the i_{th} category.

4.3 Experiments

To ensure the accuracy of the results, the same samples are used for the training and testing samples of each algorithm. 10% of the total number of samples are randomly selected as the training samples, and the remaining 90% samples are used as the testing data for SVM classification, and the average value is taken as the experimental results by running 10 times.

After subspace division of Indian Pine dataset, five subspaces are obtained as (1–36), (37–79), (80–103), (104–144), (145–200), and one band is selected in each subspace to obtain a subset of five bands. While after subspace division of university of pavia dataset, three subspaces are obtained as (1–73), (74–84), (85–103), and one band is selected in each subspace to obtain a subset of 3 bands.

In order to visualize the search performance of TLBO_QPSO, the iterative graphs of the fitness values of the optimal subset of bands searched by each algorithm are given in Fig. 5.

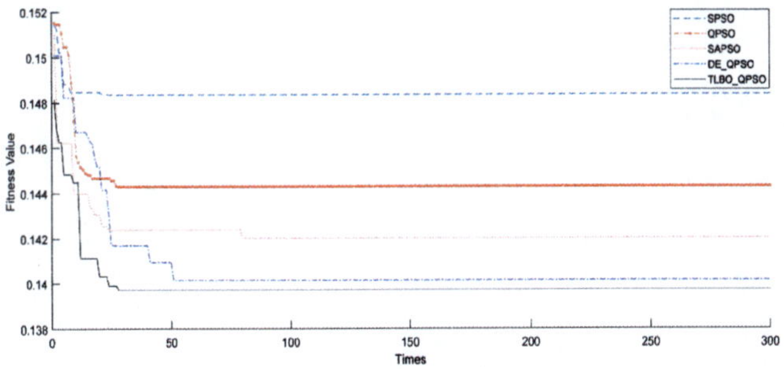

(a)

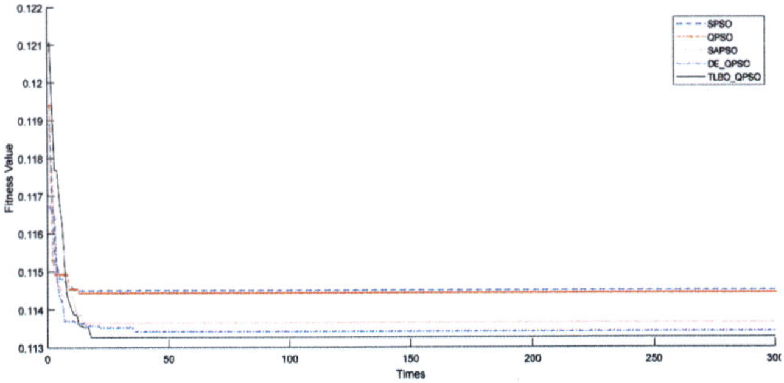

(b)

Fig. 5. (a) Iteration diagram of five algorithms for Indian Pine dataset (b) Iteration diagram of several algorithms for Pavia dataset

From the Fig. 5, it can be seen that the TLBO_QPSO algorithm converges the fastest, ensuring the diversity of the population and enhancing the global search capability while preserving the depth search capability of the QPSO algorithm. As the number of iterations increases, TLBO_QPSO can quickly search for the subset of bands for which the lowest fitness value can be obtained. The TLBO_QPSO algorithm is more efficient in terms of convergence accuracy and speed.

To verify the feature classification accuracy and robustness of the subset of bands filtered by the TLBO_QPSO algorithm, we counted the average classification accuracy of several comparison algorithms after band selection for 10 times under the same training and test sets.

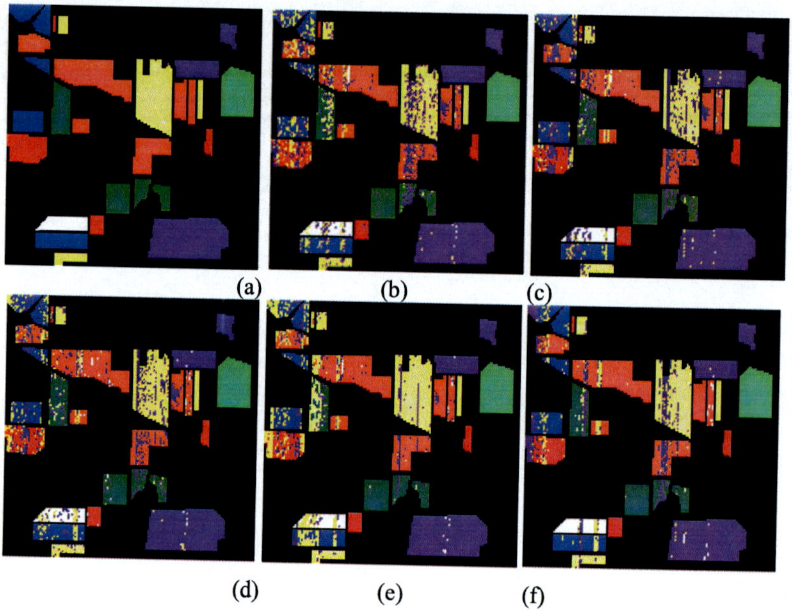

Fig. 6. Classification maps for Indian Pine dataset. (a) reference land cover map (b) SPSO (c) SAPSO (d) QPSO (e) DE_QPSO (f) TLBO_QPSO

Figure 6 and Fig. 7 show the classification effectiveness plots of several algorithms on two hyperspectral images. Tables 3 and 4 record the band combination, average correlation, OIF, and overall classification accuracy (OA) obtained by several algorithms of SPSO, SAPSO, QPSO, DE-QPSO, and TLBO_QPSO after band selection on two hyperspectral datasets using the same evaluation criteria. From the data in the table, it can be seen that the SPSO particle update capability is limited and the algorithm is prone to stagnation by falling into local optima. SPSO outperforms PSO but suffers from a lackluster state at the later stages of the iterations and a decrease in the ability to find an optimum. Compared with QPSO, TLBO_QPSO increases the population diversity, further expands the late search space, reduces the convergence speed, and avoids early convergence. This algorithm is superior to several other band selection methods in terms of both information and classification accuracy, because it retains the depth search capability of the QPSO algorithm while increasing the particle diversity to make the searched

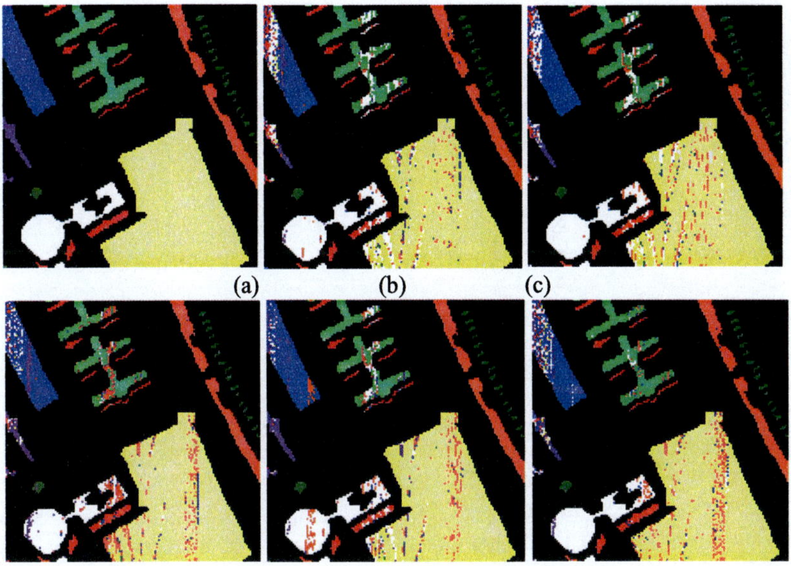

(a) (b) (c)

(d) (e) (f)

Fig. 7. Classification maps for Pavia dataset. (a) reference land cover map (b) SPSO (c) SAPSO (d) QPSO (e) DE_QPSO (f) TLBO_QPSO

bands more conducive to classification processing. From the search time, it can be seen that the time used for band selection based on TLBO_QPSO is much lower than other algorithms. It reflects the advantage of fast convergence of TLBO_QPSO, while in terms of search efficiency, the algorithm in this paper has obvious advantages, so the algorithm in this paper is more suitable for hyperspectral remote sensing image band selection.

Table 3. Experimental result Comparison (Indian Pine dataset)

Algorithms	Bands	Relevance	OIF	OA (%)	Time (s)
SPSO	(28 79 100 112 165)	1.826871	0.052065	67.46	37.53
SAPSO	(30 62 95 112 160)	1.667448	0.047629	72.08	47.09
QPSO	(30 59 90 115 164)	1.653072	0.049046	70.77	40.63
DE-QPSO	(30 37 79 113 165)	1.654474	0.055049	74.11	53.50
TLBO_QPSO	(31 37 79 115 165)	1.605289	0.055247	77.22	41.99

Table 4. Experimental result Comparison (University of Pavia dataset)

Algorithms	Bands	Relevance	OIF	OA (%)	Time (s)
SPSO	(6 84 91)	1.057379	0.077552	76.52	7.56
SAPSO	(6 84 92)	1.058931	0.078512	78.77	8.52
QPSO	(7 84 92)	1.057931	0.078512	77.75	7.32
DE-QPSO	(11 84 89)	1.062604	0.079096	79.92	9.71
TLBO_QPSO	(3 84 99)	1.045571	0.079552	81.12	7.89

5 Conclusion

Hyperspectral data have strong correlation among bands, and the band selection algorithm can reduce the dimensionality of the original data to improve the timeliness and ensure the classification accuracy. To address the problem of hyperspectral image band redundancy, this paper introduces the "learning" process of TLBO into the QPSO algorithm and proposes a TLBO-based quantum particle swarm method for hyperspectral image band selection. TLBO_QPSO retains the depth search capability of QPSO and introduces the TLBO idea to the QPSO algorithm, which ensures the population diversity and enhances the exploration ability of the optimal value of the objective function. The experimental results show that the band selection method of TLBO_QPSO algorithm converges faster than the other four band selection algorithms with guaranteed classification accuracy and is more capable of meeting practical needs, which is of certain importance for subsequent target recognition and feature classification of hyperspectral images. The optimized design of the objective function will be further studied in the later stage to improve the classification accuracy of feature data.

References

1. Chen, Z.:Research on high resolution remote sensing image classification technology. Institute of Remote Sensing Application, Beijing (2006)
2. Guyou, I., Elisseeff, A.: An introduction to variable and feature selection. J. Mach. Learn. Res. **3**, 1157–1182 (2003)
3. He, M.Y., Chang, W.J., Mei, S.H.: Research progress in feature mining of hyperspectral remote sensing data. Spacecr. Recovery Remote Sens. **34**(01), 1–12 (2013)
4. Zheng, S., Hai, Y., He, M., Wang, J.: Comparative study on vegetation extraction effect based on ZY-1 02D data. Spacecr. Recovery Remote Sens. **43**(2), 92–103 (2022)
5. Wang, L.G., Wei, F.J.: Artificial physics optimization algorithm combined band selection for hyperspectral imagery. J. Harbin Inst. Technol. **45**(9), 100–106 (2013)
6. Wang, L.G., Wei, F.J.: Band selection for hyperspectral imagery based on combination of genetic algorithm and ant colony algorithm. J. Image Graph. **18**(2), 235–242 (2013)
7. Eberhart, R.C., Kennedy, J.: A new optimizer using particle swarm theory. In: Proceedings of 6th International Symposium on Micro Machine and Human Science, pp. 39–43 (1995)
8. Eberhart, R.C., Shi, Y.: Particle swarm optimization: developments, applications and resources. In: Proceedings of Congress Evolutionary Computation, pp. 81–86 (2001)

9. Shi, Y., Eberhart, R.: A modified particle swarm optimizer. In: IEEE Conference on Evolutionary Computation, pp. 69–73 (1998)
10. Clerc, M., Kennedy, J.: The particle swarm: explosion, stability, and convergence in a multi-dimensional complex space. IEEE Trans. Evol. Comput. **6**(1), 58–73 (2002)
11. Sun, J., Feng, B., Xu, W.: Particle swarm optimization with particles having quantum behavior. In: IEEE Congress on Evolution Computation, vol. 1, pp. 325-331 (2004)
12. Sun, J., Feng, B., Xu, W.: A global search strategy of quantum-behaved particle swarm optimization. In: IEEE Conference on Cybernetics and Intelligent Systems, pp. 111–116 (2004)
13. Rao, R.V., Savsani, V.J., Vakharia, D.P.: Teaching-learning-based optimization: an optimization method for continuous non-linear large scale problems. Inf. Sci. **183**(1), 1–15 (2012)
14. Sun, J., Xu, W., Feng, B.: Adaptive parameter control for quantum-behaved particle swarm optimization on individual level. In: Systems, Man and Cybernetics, vol. 4, pp. 3049–3054 (2005)
15. Coelho, L.S.: Novel Gaussian quantum-behaved particle swarm optimiser applied to electromagnetic design. IET Sci. Meas. Technol. **1**(5), 290–294 (2007)
16. Coelho, L.D.S.: A quantum particle swarm optimizer with chaotic mutation operator. Chaos Solitons Fractals **37**(5), 1409–1418 (2008)
17. Liu, Y., Gu, Y.F., Zahng, Y.: A fast hybrid search algorithm for band selection in hyperspectral images. Opt. Tech. **33**(2), 258–261, 265 (2007)
18. Fang, W.: Best band combination of Chinese high-resolution satellite image information extraction. Spacecr. Recovery Remote Sens. **43**(2), 82–91 (2022)

High Stability Design and Verification of Hyper-spectrometer on Terrestrial Ecological Carbon Satellite

Pengfei Duan[1](✉), Weigang Wang[1], and Yilan Mao[2]

[1] Beijing Institute of Space Mechanics and Electricity, Beijing 100094, China
dpfei1949@163.com

[2] Institute of Remote Sensing Satellite, Beijing 100094, China

Abstract. The optical, mechanical and thermal stability of structure impact on the spectral performance stability of spectrometer, which ultimately affects the inverse accuracy of substance on the Earth's surface. In this paper, the design of the opto-mechanical-thermal stability of Hyper-spectrometer on Terrestrial Ecological Carbon Satellite is introduced. The verification on the ground shows that the stability of structure designed could meet the task requirements. The satellite has been launched successfully from Taiyuan Satellite Launch Center on August 4, 2022. It has been in orbit for four months (until early December 2022), and is undergoing in-orbit testing and calibration. At present, the spectral images are clear in details, rich in layers, and meet the stability requirements. Through a series of effective methods, the hyper-spectrometer achieves the requirements of spectral resolution, spectral distortion and high stability of spectrum distribution in orbit, which plays a key role in high-precision SIF spacborn inversion and makes an important contribution to the spectral accuracy of high-resolution quantitative detection satellite reaching the international advanced level.

Keywords: Spectral resolution · spectral distortion · spectrum distribution · stability · Terrestrial Ecological Carbon Satellite · Hyper-spectrometer

1 Introduction

Accurate measurement of the carbon sinks of terrestrial ecosystem is the key to achieve the dual carbon strategy of "carbon peak and carbon neutral". Satellites have great advantages in assessing the spatiotemporal changes of terrestrial ecosystem carbon sinks at global scale, Solar-Induced Chlorophyll Fluorescence (SIF) is the visible and near-infrared electromagnetic radiation emitted by chlorophyll molecules in the process of photosynthesis [1–4]. And satellite SIF remote sensing has great potential for estimating photosynthetic carbon sinks in global terrestrial ecosystem [1, 5]. ESA selected FLEX (Fluorescence Explorer) as the European "Earth Explorer" project in 2015, FLEX is the world's first SIF project dedicated to terrestrial vegetation. Originally scheduled to launch in the first half of 2022, the launch has been postponed to 2025. The spectral resolution of FLEX high resolution channel is 0.3 nm, and the corresponding spectral

H. P. Urbach and H. Jiang (Eds.): ISSOIA 2022, SPPHY 295, pp. 654–667, 2023.
https://doi.org/10.1007/978-981-99-4098-1_54

sampling interval is 0.1 nm/detector pixel. The spectral stability of the single orbit within 10 min should not exceed 1/10 of the spectral sampling interval, that is, 0.01 nm [6, 7]. The "terrestrial ecological carbon" satellite launched successfully in China was established in 2017. The hyper-spectrometer introduced in this paper is the world's first special SIF remote sensing payload in orbit. The spectral resolution is 0.3 nm, and the in-orbit stability requirement of resolution is no more than 0.015 nm [8].

SIF satellite remote sensing detection precision depends on the spectral performances of the remote sensing spectrum instrument, such as spectral resolution, spectral distortion, spectral distribution, and the stability of the related indicators above [5, 9]. As is the stability of spectrometer in orbit mainly depended on the internal factors such as the sensitivity of the optical system, the stability of the structural support, etc. and the external factors such as thermal environment in orbit, to improve the stability of the spectrometer in orbit is the basis of the application of high-precision SIF satellite. A series of optical, structural and thermal control means have been adopted to ensure the stability of the spectrometer in orbit. Many technologies are involved in the research of space optical remote sensor, such as optical, mechanical and thermal, etc., therefore the method of integrated analysis of optical, mechanical and thermal should be carried out to solve the influence of coupling loads brought to the optical precision instrument in space. There have been many studies on the integrated analysis of space optical remote sensors at home and abroad [10–13], however studies on space spectral instruments is few, and the in-orbit stability of spectral resolution, spectral distortion and spectral distributions in not sufficient. In this paper, the variation of spectral resolution, spectral distortion and spectral distribution are studied induced by the external heat flux during the imaging of the hyper-spectrometer on the Terrestrial Ecological Carbon satellite in orbit, and the stability of the index is analyzed and demonstrated. Further, the ground test shows that the stability requirements could be met by the opto-mechanical and thermal design of the spectrometer.

2 Stability Requirements in Orbit

In order to meet the detection accuracy requirements of SIF remote sensing, "Terrestrial Ecological Carbon" satellite puts forward strict stability indexes for the performance of the hyper-spectrometer, such as spectral resolution, spectral distortion and spectral distribution.

2.1 Stability Index of Spectral Resolution

The production of the Full Width at Half Maxima (FWHM) of the spectrometer's Instrument Linear Shape (ILS) and the spectral sampling interval is the spectral resolution [5, 9], as shown in Fig. 1 below.

The FWHM of ILS refers to the pixel number of the half-peak boarding of a specific spectrum on the detector image plane, and its unit is pixel. The spectral sampling interval is the wavelength broadening range of the specific spectral dispersion on the pixels of the detector in the specific field of view of the spectrometer, and its unit is nm per pixel.

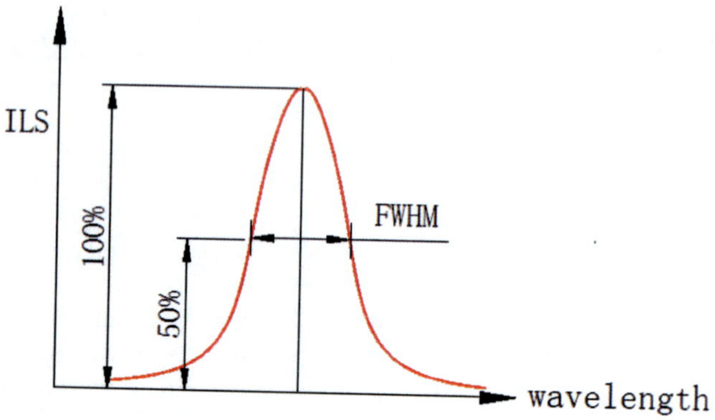

Fig. 1. Instrument linear shape and full width at half maxima

The spectral resolution $\Delta\lambda$ represents that the spectrometer could resolve two spectral lines whose wavelength difference is beyond the range of $\Delta\lambda$, which determines the chromatographic resolution power of the instrument. The in-orbit stability of the spectral resolution refers to the ability of $\Delta\lambda$ keeping its measurement constant in the space environment. The high stability of spectral resolution of the hyper-spectrometer is the basis of its in-orbit SIF detection ability. The spectral resolution index of the hyper-spectrometer is required to be no more than 0.3 nm, and the spectral resolution in orbit stability is required to be no more than 0.015 nm.

2.2 Stability Index of Spectral Distortion

The distortion of spectrometer contains two parts: Smile and Keystone [5, 9], wherein the bending of monochromatic spectral limes is called spectral direction distortion, which is denoted as Smile, meaning that when the field of view aperture (usually slit) of the spectrometer is illuminated by uniform single chromatographic light, the slit image obtained on the detector array is not completely straight, and Smile will lead to confusion between wavelengths. The spatial magnification of the spectrometer generally changes with wavelength. The distortion caused by the inconsistency of spatial magnification is called spatial distortion, which is denoted as Keystone. Keystone will lead to confusion between fields of view. The spectral directional distortion Smile and spatial directional distortion Keystone are shown in Fig. 2 below.

The high stability of spectral distortion of the hyper-spectrometer is a guarantee for the accuracy of SIF detection in orbit. The requirement of Smile and Keystone is less than 0.3 detector pixels, and the stability of the distortion is required to be less than 0.15 detector pixels.

2.3 Stability Index of Spectrum Distribution

The spectral distribution is the pixel position of the specific wavelength line responding to the image plane of the spectrometer detector. The pixel distribution position of the spectrometer detector corresponds to the wavelength of the input spectral line. The high

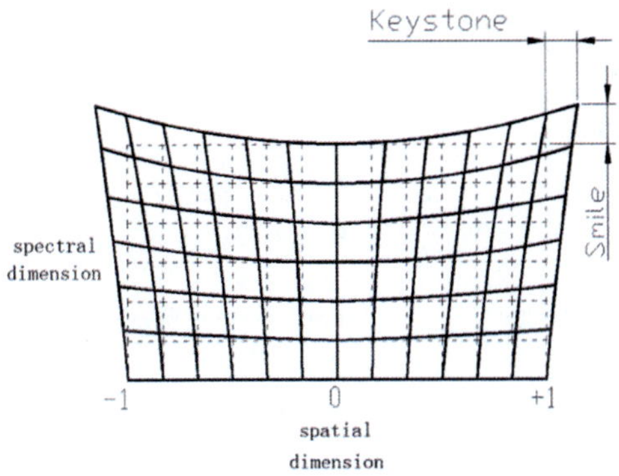

Fig. 2. Smile and Keystone

stability of the spectral distribution of the hyper-spectrometer is the basis for its in-orbit SIF quantization detection accuracy. The stability of the spectral distribution is required to be less than 0.5 detector pixels.

3 Design on Stability In-Orbit

The main body configuration of the hyper-spectrometer is shown in Fig. 3.

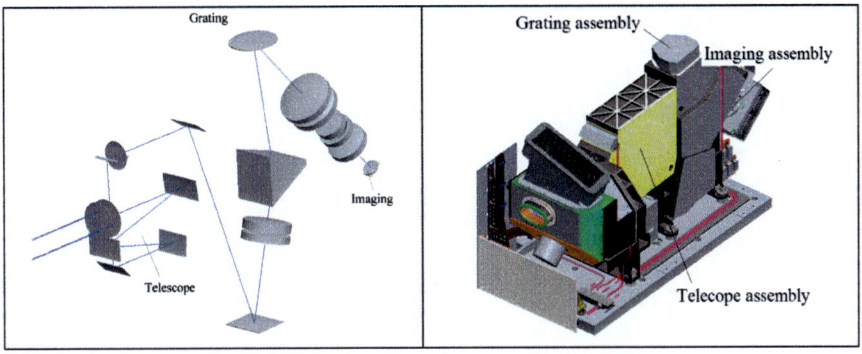

Fig. 3. Configuration of the Spectrometer

Based on the requirement of the high stability index of the hyper-spectrometer, the design of the main body is carried out from the aspects of low sensitive optical system, high stability opto-mechanical structural and high precision thermal control.

3.1 Design of Low Sensitivity Optical System

Optical system parameters are optimized to reduce the sensitivity to temperature, the results of which are as the following, the spectral resolution is 0.28 nm, Smile and

Keystone are less than 0.2 detector pixel respectively. When the operating temperature changes by ±3 °C, the spectral resolution, Smile and Keystone remain unchanged, and the spectrum distribution offset is less than ±0.05 detector pixels, which is far better than the requirements. As is shown in Table 1 below.

Table 1. Temperature sensitivity matrix

Temperature field/°C	Spectral resolution/nm	Smile/pixel	Keystone/pixel	Spectrum distribution offset/pixel
17	0.28	0.09	0.15	0.037
18	0.28	0.09	0.15	0.024
19	0.28	0.09	0.15	0.012
20 (the designed temperature)	0.28	0.09	0.15	/
21	0.28	0.09	0.15	−0.012
22	0.28	0.09	0.15	−0.024
23	0.28	0.09	0.15	−0.037

Considering the influence of the gradient temperature field of 0.1 °C, 0.2 °C and 0.3 °C from the center to the edge of the lens on the performance of the optical system, the spectral resolution, Smile and Keystone remain unchanged, and the spectrum distribution offset is within 5E−3 detector pixels, which is also far better than the requirement. As is shown in Table 2 below.

Table 2. Temperature gradient sensitivity matrix

Temperature gradient field/°C	Spectral resolution/nm	Smile/pixel	Keystone/pixel	Spectrum distribution offset/pixel
0	0.28	0.09	0.15	/
0.1	0.28	0.09	0.15	0.000
0.2	0.28	0.09	0.15	−4E−4
0.3	0.28	0.09	0.15	−5E−3

3.2 Design of High Stability Main Structure

In paper [14], a number of key technologies referring to the optical and mechanical system of space optical remote sensor are studied and a conclusion that the integrated installation design of the main structure would play an important role in removing the geometric error of the system is pointed out. The optical system of the hyper-spectrometer has designed dozens of optical components such as mirror, lens, grating, prism etc., which are large in number and complex in layout. Following the idea of integrated installation in the design of the main structure, the optical components are installed in one single structure. As is shown in Fig. 4.

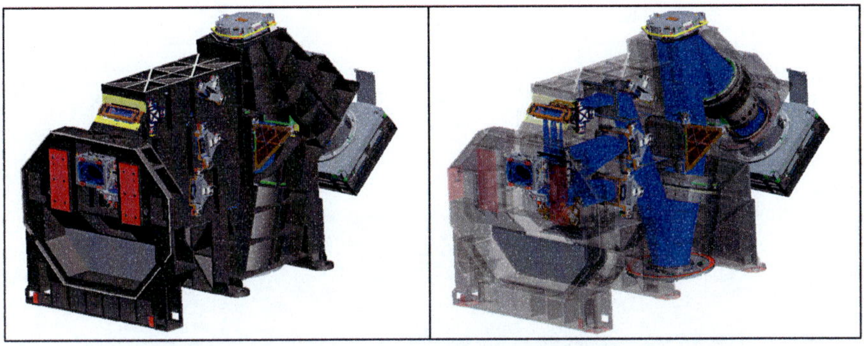

Fig. 4. Integrated design of opti-structure

The thermal deformation of the satellite structure would be induced by external heat flux in orbit, which would be transferred to the main body of the remote sensor in the form of forced displacement, which is usually 0.03–0.05 nm, and eventually the opto-mechancial structure stability would be affected. To verify the ability of the main structure to resist the forced displacement, a special test has been carried out, in which a feeler measure corresponding to the thickness of the corresponding forced displacement was heated into the installation position of the main structure to observe the change of performance indexes such as spectral resolution. The test results are seen in Table 3, which shows that the performance stability is still sufficient in the case of 0.3 mm forced displacement, indicating that the stiffness of the main body structure is sufficient to resist the influence caused by the changing thermal flux in orbit.

3.3 Design of High Precision Thermal Control

The main body of the hyper-spectrometer adopts high precision active and passive temperature control design. The main body is designed with 30 high-precision active heat control loops, and the compensation heating method is adopted to ensure the temperature stability of the structure. The whole body is warpped with multi-layer to isolate the influence of external heat flow in orbit, as shown in Fig. 5 below. Under the high precision control measures, the temperature fluctuation of the main optical components in orbit is no more than 0.2 °C. For the sensitive parts of the spectrometer, such as the grating and the prism, special thermal control measures are adopted to achieve zero fluctuation with less than 0.1 °C fluctuation in orbit.

Table 3. Test results on forced displacement

Forced displacement/mm	Variation			Spectrum distribution offset/pixel
	Spectral resolution/nm	Smile/pixel	Keystone/pixel	
0.01	0.0	0.0	0.0	0.0
0.05	0.0	0.0	0.0	0.13
0.08	0.0	0.01	0.0	0.22
0.1	0.002	0.02	0.01	0.42
0.3	0.006	0.08	0.03	0.48
0.5	0.01	0.16	0.11	0.66

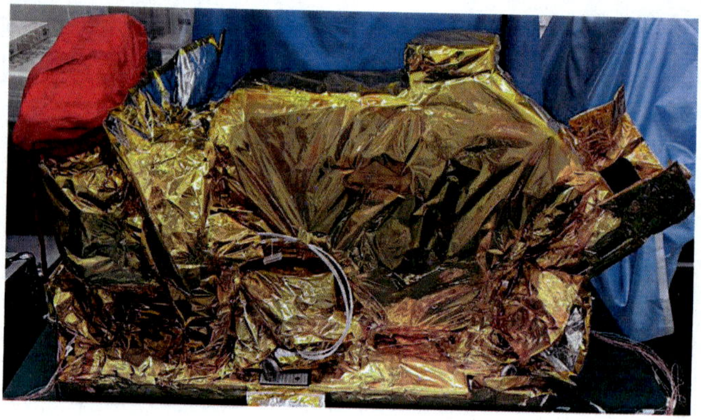

Fig. 5. Thermal multi-layer coverings on mainbody

3.4 Integrated Analysis of the System

In order to analyze whether the spectrometer could meet the requirements, the integrated analysis method is adopted to obtain the in-orbit stability of the instrument by calculating the performance changes under the condition of the in-orbit temperature field changes, which could provide the basis for the mechanical-thermal co-design and ground image correction.

The key of the integrated analysis is the transfer of data among thermal, mechanical and optical analysis models. Firstly, the 3D structure model of the spectrometer would be established, and then the thermal analysis model, mechanical analysis model and optical analysis model are established respectively. Afterwards the main body temperature field is analyzed by the thermal analysis model according to the attitude of the satellite, then the temperature field is mapped to the mechanical analysis model, who would take the mapped temperature field as load in the process of obtaining the thermal deformation field. After that the rigid displacement and high order deformation of the optical surfaces are extracted from the thermal deformation field, which are taken as inputs in the analysis of optical performance by the optical analysis model [15]. The specific process is shown in Fig. 6 below.

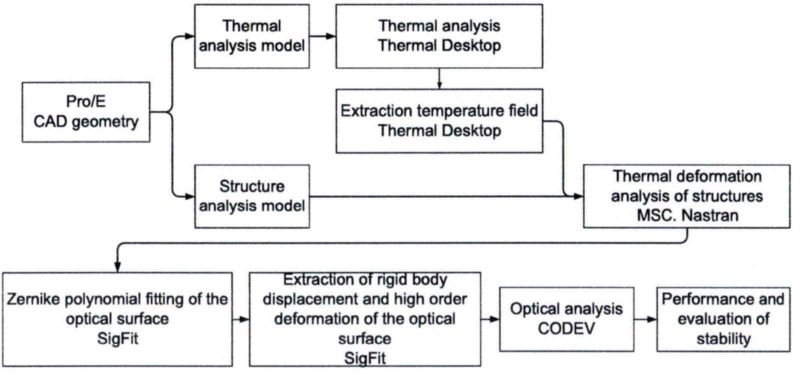

Fig. 6. Integrated analysis process

The comparition of the spectrometer temperature fields in imaging between low and high external flux cases is shown in Fig. 7, which indicates that the difference is very tiny under the temperature control method.

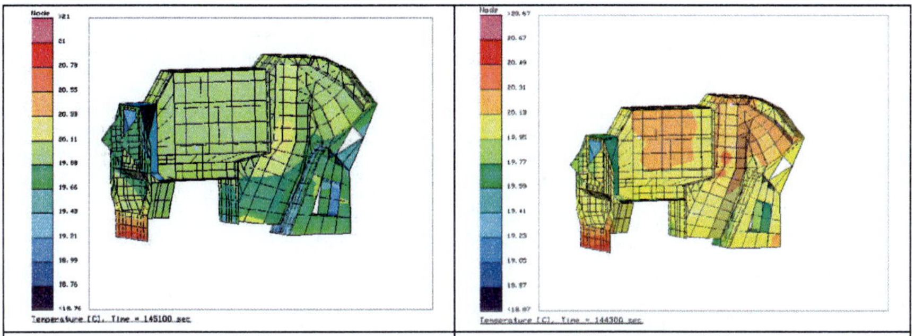

Low external flux case	High external flux case

Fig. 7. Temperature fields in low and high temperature cases

Figure 8 shows the temperature variation range of each position of the main body in orbit, taking the telescope assembly, imaging assembly and grating assembly as examples. It can be seen that the temperature of the telescope assembly and imaging assembly changes periodically under the co-action of external heat flow and temperature control measures. The temperature variation range of the telescope assembly in orbit is 0.19°C, and that of the imaging assembly is 0.17°C. Under the special control measures, the temperature variation range of the grating assembly in orbit is less than 0.1°C, almost no change.

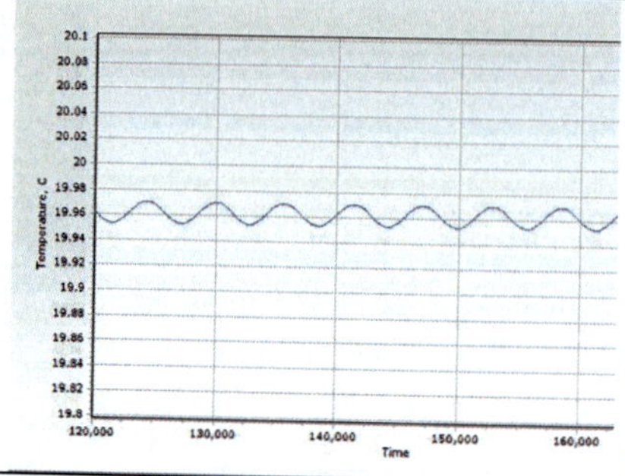

Telecope assembly in low case

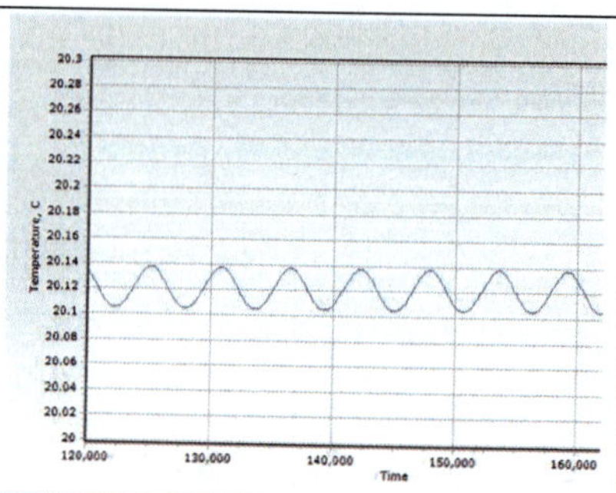

Telecope assembly in high case

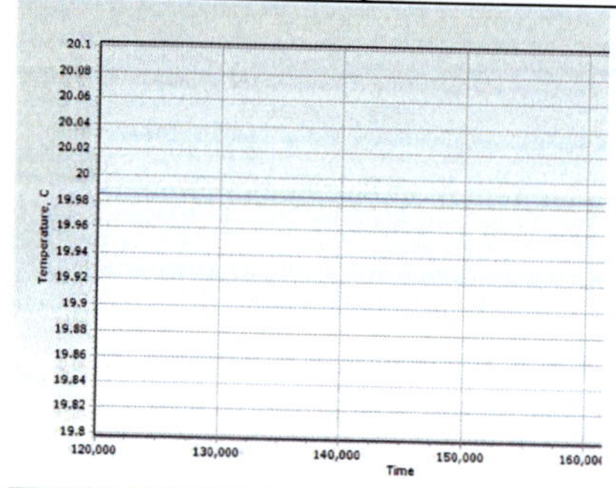

Imging assembly in low case

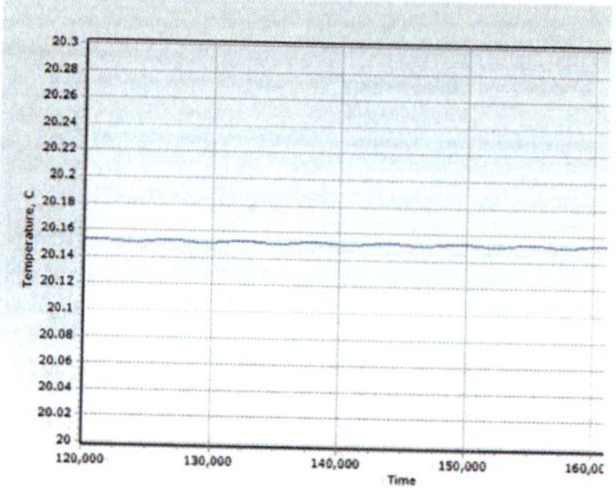

Imging assembly in high case

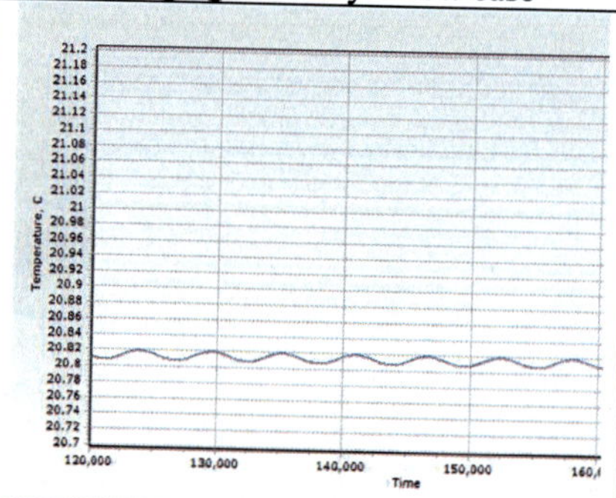

Grating assembly in low case

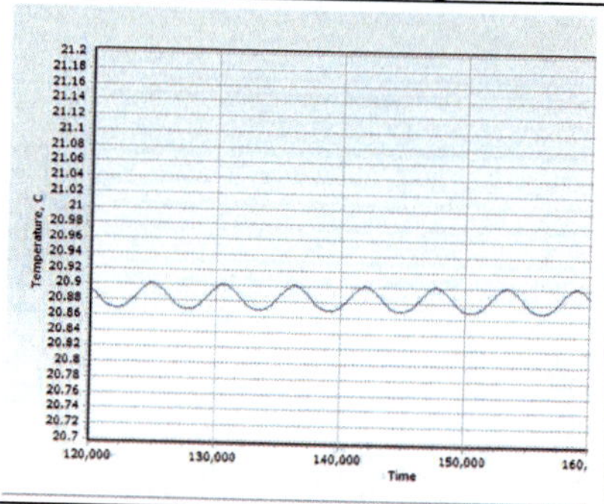

Grating assembly in high case

Fig. 8. Temperature variations of locations innor mainbody

Figure 9 shows the displacement field of the main body leading by the thermal deformation in orbit. The rigid displacement and high order deformation of the optical surfaces are obtained by the Zernike polynomial fitting of the optical surface displacements extracted from the main body displacement field. Further, the performances, such as spectral resolution, spectral distortion, and spectrum distribution etc. are calculated by the optical analysis model with the rigid displacement and high order deformation of the optical surfaces.

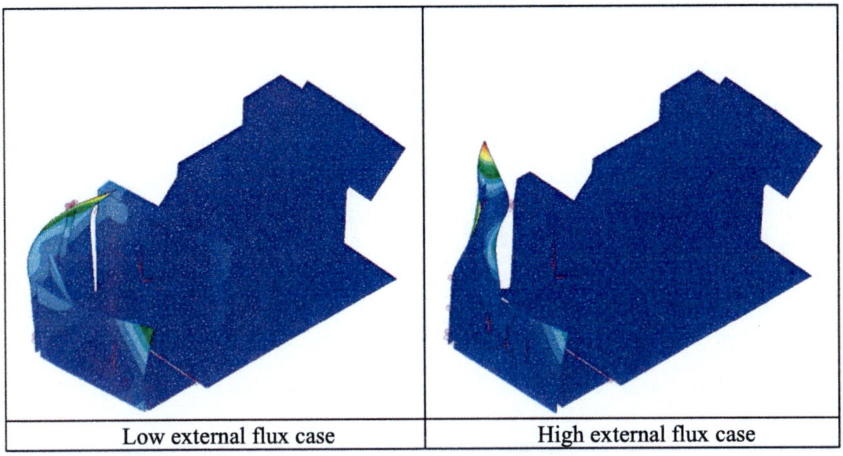

Fig. 9. Deformation plots in low and high temperature cases

The analysis results of the performance indexes of the hyper-spectrometer in various working conditions are shown in Table 4 below, and the results are in line with the requirements.

Table 4. Performances in each simulation case

Items of performance	Indicators	Design value (20 °C)	Low external flux case		High external flux case	
			Result	Conformity	Result	Conformity
Spectral resolution/nm	≤0.3	0.28	0.293	√	0.294	√
Smile/pixel	<0.3	0.09	0.16	√	0.13	√
Keystone/pixel	<0.3	0.15	0.15	√	0.19	√

The analysis results of the indexes of stability of the hyper-spectrometer in various working conditions are shown below, and the results are in line with the requirements (Table 5).

Table 5. Performance stability in each simulation case

Items of stability	Indicators	Low external flux case		High external flux case	
		Result	Conformity	Result	Conformity
Spectral resolution stability/nm	<0.015	0.013	√	0.014	√
Smile stability/pixel	<0.15	0.04	√	0.07	√
Keystone stability/pixel	<0.15	0.0	√	0.04	√
Spectrum distribution offset/pixel	<0.5	0.34	√	0.42	√

4 Ground Verification

Vacuum thermal test was carried out on the ground to verify the satisfaction of spectral indexes and in-orbit stability, and the influence of in-orbit temperature load on the spectral imaging quality. The vacuum thermal test was carried out in the vacuum simulation chamber, which could realize the space vacuum environment no higher than 1.3E−3 Pa and the space heat-sink environment no higher than 100K. The heat flows in orbit was simulated by the infrared cages and electric heaters. The test situation is shown in Fig. 10 below.

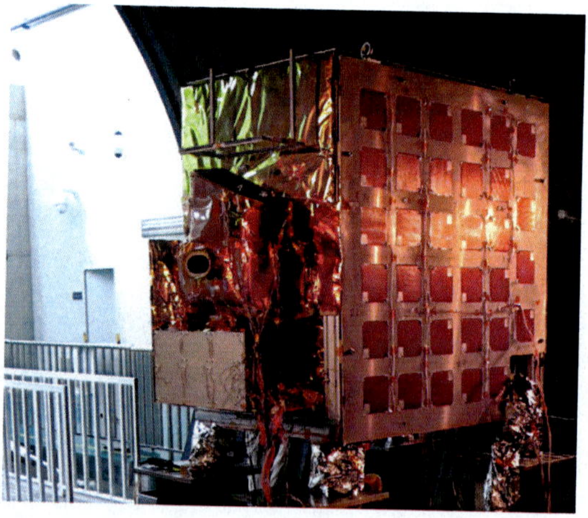

Fig. 10. Vacuum thermal test on the ground

The test has covered low and high external flux cases. The test results of the temperature, as well as the performance and stability of the instrument are shown in Table 6, which are in line with the index requirements.

Based on the ground test results, it can be seen that the indexes and stability of the hyper-spectrometer meet the requirements, and the performance is excellent.

Table 6. Test results in each test case

(a) Performances in each test case

	Temperature results/℃		
	Low external flux case	High external flux case	Difference
Telecope assembly	19.9~19.96	19.94~20.08	0.18
Imaging assembly	20.3~20.4	20.33~20.45	0.15
Grating assembly	20.9~20.96	20.94~20.98	0.08

(b) Performances in each test case

Items	Indicators	Initial case (20℃)	Conformity	Low external flux case	Conformity	High external flux case	Conformity
Spectral resolution / nm	≤ 0.3	0.248	√	0.265	√	0.270	√
Smile / pixal	< 0.3	0.25	√	0.25	√	0.23	√
Keystone / pixal	< 0.3	0.14	√	0.16	√	0.21	√

(c) Performance stability in each test case

Items on stability	Indicators	Low external flux case	Conformity	High external flux case	Conformity
Spectral resolution stability / nm	< 0.015	0.007	√	0.012	√
Smile stability / pixal	<0.15	0.0	√	0.02	√
Keystone stability / pixal	<0.15	0.02	√	0.07	√
Spectrum distribution offsets / pixal	<0.5	0.25	√	0.27	√

5 Conclusion

In this paper, performances such as the spectral resolution, spectral distortion, spectrum distribution etc. and stability of the hyper-spectrometer of Terrestrial Ecological Carbon satellite in the complex temperature field in orbit are studied and evaluated effectively by means of the integrated analysis of optical, mechanical and thermal methods. Further, the ground vacuum thermal test has been carried out to vertify the performance index and its stability in orbit. The analysis and ground test results show that the performance of the hyper-spectrometer can meet the stringent requirements of the in-orbit stability mission with excellent performance. The analytical methods and experiments of the space spectral optical remote sensor presented in this paper are of practical significance for the fabrication and verification of a class of space precision instruments with quantitative requirements.

Terrestrial Ecological Carbon satellite has been launched successfully from Taiyuan Satellite Launch Center on August 4, 2022. It has been in orbit for four months (until early December 2022), and is undergoing in-orbit testing and calibration. At present, the spectral images are clear in details, rich in layers, and meet the stability requirements. Through a series of effective methods, the hyper-spectrometer achieves the requirements of spectral resolution, spectral distortion and high stability of spectrum distribution in orbit, which plays a key role in high-precision SIF spacborn inversion and makes an important contribution to the spectral accuracy of high-resolution quantitative detection satellite reaching the international advanced level.

References

1. Tong, C., Bao, Y., Huang, Q., Wang, Y.: Progress on solar-induced chlorophyll fluorescence of satellite remote sensing. Spacecr. Recovery Remote Sens. **45**(4), 45–55 (2022). (in Chinese)
2. Zhao, F., Li, R., Verhoef, W., et al.: Reconstruction of the full spectrum of solar-induced chlorophyll fluorescence: intercomparison study for a novel method. Remote Sens. Environ. **219**, 233–246 (2018)
3. Pinto, F., Celesti, M., Acebron, K., et al.: Dynamics of sun-induced chlorophyll fluorescence and reflectance to detect stress-induced variations in canopy photosynthesis. Plant Cell Envrion. **43**, 1637–1654 (2020)
4. Zhang, Y., Guanter, L., Joiner, J., et al.: Spatially-explicit monitoring of crop photosynthetic capacity through the use of space-based chlorophyll fluorescence data. Remote Sens. Environ. **210**, 362–374 (2018)
5. Tong, Q., Zhang, B., Zheng, L.: Hyperspectral Remote Sensing: Principles, Techniques, and Applications. Higher Education Press, Beijing (2006). (in Chinese)
6. Moreno, J., et al.: Fluorescence explorer (FLEX): an optimised payload to map vegetation photosynthesis from space. In: AIAA 57th International Astronautical Congress, IAC, Valencia, Spain, 2 October 2006
7. Kraft, S., et al.: FIMAS–feasibility study of a fluorescence imaging spectrometer to be flown on a small platform in tandem with Sentinel 3. In: International Conference on Space Optics, ICSO, Rhodes Island, Greece, 4 October 2010
8. Duan, P., Wang, W.: Design and performance of the fluorescence imaging spectrometer of China. In: AIAA 68th International Astronautical Congress, IAC, Adelaide, Australia, 25 September 2017
9. Wang, J., Shu, R., Liu, Y., et al.: Introduction on the Imaging Spectroscopy. Science Press, Beijing (2011). (in Chinese)
10. Hatheway, A.E.: Overview of the finite element method in optical systems. In: Proceedings of the SPIE, vol. 1532, pp. 2–14 (1991)
11. Fu, D., Yin, C., Wu, C.: A study of thermal/structural/optical analysis of space remote sensor. J. Astronaut. **22**(3), 105–110 (2001). (in Chinese)
12. Jason, G., Jeff, L.: Collaborative design and analysis of Electro-Optical sensors. In: Proceedings of the SPIE, vol. 7427, pp. 1–12 (2009)
13. Duan, P., Li, D., Lei, W.: Collaborative design and integrated analysis of space camera. Spacecr. Eng. **23**(2), 52–57 (2014). (in Chinese)

14. Wu, Q.: Research on Key Technique of Optomechanical System Desing for Space Remote Optical Sensor. Harbin Institute of Technology, Harbin (2003). (in Chinese)
15. Li, K., Xu, P., Wang, W., et al.: Simulation analysis of hyperspectral spectrometer grating profile on spectral performance. Spacecr. Recovery Remote Sens. **39**(2), 55–62 (2018). (in Chinese)

Software-Hardware Cooperative Lightweight Research of Remote Sensing Target Detection Algorithms for Space-Borne Edge Computing

Zhang Pengcheng[1](✉), Wu Wenbo[1], Li Qiang[1,2], and Cao Chenghua[1]

[1] Beijing Institute of Space Mechanics & Electricity, Beijing 100090, China
574758761@qq.com

[2] Northwestern Polytechnic University, Xi'an 710072, China

Abstract. Real-time target detection of remote sensing image is a key technology for efficient application in the field of remote sensing. The detection rate of target detection models based on deep network is high, but this kind of models have a lot of parameters and computational load, so they are difficult to be deployed on the satellite. Lightweight optimization of such model is a problem for the current deep network target detection algorithm. Based on the YOLOv5s target detection model, this paper designs a software-hardware co-lightweight scheme. Firstly, the YOLOv5 model was improved to make it more suitable for remote sensing application, and a improved convolution kernel pruning method was used to solve the over-fitting problem of the model. Then, the model was quantified to run under the DPU of the FPGA, trained and tested on the homemade remote sensing aircraft datasets. The experimental results show that the target detection performance of the model can decline within 3% when 66% of the weights are clipped. It means a highly lightweight YOLOv5 model can be deployed to the edge device on the satellite.

Keywords: Remote sensing image · Target detection · YOLOv5 · Hardware-software · Lightweight

1 Introduction

With the continuous development of remote sensing satellites, the amount of remote sensing image data has increased dramatically. During the past few decades, the way that remote sensing satellites downlink image data to the ground station for processing and analysis has some problems such as poor real-time performance and high bandwidth requirement. Real-time on-orbit processing of remote sensing image has become an important research content [1, 2]. In the field of remote sensing image target detection, the algorithm based on convolution neural network has gradually replaced the traditional target detection algorithm based on feature extraction, and has made a breakthrough in terms of the accuracy and speed [3–5]. However, in order to improve its performance, the neural network will be designed deeper and deeper, and the amounts of parameters and calculation will increase exponentially. It is difficult to deploy the target detection

H. P. Urbach and H. Jiang (Eds.): ISSOIA 2022, SPPHY 295, pp. 668–679, 2023.
https://doi.org/10.1007/978-981-99-4098-1_55

models with excellent detection performance to the on-orbit hardware devices because of the high computational cost. According to the on-orbit hardware environment, collaborative lightweight of hardware and software for target detection model can reduce the computational and storage requirements of target detection model, which is very important [6–9].

Currently, convolution neural network is mainly accelerated by GPU. However, GPU has high power consumption, which makes it difficult to apply in the on-orbit environment. For the specific on-orbit environment, many satellites use the hardware platform of FPGA or FPGA+DSP/CPU for real-time on-orbit data processing [10–13]. These platforms have the advantage of low power consumption and high-level anti-irradiation. However, the floating-point arithmetic ability of FPGA is weak. Cooperative optimization of hardware and software for target detection algorithm can greatly enhance the efficiency of target detection model in FPGA. The deployment of deep neural network on edge devices has three major factors: the size of network weight parameters, runtime memory consumption, and calculation complexity [14]. In order to solve the problems caused by these three factors, Various methods are proposed such as pruning, quantification, low rank approximation, isomerization and so on. AGHASI [15] compresses the model by solving a convex optimization problem, suggesting that the importance of the convolution kernel is determined by the statistics of the next layer, rather than the current one. LIN [16] compresses the model by rank of the signature graph. It is believed that the signature graph with a small rank contains less information and is therefore of low importance. HU [17] proposed a strip pruning method to obtain better pruning granularity by using a convoluted core skeleton to guide model compression. TANG [18] binarizes the weights and activations of the neural network, greatly compressing the parameters and computations of the deep neural network model. CHEN [19] presents a convolution kernel pruning method based on L1 regularization method, which can effectively reduce the amounts of parameters and computation while guaranteeing the integrity of the network. It can be applied to various types of network models and is suitable for target detection model on-orbit lightweight deployment.

In this paper, a hardware-software co-lightweight scheme is designed to solve the deployment problem of target detection model in the on-orbit environment. On the premise of guaranteeing the integrity of the model structure, the target detection network is clipped, and a hardware acceleration scheme for FPGA is designed. The clipped model can run in the hardware environment of on-orbit FPGA processor. With 66% model size reduction, the target detection performance loss will decline within 3%.

2 Related Work

2.1 Remote Sensing Target Detection Models

In the application field of satellite remote sensing, the selection of target detection model is limited by the storage capacity, computing power and power consumption of on-orbit hardware devices, so the model must have the characteristics of small network parameters and low computing power requirements. In the target detection models based on deep neural network, Faster R-CNN, SSD, YOLO are widely used [20–24]. Among them, YOLO is the first one-stage target detection algorithm, and its computational complexity and real-time performance are better than two-stage target detection algorithm represented by Faster R-CNN target detection model. The improved version of YOLO algorithm, YOLOv5, absorbs the experience of different generation of YOLO algorithms and has good performance in both accuracy and speed.

YOLOv5 model is mainly composed of input layer, backbone layer, feature extraction layer, and target detection layer. Its network structure is shown in Fig. 1. The input layer uses image pre-processing techniques such as Mosaic data enhancement and adaptive anchor frame calculation, which greatly enriches the detection dataset and makes the network structure more robust. The backbone layer contains Focus unit, CBL unit, CSP unit and SPP unit. Focus unit is an improved lower sampling layer, which makes feature extraction more sufficient without loss of information. The CSP unit refers to the network structure of CSPNet [25]. First, it divides the feature mapping of the underlying layer into two parts, and then merges them through a cross-stage hierarchy, which reduces computational effort while ensuring accuracy. CBL units mainly perform convolution, batch normalization and ReLU activation functions. The SPP layer is used for maximum pooling. The feature extraction layer uses the structure of FPN [26] and PAN [27] to transfer and fuse image features by up-sampling. YOLOv5 inherits the advantages of previous versions, which is more accurate and faster in object detection, and is a high-performance generic target detection model. It is suitable to be used as the main algorithm in the on-orbit deployment scheme.

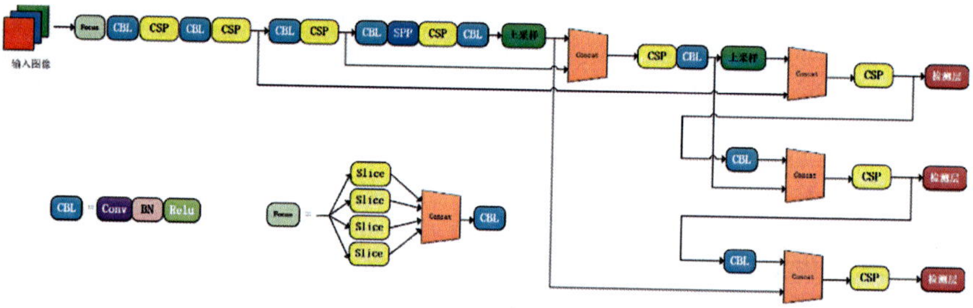

Fig. 1. YOLOv5 Model Structure.

The YOLOv5 model is divided into four models, s, m, l and x, which have the same network structure and can be resized by adjusting the parameters of depth and width. Among the four models, the YOLOv5s model has the shallowest depth, lowest accuracy and the smallest size. In this paper, YOLOv5s model is chosen as the model to be deployed. In addition, this scheme in this paper can also be applied to the other three models because these four models have the same structure.

The original YOLOv5 model uses a horizontal frame to label the information in the image. This idea is based on the premise of viewing objects from a human's horizontal perspective. However, from a top-down perspective, labeling objects with horizontal boxes will produce a lot of redundant information, which will affect model training and detection. In this paper, the data labeling method of Yolov5 model is improved by using a rotating rectangle box to label the target.

2.2 Convolution Kernel Pruning Technology

This scheme [19] introduces a scaling factor γ for each convolution kernel, the original output of the convolution kernel will be multiplied by γ as final output. The weights and scaling factors γ of the model are trained simultaneously, and the scaling factor γ is sparse regularized during training, which make the scaling factor of the unimportant convolution kernel in the network approach 0. Finally, the network is pruned according to the scaling factor, and fine-tuned the network after pruning. Specifically, the loss function of the model after training can be given by the formula (1):

$$L = \sum\nolimits_{(x,y)} 1(\mathrm{f}(x, W), y) + \lambda \sum\nolimits_{\gamma \in \Gamma} g(\gamma) \tag{1}$$

The variable L represents the loss of the model after sparse training, the variable x represents the input of the model, the variable W represents the weight of the model, the function f(x,W) represents the output of the model, the variable y represents the label of the target, the function l represents the error of the model output and label value after normal training without sparse training, the first accumulation on the right side of the formula represents the loss function of the YOLOv5 model during training, and the function g(γ) is a constraint on the scale factor γ. λ is a parameter used to balance two accumulation. This method use L1 regularization as the constraint. During the pruning process, the scaling factor γ will be used as the criterion to judge whether the convolution kernel should be pruned.

In the YOLOv5s model, a Batch Normalization layer is added after each convolution layer. Let the variables Z_{in} and Z_{out} represent the input and output of the BN layer. B Represents a set of incoming input data. Then the BN layer does the operation of formula (2):

$$z_{out} = \gamma \frac{z_{in} - \mu_B}{\sqrt{\sigma_B^2 + \epsilon}} + \beta \tag{2}$$

The scaling factor γ in the BN layer determines importance of the channel in feature selection, so the scaling factor in the BN layer can be directly used as the scaling factor γ during the training process. The advantage is that there is no additional cost to the network.

However, there are several problems in applying the convolution kernel pruning method directly to the YOLOv5 model.

(1) Clipping the convolution core directly may result in the complete elimination of an entire convolution layer, destroy the model structure, and completely lose the detection performance.
(2) There are shortcut structures in the YOLOv5 model, which also have convolution cores and BN layers. Clipping these convolution cores during pruning can severely damage the structure of the YOLOv5 model, resulting in a complete loss of detection performance.

Therefore, an improved convolution kernel pruning method is proposed in this paper, which is specially used to clip the YOLOv5 model.

3 Methodology

3.1 Improvement of Data Labeling

The original Yolov5 model uses the xyhw annotation method to label the data. (x, y) represents the coordinates of the center point of the rectangle, and (h, w) represents the width and height of the rectangle. This paper modifies the image annotation data processing part of Yolov5 model, and adopts an improved rotation frame annotation method for Yolov5 model. The annotation information of each object consists of five parameters x, y, l, s, θ. (x, y) represents the coordinates of the center point of the rectangle, l represents the length of the longer side of the rectangle, s represents the length of the shorter side of the rectangle, and θ represents the angle formed by the long side of the rectangle and the x-axis.

This design has three advantages:

(1) Compared with horizontal frame, rotating frame has less redundant information, which is more conducive to model training and detection;
(2) The four parameters x, y, l and s have the same meaning as the four parameters x, y, h and w of the original model. When modifying the label data processing module of YOLOv5 model, you only need to add the change part for the angle, without adding a large number of label conversion calculations to the model;
(3) The angle of the rotation frame can indicate the orientation of the target, which is of great significance for the relevant applications of remote sensing images (Fig. 2).

Fig. 2. Two Different Data Labeling methods.

3.2 Improvement of Convolution Kernel Pruning

(1) The scheme to avoid destroying the convolution layer structure.

By using the related method of convolution kernel pruning, a YOLOv5 model of BN layer parameter sparse distribution can be obtained through sparse training. In the pruning process, in order to prevent the systematic deletion of the convolution cores used for feature extraction, which results in under-fitting of the model in the direction of coarse-grained compression, a reasonable threshold value needs to be selected as the clipping reference.

This paper first introduces a global threshold Θ_g, which is a dynamic value whose determination method is determined by the global clipping rate N_g. When clipping, the BN layer parameters are sorted first. Then Then, the convolution kernel accounting for N_g are clipped. However, such clipping is likely to result in the systematic deletion of some of the convolutions used for feature extraction.

Therefore, this paper introduces the convolution layer security threshold Θ_i, which is a dynamic value, and its determination method is determined by the local clipping rate N_i. In the i layer of convolution neural network, only BN layer parameters below the security threshold Θ_i are subtracted to ensure the structural integrity of the convolution neural network. In practical programming, the global and security thresholds have only theoretical significance, and the pruning operation relies on the parameter N_g and N_i to compress the model and each convolution layer.

(2) The scheme to avoid destroying the shortcut structure.

YOLOv5 has a shortcut structure similar to the residual network. During pruning, it is necessary to ensure the consistency of the fusion channels of the ontology feature map and the residual feature map.

As the starting point of the shortcut, the number of output channels of the convolution layer does not directly affect the number of output channels of the residual module, and the pruning operation of the layer m is not constrained. In the shortcut structure, independent pruning of the layer m+1 and layer n causes channel characteristics to no longer match, resulting in model performance degradation after pruning. Therefore, the same results should be guaranteed. In view of the enormous performance improvement of shortcut structure in YOLOv5, the pruning results of the layer n used to generate the ontology feature map are used as the common pruning results with the shortcut structure layer m+1, and the features of the ontology feature map and the feature map i+2 are added to match the corresponding channels. This pruning method avoids destroying the shortcut structure so that the target detection ability of the model is not seriously affected (Fig. 3).

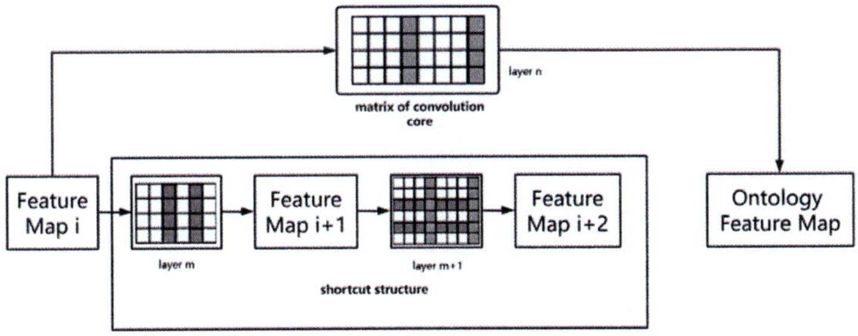

Fig. 3. Processing method of shortcut.

3.3 Optimization of FPGA deployment

In order to optimize the performance of YOLOv5 model in the space-borne field-bus environment, DPU is chosen as the computing engine of the deep neural network model in this paper. DPU is a programmable computing engine designed for better deployment of convolution neural network in the environment of the field bus. It consists of hybrid computing array module, high performance dispatcher module, instruction acquisition unit module and global memory pool module. DPU can make full use of multiplier and accumulator resources in the field bus to make the model run more efficiently in the environment of the field bus (Fig. 4).

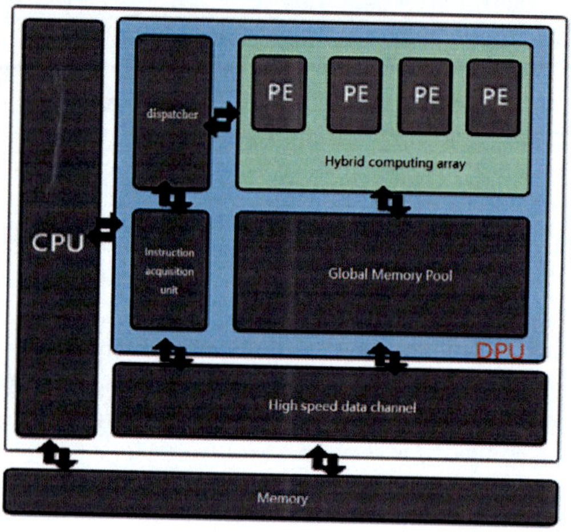

Fig. 4. DPU Architecture.

In this paper, the weight parameters in YOLOv5 model are quantified from 32-bit floating point number to 8-bit fixed point number, and a new round of training fine-tuning is made to compensate for the loss of detection accuracy caused by quantization error. The resulting 8-bit fixed-point model is then compiled to produce a lightweight model that can be run on a DPU.

3.4 Overall Collaborative Lightweight Scheme

In the specific implementation process, the YOLOv5s lightweight scheme proposed in this paper can be summarized as follows:

(1) Modify the relevant codes of the data preprocessing module and the data prediction module, and use the rotation box data annotation method to replace the horizontal box data annotation method of YOLOv5 model;
(2) Sparse the model obtained in the previous step, adjust N_g and N_i repeatedly, prune the trained model, and repeat the process several times;

(3) Quantify the model obtained in the previous step, and then conduct a normal training to obtain an 8-bit fixed-point YOLOv5s model, which is tested and evaluated in the FPGA environment (Fig. 5).

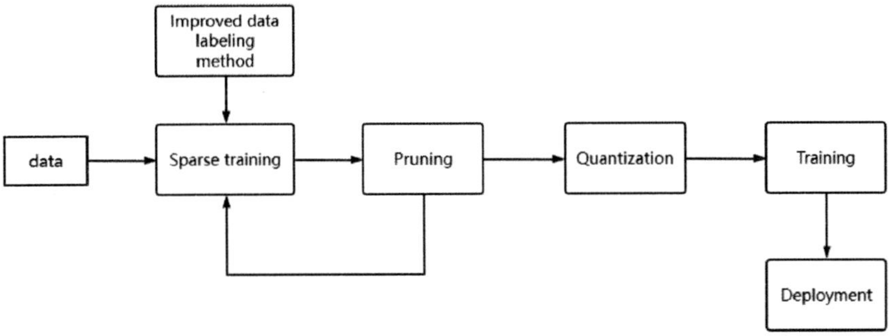

Fig. 5. Optimization Process of YOLOv5s.

4 Experiments

4.1 Experimental Setup

The model training and hardware-software co-acceleration tasks in this paper are completed on a PC with RTX3050Ti graphics card. The software environment is Pytorch1.8.2 and cuda11.1. Using the homemade remote sensing image aircraft datasets, which contains 11 types of aircraft, with a total number of targets of 9072. The training set includes 308 pictures, and the test set includes 122 pictures, all of the image are 3 channels with 4096 pixels × 4096 pixels. To speed up the training, the pictures are clipped to 3 channels with 1024 pixels × 1024 pixels, and the clipped pictures without valid target are deleted. The final training set had 2284 pictures, and the test set had 828 pictures.

The hardware used in the model deployment is the Xilinx Zcu102 development board. Zcu102 development board is a general experimental development board based on Zynq UltraScale + XCZU9EG-2FFVB1156E MPSoC. The Zcu102 development board has a 4G high-speed DDR4 SODIMM and component memory interface, an FMC extension port, thousands of megabits of serial transceivers per second, various peripheral interfaces, and a FPGA that provides a flexible prototype platform for customized design. In addition to FPGA, a quad-core Cortex-A53 and a dual-core Cortex-R5F real-time processor are integrated.

In terms of the experimental scheme, this paper sets up a comparison experiment based on the two parameters of sparse factor λ and clipping rate n, and selects the lightweight model with the best effect for the hardware acceleration. Then we test the 8-bit integer model in the custom operating system of Zcu102 development board and compare the optimization results. For compressed models, mAP, FLOPS, parameter quantities, and model file sizes are compared. Among them, mAP refers to the average accuracy rate, which is used to measure the detection accuracy of the target detection model. FLOPS represents the number of floating-point operations performed while the model is running.

4.2 Software Lightweight Results

After lightweight the YOLOv5s model according to the scheme proposed in this paper, the performance of the model is shown in Table 1. By comparison, it can be found that this scheme can greatly reduce the parameters and calculation of the model, the parameters and calculation complexity are reduced by about 50%, and the prediction rate, recall rate, mAP and other important detection indicators are within 1% of the normal training indicators, while the inference time of a single image is reduced by about 10%.

Table 1. Comparison of model parameters before and after lightweight operation.

λ	n	mAP0.5	parameter	GFLOPS	fps
0(Origin YOLOv5)	0	0.9	7525252	17.4	39.3
0.005	0.5	0.895	3542326	9.7	42.5
0.0005	0.5	0.901	3499376	9.6	44.6

The effects of different neural network optimization methods on the dataset are compared in Table 2. The slim pruning scheme is a comparative experiment on this dataset with reference to document [28]. It can be found that the method proposed in this paper based on convolution kernel pruning has certain advantages in model size and reasoning time (Fig. 6).

Table 2. Comparison of model parameters among different optimization methods.

method	Size/MB	MAP0.5	fps
YOLOv5s	15.0	0.9	39.3
Slim pruning	10.3	0.895	42.5
Kernel pruning	7.42	0.901	44.6

4.3 On-Orbit Deployment Results

After hardware acceleration of the model obtained by software lightweight according to the proposed scheme, the target detection performance of the model is shown in Table 3. By comparison, it can be found that this scheme can greatly reduce the parameters and computational load of the model, 66% of the model size are clipped, and the target detection indicators decreased by less than 3%, while the single image processing time is reduced by about 10%.

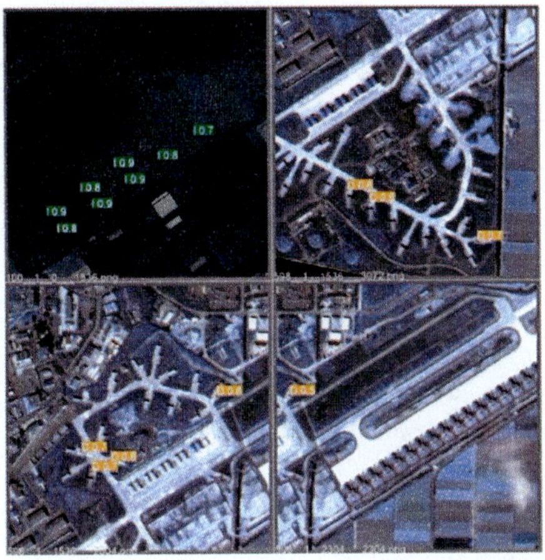

Fig. 6. Lightweight model detection effect image.

Table 3. Comparison of model performance (Model 1 is the original YOLOv5s model, Model 2 is a quantized model without software lightweight, and Model 3 is a quantized model with software lightweight).

model	size	MAP0.5	Fps in FPGA
1	15.0 MB	0.9	-
2	9.88 MB	0.885	28.0
3	5.01 MB	0.886	32.1

5 Conclusion

The rapid and intelligent target detection of massive remote sensing data on satellite edge devices is an important research topic to promote the construction of remote sensing satellite real-time service system and improve the real-time application capability of remote sensing satellites. This paper provides an optimization scheme of hardware-software cooperation for target detection real-time algorithm deployment in space-borne hardware environment. Under the premise that the model performance loss does not exceed 3%, 66% of the model size and 45% of the calculation are clipped, The experimental results show that the scheme can be effectively used in the deployment of the satellite edge device of the target detection model, and provides relevant reference for the deployment of the intelligent computing model on the space-borne hardware environment.

References

1. Qiao, K., Zhi, X.Y., Wang, D.W., et al.: Analysis and some thoughts on the development trend of the on-board intelligent information processing technology. Spacecraft Recovery Remote Sens. **42**(1), 21–27 (2021)

2. Li, W., Liu, X., Zhang, W.C., Ruan, N.J.: The application of dep learning in space-based intelligent optical remote sensing. Spacecraft Recovery Remote Sens. **41**(06), 56–65 (2020)
3. Ren, S., He, K., Girshick, R., et al.: Faster R-CNN: towards real-time object detection with region proposal networks. In: Advances in Neural Information Processing Systems, USA (2015)
4. Huang, G., Liu, Z., Van Der Maaten, L., et al.: Densely connected convolutional networks. In: Proceedings of the IEEE Conference on Computer Vision and Pattern Recognition, USA (2017)
5. Girshick, R., Donahue, J., Darrell, T., et al.: Rich feature hierarchies for accurate object detection and semantic segmentation. In: Proceedings of the IEEE Conference on Computer Vision and Pattern Recognition, USA (2014)
6. Zhang, C.X., Bao, Y.F., Ma, Z.Q., Tao, R., Li, W., Tian, Y.: Research progress on optical remote sensing object detection based on CNN. Spacecraft Recovery Remote Sens. **41**(06), 45–55 (2020)
7. Tong, Y., Zhou, H., Bian, C.J., et al.: Neural network compression method for real-time object detection on satellite. Comput. Eng. Des. **42**(6), 1713–1720 (2021)
8. Wang, S.S., Wang, M., Wang, G.Y.: Deep neural network pruning based two-stage remote sensing image object detection. J. Northeast. Univ. (Nat. Sci.) **40**(2), 174–179 (2019)
9. Li, Q., Wu, W.B., He, M.Y.: Accelerator of remote sensing image object detection based on MPSoC. Spacecraft Recovery Remote Sens. **43**(1), 58–68 (2022)
10. Li, Z.L., Wang, L.Y., Yu, J.Y., et al.: Space-borne SAR on-orbit imaging and ship target detection method. Spacecraft Eng. **27**(6), 41–47 (2018)
11. Bian, C.J.: Research on valid region on-board real-time detection and compression technology applied for optical remote sensing image. Harbin Institute of Technology, Harbin (2018)
12. Straight, S.D., Doolittle, C., Cooley, T., et al.: Tactical satellite-3 mission overview and initial lessons learned. In: Proceedings of Annual AIAA/USU Conference on Small Satellites. Washington D.C.: AIAA (2010)
13. Alessandro, G., Cody, B., Milton, D., et al.: SpaceCube v3. 0 NASA next-generation high-performance processor for science applications. In: Proceedings of Annual AIAA/USU Conference on Small Satellites. Washington D.C. AIAA (2019)
14. Zhang, P., Wu, W.B., Li, Q., Cao, C.H.: Research on lightweight of remote sensing object detection algorithm for spaceborne edge computing. Aerosp. Control Appl. **48**(05), 86–94 (2022)
15. Aghasi, A., Abdi, A., Nguyen, N., et al.: Net-trim: convex pruning of deep neural networks with performance guarantee. In: Conference on Neural Information Processing Systems. IEEE, New York (2017)
16. Lin, M., Ji, R., Wang, Y., et al.: HRank: filter pruning using high-rank feature map. In: 2020 IEEE/CVF Conference on Computer Vision & Pattern Recognition. IEEE, New York (2020)
17. Hu, Y.L., Li, Y., Zhou, Q., et al.: Structure stripe pruning for on-board object detection. Aerosp. Control Appl. **48**(05), 67–77 (2022)
18. Tang, W., Hua, G., Wang, L.: How to train a compact binary neural network with high accuracy? In: 31st AAAI Conference on Artificial Intelligence, USA (2017)
19. Chen, G., Choi, W., Yu, X., et al.: Learning efficient object detection models with knowledge distillation. In: Advances in Neural Information Processing Systems, USA (2017)
20. Cheng, Z., Lyv, J.G., Bai, Y.Q., et al.: High-resolution remote sensing image object detection algorithm combining RPN network and SSD algorithm. Sci. Survey. Mapp. **46**(4), 75–82, 99 (2021)
21. Xue, J.D., Zhu, J.J., Li, X.H., et al.: SSD model optimization for typical object detection in optical remote sensing images. Comput. Syst. Appl. **30**(10), 301–306 (2021)
22. Gong, M., Liu, Y.Y., Li, G.N.: A ship detection method for remote-sensing images based on improved YOLO-v3. Electron. Opt. Control. **27**(5), 102–107 (2020)

23. Hua, Y.R., Zhang, Z., Long, S., et al.: Remote sensing image target detection based on improved Yolo algorithm. Electron. Measur. Technol. **43**(24), 87–92 (2020)
24. Ma, Y., Shan, Y.G., Yuan, J.: Remote sensing object detection algorithm based on improved faster R-CNN. Mod. Electron. Tech. **45**(3), 58–63 (2022)
25. Wang, C.Y., Liao, H., Wu, Y.H., et al.: CSPNet: a new backbone that can enhance learning capability of CNN. In: 2020 IEEE/CVF Conference on Computer Vision and Pattern Recognition Workshops (CVPRW), USA (2020)
26. Lin, T.Y., Dollár, P., Girshick, R., et al.: Feature pyramid networks for object detection. In: Proceedings of the 2017 IEEE Conference on Computer Vision and Pattern Recognition (CVPR). Honolulu, USA, pp. 936–944 (2017)
27. Liu, S., Qi, L., Qin, H., et al.: Path aggregation network for instance segmentation. In: Proceedings of the IEEE Conference on Computer Vision and Pattern Recognition, USA (2018)
28. Zhang, P.Y., Zhong, Y.X., Li, X.Q.: SlimYOLOv3: narrower, faster and better for real-time UAV applications. In: Computer Vision and Pattern Recognition, 25 July 2019. https://arxiv.org/abs/1907.11093

Check for updates

Research on the Measuring Method of Electronic Theodolite See-Through Porthole Glass

Chunzhen Ren[1(✉)], Ziwen Wang[1], Minxing Zhou[2], Weiwei Jin[3], and Chenxu Duan[1]

[1] Beijing Institute of Spacecraft Environment Engineering, China Academy of Space Technology, Beijing 100094, China
92147000@sohu.com

[2] Institute of Telecommunication and Navigation Satellites, China Academy of Space Technology, Beijing 100094, China

[3] Beijing Institute of Spacecraft System Engineering, China Academy of Space Technology, Beijing 100094, China

Abstract. The accurate control of spacecraft in orbit is closely related to the attitude installation accuracy of equipment during the ground assembly stage, and the measurement of assembly accuracy plays an important role in the assembly process. With the continuous update and development of spacecraft, there are new requirements for the environment of equipment installation accuracy measurement. For example, in order to ensure the consistency of the installation accuracy of control equipment on the ground and the sky, some spacecraft need to simulate the in-orbit environment on the ground, that is, operating the accuracy measurement of equipment in a sealed cabin. The accuracy measurement of spacecraft mainly adopts the collimation measurement of electronic theodolite and the joint construction of multiple theodolites to determine the attitude relationship of equipment in the spacecraft's mechanical coordinate system. In the test, the collimating optical path of the electronic theodolite must be free of occlusion, otherwise, the measurement cannot be completed. In order to meet the measurement requirements of equipment in the closed cabin of spacecraft, this paper proposes a new test scheme of theodolite seeing-through porthole glass. Through the analysis of the influence of refraction accuracy produced by light and glass perspective, and through the test verification of self-developed porthole glass, the applicable conditions of theodolite seeing-through porthole glass are obtained. Finally, the problem of the installation attitude measurement of the equipment in the sealed cabin is solved. The accuracy index meets the requirements of the spacecraft design and application, and it provides a reliable solution for the installation accuracy of the equipment in the special environment of the spacecraft.

Keywords: electronic theodolite · see-through porthole glass · accuracy measurement

H. P. Urbach and H. Jiang (Eds.): ISSOIA 2022, SPPHY 295, pp. 680–687, 2023.
https://doi.org/10.1007/978-981-99-4098-1_56

1 Introduction

The measurement of spacecraft installation accuracy is an important step in the final assembly process [1]. With the development of spacecraft, the installation accuracy measurement needs in various environments. For example, in order to ensure the consistency of the installation accuracy of the equipment in the sky and theground, a spacecraft needs to measure the accuracy of the equipment in the cabin in a closed pressure-keeping state. At present, the accurate measurement of spacecraft mainly adopts the cubic mirror collimation measurement method, and the attitude relationship [2] of the equipment in the spacecraft mechanical coordinate system is determined through the joint establishment of the electronic theodolite. During the test, the alignment light path between the electronic theodolite and the measured cubic mirror must be unobstructed, otherwise, the measurement [3] cannot be completed. In order to meet the requirements of spacecraft special environment measurement, this paper puts forward a test scheme of theodolite perspective porthole glass and verifies the application feasibility of theodolite perspective porthole glass quasi-upright square mirror measurement method and the influence of measurement accuracy, thus providing a solution [4–6] for high-precision measurement of a reference mirror of equipment in a closed environment.

2 Analysis of Theodolite Measuring Method and Accuracy Influence

2.1 Measuring Principle of Theodolite

At present, a high-precision optical cube mirror or plane mirror is often used to measure the precision of spacecraft equipment, which is used to represent the optical coordinate system or a certain direction vector pointing [7, 8] of the equipment under test. As shown in Fig. 1, the electronic theodolite with self-collimation function is used to establish a joint station, such as collimation and mutual sighting, and the attitude relationship [9–11] between the tested equipment and the measurement reference coordinate system can be obtained by operation with the special measurement software for precise measurement.

2.2 Theodolite See-Through Porthole Glass Measurement Scheme

According to the pressure-maintaining state of the spacecraft's sealed cabin, two orthogonal port windows are opened at the cabin end face, and the special glass for precision measurement is installed in a sealed manner. Through the theodolite to see through the porthole glass, the normal directions of the two orthogonal surfaces of the reference mirror of the equipment in the measurement cabin are collimated to measure the installation attitude angle of the equipment. The schematic diagram is shown in Fig. 2.

2.3 Accuracy Error Analysis of Theodolite Perspective Glass

When the theodolite collimates the reference mirror in the sealed cabin, it must be measured through the glass, which will cause refraction. According to the law of refraction

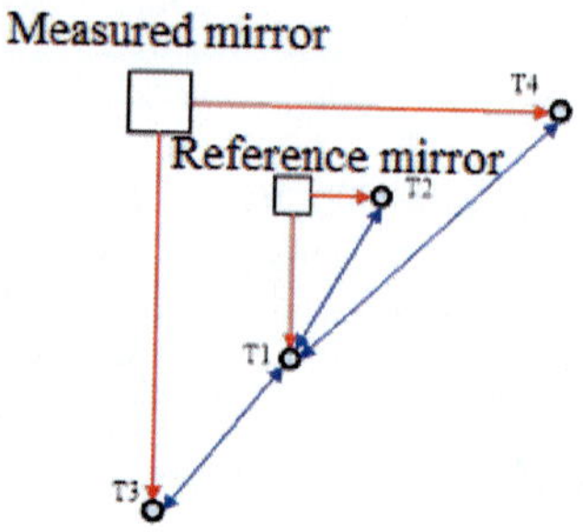

Fig. 1. Schematic diagram of theodolite collimation measurement joint station building.

Fig. 2. Schematic diagram of theodolite perspective porthole glass measurement.

of light, when light is reflected by the first medium (refractive index n_1) is injected into the second medium (refractive index n_2), on a smooth interface, part of the light is refracted after entering the second medium from the first medium. All kinds of optical instruments used to control the optical path and image are mainly based on the law of refraction and reflection of light [12]. The refraction of single-layer glass and its error influence is shown in Fig. 3.

As shown in Fig. 3, the influence of single-layer glass refraction on the direction value.

δ is:

$$\delta = \alpha_1 - \beta \tag{1}$$

When the incident angle α_1 is known, the angle β_1 is calculated as shown in formula (2):

$$tan\beta_1 = \frac{(h-d)tan\alpha_1 + d \cdot tan\alpha_2}{h} \tag{2}$$

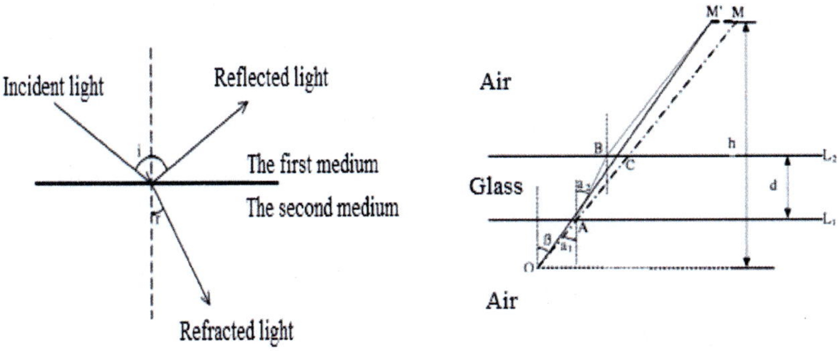

Fig. 3. Refraction of single-layer refractive glass and its error influence.

where, d is the thickness of the glass (unit mm), h is the projection length of OM on the glass normal, α_1 and α_2 satisfies the relation (3)

$$\frac{sin\alpha_1}{sin\alpha_2} = \frac{n_2}{n_1} \tag{3}$$

where n_1 is the atmospheric refractive index, take 1, n_2 is the refractive index of glass. According to the above deduction, when the glass thickness is fixed, the incident angle is α_1 is approximately inversely proportional to the distance between the measurement site and the target point, and is proportional to the β approximately.

The law of influence can be summarized as follows:

(1) The closer the incident angle is to 0°, the less influence of glass refraction on orientation is;
(2) The farther the measuring station is from the target point, the less influence of glass refraction on orientation is;
(3) When the collimation distance is fixed, the refraction increases significantly with the increase of incident angle, but the increased amplitude decreases with the increase of distance.

3 Test Verification

In this test, two portholes 1 and 2 are made of single-layer K9 glass with a thickness of 12 mm and a refractive index of 1.52. According to the influence of glass refraction error, the data in Table 1 can be obtained.

3.1 Measurement Scheme

This paper mainly compares the influence of perspective porthole on the accuracy of theodolite quasi-vertical square mirror. The test is to collimate the reference cubic mirror through theodolite, and compare with and without porthole glass shielding; Through the

Table 1. Estimation data of angle error of refraction of single-layer glass.

β	h				
	2m	4m	6m	8m	10m
0°	0.0	0.0	0.0	0.0	0.0
2°	−14.8	−7.4	−4.9	−3.7	−3.0
4°	−29.5	−14.8	−9.8	−7.4	−5.9
6°	−44.3	−22.14	−14.8	−11.1	−8.9
8°	−59.0	−29.5	−19.7	−14.7	−11.8
10°	−73.6	−36.8	−24.5	−18.4	−14.7

final data comparison and analysis, the influence of the perspective porthole glass on the accuracy of the theodolite collimating mirror is obtained, and the uncertainty of the influence of the different angles of the porthole glass on the theodolite quasi vertical square mirror is obtained.

3.2 Measurement Process

On the high precision operation platform, a precision measuring tooling cubic mirror is fixed with a glue gun; Use theodolite 1 and theodolite 2 collimate two orthogonal planes of cubic mirror respectively; The data storage results of 6 times were measured and used as the basic comparison value with porthole glass occlusion.

Install the fine survey perspective porthole 1#and porthole 2#on the fixed support. According to the different positions of portholes in Table 2, through the precise measurement of perspective porthole glass, use theodolite 1 and theodolite 2 to carry out collimation measurement on two orthogonal planes of precision measuring tooling cubic mirror, record and store the measurement data for 6 times; The mean values of final test data are shown in Table 2

3.3 Analysis of Test Data Results

Two methods are adopted to evaluate the uncertainty of test measurement in the test. First, the measurement results of the collimation measurement of the theodolite under different working conditions with a perspective porthole are obtained. The formula for calculating the average value of multiple measurement samples is shown in Eq. (4):

$$\overline{x} = \frac{\sum_{i=1}^{n}[x_i]}{n} \tag{4}$$

where $\overline{x}$ is the arithmetic average of the measured values x_i is the single observation value and n is the number of measurements. Calculate the deviation value of the arithmetic mean value of shielding with perspective portholes compared with the measured data of theodolite collimation without perspective portholes, and we can get the influence of porthole glass on theodolite collimation accuracy.

Table 2. Data results of theodolite collimating reference cubic mirror with or without porthole occlusion.

Shelter condition of porthole placement	Theodolite			
	Theodolite 1 (porthole 1 blocked)		Theodolite 2 (shielded by porthole 2)	
	horizontal angle	angle of pitch	horizontal angle	angle of pitch
Mean value without porthole	128.00589	89.99098	88.61799	90.06655
Average occlusion value of vertical middle position of porthole	128.0059	89.9911	88.6170	90.0656
Mean value of internal occlusion of horizontal translation glass diameter of porthole	128.0061	89.9906	88.6171	90.0625
Mean value of shielding within 30 in front and rear pitching direction of porthole	128.0059	89.9910	88.6187	90.0682
Mean value of rotation shielding of portholes within 30 degrees horizontally	128.0062	89.9914	88.6189	90.0665

Second, multiple measurement samples of each working condition are tested in the test measurement, as shown in Eq. (5).

$$s = \sqrt{\frac{\sum_{i=1}^{n} (x_i - \overline{x})^2}{n-1}} \tag{5}$$

where $\overline{x}$ is the arithmetic mean of the measured values, x_i is the single observation value and n is the number of measurements. Calculate the standard deviation of the sample to judge the uncertainty of the stability of the system from the dispersion degree of the sample. When the average value of multiple measurements is used as the measurement result, calculate the standard uncertainty according to the formula $u_A = s(\overline{x})$

The uncertainty under the influence of multiple factors is calculated by the formula (6) to synthesize the standard uncertainty.

$$u_c = \sqrt{\sum_{i=1}^{n} {u_i}^2} \tag{6}$$

The extended uncertainty is calculated according to the formula $U = ku_c$. Generally, the value k is 2 when the confidence level is 0.95, while k is 3 when the confidence level is 0.99.

According to the actual working conditions of the test, the vertical middle position of the porthole is least affected by the measurement accuracy of the glass shielding perspective. See Table 3 for specific comparison data.

Table 3. Measuring accuracy of theodolite passing through the vertical middle area of porthole

Porthole placement condition		Measured value of theodolite			
		Theodolite 1 (porthole 1)		Theodolite 2 (porthole 2)	
		horizontal angle	angle of pitch	horizontal angle	angle of pitch
Comparison between vertical middle position of porthole shielding and no porthole	Mean value without porthole ()	128.00589	89.99098	88.61799	90.06655
	Average occlusion value of vertical middle position of porthole ()	128.0059	89.9911	88.6170	90.0656
	Deviation value (")	0.036	0.432	3.564	3.42
The vertical middle position of the porthole is blocked	Intermediate standard deviation value (")	0.6635	0.6467	2.502	3.2089
The combined uncertainty (") of whether there is obstruction in the middle of the porthole		0.6645	0.7777	4.3545	4.690

It can be seen from Table 3 that the maximum combined uncertainty of porthole 1 is the pitch angle direction, i.e. $u_c = 0.78''$, and the expanded uncertainty is $U = 1.56''$ (k = 2); The maximum combined uncertainty of porthole 2 is its pitch angle, i.e. $u_c = 4.69''$, and the expanded uncertainty is $U = 9.38''$ $(k = 2)$ 。

4 Conclusion

(1) Through the research on the method of theodolite see-through porthole glass, it can be known that different placement states of porthole glass, that is, the size and distance of theodolite collimation incident angle, have different influences on the measurement accuracy of theodolite collimation light path.

(2) The most ideal working condition is when the incident angle between the collimating optical path of the theodolite and the porthole glass is near 0, that is, the porthole glass is perpendicular to the optical path of the theodolite.
(3) Through the test verification of self-developed porthole 1 and porthole 2. The expansion uncertainty of porthole 1 reaches 1.56″; The expansion uncertainty of porthole 2 reaches 9.38″. In the application of spacecraft sealing and pressure-keeping precision measurement, completely meets the requirement that the influence of design measurement accuracy is less than 20″.
(4) The electronic theodolite perspective porthole glass measurement method solves the problem of high-precision measurement of equipment installation posture in a closed environment, explores a new measurement scheme, and provides a brand-new measurement scheme for other non-contact such as photogrammetry and laser measurement.

References

1. Fuxiang, X., Baohua, L., Haku, H., et al.: Introduction to Satellite Engineering, pp. 1–307. Aerospace Publishing House, Beijing (2003)
2. Zaihua, Y.: An analysis of messurement technology in spacecraft assembly. Spacecraft Environ. Eng. **24**(6), 390–392 (2007)
3. Yang. L.H., Yang, X.Y., Luo, D.B., et al.: Large size coordinate measurement method based on optical plane intersection. Infrared Laser Eng. **39**(6), 1105–1109 (2010)
4. Jiang, S., Yang, L., Yu, J.: A high accuracy CCD photogrammetry system used in the vacuum cryogenic environment. Spacecraft Environ. Eng. **27**(3), 361–363 (2010). (in Chinese)
5. Lu, C.: Research and application of industrial digital photogrammetry technology in thermal vacuum deformation measurement of micro antenna. PLA Information Engineering University, Zhengzhou (2008)
6. Cheng, Z., Feng, Q., Yi, W., Li, Z., Wang, Y.: Analysis and error compensation of the effect of the protection tank of camera on the photogrammetry accuracy. Spacecraft Environ. Eng., **33**(2) (2016)
7. Wang, Z., Wu, G.: Calibration of transition matrix on cubic prisms in mapping camera and star sensor. Opt. Precis. Eng. **20**(1), 96–100 (2012). (in Chinese)
8. Zhang, J., Yang, Z., Jiao, A., et al.: Thermal deformation composite measurement experiment method of camera-star sensor combination bodies under atmospheric pressure. Nanotechnol. Precis. Eng. **12**(03), 167–175 (2014). (in Chinese)
9. Yang, Z., Li, G., Fan, B., et al.: [9] Study on collimation transfer method of cubic mirror under multiple mutual sighting conditions. Astronaut. Meter Meas. Technol. **33**(6), 10–12+30 (2013)
10. Yan, R., Wang, Y., Sun, W., Chen, M., et al.: Testing and calibrating technology in general assembly of large satellites. J. Astronaut. Metrol. Meas. **20**(3), 16–23 (2018)
11. Jianxin, L.: Spacecraft structure factors affecting equipment installation accuracy and optimization measures. Spacecraft Environ. Eng. **32**(3), 337–342 (2015)
12. Zhao, L., Wang, Q., Zhou, P., Duan, Y.: Research on correction of refraction error in displacement measurement based on vision. Instrum. Tech. Sens. **2**, 104–108 (2021)

Development and Integration of the Online Monitoring System for Early Warning of Environmental Parameters-A Case Study

Q. D. Cheng[1], Z. L. Gao[1](✉), G. F. Li[1,2], W. Z. Hu[1,2], G. L. Zeng[1,2], Y. B. Dong[1], L. Wang[1], B. Wang[1], P. P. Zhang[1], and Z. J. Hu[1]

[1] Beijing Orient Institute for Measurement and Test, Haidian, Beijing 100094, China
gzl514cast@126.com

[2] First Branch Company of Beijing HangTianHe Science Technology Development Co., Ltd., Changping, Beijing 102205, China

Abstract. In some specific scenes or environments, there are special monitoring and controlling requirements for certain environmental parameters, and the optical measurement technology plays an extremely important role in the measurement of many environmental parameters. In data centers, control of corrosive environmental parameters is one of the critical measures to maintain the equipment life and data security in the computer room. In this study, we took a typical data center as the research object, investigated its internal and external environmental quality, analyzed the possible corrosive gases and corresponding thresholds, and selected the measurement techniques and methods. Then a high-precision online monitoring system was developed and integrated, including the particle monitoring module and the corrosive gas monitoring module. The system applies high-sensitivity optical measurement technologies——the light scattering method for the particle monitoring module which can monitor $PM_{2.5}$ and PM_{10} in the air simultaneously, and photoacoustic spectroscopy and ultraviolet fluorescence for the trace corrosive gas NO_2 and SO_2 monitoring modules which can achieve ppb-level detection limits and resolutions of corrosive gas concentrations. It provides online monitoring, early-stage warning and evaluation of particles and trace corrosive gases in the internal and external environment, and has been installed in a computer room of the typical data center. In the future, along with the further development of the optical measurement technology and the continuous improvement of detection limits and measurement accuracy, such modular systems can also be applied to more clean environments or specific scenes such as operating rooms, museums, and even space stations, to monitor more contaminants online.

Keywords: Optical measurement technology · Data center · Particle · Corrosive gas · Online monitoring

H. P. Urbach and H. Jiang (Eds.): ISSOIA 2022, SPPHY 295, pp. 688–699, 2023.
https://doi.org/10.1007/978-981-99-4098-1_57

1 Introduction

1.1 The Importance of Air Quality in Specific Scenes

With the progress of science and technology and the deepening of people's understanding of pollution, the air quality in many clean environments or specific scenes is becoming more and more important. It directly affects the performance, yield, quality and reliability of products and equipment in the environment, as well as the physical and mental health of the people involved.

For instance, in museums, it is essential to continuously monitor and control indoor environmental parameters such as luminosity, temperature, humidity and corrosive gases to slow the corrosion rate and protect collections as much as possible [1, 2]; In space stations or operating rooms, light intensity, temperature and humidity, the concentration of carbon dioxide or microorganisms directly affect the physical and mental health of the personnel, even the prognosis of the operations [3–6]; In arsenals, armories or ammunition depots, environmental temperature and humidity, corrosive gases and electrostatic charges are monitored to reduce the corrosion rate and avoid deterioration and failure, to extend the storage life of firearms and ammunition, and ensure storage safety [7, 8]; Many industries, such as electronic information, semiconductor, optoelectronics, precision manufacturing, medicine and health, bioengineering and aerospace, also need to control environmental cleanliness [9].

1.2 Optical Technologies for Air Quality Monitoring

Air quality encompasses many environmental parameters. Here, we focus on the optical technologies for particulate matter and gas concentration detecting.

For particulate matters, real-time detection instruments are always based on tapered element oscillating microbalance, β radiation absorption or light scattering. Among them, the light scattering sensors have the advantages of rapidity, sensitivity, stability and small size, and have been widely used in recent years. Scattering-based optical counters (OPCs), the working principle of which is based on scattering theories such as the Mie or Ryleigh theory, can determine the number concentration of particles in multiple particle size ranges simultaneously through multiple particle size channels. If necessary, OPCs also can convert the particle number concentration to mass concentration values [10].

At present, gas concentration detection technologies, which have been widely used in atmospheric environment, industrial process, petrochemical industry or coal mine safety, mainly focus on electrochemical methods, gas chromatography and optical methods. Among them, the electrochemical method detects the gas concentration by measuring the electrical signals which are generated from the chemical reaction of the gas and electrode surface and related to the gas concentration. It is relatively mature, versatile and low-cost, but is susceptible to environment temperature, humidity and electrode surface pollution. That is, the accuracy of the electrochemical method is limited, regular calibration is required and the service life is relatively short. The gas chromatography is used to separate multi-component gas through a chromatograph to obtain the wave peaks related to the concentration, and then the concentration of each component is determined by measuring the corresponding peak area. It shows high sensitivity, high efficiency and high accuracy of detection. Nevertheless, the process of gas chromatography is complex and mainly used for laboratory measurement, i.e., it is difficult to achieve real-time on-line monitoring. The optical method is a general term for gas concentration detection using optical technologies, including chemiluminescence, ultraviolet fluorescence, tunable diode laser absorption spectroscopy (TDLAS), non-dispersive infrared (NDIR), photoacoustic spectroscopy, cavity ring-down spectroscopy (CRDS) and so on. Generally, each method has its own advantages and disadvantages, so it should be selected according to specific application scenes to meet the actual needs [11, 12].

1.3 Air Quality Requirements of Data Centers

Recent years, increasing needs for ubiquitous computing has resulted in an explosive growth of data centers around the world. Particles and corrosive gases can enter the interior of the electronic equipment to cause slow and continuous corrosion, thus affect its long-term reliability adversely. As data centers are critical infrastructures supporting cloud computing services and mobile Internet business, they have always been carefully designed and built. From the beginning of the design to the later operation and maintenance of data centers, corresponding measures such as fresh air filtering system and detection of air quality are regularly done to make the cleanliness meet the requirements [13–15].

Relevant standards or specifications have clear technical requirements for data centers. The China National Standard GB 50174-2017 *Code for design of data centers*, proposes technical requirements for environmental parameters of data centers at all grades (see Table 1). It is clear that requirements for temperature, relative humidity and particulates concentration of computer room are involved, but it has no requirements for corrosive gas [16]. The American National Standard ANSI/ISA-71.04 *Environmental Conditions for Process Measurement and Control Systems: Airborne Contaminants*, divides the corrosion severity levels of environmental pollutants into G1, G2, G3 and GX (see Table 2). Among them, corrosion is not a factor in determining equipment reliability at G1 level. Also, the standard gives the corrosive gases concentration corresponding to the reactive corrosion severity levels, which is as low as ppb (part per billion) level for G1 (see Table 3)[17]. The China National Standard GB/T 51314-2018 *Operation and maintenance standard for data center infrastructure*, proposes a concept calling *predictive maintenance*, which has become popular in recent years [18]. It means to collect and analyze data and signals through various technical means, to predict the possible consequences, and then take targeted maintenance activities. It is expected to reduce the operation and maintenance costs, and also reduce the losses caused by system failures.

In summary, air quality is extremely important to maintain the equipment life and data security for the specific scene of the data center computer room. Although there are no explicit requirements for corrosive gas in the current standards of data centers, we believe that it is highly necessary to develop a set of monitoring and early-stage warning methods and systems based on the current technologies. It is in accord with the concept of *predictive maintenance*, and could be used to monitor the corrosion-related environmental parameters. Then the monitoring data will provide references for the operation and maintenance personnel to take targeted maintenance measures. According to the existing researches, corrosivity is related to particulate matter and various corrosive

Table 1. Technical requirements for data centers at all levels.

Listings	Technical requirements			Notes
	Grade A	Grade B	Grade C	
Environmental requirements				
Ambient temperature and relative humidity of computer room (during shutdown)	5 °C ~ 45 °C, 8% ~ 80%, and the dew point temperature should not be greater than 27 °C			No condensation
Temperature change rate of computer room and auxiliary area	It should be less than 5 °C/h for magnetic tape drive unit, and less than 20 °C/h for disk driving device			
Temperature and relative humidity of auxiliary area (at startup)	18 °C ~ 28 °C, 35% ~ 75%			
Temperature and relative humidity of auxiliary area (during shutdown)	5 °C ~ 35 °C, 20% ~ 80%			
Concentration of particles in computer room	It should be less than 17600000 particles			It is the number of suspended particles (particle size ≥ 0.5 μm) per cubic meter of air
Monitoring system of environment and equipment				
Air quality	Particle concentration shall be detected		—	Offline periodic detection
Air quality	Temperature, dew point and differential pressure shall be tested		Temperature and dew point should be detected	Online monitoring or connecting parameters to the computer room environment and equipment monitoring system through data interface
Central air conditioning, fresh air system and dynamic system	Equipment operation status and filter differential pressure shall be tested			
Air conditioner for computer room	The state parameters—switch, refrigeration, heating, humidification, dehumidification, water valve opening and water flow— shall be detected The alarm parameters—temperature, relative humidity, sensor failure, com-pressor pressure, humidifier water level and air volume—shall be detected			

gases, each of which may correspond to various technologies. For specific scenes, which parameters to monitor and which technologies to choose are the key and difficult points in building a monitoring system. This study aims to make a preliminary exploration on this topic.

2 Monitoring System for a Typical Data Center

Based on the above analysis, it is clear that the control of corrosive environmental parameters in data centers is one of the critical measures to maintain the equipment life and data security in the computer room. So, we took a typical data center as the research object, investigated its internal and external environmental quality, analyzed the monitoring needs combined with relevant standards. Based on above discussion, an online monitoring and early warning system scheme was formulated, and then a

Table 2. Classification of reactive environments (1 Å = 0.1 nm = 10^{-10} m).

Level	Copper/Silver reactivity level (in angstroms)	Severity	Explanation
G1	<300 (Copper) <200 (Silver)	Mild	An environment sufficiently well controlled such that corrosion is not a factor in determining equipment reliability
G2	<1000	Moderate	An environment in which the effects of corrosion are measurable and may be a factor in determining equipment reliability
G3	<2000	Harsh	An environment in which there is a high probability that corrosive attack will occur. These harsh levels should prompt further evaluation, resulting in environmental controls or specially designed and packaged equipment
GX	≥2000	Severe	An environment in which only specially designed and packaged equipment would be expected to survive. Specifications for equipment in this class are a matter of negotiation between user and supplier

Table 3. Contaminant concentrations versus severity levels.

Severity level			G1	G2	G3	GX
			Mild	Moderate	Harsh	Sever
		Gas concentration[a]				
	Contaminant	Gas	Concentration			
Reactive species[a, b]	Group A	H_2S	<3	<10	<50	≥50
		SO_2, SO_3	<10	<100	<300	≥300
		Cl_2	<1	<2	<10	≥10
		NO_x	<50	<125	<1,250	≥1,250
	Group B[c]	HF	<1	<2	<10	≥10
		NH_3	<500	<10,000	<25,000	≥25,000
		O_3	<2	<25	<100	≥100

[a] Parts per billion by volume (ppb-v) average for test period for the gases in Groups A and B.
[b] The Group A contaminants often occur together and the reactivity levels include the synergistic effects of these contaminants
[c] The synergistic effects of Group B contaminants are not known at this time

high-precision particle and corrosive gas online monitoring system was developed and integrated. See the following for more details.

2.1 Air Quality Status

According to the China National Standard GB 3095-2012 *Ambient air quality standards*, the ambient air functions are divided into two categories: Class I areas which include natural reserves, scenic spots and other areas that need special protection; Class II areas which include residential districts, hybrid areas of commerce, traffic and residence, cultural areas, industrial areas and rural areas. The ambient air concentration limits in above two categories are shown in Table 4: Class I concentration limits apply to Class I areas, and Class II concentration limits apply to Class II areas [19].

In this study, it is considered that the data centers should be classified as *areas that need special protection* (Class I areas). So, the pollutant concentration limits in Table 4 can be used as reference for ambient air quality of our study object. In general, the overall air quality of the area where the data center is located is good. The hourly air quality data of the region in the past year were statistically analyzed, and the results are shown in Table 5.

Table 4. Concentration limits of basic items of ambient air pollutants.

No.	Item	Average time	Concentration limits		Units
			Class I	Class II	
1	Sulfur dioxide (SO_2)	Annual mean	20	60	μg/m^3
		24-h average	50	150	
		1-h average	150	500	
2	Nitrogen dioxide (NO_2)	Annual mean	40	40	
		24-h average	80	80	
		1-h average	200	200	
3	Carbon monoxide (CO)	24-h average	4	4	mg/m^3
		1-h average	10	10	
4	Ozone(O_3)	Daily maximum 8-h average	100	160	μg/m^3
		1-h average	160	200	
5	Particulate matter (PM_{10})	Annual mean	40	70	
		24-h average	50	150	
6	Particulate matter ($PM_{2.5}$)	Annual mean	15	35	
		24-h average	35	75	

Table 5. Air quality data statistics of the data center location in the past year.

Item (Units)	Mean	Maximum	Minimum	First quartile	Second quartile (Median)	Third quartile
AQI (/)[a]	58	500	12	37	50	68
CO (mg/m^3)	0.66	2.86	0.1	0.4	0.6	0.84
NO_2 (μg/m^3)	21	108	2	11	17	26
O_3 (μg/m^3)	74	244	5	47	67	93
PM_{10} (μg/m^3)	62	2246	3	29	47	75
$PM_{2.5}$ (μg/m^3)	26	317	4	14	20	32
SO_2 (μg/m^3)	11	86	2	5	8	13

[a] AQI (Air Quality Index) is a comprehensive index used to describe the degree of air cleanliness or pollution.

Although the overall air quality in the project area is good, it was found in the field survey that there are many vehicles coming and going near the data center. That means vehicle exhaust may be one of the possible pollution sources. According to the research, the components of automobile exhaust include CO, hydrocarbons, nitrogen oxides, SO_2, particulate matter, etc. Among them, nitrogen oxides and SO_2 are corrosive gases, which need special attention.

So, the operation and maintenance personnel installed copper coupons and silver coupons in the computer room for reactive corrosion monitoring and analysis, and the results are shown in Fig. 1.

It is shown that the corrosion severity level of airborne pollutants in the computer room was generally at G1-G2 level, mostly at G2, and even G3 in some periods. According to ANSI/ISA-S71.04-2013, severity level G2 means moderate, i.e., an environment in which the effects of corrosion are measurable and may be a factor in determining equipment reliability, and severity level G3 means harsh, i.e., an environment in which there is a high probability that corrosive attack will occur.

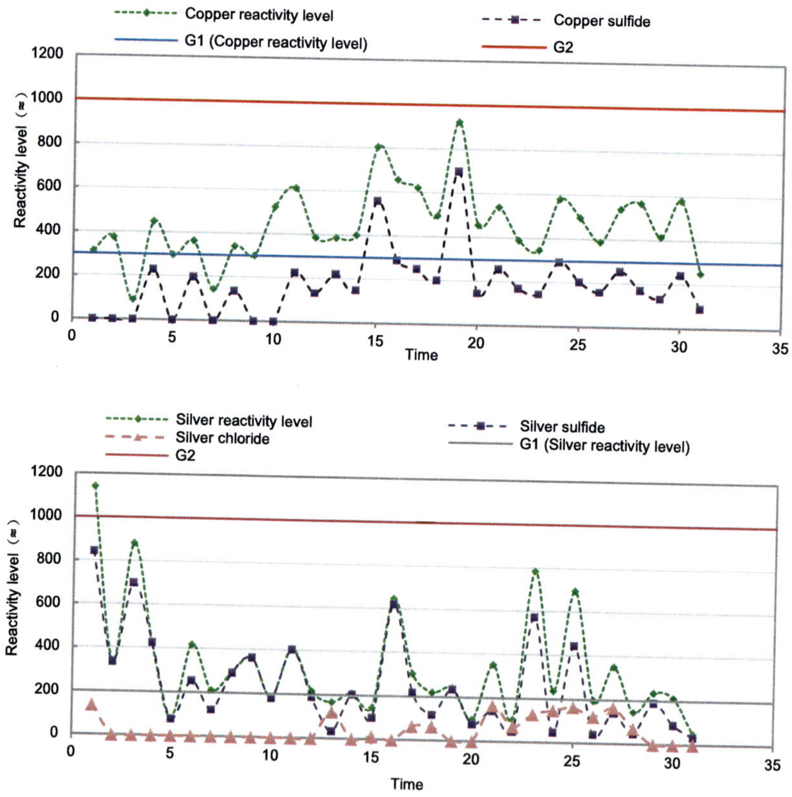

Fig. 1. Corrosion status of the computer room based on the reactivity principle.

2.2 Requirement Analysis of Monitoring

From the analysis of air quality status, it can be seen that the corrosion severity level of the computer room had not been maintained at G1, and the reactive monitoring data lagged behind. In other words, corrosion could be monitored only after it occurred, which resulted in difficult control of equipment corrosion risk. Therefore, it is necessary to monitor air quality accurately in real time, to do predictive maintenance, aiming to maintain severity level at G1 for preventive control. Specific monitoring items can be selected according to the actual situation, as detailed below.

Compared with the common sources of reactive environmental pollutants given in ANSI/ISA-71.04-2013 and the industrial conditions around the data center, it is speculated that the corrosive gases in the air of the computer room may include SO_2, NO_2, O_3, H_2S, Cl_2 and so on. From the results of copper and silver coupons in the computer room (Fig. 1), it is clear that sulfide was the main corrosion product for both copper and silver coupons. Especially for silver coupon, sulfide accounts for the majority of corrosion products. It can be inferred that sulfide corrosion was the most important corrosion mode in the computer room. According to the previous studies, SO_2 is one of the pollutants that cause sulfide corrosion. And it is reported that silver is more easily corroded by SO_2 than copper [13], which is consistent with the results of coupons in the computer room. It also adds evidence to the idea that vehicle exhaust is one of the causes of corrosion. So, considering to the main corrosion products of copper and silver coupons in the computer room and the air quality of the data center location (Fig. 1; Table 5), it is concluded that the most important corrosive gases to be controlled first should be SO_2 and NO_2.

Moreover, the particles in the atmosphere always carry gaseous pollutants and moisture which are highly corrosive. Although the air-conditioning system—Air Handling Units (AHU) has been installed, the particulate matter of high concentration in the external atmosphere will increase the burden on the filter modules. Based on comprehensive consideration, it is necessary to monitor internal and external corrosive gases and particles. We hope the monitoring results can be used to guide the operation mode of AHU, which not only conforms to the concept of *predictive maintenance* but also contributes to energy saving and environmental protection.

To sum up, the monitoring requirements can be summarized as follows: to equip a high-precision online monitoring system which is used to monitor external corrosive gases (SO_2 and NO_2) and particulate matters (PM2.5 and PM10) in real time; set the threshold of each parameter reasonably; the system will give an alarm when the value is out of limits, and the AHU system will switch its mode, so as to isolate the pollutants outdoors and maintain the corrosion severity level at G1.

Ideally, this program would effectively reduce corrosion severity level in the computer room and maintain data security, extend the service life of equipment, and save the maintenance costs. Linking with the AHU, it also can save the operating costs of the fresh air-conditioning system and reduce the replacement frequency of filtration equipment to achieve energy saving and environmental protection.

2.3 Monitoring and Early-Stage Warning System Scheme

The alarm thresholds for external environmental parameters were set preliminarily according to SO_2 and NO_2 concentrations of severity level G1 in ISA-S71.04-2013, the requirements of GB 3095-2012 on the concentration limits of particulate matters ($PM_{2.5}$ and PM_{10}) of Class I areas and the air quality of the data center location. All alarm thresholds could be adjusted according to actual needs.

When the concentration of a parameter exceeds its threshold, an alarm will be given, and the AHU system will be switched from the *mixed air mode* or *fresh air mode* to the *return air mode*. When the concentration of the parameter drops to 80% of the alarm threshold value, the AHU system will be switched back to the *mixed air mode* or *fresh air mode*, as shown in Table 6. Synchronously, the concentration of internal corrosive gases (SO_2 and NO_2) and particles shall be monitored and recorded, which will be compared with the reference concentrations of G1-GX severity levels in ISA-S71.04-2013. It should be noted that data of internal air are only used to verify the effect of the monitoring and early-stage warning of the external environment and provide a reference to adjust alarm thresholds.

Table 6. Recommended thresholds of external environmental parameters to be monitored.

Item (Units)	Alarm threshold	Switch back threshold
SO_2 (ppb)	10	8
NO_2 (ppb)	40	32
$PM_{2.5}$ ($\mu g/m^3$)	75	60
PM_{10} ($\mu g/m^3$)	150	120

In view of the environmental parameters to be monitored in this study, the existing conventional measurement methods for each of them were investigated. Reviewing the requirements of detection limits, accuracy, stability, cost and effect of each method, the optical measurement technology was finally proposed. For the particle monitoring module, we adopted the light scattering method, which can simultaneously monitor $PM_{2.5}$ and PM_{10} using one device. For the corrosive gas monitoring module, we adopted photoacoustic spectroscopy and ultraviolet fluorescence, which can achieve ppb-level detection limits and resolutions. Simultaneously, the self-calibration module was designed for corrosive gases, which could be used to calibrate equipment to ensure its accuracy. We also designed a display touch screen and an industrial computer to set parameter, monitor the status of each module, store the monitoring data and upload data to the upper computer system through the network protocol. All of the monitoring modules, self-calibration module, display touch screen and industrial computer were integrated into one cabinet to become a composite online monitoring system. The principle framework of integrated monitoring system is shown in Fig. 2. Through careful selection of the installation site and elaborate design of gas circuit, the system could alternately monitor the corrosive gas and particles in external and internal atmosphere.

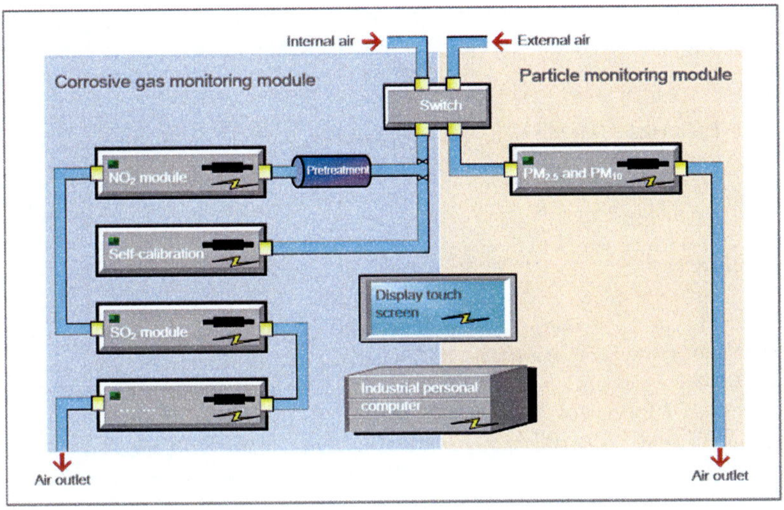

Fig. 2. Principle framework of the integrated monitoring system.

3 Conclusions and Prospects

The high-precision online monitoring system has been installed in a typical data center, running in good condition with expected measurement accuracy of each module. It can not only monitor environmental parameters online, but also carry on early-stage warning and evaluation of air quality and corrosion severity levels. Moreover, this scheme is hopeful to achieve energy saving and environmental protection through coordinated regulation with the fresh air system, which is in line with the concept of *predictive maintenance*.

However, the occurrence of corrosion is the result of a combination of multiple factors. Although the monitoring parameters and thresholds are reasonably derived and set according to the air quality data of the project area and active coupon corrosion results, certain parameters may not be identified, or the thresholds is unsuitable. In other words, the control effect of the corrosion may need further improvement and optimization. We took this into account at the beginning of the monitoring system design. For hardware, the module can be added or moved away when the gas type needs to be adjusted. For software, thresholds can also be modified and adjusted. That is, according to actual needs, the modules and alarm threshold of each parameter can be adjusted in the future. That will be conducive to maintaining the environmental cleanliness, slowing down the corrosion of equipment, and improving equipment reliability and data security.

As mentioned in the Introduction, the air quality in many clean environments or specific scenes is becoming increasingly important and the requirements are being clear and strict. Subsequently, with the further development of optical measurement technology and the continuous improvement of detection limits and measurement accuracy, such modular measurement systems can be extended to more specific environments or scenes, such as operating rooms, museums, space stations, electronics clean rooms, arsenals, armories or ammunition depots. For different scenes, the parameters to be monitored, measurement technologies and control thresholds may be various. However, special customization can be made according to the actual situation based on the concept of this study.

Acknowledgements. This research work was financially supported by 2022 Golden-Bridge Seed Funding Program of Beijing Association for Science and Technology (No. ZZ22009).

References

1. Prasanth, C.B.V., Sreekar, G., Jatin, G.V.S., Shah, J.: IoT based environment monitoring system to protect heritage artefacts. In: 5th International Conference on Electrical, Electronics, Communication, Computer Technologies and Optimization Techniques (ICEECCOT), pp. 145–150. IEEE, New York (2021)
2. Sharif-Askari, H., Abu-Hijleh, B.: Review of museums' indoor environment conditions studies and guidelines and their impact on the museums' artifacts and energy consumption. Build. Environ. **143**, 186–195 (2018)
3. Martinelli, E., et al.: Design and test of an electronic nose for monitoring the air quality in the international space station. Microgravity Sci. Technol. **19**(5–6), 60–64 (2007)
4. Bai, Z.H., Xiao, K., Yan, L.D.: Recently development of microbial corrosion of space station. Materials China **37**(1), 9–13 (2018)
5. Yamaguchi, N., et al.: Microbial monitoring of crewed habitats in space-current status and future perspectives. Microbes Environ. **29**(3), 250–260 (2014)
6. Kanamori, H., Rutala, W.A., Gergen, M.F., Weber, D.J.: Perioperative bacterial contamination from patients on contact precaution in operating room environment. Open Forum Infect. Dis. **7**(11), 1–4 (2020)
7. Thapliyal, A., Kumar, C.: Development of data acquisition console and web server using Raspberry Pi for marine platforms. Int. J. Inf. Technol. Comput. Sci. **11**, 46–53 (2016)
8. Liu, Y., et al.: Review of corrosion and anti-corrosion of ammunition metal packaging materials. Packag. Eng. **41**(9), 232–237 (2020)
9. Zhang, Z.F., et al.: Discussion on key influencing factors and solutions. For the cleanliness of clean rooms in the electronics industry. Intell. Build. Smart City **2**, 41–43 (2019)
10. Molaie, S., Lino, P.: Review of the newly developed, mobile optical sensors for real-time measurement of the atmospheric particulate matter concentration. Micromachines **12**(4), 416 (2021)
11. Wang, Q., Xu, K., Yao, C., Wang, Z., Chang, J., Ren, W.: Recent advances of power-enhanced photoacoustic spectroscopy. Chin. J. Lasers **45**(9), 0911008 (2018)
12. Wang, S.T., Wang, C.B., Pan, Z., Kong, D.M., Cheng, Q., Wang, Z.F.: Applications of optical technology in gas concentration detection. Opto-Electron. Eng. **44**(9), 862–871 (2017)
13. Kurella, A., Pathangey, B., Pukale, R., Meyyappan, K., Proctor, A.: Evaluation of corrosion films on Cu and Ag reactive monitors deployed in data center server environments. In: IEEE 60th Holm Conference on Electrical Contacts, pp. 111–115. IEEE, New Orleans, LA (2014)
14. Shah, J.M., et al.: Qualitative study of cumulative corrosion damage of information technology equipment in a data center utilizing air-side economizer operating in recommended and expanded ASHRAE envelope. J. Electron. Packag. **139**, 021002 (2017)
15. Klein, L.J., Singh, P.J., Schappert, M., Griffel, M., Hamann, H.F.: Corrosion management for data centers. In: 27th Annual IEEE Semiconductor Thermal Measurement and Management Symposium, pp. 21–26. IEEE, New York (2011)

16. Ministry of Housing and Urban-Rural Development of the People's Republic of China, General Administration of Quality Supervision, Inspection and Quarantine of the People's Republic of China.: GB 50174-2017 Code for design of data centers, pp. 39–49 (2017)
17. International Society of Automation.: ANSI/ISA-71.04-2013 Environmental Conditions for Process Measurement and Control Systems: Airborne Contaminants (2013)
18. Ministry of Housing and Urban-Rural Development of the People's Republic of China, General Administration of Quality Supervision, Inspection and Quarantine of the People's Republic of China.: GB/T 51314-2018 Operation and maintenance standard for data center infrastructure, p. 2 (2018)
19. Department for Environmental Protection, General Administration of Quality Supervision, Inspection and Quarantine of the People's Republic of China.: GB 3095-2012 Ambient air quality standards, p. 3 (2012)

Exploration and Practice of Commercial Application of AI-Empowered Remote Sensing Big Data –Taking Pine Wood Nematode Disease Space-Ground Integrated Remote Sensing Intelligent Monitoring Platform as an Example

Yao Lifan[1], Song Jiayun[1](✉), Zhang Fangfang[2], Li Bo[2], and Zhang Xinye[2]

[1] Princeton University, Princeton, NJ 08544, USA
songjy@bhqditi.com
[2] Springer Heidelberg, Tiergartenstr. 17, 69121 Heidelberg, Germany

Abstract. With the rapid development of satellite remote sensing technology, Remote sensing data acquisition methods are multi-speed, fast and time-sensitive, and remote sensing data types are becoming more and more abundant. The data volume is increasing exponentially, which has obviously shown the characteristics of "big data". Meanwhile, the rapid development of artificial intelligence technology has greatly improved the intelligent and accurate level of data analysis, which has brought new development opportunities for remote sensing big data analysis and application, and promoted the development of remote sensing application in the direction of intelligence, popularization and industrialization. Based on the current development status and actual needs of remote sensing big data application at home and abroad, this paper discusses the necessity of AI-empowered remote sensing big data application in the industry, and takes "pine wood nematode disease space-ground integrated remote sensing intelligent monitoring platform" as an example, focusing on the key technologies, main functions and application results of the platform, etc., and introduces the practice of AI-empowered remote sensing big data application in the industry in detail. Finally, it summarizes the practical exploration and experience. Finally, it looks forward to the development of AI-empowered remote sensing big data application and puts forward countermeasures and suggestions for commercial application from the dimensions of policy support, technological innovation, application mode and ecological cooperation.

Keywords: remote sensing big data · Artificial intelligence · Platform application · Commercial development · Countermeasures and suggestions

With the rapid development of satellite remote sensing technology in our country, the resolution and quality of remote sensing image have been improved continuously while the period of data acquisition and update has been shortened, which brings explosive growth to the volume of remote sensing data, remote sensing data has entered the era of big data [1]. However, the characteristics of remote sensing data diversification, real-time

H. P. Urbach and H. Jiang (Eds.): ISSOIA 2022, SPPHY 295, pp. 700–708, 2023.
https://doi.org/10.1007/978-981-99-4098-1_58

and mass, make the process of remote sensing data processing cumbersome, difficult to manage, difficult to apply, and traditional remote sensing image processing and analysis process time-consuming and labor-consuming, problems such as the need for human involvement have also been exposed. In the present intelligent era, which is represented by big data and artificial intelligence, how can we make artificial intelligence become a new driving force for the development of intelligent remote sensing applications and better release the value of remote sensing data, it is a new direction of remote sensing application.

At present, there are two main problems in the research and application of intelligent remote sensing data processing platform. On the one hand, the existing remote sensing data processing platform usually stores a large amount of remote sensing data on disk, and its performance depends on the single computer's computing power and storage power, and there is a big bottleneck in data management and calculation, it has not been able to adapt to the cloud computing, big data era of People's demand for remote sensing services, restricting the popularity of remote sensing applications. Nowadays, the distributed spatial remote sensing data service system for cloud environment has become the main supporting platform for remote sensing cloud service applications [2]. On the other hand, with the development and breakthrough of remote sensing application field in our country, a series of algorithms for remote sensing data processing have been produced, but most of these algorithms are scattered and independent, the lack of standardized data processing process, can not achieve a universal remote sensing application platform [3]. Due to the variety of satellites and sensors, the algorithms for processing different types of remote sensing data are different, and the data processing flow of existing remote sensing system is fixed, which can not meet the user's complex and changeable personalized service needs [4]. Therefore, how to make use of artificial intelligence to enable the development of remote sensing big data application industry has become a hot spot in the current commercial field.

1 Current Situation of Remote Sensing Data Processing at Home and Abroad

The traditional remote sensing image processing based on single processor is still a common method of remote sensing image processing because of its simple operation and convenience, such as ArcGIS, ENVI and other GIS software. However, the processing process relies on the manual cooperation, which is difficult to adapt to the application of batch processing of remote sensing big data, and its performance depends on the single computer's computing power and storage power. Subsequent paragraphs, however, are indented.

With the rapid development of big data and cloud computing technology, as well as the explosion and growth of remote sensing data volume, the processing platform of remote sensing image has gradually turned to the cloud. Lin et al. [5] proposed a remote sensing image processing framework based on cloud storage and distributed computing for effective management and efficient processing of geographic information data, so as to solve the limitation of hardware resources and processing time in data of a single computer. Luo Xianxian et al. [6] analyzed the background of the rapid development of deep learning technology in the field of remote sensing image processing, and in view of the intelligent application requirements of remote sensing big data, put forward the requirements of strengthening the research of algorithm theory and the intelligent construction of remote sensing big data platform. They combined deep learning intelligent algorithm with remote sensing big data processing to promote the application construction of intelligent algorithm in remote sensing image processing. Under the cloud computing environment, the remote sensing image processing system is still based on the distributed architecture, through the integration of remote sensing big data storage resources, data processing algorithms and other application platforms, the comprehensive remote sensing image processing process. At present, the design and construction of distributed intelligent remote sensing image processing system with high compatibility, high-performance computing and high scalability is still a hot spot in the research. However, for the remote sensing data processing system in the distributed cloud environment that solves the computing bottleneck, how to meet the customized application requirements of users, flexibly organize multiple algorithm models to work together, share resources and cooperate with the platform, and realize the automatic and operational processing of remote sensing images are further challenges for the remote sensing processing system.

2 Artificial Intelligence-Enabled Remote Sensing Applications

With the rapid development of artificial intelligence and satellite remote sensing technology, scholars at home and abroad have not only put forward a series of new intelligent remote sensing image processing, analysis and interpretation theories, methods and technologies. These intelligent remote sensing methods have also been widely used in resource survey, environmental monitoring, disaster monitoring, smart cities, automated analysis of agriculture, forestry and water industries, financial risk assessment and military reconnaissance. This section describes several typical application scenarios that involve complex data processing.

2.1 Application of Urban Vegetation Green Space Monitoring

Urban vegetation is an important part of urban ecosystem, so it is necessary to plan and monitor it scientifically. However, urban vegetation monitoring is faced with many problems that need to be solved. On the one hand, the types, biomass, ecological functions and greening effects of urban vegetation cover are not the same, so it is difficult to obtain the actual "green amount" data of cities through simple interpretation and analysis. Through intelligent classification technology of remote sensing images, surface vegetation types and covering areas can be identified, so as to indirectly reflect the total biomass and actual greening effects of urban vegetation. The urban vegetation green space extraction and monitoring algorithm can be used to rapidly and automatically extract the vector results of forest land change areas in remote sensing images of different phases, and the vegetation green space change map spot extraction can be quickly realized. Compared with manual methods, the efficiency can be improved by more than 100 times, providing strong technical support for intelligent monitoring and early warning of vegetation green space. On the other hand, the classification of urban vegetation is easily affected by similar features and similar spectra, resulting in vegetation leakage and misclassification, and relying on low-level features can no longer satisfy the extraction of high-quality vegetation information. In recent years, with the great success of deep learning, especially deep convolutional neural network in the field of natural image processing, it has also been extended to the analysis and processing of remote sensing images, promoting the development of deep semantic segmentation technology, which can play a good segmentation effect on object boundary contour, and which also can solve the problems of blurred segmentation boundary and noise point in the extracted vegetation area in the high-resolution remote sensing image information extraction task, and to further improve the accuracy of vegetation extraction.

2.2 Application of Urban Land Monitoring

Remote sensing provides an important means for monitoring urban land use change. Through visual interpretation of remote sensing images and automatic detection technology of building changes of remote sensing data, land use information can be obtained and monitored. Urban land monitoring is one of the main work content is "two violations of monitoring" (that is, illegal land and illegal construction), the traditional use of artificial way of inspection and then found "two violations" methods and processes, with evidence difficult, long working cycle, high cost and low efficiency, can not meet the actual business needs of "two violations" monitoring. The "two-breach monitoring" method based on big data and artificial intelligence has high accuracy and efficiency.

Specifically, based on the deep learning technology, the convolution neural network feature extraction algorithm is used to superimpose and extract the change pattern from the UAV images of the previous and previous periods, realize the automatic comparison of the pattern patterns, find the difference information of the buildings on the pattern patterns, and comprehensively determine the suspected illegal pattern patterns. At the same time, training based on massive building image data sets can repeatedly train and iterate more high-performance algorithm models that can be used for building target recognition and change monitoring, which can effectively solve the pain points and difficulties in the "two violations of monitoring" work, such as large area image extraction, interference of the season and the sun on the roof, omission of the u-net model building extraction, and the judgment criteria for suspected violations, saving labor costs and improving work efficiency.

2.3 Application of Urban Road Network Monitoring

Road is the main component of urban development, the monitoring and management of road network is an important work content of urban management. The traditional urban road network drawing and monitoring are basically carried out based on offline measurement and mapping, which is characterized by time-consuming labor, slow updating speed and low efficiency. With the rapid development of Chinese economy in recent years, the changes of road network in cities are also accelerating. Ensuring the real-time performance and high efficiency of road network monitoring is of great significance to the urban traffic management. The application of artificial intelligence technology can greatly meet the demand of high-frequency update of road network information, and carry out the monitoring, extraction, drawing, output and application of road network more quickly and more efficiently.

In terms of vehicle monitoring, based on the cluster analysis technology, the vehicle video image segmentation, contour extraction and feature detection are realized. At the same time, SVM, Bayesian classifier and BP neural network are used, and the three-layer progressive judgment model is used to classify the vehicle models. In terms of traffic flow monitoring, neural network, fuzzy logic and genetic algorithm can be used to design the intelligent control algorithm of urban road network with strong robust and adaptability, which can solve the urban traffic problems more effectively. In terms of road safety monitoring, rely on AI patrol automatic detection algorithm model, can achieve 24-h monitoring, ensure that the first found accident, congestion, violation stop road abnormal phenomenon can timely push to the management unit, overcome the manual monitoring and vehicle patrol, improve further improve the efficiency of emergency disposal and security level.

3 Space-Ground Integrated Remote Sensing Intelligent Monitoring Platform for Pine Wood Nematode Disease

"Pine wood Nematode Disease Integrated Remote Sensing Intelligent Monitoring Platform" is a whole-process monitoring platform independently developed by Qingdao Research Institute of Beijing University of Aeronautics and Astronautics and Qingdao Laoshan District Bureau of Natural Resources for precise prevention and control of pine wood nematode disease. The platform adheres to the principle of "prevention first, scientific governance, legal supervision and enhanced responsibility". By integrating UAV technology, high-resolution satellite remote sensing data and artificial intelligence deep learning algorithm, the platform builds an integrated accurate and intelligent monitoring system for pine wood nematode disease in the space and ground, which integrates "accurate monitoring", "accurate identification", "accurate assessment" and "accurate supervision". Scientific monitoring, fine management and intelligent analysis can be carried out on indicators such as "quantity of pestilentous wood", "location of pestilentous wood", "treatment progress" and "treatment efficiency", so as to provide scientific decision-making and evaluation optimization scheme for pestilentous wood treatment (see Fig. 1).

3.1 Introduction to the Platform

The platform is a scientific research industrialization achievement of the integration of the new generation of information technology, aerospace technology and natural resources protection. The key technologies are mainly interdisciplinary integration technology of RS (remote sensing) + AI (artificial intelligence) + GIS (geographic information system), one-stop integrated processing technology of big data, and high-precision monitoring algorithm model library in the field of forestry. In view of the complex conditions such as large mountain height difference, strong wind, vegetation discology in autumn and mixed varieties in pine wood nematode disease areas, aerospace remote sensing data

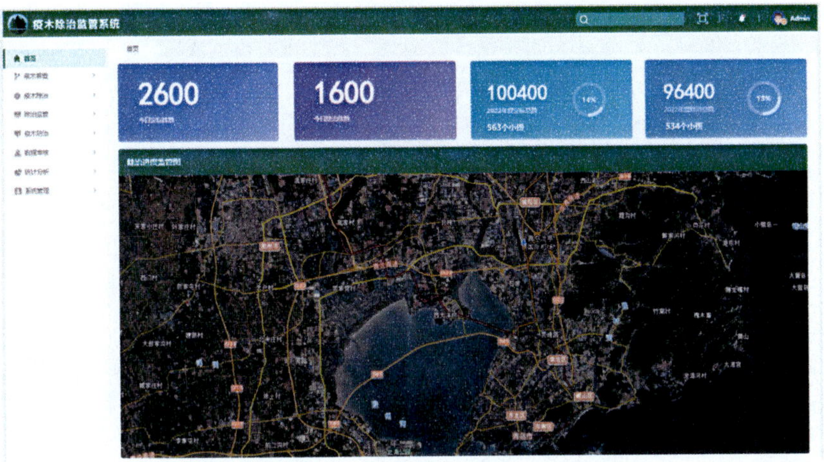

Fig. 1. Platform interface.

acquisition and fusion technology, multi-dimensional target recognition, image feature recognition and other artificial intelligence technologies, cloud visualization geographic information technology and other frontier disciplines were combined. It has independently developed and designed comprehensive solutions based on high-resolution remote sensing satellites, industrial-grade special forestry drones, and AI big data processing. It provides one-stop big data services such as aerial survey, intelligent analysis, prevention and control effect assessment, decision-making assistance, etc. It can achieve detection capacity of 50 km^2 per day, fly on the same day, and identify epidemic trees on the same day. Within 7 days, it can complete all the data processing and cloud deployment of 200 km^2. Compared with the traditional solution, the efficiency is increased by more than 5 times. The self-built special identification algorithm engine has completed millions of sample data testing, and the identification accuracy rate can reach more than 95%, so that problem trees can be found faster and users can take measures faster to timely and effectively control the epidemic (Fig. 2).

3.2 Main Features

The platform is based on remote sensing big data, with pine wood nematode monitoring as the specific application scenario, combined with deep learning algorithm, breaking a new direction for the specific application of remote sensing big data.

First, the algorithm model has high precision. on the one hand, BH-LHRAF (Large-scale High Resolution dataset for Analysis on Forestry), the largest artificial intelligence dataset in China, has 10 million sample data. It covers the year-round changing characteristics of spring, summer, autumn and winter, effectively supporting the training and reasoning of the high-performance algorithm model, and the actual test accuracy (non-theoretical value) is better than 95%. On the other hand, based on the model framework and relying on the transformation of military target detection technology, the Retinanet can independently develop AI algorithm for identifying dead and dying pine trees. The

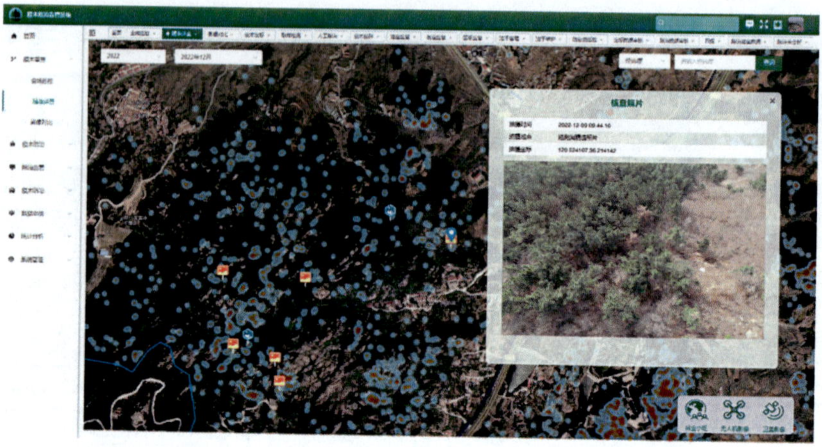

Fig. 2. Platform function: Accurate investigation.

average test accuracy rate can reach 97.22%. With the acceleration of the GPU of mobile graphics workstation, the detection time of a single aerial can be reduced to 0.46 s. The algorithm can identify the point locations of suspected dead and dying pine trees in 1 km^2 in 1 min, and the recognition accuracy is better than that of Aliyun service (Fig. 3).

Fig. 3. AI Identification of dying pine.

Second, high efficiency and low cost. On the one hand, relying on the AI middesk technology, it greatly simplifies the technical difficulty of aerospace and other professional data application. Compared with the traditional development mode of starting from scratch, the AI middesk presets a number of automatic development components and automatic processing models, greatly simplifying the complex aerospace, big data, artificial intelligence and other development processes and reducing costs. On the other hand, relying on Kubernetes and other technologies, we independently build a private cloud computing platform to realize automatic scheduling and distribution of computing power, ensure data security and rapid data calculation, and effectively support the automatic and accurate processing of the data of 300,000 mu of woodland in the whole area within 7 days, with the efficiency dozens of times higher than that of traditional means.

3.3 Effectiveness of Application

Since the platform was put into use, it has called several satellites, including the China Resources Satellite Center Gaofen-2, to take multiple photos of the forest area, covering an area of more than 400 km^2. At the same time, a special forestry flight formation (including 3 fixed-wing UAVs and 4 multi-rotor UAVs) was configured to conduct a survey of the forest area of 340,000 mu. At present, 12 times of aerial monitoring and census have been completed in the whole forest area of Laoshan District of Qingdao, with 873 unmanned aerial vehicles deployed, covering an area of 4.08 million mu. At the same time, more than 400 inspections were carried out on the far-reaching mountain area, and more than 40,000 high-definition pictures were taken.

Using artificial intelligence and big data technology to achieve accurate identification and evaluation. On the other hand, based on the big data analysis and modeling technology, the existing stock of dead and dying pine trees and the elimination progress were accurately predicted by the small class of forestry, and the prediction deviation was less than 5%, which effectively supported the coordination and deployment of relevant work.

4 Development Trend of Remote Sensing Big Data

As a typical representative of big data, remote sensing big data has made certain breakthroughs in the research of key technologies such as distributed clustered storage, high-performance computing for data-intensive applications, multi-source heterogeneous data association analysis, and geographic visualization of remote sensing big data. However, as a new research object, remote sensing big data research has just started, and there are still many problems to be further studied.

1) Multi-source information fusion in remote sensing big data environment. Data fusion is the bottleneck of big data application and intelligent decision-making process. In order to effectively utilize multi-source heterogeneous data provided by remote sensing big data, it is necessary to further study the unified expression model, association analysis model and decision rules of heterogeneous data, so as to improve the accuracy of decision making based on remote sensing data. In addition, combined with the hierarchical characteristics of remote sensing data processing, the information provided by the fusion of data level, feature level and decision level can be further comprehensively utilized to achieve the level of intelligent decision-making under complex conditions through the combination of human-computer interaction and machine learning methods.

2) Machine learning methods for remote sensing big data. The feature selection model of remote sensing big data is established based on the in-depth analysis of the physical characteristics of ground objects and their image features in images, combined with the mathematical methods of graph theory, rough set, fuzzy set, soft set, etc. By introducing the human visual cognition mechanism into the nonlinear transformation model between features, the transformation from low level image features to high level semantic features is realized. In this way, a deep learning model driven by visual mechanism is established to achieve accurate interpretation of remote sensing images. The features of feature selection, deep learning and other algorithms are deeply analyzed. Based on heterogeneous computing mode, a machine learning algorithm satisfying large-scale remote sensing data processing is built from the aspects of high-performance data storage access, memory load balancing, task scheduling optimization and so on.

References

1. Li, P.F.: Research on Key Technologies of Infrastructure in Remote Sensing Cloud Computing. Yanshan University (2010)
2. Kang, J.F.: Research on High resolution Remote sensing image storage and efficient management technology in cloud computing environment. Zhejiang University (2011)
3. Brezonik, P.L., Olmanson, L.G., Finlay, J.C., et al.: Factors affecting the measurement of CDOM by remote sensing of optically complex inland waters. Remote Sens. Environ. **157**(Sp. Iss. SI), 199–215 (2015)
4. Diao, L., Liu, R., Chen, T.: The preprocessing method of contrel point in geometric correction for UAV remot sensing image. In: International Conference of Advanced Materials & Computer Science (2016)
5. Lin, F.C., Chung, L.K., Ku, W.Y., et al.: The framework of cloud computing platform for massive remote sensing images. In: Advanced Information Networking and Applications (AINA), 2013 IEEE 27th International Conference on. IEEE (2013)
6. Luo, X.X., Zeng, W., Chen, X.Y., et al.: Research progress of deep learning method for remote sensing image processing. J. Quanzhou Norm. Univ.

Research on the Market Demand of Remote Sensing Data Application Services in Key Countries in Southeast Asia Along the Belt and Road

Luxi Chen[1(✉)], Jing Lu[1], Yonghui Pang[1], Zining Zhao[2], Yunfei Bao[1], and Cheng Jiang[1]

[1] Beijing Institute of Space Mechanics and Electricity, Beijing 100094, China
chenlxts@126.com

[2] Shanghai Electro Mechanical Engineering Institute, Shanghai 201109, China

Abstract. This paper analyzes the opportunities and risks of the countries along the Belt and Road from the perspective of international cooperation by combining the relevant information of the Belt and Road and the Six economic corridors. Combined with the application status of remote sensing data, Thailand, Malaysia and Indonesia are selected to carry out in-depth market research, and the comprehensive national conditions of the three countries were analyzed from the aspects of natural environment, political and diplomatic relations, economic and trade environment, and application demand of remote sensing data. The market demand characteristics, potential users and competitors in the three countries are analyzed through data sources, fields and patterns of remote sensing application. The results indicate that the remote sensing data application in the three countries have a wide range of demand, while facing fierce market competition. The current situation of remote sensing data application is summarized to provide reference for the Institute's remote sensing data application services to enter the international market supply chain, and the future development trend of the industry is prospected. Combining with the key technical directions of remote sensing data of the Institute, suggestions are made for data application services in key fields such as agriculture and forestry, hydrology, urban and homeland security.

Keywords: The Belt and Road · Southeast Asia · remote sensing · data application services · market demand

1 Introduction

In recent years, the Belt and Road Initiative has received increasing attention from the international community. China has pointed out that it will make full use of the existing and planned remote sensing satellite resources to carry out demonstration of space information application in countries along the Belt and Road, and actively contribute to the "Chinese solutions". With the rapid development of China's space industry, China's

H. P. Urbach and H. Jiang (Eds.): ISSOIA 2022, SPPHY 295, pp. 709–718, 2023.
https://doi.org/10.1007/978-981-99-4098-1_59

satellite manufacturing and application technology have gradually reached the international advanced level, which is a prerequisite for the construction of the Belt and Road space information corridor.

The global remote sensing services market is expected to reach USD 21.62 billion by the end of 2022. The in-depth application of remote sensing services in defense and commercial sectors is the main reason driving the growth of the remote sensing services market. In the coming years, remote sensing technology based on cloud computing and big data analytics will create more growth opportunities for the market [1].

In the context of rapid advancement of national space information corridor construction and rapid development of global remote sensing data application market, conducting market demand research will help to understand the international market demand more comprehensively and specifically, and promote the Institute's data application services to the international market gradually.

2 Comprehensive Analysis of the Belt and Road Region and Remote Sensing Data Application

2.1 Analysis of the Belt and Road Regional Situation

The Belt and Road is the world's longest economic corridor with the greatest development potential. It has also brought many opportunities and challenges for cooperation among countries [2]. The main participants of the Belt and Road Initiative are mainly Eurasian region, which can be divided into six sub-regions. On this basis, by analyzing the opportunities and risks of each sub-region and combining the characteristics of the target market demand, it is beneficial to grasp the international market entry point more accurately. The analysis of opportunities and risks for each region of the Belt and Road is shown in Table 1.

2.2 Analysis of the Current Situation of Remote Sensing Data Application

In recent years, the remote sensing satellite manufacturing industry in the world has developed rapidly, and the technical level of remote sensing data application has been greatly improved. On the one hand, the breadth and depth of remote sensing data application have been increasing, and fruitful market results have been achieved, bringing significant economic benefits to the world. On the other hand, the increasingly frequent international academic exchanges in the field of remote sensing have brought great potential and bright prospects for the development of remote sensing technology. At present, the technical level of the United States in various fields such as delivery capacity, detector development, image processing, basic theory and application is in the leading position internationally. Russia's advantages in the field of remote sensing data application are mainly reflected in its strong carrying capacity and large volume of remote sensing data. In addition, Canada, Japan, Europe and other developed countries are also relying on their own remote sensing satellite data (such as Radarsat, JERS, SPOT.) to develop data application service capabilities [7].

At present, many countries along the Belt and Road have begun to attach importance to the development of remote sensing data application technology, and part of them have

Table 1. Analysis of opportunities and risks in each region of the Belt and Road [3–6]

Regions	Opportunities	Risks
East Asia	1. China-Mongolia relations and cooperation in various fields are developing in an all-round way	1. Poor political continuity and stability; 2. China-Mongolia relations face stiff competition
Southeast Asia	1. Most countries have a positive attitude towards the Initiative; 2. Major projects led by China have further enhanced connectivity among countries	1. A few countries fear weakening of ASEAN centrality
South Asia	1. Steady progress of economic corridor construction 2. China and Indian Ocean Rim Countries Gradually Strengthen Strategic Docking	1. Some countries have fewer stable policies and less awareness of market compliance; 2. Threats and attacks by extremist forces of terrorism are frequent; 3. Extra-territorial countries are engaged in intense geopolitical games in the region
West Asia	1. As a convergence point of the Belt and Road, more results may be obtained; 2. No fundamental conflicts of interest with other countries	1. High security risk; 2. The game of Western countries in the region; 3. Lack of dispute resolution system in government
Central Asia	1. Countries have close political and economic ties with China; 2. A key strategic location as a geographical center and transportation hub in Eurasia	1. The regional security situation isn't optimistic; 2. Increased competition in regional cooperation mechanism; 3. Large regional development gap
Europe	1. Countries' willingness to cooperate continues to grow; 2. Some countries are hoping to revive their economies	1. There are differences in culture and perception of the Belt and Road; 2. Political turmoil, great power rivalries, sanctions

included it in their national development plans. Thailand, Indonesia and some other Southeast Asian countries have applied remote sensing data in agriculture, forestry, environmental monitoring, urban planning, weather forecasting, resource survey, etc. Some countries have set up full-time management agencies and technical R&D centers, and established a certain scale of professional teams, which have achieved a number of high-level achievements. It highlights the technical and market potential of the countries along the Belt and Road in the remote sensing data application industry.

2.3 Focus on Regions and Countries

This paper selects Southeast Asia as a key region for market research for the following reasons. Firstly, Southeast Asia occupies a superior and important geographical location and is a key node of the Belt and Road, linking the East Asian and European economic circles with great economic potential. Secondly, Southeast Asian countries and China have frequent economic and trade contacts, and there are many overseas Chinese, playing an important role in connectivity in the Belt and Road Initiative. Thirdly, China maintains long-term and stable diplomatic relations with ASEAN countries, therefore, the market is relatively less affected by political factors. Fourthly, the natural environment and economic development patterns of Southeast Asian countries determine their extensive demand for remote sensing data application services in major fields such as agriculture, forestry, marine, resources, disaster warning and prevention. Most of the countries themselves haven't yet formed a complete and mature data application industry chain, so they have high external demand and dependence on remote sensing data products [8].

In this paper, Thailand, Malaysia and Indonesia are selected as the key countries to carry out market research for the following reasons. Firstly, in terms of political and diplomatic relations, the three countries are politically stable in general and maintain good diplomatic relations with China. Secondly, in terms of economy, the three countries are the top three economies among the ten ASEAN countries with sound economic development momentum and frequent economic and trade exchanges with China in different industries. Thirdly, the three countries are rich in natural resources, numerous islands and developed tourism. They all have extensive demand for satellite remote sensing applications in natural resources exploration and environmental monitoring, and they also have the economic basis for cooperation in remote sensing applications; Fourthly, the countries have weak independent operation capability of remote sensing satellites, as a result, they have extensive external demand in agriculture, environment, natural resources monitoring. Fifth, the countries have established full-time administrative departments in the field of space remote sensing and carried out a series of space infrastructure (such as ground stations of remote sensing satellites) [9].

3 Comprehensive Analysis of Key Countries

3.1 Thailand

Thailand adheres to the principle of independence in its diplomacy and attaches importance to the development of good neighbourly relations with neighbouring countries. Thailand was one of the first countries to participate in the Belt and Road Initiative and has brought significant benefits to its own development and China-Thailand cooperation. Thailand has a liberal economic policy and focuses on trade liberalization.

Thailand has formulated a national development strategy in the field of remote sensing. GISTDA is the Thai research institute of remote sensing data application technology. Its mission is to receive and process different forms of satellite remote sensing data, carry out data analysis and distribution, and be responsible for the transformation and application of space and geographic information technology [10]. Thailand has a wide range of remote sensing data application demand in natural disaster prevention, such as flood, drought, typhoon and forest fire monitoring, as well as land resource survey, road planning and infrastructure construction in the field of urban planning.

3.2 Malaysia

Malaysia pursues an independent, principled and pragmatic foreign policy and maintains friendly relations with neighboring countries. Malaysia attaches importance to developing relations with major regional powers including China. Economically, China is one of the major markets for Malaysia's foreign import and export trade [11]. Malaysia is actively committed to the ASEAN Free Trade Area and economic development cooperation in the Mekong River Basin.

Malaysia has several remote sensing data application institutions, mainly including National Remote Sensing Center, UTM, ATSB. Malaysia's coastline is widely distributed, and there is a wide range of remote sensing data application demand in the field of marine environment, such as monitoring of oil spill on the sea surface, seawater temperature, coastal phytoplankton and chlorophyll.

3.3 Indonesia

Indonesia advocates a balance between major countries in its diplomacy and attaches importance to its diplomatic relations with China. Indonesia is the largest economy in ASEAN, with major industries including industry, agriculture, forestry, fishery, service industry. In recent years, Indonesia has maintained relatively fast economic growth.

Indonesia attaches importance to space development policies and strategies, and has promulgated a series of laws and regulations on satellite remote sensing to regulate activities of remote sensing data acquisition, processing, and distribution [12]. The management agency of Indonesia of remote sensing data applications is LAPAN, which is responsible for the operation of ground stations, satellite data reception and distribution. Indonesia has a wide range of remote sensing data application demand, including marine target identification, meteorological disaster monitoring, urban and rural land planning.

4 Data Sources and Fields of Remote Sensing Application in Key Countries

4.1 Data Sources for Remote Sensing Application

Thailand, Malaysia and Indonesia have a shortage of domestic remote sensing data sources and mainly rely on foreign satellite data that can be obtained for free, including Landsat, MODIS, Sentinel-2, THEOS, HJ-1.

Landsat data have long recording time, good cycle continuity and rich spectral information, which is mainly applied in water resources monitoring, agriculture and forestry monitoring, land use change monitoring [13]. MODIS is a medium resolution spectrometer, which can achieve full spectral range coverage from visible band to thermal infrared band, and it's widely used in multiple fields such as surface, atmosphere and ocean [14]. Sentinel-2 has rich feature spectral information, and is one of the best data sources for extracting vegetation information, which can provide key information about vegetation status [15]. THEOS is mainly used in the field of urban land survey, forest fire monitoring, mining, agriculture, transportation. Compared with Landsat and THEOS, HJ-1A and B satellites have a shorter revisit period (4 days) and high timeliness of data processing. They have been widely used in disaster monitoring, resource exploration, agriculture, which is a successful case of the internationalization of Chinese remote sensing data application services [16].

4.2 Disaster Monitoring and Early Warning Field

Southeast Asia is one of the natural disaster-prone regions in the world. The governments of the three countries attach great importance to the application of remote sensing data in the field of disaster monitoring and early warning [17]. Their remote sensing application patterns mainly include foreign aid, industrial cooperation and commercial services.

Foreign Aid. In 2018, an earthquake occurred in Indonesia. At the request of UNESCAP, National Remote Sensing Center of China launched the China GEO Disaster emergency response mechanism, coordinating with NSMC, CRESDA, Twenty First Century Aerospace Technology Co., LTD., Changguang Satellite Technology Co., LTD., SpaceWill Info. CO., LTD, Wuhan University and other agencies. Remote sensing data from 12 domestic and foreign satellites, including FY-3, High-resolution series, Triplesat, Jilin-1, Superview-1, Sentinel series and Landsat-8, were provided to relevant authorities, with a total of 35.61 GB of data. The satellite emergency data coverage chart of Indonesia earthquake disaster is shown in Fig. 1.

Industrial Cooperation. In 2013, GISTDA signed a cooperation agreement with Wuhan University on "Thailand Geospatial Disaster Monitoring, Assessment and Prediction System". The project mainly serves for disaster prevention and mitigation and marine environment monitoring in Thailand. The signing of this agreement helps Thailand to form a long-term, stable and reliable remote sensing disaster reduction business operation system to meet the actual demand of disaster reduction and relief in Thailand, and improve the comprehensive application capability of disaster remote sensing in Thailand. It's a typical demonstration for Thailand to comprehensively promote the development of its remote sensing data business and apply remote sensing data services to daily production and life by taking advantage of the inter-governmental industrial cooperation.

Commercial Services. The HJ-1A and B satellites ground station was jointly purchased by the Ministry of Information and Communication Technology of Thailand and Kasetsart University. The Chinese side sold the complete set of equipment and was responsible for the installation and training. The ground station receives the remote sensing image of flood disaster transmitted by HJ-1 in real time every day, which provides a strong support for disaster warning in Thailand after real-time processing and interpretation [18]. During the flood in Thailand in 2011, the Ministry of Industry and Information Technology of Thailand obtained the remote sensing data of HJ-1A and B satellites, analyzed the disaster area by using GIS technology, and distributed the results

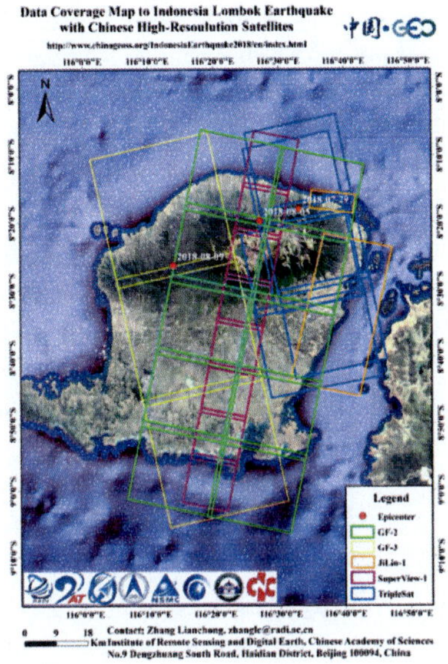

Fig. 1. Comparative map of domestic satellite emergency data coverage for earthquake disaster in Indonesia

regularly as the reference data for government decision-making. The impact of flooding in Thailand is shown in Fig. 2.

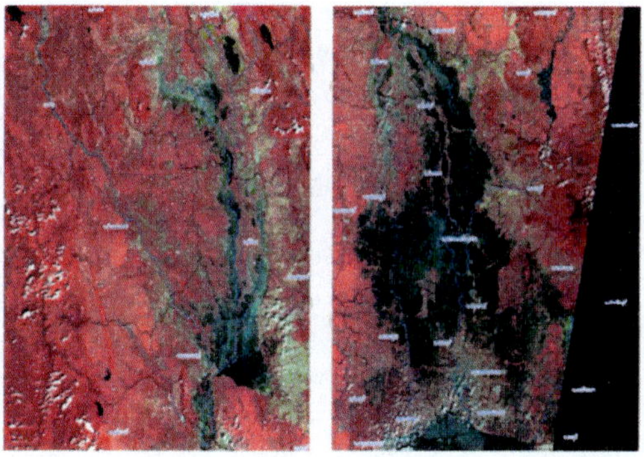

Fig. 2. Remote sensing images of flooding in Thailand

4.3 Agricultural Field

Thailand, Malaysia and Indonesia are all countries with predominantly agricultural economies. Currently, the data application patterns in agriculture and forestry mainly include government-led and commercial services.

Government-led. In order to enhance the utilization rate of farmland in Malaysia, the government has conducted long-term monitoring of the major agricultural economic belt by using HJ-1A and B satellites remote sensing data to study the land use of farmland in the East Malaysia region and using the classified data for modeling to assess the land characteristics of cash crop growing areas. The model has now been extended to other regions in Malaysia.

Business Services. In 2018, Planet signed a contract with GISTDA to provide basic image data (Planet Basemap) covering Thailand to GISTDA on a quarterly basis. These images are used by various authorities in the field of agriculture and forestry. Planet's continuous monitoring and analysis based on multivariate remote sensing data has greatly improved Thailand's geographic information acquisition and agricultural modernization management capabilities.

4.4 Resource Field

With the advancement of China High-resolution Earth Observation System (CHEOS) and space infrastructure construction, domestic remote sensing data with high spatial, high spectral and high temporal resolution began to serve the resource field of countries along the Belt and Road, such as resource survey and yield estimation, to provide effective data support for them. At present, international cooperation is the main pattern of data application in the field of resources.

In 2014, China and Indonesia signed "the Memorandum of Understanding between the China National Space Administration (CNSA) and the Indonesian Maritime Safety Coordination Agency (IMSCA) on cooperation in remote sensing ground station projects in Indonesia" [18]. The signing of the agreement laid a foundation for the two countries to use satellite data to carry out remote sensing cooperation in marine resource exploration and other fields. The installation and acceptance of the ground station have been completed.

In 2017, the "Remote Sensing Application in Water Resources Management in Northeast Thailand" project, undertaken by China, was launched. The Thai participants in the project include Khon Kaen University (KKU) and GISTDA. The project integrates China's scientific and technological achievements in water cycle remote sensing to promote the application of remote sensing science and technology in water resources management through China-Thailand cooperation. It filled the observation gap of soil moisture in Southeast Asia and provide an important ground validation for microwave remote sensing soil moisture inversion algorithm.

5 Summary and Suggestions

5.1 Summary and Prospect

This paper conducts a study on the market demand for remote sensing data applications with Thailand, Malaysia and Indonesia as the research targets, and analyzes the market demand characteristics, potential users and competitors in the three countries through different cases. The results indicate that the three countries have a large demand gap for remote sensing data application, a wide range of fields, and a large number of potential market opportunities, and governments are continuously strengthening the planning and investment in remote sensing data application. Meanwhile, developed countries have extensive and flexible business patterns in the three countries, and they have always maintained a competitive relationship with Chinese related institutions in terms of real-time monitoring and processing capabilities, observation accuracy and quantification level. Remote sensing data application has been transformed from the experimental research stage to the data production and commercial service stage. The existing remote sensing data products have met part of the application demand, but there is still room to improve the diversity of data application. The above results provide valuable reference directions for the Institute to enter the international market supply chain of remote sensing data application services.

Facing the wide range of application demand of various countries in agriculture, forestry, geology, ocean, meteorology, hydrology, environmental protection and other fields, the author believes that the future remote sensing data application services will enter a new stage of providing diversified earth observation data products and service patterns quickly and effectively, and the demand of time, space and spectral resolution of remote sensing data application services will be greatly improved. With the development of space technology, especially GIS technology, the application of remote sensing data will be increasingly extensive.

5.2 Suggestions in Key Fields

Combining the results of market research in Thailand, Malaysia and Indonesia with the technical directions of the Institute, it is suggested to take agriculture and forestry, hydrology, urban and homeland security as the key fields for market promotion.

In the field of agricultural and forestry remote sensing, the technical directions of the Institute include agricultural pest and disease monitoring, fine crop classification and crop growth monitoring, which can provide data services for the countries which are mainly in agriculture. In terms of hydrological remote sensing, the three countries have a wide range of demand for marine environment monitoring. The technical directions of the Institute include water body index extraction and remote sensing monitoring of water surface temperature, which can carry out investigations and studies on water eutrophication and monitor water quality in the target areas. In terms of urban remote sensing, the Institute's technical directions include road identification extraction and urban heat island monitoring, which can be considered to provide data services for the demand of land management and urban heat island in the three countries. For homeland security, the Institute's technology directions include ship identification and maritime rights maintenance, and therefore may consider providing data services to help maintain maritime security for the demand of other navies in effectively detecting and extracting target information.

References

1. Foresight Industry Research Institute website. https://bg.qianzhan.com/trends/detail/506/200612-e53e0480.html. Accessed 13 June 2021
2. He, M.C., Tian, B.: Difficulties in the implementation of the belt and road strategy and ideas for response - a study based on fieldwork in Central Asia, West Asia, South Asia, Southeast Asia, and Central and Eastern Europe. Acad. Front. **5**, 55–62 (2016)
3. Liu, H.M., Hu, F., Fang, K., He, G.Q., Ma, H.T., Cui, X.G.: Comprehensive assessment and prevention and control of political-economic-social risks in countries along the belt and road. Geogr. Stud. **38**(12), 2966–2984 (2019)
4. Lan, J.X.: The belt and road initiative in South Asia: progress, challenges and future. Indian Ocean Econ. Res. (4), 39–53, 139 (2017)
5. Feng, C.L.: Observations on the geopolitical landscape of South Asia and the development of regional situation under the perspective of the belt and road. South Asia Stud. **3**(1–32), 150–151 (2017)
6. Kang, L., Qi, J.: An analysis of the risks of cooperation between China and five Central Asian countries under the framework of the belt and road. Northern Economy **5**, 33–35 (2017)
7. Li, Z.X.: Research on color equalization and seamless processing of remote sensing images. Beijing University of Posts and Telecommunications (2018)
8. Li, M.J., Li, Q.R.: The process and prospect of the belt and road initiative in Southeast Asia. Bord. Ocean Stud. **4**(2), 74–88 (2019)
9. Guo, W., Li, B.M., Fang, Z., Wang, L.: Research on the construction of remote sensing satellite ground receiving station network in the belt and road. Satell. Appl. **10**, 22–25 (2016)
10. Liu, T.: Introduction to spaceflight in countries along the belt and road in Thailand. Satell. Appl. **5**, 24–30 (2016)
11. Zhou, X.M., Yao, J.: Malaysia (Country Overview of the Belt and Road), 1st edn, p. 129. Dalian Maritime University Press (2019)
12. Liu, T.: Introduction to spaceflight in Indonesia along the belt and road. Satell. Appl. **11**, 48–53 (2015)
13. Lu, G.J.: Analysis of the application potential of Landsat data in geographic country monitoring. Beijing Agric. **6**, 161–162 (2014)
14. Tang, W.J.: Application of remote sensing data based on MODIS and TM in the dynamic monitoring of agricultural drought. Zhejiang University of Agriculture and Forestry (2014)
15. Xu, F.: Study on mangrove extraction and carbon stock estimation based on Sentinel-2. Lanzhou Jiaotong University (2020)
16. Wang, Z.Z., Mongkol, R.: Data services and applications of HJ-1A&B satellites in Thailand. China Space **3**, 3–7 (2015)
17. Liu, S.C.: Current status of space technology disaster reduction applications in Indonesia and prospects for cooperation. China Disaster Reduct. **21**, 54–55 (2013)
18. Chen, X.G., Gao, H., Lin, M., Li, B.M.: Satellite applications for building dreams of the belt and road. Satell. Appl. **8**, 27–30 (2015)

Research and Application of Space Optical Payload Piggyback Technology Towards China's Future Commercial Constellations Construction

Hu Jianing[1](✉), Tian Peng[1], Wang Ke[1], and Fu Yu[2]

[1] Beijing Institute of Spacecraft System Engineering, Beijing 100094, China
hjn1202@126.com

[2] Institute of Telecommunication and Navigation Satellites, CAST, Beijing 100094, China

Abstract. With the acceleration of commercialization in the field of aerospace, the construction of commercial constellations in LEO is developing rapidly. China timely grasps the opportunity and quickly seizes the commercial aerospace market. Actively promote the development of commercial constellations system which meets the needs of supporting construction of a powerful country in aerospace. In recent years, with the successful deployment and launch of "Jilin-1", "Siwei SuperView", "China Satellite Network" and a series of multi-type commercial constellations, it makes the construction of China's commercial constellations beginning to take shape. The United States, as the bellwether in the domain of commercial aerospace, not only expedite constellation of "Starlink", a giant constellations which will be established by 42,000 satellites, but also provide additional optical payload to "Starlink" on the basis of present configuration to increase earth observation and space situational awareness. It takes full advantages of high-timeliness revisit cycle by LEO Internet mega-constellations. Moreover, the application efficiency of "Starlink" will be improved enormously. China should draw on as much as possible the experience of the advanced design concept and its construction scheme from "Starlink". This paper is based on China's future commercial constellations construction. Discovery an observation technology and a service pattern which are beneficial to the application efficiency of China's future commercial constellations by the research of space optical payload piggyback technology. These would provide a certain reference for future the research and development of commercialized space optical in China.

Keywords: Commercial Constellations · Piggyback Technology · Space Optical · Observation Service

1 Introduction

The commercial aerospace market, there are huge demands and a broad application prospects, pushes the various countries to ramp up the construction of commercial constellations system which suits their national circumstances. Especially in LEO, it becomes to the gold area because the low launch price and the huge demands from

H. P. Urbach and H. Jiang (Eds.): ISSOIA 2022, SPPHY 295, pp. 719–730, 2023.
https://doi.org/10.1007/978-981-99-4098-1_60

the commercial aerospace market. The LEO commercial constellations have advantages of optical remote sensing, space situational awareness and satellite communications by high-timeliness revisit cycle and global coverage [1, 2]. The construction of LEO commercial constellations are flourishing. A number of commercial aerospace enterprise are constantly emerging which takes the United States as representative. Especially the SpaceX, it plans to establish a LEO Internet mega-constellation by 42,000 satellites, it provides full weather, full-time and global coverage services to global users.

The year 2015 was called the first year of China's commercial aerospace, and the introduction of a series of policies marked the official takeoff of China's commercial aerospace. Compared with the United States, China's commercialization process in the aerospace field started late, but with the strong accumulation of space technology over the years, China has rapidly gained advantages in the commercial space market. At present, the rapid development of commercial optical remote sensing industry has prompted the continuous progress of China's space optical technology, which has greatly promoted the development of China's Earth observation system [3]. According to statistics, by the end of 2021, China developed and successfully launched more than 87 commercial remote sensing satellites and their test satellites. As of October 2022, the number of commercial remote sensing satellites in orbit in China exceeded 140. In addition, the field of communication satellites is highly commercialized, and in the future, China will be committed to the construction of LEO Internet commercial constellations. Information shows that in 2021, China launched a total of 36 commercial satellites. By the end of 2021, China had a total of 179 commercial satellites in orbit (excluding test satellites). 2022, China is expected to launch more than 50 commercial satellites.

With the number of commercial satellite launches rising year by year, the process of commercial constellation system construction has been accelerating. In order to meet the market development and application needs, China should explore a kind of observation technology and service mode that can help enhance the application efficiency of commercial constellations. Transform the satellite development mode from the payload development capability as the starting point to the application service as the traction point. By exploring different application patterns to promote the continuous progress of China's space optical technology, explore the means to enhance the application effectiveness of China's commercial constellations, and maintain the high competitiveness of China's commercial aerospace in the international market. In the future, commercial optical observation technology and services will definitely become the driving force of China's commercial constellation construction (Fig. 1).

2 Overview of China's Commercial Constellation Development

China's commercial constellations have maintained a rapid development trend, with a series of multi-type commercial constellations such as "Jilin-1", "Siwei SuperView" and "China Satellite Network" successfully deployed to accelerate the construction of China's commercial constellation map. The construction of China's commercial constellation map, promote the rapid rise of China's commercial constellations in the global field, further enhance China's commercial application capabilities in the field of aerospace, and expand the service pattern of commercial constellations (Table 1).

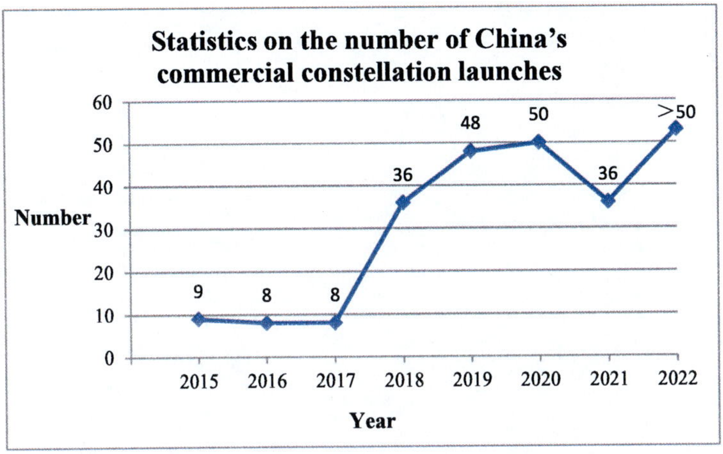

Fig. 1. Statistics on the Number of China's Commercial Constellation launches (2015—2022) [4–7]

Table 1. Part of China's Commercial Constellation Plans

Enterprise	Constellations	Field	Number
CASC	Hongyan	Communication	300
CASIC	Xingyun	Communication	80
CASIC	Hongyun	Communication	156
Chang Guang	Jilin-1	Remote Sensing	138
China Siwei	Siwei Superview	Remote Sensing	28
Galaxy Space	Galaxy	Communication	> 1000
CSCN	China Satellite Network	Communication	12992

(1) "Jilin-1"

"Jilin-1" constellation is the core project under construction by Chang Guang Satellite Technology Company, which forms the largest commercial remote sensing satellite constellation in China. As of August 2022, Chang Guang has successfully launched 70 "Jilin-1" satellites into space through 21 launches, marking the successful completion of the first phase of the "Jilin-1" constellation. At this stage, the 70 "Jilin-1" satellites in orbit can revisit any location in the world 23–25 times a day, with the ability to cover the world twice a year and the whole China six times a year. Chang Guang plans to complete the launch of all 138 satellites in the first phase of the "Jilin-1" constellation project by the end of 2023, and the number of "Jilin-1" satellites in orbit will exceed 300 by the end of 2025 (Fig. 2).

(2) "Siwei SuperView"

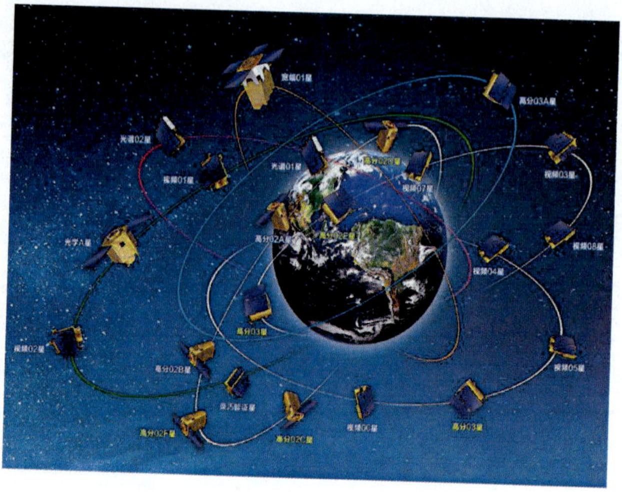

Fig. 2. "Jilin-1"

China Siwei Company plans to fully build China's new generation commercial remote sensing satellite system by 2025. At present, with the successful launch of "Siwei SuperView" No.1 01/02 and "Siwei SuperView" No.2 01/02 satellites, it marks the first phase of the system construction successfully. After the completion of the system construction, the system will have a high revisit capability of 25 times per day globally, a high acquisition capability of more than 30 million km2 per day, and an emergency delivery capability within 2 h. The space segment plans to launch "16 + 4 + 8" 28 satellites, including 16 optical satellites with 0.5 m resolution, 4 wide optical satellites with 0.7 m resolution, 8 SAR satellites with 1m resolution, and the simultaneous construction of the matching "Siwei cloud" based ground segment. Later, according to the market development can be expanded to 56 satellites (Fig. 3).

Fig. 3. "Siwei SuperView"

(3) "Galaxy"

Since its establishment in 2016, Galaxy Space has been committed to building the world's leading LEO broadband communication satellite constellation through the development of low-cost, high-performance 5G satellites, and establishing a worldwide converged 5G communication network. To this end, Galaxy Space has planned to build the 5G standard "Galaxy" LEO broadband satellite constellation, which consists of thousands of self-developed 5G satellites, forming a network constellation in LEO of about 1200 km. At present, the "Galaxy" constellation consists of the world's first Q/V-band LEO broadband satellite and six LEO broadband communication satellites to successfully build a star-ground convergence 5G test network called "Little Spider Web". Galaxy Space is developing a new generation of stackable flat-panel satellites, scheduled for launch in 2023 (Fig. 4).

Fig. 4. "Galaxy" 6 satellites in orbit

(4) "China Satellite Network"

In September 2020, China submitted an application to the International Telecommunication Union for a giant constellation of 12,992 satellites in LEO, and in April 2021, China Satellite Network Group Company was established to lead the "China Satellite Network" constellation program. The "China Satellite Network" constellation will integrate several LEO constellation programs of China's two major aerospace groups, such as "Hong Yan", "Xing Yun", "Hong Yun". The "Galaxy" constellation will also be part of the "China Satellite Network" constellation. The "China Satellite Network" constellation consists of two constellations, codenamed "GW-A59" and "GW-2" respectively. "GW-A59 consists of three sub-constellations with 6,080 satellites. "GW-2" consists of 4 sub-constellations with 6,912 satellites (Table 2).

Global commercial space development has long been in the fast lane, especially in the LEO region, competition is particularly fierce. In order to quickly seize the commercial space market in the LEO region, China is accelerating the construction of a commercial constellation system that meets the needs of China's aerospace power construction.

Table 2. The Construction Planning of "China Satellite Network"

Cobstellation	Sub	Altitude/km	Inclination/°	Orbit Planes	Number of Per Orbit Plane	Number
GW-A59	1	480	85	16	30	480
	2	2,000	50	40	50	2,000
	3	3,600	55	60	60	3,600
					Sum	6,080
GW-2	1	1,728	30	48	36	1,728
	2	1,728	40	48	36	1,728
	3	1,728	50	48	36	1,728
	4	1,728	60	48	36	1,728
					Sum	6,912
					Sum	12,992

Admittedly, China's LEO commercial constellations are taking shape, with remote sensing and communications as representatives, and several LEO commercial constellations are being planned and constructed, and some of them have successfully completed the first phase of construction tasks. However, compared with foreign advanced commercial constellation construction process is still in the catch-up stage. China's commercial constellation construction should further expand the constellation scale, but also to foreign advanced commercial constellation as a reference, learn from its construction ideas; to commercial constellation capacity as a point of enhancement, focus on commercial constellation application effectiveness; to market demand as a starting point, to explore different service patterns.

3 Analysis of "Starlink" Piggyback Situation

Since SpaceX proposed the "Starlink" constellation in 2015 (as of October 28th, 2022), 65 batches of 3558 "Starlink" have been launched. Currently, there are three phases of the "Starlink" constellation, which is planned to be a giant constellation of about 42,000 satellites for the LEO Internet. The first phase of the "Starlink" constellation contains 4408 satellites deployed in 190 orbital planes with orbital altitudes ranging from 540 km–570 km; the second phase contains 7518 satellites with orbital altitudes ranging from 330 km–350 km The "Starlink" constellation will form a space-based global communication system with global coverage, high speed, high capacity, high adaptability and low latency, aiming to provide global high-speed broadband access services by LEO giant constellation (Fig. 5).

In April 2020, SpaceX applied to adjust the orbital altitude and orbital inclination of the satellites in the first phase of the "Starlink" constellation. Among them, the fourth and fifth sub-constellation of "Starlink" orbits were adjusted from 1275 km, 81° and 1325 km 70° to 560 km, 97.6°, a total of 520 satellites. The adjusted orbits coincide with

Fig. 5. "Starlink"

the SSO. According to speculation, the "Starlink" constellation is not only used as a LEO Internet constellation, but also may have the capability of remote sensing constellation.

In October 2020, SpaceX was awarded a contract by the Space Development Agency to build four ballistic missile and hypersonic missile detection and tracking satellites. SpaceX has proposed a satellite design based on the "Starlink" platform that includes a "wide-field" overhead persistent infrared payload capable of detecting and tracking advanced missile threats from LEO. The four satellites are scheduled to be launched in 2022.

Therefore, SpaceX will piggyback optical payloads on the "Starlink" satellite. In the future, it is not ruled out to continue to equip the "Starlink" with optical payloads such as remote sensing and space target detection, so as to complete high-density communication missions and at the same time have the ability to observe the earth and space situational awareness. With its super large-scale deployment, the "Starlink" satellites can achieve global time-efficient revisit, and if they piggyback optical payloads, they can achieve global uninterrupted monitoring, which will greatly enhance the application effectiveness of the "Starlink" constellation. Through the analysis of "Starlink" constellation construction plan and piggyback situation, we can see that in addition to accelerating the construction of commercial constellation system in line with China's national conditions, we should also dig deeper into the commercial constellation capacity growth points around applications and services, and explore multiple service modes of commercial constellation. Therefore, China's future commercial constellation development can fully learn from the U.S. "star chain" constellation construction ideas, in gradually expanding the scale of commercial constellations on the basis of further enhance the effectiveness of commercial constellation applications. Based on this, we should carry out research on standardized piggyback interfaces, light and small space optical payloads, and commercialized service patterns.

4 Research and Application of Space Optical Payload Piggyback Technology

4.1 Piggyback Technology of Commercial Constellation

According to the above analysis, by piggyback optical payloads, "Starlink" have the capability of earth observation and space situational awareness while completing the communication function. As the scale of China's commercial constellation gradually expands, the number of commercial satellites in orbit increases year by year. In order to further enhance the application efficiency of commercial constellation and expand the commercial service patter, the extra weight and space of commercial satellites can be used to piggyback standardized optical payloads. However, there are many kinds of interfaces between optical payloads and commercial satellites, and there is a lack of unified design standards and piggyback schemes, so it is necessary to carry out research on piggyback technology for commercial constellations.

For China's future commercial constellation construction planning, through the task analysis of commercial satellites piggyback optical payloads, sort out the optical payloads with the ability to piggyback into orbit with multiple types of commercial satellites. The analysis will focus on the weight, volume, launch envelope, launch mode, orbital conditions and other factors that may affect the piggyback of existing China's commercial satellites, and standardize the design of the piggyback interface for multiple types of commercial satellites, so as to reduce the weight and volume of mechanical, electrical and thermal control interfaces as much as possible, as well as the space and power consumption requirements of optical payloads on commercial satellites, and finally form a standardized interface system for commercial satellite piggyback payloads (Table 3).

Table 3. Piggyback Capability Index

Piggyback Capability	Parameter Index	Value
Piggyback Limit	Weight	30 kg
	Size	800 mm × 800 mm × 500 mm
	Power	20 W
Piggyback Device	Weight	1 kg
	Size	φ140 mm × 120 mm

4.2 Integrated Detection and Imaging Technology

Light and small space optical payloads are highly sought after in the commercial space field because of their, light mass, small size and low cost. Compared with traditional large space optical payloads, light and small space optical payloads exchange their low development cost for high mission cost performance. It is especially widely used in micro-nano satellites, which generally refer to satellites weighing 1–100 kg. With the continuous development of aerospace technology, the resolution of micro-nano optical remote sensing satellites has been greatly improved, and the resolution of 50 kg class micro-nano satellite is better than 1 m. Therefore, the remaining capacity of commercial constellations piggyback light and small space optical payloads will undoubtedly enhance the application effectiveness of commercial constellations. The booming development of micro-nano satellites and diversified user needs have led to the development of new technologies for light and small space optical payloads, such as lightweight design of optical payloads, integrated platform and payload design, and integrated detection and imaging technologies [8, 9].

The fact that most space optical payloads do not have both detection and imaging capabilities is related to the space observation mission. Usually, remote sensing satellites or space-based surveillance satellites, etc. only need to accomplish a single type mission. In order to fully utilize the on-board resources and maximize the effectiveness of commercial constellation applications, the observation technology and service pattern of space optical payloads are further explored, and research on the integrated technology of detection and imaging of space optical payloads is carried out to complete detection and imaging missions for key targets. Since there are certain differences in the optical systems of the detection and imaging parts, they can be realized by methods such as split-optical path detection and imaging or free-form surface technology [10, 11].

4.3 Space Optical Payload Application Model

(1) Space debris detection services

Currently, there are more than 17,000 space targets in orbit cataloged in the LEO region, including spacecraft in orbit that are functioning, spacecraft that are not functioning, space debris, etc. According to statistics, there are more than 500,000 space debris with a diameter of 1 cm or more. More than 80% of space debris is concentrated in the LEO, and more than 60% of space debris is distributed in the LEO below 1200 km and in GEO. The increasing amount of space debris in the LEO will pose a great risk to the safe operation of spacecraft. China's space debris detection mainly relies on ground-based systems, and due to the limitation of ground-based station deployment, China's ground-based space debris detection area can only cover the space over the national territory, and the timeliness is low. Therefore, piggyback the integrated optical payload of detection and imaging on our commercial constellation is beneficial to the safe operation of our commercial constellation in orbit, and the joint application with ground-based space debris surveillance system is an extension and supplement to our space debris detection capability, further enhancing the coverage, timeliness and cataloging accuracy of our space debris detection [12–14] (Fig. 6).

(2) Earth observation services.

China's future commercial constellation construction planning mainly composed of remote sensing and communications constellations. Among them, the giant constellation represented by the "China Satellite Network" constellation is the largest in scale. It is a giant constellation of LEO Internet consisting of 12,992 satellites in 308 orbital planes. Therefore, China can fully learn from the U.S. "Starlink"

Fig. 6. Space Debris Detection

constellation construction idea, and install optical payloads on the ground in communication satellites, so that the giant constellation in low orbit has the capability of earth observation or space debris detection. Using the global time-efficient revisit capability achieved by the super large-scale deployment of the giant constellation's multiple orbital surfaces, the detection and imaging integrated optical payloads can be piggyback to provide users with all-weather earth observation services around the globe. In addition, the detection and imaging integrated optical payloads can also be used to complete high-time-effective space debris detection missions in LEO. The joint application of the high time-effective revisit capability of the giant constellation of the LEO Internet, the integrated detection and imaging capability of the optical payload, and the high-resolution imaging capability of the commercial remote sensing constellation is conducive to improving the application efficiency of China's commercial constellations, exploring new observation technologies, and providing users with multiple service modes (Fig. 7).

(3) In-orbit Maintenance Support

China's satellite development process reliability and quality requirements are high, the probability of failure in orbit is low, but once the failure is very costly. Timely troubleshooting for satellite in orbit can improve the satellite in orbit reliability and extend the satellite in orbit life. Satellite in-orbit failure is mainly composed of payload, control, power supply and distribution, propulsion and other aspects, but in some cases only through telemetry data and other ground analysis means can not accurately determine the cause of in-orbit failure in a timely manner. Such as facing the satellite into orbit after the loss of the satellite power loss, the satellite abnormal high-speed spin, space debris collision and other sudden failure or anomaly, through its natural rendezvous of commercial satellites piggyback optical payloads for closer opportunity to detect / imaging, and through the imaging and detection data quickly down to the ground, for the ground fault interpretation and troubleshooting plans to provide effective measurement information. Through the commercial constellation piggyback detection and imaging integrated optical payloads, timely access to accurate orbital information and high-resolution images of satellites, the state of health

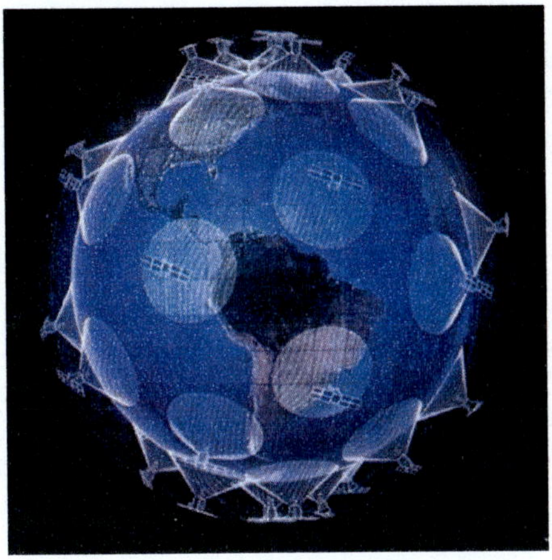

Fig. 7. Earth Observation Services

of our satellites in orbit on a regular basis to grasp, to provide effective support for the realization of high time effective in-orbit maintenance (Fig. 8).

Fig. 8. In-orbit Maintenance Support

5 Summary

China's commercial constellation construction is not yet mature, compared with the United States and other foreign advanced commercial constellations there are still some gaps. However, with the commercialization of China's space field, a series of multi-type commercial constellations are springing up, and the construction of commercial constellations in China is in a rapid development stage. For the future construction of commercial constellations in China, on the one hand, we should combine our national conditions, plan and improve the construction of commercial constellation system, and seize the vast market in the world commercial space field; on the other hand, we should focus on improving the application efficiency of commercial constellations, combining with user needs, digging deeper into the service model, and leading the world commercial constellation application market.

References

1. Qi, Y., et al.: Exploration practice and development suggestions for commercial remote sensing satellites of China. Spacecr. Eng. **30**(6), 188–194 (2021)
2. Ruan, Y., Hu, M., Yun, C.: Advances and prospects of the configuration design and control research of the LEO mega-constellations. Chin. Space Sci. Technol. **42**(1), 1–15 (2022)
3. Wang Xiaoyong, F.: Development and prospect of space optical technology. Spacecr. Recovery & Remote Sens. **39**(4), 79–86 (2018)
4. Sun, Q., Liu, J.: Global commercial launch market overview 2015. Aerosp. China (6), 40–43 (2016)
5. Liu, J., He, Y.: Global launch services market analysis 2016. Aerosp. China (6), 60–63 (2017)
6. Hao, L.: New breakthrough in China's commercial launches in 2019. Aerosp. China (1), 40–42 (2020)
7. Liu, J., Xu, Z., Yang, X., Li, H.: Review of China's commercial space activities in 2020. Satell. Appl. (5), 29–33 (2021)
8. Fu, D., et al.: The opportunities and challenges in optical payload of micro-nano satellite. Spacecr. Recovery Remote Sens. **39**(4), 64–69 (2018)
9. Sha, Z., Yu, Y., Jiao, J.: The latest development of micro-nano remote-sensing payload technology abroad. Spacecr. Recovery Remote Sens. **42**(5), 39–48 (2021)
10. Jiang, H., Fu, Q., Zhang, Y., Jiang, L.: Discussion of the laser ranging with polarization spectral imaging observations and communication technology for space debris. Infrared Laser Eng. **45**(4), 2–7 (2016)
11. Ni, D.: An Integrated Optical System for Detection and Imaging Based on Freeform Surface. Xi'an Institute of Optics & Precision Mechanics, Chinese Academy of Sciences (2019)
12. Hu, T. Li, H., Pan, T., He, D., Zhuang, H.: Configuration scheme and effectiveness analysis of space-based orbital debris surveillance system. Space Debris Res. **17**(3), 8–12 (2017)
13. Feng, H., Tian, B., Zhang, X.: Analysis on influence of giant commercial constellations on orbit resources. Space Debris Res. **21**(2), 40–45 (2021)
14. Zhang, P., Sang, J., Pan, T., Li, H.: Initial orbit determination method based on range searching for LEO space debris. Spacecr. Eng. **26**(2), 22–28 (2017)

Author Index

H. P. Urbach and H. Jiang (Eds.): ISSOIA 2022, SPPHY 295, pp. 731–734, 2023.
https://doi.org/10.1007/978-981-99-4098-1

Z